Modeling and Optimization of Interdependent Energy Infrastructures

Wei Wei • Jianhui Wang

Modeling and Optimization of Interdependent Energy Infrastructures

Springer

Wei Wei
State Key Laboratory of Power Systems
Department of Electrical Engineering
Beijing, China

Jianhui Wang
Department of Electrical and Computer Engineering
Southern Methodist University
Dallas, TX, USA

ISBN 978-3-030-25960-0 ISBN 978-3-030-25958-7 (eBook)
https://doi.org/10.1007/978-3-030-25958-7

This Springer imprint is published by the registered company Springer Nature Switzerland AG.
The registered company address is: Gewerbestrasse 11, 6330 Cham, Switzerland

To our families.

Preface

Motivation of This Book

The everlasting consumption of fossil fuels with limited reserves amid climate change and environmental pollution arises public awareness of sustainable development, which calls for technology innovations as well as policy reformations to encourage energy saving, enhance system efficiency, and harness green resources. Nowadays, our energy supply infrastructures are experiencing rapid changes in two aspects. First, renewable energies are utilized at various levels of our energy systems. This initiative alleviates the dependence on traditional fossil fuels and remarkably reduces carbon emissions. However, their volatility could impose significant challenges on real-time power balancing of the electric power grid and may even wreck system security. Second, system integration has become a prevalent issue so as to unleash synergetic potentials across multiple energy carriers. The scalable storage of heat makes it an attractive medium for renewable energy storage; the relatively low carbon impact of natural gas, its large reserve in shale rocks, the fast-response capability of gas-fired power generation units, and the growing maturity of power-to-gas technology will underpin a nexus of renewables, natural gas, and electricity in the coming decades. Moreover, the bidirectional information and energy flows allow electric vehicles to play an increasingly influential role in future power grids. The complementary natures of different energy resources precipitate the integration of traditionally independent energy systems, such as the power system, natural gas system, heating system, and even the transportation system, creating energy flow interdependence across involved physical systems.

The transition to a cleaner and more sustainable energy industry is affecting the current structure of energy systems ranging from the national, state, and city scales down to the distribution, residential, and building levels. However, this ambitious plan necessitates brand-new regulatory frameworks, strategic economic incentives, innovative business models, and active participation of consumers as responsive demands.

With the progress of energy system integration, decision-making problems for all involved stakeholders are becoming more challenging, complicated, and interactive. We perceive the necessity of disseminating fundamental methods in energy system modeling as well as advanced theory and methodology in mathematical programming which could address these problems faced by the broad energy research community.

Organization of This Book

This book will cover some promising and active research topics and will comprehensively address the modeling and optimization issues in the analysis, management, and operation of interdependent energy systems from technical, economic, and regulatory perspectives. Based on thorough mathematical methodologies, we can make best use of potential synergetic effects and achieve desired goals. This book is organized as follows:

Chapter 1 provides an overview on the functionality of several representative physical systems which will be discussed in this book, including the electric power grid, natural gas network, district heating network, and urban transportation network. The complementary natural and synergetic effects among different energy resources are revealed. The advantages, benefits, and challenges lying ahead, brought by energy system integration, are introduced. The remaining chapters are dedicated to special forms of integrated energy systems.

Chapter 2 discusses the power system with renewable power integration. The main purpose of this chapter is to tackle the non-convexity of power flow equations as well as uncertainties brought by renewable integration. In view of the fundamental role of power flow in power system analysis, we give a thorough tutorial on various power flow models in Sect. 2.2, including AC power flow, DC power flow, branch flow, and its linear approximation. Then, we proceed to the optimal power flow problem and its variations, which actually come down to optimize a specific objective function subject to power flow constraints. Convex relaxation methods are employed to cope with the intrinsic non-convexity. We propose a sequential convex optimization method to recover a high-quality solution when the convex relaxation is inexact, which usually happens in the maximum loadability problem and the multi-objective optimal power flow problem. Next, we commit efforts to cope with uncertainties resulting from renewable integration. Depending on the available information on the uncertainty, we apply two kinds of robust optimization approaches to the hourly ahead energy and reserve dispatch problem and the day-ahead unit commitment problem, aiming at improving system reliability at the minimum cost. We propose the concept of dispatchable region, which exactly characterizes how much uncertainty can be dealt with in real-time operation with a fixed energy and reserve dispatch strategy and propose an integer programming approach to generate the boundaries of the dispatchable region. This concept inspires many interesting applications. For example, control and optimization of the dispatchable region, as well as data-driven reliability assessment without probability distribution, are elaborated in this chapter.

Chapter 3 is devoted to the integrated gas-power system. Such integration is usually seen at the inter-area transmission level. In Sect. 3.2, we provide the mathematical model of a natural gas pipeline network and summarize two approximations for the gas flow propagation equation, which is a partial differential equation. The steady-state approximation is known as the Weymouth equation, and the quasi steady-state approximation is usually referred to as the line pack model. Both of them render algebraic equations. In the rest of this chapter, convex optimization methods for gas-power flow, equilibria of coupled gas-power markets, and resilient operation of gas-power systems are discussed.

Chapter 4 is devoted to the integrated heat-power system. Such integration is usually seen at the distribution level, as long-distance delivery of heat is not economically efficient. In Sect. 4.2, we introduce the mathematic model of a district heating pipeline network, called the hydraulic-thermal model. We reveal the decomposition structure of the optimal hydraulic-thermal flow problem and propose a method to determine the near optimal hydraulic conditions in the system, such that the thermal state optimization gives rise to a convex program with linear constraints. In the rest of this chapter, market-based decentralized operation schemes for the heat-power system, equilibria of coupled heat-power markets, strategic behaviors of energy hubs in electricity and heating markets, and capacity planning of energy hubs are presented.

Chapter 5 is devoted to the city-sized transportation network and power distribution network which are coupled by electric vehicles and on-road fast-charging stations. In Sect. 5.2, we present the mathematical descriptions of two traffic user equilibrium models, which determine the steady-state vehicular flow distributions in the transportation network. Different from other energy flow problems, in which the system is controlled by a central authority, the equilibrium in a transportation network is a spontaneous outcome of individual rationalities of travelers. A practical algorithm is introduced to solve large-scale instances by dynamically generating active paths. In the rest of this chapter, important issues covering the topics in network flow equilibria and optimization, system operation with uncertain traffic demands, vulnerability analysis, and capacity expansion planning are studied.

Given the interdependence across multiple energy systems, a central issue encountered in energy markets is the transaction policy. Imitating the pool-based power market, we generalize the marginal cost-based energy pricing scheme to the companion energy systems in Chaps. 3–5 and propose a bilateral trading framework for the integrated energy systems. Because the energy price should be retrieved from the dual variables from the corresponding market clearing problems, the analysis of strategic behaviors of market participants and calculation of equilibrium points can be challenging. We establish fundamental computation tools for such interactive decision-making problems based on sophisticated optimization theories and methodologies.

The formal contents of this book center on engineering optimization problems of the interdependent energy systems. Due to their complexity, it is usually difficult to clearly describe both the specific problem formulation and the mathematical substance simultaneously, although we have tried our best to do so. In this book, engineering problems in practice are set up in the main body, namely, Chaps. 2–5. Four advanced optimization techniques which are frequently used throughout the main body are comprehensively tutored in the appendices, where optimization problems are cast as compact matrix form in order to highlight the mathematical substance. The majority of materials in this part are concentrated from the operational research literature and presented in a more systematic and application-oriented manner after our filtering and refinement. Particular tricks discovered in our own research can be quite useful for certain problems. Comprehensive surveys on the state-of-the-art development in respective fields are provided at the end of each chapter. As such, we hope the readers can get clear landscape and concise summaries of these methods without being trapped in tedious symbols and notations for specific applications or abstruse mathematical theory. The minimal background knowledge required to understand these contents is tantamount to the level of an introductory course on mathematical programming or operational research.

Appendix A reviews basic concepts in convex optimization. The most attractive features of convex optimization problems are their polynomial-time solvability and zero duality gap under mild conditions. The linear programming and its duality theory are extended to conic linear programming in this chapter. This comparative description helps readers who are not familiar with convex optimization develop intuitive understanding and necessary skills for applying conic linear programming and duality techniques in their problems, such as the semidefinite programming and second-order cone programming encountered in the optimal power flow studies. We pay special attention to the most general quadratic programs which include non-convex functions in constraints or the objective function or both. Such a problem frequently arises in power flow optimization, robust optimization, and equilibrium problems. Particularly, it is shown that when the constraints are linear, a quadratic program with a non-convex objective function can be equivalently formulated via a mixed-integer linear program without exploiting approximation.

Appendix B summarizes linearization techniques in mixed-integer linear programs. It is common knowledge that real-world optimization problems are usually non-convex and a non-convex program is generally NP-hard to solve globally. Nevertheless, the progress in state-of-the-art solvers makes it possible for solving a large-scale mixed-integer linear program in reasonable time, although it is also NP-hard in the worst case. More importantly, many non-convex programs can be approximated via mixed-integer linear programs, which can be solved in reasonable time. In this chapter, we summarize various formulation tricks which can convert non-convex terms into linear expressions with integer variables, offering a viable option to approximately solve particular classes of hard optimization problems (globally) via mixed-integer linear programs. These techniques are particularly useful for dealing with univariate nonlinear functions, bilinear terms, and logical (disjunctive) constraints.

Appendix C illuminates the basics of robust optimization approaches, categorized by the descriptions of uncertainty (whether distributional information is taken into account) and the way of mitigating uncertainty (whether a recourse action is allowed). It is demonstrated that for a fairly broad class of uncertain set, a robust optimization problem admits a tractable counterpart, which can be efficiently solved. In other cases, the problem can be solved in a decomposition fashion, and the difficulty rests in robust feasibility check or the determination of the worst-case scenario for a given strategy, which can be formulated as a max-min problem. The development of robust optimization theory and methodology has already put it into application on real engineering problems, which is desired by the electricity sector due to the volatility of ubiquitous renewable generation. This method is attracting research attention at an unprecedented level.

Appendix D sheds lights on equilibrium problems. For single-level equilibrium problems, we discuss the standard Nash game and the generalized Nash game. We show how these problems that entail simultaneous decision-making can be solved in a distributed manner via a best response iteration algorithm. As special cases of the standard Nash game, matrix game and potential game are also elucidated with more dedicated treatments. For bilevel equilibrium problems, we elaborate bilevel program (single-leader-single-follower Stackelberg game), mathematical program with equilibrium constraints (single-leader-multi-follower Stackelberg game), and equilibrium program with equilibrium constraints (multi-leader-multi-follower Stackelberg game). We set forth solution methodologies for these challenging problems mainly based on mixed-integer linear programming techniques. Specifically, the linear complementarity problem is discussed which serves as the basis for the equivalent reformulation of bilevel games. In addition, we discuss several useful market models abstracted from our own research which are general enough to encompass a wide spectrum of practical energy market problems. We believe these models help practitioners understand the mathematical substance behind complex equations and apply related theory more conveniently.

Although a lot of mathematical theorems are involved in the appendices of this book, we avoid mathematical formality and do not focus on proofs of theoretical results. Instead, we give their implications on algorithm development and still maintain sufficient mathematical rigor for readers to understand the elegance of optimization.

From the application perspective, mathematical tricks performed in linearization, convexification, and robustification, among other reformulation techniques that help solve a complicated decision-making problem, are rather scrappy and fragmentary. The recent upsurge of research on robust and distributionally robust optimization makes it difficult for beginners to quickly grasp the essence of robust optimization. We feel the necessity of something like a dictionary which encapsulates all useful knowledge along each direction in a unified and concise architecture. This motivates the appendix part, and we believe it will make the book widely accessible and helpful for practitioners in their research.

Background Knowledge Required to Read This Book

The required background knowledge is an undergraduate level of understandings on linear algebra and power systems, basic knowledge on optimization (e.g., undergraduate-level courses on linear programming, mathematical programming, or operational research), as well as probability theory.

Potential Readers and Usage of This Book

This book addresses emerging and hot topics in the field of power and energy engineering research using advanced optimization methods. Thus, the book may be of interest to researchers in engineering, operations research, and government authorities who are primarily working in system planning and operation, energy system economics, markets, and policies.

We hope that this book will be useful as the reference of advanced courses within the professions of energy, industrial engineering, and operations research. This book also serves as an application-oriented tutorial on several important mathematical programming techniques, including convex optimization, mixed-integer linear programming, robust optimization, and equilibrium programming.

Beijing, China
Dallas, TX, USA
January 2019

Wei Wei
Jianhui Wang

Acknowledgments

Wei Wei would like to acknowledge the financial support from the National Natural Science Foundation of China under Grant 51807101, Grant 51621065, and Grant U1766203.

Contents

Acronyms

The abbreviations and universal symbols used throughout this book are summarized as follows for quick reference. Others are defined after their first appearance in case of need.

Abbreviations

AC	Alternating Current
ADMM	Alternating Direction Method of Multipliers
ARO	Adjustable Robust Optimization
BFM	Branch Flow Model
BLP	Bilinear Program
CAES	Compressed Air Energy Storage
CCG	Constraint-and-Column Generation
CCP	Convex-Concave Procedure
CHP	Combined Heat and Power
CLMP	Continuous Locational Marginal Price
COP	Coefficient of Performance
CSP	Concentrated Solar Power
CVaR	Conditional Value at Risk
DC	Direct Current
DHN	District Heating Network
DR	Demand Response
DRO	Distributionally Robust Optimization
DR-SO	Data-Driven Robust Stochastic Optimization
EPEC	Equilibrium Program with Equilibrium Constraints
ESU	Electricity Storage Unit
EV	Electric Vehicles
FCS	Fast-Charging Station
GCI	Generalized Chebyshev Inequality

GGI	Generalized Gauss Inequality
GPI	Generalized Probability Inequality
GNE	Generalized Nash Equilibrium
GNEP	Generalized Nash Equilibrium Problem
HP	Heat Pump
KKT	Karush-Kuhn-Tucker
KL	Kullback-Leibler
LCP	Linear Complementarity Program
LMEP	Locational Marginal Electricity Price
LMGP	Locational Marginal Gas Price
LMI	Linear Matrix Inequality
LMP	Locational Marginal Price
LP	Linear Program
LPCC	Linear Program with Complementarity Constraints
MILP	Mixed-Integer Linear Program
MINLP	Mixed-Integer Nonlinear Program
MISOCP	Mixed-Integer Second-Order Cone Program
MPCC	Mathematical Program with Complementarity Constraints
MPEC	Mathematical Program with Equilibrium Constraints
NBI	Normal Boundary Intersection
NCP	Nonlinear Complementarity Program
NE	Nash Equilibrium
NEP	Nash Equilibrium Problem
NLP	Nonlinear Program
O-D	Origin-Destination
OGF	Optimal Gas Flow
OGPF	Optimal Gas-Power Flow
OHTF	Optimal Hydraulic-Thermal Flow
OPF	Optimal Power Flow
OTPF	Optimal Traffic-Power Flow
P2G	Power-to-Gas
PDE	Partial Differential Equation
PDF	Probability Distribution Function
PDN	Power Distribution Network
PSD	Positive Semidefinite
PTDF	Power Transfer Distribution Factor
PV	Photovoltaics
PWL	Piece-Wise Linear
QCQP	Quadratically Constrained Quadratic Program
QP	Quadratic Program
RLT	Reformulation-Linearization Technique
RO	Robust Optimization
ROD	Radius of Dispatchable Region
SAA	Sampling Average Approximation
SDP	Semidefinite Program

SO	Stochastic Optimization
SOC	Second-Order Cone
SoC	State of Charge
SOCP	Second-Order Cone Program
SOS	Sum of Square
SOS2	Special-Ordered Sets of Type 2
SQP	Sequential Quadratic Programming
TAP	Traffic Assignment
TN	Transportation Network
TSU	Thermal Storage Unit
UC	Unit Commitment
UE	User Equilibrium
V2G	Vehicle to Grid
VaR	Value at Risk
VE	Variational Equilibrium
VI	Variational Inequality

Symbols and Functions

$\|A\|$	The number of elements (Cardinality) of set A
$\lceil A \rceil$	The minimal integer that is greater than or equal to the real scalar A (rounding up towards plus infinity)
$\lfloor A \rfloor$	The maximal integer that is smaller than or equal to the real scalar A (rounding down towards minus infinity)
conv(A)	Convex hull of set A
$\mathbb{E}_P(\cdot)$	Expectation over probability distribution P
int(A)	The interior of set A
Pr(A)	Probability of event A
Tr(A)	Trace of matrix A
vert(A)	Vertices of polytope A
$\mathbb{B}^n$	Set of n-dimensional vectors with binary entries, also denoted as $\{0, 1\}^n$
$\mathbb{C}^n_+$	Set of $n \times n$ copositive matrices (copositive cone)
$\mathbb{R}^n$	Set of n-dimensional vectors with real entries.
$\mathbb{R}^n_+$	Set of n-dimensional vectors with non-negative real entries
$\mathbb{R}^n_{++}$	Set of n-dimensional vectors with strictly positive real entries
$\mathbb{L}^{n+1}_C$	Standard second-order cone (Lorentz cone or ice-cream cone) in $\mathbb{R}^{n+1}$
$\mathbb{M}^{m \times n}$	Set of $m \times n$ matrices
$\mathbb{S}^n$	Set of $n \times n$ symmetric matrices
$\mathbb{S}^n_+$	Set of $n \times n$ positive semidefinite (symmetric) matrices
$\mathbb{S}^n_{++}$	Set of $n \times n$ positive definite (symmetric) matrices

Chapter 1
Introduction

The desire for a secure, efficient, and sustainable energy supply is calling for significant changes in our energy systems. The high share of renewable energy resources is paving the way towards low-carbon electricity production. Furthermore, the cutting-edge technologies in a vast array of engineering disciplines allow building physical linkages between different energy systems, such as electricity, district heating, natural gas, and urban transportation. Such an initiative has broken down the barriers among traditionally isolated sectors, so as to create additional flexibility enabled by the complementary nature and synergetic potentials of heterogeneous resources. As a result, the whole energy supply chain in modern society is undergoing a rapid transition to a highly integrated energy system. During this transition, there are many emerging technical and political challenges to be addressed, which require interdisciplinary research on innovative optimization models and analytical methods. This section introduces necessary backgrounds in this mushrooming research area.

1.1 Energy Resources

1. Primary Energy
A primary energy resource refers to the substance existing in nature which is directly consumed or can be converted into other desired forms. Primary energy can be non-renewable or renewable.

A non-renewable resource is the one which is not able to renew itself at a sufficient rate for sustainable utilization in a meaningful human timescale. Fossil fuels, such as coal, petroleum, and natural gas, are non-renewable.

Coal is an abundant primary energy resource and has been used throughout human history, primarily in the production of heat and electricity, as well as other industrial purposes, such as refining chemical materials. Coal has also accounted for

W. Wei, J. Wang, *Modeling and Optimization of Interdependent Energy Infrastructures*, https://doi.org/10.1007/978-3-030-25958-7_1

a large share of carbon dioxide emissions. The usage of coal in energy production and its other consumptions are associated with environmental impacts amid climate change concerns.

Crude oil is a naturally occurring, viscous liquid found beneath the Earth's surface. Petroleum covers both unprocessed crude oil and chemical products that are refined from the crude oil, and is another kind of primary energy resource.

Natural gas is a naturally occurring gas mixture consisting primarily of methane and including varying amounts of other higher alkanes. Natural gas is commonly used for heating, electricity generation, and driving vehicles as the fuel.

Renewable energies refer to energy resources which are naturally replenished in a meaningful human timescale. Renewable energies are often used in heat and electricity generation.

Solar energy is an important source of renewable energy. It originates from the radiant light and heat from the Sun. The energy is captured and converted to electricity either directly using the photovoltaic effect or indirectly using concentrated solar power generation system [1]. Such a system focus the sunlight in a large area into a small beam using mirrors, and then the light is converted to heat, which drives a turbine to produce electricity. It is estimated that solar energy would become the world's largest source of renewable generation by the end of 2050 [2]. By then, most solar power generation would be installed in China, USA, and India.

Wind power is another plentiful renewable energy resource. When air flow blows wind turbines, kinetic energy is transformed into electricity. Onshore wind is convenient to exploit, economically competitive with coal and gas energies. Offshore wind is stronger and less volatile than on land, but the investment and maintenance costs are considerably higher. By the end of 2015, more than 80 countries around the world are using wind power for electricity generation. The global wind power capacity has expanded to 432.9 GW, and wind power generation has reached 186.3 TWh, accounting for 3.3% of total electricity production [3].

Hydropower is converted from the energy of falling water or fast running water, which can be harnessed for producing electricity or actuating mechanical devices. Historical hydropower stations require constructing large dams and reservoirs. Run-of-the-river hydropower plants derive kinetic energy from running water without the creation of large reservoirs.

Geothermal energy originates from the radioactive decay of minerals. It continuously drives thermal energy from the core to the surface of our planet, which can be harnessed for heating and other purposes.

Biofuel can be converted from biological substances derived from living or recently living organisms, such as plants or plant-derived materials. Biofuel can be burnt to produce heat and generate electricity. However, pollutants such as sulfurous oxides, nitrous oxides, and particulate matters are companied with the combustion of biofuels.

The primary energy demands in major countries and regions by 2040 have been shown in Fig. 1.1. It is seen that till the middle of this century, coal, oil, and natural gas remain the dominant primary energy resources, which are converted to fuels or used to produce electricity and heat. Among the end-user sectors, industry,

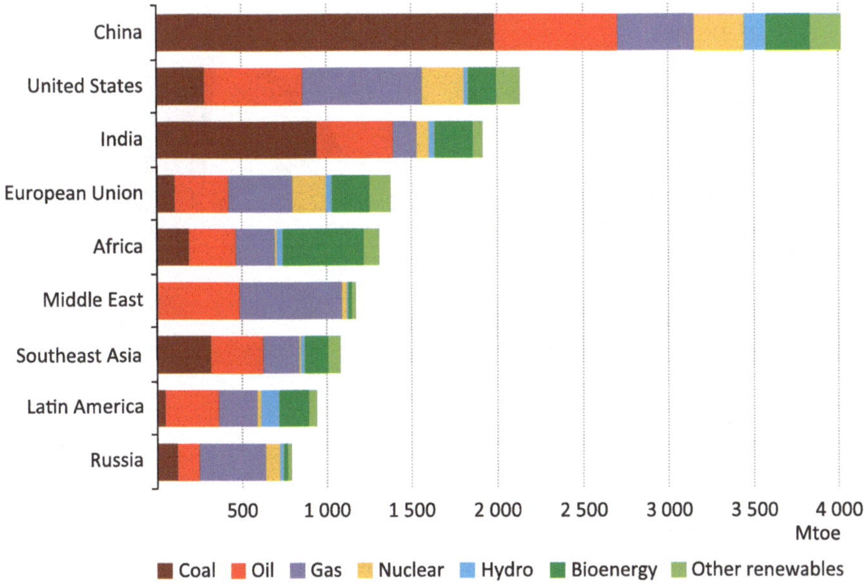

Fig. 1.1 Primary energy consumptions in selected regions by 2040 [4]

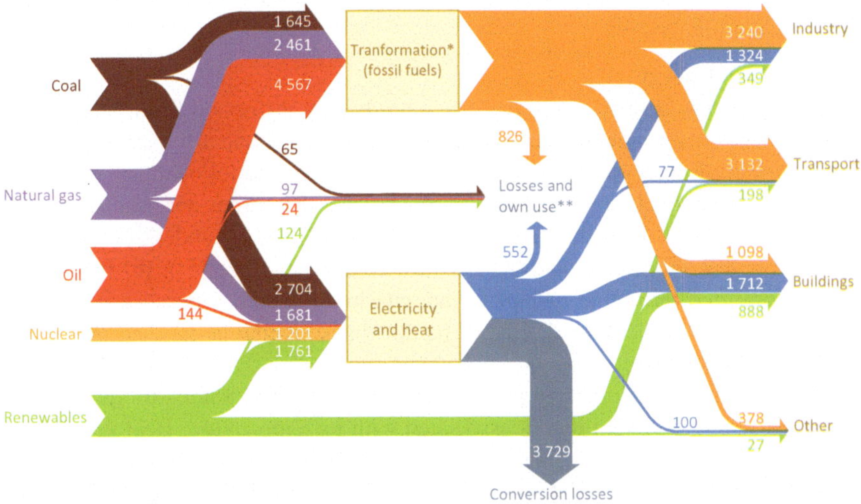

Fig. 1.2 Energy conversion flows in 2040 (Mtoe) [4]

transportation, and buildings are the biggest ones, see Fig. 1.2 for a summary. The fractions of electricity produced by different energy resources are given in Fig. 1.3. Currently, coal is still the largest source of electric power generation worldwide, contributing almost twice production volume as much as the second-largest one,

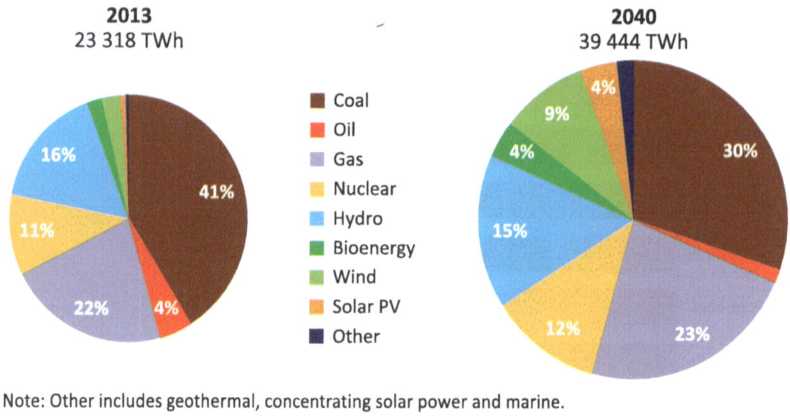

Fig. 1.3 Electricity generation by different resources [4]

the natural gas. By 2040, the share of coal in the electricity sector will decrease significantly. Renewable generation capacities are predicted to expand rapidly in the next two decades, and it is expected to become the largest source of electricity supply (including hydropower, contributing approximately 46% of total renewable-based electricity generation) by 2040 which accounts for more than one third of the world's electricity supply.

2. Secondary Energy

Secondary energy is in certain form which has been transformed from the primary energy. Electricity is a good example of secondary energy, and also one of the most commonly used energy carriers. It is advantageous for the following reasons: (a) it can be transformed from numerous primary energy sources and transmitted to where it is needed through a network of overhead lines or underground cables (wireless power transfer is possible but restricted by distance [5]). (b) Electricity can be converted into other energy forms with high efficiency because it has low entropy (is highly ordered) [6]. One of its drawbacks is that large-scale storage of electricity is expensive.

Thermal energy, another important secondary energy, refers to the energy contained within a system that is associated with its temperature. Heat is the flow of thermal energy. Thermal energy can be delivered through working fluid (such as water) in a pipeline network. In complementarity with electrical energy, thermal energy is easy to store but difficult to deliver for a long distance.

Electrical and thermal demands are the most prevalent energy demands in the daily life of modern society. A majority fraction of primary energy resources are consumed to produce electricity and heat, as demonstrated in Fig. 1.2.

1.2 Physical Infrastructures

Energy resources are produced, transited, and distributed in different infrastructures subject to their individual physical laws. This section provides descriptive introductions on some representative ones which will be discussed in this book.

1.2.1 Electric Power Systems

An electric power system (or power grid) consists of networked electrical components for the purpose of electricity generation, transmission, distribution, and consumption. A sketch is illustrated in Fig. 1.4. According to the functionality, voltage level, and spanned area, a power system can be broadly divided into transmission network and distribution network.

Transmission networks consist of power plants and high-voltage transmission links between power stations and load centers. Power stations located near a coal mine or a reservoir as well as renewable generation centers are often far away from heavily populated areas. They are large in capacity to take advantage of the scale effect. The output of a power station with a voltage of 11–35 kV is injected into the power transmission network through a transformer, after which the voltage level is stepped up to 220–1000 kV (The 1000 kV ultra-high-voltage transmission line has been put into operation in China [8]). The transmission network will move electrical power across hundreds or even thousands of kilometers from the power plants to load centers (usually metropolitan areas).

Upon arrival at a substation, the voltage will be stepped down from a transmission level to a distribution level, usually 2–35 kV, and the electric power enters a distribution network, which carries electricity from the substation to individual consumers. Finally, when reaching service locations, the voltage is stepped down again to the desired values, say, 220–380 V. The service area of a distribution network is usually a whole city.

Besides their voltage levels and service areas, transmission networks and distribution networks are different in several ways. The former one is meshed in topology, and the delivered power in each of the three phases is balanced. Moreover, transmission lines have very small resistance to reactance ratios, as a result, the active power losses in the transmission network can be neglected, and nodal bus voltage is mainly affected by the delivery of reactive power. Equipped with adequate reactive power compensation devices, the bus voltages can be well maintained close to the reference value. On the contrary, the distribution network is intentionally operated with tree topology due to relay protection related issues. In addition, the load supplied by each phase could be unbalanced. The resistance and reactance of a distribution line are numerically comparable, so both of active and reactive power delivery would have notable impacts on bus voltage magnitudes, and line losses can be higher. With the proliferation of distributed renewable generation and

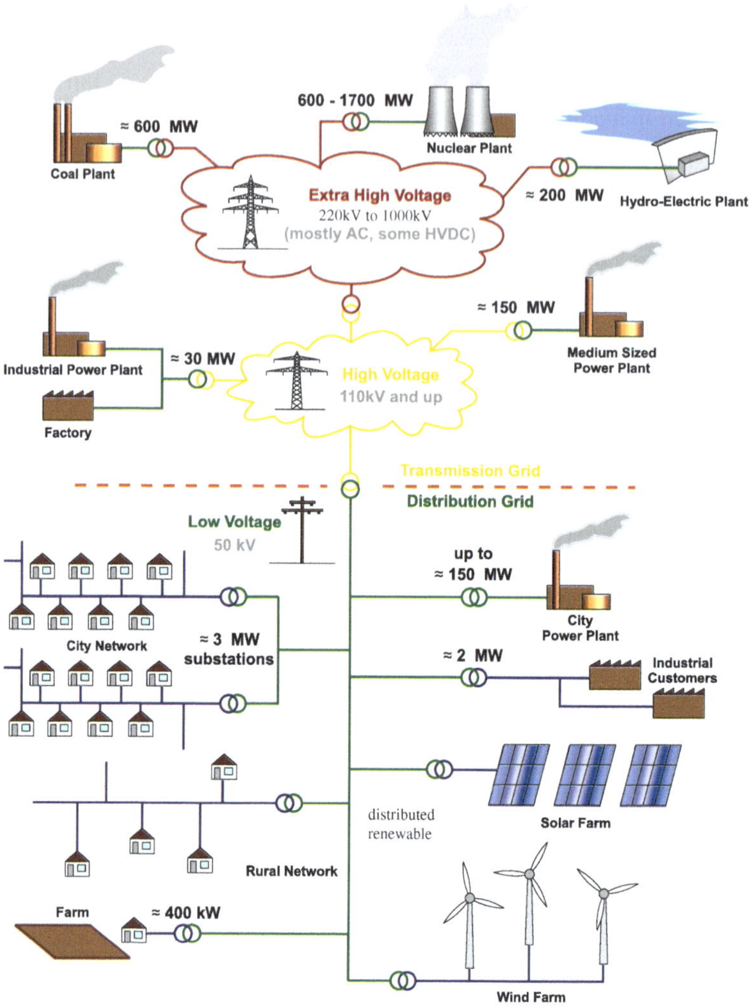

Fig. 1.4 A sketch of the electric power system [7]

microgrids, traditional demands are becoming more active and responsive, and the infrastructures of distribution networks are also undergoing a rapid transition, such as meshed topology and bi-directional power flows. New operation strategies and market incentives are proposed to facilitate more efficient use of energy. See [9–11] for more surveys.

The steady-state operating status of a transmission or distribution network is known as the (alternating current) power flow or load flow solution, which includes voltages and injections at every bus as well as active and reactive power flows through every branch. In the power flow analysis, the network is treated as a circuit, and power flow equations are constructed in accordance with Kirchhoff's laws and

Ohm's law of circuits. Although the network is linear, however, the given demands render the production of voltage and current, so power flow calculation boils down to solving nonlinear equations. Power flow is the most fundamental problem in power system operation, which will be detailed in Chap. 2.

Specialized power systems, such as those in aircrafts, electric ships, microgrids, and some of transmission lines, may rely upon direct current power delivery. They are not the main focus of this book, but can be active research topics. See [12–14] for reviews on direct current power system technologies.

1.2.2 Natural Gas System

In most countries or regions, the gas wells where natural gas is mined are distant from populated areas where it will be consumed. Therefore, natural gas resources are transported from its production sites to load centers either through a pipeline network or by trains and ships in its liquefied form. This book will mainly focus on the former approach.

Similar to the electric power system, natural gas pipeline networks can be distinguished into transportation networks and distribution networks according to their functionalities and gas pressures. Gas transportation networks are filled with high pressure gas and deliver the resource thousands of kilometers from gas wells to where it is needed for consumption or storage. Gas distribution networks are operated under low pressures. They distribute natural gas delivered from the transmission network to end consumers, e.g., city gas stations, industries, households, underground storage facilities, etc.

When gas moves through a pipeline, its pressure drops due to friction. Thus, maintaining constant gas flow requires a certain pressure gradient along pipes. The compressor is an indispensable component in natural gas systems to step up gas pressures in the pipelines. Compressors are installed commonly at 50–100 mile intervals. The operation of a compressor consumes certain amounts of energy, either gas or electricity. Proper placement of compressors could significantly decrease the operation cost and improve the reliability of gas supply. Valves are used to isolate faulted branches and protect the system under abnormal conditions. A sketch of a gas pipeline network is illustrated in Fig. 1.5.

Gas flow analysis is to determine the mass flow rate in every pipeline and gas pressure at every node. Such an analysis allows prediction of the behavior of gas network systems in response to different load conditions, and can be used to guide real system design and operation informatively. Steady-state analysis assumes the flow status does not change over time, and the network model is described by a set of nonlinear equations derived from Kirchhoff's laws as well as pressure loss equations. In transient flow analysis, the network model is described by partial differential equations, and gas flows are functions of time. This book will be primarily focused on the steady-state analysis. The transient behavior will be

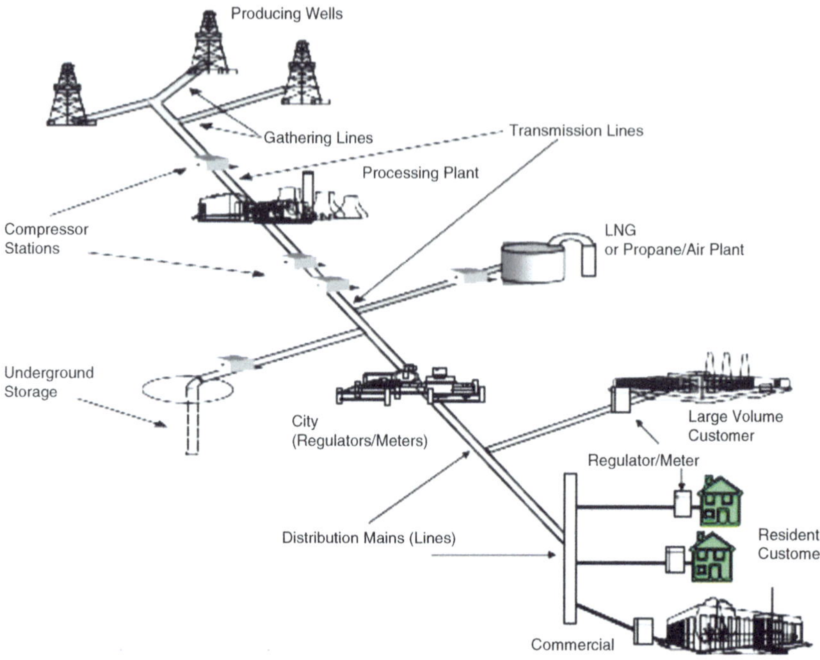

Fig. 1.5 A sketch of the natural gas system [15]

simplified and approximated by algebraic equations for ease of analysis. The gas flow model in the pipeline network will be portrayed in Chap. 3.

1.2.3 District Heating System

A district heating system is an infrastructure for delivering thermal energy from heat sources to loads via a pipeline network carrying steam or hot water. The history of district heating systems is reviewed in [16]. A sketch of the district heating system is outlined in Fig. 1.6.

The heat sources can be combined heat-power units, heat pumps, electric boilers, etc. Future district heating will utilize surplus sources from various industrial processes, geothermal and solar thermal plants.

A district heating network is comprised of symmetric supply pipes and return pipes. On the supply side, water is heated by the source with a temperature between 70 °C and 90 °C; on the return side, water flows back to the source with a temperature between 30 °C and 70 °C. The heat is transferred to the consumer through a heat exchanger. A circulation pump driven by electricity is located near the source to maintain a certain pressure between supply and return pipes, so as to direct the water flow in the network.

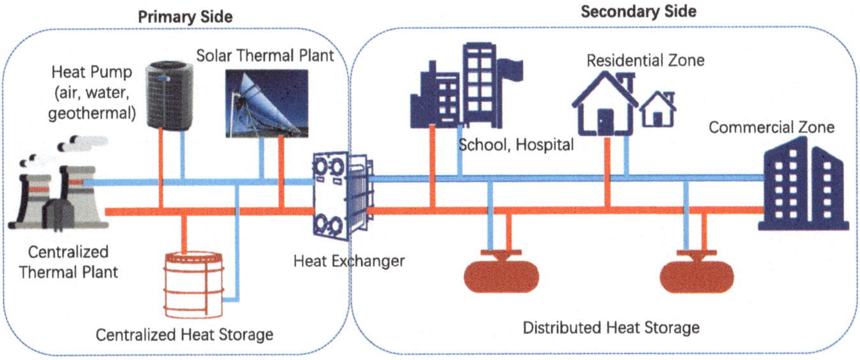

Fig. 1.6 A sketch of the district heating system

Hydraulic and thermal analysis should be conducted to calculate the operating status of the district heating network. The hydraulic analysis is to determine the mass flow rates and pressure drops in the individual pipes in the network. The hydraulic model includes flow continuity condition (in analogy to Kirchhoff's current law), loop pressure condition (in analogy to Kirchhoff's voltage law), and head loss equation (in analogy to Ohm's law). The thermal analysis is to determine the supply and return temperatures at each node due to heat exchange and dissipativity (net losses). Certainly, hydraulic and thermal conditions are tightly coupled. Hydraulic and thermal models of the district heating network will be detailed in Chap. 4.

1.2.4 Urban Transportation System

Transportation is one of the largest energy consumption sectors, accounting for about one third of the total energy demands. Road transport consumes approximately 70% of the energy used by the global transportation system [17]. Unlike three systems discussed above, a transportation system neither directly produces primary or secondary energy resources nor distributes them via networked components. However, it can be useful to investigate the geographical distribution of fuel demands so as to make wise urban planning decisions. In addition, the proliferation of electric vehicles and wide deployment of on-road charging facilities are creating notable interplays across the transportation system and the power distribution system, and precipitating electrification of the transportation sector. To study the fuel and electricity demands in the transportation system, it is imperative to characterize the traffic flow distribution, which is known as the traffic assignment problem.

The traffic flow pattern over a congested urban transportation network depends on two interactive factors. On the one hand, every motorist attempts to choose the route with minimal expense. It is believed that the travel time is often the primary concern of drivers. On the other hand, the travel expense of a particular traveler may also depend on the route choices of others. Take the travel time for example, the fastest route corresponds to the shortest route without congestion. However, an urban street has limited capacity compared with the traffic flow it carries, and the travel time increases when the traffic flow exceeds a certain limit due to congestion. In this regard, traveling on the shortest path may not receive the shortest travel time. Therefore, it is not sufficient to determine the network flow pattern using only geographical information. If the travel costs, such as congestion tolls and payments for battery recharging, are taken into account, the problem would become more complicated.

In summary, the force which directs the traffic flow over a transportation network to reach a reasonable steady-state pattern is the rationalities of road users. If the travel expenses of individual users cannot be further reduced by altering the route choice unilaterally, the traffic flow will reach a stable pattern. Mathematical descriptions and computation methods of this user equilibrium pattern will be elaborated in Chap. 5.

1.3 Transition to the Integrated Energy System

The growing demands for energy resources amid strict carbon dioxide reduction targets require revolutionary changes in the energy sector all over the world. The past decades have witnessed a continuous transition from conventional fossil fuel dominated power generation to hybrid energy production with low-carbon facilities or even zero-carbon renewable energies in the power supply infrastructure. Wind turbines and photovoltaic (PV) panels offer variable power which is difficult to predict accurately. Therefore, they are used in complementarity with other energy sources (e.g., hydropower) as well as storage devices to offer a smooth power supply. As the penetration level of volatile resources increases, additional measures such as flexible generation and demand response programs are in great need to provide a secure balance between supply and demand.

In addition to the measures taken locally in the electric power generation, a deep integration of available energy forms could be a promising option for high-efficiency energy utilization. For example, coupling the electrical power grid, natural gas network, district heating/cooling network, and river network gives a more flexible and economical energy system. The main target of such interdependent networks is to provide efficient and flexible energy supply while achieving emission reduction by utilizing the synergies of different energy forms and optimizing their interactions in operation. Excellent overviews on the benefits, challenges, opportunities, and research directions can be found in [18, 19]. A conceptual sketch of an integrated energy system is illustrated in Fig. 1.7. It highlights the tightly interdependent energy flows intertwined at the coupling components, and demonstrates the central role of electricity as the core energy carrier.

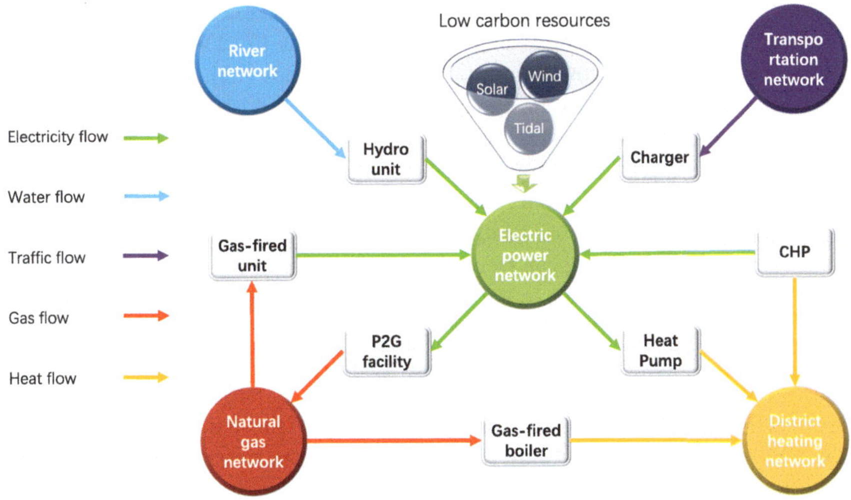

Fig. 1.7 Illustration of system-level interdependency

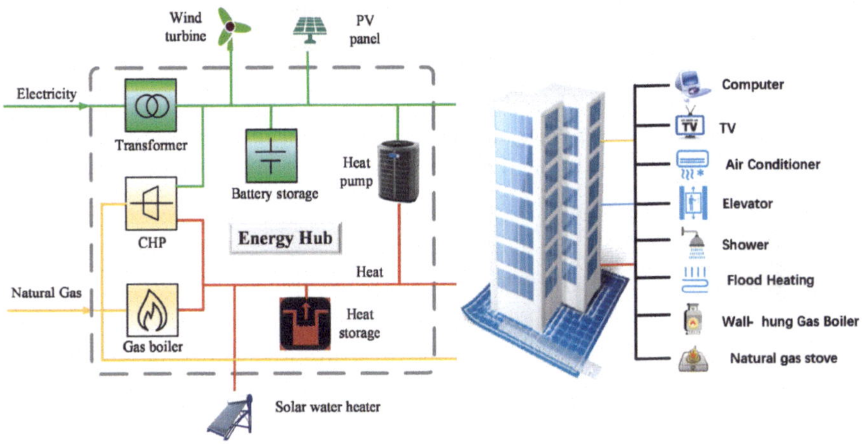

Fig. 1.8 Energy hub at the demand side

Apart from the system-level integration, different energy carriers, such as electricity, natural gas, and heat, can be collected in a so-called energy hub [20], which is portrayed in Fig. 1.8, and re-directed to supply household demands at the demand side. In the hub station, natural gas can be used by a combined heat and power (CHP) unit and a boiler to produce electricity and heat. Electricity to heat conversion is enabled by a heat pump. Electricity to gas is also possible with the installation of power-to-gas (P2G) equipment. Such an integrated facility can be described by a multi-input multi-output transfer matrix in an optimal energy flow study [21].

The major motivation for this combination is the tie between electricity and heat demands. Take Europe as an example, in 2009, about 175 TWh of electricity is used to cover 7% of the total domestic heating demands [22]. This correlation is expected to be intensified in the future. On the one hand, state-of-the-art heating devices, heat pumps, have coefficients of performance around $3 \sim 4$, which means an up to 66–75% electricity saving for generating the same volume of heat, making electrification of heat provision a competitive option. On the other hand, shifting to renewable energy based electric heating can help realize decarbonization in heating supply. Since thermal energy can be easily stored, the volatility of renewable power would not significantly challenge the operation of a heating system. In other words, the heating system can play the role of energy storage to mitigate fluctuations of renewable output. However, thermal energy is expensive to transmit across a long distance, so a district heating network is typically not very large, and heat-electricity integration is usually seen at the distribution level or on the demand side.

Gas-electricity integration may also occur at the transmission level. The interconnected electric power system and inter-state natural gas transmission network are comparable in their scales. During the past decade, the natural gas price has declined dramatically due to the shale rock revolution [23], leading to the wide deployment of gas-fired units. Meanwhile, thanks to the emerging P2G technology [24], excessive electricity can be used to produce gases, such as hydrogen and methane, which can be injected back into the gas network or stored in existing tanks. The gas-fired units and P2G facilities greatly enhance the ability to compensate renewable generation uncertainty, and also create stronger interdependency across gas and electricity systems through bi-directional energy conversions.

1.4 Benefits and Challenges

There is an increasing recognition that the diversity of energy reliance and deep integration of energy systems will lead to vast benefits [25]. Important implications are summarized as follows.

1. Improving System Operating Flexibility

Integrated energy systems produce, deliver, and distribute different forms of energy through multiple channels. Coordinated operation of multiple resources adds additional freedom to satisfy multiple demands without violating security constraints. From a mathematical perspective, the feasible operating region of the integrated energy networks with multiple carriers is larger than the Cartesian product of the individual feasible regions of the constitutors. As a result, when projected back to the strategy spaces of individual subsystems, the feasible region of each subsystem is enlarged, so the operation becomes more flexible. Attempts which use geometric approaches to characterize flexibility of the integrated energy system are found in [26, 27].

Such flexibility provides effective means to tackle congestions and other security issues in heavily loaded systems and may even postpone network reinforcements which are usually costly. Image a power transmission line connecting a power station and a city, which is congested and needs upgrade. If an existing gas pipeline is available to use, a cost-saving option to resolve the congestion problem is to build a P2G facility at the power station and a gas-fired unit near the city. The fuel is delivered from the gas pipeline. This option is attractive for renewable energy harnessing because gas is compressible and the linepack effect naturally provides certain storage capacity without additional investments.

2. Enhancing the Overall Energy Efficiency

One primary driver for the improvement on energy efficiency is the co-generation of multiple resources. Take heat and power co-generation systems as an example, such a combination has been widely acknowledged to have high energy and exergy efficiency compared to the isolated heat and power production [28]. Figure 1.9 shows an example taken from [25] to elucidate energy saving and carbon emissions reduction in a specified scenario. The respective input energy and carbon dioxide emission when supplying electricity and heat from a co-generation system and independent production of electricity and heat via a thermal power plant and a gas-fired boiler are compared.

In the separate generation mode, the efficiency is low, because a large portion (45% in this example) of the fuel energy is emitted to the ambient through the condenser cooling systems. In the co-generation mode, the useless energy in electricity production is captured and used by the heat generation system, for the purpose of water or space heating. This example shows 18.7% fuel energy saving by

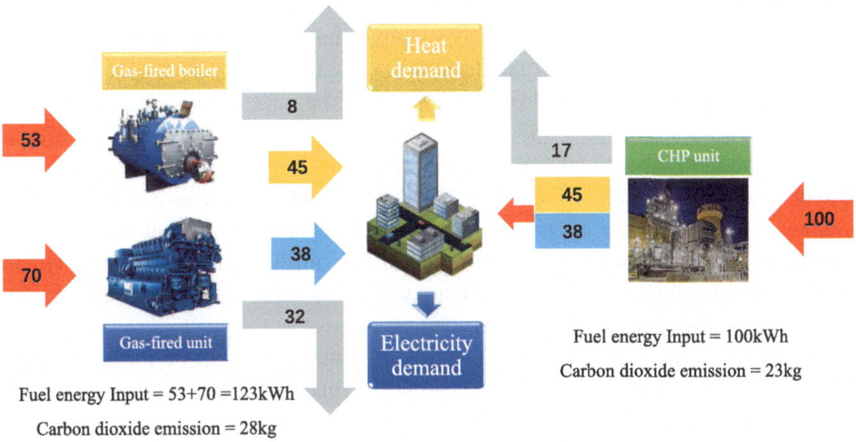

Fig. 1.9 Comparison of energy consumptions and carbon emissions of independent generation and combined heat-power generation

the co-generation system to meet the same electricity and heat loads. Furthermore, the carbon dioxide emission is also declined by 17.8%.

Multi-generation such as combined cooling, heat, and power generation has the potential for further increasing the energy efficiency, according to the energetic and exergetic analysis in [29]. In many situations, the occurrence of heat and cooling demands in different seasons may hide the advantage of the tri-generation system mentioned above. Nevertheless, the connection of local energy networks as well as the seasonal heat storage technology [30, 31] may create adequate energy demands scattered over wide geographical regions and long time horizons.

3. Boosting the Usage of Renewable Energies
Wind and solar power is volatile, and backup capacity is needed to balance real-time discrepancy between supply and demand in real time. Penetration levels of renewables and economical benefits may be hindered by intensive investments on large-scale energy storage facilities. When the security of power system is threatened, the operator will impose an upper bound on the production level of renewable power, and the excessive generation will be curtailed.

There is a great potential to increase the utilization of renewable energy by exploiting the storage/regulation capacity offered by energy system integration. Excessive power is used to produce gas and heat which are directly consumed or can be easily stored, and the utilization rates of wind and solar generation assets increase remarkably. It is unnecessary to convert them back to electricity due to efficiency loss. Such technologies are termed power-to-X (P2X) in recent literature [32, 33].

4. Increasing Reliability and Resilience of the Entire Network
To guarantee desired functionality in certain abnormal conditions, any physical infrastructures are built with a certain level of redundancy to avoid potential system failures. As for the power system, it is required that the system should remain in operation after the failure of a single network element, such as a generator or a transmission line. The integration of energy systems with multiple carriers offers additional choices to supply the demand when unexpected contingencies in one system occur, and hence increases the reliability and resilience of the entire network if managed properly [34, 35]. However, when the loads shift from one network to another, the latter will be operated in a more stressed condition. In this regard, the prominent interdependency across different energy infrastructures may also create the risk of cascading failures. This phenomenon has been reported in [36–38] for more general interconnected systems.

Despite the benefits mentioned above, the integration of multiple physical systems with different energy carriers also poses fundamental challenges to implement the coordinated design and operation of the interdependent energy networks. Important difficulties are summarized here.

1. High Interdependency Amid Heterogeneous Energy Flows
The increasing interactions among different energy systems with heterogeneous carriers allow making full use of complementarities of resources. However, the first step to realize an efficient operation is to understand the interdependency brought by heterogeneous energy flows which are governed by different physical laws and intertwine through conversion devices. Managing such interactive systems requires elaborated actions to maintain security and reach a social optimum. Therefore, research and project development will call for knowledge in different engineering disciplines, such as thermodynamics, fluid dynamics, and power system analysis.

2. Increased Complexity in Cyber and Physical Infrastructures
Integration of multiple energy systems not only adds complexity in the physical infrastructure, but also entails the cutting-edge technologies in communication and internet, as well as powerful hardware for managing big data, creating a so-called cyberphysical system [39]. The interdependency between data flow and energy flow will be complicated, even only in the electric power system [40]. Innovative models and methods are required to analyze the cyberphysical security and resilience of the networked energy infrastructures with multiple carriers. Nevertheless, cyberphysical related issue is not the main focus of this book.

3. Institutional Barriers for Convenient Market Organization and Energy Trading
The regulatory policies in the involved systems differ dramatically due to the physical features of their respective energy carriers, resulting in a fragmented operational framework in which no single stakeholder is responsible for the improvement of the overall system performances. The market organizations can be extremely different, too. For example, the current electricity market and natural gas market are cleared at different times with different frequencies. Synchronizing issues are discussed in [41]. Distinctions also exist in the pricing policies. The electricity price is set in accordance with the locational marginal price [42] extracted from the Lagrangian dual multipliers of an optimal power flow problem, while the natural gas and heating prices are mainly driven by supply and demand fundamentals and remain unchanged in the time scale of power system generation scheduling, usually 1 day. The similarities between the deregulated electricity market and heat market as well as marginal pricing of heat are discussed in [43]. Researchers also suggest exergy-based heat pricing schemes [44]. A thorough review on heat pricing approaches in district heating systems can be found in [45]. In summary, new business modes that promote collaboration among applicable authorities without dramatically increasing operation complexity need further exploration. Only by breaking the institutional barrier and creating appropriate business models for convenient energy trading can the potential benefits of system integration be fully realized.

1.5 Focus of This Book

Given the tendency towards a more integrated energy industry, significant interests and attention have been attracted from academics, industry, and policy makers around the world to resolve the theoretical, technical, and marketing issues to leverage synergies in different energy forms and avoid any negative impacts. Among the vast problems to be resolved, this book mainly focuses on the steady-state operational issues, including the following

1. Modeling Network Energy Flows
The energy flows describe the steady-state operating condition of a system which does not change over time. It underlies many applications such as capacity expansion planning, economic operation, and security assessment. The individual network flow models of the electrical power system, natural gas pipeline system, district heating system, and urban transportation system are introduced in the second sections in Chaps. 2–5, respectively.

2. Interdependency Analysis and Optimal System Operation
Integrated modeling of coupled energy flows is the main theme of this book. The interdependencies investigated in this book can be categorized by the involved systems. Electricity and gas coupling is addressed in Chap. 3; Electricity and district heating coupling is discussed in Chap. 4; Electricity and transportation coupling is detailed in Chap. 5. The optimal operation of coupled networks in each category is elaborated in the respective chapters based upon the developed optimal energy flow models, combined with state-of-the-art optimization methods.

3. Market Design and Equilibrium Computation
Due to the presence of strategic interactions among different stakeholders and energy flows, there is a clear need for innovation in trading policy design and developing analytic methods to evaluate market equilibrium points. This book investigates interdependent markets with locational marginal energy pricing-based transactions. Market equilibrium models and computational approaches for the three categories of coupled networks are developed in Chaps. 3–5, respectively.

4. Advanced Optimization Methods for Integrated Energy Systems
Besides above emerging issues, another purpose of this book is to disseminate advanced optimization methods which facilitate resolving broader classes of decision-making problems in the integrated energy system. We provide comprehensive and thorough tutorials on four widely used approaches in Appendices A–D. Convex optimization allows solving large-scale problems reliably and efficiently; Mixed-integer linear reformulation provides a systematic mean to globally solve certain classes of non-convex programs; Robust optimization deals with uncertainties; and equilibrium program models complex interactions among multiple stakeholders.

We provide rigorous mathematical foundations for these useful modeling techniques and explain how to solve them leveraging off-the-shelf solvers. To complement previous chapters, we use compact formations in appendices which are much easier to follow. In this way, we hope this book will be able to close the gap between theorists and practitioners in the research field of energy system optimization.

1.6 Summary

Management, operation, and marketization of the highly-integrated multi-carrier energy system entail sophisticated optimization models which will simultaneously capture the interaction and competition among its components including the electric power, natural gas, district heating, and urban transportation networks. This book will develop advanced computational methodologies to address the operational and marketing issues in interdependent energy networks.

References

1. Zhang, H.L., Baeyens, J., Degrève, J., Cacères, G.: Concentrated solar power plants: review and design methodology. Renew. Sust. Energy Rev. **22**, 466–481 (2013)
2. International Energy Agency: Technology roadmap: solar photovoltaic energy. Tech. Rep. Available at: http://www.iea.org/publications/freepublications/publication/
3. Global Wind Energy Council: Global wind report: annual market update. Tech. Rep. Available at: https://www.gwec.net/wp-content/uploads/vip/GWEC-Global-Wind-2015-Report_April-2016_19_04.pdf (2015)
4. International Energy Agency: World Energy Outlook. Tech. Rep. (2015)
5. Hui, S., Zhong, W., Lee, C.K.: A critical review of recent progress in mid-range wireless power transfer. IEEE Trans. Power Electron. **29**(9), 4500–4511 (2014)
6. Dincer, I., Cengel, Y.A.: Energy, entropy and exergy concepts and their roles in thermal engineering. Entropy **3**(3), 116–149 (2001)
7. https://en.wikipedia.org/wiki/Electrical_grid
8. https://en.wikipedia.org/wiki/Ultra-high-voltage_electricity_transmission_in_China
9. Palizban, O., Kauhaniemi, K., Guerrero, J.M.: Microgrids in active network management-Part I: hierarchical control, energy storage, virtual power plants, and market participation. Renew. Sust. Energy Rev. **36**, 428–439 (2014)
10. Palizban, O., Kauhaniemi, K., Guerrero, J.M.: Microgrids in active network management-Part II: system operation, power quality and protection. Renew. Sust. Energy Rev. **36**: 440–451 (2014)
11. Ruiz-Romero, S., Colmenar-Santos, A., Mur-Pérez F., Lòpez-Rey, A.: Integration of distributed generation in the power distribution network: the need for smart grid control systems, communication and equipment for a smart city-use cases. Renew. Sust. Energy Rev. **38**, 223–234 (2014)
12. Justo, J., Mwasilu, F., Lee, J., Jung, J.: AC-microgrids versus DC-microgrids with distributed energy resources: a review. Renew. Sust. Energy Rev. **24**, 387–405 (2013)
13. Planas, E., Andreu, J., Gárate J., de Alegria, I., Ibarra, E.: AC and DC technology in microgrids: a review. Renew. Sust. Energy Rev. **43**, 726–749 (2015)

14. Salomonsson, D., Sannino, A.: Low-voltage DC distribution system for commercial power systems with sensitive electronic loads. IEEE Trans. Power Deliv. **22**(3), 1620–1627 (2007)
15. Onwukwe, M., Duru, U., Ikpeka, P. Natural gas utilization through domestic gas distribution in Nigeria. Journal of Advanced Research in Petroleum Technology & Management, 2015, **1**(1), 1–7.
16. Lund, H., Werner, S., Wiltshire, R., Svendsen, S., Eric Thorsen, J., Hvelplund, F., Vad Mathiesen, B.: 4th generation district heating (4GDH): integrating smart thermal grids into future sustainable energy systems. Energy **68**, 1–11 (2014)
17. Urban Transport and Efficiency. Tech. Rep. Available at: http://www2.giz.de/wbf/4tDx9kw63gma/SUT_module5h.pdf
18. Wu, J., Yan, J., Desideri, U., et al.: Synergies between energy supply networks. Appl. Energy **192**, 263–267 (2017)
19. Abeysekera, M., Wu, J., Jenkins, N.: Integrated energy systems: an overview of benefits, analysis methods, research gaps and opportunities. Tech. Rep. Available at http://www.hubnet.org.uk/filebyid/791/InteEnergySystems.pdf
20. Geidl, M., Koeppel, G., Favre-Perrod, P., Klockl, B., Andersson, G., Frohlich, K.: Energy hubs for the future. IEEE Power Energy Manag. **5**(1), 24–30 (2007)
21. Geidl, M., Andersson, G.: Optimal power flow of multiple energy carriers. IEEE Trans. Power Syst. **22**(1), 145–155 (2007)
22. Roadmap 2050: A practical guide to a prosperous low-carbon Europe. Tech. Rep. European Climate Foundation (2010)
23. Rognerud, E.: The impact of the shale gas revolution. Tech. Rep. Available at: http://www.wgei.org/wp-content/uploads/2015/11/The-shale-gas-revolution_Eli-W-Rognerud-Aug-2015.pdf
24. Schiebahn, S., Grube, T., Robinius, M., et al.: Power to gas: technological overview, systems analysis and economic assessment for a case study in Germany. Int. J. Hydrogen Energy **40**(12), 4285–4294 (2015)
25. Abeysekera, M.: Combined analysis of coupled energy networks. Doctor Dissertation, Cardiff University (2016)
26. Pan, Z., Guo, Q., Sun, H.: Feasible region method based integrated heat and electricity dispatch considering building thermal inertia. Appl. Energy **192**, 395–407 (2017)
27. Zhao, L., Zhang, W., Hao, H., Kalsi, K.: A geometric approach to aggregate flexibility modeling of thermostatically controlled loads. IEEE Trans. Power Syst. **32**(6), 4721–4731 (2017)
28. Horlock, J.H.: Cogeneration: Combined Heat and Power, Thermodynamics and Economics. Pergamon, Oxford (1987)
29. Cho, H., Smith, A.D., Mago, P.: Combined cooling, heating and power: a review of performance improvement and optimization. Appl. Energy **136**, 168–185 (2014)
30. Xu, J., Wang, R.Z., Li, Y.: A review of available technologies for seasonal thermal energy storage. Solar Energy **103**, 610–638 (2014)
31. Hesaraki, A., Holmberg, S., Haghighat, F.: Seasonal thermal energy storage with heat pumps and low temperatures in building projects-a comparative review. Renew. Sust. Energy Rev. **43**, 1199–1213 (2015)
32. Lund, P.D., Lindgren, J., Mikkola, J., Salpakari, J.: Review of energy system flexibility measures to enable high levels of variable renewable electricity. Renew. Sust. Energy Rev. **45**, 785–807 (2015)
33. Flexibility concepts for the German power supply in 2050: ensuring stability in the age of renewable energies. Tech. Rep. Berlin, Germany (2016)
34. Fu, X., Guo, Q., Sun, H., Zhang, X., Wang, L.: Estimation of the failure probability of an integrated energy system based on the first order reliability method. Energy **134**, 1068–1078 (2017)
35. Manshadi, S.D., Khodayar, M.E.: Resilient operation of multiple energy carrier microgrids. IEEE Trans. Smart Grid **6**(5), 2283–2292 (2015)
36. Buldyrev, S.V., Parshani, R., Paul, G., Stanley, H.E., Havlin, S.: Catastrophic cascade of failures in interdependent networks. Nature **464**, 1025–1028 (2010)

37. Rinaldi, S.M., Peerenboom, J.P., Kelly, T.K.: Identifying, understanding, and analyzing critical infrastructure interdependencies. IEEE Control Syst. **21**(6), 11–25 (2001)
38. Little, R.G.: Controlling cascading failure: understanding the vulnerabilities of interconnected infrastructures. J. Urban Technol. **9**(1), 109–123 (2002)
39. Sandberg, H., Amin, S., Johansson, K.H.: Cyberphysical security in networked control systems: an introduction to the issue. IEEE Control Syst. **35**(1), 20–23 (2015)
40. Ashok, A., Govindarasu, M., Wang, J.: Cyber-physical attack-resilient wide-area monitoring, protection, and control for the power grid. Proc. IEEE **105**(7), 1389–1407 (2017)
41. Weigand, P., Lander, G., Malme, R.: Synchronizing natural gas and power market: a series of proposed solutions. Tech. Rep. (2013)
42. Ott, A.L.: Experience with PJM market operation, system design, and implementation. IEEE Trans. Power Syst. **18**(2), 528–534 (2003)
43. Gebremedhin, A., Moshfegh, B.: Modelling and optimization of district heating and industrial energy system-an approach to a locally deregulated heat market. Int. J. Energy Res. **28**(5), 411–422 (2004)
44. Poredoš, A., Kitanovski, A.: Exergy loss as a basis for the price of thermal energy. Energy Convers. Manag. **43**(16), 2163–2173 (2002)
45. Li, H., Sun, Q., Zhang, Q., Wallin, F.: A review of the pricing mechanisms for district heating systems. Renew. Sust. Energy Rev. **42**, 56–65 (2015)

Chapter 2
Electric Power System with Renewable Generation

2.1 Introduction

Electricity is ubiquitously used in the modern society. Its availability depends on the reliable operation of the most complicated man-made creation: the electric power system, where electrical energy is produced by generators, transported through transmission lines, and distributed to load centers. The system demand varies over time, and the total generation must be maintained equal to the demand on a moment-by-moment basis. If the balance is lost, the frequency of alternating current (AC) power delivery deviates from the reference value, and the quality and security of power supply is compromised, because most components in a power system can only bear a narrow range of frequency deviation.

In power system operation, load balancing is realized through a meticulous generation schedule in multiple time scales: day-ahead balancing through unit commitment; hourly-ahead balancing through economic dispatch; balancing across 5 min through automatic generation control, balancing across seconds through transient control of generator governors, and the remaining discrepancy translates into system frequency fluctuations. Dispatchable power plants, such as thermal and hydro units, account for a substantial share in traditional power systems.

During the past decades, the advantages of renewable generation as clean and low-cost energy resources have inspired the dramatic integration of wind and solar energy into power systems. Unlike conventional units whose generation capacity is constant, the maximum output of wind turbines and PV panels is affected by the wind speed and solar irradiance intensity, which are beyond the control of system operators, and thus challenges the operation of power systems. If there is not adequate backup of balancing capability, it may become inevitable to curtail excessive renewable energy or shed a fraction of load to ensure a secure operation.

Many countries allow trading electricity in power markets. Energy transactions are determined in various short-term and long-term trading plans. For example, the day-ahead market determines the unit commitment of coal-fired units due to their

W. Wei, J. Wang, *Modeling and Optimization of Interdependent Energy Infrastructures*, https://doi.org/10.1007/978-3-030-25958-7_2

slow response. The intra-day markets allow energy/reserve trading one or a few hours before energy delivery. The real-time market provides the last opportunity to narrow discrepancy between generation and consumption half or one hour before energy delivery. With time rolling on, renewable producers are becoming more certain about their production levels, which would generally reduce frequency volatility at the real-time stage.

To realize the most efficient utilization of generation assets while guaranteeing operating security, optimization methods are applied to a wide range of problems covering time frames ranging from seconds to years. However, most conventional models are based on exact load and renewable power forecast and may fail to provide useful strategies without accurate information on system state. The operation and economics of an electric power system with increasing penetrations of volatile renewable energy resources are significantly different from those with only deterministic and controllable facilities. To accommodate high shares of variable generation, flexible backup resources are indispensable to compensate for the real-time mismatch between generation and load, which introduces additional costs.

This chapter will mainly focus on the operational-level issues in electric power systems with stochastic renewable generation in the time frame of 1 day, and elucidate models and approaches for three most fundamental problems in power system operation. The first one is the optimal power flow (OPF) problem, which is solved every few minutes with real-time measurements on system loads and renewable generations, and determines steady-state distribution of bus voltages and line power flows under the most cost-effective energy production pattern. The second one is the economic dispatch, which is solved every one or a few hours and provides generation and reserve schedules for the upcoming periods. The third one is the unit commitment (UC), which is used to clear a day-ahead power market and determine on-off status of generators in each time slot. Volatility and intermittency of renewable generation are considered in the latter two problems. Their connection is shown in Fig. 2.1.

In fact, even in the deterministic setting, optimization problems associated with large-scale power networks are challenging to solve. One reason rests on the fact

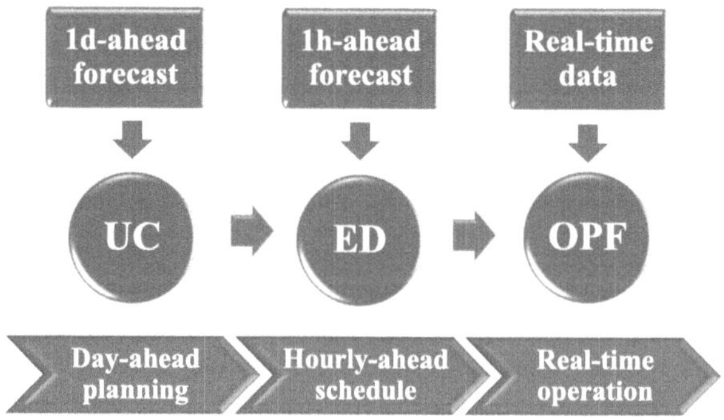

Fig. 2.1 Fundamental issues in power system daily operation

that power flow equations for the power transmission network are characterized by quadratic equalities which are non-convex and make it difficult to find a global optimal solution in polynomial time. Another is that the deterministic equivalence of optimization problems involving uncertainties has much larger sizes and high complexity than the original one. To cope with the volatile output of renewable units in a data-driven (distribution-free) and more tractable manner, most approaches discussed in this chapter rely on the theory of robust optimization. These techniques and methods help system operators make better decisions and manage their systems more reliably and efficiently under increased system scales, renewable volatility, and economic/environmental pressures.

This chapter is organized as follows. Basic power flow equations in the bus injection format and branch flow format as well as their linear approximations are introduced in Sect. 2.2, and will be used throughout the book. The convex relaxation methods for the ACOPF problem and three variations are discussed in Sect. 2.3. Because the OPF problem corresponds to the real-time operation, and renewable output has already been observed or can be predicted accurately, no uncertainty is taken into account. Co-optimization of energy and reserve dispatch, a special economic dispatch problem under uncertainty, is addressed via traditional robust optimization and distributionally robust optimization methods in Sect. 2.4. This approach provides robust dispatch strategies against target uncertainty and ambiguity sets. Dispatchability is a reverse problem: given the generation schedule and reserve offer, how much uncertainty can be dealt with. The theory of describing and optimizing renewable power dispatchable regions in joint energy and reserve dispatch is thoroughly developed in Sect. 2.5. Unit commitment is studied in Sect. 2.6, in which the traditional robust optimization is applied to ensure the invariance of UC decisions and network security under wind generation uncertainty. Conclusions are drawn in Sect. 2.7. Further readings are suggested in Sect. 2.8.

2.2 Power Flow Model

Power flow is also known as load flow. It is a steady-state network solution which depicts bus voltages as well as line power flows in response to a given set of loads and generations. As all lines are described by constant impedances, the bus voltage and nodal current injection have a linear dependence. However, the power flow equations are nonlinear because it is the nodal power injections, rather than the current, are specified, and the relationship between voltage (or current) and power is nonlinear. Power flow equations play an essential role in power system operation, and can be found in any elementary textbook of power system analysis, such as [1] and [2]. An open-source toolbox for calculating power flow is available at [3]. Although power flow models and algorithms are common knowledge among power system researchers, to make this book self-contained, in this section we introduce four pervasive power flow models which will be frequently used in the rest of this book.

2.2.1 Bus Injection Model

1. AC Power Flow Formulation

Symbols and notations used in this formulation will be defined first.

e_i	Real part of complex voltage at bus i, $e = [e_1, e_2, \ldots, e_n]^T$
f_i	Imaginary part of complex voltage at bus i, $f = [f_1, f_2, \ldots, f_n]^T$
i	Unit imaginary number, $\mathrm{i} = \sqrt{-1}$
$\dot{I}_i$	Complex injection current at bus i, $\dot{I} = [\dot{I}_1, \dot{I}_2, \ldots, \dot{I}_n]^T$
E_L	The set of lines
E_N	The set of buses with cardinality $\lvert E_N \rvert = n$
p_i^g	Active power generation at bus i, $p^g = [p_1^g, p_2^g, \ldots, p_n^g]^T$
p_i^d	Active power demand at bus i, $p^d = [p_1^d, p_2^d, \ldots, p_n^d]^T$
P_i^N	Net active power injection at bus i, $P^N = [P_1^N, P_2^N, \ldots, P_n^N]^T$
q_i^g	Reactive power generation at bus i, $q^g = [q_1^g, q_2^g, \ldots, q_n^g]^T$
q_i^d	Reactive power demand at bus i, $q^d = [q_1^d, q_2^d, \ldots, q_n^d]^T$
Q_i^N	Net reactive power injection at bus i, $Q^N = [Q_1^N, Q_2^N, \ldots, Q_n^N]^T$
S_i^N	Net complex power injection at bus i, $S^N = [S_1^N, S_2^N, \ldots, S_n^N]^T$
$\dot{V}_i$	Complex voltage at bus i, $\dot{V}_i = e_i + \mathrm{i} f_i$, $\dot{V} = [\dot{V}_1, \dot{V}_2, \ldots, \dot{V}_n]^T$
V_i	Magnitude of complex bus voltage $\dot{V}_i$
Y	Bus admittance matrix, $Y \in \mathbb{C}^{n \times n}$, its diagonal entry Y_{ii} is the sum of all branch admittances connecting to bus i, and off-diagonal element Y_{ij} equals the negative of branch admittance linking buses i and j, or zero if there is no connection line between buses i and j. Matrices $G = \mathrm{Re}(Y)$, and $B = \mathrm{Im}(Y)$.

For a general network with n buses, according to basic principles of circuits we have $Y\dot{V} = \dot{I}$, or in an element-wise form

$$
\begin{aligned}
& Y_{11}\dot{V}_1 + Y_{12}\dot{V}_2 + \cdots + Y_{1n}\dot{V}_n = \dot{I}_1 \\
& Y_{21}\dot{V}_1 + Y_{22}\dot{V}_2 + \cdots + Y_{2n}\dot{V}_n = \dot{I}_2 \\
& \qquad \cdots\cdots \\
& Y_{n1}\dot{V}_1 + Y_{n2}\dot{V}_2 + \cdots + Y_{nn}\dot{V}_n = \dot{I}_n
\end{aligned}
\tag{2.1}
$$

Giving a net injection vector $S^N = P^N + iQ^N$, the current variable can be eliminated by substituting

$$\dot{I}_i = \frac{P_i^N - \mathrm{i}Q_i^N}{\hat{V}_i}, \ i = 1, 2, \ldots, n \tag{2.2}$$

into (2.1), resulting in

$$P_i^N - \mathrm{i}Q_i^N = \hat{V}_i \sum_{j=1}^{n} Y_{ij}\dot{V}_j, \ i = 1, 2, \ldots, n \tag{2.3}$$

In rectangular coordinates, the complex voltage of bus i can be expressed by $\dot{V}_i = e_i + \mathrm{i} f_i$. Equaling the real and imaginary parts on both sides of (2.3) respectively, we arrive at two equations for each bus as follows:

$$P_i^N = \sum_{j=1}^{n} [G_{ij}(e_i e_j + f_i f_j) - B_{ij}(e_i f_j - e_j f_i)],\ i = 1, 2, \ldots, n \tag{2.4a}$$

$$Q_i^N = \sum_{j=1}^{n} [-B_{ij}(e_i e_j + f_i f_j) - G_{ij}(e_i f_j - e_j f_i)],\ i = 1, 2, \ldots, n \tag{2.4b}$$

where P_i^N, Q_i^N, e_i, and f_i at each bus are to be determined. To solve the power flow equations (2.4), two variables at each bus should be given. According to the power system operating conditions, the buses can be classified into three types, and two of the four variables are fixed in each type.

1. Slack bus ($V\theta$ bus): Some observations inspire the definition of slack bus. First, the total loss on transmission lines is not known in advance, therefore the nodal power injection of at least one bus should be adjustable to balance the generation and demand. Second, it is necessary to specify a reference angle in order to define real and imaginary parts of bus voltages. Mathematically, the voltage magnitude and angle at the slack bus are specified, i.e., $e_i = V$ and $\theta_i = 0$. Physically, a large generator will be placed at the slack bus to maintain a constant voltage magnitude and compensate discrepancy between generation and load (plus losses). The slack bus is unique in the traditional power flow model. In practice, multiple slack buses could exist, so that frequency-regulation generators jointly undertake the power mismatch according to some protocols, such as the unit participation factor rule. Please refer to [4, 5] for more information on power flow models with distributed slack bus.
2. PQ bus: This type of bus usually connects to a load ($p_i^g = 0, q_i^g = 0$). As its name implies, the active and reactive power injection (demand) p_i^d and q_i^d are given at a PQ bus, and the bus voltage is unknown.
3. PV bus: This type of bus usually connects to a generator ($p_i^d = 0, q_i^d = 0$), whose square voltage magnitude $V_i^2 = e_i^2 + f_i^2$ is regulated as a constant through adjusting reactive power, and the active power output p_i^d is set according to an energy transaction contract.

Given above definitions, there are $2n$ independent variables in $2n$ equations, and the power flow problem can be solved by a number of traditional numerical methods, such as the standard Newton's method [6, 7], the fast-decoupled method [8, 9] and its variations [10, 11]. In fact, all variables are physically adjustable. For example, optimal power flow dispatches generator output p_i^g, q_i^g and regulates bus voltage V_i; demand response optimizes load consumptions p_i^d and q_i^d. The same applies to the branch flow model presented in Sect. 2.2.2.

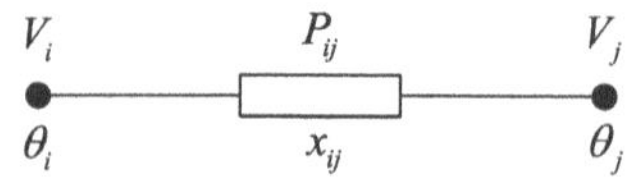

Fig. 2.2 Active power delivery in an AC transmission line

2. DC Power Flow Formulation

AC power flow models provide an accurate operating status of the power system. Prevailing power flow algorithms depend on an initial point and are iterative, making it difficult to consider AC power flow equations in higher level optimization problems. In practice, the production cost of a generator largely depends on its active power output. Restricting attention to active power is especially attractive when the computation efficiency is a major concern, such as in the unit commitment problem and power market clearing. In such circumstances, AC power flow equations can be approximated by a set of linear equations, which is called the direct current (DC) power flow model.

To explain DC power flow, consider an AC transmission line between buses i and j shown in Fig. 2.2 with reactance x_{ij}. The voltage magnitudes are V_i / V_j, and voltage angles are θ_i / θ_j, respectively; $\dot{V}_i$ and $\hat{V}_i$ stand for the complex bus voltage and its conjugate. According to basic power system knowledge, the active power delivered from bus i to bus j is

$$P_{ij} = \Re \left\{ \frac{\dot{V}_i(\hat{V}_i - \hat{V}_j)}{-\mathrm{i}x_{ij}} \right\} = \frac{V_i V_j \sin(\theta_i - \theta_j)}{x_{ij}} \tag{2.5}$$

DC power flow model makes the following assumptions:

1. Voltage magnitudes are equal to 1.0 at all buses in the system.
2. The angle difference at both sides of a branch is very small, such that

$$\sin(\theta_i - \theta_j) \approx \theta_i - \theta_j \tag{2.6}$$

3. Ignore all ground branches and resistances, and construct matrix B' as

$$B'_{ij} = -\frac{1}{x_{ij}}, \; B'_{ii} = -\sum_{j \neq i} B'_{ij}$$

Under these assumptions, (2.5) is simplified as

$$P_{ij} = \frac{\theta_i - \theta_j}{x_{ij}} \tag{2.7}$$

and the DC power flow model comes down to a set of linear equations

$$P^N = B'\theta, \; \theta_n = 0 \tag{2.8}$$

Given the solution of θ in Eq. (2.8), active power flow in each line can be calculated from (2.7). This characterization is very similar to a DC circuit, in which the branch resistance, bus voltage, and branch current are proportional to the branch reactance, voltage angle, and branch active power flow of the AC network. This is how the term "DC power flow" comes out.

Since branch resistance is neglected, the power delivery in Fig. 2.2 is lossless. This approximation is often satisfactory for high-voltage power transmission networks, and may be less accurate for power distribution networks. Furthermore, the DC power flow model does not capture reactive power and voltage behavior, it cannot be used in reactive power related applications, such as capacitor placement, voltage stability, volt/var control. If the power system is highly stressed, or the resistance to reactance ratio of each branch is high, the accuracy of the DC power flow model might be questionable. Further discussions on variant DC power flow models can be found in [12] and the references therein. Their accuracies largely depend on network data, load levels, and may vary significantly in different cases. An improved linear approximation of BIM is proposed in [13], in which reactive power and voltage magnitudes are taken into account.

To get a picture for the accuracy of DC power flow solution, an illustrative example based on a 4-bus system with two generators and four transmission lines is provided. System topology is shown in Fig. 2.3, and the data set can be found in file "case4gs" in Matpower toolbox. The r/x ratio is 0.2. AC power flow and DC power flow equations are solved by build-in solvers offered by Matpower. For clarity, we omit bus injection power and exhibit complex bus voltage and line power flow in Fig. 2.3. It can be observed that the DC model provides a good approximation for line active power flow.

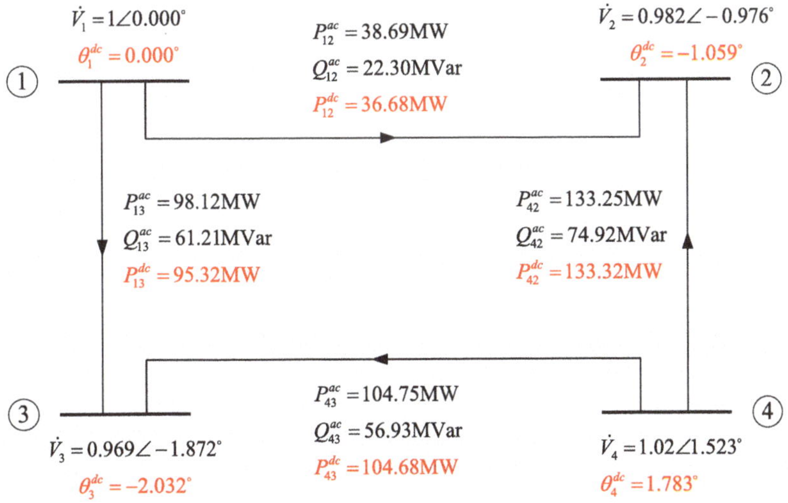

Fig. 2.3 Comparison of AC and DC power flow results

By solving Eq. (2.8), bus angle variables can be eliminated from the DC power flow model, and line power flow P_{ij} in (2.7) can be expressed via linear functions of net power injection in a compact form

$$P_{Line} = \text{PTDF} \cdot p^N \tag{2.9}$$

where PTDF is the so-called power transfer distribution factor, which indicates the incremental change of active power in transmission lines caused by the unit change of nodal injection. Although (2.9) incorporates fewer variables than (2.7)–(2.8) however, the coefficient matrices in the latter formulation enjoy a sparser structure. Which one leads to better computational performance depends on real data.

3. Jabr's Formulation via Variable Transformation

Quadratic equalities are non-convex thus are difficult to tackle in optimization problems. The quadratic terms in power flow equations exhibit an appropriate form. By deploying a suitable variable transformation, most of the power flow equations become linear, and non-convexity is packaged in a few constraints. To see this, write (2.4) in the following form

$$P_i^N = G_{ii}(e_i^2 + f_i^2) + \sum_{j \in c(i)} [G_{ij}(e_i e_j + f_i f_j) - B_{ij}(e_i f_j - e_j f_i)], \ \forall i \tag{2.10a}$$

$$Q_i^N = -B_{ii}(e_i^2 + f_i^2) + \sum_{j \in c(i)} [-B_{ij}(e_i e_j + f_i f_j) - G_{ij}(e_i f_j - e_j f_i)], \ \forall i \tag{2.10b}$$

where $c(i)$ denotes the set of buses that connects to bus i through some transmission line. Define the following variable transformation

$$c_{ii} = e_i^2 + f_i^2, \quad c_{ij} = e_i e_j + f_i f_j, \quad s_{ij} = e_i f_j - e_j f_i \tag{2.11a}$$

It is easy to verify the relation

$$c_{ij} = c_{ji}, \ s_{ij} = -s_{ji} \tag{2.11b}$$

$$c_{ij}^2 + s_{ij}^2 = c_{ii} c_{jj} \tag{2.11c}$$

Suppose the network consists of N_B buses and N_L lines. The original BIM include $2N_B$ equalities. Transformation (2.11a) introduces $N_B + 4N_L$ new variables, and (2.11b)–(2.11c) bring $3N_L$ equalities, so the total number of independent variables in (2.11) is $N_B + N_L$. For radial networks, $N_B = N_L$ holds true, therefore, there are actually $2N_B$ unknown variables, and the power flow equations are well defined under transformation (2.11).

For meshed networks, $N_L > N_B$ holds due to the existence of cycles, the new power flow equations will be under-determined. By introducing angle variables θ_i, $i = 1, 2, \dots, N_B$ for each bus and noticing the fact

$$e_i e_j + f_i f_j = V_i V_j \cos(\theta_i - \theta_j) = c_{ij}, \ \forall(i, j)$$
$$e_i f_j - e_j f_i = V_i V_j \sin(\theta_i - \theta_j) = s_{ij}, \ \forall(i, j)$$

additional N_L linkages can be built by

$$\tan(\theta_i - \theta_j) = \frac{s_{ij}}{c_{ij}}, \ \forall(i, j) \tag{2.12}$$

which balances the number of variables and equations in the power flow model. The new power flow model is summarized as follows:

$$p_i^g - p_i^d = G_{ii} c_{ii} + \sum_{j \in c(i)} (G_{ij} c_{ij} - B_{ij} s_{ij}), \ \forall i \tag{2.13a}$$

$$q_i^g - q_i^d = -B_{ii} c_{ii} + \sum_{j \in c(i)} (-B_{ij} c_{ij} - G_{ij} s_{ij}), \ \forall i \tag{2.13b}$$

$$c_{ij} = c_{ji}, \ s_{ij} = -s_{ji}, \ c_{ij}^2 + s_{ij}^2 = c_{ii} c_{jj}, \ \forall(i, j) \tag{2.13c}$$

$$\tan(\theta_i - \theta_j) = s_{ij}/c_{ij}, \ \forall(i, j) \tag{2.13d}$$

In (2.13), there are $2N_B + 4N_L$ variables and the same number of equations.

This formulation is developed by Jabr and Esposito in [14–16] for load flow calculation. For radial networks, (2.13a)–(2.13c) constitute a well-defined equation set. Except the last equality in (2.13c), remaining equations are all linear. Complex bus voltages can be recovered from the inverse transformation of (2.11), and angle variables are instantly available after the voltage is obtained.

2.2.2 Branch Flow Model

1. AC Power Flow Formulation

Symbols and notations used in this section will be defined first.

Sets and Indices

E_L Set of lines

E_N Set of buses except the slack bus

$c(i)$ Child buses of bus i, empty set for terminal buses

l, i, j Triplet (l, i, j) with superscript l and subscripts i and j is used to index line $l \in E_L$ whose head (from) bus is $i \in E_N$ and tail (to) bus is $j \in E_N$; $i = 0$ stands for the slack bus

Parameters and Variables

I_{ij}^l	Square magnitude of current in line l
p_i^d	Fixed active power demand at bus i
p_i^g	Active power generation at bus i
P_{ij}^l	Active power flow in line l
q_i^d	Fixed reactive power demand at bus i
q_i^g	Reactive power generation at bus i
Q_{ij}^l	Reactive power flow in line l
r_{ij}^l	Resistance of line l
S_l	Apparent power flow capacity of line l
x_{ij}^l	Reactance of line l
z_{ij}^l	Impedance of line l, $(z_{ij}^l)^2 = (r_{ij}^l)^2 + (x_{ij}^l)^2$
U_i	Square magnitude of voltage V_i at bus i

Branch flow model (BFM) is originally proposed for studying distribution networks [17–19], which are intentionally operated with a tree topology. BFM incorporates a set of independent variables different from those in BIM. As its name implies, the active and reactive power flows in each line will replace voltage angle at each bus and become variables in BFM; bus voltage magnitudes remain variables in BFM, as they are important indicators of power quality and closely monitored in distribution systems. In what follows, we provide a detailed explanation for the forward voltage drop equability introduced in [17–19].

Consider a branch with impedance $z_{ij} = r_{ij} + \mathrm{j}x_{ij}$ between buses i and j, where the complex voltages are given by $\dot{V}_i = V_i \angle \theta_i$ and $\dot{V}_j = V_j \angle \theta_j$, respectively. The apparent power $\dot{S}_{ij}$ delivered at the head bus i can be expressed as

$$\dot{S}_{ij} = P_{ij} + \mathrm{j}Q_{ij} = \frac{\dot{V}_i(\hat{V}_i - \hat{V}_j)}{r_{ij} - \mathrm{j}x_{ij}} \tag{2.14}$$

Because $\dot{V}_i\hat{V}_i = V_i^2$, $\dot{V}_i\hat{V}_j = V_iV_j\angle\theta_{ij}$, where $\theta_{ij} = \theta_i - \theta_j$, multiplying both sides of (2.14) with $r_{ij} - \mathrm{j}x_{ij}$ yields

$$P_{ij}r_{ij} + Q_{ij}x_{ij} + \mathrm{j}(Q_{ij}r_{ij} - P_{ij}x_{ij}) = V_i^2 - V_iV_j(\cos\theta_{ij} + \mathrm{j}\sin\theta_{ij}) \tag{2.15}$$

Equaling real and imaginary parts of both sides of (2.15) we have

$$\begin{aligned} V_iV_j\cos\theta_{ij} &= V_i^2 - (P_{ij}r_{ij} + Q_{ij}x_{ij}) \\ V_iV_j\sin\theta_{ij} &= P_{ij}x_{ij} - Q_{ij}r_{ij} \end{aligned}$$

Eliminating θ_{ij} gives

$$V_i^2V_j^2 = V_i^4 - 2V_i^2(P_{ij}r_{ij} + Q_{ij}x_{ij}) + (P_{ij}r_{ij} + Q_{ij}x_{ij})^2 + (P_{ij}x_{ij} - Q_{ij}r_{ij})^2 \tag{2.16}$$

Since $z_{ij}^2 = r_{ij}^2 + x_{ij}^2$, (2.16) is equivalent to

$$V_j^2 = V_i^2 - 2(P_{ij}r_{ij} + Q_{ij}x_{ij}) + z_{ij}^2 \frac{(P_{ij}^2 + Q_{ij}^2)}{V_i^2} \tag{2.17}$$

Please bear in mind that $\dot{S}_{ij} \neq \dot{S}_{ji}$ due to line loss. In Eq. (2.17), the voltage magnitude at the tail bus is expressed in terms of the voltage and apparent power at the head bus. It is called the forward voltage drop equation.

We will restrict our discussion to radial networks with n buses. Because there is no loop, the number of branches is equal to the number of buses (except the slack bus), i.e., $|E_L| = |E_N| = n$. The typical connection of branches in a radial network is depicted in Fig. 2.4, where the nodal injection power $\dot{S}_i^N = P_i^N + \mathrm{i}Q_i^N$, $P_i^N = p_i^g - p_i^d$, $Q_i^N = q_i^g - q_i^d$. Taking the power balancing condition for every bus into consideration, the network power flow can be described by a set of recursive equations

$$P_{ij}^l - r_{ij}^l \frac{(P_{ij}^{l2} + Q_{ij}^{l2})}{V_i^2} + P_j^N = \sum_{k \in c(j)} P_{jk}^l, \ \forall j \in E_N \tag{2.18a}$$

$$Q_{ij}^l - x_{ij}^l \frac{(P_{ij}^{l2} + Q_{ij}^{l2})}{V_i^2} + Q_j^N = \sum_{k \in c(j)} Q_{jk}^l, \ \forall j \in E_N \tag{2.18b}$$

$$V_j^2 = V_i^2 - 2(P_{ij}^l r_{ij}^l + Q_{ij}^l x_{ij}^l) + (z_{ij}^l)^2 \frac{(P_{ij}^{l2} + Q_{ij}^{l2})}{V_i^2}, \ \forall l \in E_L \tag{2.18c}$$

where constraints (2.18a) and (2.18b) are nodal active power and reactive power balancing conditions; According to the reference direction of power flow shown in Fig. 2.4, the left-hand (right-hand) side gathers total active and reactive power injected (withdrawn) in (from) bus j; (2.18c) describes voltage drop on each line. The voltage magnitude V_0 at the slack bus is a constant.

In BFM, the net injection power P_i^N, Q_i^N and voltage magnitude V_i at each bus, as well as power flow P_{ij}^l and Q_{ij}^l in each line, totally $5n$ variables, are to be determined from equation set (2.18) which includes $3n$ equalities. Similar to BIM, the buses are classified into three types, depending on which two of the three

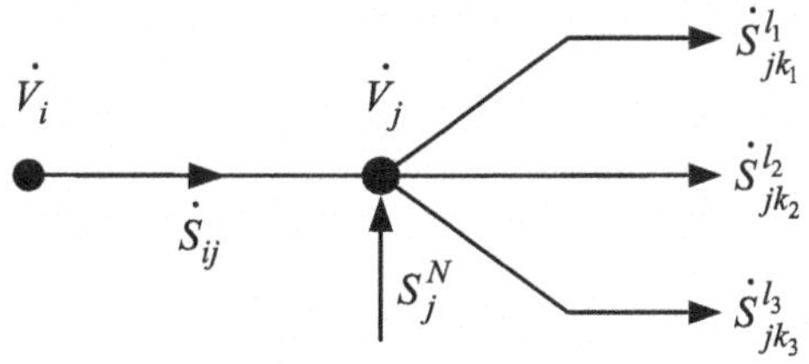

Fig. 2.4 Typical connection of branches in a radial network

variables P_i^N, Q_i^N, and V_i at each bus are specified. For PQ buses, the net injection power $\dot{S}_i^N = P_i^N + \mathrm{i}Q_i^N$ is given; For PV buses, the active power injection P_i^N and voltage magnitude V_i are given. In distribution networks, generators and loads can connect to the same bus. The bus type depends on whether the voltage magnitude is regulated as a constant. The voltage magnitude V_0 at the slack bus is known, the apparent power injected from the slack bus is given by

$$P_0^N + \mathrm{i}Q_0^N = \sum_{k \in c(0)} (P_{0k}^l + \mathrm{i}Q_{0k}^l) \tag{2.19}$$

BFM has a special recursive structure, inspiring a forward-backward sweep load flow algorithm [17], the most representative method for distribution load flow problems. A comprehensive review on this algorithm can be found in [20].

2. Linearized BFM

For simplification, the DC power flow model can be applied. However, for distribution networks where the branch r/x value can be significantly larger than that in a transmission network, the accuracy of DC model may not be satisfactory. One reason is that bus voltages usually scatter in a wider range rather than close to 1 p.u. The other reason is that both active and reactive power flows influence bus voltages due to the large r/x ratio, but reactive power is neglected in the DC model. Here we introduce a linearized BFM that is superior to the DC model for radial distribution networks.

Assuming that the line loss accounts for a small portion of the total demand, typically less than 5% in 10 kV voltage level and 3% in 35 kV voltage level, the lossy terms in BFM (2.18) can be ignored, giving rise to a linear BFM

$$P_{ij}^l + P_j^N = \sum_{k \in c(j)} P_{jk}^l, \ \forall j \in E_N \tag{2.20a}$$

$$Q_{ij}^l + Q_j^N = \sum_{k \in c(j)} Q_{jk}^l, \ \forall j \in E_N \tag{2.20b}$$

$$U_j = U_i - 2(P_{ij}^l r_{ij}^l + Q_{ij}^l x_{ij}^l), \ \forall l \in E_L \tag{2.20c}$$

where $U_j = V_j^2$ and $U_i = V_i^2$. Equations (2.20a)–(2.20c) are linear equations. There is an alternative formulation for the voltage drop equation. Since the bus voltages are close to that at the reference bus, $(V_i - V_0)^2 \approx 0$, indicating $V_j^2 = 2V_0V_j - V_0^2$. Substituting it into (2.20c) yields

$$V_j = V_i - \frac{(P_{ij}^l r_{ij}^l + Q_{ij}^l x_{ij}^l)}{V_0}, \ \forall l \in E_L \tag{2.20d}$$

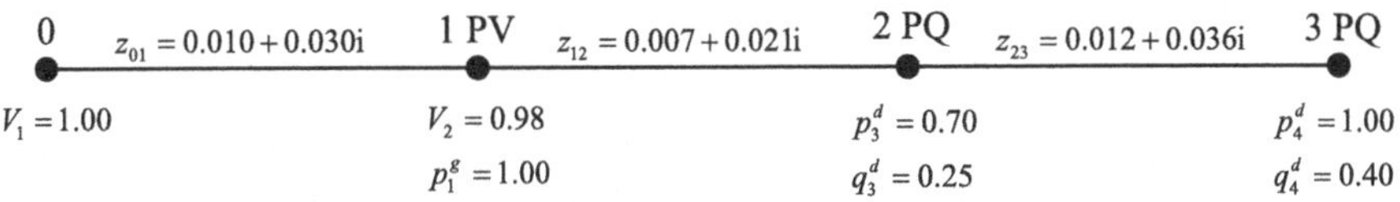

Fig. 2.5 A test radial network

Table 2.1 Comparison of AC power flow model and linearized BFM

	V_i			P^l_{ij}		Q^l_{ij}	
Node	AC	BFM-L	Line	AC	BFM-L	AC	BFM-L
1	0.980	0.980	0–1	75.03	70.00	42.22	43.33
2	0.951	0.954	1–2	174.29	170.00	77.88	65.00
3	0.922	0.928	2–3	101.64	100.00	44.91	40.00

which is also linear. The linearized BFM consisting of (2.20a), (2.20b), and (2.20d) will be used later. This model has been justified in a number of distribution system studies, such as [21–24].

To get an outline for the accuracy of the linearized BFM, an illustrative example based on a 4-bus system is provided. System topology and data are shown in Fig. 2.5. Results of AC power flow model and linearized BFM are listed in Table 2.1. We can observe that the linearized BFM gives reasonably accurate voltage magnitudes at the PQ buses, because reactive power and line resistance are taken into account. Since active net injections at all individual buses are given, line active power flows in the linearized BFM are instantly available according to the Kirchhoff's Current Law. In this case, active power losses account for 2.94% of the total system demand. The reactive power flow in line 0–1 depends not only on the system demands, but also on the voltage at buses 0 and 1 as well as active power flow in line 0–1, according to (2.20d). For this reason, the reactive power flow offered by the linearized BFM is less accurate than the active power flow.

Remark 2.1 Comparison of the linearized BFM and the DC power flow model. Although both models ignore network losses, the latter assumes fixed voltage magnitudes and completely neglects reactive power, so it is unable to model PV buses and proactive voltage regulation through reactive power control, whereas the former accounts for both active and reactive power, and consider line resistance in the voltage drop equation (2.20d). This is particularly important for distribution networks where line resistance is not negligible, and voltage magnitudes can be significantly affected by both of active and reactive power flows. Because the network has a tree topology and the network injection P_i^N for every bus is given, both models offer the same results for line active power flow P_{ij}^l which can be recursively determined from (2.20a), and is independent of the voltage magnitude. The linearized BFM also determines reactive injection Q_i^N at PV buses and voltage magnitude V_i at PQ buses simultaneously. Nevertheless, BFM only applies to radial networks. For meshed networks, because the bus angle is neglected, BFM may give incorrect results. This pitfall does not exist in the DC power flow model.

Remark 2.2 Comparison of BFM and BIM. From the analysis of voltage drop equation (2.17), we can image that BIM and BFM are equivalent for radial networks: their solutions have a one-to-one correspondence. BIM formulates complex bus voltages in rectangle coordinates or polar coordinates. In rectangle coordinates, power flow equations render quadratic equalities which is convenient for SDP based convex relaxation. In polar coordinates, sine and cosine functions appear, which are not compatible with the majority of convex relaxation methods. So we omit the polar formulation in this section. Primary interests of BIM originate from applications in meshed networks. For radial networks, BFM is equivalent to BIM, and directly handles bus voltage magnitude and line power flows, which are easily measurable (voltage angle is difficult to measure without phasor measurement units). In addition, power flow capacity constraints in BFM are simply lower or upper bounds of decision variables, and hence convenient for further manipulation in sophisticated optimization models, such as a bilevel program or a robust optimization problem.

In the following chapters, we will pertinently choose one of the four models to represent the electric power system depending on which one is more convenient for the problem at hand.

2.3 OPF and Its Variations

OPF is a mathematical program that seeks to optimize a given objective function, such as the generation cost or the network loss, subject to under-determined power flow equations and technical limits of individual components. Voltage and injection power at each generator bus are adjustable so as to improve the objective. The original form of OPF can be traced back to 1962 [25]. Its formulation has changed little over half a century. Different OPF formulations are reviewed in [26]. OPF is the basis of power market clearing and underlies many applications such as unit commitment, economic dispatch, volt/var control, to name just a few. With different requirements on model accuracy and complexity, all formulations discussed in Sect. 2.2 can serve as power flow constraints, for example, in unit commitment and economic dispatch problems, where generation cost and active power are the main concerns, the DC power flow model and linearized BFM are often used for transmission networks and distribution networks, respectively; in volt/var control problems where bus voltage and line reactive power become the leading characteristics, the bus injection model in AC format can be used for transmission networks, and two BFMs are eligible for distribution networks.

As what has been shown in Sect. 2.2, AC power flow equations include quadratic equalities, and hence ACOPF gives rise to a QCQP, which is non-convex and NP-hard. In general, the global optimal solution cannot be found in polynomial time. ACOPF problems can be solved locally by a number of general-purpose NLP techniques, such as the interior point algorithm [27–29] and the sequential quadratic programming algorithm [30–32]. A comprehensive survey of OPF algorithms can be found in [33] (for deterministic ones) and [34] (for non-deterministic ones), with different compromises on optimality and computational efficiency.

In view of the potential economic benefit, there has been sustainable interest on computing the global OPF solution. In recent years, important progresses have been reported based on convex relaxation methods. Two most successful approaches are SDP relaxation [35, 36] and SOCP relaxation [37]. Convex relaxation models could provide a lower bound for the objective value at optimum, and may even globally solve an OPF problem if the relaxation is exact.

This section will first introduce basic formulation of ACOPF in bus injection format and branch flow format as well as their convex relaxation models. For radial networks and the branch flow based model, we propose an iterative convexification procedure to refine a feasible or an optimal solution if the original convex relaxation is not exact. Moreover, some operational problems that involve optimizing special targets over the feasible region defined by power flow equations can be regarded as variants of the OPF problem. Three representative ones are addressed in this section, including the maximal loadability problem, the bi-objective OPF problem, and the equilibrium of a demand response market. In the former two variants, existing convex relaxation models are generally inexact. In the last one, system demands respond to the locational marginal electricity price. Discussions on these variant problems are restricted to radial networks based on the proposed iterative procedure. Nevertheless, we give supplementary discussions on how the proposed method can be extended to meshed networks. Materials in this section come from [38, 39].

2.3.1 Cost-Minimizing OPF

ACOPF in bus-injection format (OPF-BIM for short) and branch flow format (OPF-BFM for short) exhibit different mathematic structure: OPF-BIM gives rise to a QCQP, which allows a clear matrix representation and SDP relaxation; OPF-BFM can boil down to optimization with a convex objective over the intersection of a polyhedron and boundaries of rotated SOCs, which leads to a convenient SOCP relaxation. Details are provided as follows.

1. OPF-BIM and SDP Relaxation

For notation brevity, introduce the standard basis vectors in $\mathbb{R}^n$ as $b_1, b_2, \ldots, b_n$, for each basis vector b_k, $k = 1, 2, \ldots, n$, the k-th element of b_k is 1, and the remaining entries are 0. By replacing transmission lines and transformers with their Π equivalent models, the power grid comes down to an equivalent circuit consisting of resistances, inductances, and capacitances. Let $\bar{y}^l_{ij}$ be the admittance of the shunt components at bus i, y^l_{ij} the admittance of line l connecting buses i and j. The nodal admittance matrix of the equivalent circuit model is Y, whose off-diagonal element is equal to $-y^l_{ij}$ and diagonal element is given by $\bar{y}^l_{ii} + \sum_{j \in c(i)} y^l_{ij}$.

Define auxiliary matrices

$$Y^N_k = b_k b^T_k Y, \quad Y^L_{ij} = (\bar{y}^l_{ij} + y^l_{ij}) b_i b^T_i - y^l_{ij} b_i b^T_j$$

$$M_k = \begin{bmatrix} b_k b^T_k & 0 \\ 0 & b_k b^T_k \end{bmatrix}, \; M_{ij} = \frac{1}{2} \begin{bmatrix} (b_i - b_j)(b_i - b_j)^T & 0 \\ 0 & (b_i - b_j)(b_i - b_j)^T \end{bmatrix}$$

$$Z_k = \frac{1}{2} \begin{bmatrix} \Re\{Y^N_k + (Y^N_k)^T\} & \Im\{(Y^N_k)^T - Y^N_k\} \\ \Im\{Y^N_k - (Y^N_k)^T\} & \Re\{Y^N_k + (Y^N_k)^T\} \end{bmatrix} \in \mathbb{R}^{2n \times 2n}$$

$$\bar{Z}_k = -\frac{1}{2} \begin{bmatrix} \Im\{Y^N_k + (Y^N_k)^T\} & \Re\{Y^N_k - (Y^N_k)^T\} \\ \Re\{(Y^N_k)^T - Y^N_k\} & \Im\{Y^N_k + (Y^N_k)^T\} \end{bmatrix} \in \mathbb{R}^{2n \times 2n}$$

$$Z^L_{ij} = \frac{1}{2} \begin{bmatrix} \Re\{Y^L_{ij} + (Y^L_{ij})^T\} & \Im\{(Y^L_{ij})^T - Y^L_{ij}\} \\ \Im\{Y^L_{ij} - (Y^L_{ij})^T\} & \Re\{Y^L_{ij} + (Y^L_{ij})^T\} \end{bmatrix} \in \mathbb{R}^{2n \times 2n}$$

$$\bar{Z}^L_{ij} = -\frac{1}{2} \begin{bmatrix} \Im\{Y^L_{ij} + (Y^L_{ij})^T\} & \Re\{Y^L_{ij} - (Y^L_{ij})^T\} \\ \Re\{(Y^L_{ij})^T - Y^L_{ij}\} & \Im\{Y^L_{ij} + (Y^L_{ij})^T\} \end{bmatrix} \in \mathbb{R}^{2n \times 2n}$$

where operator $\Re\{\cdot\}/\Im\{\cdot\}$ represents the (element-wise) real/imaginary part of a complex input (scalar, vector, or matrix). The normalized vector x of decision variables is

$$x = [\Re\{\dot{V}^T\}, \Im\{\dot{V}^T\}]^T$$

Using above notations, the following equations hold [36]:

$$P^N_k = x^T Z_k x, \; \forall k \tag{2.21a}$$

$$Q^N_k = x^T \bar{Z}_k x, \; \forall k \tag{2.21b}$$

$$V^2_k = x^T M_k x, \; \forall k \tag{2.21c}$$

$$|\dot{S}^l_{ij}|^2 = (x^T Z^L_{ij} x)^2 + (x^T \bar{Z}^L_{ij} x)^2 \tag{2.21d}$$

where (2.21a) and (2.21b) correspond to power flow equations (2.4a) and (2.4b), respectively; (2.21c) is simply the definition of voltage magnitude; (2.21d) is the square magnitude of apparent power flow at the head bus of line l.

OPF-BIM can be written as the following QCQP

$$
\begin{aligned}
\min \quad & \sum_k \left[a_k^g (p_k^g)^2 + b_k^g p_k^g\right] \\
\text{s.t.} \quad & p_k^g = x^T Z_k x + p_k^d,\ p_k^{gn} \le p_k^g \le p_k^{gm},\ \forall k \\
& q_k^g = x^T \bar{Z}_k x + q_k^d,\ q_k^{gn} \le q_k^g \le q_k^{gm},\ \forall k \\
& (V_k^n)^2 \le x^T M_k x \le (V_k^m)^2,\ \forall k \\
& (x^T Z_{ij}^L x)^2 + (x^T \bar{Z}_{ij}^L x)^2 \le (S_l^m)^2,\ \forall l
\end{aligned}
\tag{2.22}
$$

where a_k^g and b_k^g are production cost coefficients, p_k^{gn}/p_k^{gm}, q_k^{gn}/q_k^{gm}, and V_k^n/V_k^m are the upper/lower bound of active power, reactive power, and voltage magnitude at bus k; the power limits are 0 for buses without a generator; S_l^m is the power flow capacity of line l.

It is clear that (2.22) is non-convex due to the presence of quadratic terms in power balancing equalities and voltage lower bound inequalities. A common idea is to lift the problem to a higher dimensional space by introducing auxiliary variables. For example, a quadratic term can be linearized via $x^T Dx = \mathrm{Tr}(Dxx^T) = \mathrm{Tr}(DX)$, where D is a symmetric matrix, $\mathrm{Tr}(\cdot)$ is the matrix trace operator, $X = xx^T$ is the lifting variable. Applying this trick to (2.22) results in

$$\min \quad \sum_k f_k^g(X) \tag{2.23a}$$

$$\text{s.t.} \quad p_k^{gn} \le \mathrm{Tr}(Z_k X) + p_k^d \le p_k^{gm},\ \forall k \tag{2.23b}$$

$$q_k^{gn} \le \mathrm{Tr}(\bar{Z}_k X) + q_k^d \le q_k^{gm},\ \forall k \tag{2.23c}$$

$$(V_k^n)^2 \le \mathrm{Tr}(M_k X) \le (V_k^m)^2,\ \forall k \tag{2.23d}$$

$$\mathrm{Tr}(Z_{ij}^L X)^2 + \mathrm{Tr}(\bar{Z}_{ij}^L X)^2 \le (S_l^m)^2,\ \forall l \tag{2.23e}$$

$$\mathrm{rank}(X) = 1 \tag{2.23f}$$

where $f_k^g(X) = a_k^g \left[\mathrm{Tr}(Z_k X) + p_k^d\right]^2 + b_k^g \left[\mathrm{Tr}(Z_k X) + p_k^d\right]$ is the generation cost at bus k. Constraint (2.23f) ensures that the lifting variable X has a rank-1 decomposition $X = xx^T$, and x will be the solution of OPF-BIM. In this way, non-convexity is encapsulated in the rank-1 constraint (2.23f). For an arbitrary vector v with a compatible dimension, $v^T X v = v^T xx^T v = (v^T x)^2 \ge 0$, so X must be a positive semidefinite (PSD) matrix.

The line flow capacity constraint (2.23e) and the objective function (2.23a) remain nonlinear. Nevertheless, they are convex. By applying Schur complement theorem, convex quadratic constraint (2.23e) can be reformulated as a linear matrix inequality (LMI)

$$\begin{bmatrix} (S_l^m)^2 & \text{Tr}(Z_{ij}^l X) & \text{Tr}(\bar{Z}_{ij}^l X) \\ \text{Tr}(Z_{ij}^l X) & 1 & 0 \\ \text{Tr}(\bar{Z}_{ij}^X) & 0 & 1 \end{bmatrix} \succeq 0, \ \forall l \tag{2.24}$$

Following a similar procedure, inequality $\beta_k \geq f_k^g(X)$ has an LMI form as

$$\begin{bmatrix} \beta_k - b_k^g \left[\text{Tr}(Z_k X) + p_k^d\right] & \sqrt{a_k^g}\left[\text{Tr}(Z_k X) + p_k^d\right] \\ \sqrt{a_k^g}\left[\text{Tr}(Z_k X) + p_k^d\right] & 1 \end{bmatrix} \succeq 0, \ \forall k \tag{2.25}$$

Owing to above discussions, one can transform OPF-BIM formalized in (2.23) into an SDP

$$\min \left\{ \sum_k \beta_k \,\middle|\, \text{(2.23b)–(2.23d), (2.24), (2.25)}, X \succeq 0 \right\} \tag{2.26}$$

SDP (2.26) is in fact a relaxation of (2.23), because $X \succeq 0$ is only a necessary condition of (2.23f), and the feasible region of (2.23) is enlarged by this replacement. If (2.26) has a rank-1 solution X^*, then the decomposition $X^* = x^*(x^*)^T$ gives the optimal solution x^* for OPF-BIM; otherwise, the optimal value of (2.26) generally provides a lower bound for the optimum of (2.23).

From a theoretical point of view, establishing sufficient conditions that guarantee an exact SDP relaxation has received substantial attention. A number of guarantees have been reported in the literature. The pioneer work in [36] finds that the SDP relaxation will be exact for IEEE benchmark systems after adding a small resistance to every purely inductive transformer branch, and opens up a promising research direction in OPF studies. It is proven that SDP relaxation will be exact with allowance of load over satisfaction and utilization of virtual phase shifters [40, 41], or the network is radial [42]. Readers who are interested in more conditions and in-depth discussions on the exactness of the SDP relaxation may find extensive information in [43–49], which are beyond the scope of this section. At the end of this chapter, we recommend relevant literature on general convex relaxation methods for QCQPs. In Appendix A.3, we give a clear description on the SDP relaxation technique for QCQPs.

If the optimizer of (2.26) has a rank higher than 1, a physically meaningful power flow solution cannot be recovered. In such a circumstance, one may turn to seek a feasible rank-1 solution while sacrificing the global optimality guarantee. In order to enforce a rank-1 matrix solution, it is proposed to append rank-related penalty terms in the objective function [45, 46]. Computational results show that the rank penalty method is very inspiring to identify a high quality solution. Another approach to construct a tight convex relaxation is to employ the moment method from polynomial optimization [50], which solves a sequence of SDPs with increasing problem sizes, as discussed in [51–54]. In theory, the global OPF solution can be recovered with the order of moment relaxation approaching infinity. However, this method may suffer from a high computational overhead due to the dramatic growing SDP sizes when the order of moment relaxation becomes higher. There have been research efforts on globally solving the OPF problem by using branch-and-bound type methods, such as those in [55, 56], while abandoning a polynomial-time complexity guarantee. To the best of our knowledge, how to recover a high-quality solution within affordable computational expense if the SDP relaxation is inexact remains an open problem. Next, we will discuss this problem starting from the SOCP relaxation of OPF-BFM.

2. OPF-BFM and SOCP Relaxation

Recall BFM (2.18), if we introduce new variables U_i and I^l_{ij} which represent squares of voltage magnitude and current magnitude as

$$U_i = V_i^2,\ \forall i \in E_N \tag{2.27a}$$

$$I^l_{ij} = \frac{(P^l_{ij})^2 + (Q^l_{ij})^2}{V_i^2},\ \forall l \in E_L \tag{2.27b}$$

BFM (2.18) becomes

$$P^l_{ij} - r^l_{ij} I^l_{ij} + p^g_j = p^d_j + \sum_{k \in c(j)} P^l_{jk},\ \forall j \in E_N \tag{2.28a}$$

$$Q^l_{ij} - x^l_{ij} I^l_{ij} + q^g_j = q^d_j + \sum_{k \in c(j)} Q^l_{jk},\ \forall j \in E_N \tag{2.28b}$$

$$U_j = U_i - 2(P^l_{ij} r^l_{ij} + Q^l_{ij} x^l_{ij}) + (z^l_{ij})^2 I^l_{ij},\ \forall l \in E_L \tag{2.28c}$$

$$U_i I^l_{ij} = (P^l_{ij})^2 + (Q^l_{ij})^2,\ \forall l \in E_L \tag{2.28d}$$

Compared with (2.18a)–(2.18c), (2.28a)–(2.28c) are linear equalities, and non-convexity is encapsulated in (2.28d), which defines the apparent power injected at the head bus of each line. Component security boundaries can be gathered as

$$p^{gn}_j \le p^g_j \le p^{gm}_j,\ q^{gn}_j \le q^g_j \le q^{gm}_j,\ \forall j \in E_N \tag{2.29a}$$

$$U^n_j \le U_j \le U^m_j,\ \forall j \in E_N \tag{2.29b}$$

$$P^l_{ij}, Q^l_{ij} \ge 0,\ \sqrt{(P^l_{ij})^2 + (Q^l_{ij})^2} \le S_l,\ \forall l \in E_L \tag{2.29c}$$

including generation capacity ranges (2.29a), bus voltage intervals (2.29b), and line power flow limits (2.29c). Some of the boundary constraints may not be necessary for a practical OPF problem. Undesired ones can be removed from the model with little impact on the discussion in this section. For instance, the positivity of line flow in (2.29c) aims to capture the relay protection requirement of traditional distribution networks, but may no longer be necessary for the emerging active distribution network endowed with smart grid features. Define

$$\text{Cons-BFM} = \{(2.28a) - (2.28d)\}$$

$$\text{Cons-BND} = \{(2.29a) - (2.29c)\}$$

OPF-BFM can be written as the following optimization problem

$$\begin{aligned} \min \ & F_C = \sum_{j \in E_N} f_j^g(p_j^g) + \rho \sum_{j \in \pi(0)} P_{0j}^l \\ \text{s.t.} \ & \text{Cons-BFM, Cons-BND} \end{aligned} \tag{2.30}$$

In objective function F_C, $f_j^g(p_j^g) = a_j^g(p_j^g)^2 + b_j^g p_j^g$ represents the cost function of generator at bus k; P_{0j}^l stands for the active power delivered through each branch connected to the slack bus; we assume that electricity can be purchased from a higher level power market at a contract price ρ and delivered via the slack bus, so the second term is the transaction cost paid to the power market. If the slack bus is connected to a generator, the second term can be replaced by a conventional quadratic cost function $f_0^g(p_0^g)$ where $p_0^g = \sum_{j \in \pi(0)} P_{0j}^l$. Without particular mention, we assume throughout this section that the OPF problem (2.30) is feasible and square voltage magnitudes at all buses are strictly positive.

The feasible region of OPF-BFM (2.30) is the intersection of a polyhedron with the boundaries of rotated second-order cones. To obtain a convex model, it is natural to replace (2.28d) with an inequality

$$U_i I_{ij}^l \geq (P_{ij}^l)^2 + (Q_{ij}^l)^2, \ \forall l \in E_L$$

whose canonical SOC form is given by

$$\left\| \begin{matrix} 2P_{ij}^l \\ 2Q_{ij}^l \\ I_{ij}^l - U_i \end{matrix} \right\|_2 \leq I_{ij}^l + U_i, \ \forall l \in E_L \tag{2.31}$$

Define the convex feasible region

$$\text{Cons-BFM-Cr} = \{(2.28a)\text{–}(2.28c), (2.31)\}$$

The SOCP relaxation of OPF-BFM (OPF-BFM-Cr for short) is defined by

$$\min \{F_C \mid \text{Cons-BFM-Cr, Cons-BND}\} \tag{2.32}$$

It is proven that the SOCP relaxation is equivalent to the SDP relaxation for OPF problems over radial networks [44], therefore the strength and limitation of SDP relaxation also apply to the SOCP relaxation of OPF-BFM. However, some sufficient conditions for exact SOCP relaxation provide deeper insights on power flows over acyclic distribution networks, such as those in [57, 58], which are difficult to be derived from OPF-BIM and its SDP relaxation. A sufficient condition for the exactness of OPF-BFM-Cr when there is no simultaneous reverse active and reactive power flow is proposed in [59].

3. SOCP Relaxation for Jabr's Formulation

Recall Jabr's formulation (2.13) for BIM and corresponding variable transformation (2.11), define generator capacities

$$p_i^{gn} \le p_i^g \le p_i^{gm},\ q_i^{gn} \le q_i^g \le q_i^{gm},\ \forall i \tag{2.33}$$

and bus voltage magnitude boundaries

$$(V_i^n)^2 \le c_{ii} \le (V_i^m)^2,\ \forall i \tag{2.34}$$

Line power flow in original variables has the form of

$$\begin{aligned}&\left[-G_{ij}(e_i^2+f_i^2)+G_{ij}(e_ie_j+f_if_j)-B_{ij}(e_if_j-e_jf_i)\right]^2\\&+\left[B_{ij}(e_i^2+f_i^2)-B_{ij}(e_ie_j+f_if_j)-G_{ij}(e_if_j-e_jf_i)\right]^2 \le S_l^2\end{aligned}$$

Using new variables in Jabr's formulation, line flow limits come down to

$$\left[-G_{ij}c_{ii}+G_{ij}c_{ij}-B_{ij}s_{ij}\right]^2+\left[B_{ij}c_{ii}-B_{ij}c_{ij}-G_{ij}s_{ij}\right]^2 \le S_l^2 \tag{2.35}$$

which is an SOC and does not destroy model convexity.

Now we can build the following OPF problem:

$$\min\left\{\sum_i f_i^g(p_i^g)\ \middle|\ \text{Cons-BIM-VT, Cons-BND-VT}\right\} \tag{2.36}$$

where Cons-BIM-VT includes constraints (2.13a)–(2.13c), and Cons-BND-VT packages (2.33)–(2.35).

Problem (2.36) is non-convex due to the quadratic equality in (2.13c). For radial networks, (2.13d) is redundant thus can be omitted. The feasible region of (2.36) is the intersection of a polytope with the boundaries of rotated SOCs. To convexify model (2.36), one can replace

$$c_{ij}^2 + s_{ij}^2 = c_{ii}c_{jj}$$

with the following SOC

$$c_{ij}^2 + s_{ij}^2 \leq c_{ii}c_{jj}$$

which is similar to what we did in (2.31). The convex relaxation of (2.36) is thoroughly studied in [60, 61]. The SOCP relaxation is exact under mild conditions. However, if not, valid inequalities, such as McCormick envelops, can provide stronger relaxation, but are still unable to guarantee exactness in general cases.

In view of similarities in (2.30) and (2.36) for radial networks, the solution recovery procedure discussed later applies to both of them if the original SOCP relaxation is inexact. One possible drawback of (2.36) may be it is not straightforward to monitor and manipulate the original power flow variables. To recover the physical power flow status, reverse transformation is needed, which is nonlinear.

Remark SOCP relaxation models are more tractable than the SDP relaxation model from the computational perspective. For one reason, the interior point algorithm for SOCP is faster than that for SDP with a similar problem size; For another, the number of decision variables in (2.32) and (2.36) grows linearly with respect to the network scales. The dimension of matrix variable X in (2.26) is $2N_B \times 2N_B$, which grows with a quadratic rate with respect to the network scales. Because both relaxations have the same tightness for radial networks, the BFM and SOCP relaxation are preferred if the network has a tree topology; otherwise, the BIM and SDP relaxation can be considered.

4. Solution Recovery When the Convex Relaxation is Inexact

As mentioned before, previous research efforts have been paid to the sufficient conditions which guarantee an exact convex relaxation. As revealed in the literature, the exactness condition for general meshed networks largely depends on the choice of the objective function and may be sensitive to system data. If the relaxed model is not exact, its solution is not feasible for the original OPF problem and has no clear physical meaning. Possible remedies to improve the solution or even solve the problem globally are summarized as follows:

(a) Valid inequality [60, 61]
(b) Rank penalty [45, 46]
(c) Moment relaxation [51–54]
(d) Branch and bound [55, 56]

All of them have their own advantages and drawbacks. For example, valid inequalities do enhance the tightness of relaxation and reduce the relaxation gap, but still fail to recover the global optimal solution in many cases. The moment method can offer the global optimum but may encounter computational overhead when the order of relaxation goes higher.

Although the convex relaxation is proved to be exact for radial networks under relative mild conditions. However, there are still a lot of factors that will challenge these conditions, especially those on the objective function. In the cost-minimum OPF, if the cost function is not strictly increasing in the nodal active power injection (this situation is not rare because renewable units can have zero production costs), the exactness may no longer be guaranteed. Moreover, if non-cost functions should be optimized, the objective may appear to be non-monotonic, which also jeopardizes the exactness.

Here we develop a versatile computational framework for broader classes of OPF problems over radial networks motivated by the method in [62]. The non-convex branch flow equation is treated as constraints involving a difference of two convex functions. The concave part is then replaced by its linear approximation, and is updated in each iteration. Starting from an initial point offered by an inexact SOCP relaxation, this approach solves a sequence of convexified penalization problems, and recovers a feasible power flow solution, which usually appears to be very close, if not identical, to the global optimal one. It removes the need of an initial guess, which is required by the majority of prevalent local NLP algorithms, and the need of a prior exactness guarantee, which is desired by existing convex relaxation methods. It reduces the computational overhead of branch-and-bound type methods, and provides a high-quality solution within reasonable computational resources. Some notations that will be frequently used are explained as follows:

DC decomposition	Express a function as the difference of two convex functions
DC function	A function that has a DC decomposition
DC inequality	$d(x) \le 0$ where $d(x)$ is a DC function
DC representable	Item that can be represented by DC functions or inequalities
DC program	A mathematical program whose objective and constraints contain DC functions and DC inequalities
DCP	Abbreviation for DC program
DC formulation	Formulate a target using DC functions

Our primary problem is OPF-BFM (2.30) in which the branch flow equality (2.28d) is non-convex. Let x be the vector of decision variables in compact form. Define two convex quadratic functions

$$f_l(x) = (U_i + I_{ij}^l)^2, \ \forall l \in E_L \tag{2.37a}$$

$$g_l(x) = (U_i - I_{ij}^l)^2 + (2P_{ij}^l)^2 + (2Q_{ij}^l)^2, \ \forall l \in E_L \tag{2.37b}$$

equality (2.28d) can be replaced by two opposite inequalities

$$f_l(x) - g_l(x) \geq 0, \ \forall l \in E_L \tag{2.38a}$$

$$f_l(x) - g_l(x) \leq 0, \ \forall l \in E_L \tag{2.38b}$$

The first one in (2.38a) is merely the SOC inequality (2.31), while the second one in (2.38b) is a DC inequality. OPF-BFM (2.30) can be now expressed via a DCP

$$\min \{F_C \mid \text{Cons-BFM-Cr, Cons-BND, (2.38b)}\} \tag{2.39}$$

DC constraints are non-convex and do not alter problem complexity. Nevertheless, the special structure of DC inequality helps develop algorithms with good convergence properties based on convex optimization.

The proposed method, which is referred to as CCP-OPF (CCP is the abbreviation for convex-concave procedure) consists of three major steps: an initial SOCP relaxation, a feasibility recovery procedure, and an optimality recovery procedure. The first step refers to the well-known SOCP relaxation. If the relaxation is not exact, the infeasible solution is passed to the feasibility procedure, which produces a feasible power flow solution in the second step. Finally, an adjacent optimal solution is refined by the optimality recovery procedure in the third step, in case the solution offered in the second step is not optimal. The flowchart of CCP-OPF is shown in Algorithms 2.1 and 2.2.

Algorithm 2.1 CCP-OPF

1: Solve the SOCP relaxation model (2.32), the optimal solution is x^*. Evaluate the relaxation gap at x^* defined by

$$\text{Gap}_r(x^*) = \sum_l r_{ij}^l \left[I_{ij}^{l*} U_i^* - (P_{ij}^{l*})^2 - (Q_{ij}^{l*})^2 \right] \tag{2.40}$$

If $\text{Gap}_r(x^*) \leq \varepsilon$, where ε is a pre-specified tolerance, terminate and report the optimal solution x^* and the optimal value v^*.

2: Perform feasibility recovery procedure (summarized in Algorithm 2.2) with initial point x^*, and find a feasible power flow solution x^F.

3: Perform optimality recovery procedure (solve problem (2.39) via local NLP methods) with initial point x^F, and refine an optimal solution x^S.

In the feasibility recovery procedure, DC inequality (2.38b) which is ignored in the initial SOCP relaxation model (2.32) is convexified by linearizing concave terms and adding slack variables. Moreover, the total constraint violation is penalized in the objective function. The flowchart of the feasibility recovery procedure is summarized in Algorithm 2.2.

Algorithm 2.2 Feasibility recovery procedure

1: Initiating a penalty coefficient $\rho^1 > 0$, a penalty growth rate $\tau > 1$, a penalty upper bound ρ_M, a convergence tolerance $\varepsilon > 0$. Let the iteration index $k = 1$, and the initial point $x^1 = x^*$ (passed from step 1 of Algorithm 2.1).
2: Form linear approximation of $g_l(x)$ at x^k

$$\bar{g}_l(x, x^k) = g_l(x^k) + \nabla g_l(x^k)^T (x - x^k), \forall l \in E_L \tag{2.41}$$

and solve the following penalized problem

$$\begin{aligned} \min \quad & v(x, s) = F(x) + \rho^k \sum\nolimits_l s_l \\ \text{s.t.} \quad & x \in X, \; s_l \ge 0, \; \forall l \in L \\ & f_l(x) - \bar{g}_l(x, x^k) \le s_l, \; \forall l \in L \end{aligned} \tag{2.42}$$

the optimal solution is (x^{k+1}, s^{k+1}), and the optimal value is v^{k+1}.
3: Evaluate the relaxation gap at x^{k+1} according to (2.40), if $\text{Gap}_r(x^{k+1}) \le \varepsilon$, terminate and report $x^F = x^{k+1}$; otherwise; update $\rho^{k+1} = \min\{\tau\rho^k, \rho_M\}$, $k \leftarrow k + 1$, and go to step 2.

Its convergence is explained as follows.

If $\text{Gap}(x^k)$ becomes small enough before ρ^k reaches ρ_M, the convergence criterion has already been reached. Otherwise, the penalty parameter ρ^k will be fixed at ρ_M and the objective function of (2.42) does not change in the subsequent iterations. Next we consider the latter case, and demonstrate that the objective value $\{v(x^k, s^k)\}_k$ generated in step 2 monotonically decreases when $k > \log_\tau(\rho_M/\rho^1)$. Two basic facts on convex function $g_l(x)$ and its first-order Taylor expansion $\bar{g}_l(x, x^k)$ play an important role in the discussion.

F1: According to (2.41), $\bar{g}_l(x^k, x^k) = g_l(x^k)$.
F2: According to the property of convex functions, for an arbitrarily given x^k, $g_l(x) \ge \bar{g}_l(x, x^k)$, $\forall x$ holds.

The relaxation gap of line l in iteration k is defined as

$$\epsilon_l^k = f_l(x^k) - g_l(x^k) \tag{2.43}$$

and vector $\epsilon^k = [\epsilon_l^k]$, $\forall l$ will be frequently used.

The definition of ϵ_l^k and F1 implies that $f_l(x^k) - g_l(x^k) = f_l(x^k) - \bar{g}_l(x^k, x^k) \le \epsilon^k$, $\forall l$ is satisfied, therefore (x^k, ϵ^k) is a feasible solution of (2.42) in iteration k. Denote by (x^{k+1}, s^{k+1}) the optimizer of (2.42) in iteration k, we can assert that

$$v(x^{k+1}, s^{k+1}) \le v(x^k, \epsilon^k) \tag{2.44}$$

because $v(x, s)$ is minimized in that iteration, and the objective at optimal solution (x^{k+1}, s^{k+1}) deserves a value no greater than that corresponds to the feasible solution (x^k, ϵ^k). In addition, according to F2, we have

$$\epsilon_l^{k+1} = f_l(x^{k+1}) - g_l(x^{k+1}) \leq f_l(x^{k+1}) - \bar{g}_l(x^{k+1}, x^k) \leq s_l^{k+1}, \ \forall l \in E_L$$

The last inequality is a constraint of (2.42) which the optimal solution (x^{k+1}, s^{k+1}) must satisfy. For the same reason, $\epsilon_l^k \leq s_l^k, \forall l$. Because $v(x, s)$ is strictly increasing in its second input, we arrive at

$$v(x^k, \epsilon^k) \leq v(x^k, s^k) \tag{2.45}$$

Combining (2.44) and (2.45) we conclude the following monotonic relation

$$v(x^{k+1}, s^{k+1}) \leq v(x^k, s^k) \tag{2.46}$$

If one or more of the following conditions are met:

(a) $v(x^{k+1}, s^{k+1}) < v(x^k, \epsilon^k)$
(b) $U_i^{k+1} - I_{ij}^{l(k+1)} \neq U_i^k - I_{ij}^{lk}$
(c) $P_{ij}^{l(k+1)} \neq P_{ij}^{lk}$
(d) $Q_{ij}^{l(k+1)} \neq Q_{ij}^{lk}$

strict inequality holds in (2.46), which is usually met in practical OPF problems, therefore, the sequence $\{v(x^k, s^k)\}_k$ is decreasing, indicating that as long as the convergence criterion is not satisfied, $v(x, s)$ can be strictly improved in the next iteration. Since v has a finite optimum, and problem (2.42) is bounded below, Algorithm 2.2 converges with $v(x^k, s^k)$ approaching constant values. Because OPF problems with network losses usually has a unique optimal solution [43], there is a one-to-one mapping between the optimal solution and the optimal value, so (x^k, s^k) also converges.

To ensure slack variables s^k converge to 0, the value of ρ_M should be greater than the exact penalty parameter ρ_M^* [62]. According to the exact penalty function theory [63, 64], if OPF problem (2.39) is feasible and certain constraint qualification holds, there will be a finite penalty parameter ρ_M^*, such that for any $\rho^* \geq \rho_M^*$, the following penalized problem

$$\min_{x \in X} \ F(x) + \rho^* \sum\nolimits_l [f_l(x) - g_l(x)]$$

has the same optimal solution as (2.39).

In theory, if ρ_M is not big enough, problem (2.42) may give $x^{k+1} = x^k$ before s^k approaches 0 due to the lack of enough penalty. In such circumstance, v^k (ϵ^k) cannot be improved (reduced) any more, and Algorithm 2.2 may fail to converge.

Nevertheless, numerical tests suggest that Algorithm 2.2 works well with penalty parameters in a fairly wide range.

Owing to the convergence criterion used in Algorithm 2.2, the solution x^F offered by the feasibility recovery procedure may not be optimal. Thus we incorporate optimality recovery procedure to refine an adjacent optimal solution. Any traditional OPF methods or NLP solvers are eligible for this task, such as the interior point method [27–29], and the sequential quadratic programming method [30–32]. Nonetheless, if we delicately choose penalty parameters in Algorithm 2.2, it will directly procure a solution that is very close to, if not identical to, the global optimal one. Additional observations are given below.

1. From our numerical experiences, to recover the global optimal solution, it is very useful to use a small initial value of ρ^1, and gradually increase it with a moderate growth rate τ. First, temporarily constraint violations may allow x^k to move to a more favorable region in which the optimal solution stays. Second, light penalty does not cause dramatic change of optimal solutions in two successive iterations. Since $\Delta x^k = |x^{k+1} - x^k|$ is small, $\bar{g}_l(x, x^k)$ can provide relative accurate approximation for $g_l(x)$ in k-th iteration. Note that constraint violation will be eliminated by the growing penalty parameter ρ^k. In consequence, Algorithm 2.2 will spend more iterations before it can converge.
2. Despite the existence of an exact penalty parameter ρ_M^*, its value can scarcely be determined in advance. In fact, we do not need its exact value. We can prudently choose a relatively large value for ρ_M. According to our experience, Algorithm 2.2 will converge even when ρ^k is much smaller than ρ_M. One reason is that $\text{Gap}_r(x^k) = \sum_l r_{ij}^l \epsilon_l^k / 4$ is actually quantified by ϵ^k. According to the previous analysis, ϵ^k is generally strictly smaller than s^k, and thus the convergence criterion could be met before all slack variables approach to 0. In such a circumstance, little can be said on the monotonicity of the optimal value sequence v^k. Experimental results suggest that $\{F(x^k)\}_k$ will be increasing and $\{\text{Gap}_r(x^k)\}_k$ will be decreasing. However, rigorous mathematical proof is nontrivial, because the feasible region of (2.42) which depends on x^k changes in every iteration, which is different from the situation in the traditional penalty function theory.
3. Failure of convergence may occur in two cases: (a) When the original OPF problem (2.39) is infeasible. In such a situation, the exact penalty parameter ρ_M^* does not exist; (b) If the initial relaxation is very poor or an arbitrary initiation is used. In such a situation, Algorithm 2.2 may converge very slowly. Possible remedy would be using valid inequalities [60, 61] to strengthen the initial relaxation.
4. In theory, CCP needs an initial point to perform linear approximation in (2.41). In Algorithm 2.2, the initial value is offered by OPF-BFM-Cr (2.32). In this regard, Algorithm 2.1 makes no reference to any heuristic initial guess.

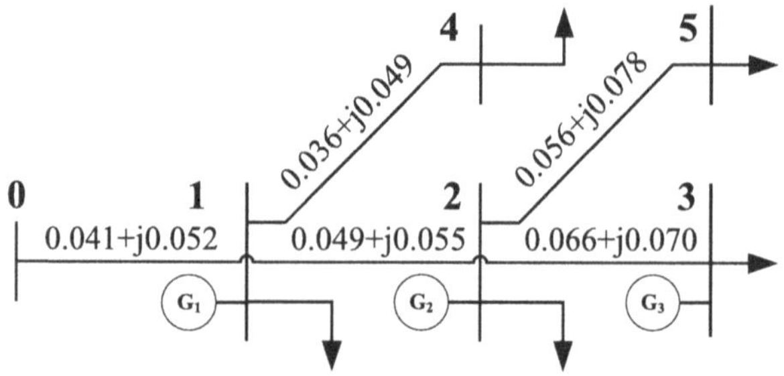

Fig. 2.6 Topology and line data of the 6-bus system

Table 2.2 Parameters of generators (in p.u.)

Unit	a_i^g	b_i^g	p_i^{gm}	q_i^{gm}
G_1	1.58	7.62	1.5	1.0
G_2	2.13	6.10	1.5	0.5
G_3	1.09	8.85	1.5	0.5

A 6-bus system is used to illustrate the performance of Algorithms 2.1 and 2.2. Network topology and line data are given in Fig. 2.6. Generator data is provided in Table 2.2, $p_i^{gn} = 0$, $q_i^{gn} = -q_i^{gm}$, $\forall i \in E_B$. Active and reactive nodal power demands are $p_i^d = 0.6$, $q_i^d = 0.2$, $\forall i \in E_B$. Voltage magnitude limits are $U_i^n = 0.90$, $U_i^m = 1.05$, $U_0 = 1.05$. Parameters and test results are given in p.u. This example is devoted to exhibiting the performance of the proposed method in terms of convergence rate and solution quality, when the initial SOCP relaxation is not exact. The scalability and efficiency will be demonstrated on practically sized power systems in Sects. 2.3.2 and 2.3.3. A main consideration is that it is usually more difficult to recover a feasible or optimal solution in the forthcoming two variant OPF problems. In this test, Algorithm 2.2 converges and the solution is already very close to the global optimal one. To highlight this feature, optimality recovery procedure is not deployed.

Case 1 The objective is to minimize the production cost F_C defined in (2.30). Three scenarios are under investigation. In scenario i, $i = 1, 2, 3$, only the cost parameters a_i^g and b_i^g take values in accordance with those in Table 2.2, and others are intentionally set to 0. In this way, the objective F_C is not strictly increasing in every p_i^g. This situation is not contrived, because distributed renewable generation units, which are not rare in active distribution networks, usually have zero marginal production costs. Given penalty parameters $\rho^1 = 10^{-3}$, $\tau = 2$, and convergence tolerance $\varepsilon = 10^{-6}$ in Algorithm 2.2, results are provided in Table 2.3. The second column is the initial relaxation gap $\text{Gap}_r(x^0)$. We can see that SOCP relaxations in the former two scenarios are inexact. Feasibility recovery procedure closes the relaxation gap within the pre-specified threshold in one or two iterations, and identifies the global optimal solution in all three scenarios. Global optimality is verified by comparing the optimal values with those reported by BARON, a global optimization solver for NLPs using spatial branch-and-bound algorithm.

Table 2.3 Performance of CCP-OPF method in Case 1

Scenario	$\text{Gap}_r(x^0)$	$\text{Gap}_r(x^k)$	Iter.	F_C	BARON
1	0.0712	4.92×10^{-7}	2	10.015	10.015
2	0.0382	1.45×10^{-7}	1	7.3265	7.3265
3	4.26×10^{-9}	4.26×10^{-9}	0	0.5321	0.5321

Table 2.4 Performance of CCP-OPF method in Case 2 with $\rho^1 = 10^{-3}$

λ (10^{-4})	$\text{Gap}_r(x^0)$	$\text{Gap}_r(x^k)$	Δx^k
0	0.0576	1.67×10^{-8}	1.9340
0.5	0.0280	4.87×10^{-9}	3.1460
1.0	0.0259	4.29×10^{-9}	3.0868
1.5	0.0238	3.98×10^{-9}	3.0296
2.0	0.0217	3.93×10^{-9}	2.9730
2.5	0.0038	9.29×10^{-9}	2.4228
3.0	0.00001	4.78×10^{-8}	2.2933

Table 2.5 Optimality gap in Case 2 with $\rho^1 = 10^{-3}$

λ (10^{-4})	$F(x^*)$	BARON	Gap_F
0	0.0078	0.0077	1.29%
0.5	0.0093	0.0090	3.33%
1.0	0.0106	0.0104	1.92%
1.5	0.0120	0.0117	2.48%
2.0	0.0134	0.0130	3.07%
2.5	0.0148	0.0144	2.78%
3.0	0.0161	0.0157	2.49%

Case 2 All generator cost data take values as those shown in Table 2.2. In distribution system operation, one task is to maintain the voltage magnitude at every bus close to its reference value, yielding the following objective

$$F_V = \sum_i (U_i - U_i^r)^2 \tag{2.47}$$

where the voltage reference values are selected as $U_i^r = 1.00, \forall i \in E_B$. Although F_V is convex, it is not monotonic. Thus convex relaxation for pertinent OPF problems is generally inexact if F_V appears in their objective(s). In this test, the final objective is to minimize the weighted-sum of F_C and F_V, i.e.,

$$F(x) = \lambda F_C(x) + (1 - \lambda) F_V(x) \tag{2.48}$$

This problem is closely related to the bi-objective OPF problem discussed in Sect. 2.3.3. Let $\rho^1 = 10^{-3}$, $\tau = 2$, and change the weight parameter λ from 0 to 3.0×10^{-4}. Results are provided in Tables 2.4 and 2.5.

Because F_V is not strictly monotonic, the SOCP relaxation is inexact when $\lambda \leq 3 \times 10^{-4}$. Feasibility recovery procedure closes the gap within one iteration in all cases, and identifies a feasible power flow solution. To examine the solution quality, the optimal values offered by BARON are shown in the third column of Table 2.5. We can observe that when $\rho^1 = 10^{-3}$, the optimal values $F(x^*)$ offered by Algorithm 2.2 are slightly higher than those $F(x^B)$ offered by BARON, the optimality gap

$$\text{Gap}_F = \frac{F(x^*) - F(x^B)}{F(x^B)} \times 100\% \tag{2.49}$$

shown in the last column of Table 2.5 varies from 1% to 3%. This is because the changes in optimal solutions $\Delta x^k = |x^k - x^{k-1}|$ shown in the last column of Table 2.4 are relatively large, thus the linear approximation $\bar{g}(x, x^k)$ is not accurate enough.

We further reduce the value of ρ^1 and investigate the performance of FRP again. Results are shown in Table 2.6. We can see that when a smaller ρ^1 is adopted, the optimal value offered by Algorithm 2.2 declines to the global optimum, and Δx^k in the last iteration decreases as well, while the number of iterations increases at the same time. In view of this, there should be a trade-off between optimality and efficiency. In this case, $\rho^1 = 10^{-4}$ seems to be a good choice. Computation time is not provided in this case because the system is small, but will be elucidated in the subsequent sections.

When $\lambda = 0$ and $\rho^1 = 10^{-5}$, the values of relaxation gap $\text{Gap}_r(x^k)$ defined in (2.40), optimality gap $\text{Gap}_F(x^k)$ defined in (2.49), and objective function $F(x^k)$ defined in (2.48) generated in Algorithm 2.2 are plotted in Fig. 2.7, showing that the two gaps decrease to 0, and the objective value approaches the global optimum.

When $\lambda > 3.1 \times 10^{-4}$, the SOCP relaxation is exact, and the optimal value does not change with respect to λ. In fact, when λ varies from 0 to 3.1×10^{-4}, the optimal solutions constitute the Pareto front of a bi-objective OPF problem with F_C and F_V being the objectives. However, it is not easy to derive a convincing λ because F_C and F_V exhibit different orders of magnitudes.

Table 2.6 Performance of CCP-OPF in Case 2 with other ρ^1

λ	$\rho^1 = 10^{-4}$			$\rho^1 = 10^{-5}$		
10^{-4}	Iter.	$F(x^*)$	Δx^k	Iter.	$F(x^*)$	Δx^k
0	1	0.0077	1.6450	4	0.0077	0.0192
0.5	1	0.0090	2.6629	4	0.0090	0.1922
1.0	1	0.0104	2.4598	3	0.0104	0.4348
1.5	2	0.0118	0.2934	3	0.0117	0.4220
2.0	1	0.0131	2.1223	3	0.0130	0.4778
2.5	1	0.0145	1.4825	3	0.0144	0.4784
3.0	1	0.0158	1.2767	2	0.0157	0.2439

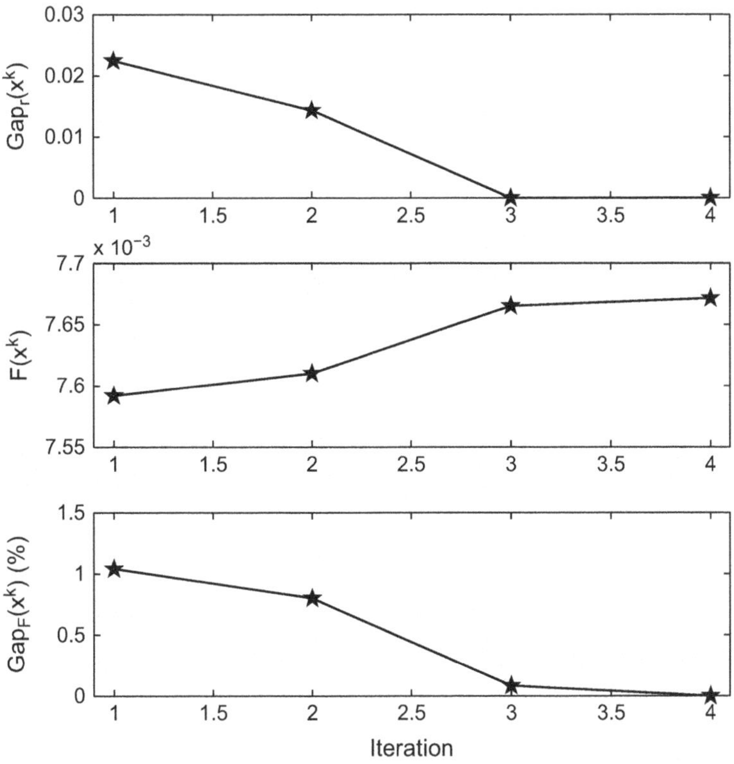

Fig. 2.7 Convergence performance of FRP when $\lambda = 0$ and $\rho^1 = 10^{-5}$

5. Extensions for Meshed Networks

1. Lifted Formulation

In the SDP relaxation model (2.26) for OPF-BIM, exactness is certified if the optimal solution X is a rank-1 matrix. To build a connection between the SDP method and CCP-OPF method, we investigate a DC formulation for the constraint $\text{rank}(X) = 1$. Consider the rank-1 factorization $X = xx^T$ in an element-wise form

$$X_{ij} = x_i x_j, \ \forall i, j$$

resulting in

$$X_{ij} X_{ji} = x_i^2 x_j^2 = X_{ij}^2 = X_{ii} X_{jj}, \ \forall i, j$$

The last equality can be replaced by opposite inequalities

$$X_{ij}^2 \le X_{ii} X_{jj}, \ \ X_{ij}^2 \ge X_{ii} X_{jj}, \ \forall i, j$$

The former inequality is redundant to LMI $X \succeq 0$. The latter one is in fact a DC inequality $f_{ij}(X) - g_{ij}(X) \le 0, \forall i, j$, where

$$f_{ij}(X) = (X_{ii} + X_{jj})^2, \ \forall i, j$$

$$g_{ij}(X) = (X_{ii} - X_{jj})^2 + 4X_{ij}^2, \ \forall i, j$$

These constraints can be handled by a framework similar to the one presented in Algorithm 2.2. This formulation eliminates the original OPF variable x and works with the lifted variable X, so is called the lifted formulation.

2. Extended Formulation

In the SDP relaxation, we actually use $X \succeq 0$ to approximate quadratic equality $X = xx^T$, which can be rewritten as two opposite LMIs

$$X - xx^T \succeq 0$$

$$xx^T - X \succeq 0$$

Applying Schur complement theorem, the first one actually boils down to

$$\begin{bmatrix} 1 & x^T \\ x & X \end{bmatrix} \succeq 0 \tag{2.50}$$

which is also an LMI. The second one is non-convex. We seek for simpler equivalent condition. Let b_i, $i = 1, \ldots, n$ be a set of orthogonal basis in $\mathbb{R}^n$, and the following condition holds

$$\left\langle X, b_i b_i^T \right\rangle \le (b_i^T x)^2, \ i = 1, \ldots, n \tag{2.51}$$

Then for any vector $v = \lambda_1 b_1 + \cdots + \lambda_n x_n$, we have

$$v^T (xx^T - X)v = \sum_{i=1}^{n} \lambda_i^2 \left[(b_i^T x)^2 - \left\langle X, b_i b_i^T \right\rangle \right] \ge 0$$

So $xx^T - X \succeq 0$ holds.

Furthermore, let's assume b_i, $i = 1, \ldots, n$ constitute the standard orthogonal basis in $\mathbb{R}^n$, in such situation, condition (2.51) degenerates into

$$X_{ii} - x_i^2 \le 0, \ i = 1, \ldots, n \tag{2.52}$$

which is clearly a DC inequality, and can be simply handled by a framework similar to the one presented in Algorithm 2.2. This formulation works with both the original OPF variable x and the lifted variable X, and the matrix variable in LMI (2.50) adds an additional row and column to X, so it is called the extended formulation, whose computation complexity is not significantly changed. In addition, valid inequalities can be considered since x is explicitly modeled. See Appendix A.3.1 for more details. This technique has been used in [65] to develop disjunctive cuts and integer programming models, which provides another way to solve a non-convex quadratic program.

Constraint (2.52) imposes an upper bound on the diagonal elements of X. In other words, it prevents X_{ii} from being too large. To remove the non-convex constraint, a heuristic method would be adding a penalty term $\rho \text{Tr}[\Lambda X]$ in the objective function, where Λ is a constant diagonal matrix, and ρ is the penalty parameter. Certainly, the choices of these parameters influence the OPF results. This technique has been discussed in [66]. For OPF problems, such a penalty term reflects the weighted sum of voltage magnitude squares. In power system operation, the bus voltage is tightly monitored, which may restrict the performance of this trick. From another perspective, (2.52) prefers larger off-diagonal elements of X. It is noticed in [46] that the decrease in system reactive power generation and line losses leads to the increase of off-diagonal entries of X, so helps to recover a low rank solution.

3. Other Possible Means

Another choice for OPF over meshed networks is the extended conic quadratic model (2.36) (with angle constraints) or the cycle-based model in [61], in which power flow equations in bus-injection format are converted to linear equalities, rotated conic quadratic equalities, and arctangent equalities (or cycling polynomials) via variable transformation. If an appropriate DC decomposition for the arctangent function (or the cycling polynomial) is available, CCP-OPF method can be extended for coping with meshed networks.

Please bear in mind that the suggested framework requires solving SDPs repeatedly, which may not be efficient for large-scale systems, in spite of potential acceleration enabled by exploiting the special sparsity pattern. Nonetheless, these discussions may encourage new research that will leverage the computation superiority of convex optimization by exploring more appropriate DC formulation for the non-convex constraints.

The 2-bus system in [67] is studied to illustrate the performance of method suggested above. It is well-known that the SDP relaxation is not exact for this system when the voltage upper bound parameter $V_2^m \in [0.98, 1.03]$ p.u., although the system is not meshed in topology. The exact global optimums offered by BARON are shown in the last column of Table 2.7. The optimal values offered by SDP relaxation model (2.26) are listed in the second column, showing an increasing optimality gap with growing V_2^m. Using the lifted formulation, we deploy an iterative method similar to Algorithm 2.2 with $\rho^1 = 200$, $\tau = 1.5$. It successfully recovers the global solution after a few number of iterations. When $V_2^m = 0.989$, the vector $\varsigma(X)$ which consists of singular values of the optimal solution X in each iteration is provided in Table 2.8. Since rank(X) is equal to the number of non-zero elements in $\varsigma(X)$, the initial SDP relaxation gives a rank-3 solution, and the

Table 2.7 Computation performance with $\rho^1 = 200$, $\tau = 1.5$

	SDP-CCP			
V_2^m	Obj-0	Obj-n	Iter.	BARON
0.976	905.76	905.76	0	905.76
0.983	903.12	905.73	3	905.73
0.989	900.84	905.73	4	905.73
0.996	898.17	905.73	7	905.73
1.002	895.86	905.73	12	905.73

Table 2.8 $\varsigma(X)$ in each iteration when $V_2^m = 0.989$

Iter. 0	Iter. 1	Iter. 2	Iter. 3	Iter. 4
1.8783	1.8794	1.8763	1.8675	1.8552
0.0020	0.0012	0.0011	0.0007	0.0000
0.0003	0.0000	0.0000	0.0000	0.0000
0.0000	0.0000	0.0000	0.0000	0.0000

final optimal solution is indeed rank-1. It is also observed that when the initial SDP relaxation becomes weaker, more iterations are needed to recover a rank-1 solution.

2.3.2 *Maximum Loadability Problem*

1. Brief Introduction

Loadability refers to the ability of power systems to maintain a feasible power flow status and reliable power delivery when the demand varies. Insolvability of power flow equations is closely related to voltage instability, which is one of the major threats to a secure operation of distribution networks.

Traditional approaches for calculating maximum loadability or voltage stability margin can be categorized into continuation-based ones and optimization-based ones. The former methods solve power flow equations repeatedly along a load increasing direction till a bifurcation point is found, such as those in [68–70]. The latter ones directly solve an optimization problem which maximizes the served load along a given direction subject to the solvability of power flow equations, such as those in [71, 72]. In this regard, maximum loadability is a variant of OPF problem. It is pointed out in [72] that optimization-based approaches are more flexible in modeling various regulation measures and operating constraints, and could usually be more effective than continuation-based ones in maximum loadability calculation.

Although existing continuation power flow techniques are mature and tractable for large-scale instances, they may encounter challenges in determining a proper initial point that satisfies all network and equipment operating constraints, which may be difficult to acquire when an initial feasible power flow point is not clear. In addition, continuation approaches need special treatment for updating the search step around the bifurcation point. Optimization-based approaches are convenient in model setup, but NLP solvers may not be very robust due to the non-convexity of power flow equations.

To overcome these difficulties, convex relaxation models are investigated in [73] and [74], for detecting power flow insolvability. SDP and SOCP belong to the category of convex optimization, and can be solved reliably and efficiently. In general, these two models can estimate an upper bound on the loadability margin without the need of a manually supplied start point. Unfortunately, although the exactness of SDP and SOCP relaxations can be guaranteed in cost-minimum OPF problems under some technical conditions, maximum loadability applications may fail to meet those conditions [73, 74]. As a result, the outcomes offered by these convex relaxation models are usually inexact and can be over-optimistic.

Motivated by [73, 74], this chapter proposes a convex optimization based approach to compute loadability margin of radial networks, which is formulated as a special OPF problem, which can be solved by Algorithm 2.1. Different from SDP and SOCP relaxation models in [73] and [74], Algorithm 2.1 generates an exact loadability margin.

2. Mathematic Model

With notations defined in Sect. 2.3.1, maximum loadability problem of radial networks can be expressed as a special OPF problem as follows:

$$\begin{aligned} \max_{x,\eta} \quad & \eta \\ \text{s.t.} \quad & \text{Cons-BFM, Cons-BND} \\ & p^d = p^{d0} + \eta \Delta p \\ & q^d = q^{d0} + \eta \Delta q \end{aligned} \tag{2.53}$$

where active (reactive) power demands p^{d0} (q^{d0}) correspond to a certain load scenario; Δp and Δq represent load incremental directions. Such parameters should be specified by the system operator. As a straightforward application of the proposed OPF method, we are not aiming to develop a comprehensive model that captures every detail in voltage stability or continuation power flow calculation. In case of need, generators can be treated as ideal voltage sources [73]; to consider reactive power limits of generators, complementarity constraints are proposed in [72] to model the logic of bus type switching. Problem (2.53) is a variant of OPF problem (2.30), in view of their similarities in constraints. Because the objective function fails to meet conditions which ensure a tight SOCP relaxation, Algorithm 2.1 is applied to solve problem (2.53), and the exact loadability margin can be recovered.

3. Case Study

Besides the 6-bus system investigated in Sect. 2.3.1, several practically-sized distribution systems are also tested in this section. The data sets are available in [75] and [76]. Computation is performed on a laptop with Intel i5-3210M CPU and

4 GB memory. All test systems are named in the form of "DTN-X" where X is the number of buses, for example, DTN-6 refers to the previous 6-bus system. In problem (2.53), the nodal active and reactive power demands are chosen as $p^d = \eta p^{d0}$, $q^d = \eta q^{d0}$, where p^{d0} and q^{d0} are constants given by system load data, and η is the loadability index, which is to be maximized. Parameters of Algorithm 2.2 are chosen as $\rho^1 = 10^{-4}$, $\tau = 2$, and $\varepsilon = 10^{-6}$. To compare Algorithm 2.1 with NLP solvers, problem (2.53) is also solved by KNITRO [77] and BARON [78]. Results are provided in Tables 2.9 and 2.10.

We can observe that KNITRO fails to solve problem (2.53) associated with DTN-30, DTN-69, and DTN-861, while BARON only successes in DTN-6, DTN-14, and DTN-33, due to the increasing problem sizes. In other cases, BARON does not converge within the default time limit. Algorithm 2.1 successfully solves all instances, verifying its robustness and scalability. It is also confirmed that Algorithm 2.1 offers consistent loadability margins with those provided by KNITRO and BARON (if both of them work well), and is more efficient. Finally, it can be seen from the number of iterations in Table 2.9 that SOCP relaxation performed on problem (2.53) is generally not exact. Algorithm 2.2 manages to recover the feasible, and also optimal, solution in a few number of iterations.

Table 2.9 Computation performance of Algorithm 2.1

System	η	$\text{Gap}_r(x^k)$	Iter.	Time (s)
DTN-6	3.1488	4.27×10^{-9}	1	0.44
DTN-14	0.5265	2.31×10^{-8}	4	0.63
DTN-30	1.5586	1.64×10^{-8}	1	0.85
DTN-33	4.1115	5.88×10^{-7}	3	1.16
DTN-57	3.1545	3.65×10^{-7}	3	1.55
DTN-69	4.1737	1.75×10^{-7}	9	6.03
DTN-123	2.4532	2.73×10^{-10}	1	2.04
DTN-861	7.9859	4.04×10^{-7}	2	28.3

Table 2.10 Computation performances of NLP solvers

	BARON		KNITRO	
System	η	Time (s)	η	Time (s)
DTN-6	3.1488	3.72	3.1488	1.52
DTN-14	0.5265	14.9	0.5265	1.42
DTN-30	–	–	Fail	Fail
DTN-33	4.1114	332	4.1115	2.40
DTN-57	–	–	3.1545	2.56
DTN-69	–	–	Fail	Fail
DTN-123	–	–	2.4532	8.32
DTN-861	–	–	Fail	Fail

2.3.3 Bi-objective OPF Problem

1. Brief Introduction

Generation dispatch of power systems may involve multiple objectives, such as reducing fuel cost, carbon emission, net losses, or improving stability, voltage profile, etc. In many cases, several objectives should be coordinated or compromised, giving rise to a multi-objective OPF problem. Methods for general multi-objective optimization problems can be divided into two classes. One of them procures only one Pareto solution that compromises all objectives through a scalarized model, including the goal programming method, the weighted-sum method, and the ε-constrained method, see [79, 80] for more comprehensive surveys. A common difficulty for these methods would be how to determine convincing weights or parameters for building a single-objective optimization model that properly balances all targets. The other provides the entire Pareto front or evenly distributed Pareto points, such as the normal boundary intersection (NBI) method [81], and evolutionary algorithms [82]. The user selects a final decision according to his own preference.

This section proposes a non-parametric scalarization model for bi-objective OPF problems, which can be convexified and solved by Algorithm 2.1. The Pareto front can be computed, in case of need, via Algorithm 2.1 embedded in the ε-constraint method (with multiple values of ε which produce a well-distributed Pareto front) or the NBI method.

2. Mathematic Model

With notations defined in Sect. 2.3.1, a bi-objective OPF problem can be expressed as follows:

$$
\begin{aligned}
\min \quad & \{F_1(x), F_2(x)\} \\
\text{s.t.} \quad & \text{Cons-BFM} \\
& \text{Cons-BND}
\end{aligned}
\tag{2.54}
$$

where objective functions $F_1(x)$ and $F_2(x)$ are assumed to be convex. Let x_1^* (x_2^*) denote the unilateral optimal solution if only $F_1(x)$ ($F_2(x)$) is optimized, and $F_1^n = F_1(x_1^*)$, $F_1^m = F_1(x_2^*)$, $F_2^n = F_2(x_2^*)$, $F_2^m = F_2(x_1^*)$, then a compromising solution x^* of problem (2.54) can be determined by solving the following optimization problem with a scalarized objective:

$$
\begin{aligned}
\max \quad & \left(F_1^m - F_1(x)\right)\left(F_2^m - F_2(x)\right) \\
\text{s.t.} \quad & \text{Cons-BFM, Cons-BND}
\end{aligned}
\tag{2.55}
$$

Its optimal solution x^* has two attractive properties:

1. x^* keeps invariant even if either objective is multiplied by a positive scalar.
2. x^* is non-dominated (Pareto optimal).

The first property is clear. For example, if $F_1(x)$ becomes $\sigma F_1(x)$, where $\sigma > 0$ is a constant, then both F_1^m and the objective of (2.55) will be multiplied by σ, and the optimal solution remains the same. This property implies that x^* is independent of the relative ratio between the two objectives.

To see the second one, suppose x^* is not a Pareto solution, and can be dominated by x^{**}, then it must satisfy $F_1(x^{**}) \le F_1(x^*)$, $F_2(x^{**}) \le F_2(x^*)$, and at least one of them holds as a strict inequality, so the objective function of (2.55) receives a larger value at x^{**} than at x^*, which is in contradiction with the assumption that x^* is an optimal solution of problem (2.55).

To convexify the objective function, we introduce auxiliary variables t_1, t_2, and t_b, and then build the problem

$$\begin{aligned} \max \ & t_b \\ s.t. \ & \text{Cons-BFM, Cons-BND} \\ & t_1 \le F_1^m - F_1(x) \\ & t_2 \le F_2^m - F_2(x) \\ & t_1 \ge 0, \ t_2 \ge 0, \ t_1 t_2 \ge t_b^2 \end{aligned} \tag{2.56}$$

Problem (2.56) and problem (2.55) share the same optimal solution in variable x, and $\sqrt{(F_1^m - F_1(x^*))(F_2^m - F_2(x^*))} = t_b^*$ holds at optimum. The last hyperbolic inequality of (2.56) can be replaced with an SOC inequality

$$\left\| \begin{matrix} 2t_b \\ t_1 - t_2 \end{matrix} \right\| \le t_1 + t_2$$

Because $F_1(x)$ and $F_2(x)$ are convex, only the non-convexity only rests in the branch flow equality in Cons-BFM. In this regard, problem (2.56) is another variant of OPF problem (2.30), and can be solved by Algorithm 2.1.

If the goal of a bi-objective OPF problem is to find a set of uniformly distributed Pareto points, rather than only one non-dominated solution, the ε-constraint method and NBI method will be good choices for this task. The former approach solves the following problem with some values of ε that are delicately sampled from interval $[F_2^l, F_2^m]$

$$\min \ \{F_1(x) : \text{Cons-BFM, Cons-BND}, \ F_2(x) \le \varepsilon\} \tag{2.57}$$

The last inequality is convex. Problem (2.57) with a given ε is similar to problem (2.30), and can be solved by Algorithm 2.1.

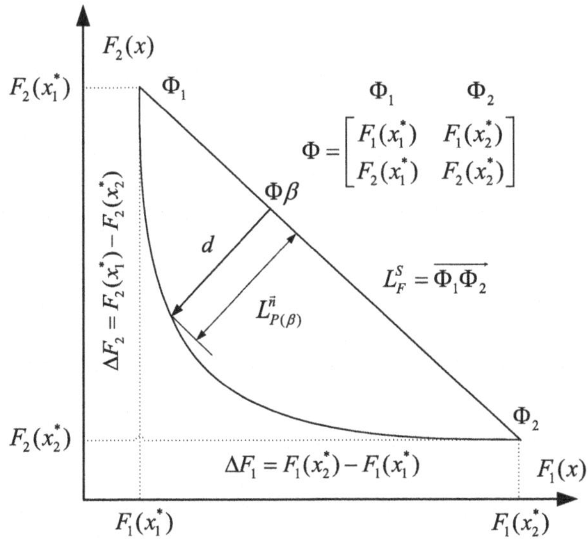

Fig. 2.8 Illustration of NBI method

The NBI method has been applied to the multi-objective OPF problem in NLP form in [83]. The bi-objective case is elucidated in Fig. 2.8. Two extreme points of the Pareto front are

$$\Phi_1 = \begin{bmatrix} F_1(x_1^*) \\ F_2(x_1^*) \end{bmatrix}, \quad \Phi_2 = \begin{bmatrix} F_1(x_2^*) \\ F_2(x_2^*) \end{bmatrix}$$

Define matrix $\Phi = [\Phi_1, \Phi_2]$, then any point on line segment L_F^S connecting Φ_1 and Φ_2 can be expressed via

$$P(\beta) = \Phi\beta, \ \beta = [\beta_1, \beta_2], \ 0 \le \beta_1, \beta_2 \le 1, \ \beta_1 + \beta_2 = 1$$

where parameter β controls the location of point $P(\beta)$.

Let $\mathbf{n}$ be a unit vector that is perpendicular to L_F^S, which can be calculated as

$$\mathbf{n} = \frac{-[\Delta F_1, \Delta F_2]^T}{\sqrt{\Delta F_1^2 + \Delta F_2^2}}$$

where $\Delta F_1 = F_1^m - F_1^n > 0$, $\Delta F_2 = F_2^m - F_2^n > 0$.

Provided with well-distributed initiating points $P(\beta_m)$, $m = 1, 2, \ldots, M$ on L_F^S, the key step performed in NBI method is to detect a Pareto point on line $L_{P(\beta_m)}^{\mathbf{n}}$ which passes through point $P(\beta_m)$ and is orthogonal to L_F^S. This task can be implemented by solving the following problem [81, 83]

$$
\begin{aligned}
& \max_{x,d} \ d \\
& s.t. \ \text{Cons-BFM}, \ \text{Cons-BND} \\
& \Phi\beta_m + d\mathbf{n} = \begin{bmatrix} F_1(x) \\ F_2(x) \end{bmatrix}
\end{aligned}
\tag{2.58}
$$

If both $F_1(x)$ and $F_2(x)$ are linear, the last constraint is also linear. The Pareto point detection problem (2.58) is similar to the maximum loadability problem (2.53), and the optimal solution x_m^* can be solved by Algorithm 2.1. The set of points $[F_1(x_m^*), F_2(x_m^*)]^T$, $m = 1, 2, \ldots$ are almost evenly distributed on the Pareto front. The last equality can be relaxed as convex inequality

$$
\Phi\beta_m + d\mathbf{n} \geq \begin{bmatrix} F_1(x) \\ F_2(x) \end{bmatrix}
$$

when $F_1(x)$ and $F_2(x)$ are nonlinear but convex functions. The geometric meaning of this inequality can be explained through Fig. 2.8. For a given number d, $\Phi\beta_m + d\mathbf{n}$ defines a point, which still resides in the first quadrant if the origin moves to the new point $(F_1(x), F_2(x))$. Since d is to be maximized, the point $\Phi\beta_m + d\mathbf{n}$ will touch the Pareto front at the optimal solution, and thus equality holds.

3. Extension to Incorporating More Objectives

We assume there are totally K convex objective functions, which can be approximated via PWL functions using the convex combination method presented in Appendix B.1.1; the worst outcome of objective k when the remaining ones are optimized is denoted by F_k^m. Similar to the bi-objective case, the scalarized problem can be formulated as

$$
\begin{aligned}
& \max \ \prod_k \left(F_k^m - c_k^T x \right) \\
& \text{s.t. Cons-BFM, Cons-BND}
\end{aligned}
\tag{2.59}
$$

where $c_k^T x$ is the linearized objective function. Problem (2.59) cannot be convexified as (2.55). Nevertheless, maximizing $\prod_k \left(F_k^m - c_k^T x \right)$ is equivalent to maximizing its logarithm, because a logarithm function is monotonically increasing, and every production term is linear in x. In view of this, we can resort to solving the following problem

$$
\begin{aligned}
& \max \ \sum_k \ln \left(F_k^m - c_k^T x \right) \\
& \text{s.t. Cons-BFM, Cons-BND}
\end{aligned}
\tag{2.60}
$$

in which the objective function is strictly concave, and non-convexity only exists in Cons-BFM. Problem (2.60) can be solved by Algorithm 2.1.

4. A Tailored Feasibility Cut

We observe that in the penalized problem (2.42), if we replace the feasibility cut with $f_l(U^k_{i,t}, I^l_{ijt}) - \bar{g}_l(x, x^k) \le s_l$, or more precisely, fix the value of voltage variables at the optimal solution in the previous step, then Algorithm 2.2 may exhibit a better performance in voltage regulation problems with a non-monotonic objective function in voltage magnitude like (2.47). The motivation is that: the inexactness of convex relaxation is caused by the over-high value of bus voltage whenever this is preferred by the optimization target. Since we keep it constant in the feasibility cut, the line current cannot be too high as larger currents lead to higher network losses. Nevertheless, a rigorous proof on convergence is non-trivial.

5. Case Study

Testing systems in Sect. 2.3.2 are used again for studying the bi-objective OPF problem. Production cost $F_C(x)$ in (2.30) and voltage profile $F_V(x)$ in (2.47) are considered for optimization. The computation environment is the same as that in Sect. 2.3.2. Bound parameters F^l_C, F^m_C, F^l_V, and F^m_V of both objective functions are given in Table 2.11. The base value of cost is \$1000.

The bi-objective OPF problem associated with DTN-6 is studied first. The Pareto front is computed by using the ε-constraint method, and shown in Fig. 2.9. The trade-off solution retrieved from problem (2.56) is plotted in the same figure, located in the "middle" of Pareto front without a manually supplied parameter or particular normalization process. This is an attractive feature because the values of different objectives usually exhibit different orders of magnitudes, and the optimal solution could be extremely sensitive to the normalization parameter.

Results of remaining systems are shown in Tables 2.12, 2.13 and 2.14, where x^*_b is the tradeoff solution computed from (2.56), $B_V(x^*_b) = [F^m_C - F_C(x^*_b)][F^m_V - F_V(x^*_b)]$ is the optimal value of (2.55). Algorithm 2.1 is applied to solve these bi-objective OPF problems in form of (2.56), and successes in all instances within a few number of iterations. The computation time for DTN-861 is a little bit longer, but still acceptable. For NLP solvers, BARON manages to solve most of these instances, as for DTN-861, it fails to converge within the default time limit. Results shown

Table 2.11 Bound parameters of F_C and F_V

System	F^l_C	F^m_C	F^l_V	F^m_V
DTN-6	7.3265	15.8204	0.0077	0.0137
DTN-14	8.7638	10.328	0.0280	0.0745
DTN-30	11.785	13.685	0.0984	0.5950
DTN-33	35.5128	36.4881	0.2710	0.4379
DTN-57	12.410	13.860	0.0231	0.0535
DTN-69	46.2022	49.8723	0.3356	0.7408
DTN-123	24.9816	26.2172	0.1647	0.4568
DTN-861	26.3904	87.3602	0.9971	3.1522

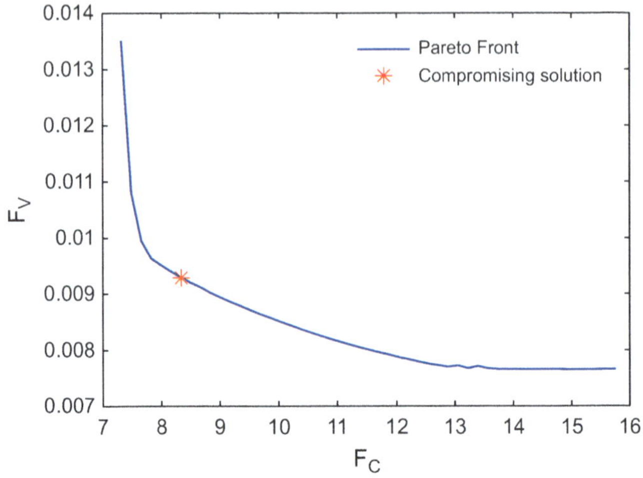

Fig. 2.9 Pareto front and trade-off solution of the 6-bus system

Table 2.12 Bi-objective OPF results offered by OPF-CCP

System	Iter.	Time(s)	$F_C(x_b^*)$	$F_V(x_b^*)$	$B_V(x_b^*)$
DTN-6	3	0.34	8.3540	0.0093	0.0326
DTN-14	7	2.79	9.2468	0.0377	0.0436
DTN-30	1	1.12	12.369	0.2687	0.4294
DTN-33	1	0.99	35.8610	0.3137	0.0779
DTN-57	5	8.41	13.021	0.0361	0.0146
DTN-69	2	1.34	46.9372	0.3664	1.0988
DTN-123	3	8.77	25.3540	0.2492	0.1792
DTN-861	9	98.7	26.8086	1.3582	108.63

Table 2.13 Bi-objective OPF results offered by BARON

System	$F_C^b(x_b^*)$	$F_V^b(x_b^*)$	$B_V(x_b^*)$	Time(s)
DTN-6	8.3174	0.0093	0.0329	239
DTN-14	9.2470	0.0377	0.0436	12
DTN-30	12.369	0.2686	0.4295	57
DTN-33	35.8617	0.3136	0.0779	114
DTN-57	13.021	0.0361	0.0146	132
DTN-69	46.9341	0.3668	1.0988	170
DTN-123	25.3545	0.2485	0.1797	66
DTN-861	–	–	–	–

in Tables 2.12 and 2.13 corroborate that those solutions offered by Algorithm 2.1 are indeed very close to the global optimal ones. Nevertheless, Algorithm 2.1 is much more efficient. We find that KNITRO is not robust in solving bi-objective OPF problems, and switch to IPOPT [84], which is implemented based on the primal-dual interior point method. The same instances are solved again. Results are provided in Table 2.14. In these tests, IPOPT fails in problems of DTN-69 and

Table 2.14 Bi-objective OPF results offered by IPOPT

System	$F_C^b(x_b^*)$	$F_V^b(x_b^*)$	$B_V(x_b^*)$	Time(s)
DTN-6	8.3645	0.0093	0.0329	3.02
DTN-14	9.2465	0.0377	0.0436	2.71
DTN-30	12.368	0.2688	0.4295	1.64
DTN-33	35.8609	0.3137	0.0779	2.26
DTN-57	Non-opt	Non-opt	Non-opt	Non-opt
DTN-69	Fail	Fail	Fail	Fail
DTN-123	25.3541	0.2486	0.1797	7.32
DTN-861	Fail	Fail	Fail	Fail

DTN-861, and does not offer an (local) optimal solution for DTN-57 with default settings. Although the computational efficiency of IPOPT is comparable with CCP-OPF, especially for medium-scale systems, the latter appears to be more robust, because it only involves solving SOCPs.

2.3.4 OPF with Elastic Demands

1. Brief Introduction

Time-varying electricity prices may encourage consumers to reduce or shift their demands during peak hours. Demand response (DR) is just such a financial incentive. Typical programs include time-of-use pricing, peak pricing, real-time pricing, and direct load control with subsidies. Research shows that the real-time pricing would be advantageous from the system operation perspective, as it carries instantaneous information on the operating status. With the opening of distribution energy markets, the locational marginal price (LMP) will be a candidate pricing policy in DR. We recommend literature [85–88] which shed light on distribution markets and distribution LMP.

In the presence of DR, generation dispatch and load adjustment become interactive: the amount of elastic demand keeps changing in response to LMPs, and the changes in nodal load levels will further influence the OPF solution and in turn impact LMPs. Such dynamics will bring equilibrium and stability issues in DR markets. Two fundamental questions are: Will there be an invariant state in such a DR market? Will the load and price reach a fixed point after a small disturbance? The first one refers to the existence of at least one market equilibrium; the second one is related to the stability of the equilibrium, if one exists.

The state-space model of market dynamics is proposed in [89], which is used in stability assessment and market design in [90, 91]. Dynamics of supply, demand, and electricity price in a retail market is discussed in [92]. The similar problem is revisited using a static model in [93]: convergence of a uniform marginal pricing market with an aggregated elastic load is investigated, where convergence means the price and demand series could approach constant values if generators and loads update their strategies sequentially. Since only generator capacity and power balancing condition are considered in market clearing, an analytical criterion can be derived using contractive mapping theory.

This section extends the problem in [93] to a distribution energy market with LMP based real-time pricing and elastic demands. In the market level, the operator receives demand bidding and clears the market according to an ACOPF; LMPs are determined from the Lagrangian dual multipliers associated with active power balancing equalities. In the demand level, responsive consumers receive LMPs and update their demand bidding. Two levels constitute a closed-loop feedback process, which can be viewed as an OPF with demand elasticity. Because line congestions are modeled, it is difficult to derive analytical conditions that ensure the existence and stability of the market equilibrium. We provide a heuristic criterion that depends on very simple data, and suggest an iterative algorithm to identify the market equilibrium based on Algorithm 2.1.

2. Mathematic Model and an Iterative Algorithm

With notations defined in Sect. 2.3.1, the OPF problem of DR market can be formulated as follows:

$$
\begin{aligned}
\min \quad & F_C \\
\text{s.t.} \quad & \text{Cons-BFM}(\lambda),\ \text{Cons-BND} \\
& p_j^d = p_j^{d0} + p_j^{dr},\ \forall j \in E_B \\
& p_j^{dr} = \begin{cases} p_{jm}^{dr}, & \text{if } \lambda_j < \lambda_j^n; \\ p_j^L(\lambda_j), & \text{if } \lambda_j \in [\lambda_j^n, \lambda_j^m] \\ p_{jn}^{dr}, & \text{if } \lambda_j > \lambda_j^m; \end{cases},\ \forall j \in E_B
\end{aligned}
\tag{2.61}
$$

where F_C, Cons-BFM, and Cons-BND have been defined in problem (2.30), active power demand p_j^d consists of a fixed part p_j^{d0} and a price-sensitive part p_j^{dr} which is a PWL function in LMP λ_j, and the linear part is given by

$$
p_j^L(\lambda_j) = p_{jm}^{dr} - \frac{\lambda_j - \lambda_j^n}{\lambda_j^m - \lambda_j^n}(p_{jm}^{dr} - p_{jn}^{dr}) \tag{2.62}
$$

where λ_j^n and λ_j^m are threshold prices, p_{jn}^{dr} and p_{jm}^{dr} are minimal and maximal values of the elastic demand. Other continuous nonlinear functions, such as those investigated in [93], can be used in (2.62) as well. We make no special requirements on the DR function, except continuity. Discontinuity may influence the existence of market equilibria.

Different from any OPF problems discussed in previous sections, the amount of responsive loads in (2.62) depends on the dual variable λ at an optimal solution, which is unknown before the OPF problem is solved; meanwhile, the spatial distribution of demands in turn affects the OPF solution and LMP. Such an interplay constitutes a closed-loop feedback mechanism. A fixed-point iterative procedure for OPF problem (2.61) with demand elasticity is shown in Algorithm 2.3.

Additional remarks are provided as follows.

Algorithm 2.3 Fixed-point iteration

1: Choose $\varepsilon > 0$, and a maximal number of iterations K_{it}^M. Set $k = 1$, demand $p^d(0) = p^{d0}$, $p^d(1) = p^{d0} + p_m^{dr}$.
2: Solve problem (2.61) with fixed $p^d(k)$, and retrieve LMP vector λ.
3: If $\|p^d(k) - p^d(k-1)\|_2 / \|p_m^{dr}\|_2 \le \varepsilon$, return the OPF solution and LMP; else if $k = K_{it}^M$, report that the algorithm fails to converge; else update p^{dr} using the obtained LMP according to (2.61) and (2.62), $k = k + 1$, $p^d(k) = p^{d0} + p^{dr}$, and go to step 2.

---

1. Retrieving the LMP

OPF problem (2.61) with a fixed demand can be solved by Algorithm 2.1, and LMPs can be calculated from sensitivity analysis, or can be recovered by solving the KKT optimality condition of problem (2.61). Since the optimal solution is available at hand, KKT conditions give rise to linear equations, which is readily solvable. If the SOCP relaxation is exact and certain constraint quantification is met, the duality gap is zero, and LMP can be extracted from the variables in the dual problem of the SOCP relaxation model. The primal and dual form of SOCPs can be found in Appendix A.2.3.

2. Convergence Guarantee

If the OPF solution at any fixed demands satisfies the second-order sufficient condition, and no active inequality in complementarity and slackness conditions is degenerate, then LMP is uniquely determined by a mapping $\lambda = \Phi(p^{dr})$. Meanwhile, the PWL function in (2.61) denotes another one-to-one mapping $p^{dr} = \Theta(\lambda)$. Thus, the elastic demand p^{dr} at an equilibrium state satisfies:

$$p^{dr} = \Theta(\Phi(p^{dr})) \tag{2.63}$$

Because all possible p^{dr} constitutes a hypercube region P_D^R which is convex and compact, Brouwer fixed-point theorem [94] says that a fixed point of (2.63) exists if $\Theta(\Phi(\cdot))$ is continuous; moreover, if $\Theta(\Phi(\cdot))$ is contractive, Banach contraction principle ensures that the fixed point is unique and stable [94].

In problem (2.61), unit production costs are convex quadratic functions, the uniqueness of LMP can be guaranteed for any given p^{dr}, and continuity can be expected when P_D^R is not too large. A solution under demand elasticity exists as long as (2.61) is feasible for all $p^{dr} \in P_D^R$. The existence of a solution can be understood from a geometric perspective: at bus i, λ_i is increasing in p_i^{dr}, while p_i^{dr} is decreasing in λ_i, so the supply curve and demand curve must have an intersection.

Given the interactive feature, a solution of (2.61) is called the market equilibrium, which is stable if $\Theta(\Phi(\cdot))$ is contractive. In unstable cases, Algorithm 2.3 may fail to converge [93]. To validate a contraction, the spectral radius of sensitivity matrix $J = [\partial\lambda_i/\partial p_j^{dr}]$, $\forall i, j$ should be less than 1 for all $p^{dr} \in P_D^R$. However, this condition is difficult to check, because J does not have a closed-form expression, although it can be numerically evaluated using the sensitivity method in [95]. In fact, even if the LMP is not continuous, a stable solution may still exist.

Here we provide a heuristic but simple criterion to judge the convergence of Algorithm 2.3 without continuity and contraction assumption. Geometrically, the equilibria of (2.61) is stable if the sensitivities of supply and demand curves satisfy

$$\left|\frac{\partial\lambda_i(p^{dr})}{\partial p_j^{dr}}\right|\left|\frac{\mathrm{d}p_j^{dr}(\lambda_j)}{\mathrm{d}\lambda_j}\right| < 1,\ \forall i, j \in E_B \tag{2.64}$$

According to (2.61), if $\lambda_j \notin [\lambda_j^n, \lambda_j^m]$, the second derivative is 0 and (2.64) is naturally met. The marginal cost of generator is $2a_i p_i^g + b_i$. Suppose the load increment is fully supplied by the unit with the largest a_i, the first sensitivity is bounded by $2a_i$. When the elastic demand varies in $[p_{jn}^{dr}, p_{jm}^{dr}]$, the sensitivity to price is a constant $(p_{jm}^{dr} - p_{jn}^{dr})/(\lambda_j^m - \lambda_j^n)$. As a result, the stability condition (2.64) evolves into

$$2a_m(p_{jm}^{dr} - p_{jn}^{dr}) < \lambda_j^m - \lambda_j^n, \forall j \in B \tag{2.65}$$

where $a_m = \max_i\{a_i\}, \forall i$. Condition (2.65) is easy to check because it is described by simple data and is independent of the OPF solution. If the LMP is continuous, (2.65) will be a sufficient condition, because we consider the sensitivity in the worst case. Otherwise, the market equilibrium may not exist because the supply and demand curves may have no intersection. If a nonlinear demand function is used, the sensitivity $|\mathrm{d}p_j^{dr}(\lambda_j)/\mathrm{d}\lambda_j|$ in (2.64) can be replaced by its upper bound in $[\lambda_j^n, \lambda_j^m]$. Otherwise, certification of (2.64) with $|\mathrm{d}p_j^{dr}(\lambda_j)/\mathrm{d}\lambda_j|$ evaluated at a given point provides a local stability guarantee for the market equilibrium. The proposed framework can be extended to meshed networks by employing a BIM and a suitable OPF algorithm, such as the SDP relaxation method in Sect. 2.3.1.

3. Case Study

We present numerical results on a 69-bus test system. Network data are available in [75]. Generators are identical whose parameters are given by: $p_j^{gn} = 0$, $p_j^{gm} = 2$, $q_j^{gn} = -1$, $q_j^{gm} = 1$; the fuel cost function is $200(p_j^g)^2 + 800p_j^g$. Contract price with the main grid is $\rho = 800$\$/MWh. Elastic load data are: $p_{jn}^{dr} = 0.1$, $\lambda_j^n = 900$\$/MWh, $\lambda_j^m = 1100$\$/MWh. In order to validate (2.65), p_{jm}^{dr} is parameterized in $\Delta \in [0, 1]$ as $p_{jm}^{dr} = p_{jn}^{dr} + \Delta(p_{jM}^{dr} - p_{jn}^{dr})$, where $p_{jM}^{dr} = (\lambda_j^m - \lambda_j^n)/2a_m + p_{jn}^{dr}$. In Algorithm 2.3, we set $\varepsilon = 10^{-3}$, $K_{it}^M = 200$.

Table 2.15 Computation time under different values of Δ

Δ	0.40	0.50	0.60	0.70	0.80	0.90	1.00
Iteration	7	9	12	18	28	59	Fail
Time (s)	8.47	10.1	13.4	19.3	33.3	83.7	Fail

Results with different values of Δ are provided in Table 2.15. According to the definition of p_{im}^{dr}, condition (2.65) translates into $\Delta < 1$. From Table 2.15 we can see that the number of iterations increases with Δ growing; when $\Delta = 1$, Algorithm 2.3 fails to converge in 200 iterations, although we observe that the value of $\|p^{dr} - p_0^{dr}\|_2/\|p_m^{dr}\|_2$ still decreases gradually. At the equilibrium with $\Delta = 0.90$, LMPs at 58 buses still reside in the interval [900, 1100]\$/MWh. Numeric studies on the 246-bus system, 615-bus system, and 861-bus system whose data are provided at the same website convey a similar message: Algorithm 2.3 converges faster when Δ is smaller.

2.4 Robust Energy and Reserve Dispatch

The advantages of renewable generations as clean and cheap energy resources have inspired the dramatic integration of wind and solar energy into power systems during the past decade. However, the uncertain behavior and low controllability of such variable energy resources also bring challenges to power system operation. As a result, spinning reserves are increasingly involved in order to compensate real-time mismatch between generation and load, introducing additional expenditures. So sophisticated decision-making tools which actively account for uncertainties are in great need.

The volatility of renewable generation is usually modeled through probability distribution functions (PDFs), such as Gamma distribution, Weibull distribution, Rayleigh distribution, to name just a few [96]. Gaussian distribution is often used to describe the wind power forecast error in short-term generation scheduling problems. Given the PDFs of renewable power, the economic dispatch and spinning reserve can be scheduled in the framework of stochastic optimization. Most stochastic approaches minimize the expected cost in their objective functions. In order to acquire an accurate enough PDF that reflects the uncertain behavior of actual renewable generation, sufficient historical data are required but not always available at hand. The adjustable robust optimization raises an appealing decision-making paradigm for power system generation scheduling problems without exploiting the PDFs, which is employed in this section to develop an adaptive robust energy and reserve dispatch method (A-RERD for short). Robust optimization based approaches minimize the cost in the worst-case scenario. According to current researches, both stochastic optimization and robust optimization provide powerful decision-making tools for power system optimization problems under uncertainties.

If accurate PDFs are available, the former one is usually preferred as it produces economically efficient solutions from the statistical point of view; otherwise, the latter one is a good alternative as it only requires the forecast and can protect the system against a pre-defined uncertainty set, although it is likely to be more conservative.

Inspired by recent advances in the distributionally robust optimization [97, 98], this section intends to develop a comprehensive method that combines the advantages of stochastic and robust optimization approaches. In particular, we propose a two-stage distributionally robust optimization model for the energy and reserve dispatch (D-RERD for short). The first-stage decides the set point of generators and renewable power plants, as well as the spinning reserve capacity preserved in each generator; In the second stage, corrective actions, including the redispatch of generators, renewable power spillage, and load shedding, are deployed in real-time dispatch in response to the realization of renewable power. In the proposed model, the uncertainty is described via an ambiguous PDF with known expectation and covariance matrix. The expected cost of corrective actions under the worst-case renewable power distribution is minimized. Compared with the stochastic optimization method, D-RERD does not require the exact PDFs, which may be difficult to retrieve; compared with the robust optimization method, D-RERD accounts for the distribution property and makes full use of the variances information. D-RERD is equivalent to an SDP and can be solved by a delayed constraint generation algorithm in a tractable manner. Materials in this section come from [99, 100].

Main symbols and notations used throughout this section are defined below for quick reference. Others are clarified after their first appearances as required.

Indices

i	Index of generators.
j	Index of renewable power plants.
q	Index of loads.
s	Index of scenarios.

Parameters

a_i, b_i	Energy production cost coefficients of generator i.
C_j^V	Installed capacity of renewable power plant j, $C^V = \{C_j^V\}, \forall j$.
d_i^{g+}	Up-regulation cost coefficient of generator i.
d_i^{g-}	Down-regulation cost coefficient of generator i.
d_i^{r+}	Up-reserve cost coefficient of generator i.
d_i^{r-}	Down-reserve cost coefficient of generator i.
F_l	Power flow capacity of transmission line l.
P_i^l	Minimal output of generator i.
P_i^u	Maximal output of generator i.
p_q	Power demand of load q.
N_E	Number of extreme points.
N_G	Number of generators.

N_Q Number of loads.
N_W Number of renewable power plants.
Pr_s Probability of scenario s.
R_i^+ Ramp-up limit of generator i.
R_i^- Ramp-down limit of generator i.
Δt Time duration of the current dispatch interval.
w_j^e Maximum output forecast of renewable power plant j, vector $w^e = \{w_j^e\}, \forall j$.
w_j^g Actual generation capability of renewable power plant j in the deterministic formulation, vector $w^g = \{w_j^g\}, \forall j$.
w_j^l Minimal generation capability of renewable power plant j .
w_j^u Maximal generation capability of renewable power plant j .
w_j^h Half of the forecast interval, $w_j^h = 0.5(w_j^u - w_j^l), \forall j$
π_{il} PTDF from generator i to line l.
π_{jl} PTDF from renewable power plant j to line l.
π_{ql} PTDF from load q to line l.
Vr_j Variance of the maximum output forecast error at a single plant.
Σ Covariance matrix of the maximum output forecast error.
Θ Second-order moment matrix, $\Theta = \Sigma + w^e(w^e)^T$.
ρ_j Generation curtailment cost at renewable power plant j.
ρ_q Load-shedding cost of load q.

First-Stage Variables

p_i^f Scheduled output of generator i, vector $p^f = \{p_i^f\}, \forall i$.
r_i^+ Up-spinning reserve capacity offered by generator i, $r^+ = \{r_i^+\}, \forall i$.
r_i^- Down-spinning reserve capacity offered by generator i, $r^- = \{r_i^-\}, \forall i$.
w_j^f Scheduled output of renewable power plant j, $w^f = \{w_j^f\}, \forall j$.
x First-stage decision in the compact form, $x = \{p^f, r^+, r^-, w^f\}$.

Second-Stage Variables

p_i^+ Up-regulation power deployed by generator i in response to the actual renewable power generation, vector $p^+ = \{p_i^+\}, \forall i$.
p_i^- Down-regulation power deployed by generator i in response to the actual renewable power generation, vector $p^- = \{p_i^-\}, \forall i$.
p_q^l Demand curtailment of load q in response to the actual renewable power generation, vector $p^l = \{p_q^l\}, \forall q$.
w_j^c Renewable power spillage in plant j, $w^c = \{w_j^c\}, \forall j$
$\tilde{w}_j^g$ Random generation capability of renewable power plant j in the second stage of D-RERD, vector $\tilde{w}^g = \{\tilde{w}_j^g\}, \forall j$.
$\bar{w}_j^g$ Uncertain generation capability of renewable power plant j in the second stage of A-RERD, vector $\bar{w}^g = \{\bar{w}_j^g\}, \forall j$.
y Second-stage decision in the compact form, $y = \{p^+, p^-, p^l, w^c\}$.

H Dual variable (a matrix) used in D-RERD.
h Dual variable (a vector) used in D-RERD.
h_0 Dual variable (a scalar) used in D-RERD.
u Dual variable of the operating constraints.

Sets

W Uncertainty set used in A-RERD.
Ω Ambiguity set of PDFs used in D-RERD.
X Feasible set of the first-stage decision x.
U Feasible set of dual variable u.
B^V Hypercube of renewable power output, $B^V = \{w^g \mid 0 \le w^g \le C^V\}$.

2.4.1 Deterministic Reserve Provision and Deployment

1. The First-Stage Problem

In order to cover renewable power fluctuation in short term dispatch, generators have to preserve a certain amount of spinning reserve capacity. The joint optimization of energy production and reserve provision can be implemented by incorporating reserve constraint and regulation cost in the OPF problem as follows:

$$\min \sum_{i=1}^{N_G} \left[a_i^2 (p_i^f)^2 + b_i p_i^f + d_i^{r+} r_i^+ + d_i^{r-} r_i^- \right] + \sum_{j=1}^{N_W} \rho_j (w_j^e - w_j^f) \tag{2.66a}$$

$$\text{s.t. } P_i^l \le p_i^f - r_i^-, \quad p_i^f + r_i^+ \le P_i^u, \forall i \tag{2.66b}$$

$$\sum_{i=1}^{N_G} p_i^f + \sum_{j=1}^{N_W} w_i^f = \sum_{q=1}^{N_Q} p_q \tag{2.66c}$$

$$-F_l \le \sum_{i=1}^{N_G} \pi_{il} p_i^f + \sum_{j=1}^{N_W} \pi_{jl} w_j^f - \sum_{q=1}^{N_q} \pi_{ql} p_q \le F_l, \quad \forall l \tag{2.66d}$$

$$0 \le r_i^+ \le R_i^+ \Delta t, \; 0 \le r_i^- \le R_i^- \Delta t, \; \forall i, \; 0 \le w_j^f \le w_j^e, \forall j \tag{2.66e}$$

where objective (2.66a) is the total cost, including the generation cost, the spinning reserve cost, and spillage cost of renewable power. Constraint (2.66b) is the generation capacity limitation considering spinning reserve offers; constraints (2.66c) and (2.66d) are the power balancing condition and power flow restriction of transmission lines with respect to the renewable generation forecast, where PTDF parameters can be derived from the DC power flow model presented in Sect. 2.2.1 by eliminating voltage angle variables; constraint (2.66e) enforces that the spinning reserve capacity provided by each generator cannot exceed its ramping limit in the considered dispatch interval, and the scheduled output of renewable power plant should neither become negative nor exceed its predicted generation capability. Load shedding is not allowed at this time, but will be the last resort to maintain power balancing at the real-time dispatch stage. Since the convex quadratic terms in (2.66a) can be approximated by PWL functions, such as the method in [101] and the convex combination method in Appendix B.1.1, problem (2.66) can be arranged into a compact linear form without loss of generality

$$\min_{x \in X} c^T x \tag{2.67}$$

Problem (2.67) is associated with renewable generation forecast w^e. It is called the first-stage problem, and x the first-stage decision. Feasible set X corresponds to constraints (2.66b)–(2.66e) which only involves the first-stage decision. If the load and renewable predictions are accurate and no uncertainty is considered, the reserve capacity will be 0 in the optimal solution.

2. The Second-Stage Problem

After the first-stage decision is deployed, the actual renewable power generation w^g may deviate from w^e, and corrective actions $y = \{p^+, p^-, p^l, w^c\}$ are needed to restore operating constraints given the discrepancy $\Delta w = w^g - w^e$, giving rise to the following optimization problem:

$$\min \quad \sum_{i=1}^{N_G} (d_i^{g+} p_i^+ + d_i^{g-} p_i^-) + \sum_{j=1}^{N_W} \rho_j w_j^c + \sum_{q=1}^{N_Q} \rho_q p_q^l \tag{2.68a}$$

$$\text{s.t.} \quad \sum_{q=1}^{N_q} (p_q - p_q^l) = \sum_{j=1}^{N_W} (w_j^g - w_j^c) + \sum_{i=1}^{N_G} (p_i^f + p_i^+ - p_i^-) \tag{2.68b}$$

$$\begin{aligned} -F_l \le \sum_{i=1}^{N_G} \pi_{il}(p_i^f + p_i^+ - p_i^-) + \sum_{j=1}^{N_W} \pi_{jl}(w_j^g - w_j^c) \\ - \sum_{q=1}^{N_Q} \pi_{ql}(p_q - p_q^l) \le F_l, \ \forall l \end{aligned} \tag{2.68c}$$

$$0 \le p_i^+ \le r_i^+, \ \forall i, \quad 0 \le p_i^- \le r_i^-, \ \forall i \tag{2.68d}$$

$$0 \le w_j^c \le w_j^g, \ \forall j, \quad 0 \le p_q^l \le p_q, \ \forall q \tag{2.68e}$$

where the objective (2.68a) minimizes the cost of corrective actions in order to recover the operating constraints, including the re-dispatch cost of generators, the renewable power spillage cost, and the load shedding cost; constraint (2.68b) represents the power balancing condition; constraint (2.68c) is the security limitation of transmission lines; constraint (2.68d) stipulates the real-time adjustment of generators within its spinning reserve capacity preserved in the first stage; constraint (2.68e) sets the bounds on the amount of renewable power spillage and load shedding, indicating that the curtailed power or load could neither become negative nor exceed their maximal values. The compact form of LP (2.68) with given x and w^g is shown below

$$Q(x, w^g) = \min_{y \in Y(x, w^g)} f^T y \tag{2.69a}$$

$$Y(x, w^g) = \{y \mid By \le b - Ax - Cw^g\} \tag{2.69b}$$

where matrices A, B, C and vector b correspond to the coefficients in constraints (2.68b)–(2.68e), vector f corresponds to the coefficients in objective (2.68a). Function $Q(x, w^g)$ denotes the optimal value of LP (2.69) as a function of x and w^g.

Problem (2.69) is linked with the actual value of renewable generation w^g. It is called the second-stage problem, and y the second-stage decision. Feasible set $Y(x, w^g)$ corresponds to (2.68b)–(2.68e) in which the first-stage decision x has been fixed and the value of w^g has been observed or can be predicted accurately.

Clearly, two problems in both stages are tightly coupled. If we schedule more spinning reserves in the first stage, the risk in the second stage declines; and vice versa. Since w^g is not known exactly in the first stage, different descriptions of the randomness of w^g are adopted in various decision-making tools. In what follows, we introduce two emerging approaches based on robust optimization.

2.4.2 Scenario Set Based Robust Formulation

In this formulation, all possible values of w^g are restricted in a given set

$$W = \left\{ \bar{w}^g \;\middle|\; \begin{array}{l} w^l_j \le \bar{w}^g_j \le w^u_j, \forall j \\ \displaystyle\sum_{j=1}^{N_W} \frac{|\bar{w}^g_j - w^e_j|}{w^h_j} \le \Gamma \end{array} \right\} \tag{2.70}$$

where the first constraint enforces $\bar{w}^g_j$ within the interval $[w^l_j, w^u_j]$, the second constraint restricts the total deviation of the actual renewable generation from the forecast, parameter Γ is called the budget of uncertainty. Without loss of generality, we assume W is a polytope. When Γ is an integer, all extreme points of W constitute a cardinality constrained uncertainty set, which can be represented by binary variables. The elements in W are equally treated, no dispersion effect is taken into account. This assumption greatly simplifies data acquisition process, because only the forecast and range information is needed, whereas fitting a PDF requires extensive historic data.

Based on this description, the energy and reserve dispatch problem can be formulated as

$$\min_{x \in X} \left\{ c^T x + \max_{\bar{w}^g \in W} Q(x, \bar{w}^g) \right\} \tag{2.71}$$

In problem (2.71), the total cost in the worst-case scenario is optimized. Since $Q(x, \bar{w}^g)$ is the optimum of an LP, (2.71) is a trilevel program. Theory and algorithm for such kind of robust optimization problems are dedicatedly discussed in Appendix C.2. Since the worst-case scenario rarely happens in reality, the solution of (2.71) may be pessimistic. Nevertheless, it could protect the system against any disturbance resides in W and is acceptable in power system operation, because the cost of system failure is extremely high.

In contrast to the uncertainty set description, the randomness of renewable generation is characterized by the PDF in stochastic optimization models, in which the expected cost is optimized as

$$\min_{x \in X} \left\{ c^T x + \mathbb{E}_P Q(x, \tilde{w}^g) \right\} \tag{2.72}$$

where $\mathbb{E}$ is the expectation operator, and P is the PDF of $\tilde{w}$. In view that $\mathbb{E}_P Q(x, \tilde{w}^g)$ does not have a closed-form expression, a typical mean is called the sampling average approximation (SAA), in which a set of scenarios $\{w^s\}, s = 1, 2, \ldots$ are sampled from the PDF of $\tilde{w}^g$ associated with a probability $\Pr_s$, and the expectation can be approximated by the weighted-sum function, yielding

$$\min_{x \in X, y^s \in Y(x, w^s), \forall s} \left\{ c^T x + \sum_s f^T y^s \right\} \tag{2.73}$$

Because extreme conditions usually have low probabilities, formulation (2.72) could produce economically efficient solutions from a statistical perspective.

We assume the exact PDF of renewable generation is unavailable due to the lack of enough historical data, so we will not go into depth about stochastic optimization model (2.72) and (2.73). Instead, we will consider what if we know some yet incomplete information on the PDF of the renewable generation.

2.4.3 Ambiguity Set Based Robust Formulation

Combining the idea of stochastic optimization and robust optimization, we assume that the PDF of renewable generation is not known exactly and belongs to a functional uncertainty set defined as

$$\Omega = \left\{ P \left| \begin{aligned} & \Pr[\tilde{w}^g \in \mathbb{R}^{N_W}] = 1 \\ & \mathbb{E}_P[\tilde{w}^g] = w^e \\ & \mathbb{E}_P[\tilde{w}^g (\tilde{w}^g)^T] = \Sigma + w^e (w^e)^T \end{aligned} \right. \right\} \tag{2.74}$$

which is also called an ambiguity set. It defines the joint PDF of uncertain data, and is significantly different from the uncertainty set which restricts the values of uncertain data. From the mathematic modeling perspective, the dimension of Ω is infinite. All candidate PDFs in set Ω share the same expectation w^e and covariance matrix Σ. Based on this ambiguity set, the energy and reserve dispatch problem can be formulated as

$$\min_{x \in X} \left\{ c^T x + \max_{P \in \Omega} \mathbb{E}_P[Q(x, \tilde{w}^g)] \right\} \tag{2.75}$$

In problem (2.75), ambiguity set Ω plays the role as the uncertainty set W does in problem (2.71), and the expected cost in the worst-case distribution is optimized.

Formulation (2.75) is different from (2.72) because the PDF P is not known exactly except the forecast w^e and the covariance matrix Σ. The former is more pessimistic than the latter, because the specific PDF used in (2.72) is merely a candidate in the ambiguity set Ω, therefore the optimal value of (2.72) should be a lower bound of the robust formulation (2.75). Since (2.75) copes with the worst distribution, its optimal value would be less sensitive with respect to the PDF perturbation, although may be more conservative from an economical point of view.

Formulation (2.75) also differs from (2.71). Its advantages rest on two observations. First, the conservativeness of (2.71) largely depends on the choice of uncertainty set W. The trade-off is usually manually decided via selecting a budget of uncertainty Γ, while all information used in (2.75) can be derived from actual data. Second, the renewable power is still a random variable described by a PDF, thus the dispersion property is explicitly taken into account, and the objective function in (2.75) is an expectation reflecting the statistical performance, rather than that in (2.71) associated with only a single worst-case scenario. As the variance is fixed, a scenario that deviates far away from the forecast would have a low probability. Moreover, it is usually important to tackle the risk caused by the "tail effect" in probability theory, which indicates that an occurrence of renewable generation that is far away from the forecast may induce heavy losses of the system in spite of its low probability. Such phenomenon is naturally modeled in (2.75) but neglected in (2.71), because probability information is ignored in uncertainty set W. If the dispatch strategy of (2.71) should protect the system against all rare events, the operational cost could be extremely high and unacceptable in practice.

In summary, formulation (2.75) inherits the advantages of stochastic optimization and robust optimization. Theory and algorithm for such kind of robust optimization problems are dedicatedly discussed in Appendix C.3.2.

2.4.4 Case Study

A-RERD (2.71) and D-RERD (2.75) are tested on the IEEE 118-bus system, whose data are available at [102]. This system possesses 54 generating units and 186 transmission lines. In the considered dispatch interval, the total demand is 5500 MW. The reserve cost coefficients d_i^{r+}/d_i^{r-} of each unit are assumed to be 10% of its cost coefficient b_i. The regulation cost coefficients d_i^{g+}/d_i^{g-} of each generator are assumed to be the per-unit production cost at the nominal operating point, i.e. $d_i^{g+} = d_i^{g-} = a_i P_i^u + b_i + c_i/P_i^u$. The ramping limits R_i^+/R_i^- of each generator is assumed to be 40% of its maximal output P_i^u. Nine wind farms are connected to the system at bus #70, #12, #17 (Area 1), #49, #59, #77, #80 (Area 2), #100, #92 (Area 3). Their predicted generation level is 100 MW. The wind power curtailment cost is 5\$/MWh. The load shedding cost is 500\$/MWh. Other parameter without particular announcement is the same as those provided online. SDPs are solved by MOSEK [103], CQPs and LPs are solved by CPLEX [104].

1. Model of Uncertainty

In A-RERD, the uncertainty set is defined in (2.70). In order to identify a proper volume that is neither too big nor too small, we determine the parameters according to [105]. The bound parameter w_j^l and w_j^u can be selected as

$$w_j^l = w_j^e - z(\alpha)\mathrm{Vr}_j, \quad w_j^u = w_j^e + z(\alpha)\mathrm{Vr}_j$$

where $z(\alpha)$ represents the α confidence level for the standard Gaussian distribution. In our tests, $\alpha = 99.9\%$ is adopted to guarantee the probability $\Pr[w^l \le w \le w^u]$ is high. Because there are 18 inequalities, $\Pr[w^l \le w \le w^u] = 0.999^{18} = 98.2\%$. According to [106], the root mean square error of the hourly-ahead forecast is around 10% of the predicted output. In our tests, $8\% \le \sqrt{\mathrm{Vr}}/w^e \le 14\%$ is investigated. The corresponding lower bound w^l and upper bound w^u in (2.70) are shown in Table 2.16. As for the budget of uncertainty Γ, if the number of wind farms n is large enough, it can be selected as $\Gamma \sim O(\sqrt{n})$. In our tests, we increase it from 6 to 9 (equivalent to $2\sqrt{N_W} \le \Gamma \le 3\sqrt{N_W}$). Please bear in mind that A-RERD itself does not make any assumption on the PDF of wind generation. All parameters can be selected differently in compliance with specific preferences.

In D-RERD, the PDF of wind generation are restricted in the ambiguity set Ω described in (2.74). We further assume the forecast error is independent so that the covariance matrix Σ is diagonal and shown below

Table 2.16 Confident intervals of A-RERD

$\sqrt{\mathrm{Vr}}/w^e$	w^l (MW)	w^u (MW)
8%	75.28	124.72
10%	69.10	130.90
12%	62.92	137.08
14%	56.74	143.26

$$\Sigma = \begin{bmatrix} \mathrm{Vr}_1 & & \\ & \ddots & \\ & & \mathrm{Vr}_{N_W} \end{bmatrix}$$

Nevertheless, in practical usage, the covariance matrix Σ should be calculated from actual data, correlation is possible and not a limitation of our method.

2. Results

We compare the performance of A-RERD and D-RERD from four aspects:

1. The computational efficiency.
2. The respective total cost and the second-stage cost.
3. The expected total cost when the forecast error follows a specific distribution.
4. The total cost of D-RERD in the worst-case scenario of A-RERD.

D-RERD is solved by Algorithm C.5 in Appendix C.3.2. In Algorithm C.5, for the bi-convex program (C.117), we use the following $N_W + 1$ points as the initial points

$$w^g = 0, \; w^g = C_j^V e_j^{N_W}, \; j = 1, 2, \ldots, N_W$$

where C_j^V is the installed capacity of wind farm j, and $e_j^{N_W}$ is the j-th column of an $N_W \times N_W$ identity matrix.

A-RERD is solved by Algorithm C.2 in Appendix C.2.3. Since the uncertainty set W is a polyhedron, the bilinear subproblems (C.59) and (C.60) are solved in the equivalent MILP reformulation (C.63). The computation times of D-RERD and A-RERD are compared in Table 2.17, suggesting that their computational complexities are generally comparable, despite that the latter can be solved faster when Γ is small. This result is inspiring because D-RERD is more close to stochastic optimization models that accounts for the average performance over all possible scenarios, however, SAA models are more challenging to solve than robust optimization models. The reason is that the second-stage problem of D-RERD is considered in its dual form, which enjoys lower complexity.

Optimal valuęs of D-RERD and A-RERD under different forecast accuracies are provided in Fig. 2.10, showing that the total cost of A-RERD is higher than that of

Table 2.17 Comparison of computational times

		A-RERD (s)			
$\sqrt{\mathrm{Vr}}/w^e$	D-RERD (s)	$\Gamma = 6$	$\Gamma = 7$	$\Gamma = 8$	$\Gamma = 9$
8%	12.9	7.53	8.51	9.23	12.7
10%	13.4	8.11	9.45	11.0	14.9
12%	12.8	8.56	9.16	10.5	13.5
14%	13.1	8.93	14.1	14.0	10.4

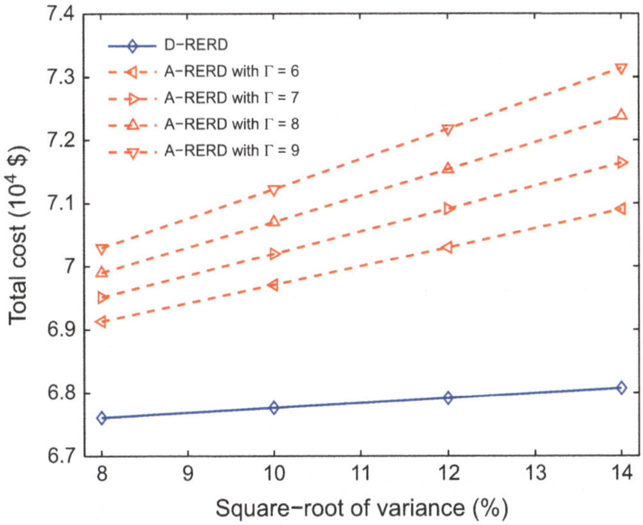

Fig. 2.10 Comparison of the total costs of A-RERD and D-RERD

D-RERD, due to the different criterions used in the second stage. Moreover, when the wind generation forecast becomes less accurate ($\sqrt{\text{Vr}}/w^e$ becomes larger), the cost of D-RERD grows slower than that of A-RERD. This is also understandable, as D-RERD considers the expected cost in the second stage. Even in the worst-case distribution, most scenarios still appear to be near to the forecast due to the variance constraint in (2.74). If the majority of scenarios leave far away from the forecast, the moment constraint will be violated.

We compare the second-stage cost of D-RERD in the worst-case distribution with that in the Gaussian distribution. The former can be computed by solving SDP (C.113), and the latter is obtained from Monte Carlo simulation, which proceeds as follows:

1. Solve D-RERD and fix the first-stage decision;
2. By assuming the forecast error follows Gaussian distribution with the provided mean and variance, generate 10,000 samples of w^g. Solve the deterministic second-stage problem (2.68) for each scenario.
3. Calculate the mean of the second-stage cost.

Wind generation may follow other distributions in practice. Nevertheless, Gaussian distribution is not used for making decisions or computing the true expectation in practice, it is only used in Monte Carlo simulation for two purposes:

1. Demonstrating the gap between the expected cost in the worst-case distribution and a particular distribution, or how conservative the D-RERD may appear to be.
2. Evaluating the performances of different dispatch strategies offered by D-RERD and A-RERD.

Results are provided in Fig. 2.11, demonstrating the cost in the Gaussian distribution is about half of that in the worst-case distribution, which indicates that the Gaussian distribution is not the worst one. For comparison, the second-stage costs of A-RERD in the worst-case scenario under different budget Γ are shown in Fig. 2.12, which are much larger than that of D-RERD, because they are associated with the worst-case scenario.

We compare the expected total cost of D-RERD and A-RERD under Gaussian distribution. We first acquire the expected second-stage cost of D-RERD and

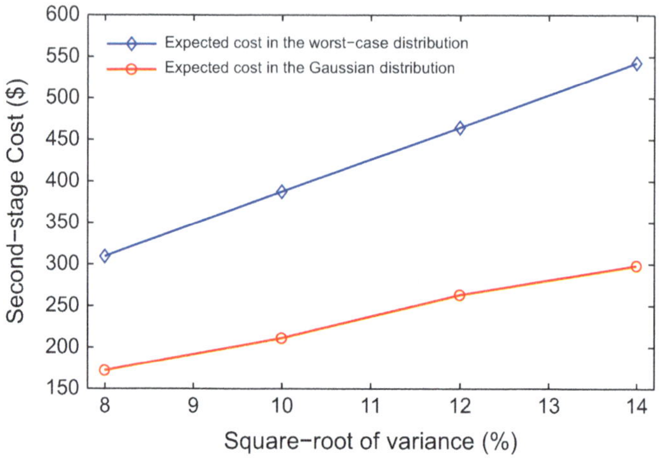

Fig. 2.11 $\mathbb{E}_P[Q(x, \tilde{w}^g)]$ in the worst-case distribution and Gaussian distribution

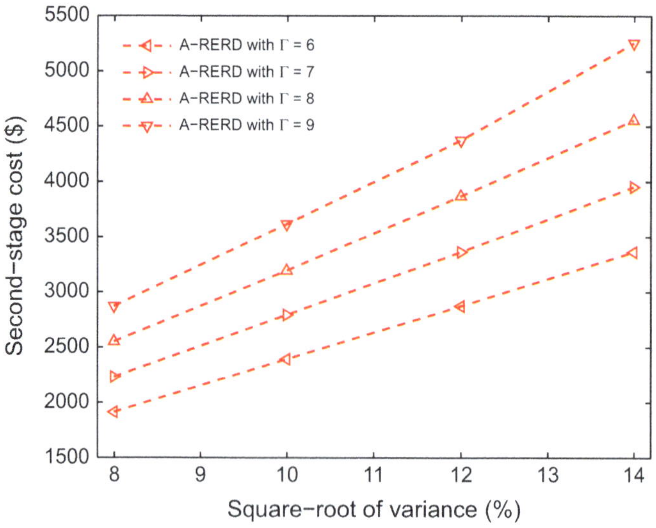

Fig. 2.12 Second-stage cost of A-RERD under different budget Γ

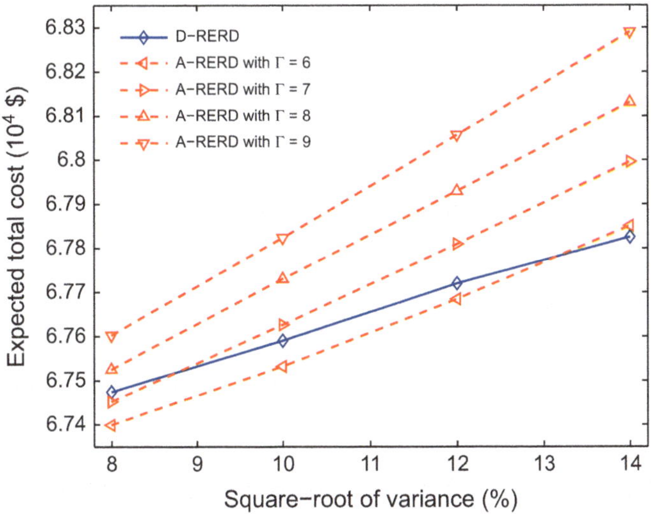

Fig. 2.13 Comparison of the expected costs of A-RERD and D-RERD

A-RERD via Monte Carlo simulation, and then add them with the corresponding first-stage cost. The results are shown in Fig. 2.13, indicating that A-RERD is not necessarily always conservative than D-RERD. When Γ is small, the former may have a lower expected total cost. With the forecast being less accurate, the latter is likely to have a lower expected total cost. It is worth mentioning that this is partly because of the data set we used, where the reserve cost and regulation cost is in proportion to the marginal cost of production. If the simulation was setup in the opposite way such that units with high marginal cost, e.g., gas-fired units, have low reserve cost and regulation cost, and slower units with low marginal cost have high reserve cost and regulation cost, A-RERD will prefer to dispatch the former ones, since the second-stage cost will be smaller when reserves are fully deployed in the worst-case scenario. This would be suboptimal in terms of expectation, resulting in worse performance of A-RERD compared to D-RERD. Another reason rests on the Gaussian distribution we assumed while selecting the bound parameter of the uncertainty set. If another distribution is adopted, A-RERD may appear to be more pessimistic (due to slower decay rate of PDF and larger uncertainty set).

Compared with Fig. 2.10, the expected total cost of A-RERD is much less than the total cost in the worst-case scenario. In practice, the worst-case scenario is unlikely to happen, the realization of uncertainty is usually near to the forecast. In this regard, the A-RERD may not be that conservative than it was expected to be. The most important task in A-RERD is to derive the convincing parameters of the uncertainty set, especially the budget of uncertainty Γ, which accounts for correlation. In contrast, there is no manually supplied parameter in D-RERD. Meanwhile, it performs well in a wide range of the forecast accuracy.

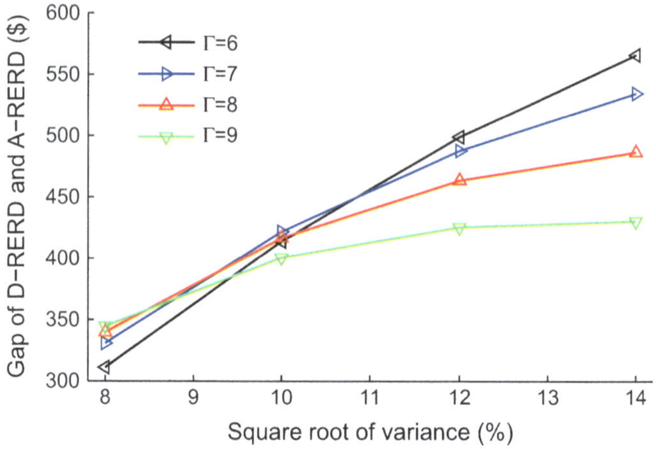

Fig. 2.14 Comparison of the costs of D-RERD and A-RERD in the worst-case scenario

The gap between the optimal total costs of D-RERD and A-RERD in the worst-case scenario is shown in Fig. 2.14. The total cost of A-RERD is the optimal value of problem (2.71), i.e., $C_T^A = c^T x_A^* + Q(x_A^*, w_A^*) = c^T x_A^* + f^T y_A^*$. The total cost of D-RERD C_T^D is procured in the following way: fix the optimal solution x_D^* in the first stage, and the second-stage decision y_D^* is obtained from solving problem (2.69) with the worst-case scenario w_A^* identified by A-RERD. In this way, the total cost of D-RERD is $C_T^D = c^T x_D^* + Q(x_D^*, w_A^*) = c^T x_D^* + f^T y_D^*$, which must be larger than C_T^A, because both stages of A-RERD are optimized for w_A^*. This assertion is certified by Fig. 2.14, from which we can see that the gap $C_T^D - C_T^A$ is always positive. Figure 2.14 also illustrates the following two facts: first, for a fixed budget Γ, with the forecast becoming less accurate, the gap grows larger, as the worst-case scenario w_A^* leaves farther from the forecast, and the results of D-RERD become less optimized for w_A^*; second, with the budget Γ growing, the gap is becoming less sensitive to the accuracy of forecast. These results demonstrate that deploying the strategy of D-RERD will not cause disastrous contingencies even under the worst-case scenario, although may incur a higher but acceptable cost in this rare event.

Finally, the reserve capacity scheduled by D-RERD and A-RERD are given in Fig. 2.15. In this case, only up-reserve capacity is scheduled and down-reserve capacity is 0 because the wind curtailment cost is lower than the down-regulation cost. As a result, excessive wind generation will be curtailed rather than deploying down-reserve capacity. The situation may change when both the wind power and reserve capacity participate in a joint energy and reserve market, where their respective cost is comparable. From Fig. 2.15, it seems that D-RERD will dispatch more reserve capacity than A-RERD, because of the phenomenon called "tail effect," i.e. in the worst-case distribution, a larger share of wind power occurrence tends to leave as far away from the forecast as possible, subjecting to the constraints in (2.74). A-RERD is difficult to model such a phenomenon because the uncertainty

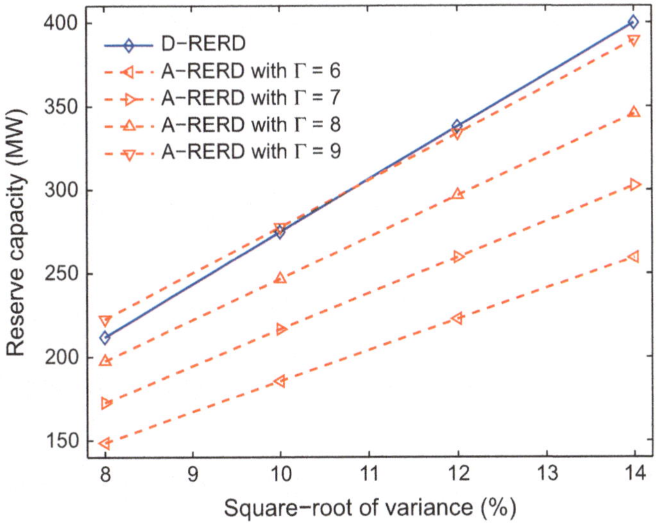

Fig. 2.15 Reserve capacity scheduled by D-RERD and A-RERD

set W is a bounded polytope that does not involve any probability information. Despite the fact that more reserve capacity is scheduled, we have already known from Fig. 2.13 that the total expected cost of D-RERD is still lower than that of A-RERD in most cases. Moreover, the optimal value of D-RERD provides an upper bound of the expected total cost when the wind generation follows the PDFs in set Ω.

2.5 Dispatchable Region: Characterization and Optimization

A-RERD in Sect. 2.4 discusses how to schedule generators with an estimated region of renewable power volatility. This section investigates the inverse problem: how much uncertainty can be dealt with under fixed generation and reserve schedules? The answer is quantified by a set in the nodal injection space, which is called the dispatchable region. We establish a systematic theory of dispatchable region and methodologies to help power system operation in four subsections: The mathematical definition and property of dispatchable region is given in Sect. 2.5.1; a cutting plane method that dynamically generates boundaries of the dispatchable region is developed in Sect. 2.5.2; a data-driven method to assess the probability of DC power flow insolvability based on the dispatchable region is proposed in Sect. 2.5.3; finally, the optimal control of dispatchable region through an affine dispatch policy is discussed in Sect. 2.5.4. Materials in this section come from [107–110].

2.5.1 Definition and Property of the Dispatchable Region

Most symbols and notations are inherited from Sect. 2.4. Suppose the output p, reserve level r^+, r^- of generation units, and the current output w^e from renewable power plants are known. The output level of renewable units may deviate from w^e; the deviation or forecast error is Δw. Once the actual output $w = w^e + \Delta w$ is observed, corrective actions $\{p^+, p^-\}$ are deployed to recover operating constraints; after that the generator output vector is changed to $p + p^+ - p^-$. Due to the limited ramping capability and the cost of reserve deployment, the system cannot accommodate arbitrarily large renewable variations. The task is to identify the largest set W, such that for any given $w = w^e + \Delta w$, $\Delta w \in W$, there is at least one valid corrective action $\{p^+, p^-\}$ which procures a feasible DC power flow solution in the following constraint set

$$\sum_{i=1}^{N_G}(p_i + p_i^+ - p_i^-) + \sum_{j=1}^{N_W}(w_j^e + \Delta w_j) = \sum_{q=1}^{N_Q} p_q \tag{2.76a}$$

$$-F_l \le \sum_{i=1}^{N_G}\pi_{il}(p_i + p_i^+ - p_i^-) + \sum_{j=1}^{N_W}\pi_{jl}(w_j^e + \Delta w_j) - \sum_{q=1}^{N_Q}\pi_{ql}p_q \le F_l,\ \forall l \tag{2.76b}$$

$$P_i^l \le p_i + p_i^+ - p_i^- \le P_i^u,\ \forall i \tag{2.76c}$$

$$0 \le p_i^+ \le r_i^+,\ 0 \le p_i^- \le r_i^-,\ \forall i \tag{2.76d}$$

$$\sum_{i=1}^{N_G}(b_i^+ p_i^+ + b_i^- p_i^-) \le C^R \tag{2.76e}$$

where constraints (2.76a)–(2.76c) are power balancing condition, power flow limits of transmission lines, and generation capacity, respectively; constraint (2.76d) stipulates that the regulation power is subjected to the reserve level, which depends on the ramping speed as $r_i^+ \le R_i^+\Delta t$, $r_i^- \le R_i^-\Delta t$, where R_i^+, R_i^- are the maximum upward and downward regulation speed of unit i, Δt is the duration of dispatch interval, or can be purchased from the reserve market; budget constraint (2.76e) restricts the total cost of re-dispatch, which couples the regulation power of all units.

To acquire the optimal corrective actions, one could minimize the cost $C^R = \sum_i (b_i^+ p_i^+ + b_i^- p_i^-)$ subject to constraints (2.76a)–(2.76d). In the dispatchable region study, we aim to examine the existence of corrective actions $\{p^+, p^-\}$ rather than acquiring a strategic solution. So the cost is considered as an available budget in (2.76e). The current operating strategies $\{p, r^+, r^-, w^e\}$ are known parameters in constraint set (2.76). They act as the input of dispatchable region. How to produce the optimal dispatch strategy is not the main concern of this section. For the purpose of clarity, renewable spillage and load shedding are neglected, although there is no difficulty to model such issues.

Constraints (2.76a)–(2.76e) can be written in a compact form

$$Ax + By + C(w^e + \Delta w) \leq b^0$$

where $x = [p, r^+, r^-]$ is given parameter; variable $y = [p^+, p^-]$ can be adjusted in response to the value of Δw; matrices A, B, C and vector b^0 correspond to coefficients in (2.76a)–(2.76e). Define the feasible set of y under given x and Δw

$$Y(x, \Delta w) = \{y \mid By \leq b - Ax - C\Delta w\} \tag{2.77}$$

where vector $b = b^0 - Cw^e$. The compact form (2.77) brings up the definition of dispatchable region as follows:

Definition 2.1 The dispatchable region W_R^D is a set comprised of all Δw such that (2.76) is feasible, or mathematically

$$W_R^D = \{\Delta w \mid Y(x, \Delta w) \neq \emptyset\} \tag{2.78}$$

According to Definition 2.1, $\Delta w \in W_R^D$ is a necessary and sufficient condition for DC power flow solvability. In view of this, W_R^D acts as a measure on the flexibility of the system under the given dispatch strategy x, or can be regarded as an extension of the crucial concept "security region" in the uncertainty space. It informs the operator exactly how much uncertainty the system can deal with. It is interesting to investigate the geometry property of dispatchable region.

Proposition 2.1 *The dispatchable region is a bounded polytope.*

This assertion is easy to understand. Consider the set

$$\Theta = \{(y, \Delta w) | By + C\Delta w \leq b - Ax\}$$

It is a bounded polytope because all variables have physical limits. According to Definition 2.1, the dispatchable region is the projection of Θ on Δw-subspace, i.e.

$$W_R^D = \prod\nolimits_{\Delta w}(\Theta) = \{\Delta w \mid \exists y : By + C\Delta w \leq b - Ax\} \tag{2.79}$$

where Π is the projection operator. Therefore, the dispatchable region is also a bounded polytope as projection is a linear mapping.

Some popular projection algorithms in computational geometry, such as the Fourier-Motzkin elimination method, the extreme point method, and the convex hull method discussed in [111], can be exploited to compute the dispatchable region using its projected form (2.79). The Multi-Parametric Toolbox [112] also offers build-in function for polyhedral projection. However, projection algorithms may be less efficient for large problems. To see this, the Fourier-Motzkin elimination method creates $O(m^2)$ (where m is the total number of linear inequalities in Θ) new inequalities after eliminating one variable, so the calculation quickly becomes unmanageable after a few elimination steps. The extreme point method relies on vertex enumeration, whose complexity may grow exponentially with respect to the dimension of Θ. The convex hull method may be the most efficient one among the three from a computational point of view. However, it still requires a facet enumeration in each iteration, which is less efficient for large-scale instances. In the next subsection, we develop a special cutting plane method for computing the dispatchable region.

2.5.2 Computing the Dispatchable Region

As stated in Proposition 2.1, the dispatchable region is a polytope characterized by $W_R^D = \{\Delta w | H\Delta w \le h\}$. We elucidate how to retrieve matrix H and vector h from the coefficient matrices in (2.79) and provide a closed form of the dispatchable region.

Proposition 2.2 *The dispatchable region can be expressed as*

$$W_R^D = \{\Delta w \mid u^T C \Delta w \ge u^T (b - Ax), \forall u \in \text{vert}(U)\} \tag{2.80}$$

where $U = \{u \mid B^T u = 0, -1 \le u \le 0\}$ *is a bounded polytope, and* $\text{vert}(U)$ *represents all the vertices of* U.

Proof For fixed x and Δw, consider the following feasibility detection problem

$$\begin{aligned} \min \quad & 1^T s^+ + 1^T s^- \\ \text{s.t.} \quad & By + Is^+ - Is^- \le b - Ax - C\Delta w \\ & s^+ \ge 0, \; s^- \ge 0 \end{aligned} \tag{2.81}$$

where s^+ and s^- are non-negative slack variables, 1^T and I are the unit vector and identity matrix with compatible dimensions, respectively. Equation (2.81) is always feasible and has a finite minimum. The dual problem of LP (2.81) is

$$\begin{aligned} \max \quad & u^T (b - Ap - C\Delta w) \\ \text{s.t.} \quad & B^T u = 0, \; u \le 0 \\ & -1 \le u \le 1 \end{aligned} \tag{2.82}$$

where u is the dual variable. Set $Y(p, \Delta w) \ne \emptyset$ if and only if the optimal value of (2.81) is 0. Because strong duality always holds for LPs, the optimal value of (2.82) is also 0, implying

$$u^T(b - Ap - C\Delta w) \le 0, \ \forall u \in U$$

where $U = \{u | B^T u = 0, -1 \le u \le 0\}$ is the feasible set of the dual variable u. If we consider Δw as a controllable parameter in the condition above, we can claim that W_R^D has the following polyhedral representation by noticing that the optimal solution of LPs can always be found at one of its vertices

$$\begin{aligned} W_R^D &= \{\Delta w \mid u^T C \Delta w \ge u^T(b - Ax), \forall u \in U\} \\ &= \{\Delta w \mid u^T C \Delta w \ge u^T(b - Ax), \forall u \in \text{vert}(U)\} \end{aligned}$$

This completes the proof. ■

Proposition 2.2 gives an explicit polyhedral form of W_R^D based on vertex enumeration. Some further discussions are provided.

1. The polytope U only depends on the matrix B, which depends on the parameters of generators and the transmission network, and is independent of operating conditions, so is constant for a given system. For some small-scale power systems, the vertices of polytope U can be computed off-line, then W_R^D can be directly constructed according to Proposition 2.2 after receiving the current generation dispatch p and VERs' output w^e (contained in the vector $b = b^0 - Cw^e$).
2. In fact, in the vertex enumeration based formulation (2.80), most linear inequalities in W_R^D will be redundant. The method proposed in [113] can be applied to remove redundant constraints through solving an auxiliary LP.
3. Due to the difficulty and complexity of vertex enumeration [114], it is usually impossible to enumerate all vertices of set U even for medium scale power systems. In this section, an algorithm will be proposed to identify binding vertices in U and generate the boundaries of W^{RTD} iteratively.

To this end, we first consider a problem: given a polytope W, certify $W \subseteq W_R^D$, or identify a point $w^* \in W$ and $w \notin W_R^D$, which is called a separation oracle. According to Proposition 2.2, if $W \subseteq W_R^D$ is true, there must be

$$u^T C \Delta w \ge u^T(b - Ax), \ \forall u \in U, \ \forall \Delta w \in W \tag{2.83}$$

Recall Definition 2.1, W_R^D is the largest set that makes (2.83) hold, indicating that if $\exists \Delta w^* \notin W_R^D$, there must be some vertex $u^* \in U$ such that

$$(u^*)^T(b - Ax - C\Delta w^*) > 0 \tag{2.84}$$

which brings up the definition of critical vertex.

Definition 2.2 (Critical Vertex) An element $u^* \in U$ is called a critical vertex if $\exists \Delta w^* \notin W_R^D$ such that (2.84) is met.

Given $w^* \notin W_R^D$ and its corresponding critical vertex $u^* \in U$, the following hyperplane strictly separates Δw^* from W_R^D

$$(u^*)^T C \Delta w = (u^*)^T (b - Ax) \tag{2.85}$$

Hyperplane (2.85) will not remove any point inside W_R^D. More precisely,

$$\left\{ \Delta w \,\middle|\, (u^*)^T C \Delta w < (u^*)^T (b - Ax) \right\} \cap W_R^D = \emptyset$$

because of (2.80).

In view of analysis above, our strategy for computing W_R^D is to create a large enough initial box set W^B such that $W_R^D \subseteq W^B$. Then remove $\Delta w^* \notin W_R^D$ from W^B by identifying critical vertices and creating hyperplanes, until $W^B \subseteq W_R^D$ is certified, therefore $W^B = W^{RTD}$ can be found. This procedure involves the following important problem:

Separation Problem Given a polytope W, validate (2.83) or find a pair $(u^*, \Delta w)$ which satisfies (2.84).

The separation problem can boil down to a bilinear program

$$\begin{aligned} R(x) = \max \quad & u^T (b - Ax - C\Delta w) \\ \text{s.t.} \quad & u \in U, \ \Delta w \in W \end{aligned} \tag{2.86}$$

Since $u = 0$ is always feasible, the optimal value $R(x)$ must be non-negative. If $R(x) = 0$, then $\forall \Delta w \in W$, $Y(x, \Delta w) \neq \emptyset$, or equivalently, (2.83) and $W \subseteq W_R^D$ hold true for sets U and W. Otherwise, if $R(x) > 0$, i.e., (2.84) is met at the pair $(u^*, \Delta w^*)$, the optimal solution Δw^* does not belong to W_R^D, and u^* is a corresponding critical vertex. Problem (2.86) is non-convex. Two methods for solving this problem are given in Appendix C.2.3. The first one reformulates (2.86) as the equivalent MILP (C.63), which can be handled by off-the-shelf solvers. The second one is the mountain climbing algorithm which solves LPs iteratively and summarized in Algorithm C.1. There is no provable guarantee for finding the global optimal solution. In what follows, we assume problem (2.86) can be solved without difficulties.

The algorithm of computing the dispatchable region is formally provided in Algorithm 2.4, which will terminate in a finite number of iterations because polytope U has finite vertices. Several remarks are given.

Algorithm 2.4 Adaptive vertex generation

1: Select a large enough hypercube $W^B = \{\Delta w \mid H\Delta w \leq h\}$ and tolerance $\delta > 0$;
2: Solve problem (2.86) with the current W^B, the optimal solution is $(u^*, \Delta w^*)$, and the optimal value is $R(x)$.
3: If $R(x) \leq \delta$, terminate, report $W_R^D = W^B$; else add the following constraint to W^B, update matrix H and vector h, and go to step 2.

$$(u^*)^T C\Delta w \geq (u^*)^T (b - Ax) \tag{2.87}$$

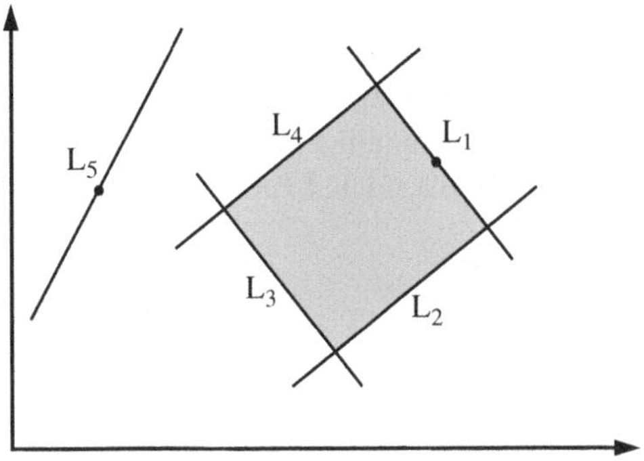

Fig. 2.16 Illustration of a redundant constraint

1. Instead of solving problem (2.86) rigorously, any $(u^*, \Delta w^*)$ pair found during computation that satisfies (2.84) is eligible to create the cutting plane (2.87). This trick can reduce the computation burden of globally solving (2.86).
2. Because the arbitrary initiation W^B could be very large, we can use the mountain climbing method to solve (2.86) in step 2 due to its high efficiency. If mountain climbing finds no solution with an objective value greater than δ, we switch to solving MILP (C.63) and continue.
3. The discussed method is able to incorporate load variations as well. However, with the dimension of the uncertainty space growing, computing W_R^D will become more challenging, as problem (2.86) will involve more variables, and Algorithm 2.4 will spend more iterations.
4. Because the initial set W^B is larger than W_R^D, redundant constraints may exist in the result of Algorithm 2.4, which can be removed by using the method in [113], which is briefly introduced here.

Consider a polyhedral set expressed by $\{x|Ax \leq b\}$ or $\{x|a_i^T x \leq b_i, i = 1, \ldots, m\}$. A constraint is redundant if removing it from the set does not affect the feasible region. Taking Fig. 2.16 for example, inequalities defined by L_1–L_4 are umbrella (non-redundant) constraints following the terminology in [113], and L_5 gives a redundant constraint. Please keep in mind that redundancy is independent of the objective function and the optimal solution, whereas active (binding) constraints are a subset of umbrella constraints depending on the optimal solution. For a non-redundant constraint, there is at least one point which makes it active and satisfies all remaining constraints, e.g., the point on L_1 in Fig. 2.16. Also, for a redundant constraint, there is no such a point; the point on L_5 violates constraints generated by L_3 and L_4. A linear programming method is suggested in [113] to determine whether a constraint is redundant or not based on above observations. Specifically, for the i-th constraint, we solve

$$\begin{aligned} \min \ & s \\ \text{s.t.} \ & Ax \leq b, \ s \geq 0 \\ & a_i^T x + s \geq b_i \end{aligned}$$

If the optimal value s^* is equal to zero, the optimal solution x^* activates the i-th constraint and complies with all the other constraints; otherwise, if s^* is strictly positive, we cannot find such a point, and the constraint is redundant.The procedure can be done in a single LP, if desired.

$$\begin{aligned} \min \ & \sum\nolimits_{j=1}^{m} s_j \\ \text{s.t.} \ & Ax_j \leq b, \ s_j \geq 0, \ \forall j \\ & a_j^T x_j + s_j \geq b_j, \ \forall j \end{aligned}$$

Constraints associated with $s_j^* > 0$ are redundant.

Dispatchable region contains useful information on system operating security. Several implications are revealed.

1. Vulnerability Assessment

Given a forecast error $\Delta w \notin W_R^D$, DC power flow is infeasible, there must be one or more operating constraints violated, due to the insufficient reserve backup or ramping rates, or the lack of transmission capacities, or the shortage of cost, or the combination of them. The terminology "vulnerability" means the factors which prevent the system from being restored. To identify such factors, it is necessary to pick up the violated inequalities in W_R^D such that

$$(u^i)^T C \Delta w < h_i, \ i \in I_V$$

where I_V corresponds to the indices of inequalities that are violated due to Δw. Each non-zero element in u^i, $i \in I_V$ suggests a binding constraint in (2.76), which indicates that the associated resource will be used up in re-dispatch stage, thus the corresponding element is vulnerable to the deviation pattern Δw. The magnitudes of these elements may naturally give a ranking of vulnerable components.

2. Security Assessment

Given a forecast error $\Delta w \in W^{RTD}$, the corrective action $y \in Y(x, \Delta w)$ can restore the power system in re-dispatch. From a practical point of view, the minimal distance d from Δw to the boundaries of W_R^D is desired, reflecting the security margin to infeasibility. The distance can be explicitly calculated by

$$d = \min\left\{\frac{|H^i(\Delta w - \Delta w^{bi})|}{\sqrt{H^i(H^i)^T}}, \forall i\right\} \tag{2.88}$$

where H^i is the i-th row of matrix H, h_i is the i-th element of vector h, Δw^{bi} is an arbitrary point on the i-th hyperplane, satisfying $H^i \Delta w^{bi} = h_i$, so h does not appear in (2.88) explicitly. Because the expression of W_R^D is known at this stage, it's quite easy to acquire such a point.

It is interesting to study how reserve should be scheduled in order to increase the security margin d. However, because W_R^D does not has a closed form, it is difficult to quantify the impact of generation schedule on the dispatchable region. In Sect. 2.5.4, we will investigate how to control and optimize the shape of dispatchable region under an affine dispatch policy.

3. Reliability Assessment
If the probability distribution of renewable power is available, one can test the probability of power flow insolvability $\Pr[\Delta w \notin W_R^D]$ from Monte Carlo simulation without solving DC power flow equations, and thus can be implemented online. Moreover, the system reliability level under different cost C^R can be easily examined, providing a reference for operators with different risk preferences and economic interests.

In Sect. 2.5.3, we will develop a data-driven method to assess system reliability without an exact probability distribution function, which will be further used in Sect. 2.5.4 to optimize the shape of the dispatchable region.

The proposed method is applied to two testing system to illustrate the concept of dispatchable region and validate Algorithm 2.4.

1. Modified PJM 5-Bus System
System topology is shown in Fig. 2.17. It includes three thermal units at Buses A, D, and E. Wind farm W_1 with 150 MW output is connected to the grid at bus D. Wind farm W_2 with 100 MW output is connected to the grid at bus A. In the considered dispatch interval, the wind power is assumed to vary from 90%–110% of its current

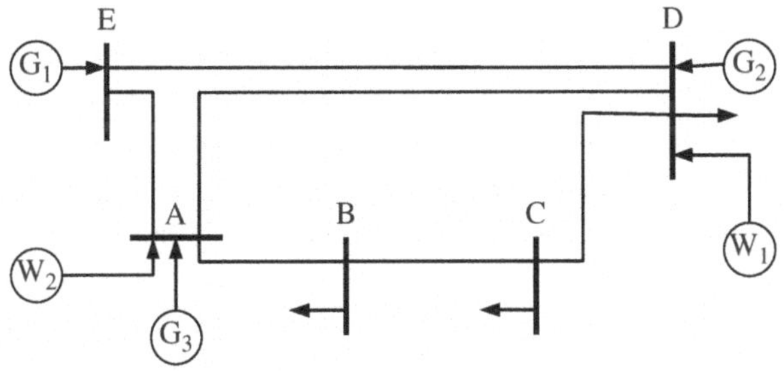

Fig. 2.17 Diagram of the PJM 5-bus system

Table 2.18 Generator data

Unit	$[P_i^l, P_i^u]$ MW	Generation cost CNY/MWh	Reserve cost CNY/MWh	Ramping limit MW/h
G_1	[150, 400]	200	300	40
G_2	[200, 500]	300	450	50
G_3	[250, 600]	360	520	60

Table 2.19 Transmission line data

Line	From	To	Reactance (p.u.)	Capacity (MW)
L1	A	B	0.0281	600
L2	A	D	0.0304	300
L3	A	E	0.0064	200
L4	B	C	0.0108	100
L5	C	D	0.0297	401
L6	D	E	0.0297	300

output. Deterministic demands at Buses B, C, and D are 550 MW, 450 MW, and 350 MW, respectively. Generator and line data are shown in Tables 2.18 and 2.19.

Case 1 (Traditional Dispatch) The total reserve capacity is $R_T = 25$ MW, which is equal to the range of wind power uncertainty. The reserve capacity offered by each generator is proportional to its capacity, i.e.:

$$r_i = R_T P_i^u / \sum_i P_i^u, \ \forall i \tag{2.89}$$

Then the generation dispatch is determined from problem (2.90),

$$\begin{aligned} \min \ & \sum_i (a_i p_i^2 + b_i p_i) \\ \text{s.t. } & P_i^l + r_i \le p_i \le P_i^u - r_i, \forall i, \ \sum_i p_i + \sum_j w_j^e = \sum_q p_q \\ & - F_l \le \sum_i \pi_{il} p_i + \sum_j \pi_{jl} w_j^e - \sum_q \pi_{ql} p_q \le F_l, \ \forall l \end{aligned} \tag{2.90}$$

The generation dispatch, reserve allocation, and line power flow are shown in Fig. 2.18. The dispatchable region is depicted through Fig. 2.19, which is characterized by eight linear inequalities and can be rewritten as:

$$H^{TED} \begin{bmatrix} \Delta w_1 \\ \Delta w_2 \end{bmatrix} \le h^{TED}$$

where

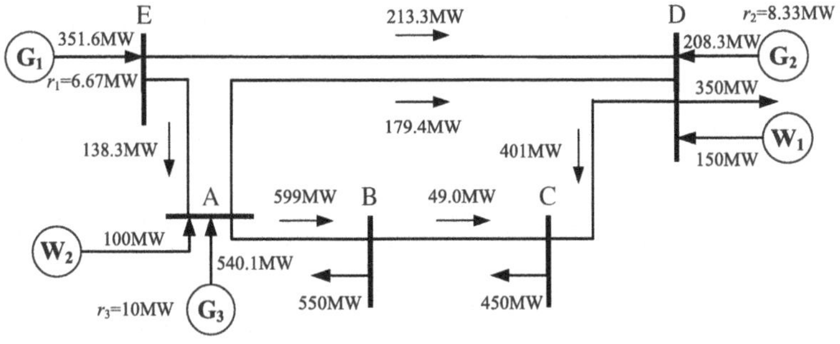

Fig. 2.18 Generation schedule and DC power flow under traditional dispatch

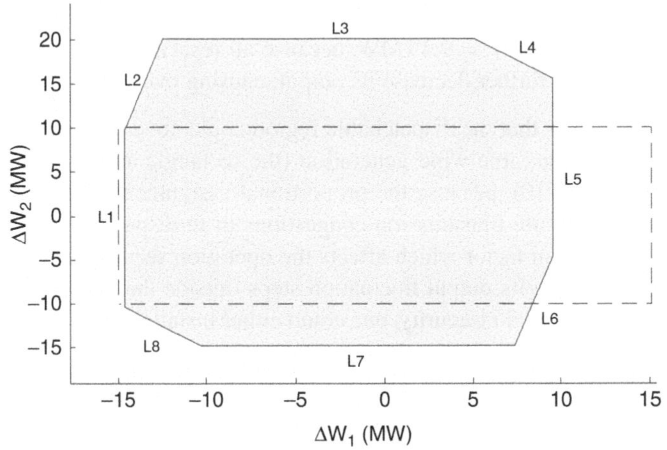

Fig. 2.19 Dispatchable region under traditional generation schedule

$$H^{TED} = \begin{bmatrix} -1.0000 & -0.1595 & 0.0000 & 1.0000 & 1.0000 & 0.1595 & 0.0000 & -1.0000 \\ 0.0000 & 0.0344 & 1.0000 & 1.0000 & 0.0000 & -0.0344 & -1.0000 & -1.0000 \end{bmatrix}^T$$

$$h^{TED} = \begin{bmatrix} 14.6500 & 2.6707 & 20.6300 & 25.0000 & 9.4900 & 1.6717 & 14.6600 & 25.0000 \end{bmatrix}^T$$

Among these boundaries, L4 and L8 are caused by the lack of total reserve capacity, while others are caused by transmission congestions. Details are shown in Table 2.20. Take L1 for example: when $\Delta w_1 = -14.65$ MW and $\Delta w_2 = 0$, the output of G_2 will increase 8.33 MW. In order to maintain the active power through Line AB within 600 MW, G_1's output has to increase 6.67 MW, and G_3's output has to decrease 0.35 MW. If $\Delta w_1 < 14.65$ MW, because all reserve capacity in G_1 and G_2 is used up, G_3 has to further increase its output, causing congestion in Line AB. Take L5 for another example: when $\Delta w_1 = 9.49$ MW and $\Delta w_2 = 0$, G_2's output will decrease 8.33 MW. In order to maintain the active power through Line CD within 401 MW, G_1's output has to decrease 6.67 MW, and G_3's output has to

Table 2.20 Inducements of boundaries in Case 1

Boundary	Inducement	Reaction from generators		
		G_1	G_2	G_3
L1	Flow limit of Line AB	↑	↑	Slack
L2	Flow limit of Line AB	↓	↑	Slack
L3	Flow limit of Line AB	↓	Slack	↓
L4	Lack of reserve	↑	↑	↑
L5	Flow limit of Line CD	↓	↓	Slack
L6	Flow limit of Line CD	Slack	↓	↑
L7	Flow limit of Line CD	↑	Slack	↑
L8	Lack of reserve	↓	↓	↓

increase 5.51 MW. If $\Delta w_1 > 9.49$ MW, because all reserve capacity in G_1 and G_2 is used up, G_3 has to further decrease its output, causing over flow in Line CD.

Figure 2.19 shows that the dispatchable region could cover the majority, but not the entire range of possible wind generation (the rectangle area: $-15 \leq \Delta w_1 \leq 15$, $-10 \leq \Delta w_2 \leq 10$), because the proportional assignment of reserve capacity does not consider future transmission congestions in re-dispatch. Figure 2.19 also indicates that the main factor which affects the operation security is the wind farm W_1 at Bus D, because its output fluctuation steps outside the dispatchable region. To eliminate the potential insecurity, one could either install storage devices in wind farm W_1, or modify the dispatch strategy.

Case 2 (Robust Dispatch) A-RERD method presented in Sect. 2.4 is used to determine the generation and reserve schedule. The uncertainty set of wind generation is $W = \{\Delta w | -15 \leq \Delta w_1 \leq 15, -10 \leq \Delta w_2 \leq 10\}$. Wind power spillage and load shedding are prohibited. The generation dispatch, reserve allocation, and line power flow are shown through Fig. 2.20. The dispatchable region is plotted in Fig. 2.21, which is characterized by six linear inequalities and can be rewritten as:

$$H^{RED} \begin{bmatrix} \Delta w_1 \\ \Delta w_2 \end{bmatrix} \leq h^{RED}$$

where

$$H^{RED} = \begin{bmatrix} -1 & 0 & 1 & 1 & 0 & -1 \\ 0 & 1 & 1 & 0 & -1 & -1 \end{bmatrix}^T$$

$$h^{RED} = \begin{bmatrix} 17.10 & 11.63 & 25.00 & 17.10 & 11.63 & 25.00 \end{bmatrix}^T$$

Among these boundaries, L3 and L6 are caused by the lack of reserve capacity, while others are caused by transmission congestions. Details are shown in Table 2.21. Similar analysis can be conducted as that in the previous case. The difference is that the dispatchable region under robust generation schedule covers

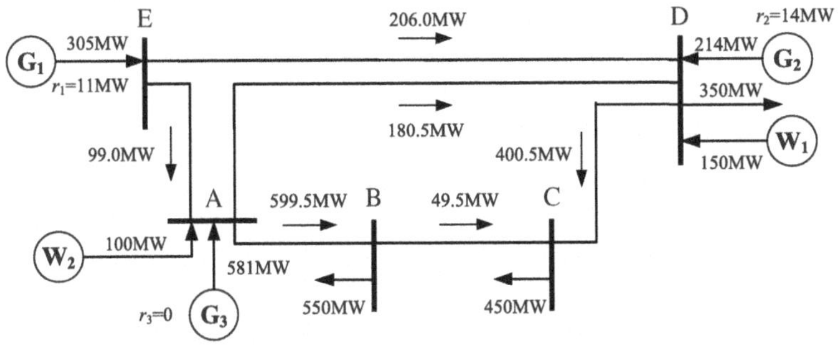

Fig. 2.20 Generation schedule and DC power flow under robust dispatch

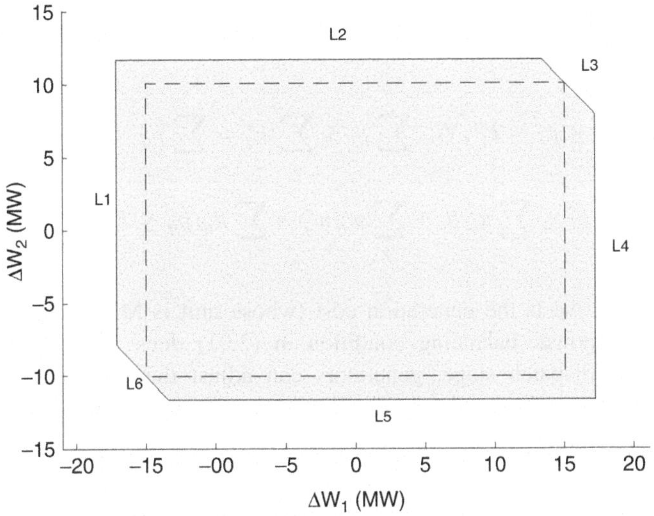

Fig. 2.21 Dispatchable region under robust generation schedule

the entire uncertainty set of wind generation using the same amount of reserve capacity, because the A-RERD method considers operating constraints in re-dispatch stage. Figure 2.21 shows the strictly inclusive relation $W \subset W_R^D$. This may explain why robust strategies are generally regarded as reliable but conservative.

2. IEEE 118-Bus System

System data are available in [102]. There are 54 thermal units and 186 transmission lines in this system. The total demand is 5500 MW. The up/down regulation cost coefficient b_i^+/b_i^- of each generator is assumed to be 10% of its production cost coefficient b_i. The up/down ramping limit R_i^+/R_i^- of each generator is assumed to be 25% of its capacity. Virtual wind farms are connected. Dispatchable region of this system with different numbers of wind farms are computed. The following

Table 2.21 Inducements of boundaries in Case 2

Boundary	Inducement	Reaction from generators	
		G_1	G_2
L1	Flow limit of Line AB	Slack	↑
L2	Flow limit of Line AB	↓	Slack
L3	Lack of reserve	↑	↑
L4	Flow limit of Line CD	Slack	↓
L5	Flow limit of Line CD	↑	Slack
L6	Lack of reserve	↓	↓

economic dispatch problem is solved to determine the generator output p

$$\begin{aligned} \min \ & \sum_i (a_i p_i^2 + b_i p_i) \\ \text{s.t.} \ & P_i^l \le p_i \le P_i^u, \forall i, \ \sum_i p_i + \sum_j w_j^e = \sum_q p_q \\ & -F_l \le \sum_i \pi_{il} p_i + \sum_j \pi_{jl} w_j^e - \sum_q \pi_{ql} p_q \le F_l, \ \forall l \end{aligned} \tag{2.91}$$

where the objective is the generation cost (whose unit is MBtu in the data file). Unlike (2.90), power balancing condition in (2.91) does not consider reserve capacity. In re-dispatch stage, generators can adjust their output at maximum ramping rates, i.e., $r_i^+ = R_i^+ \Delta t$, $r_i^- = R_i^- \Delta t$, $\Delta t = 1$ are used in (2.76d).

Two wind farms are connected at bus #70 in Zone 1 and bus #49 in Zone 2. Their respective current output is equal to 350 MW. Problem (2.91) is solved to retrieve unit output, and then Algorithm 2.4 is applied to compute the dispatchable region W_R^D under different values of C^R. The computation time is less than 5 s in all these tests. Results are shown in Fig. 2.22.

For each case, 1000 uniformly distributed samples Δw^s are generated and the feasibility of (2.76) is tested. It is found that (2.76) has at least one feasible solution if and only if $\Delta w^s \in W_R^D$, which complies with Definition 2.1 and thus validates Algorithm 2.4. The impact of parameter C^R on W_R^D can be observed from Fig. 2.22, which shows that W_R^D grows larger with C^R increasing. However, when C^R is sufficiently large, the cost constraint (2.76e) never becomes binding, and the solvability of DC power flow will mainly depend on the ramping limits of generators and power flow capacities of transmission lines, and C^R becomes less important.

One more wind farm is connected at bus #100 in Zone 3. The output of the three wind farms is equal to 250 MW. The dispatchable regions with different values of C^R are illustrated through Figs. 2.23, 2.24 and 2.25, from which we can see clearly that W_R^D consists of hyperplanes and is polyhedral. Similar to Fig. 2.22, W_R^D grows larger with C^R increasing.

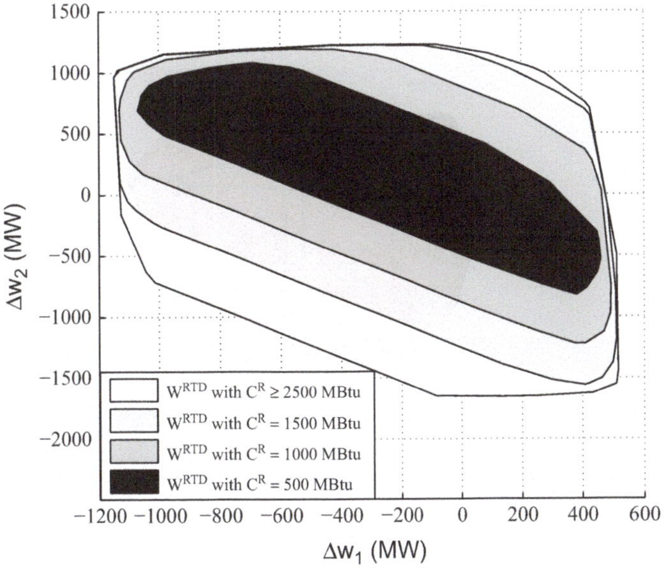

Fig. 2.22 W_R^D of two wind farms under different C^R

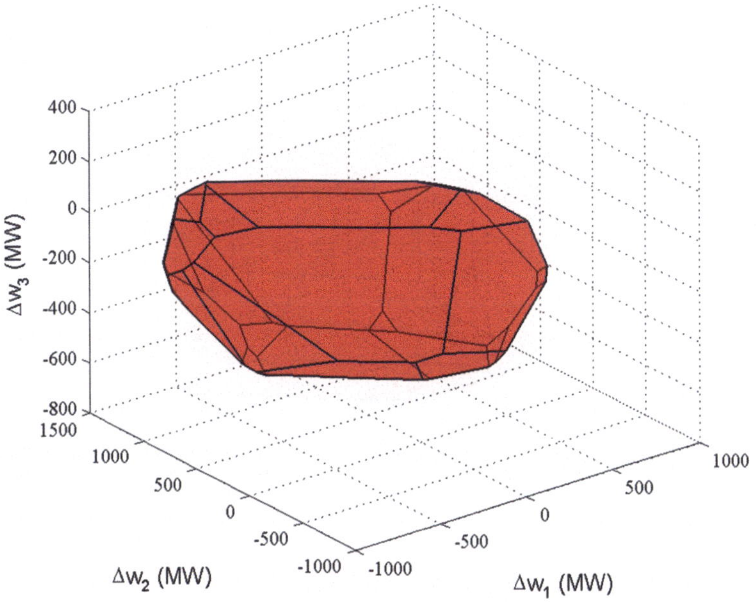

Fig. 2.23 W_R^D with three wind farms and C^R=200 MBtu

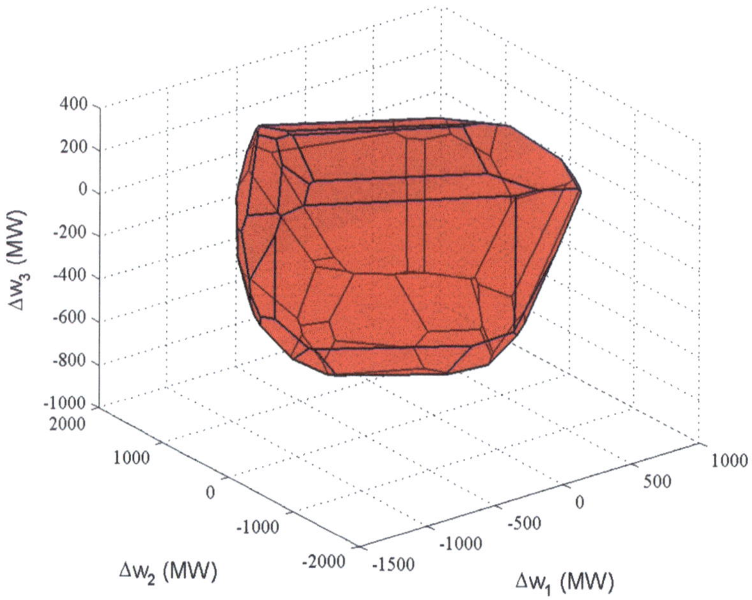

Fig. 2.24 W_R^D with three wind farms and C^R=600 MBtu

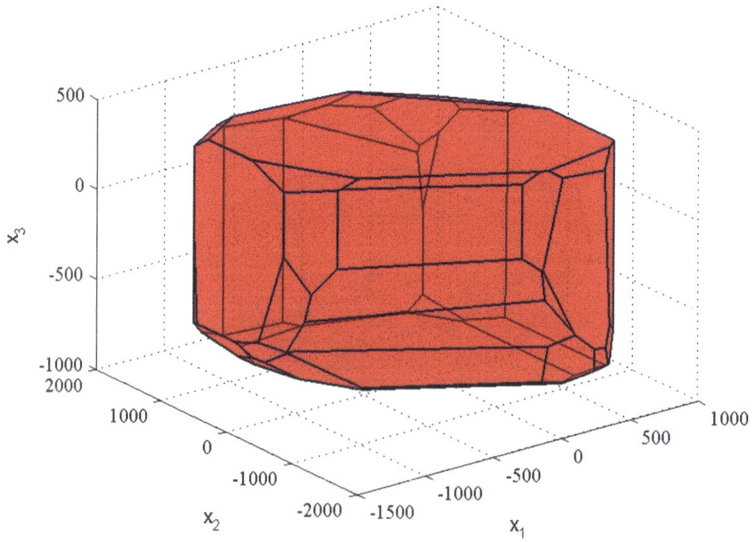

Fig. 2.25 W_R^D with three wind farms and C^R=2500 MBtu

These case studies demonstrate that dispatchable region provides both analytical tools as well as visualization paradigm to study the ability of power systems to accommodate uncertain renewable power generations.

Finally, we would like to highlight that the proposed method is especially suitable for power transmission systems, because wind and solar generation is centrally integrated, the dimension of uncertainty space is not very high, and the MILP in step 2 of Algorithm 2.4 can be solved efficiently; in addition, the DC power flow model provides satisfactory accuracy for such networks. In distribution networks, line resistance, reactive power control, and voltage profile should be modeled.

The dispatchable region concept has been generalized to distribution system in [115]. Nevertheless, using the linearized BFM, in Sect. 2.2, the proposed Algorithm 2.4 can be used to access the dispatchable region of radial distribution networks with reactive power and voltage constraints. When the AC power flow model is considered, the dispatchable region can be defined in the similar way. However, the exact region is difficult to acquire. The methods in [116, 117] offer viable options to compute conservative (inner) approximations.

2.5.3 Probability of DC Power Flow Infeasibility

In many power system applications, the variable nature of renewable generation is described by a PDF, for instance, the stochastic programming model of generation scheduling problem and the Monte Carlo simulation require the PDF of wind/solar power for scenario generation. However, the exact PDF may be difficult to acquire. One can only seek approximations that best fit to the historical data. Without a PDF, Monte Carlo simulation can hardly be performed. Nevertheless, the historical data, even if not enough to generate an exact PDF, may have further implications on some probability guarantee.

The simplest case of our problem is illustrated in Fig. 2.26. The system consists of a conventional generator, a wind farm, and a load with 400 MW deterministic demand. The current output of generator and wind farm is equal to 200 MW, respectively. The wind generation w is stochastic and its PDF is unknown. We only have its forecast ($\mathbb{E}[w] = 200$ MW, first-order moment) and variance ($\text{Var}[w] = \sigma^2$, second-order center moment) in the next 1 h. If the wind generation deviates from 200 MW, the generator will respond to the forecast error by adjusting its output so as to balance the real-time mismatch between generation and demand. Because the ramping speed of the generator is limited, say 100 MW/h, the power balancing condition cannot be maintained if $|w - 200| > 100$. The probability of infeasible

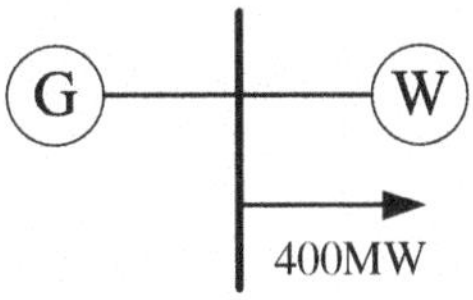

Fig. 2.26 A single-bus system

dispatch depends on the forecast accuracy. According to Chebyshev inequality, an upper bound of the probability of infeasibility is given by

$$\Pr(|w - 200| > 100) \leq \sigma^2/100^2$$

This inequality holds true regardless of the distribution of random variable w. In this way, we get an estimation on the probability of an infeasible operation without a clear PDF. However, if there are two or more uncertain sources, Chebyshev inequality no longer works. Another shortcoming is that the estimation offered by Chebyshev inequality is often very conservative.

In this section, we study the similar problem in power transmission networks with multiple wind farms. Recent advances in [118–120] raise optimization based approaches to compute the probability of certain event with an ambiguous PDF. Combining the methodology of dispatchable region in Sects. 2.5.1 and 2.5.2, and the generalized probability inequality theory in [119], we propose a data-driven and convex optimization based method to quantify the probability of infeasible operation caused by renewable volatility. The required information is a finite sequence of moments, instead of the exact PDF of renewable power. By assuming the dispatchable region is available, the problem is considered as a multi-dimensional generalized probability inequality. We propose two methods to assess its solution. One is based on SDP and the other is based on LP. We also discuss how to use the generalized Chebyshev inequality (GCI) [118] and the generalized Gauss inequality (GGI) [120] to accomplish this task if only the first and second order moments are available. Both of the two methods come down to solving SDPs.

1. Notations

For the purpose of notation brevity and clarity, let $\kappa = [k_1, k_2, \ldots, k_{N_W}]$ be a vector index with N_W entries, where N_W is the number of wind farms, $k_j \in \mathbb{Z}^+, j = 1, 2, \ldots, N_W$; Index set $J_k = \{\kappa | 1^T \kappa = k\}$ will be used to define k-th order moment; the notation $w^\kappa = w_1^{k_1} w_2^{k_2} \ldots w_{N_W}^{k_{N_W}}$, where $w = [w_1, w_2, \ldots w_{N_W}]^T$, will be frequently used; The k-th order moment of renewable generation is the sequence $\{\sigma_\kappa = \mathbb{E}[w^\kappa]\}, \forall \kappa \in J_k$, where $\mathbb{E}$ is the expectation operator; If $w_j, \forall j$ is independent, the moment $\sigma_\kappa = E[w_1^{k_1}]E[w_2^{k_2}] \ldots E[w_J^{k_J}], \forall \kappa \in J_k$ can be easily retrieved from the local data of each wind farm, otherwise, the sampled data should be synchronized in the presence of correlation; index set $J(K) = \bigcup_{k=0}^{K} J_k$.

The dispatchable region can be computed using Algorithm 2.4 in Sect. 2.5.2 and represented in the forecast error space as

$$W_R^D = \{\Delta w \in \mathbb{R}^{N_W} | H_1 \Delta w \leq h_1\} \tag{2.92a}$$

or in the renewable power space as

$$W_R^D = \{w \in \mathbb{R}^{N_W} | H_2 w \leq h_2\} \tag{2.92b}$$

where $H_2 = H_1$, $h_2 = h_1 + H_1 w^e$, w^e is the renewable generation forecast, or in the unit value space as

$$W_R^D = \{\bar{w} \in \mathbb{R}^{N_W} | H_3 \bar{w} \le h_3\} \tag{2.92c}$$

where $H_3 = H_2 D(w^B)$, $h_3 = h_2$, $D(w^B)$ is a diagonal matrix whose non-zero elements are the base values of renewable plants. In the following context, dispatchable region refers to (2.92b) and the subscript is omitted, i.e. $W_R^D = \{w \in \mathbb{R}^{N_W} | Hw \le h\}$,

In this section, we also use the non-dispatchable region

$$\bar{W}_R^D = \{w \mid w \notin W_R^D\} \tag{2.93}$$

which is complementary to W_R^D on $\mathbb{R}^{N_W}$ and consists of all renewable generation scenarios that will cause infeasibility. Strictly speaking, the non-dispatchable region $\bar{W}_R^D$ does not contain the boundaries of W_R^D and is an open set. Nonetheless, in the following context we will include the boundary of W_R^D in $\bar{W}_R^D$ to facilitate developing tractable reformulations. This will not influence the result because the boundary set ∂W_R^D has zero measure on $\mathbb{R}^{N_W}$.

2. Formulation of the Probability Estimation Problem

Suppose the PDF $f(w)$ of renewable generation is not known exactly. Nevertheless, we can compute the moments of w up to some order from statistic analysis, and the PDF should provide moments consistent with those recovered from actual data. In this regard, $f(w)$ belongs to the ambiguity set Ω_M^K subject to moment constraints up to K-th order

$$\Omega_M^K = \left\{ f(w) \left| \begin{array}{c} \displaystyle\int_{w \in E_B^W} w^\kappa f(w) \mathrm{d}w = \sigma_\kappa, \forall \kappa \in J(K) \\ f(w) \ge 0, \ \forall w \in \mathbb{R}^{N_W} \end{array} \right. \right\} \tag{2.94}$$

where $E_B^W = \{w \in \mathbb{R}^{N_W} \mid 0 \le w \le C^W\}$, C_j^W is the capacity of j-th renewable plant. Because the output of plant j can neither become negative nor exceed its capacity, we restrict the integration over the hypercube E_B^W. We impose non-negativity on $f(w)$, and define $\sigma_{0,\ldots,0} = 1$ in order to make sure that $\int_{w \in E_B^W} f(w) \mathrm{d}w = 1$. Remaining constraints in the first equation define each order moment of w.

We aim to estimate how likely the actual renewable power w would step outside W_R^D, which will lead to insolvable DC power flow or infeasible system operation. This brings up the following problem

$$Z_P = \max \ \Pr[w \in \bar{W}_R^D] = \max_{f(w) \in \Omega_M^K} \int_{w \in \bar{W}^D} f(w) \mathrm{d}w \tag{2.95}$$

In problem (2.95), the decision variables are the values of $f(w)$ over all possible $w \in \mathbb{R}^{N_W}$, so there are infinitely many decision variables, and problem (2.95) is an infinite-dimensional LP. It maximizes the probability that w steps outside the dispatchable region W_R^D over all candidate PDFs. The optimal solution of problem (2.95) gives the worst-case PDF. However, it is difficult to solve (2.95) directly.

Associating a dual variable y_κ with each moment constraint in set Ω_M^K, we can derive the dual form of problem (2.95) in the spirit of the duality theory of conic LP following the method in [119]

$$Z_D = \min_{y_\kappa} \sum_{\kappa \in J(K)} y_\kappa \sigma_\kappa \tag{2.96a}$$

$$\text{s.t. } g(w) \geq 1, \ \forall w \in \bar{W}^D \tag{2.96b}$$

$$g(w) \geq 0, \ \forall w \in E_B^W \tag{2.96c}$$

where $g(w) = \sum_{\kappa \in J(K)} y_\kappa w^\kappa$ is a polynomial in w. The valid decision variables of problem (2.96) are coefficients $y_\kappa, \forall \kappa$. The argument w of polynomial $g(w)$ plays the role of parameter. To see this, consider a particular w^0, substitute it into (2.96b), then $g(w^0) \geq 1$ gives rise to a linear inequality in variables $y_\kappa, \forall \kappa$; (2.96b) requires $g(w) \geq 1$ should hold for all possible values of w in $\bar{W}^D$, imposing certain feasible region on $y_\kappa, \forall \kappa$. Equation (2.96c) has the same interpretation. In view of this, unlike the primal problem (2.95) that has infinite decision variables, the dual problem (2.96) has finite variables and an infinite number of linear constraints; in other words, we are optimizing objective (2.96a) over the coefficients y_κ of the polynomial $g(w)$ that satisfies (2.96b) and (2.96c), giving rise to a semi-infinite LP. It is clear that the optimum Z_P and Z_D depends on the generation dispatch, which influences the shape and size of $W^D/\bar{W}^D$.

Assumption 2.1 *The moment vector $\bar{\sigma} = \{\sigma_\kappa\}, \forall \kappa \in J(K)$ is an interior point of the set of feasible moment vectors.*

A moment vector is said to be feasible if there exists a PDF such that

$$\sigma_\kappa = E[w_1^{k_1} w_2^{k_2} \ldots w_{N_W}^{k_{N_W}}], \ \forall \kappa \in J(K)$$

Assumption 2.1 means that all moment vectors in a small neighborhood of $\bar{\sigma}$ are feasible. For more information about the feasibility of moment constraints, please consult [119] and [50]. We assume Assumption 2.1 always holds for the given data from an engineering perspective.

Proposition 2.3 ([50, 119]) *If Assumption 2.1 is satisfied, then strong duality holds for problems (2.95) and (2.96), i.e., $Z_P = Z_D$.*

Next, we develop tractable reformulations for dual problem (2.96). We will mainly focus on how to eliminate variable w in constraints (2.96b) and (2.96c) which actually restrict the coefficients of polynomial $g(w)$. To this end, we first introduce the notation of sum-of-squares (SOS) polynomial which is used to certify non-negativity.

Definition 2.3 $g(w)$ is an SOS polynomial, if it can be decomposed as $g(w) = \sum_i h_i(w)^2$, where $h_i(w)$ are polynomials.

Let $\Sigma(w)$ be the set of all SOS polynomials. Clearly, if a polynomial $g(w) \in \Sigma(w)$, it must be nonnegative on $\mathbb{R}^{N_W}$.

3. SDP Reformulation Based Upper Bounding

a. Reformulation of Constraint (2.96c)
This constraint imposes non-negativity of $g(w)$ over the hypercube E_B^W, which is a subset of $\mathbb{R}^{N_W}$. We can adopt the paradigm of positivstellensatz. By introducing multiplier variables, we have the following proposition.

Proposition 2.4 *$g(w) \geq 0$, $\forall w \in E_B^W$ holds if there exists multipliers $\lambda^l \geq 0$, $\lambda^u \geq 0$, such that the polynomial $l_E(w) = w^T \lambda^l + (C^W - w)^T \lambda^u$ satisfies*

$$g(w) - l_E(w) \in \Sigma(w) \tag{2.97}$$

This proposition is easy to understand. Denote by $g(w) - l_E(w) = p_0(w)$. For any $w \in E_B^W$, $0 \leq w \leq C_i^W$, $\forall i$ holds, so $l_E(w)$ must be non-negative as $\lambda^l \geq 0$, $\lambda^u \geq 0$. Moreover, $p_0(w) \in \Sigma(w) \geq 0$, thus $g(w) = p_0(w) + l_E(w) \geq 0$. Because the order of an SOS polynomial must be even, we need to know the even order moments to guarantee polynomial $g(w)$ has an even order.

From the analysis above, constraint (2.96c) can be approximated by the following SOS constraint

$$g(w) - l_E(w) = p_0(w) \in \Sigma(w),\ \lambda^l, \lambda^u \geq 0 \tag{2.98}$$

Constraint (2.98) is a sufficient condition of (2.96c), this approximation may lead to a larger optimum Z_D.

In fact, the argument w of each polynomial can be eliminated by introducing a PSD matrix variable Q. To this end, let $\mathrm{w}^{(d)}$ be the vector consisting of all monomials in variables $\{w_j\}$, $\forall j$ with the highest degree of d, for example, if $N_W = 2, d = 2$ then $\mathrm{w}^{(d)} = [1, w_1, w_2, w_1^2, w_1 w_2, w_2^2]^T$.

Proposition 2.5 ([119]) *Polynomial $p_0(w)$ of degree $2d$ is SOS if and only if $\exists Q \succeq 0$, such that $p_0(w) = \mathrm{w}^{(d)T} Q \mathrm{w}^{(d)}$.*

This proposition is easy to understand. For necessity, since $Q \succeq 0$, it admits a Cholesky decomposition $Q = H^T H$, and thus

$$p_0(w) = (H\mathrm{w}^{(d)})^T (H\mathrm{w}^{(d)}) = \sum_i (H_i \mathrm{w}^{(d)})^2$$

where H_i, the i-th row of matrix H, is a constant vector, and $H_i \mathrm{w}^{(d)}$ is a polynomial, so $p_0(w)$ is expressed as an SOS form.

For sufficiency, suppose that $p_0(w)$ of degree $2d$ is SOS and can be represented via $p_0(w) = \sum_i [h_i(w)]^2$. The degree of $h_i(w)$ is no greater than d with a coefficient vector $\mathbf{h}_i$ such that $h_i(w) = \mathbf{h}_i^T \mathrm{w}^{(d)}$. Hence

$$p_0(w) = \sum_i (\mathrm{w}^{(d)})^T \mathbf{h}_i \mathbf{h}_i^T \mathrm{w}^{(d)} = (\mathrm{w}^{(d)})^T Q \mathrm{w}^{(d)}$$

where $Q = \sum_i \mathbf{h}_i \mathbf{h}_i^T \succeq 0$ (because each $\mathbf{h}_i \mathbf{h}_i^T$ is PSD).

In view of Proposition 2.5, we can write (2.98) as explicit LMIs in variable y_κ, Q and λ^l, λ^u by equating the coefficients of each monomial on both sides. In this way, the enumeration of w over hypercube E_B^W in Eq. (2.96c) is eliminated.

When SOS constraint (2.98) is added in an optimization problem, the valid decision variables are coefficients y_κ of the polynomial $g(w)$ as well as multipliers λ^l, λ^u.

b. Reformulation of Constraint (2.96b)

Different from constraint (2.96c), the non-dispatchable region $\bar{W}^D$ is not convex and the positivstellensatz condition does not apply. Nevertheless, the dispatchable region W^D consists of linear inequalities, which allows us to represent $\bar{W}^D$ via the union of half-spaces, and develop equivalent LMI constraints for constraint (2.96b).

Since $W_R^D = \bigcap_i L_i^H$, where the half-space $L_i^H = \{w | H_i w \le h_i\}$, H_i is the i-th row of matrix H, h_i is the i-th element of vector h, the non-dispatchable region $\bar{W}^D = \bigcup_i \bar{L}_i^H$; the set $\bar{L}_i^H = \{w | H_i w \ge h_i\}$ is still a half-space (as mentioned before, we intentionally require $\bar{W}^D$ include the boundary of W^D). Therefore, constraint (2.96b) is equivalent to the following condition

$$g(w) \ge 1, \forall w \in \bigcup_i \bar{L}_i^H$$

It can be split into individual constraints with simple regions on w

$$g(w) \ge 1, \forall w \in \bar{L}_i^H, \forall i, \tag{2.99}$$

For each half-space $\bar{L}_i^H$, we have a reformulation similar to (2.97). As a result, we arrive at the following proposition.

Proposition 2.6 $g(w) \ge 1$, $\forall w \in \bar{W}^D$ *if there exists a multiplier vector* $\rho > 0$ *and polynomials* $l_i^W(w) = \rho_i(H_i w - h_i)$, $\forall i$, *where* ρ_i *is the* i*-th element of* ρ, *such that*

$$g(w) - l_i^W(w) - 1 \in \Sigma(w), \forall i \tag{2.100}$$

Similar to (2.98), constraint (2.96b) can be approximated by

$$g(w) - l_i^W(w) - 1 \in \Sigma(w), \ \rho_i \ge 0, \ \forall i \tag{2.101}$$

In view of Proposition 2.5, constraint (2.101) is equivalent to a set of LMIs in variables y_κ, ρ and additional matrix variables. The enumeration of w over set $\bar{W}^D$ in (2.96b) is eliminated. Constraint (2.101) is a sufficient condition of constraint (2.96b). Replacing (2.96b) with (2.101) may also lead to a larger dual optimum Z_D.

c. The Overall SDP

Previous outcomes allow us to reformulate the dual problem (2.96) using SOS constraints. The overall formulation is summarized as follows:

$$\begin{aligned} Z_D^S = \min \ & \sum_{\kappa \in J_K} y_\kappa \sigma_\kappa \\ \text{s.t. } & \lambda^l \geq 0,\ \lambda^u \geq 0,\ \rho \geq 0 \\ & g(w) - w^T \lambda^l - (C^W - w)^T \lambda^u \in \Sigma(w) \\ & g(w) - \rho_i (H_i w - h_i) - 1 \in \Sigma(w), \forall i \end{aligned} \tag{2.102}$$

Because SOS constraints imply LMI constraints, problem (2.102) is an SDP. In our implementation, problem (13) is built using the SOS module in YALMIP toolbox [121], and automatically transformed into its equivalent SDP without variable w. Several discussions are provided.

1. As mentioned before, Propositions 2.4 and 2.6 are sufficient conditions for constraint (2.96c) and constraint (2.96b). In general, the optimal value Z_D^S of problem (2.102) yields an upper bound of Z_D, even in the worst-case distribution.
2. Although we can solve problem (2.102) via its equivalent SDP and obtain the optimal value as an estimation on the probability of an infeasible operation, it is still difficult to recover the corresponding optimal solution of the primal problem (2.95), which means we do not have the worst-case distribution. Nevertheless, as long as the system operator knows the likelihood of infeasibility in the worst case, the corresponding PDF usually becomes less important, and the actual probability must be smaller.

4. LP Relaxation Based Lower Bounding

As problem (2.96) is a semi-infinite LP in variable y_κ, it is natural to use finite samples w^s, $s = 1, 2, \ldots$ of $\bar{W}_R^D$ and E_B^W to construct the following LP relaxation

$$\begin{aligned} Z_D^L = \min \ & \sum_{\kappa \in J_K} y_\kappa \sigma_\kappa \\ \text{s.t. } & g(w^s) \geq 1, \text{ if } w^s \in \bar{W}_R^D, \forall s \\ & g(w^s) \geq 0, \text{ if } w^s \in E_B^W, \forall s \end{aligned} \tag{2.103}$$

As the constraints in (2.103) only pick up a finite subset from infinitely many inequalities in (2.96b) and (2.96c), the optimal value Z_D^L provides a lower bound of Z_D. Adding more samples in (2.103) can generally tighten the relaxation and reduce the gap. Several discussions are provided as follows.

1. One possible way to generate the samples w^s, $s = 1, 2, \ldots$ is to use the "grid points," which refers to the extreme points of smaller hypercubes E_i^S, such that $E_i^S \bigcap E_j^S = \emptyset, \forall i \neq j$ and $\bigcup_i E_i^S = E_B^W$. As the dispatchable region W^D is available, testing whether a grid point w_E^G belongs to W^D yields checking the validness of the linear inequality set $Hw_E^G \le h$, which only involves algebraic operation. This motivation is highly appropriate if there are only two or three wind farms, so the number of grid points will not be prohibitively high.
2. When AC power flow is considered, if we can get an inner polyhedral approximation of the dispatchable region, such as the method in [117], we can still apply the proposed method, although the result could be rather pessimistic.
3. The relationship among the primal problem (2.95), the dual problem (2.96), SDP (2.102) and LP (2.103) is shown in Fig. 2.27, their optimums satisfy $Z_D^S \ge Z_P = Z_D \ge Z_D^L$. Empirical results show that the gap $Z_D^S - Z_D^L$ tends to zero when more samples are used and the linear relaxation becomes tighter. This indicates the relation $Z_D^S = Z_D$, which means that SDP (2.102) usually provides a good estimation of Z_D.
4. An important implication of the LP relaxation model rises in the situation when the dispatchable region W_R^D is not available, therefore, the SDP reformulation, GCI and GGI are not applicable. Nevertheless, we can still set up LP (2.103) and obtain a probability of infeasibility. In such circumstance, we need to solve an additional LP for each grid point w_E^G, in order to examine whether it belongs to W_R^D. In view of this, the computational efficiency may fail to satisfy the requirement of online application.
5. When the dimension of uncertainty grows, the number of grid points increases exponentially, so does the number of constraints in LP (2.103). Trade-off should be made between the number of sampled points and the precision of approximation. In the case that the precision of grid points based approximation is not satisfactory, one way to enhance the LP relaxation is to check constraint violation and update sets $\bar{W}_{RB}^D$ and E_{BS}^W dynamically, yielding the following two steps

   Step 1: solve the master problem LP (2.103) with current $\bar{W}_R^D$ or E_R^W, the optimal solution is y_κ^*.

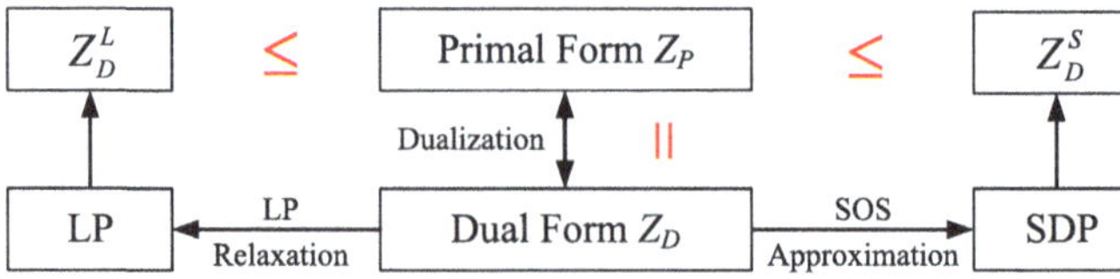

Fig. 2.27 Relations among different methods

Step 2: solve the following two subproblems with $y_\kappa = y_\kappa^*$

$$g_1^* = \min_{w \in \bar{W}^D} g(w) \tag{2.104a}$$

$$g_2^* = \min_{w \in E_B^W} g(w) \tag{2.104b}$$

The optimal solution is w^*. If $g_1^* < 1$ or $g_2^* < 0$, add w^* into $\bar{W}_R^D$ or E_R^W, and then return Step 1. The procedure continues until constraints (2.96b) and (2.96c) are met.

In view of the complexity of $\bar{W}^D$, problem (2.104a) can be replaced by a set of simpler problems

$$g_1^* = \min\{g_{1i}^*, \forall i\}, \; g_{1i}^* = \min_{w \in \bar{L}_i^H} g(w) \tag{2.105}$$

Because E_B^W is a hypercube and $\bar{L}_i^H$ is a hyperplane, these subproblems are polynomial optimizations with linear constraints. Interested readers are referred to [50] for a comprehensive discussion on polynomial programs. If a local optimum is acceptable, one can use any NLP algorithm to solve subproblems. Because the constraint set is simple, the computation can be very efficient. Because only a single sample is generated in each iteration, this method only works in theory, but it provides some guidelines if we can generate a set of samples dynamically using the latest information offered by (2.104).

Data acquirement and computation related issues of the overall methodology are discussed as follows.

1. The moments are the only information required in the ambiguity set $\Omega_{\mathrm{M}}^{\mathrm{K}}$. Compared with an accurate PDF, moments can be directly computed from actual data. Estimating the first- and second-order moments of stochastic wind generation has been well explored in power system applications. The first-order moment pertains to the wind power forecast, which has been extensively studied [106]; The second-order moment, or the covariance matrix of wind generation, is elaborated in [122–124]; higher-order moments are investigated in [125].
2. If the point estimation of moment cannot meet desired accuracy, we can alternatively exploit the following ambiguity set with interval moment estimation

$$\Omega_M^K = \left\{ f(w) \left| \begin{array}{l} \displaystyle\int_{w \in E_B^W} w^\kappa f(w) \mathrm{d}w \le \sigma_\kappa^m, \forall \kappa \in J(K) \\ \displaystyle\int_{w \in E_B^W} w^\kappa f(w) \mathrm{d}w \ge \sigma_\kappa^l, \forall \kappa \in J(K) \\ f(w) \ge 0, \; \forall w \in \mathbb{R}^{N_W} \end{array} \right. \right\}$$

where σ_κ^l and σ_κ^m is the lower bound and upper bound of k-th order moment σ_κ, $\sigma_0^m = \sigma_0^l = 1$, and then the corresponding dual problem will become

$$
\begin{aligned}
Z_D = \min_{y_\kappa} \quad & \sum_{\kappa \in J(K)} (y_\kappa^m \sigma_\kappa^m + y_\kappa^l \sigma_\kappa^l) \\
\text{s.t.} \quad & g(w) \geq 1, \ \forall w \in \bar{W}^D \\
& g(w) \geq 0, \ \forall w \in E_B^W \\
& y_\kappa^m \geq 0, \ y_\kappa^l \leq 0, \ \forall \kappa \in J(K)
\end{aligned}
$$

where the polynomial $g(w) = \sum_{\kappa \in J(K)} (y_\kappa^m + y_\kappa^l) w^\kappa$. An SDP reformulation can be derived in a similar way.

3. The proposed method relies on solving SDP, which is a convex optimization problem and admits efficient computation using commercial software. However, difficulty arises when the number of wind farms grows and the orders of moments become high. This is because in Proposition 2.5, the dimension of vector $\mathrm{w}^{(d)}$ is $C_d^{N_W+d}$, where C_m^n is the combination number of choosing m items from n items, and $d = K/2$ is the highest degree of monomials in $\mathrm{w}^{(d)}$, so the dimension of the PSD matrix Q will be $C_d^{N_W+d} \times C_d^{N_W+d}$, which grows quickly with N_W and d increasing, but is independent of the power system model. In this regard, this method is restricted to bulk power systems with centralized renewable integration.
4. Because the values of moments exhibit different orders of magnitudes, problems (2.102) and (2.103) may be ill-conditioned. In order to reduce the conditional number, one possible way is to divide the output data w by a constant, such that w_j varies around 1. We can choose different base values for different wind farms. Consequently, the dispatchable region is given by (2.92c). Another way is to use the center moment, and the dispatchable region is given by (2.92a). Empirical studies suggest that the former one is usually more effective.

If only the first and second order moments are available, the GCI approach in [118] and the GGI approach in [120] can provide probability estimation on infeasible operation given the dispatchable region, which are briefly introduced below. We will define some notations which are more convenient for GCI and GGI. The dispatchable region in the coordinate of forecast error Δw is expressed as

$$
W_R^D = \{\Delta w \mid a_i^T \Delta w \leq b_i, \ i = 1, \ldots, k\} \tag{2.106}
$$

The first-order moment of Δw is 0, and the second-order moment of Δw is the covariance matrix Θ.

4. Generalized Chebyshev Inequality Approach

GCI [118] extends the traditional Chebyshev inequality to a multivariate setting and provides the probability of a random point stepping outside a given polytope under the worst distribution in the Chebyshev ambiguity set Ω_C described by

$$\Omega_C = \left\{ f(\Delta w) \left| \begin{array}{c} \int_{\mathbb{R}^{N_W}} f(\Delta w) \mathrm{d}w = 1, \ f(\Delta w) \geq 0 \\ \int_{\mathbb{R}^{N_W}} \Delta w f(\Delta w) \mathrm{d}w = 0 \\ \int_{\mathbb{R}^{N_W}} \Delta w \Delta w^T f(\Delta w) \mathrm{d}w = \Theta \end{array} \right. \right\} \tag{2.107}$$

GCI can boil down to a single SDP [118]

$$\begin{aligned} \max_{f(w) \in \Omega_C} \quad & \Pr[w \in \bar{W}_R^D] = \max \sum_{i=1}^{k} \lambda_i \\ \text{s.t.} \quad & z_i \in \mathbb{R}^{N_W}, \ Z_i \in \mathbb{S}^{N_W}, \forall i = 1, \ldots, k \\ & \lambda_i \in \mathbb{R}, \ a_i^T z_i \geq b_i \lambda_i, \ \forall i = 1, \ldots, k \\ & \begin{pmatrix} Z_i & z_i \\ z_i^T & \lambda_i \end{pmatrix} \succeq 0, \ \forall i = 1, \ldots, k \\ & \begin{pmatrix} \Theta & 0 \\ 0 & 1 \end{pmatrix} \succeq \sum_{i=1}^{k} \begin{pmatrix} Z_i & z_i \\ z_i^T & \lambda_i \end{pmatrix} \end{aligned} \tag{2.108}$$

However, GCI often appears to be very pessimistic. The reason is that the worst-case distribution has only few discretization points [118], instead of a continuous PDF.

5. Generalized Gauss Inequality Approach
To alleviate the conservativeness of GCI without sacrificing the convexity, GGI is proposed in [120] by exploring the unimodality which excludes the discrete distributions from the candidate PDFs, so the conservatism can be reduced remarkably. Intuitively, the PDF of a unimodal distribution has only one maximum, or in other words, the PDF is non-increasing when leaving from the maximum. For rigorous mathematical definition of unimodality, please refer to [120]. Although the PDF of wind speed over a long time period may be multi-modal, the PDF of short-term wind power forecast error often appears to be unimodal around its prediction, but not necessarily symmetric.

For the univariate case, Gauss proved that the Chebyshev bound can be improved as follows when the PDF is known to be unimodal [120]

$$\Pr[|\xi - \mu| \geq k\sigma] \leq \begin{cases} 4/9k^2, & \text{if } k > 2/\sqrt{3}; \\ 1 - k/\sqrt{3}, & \text{otherwise}; \end{cases} \tag{2.109}$$

When Gauss inequality (2.109) is applied to the case at the beginning of this section, the probability will typically be reduced to 4/9 of that provided by Chebyshev inequality. Reference [120] generalizes above result to incorporate multiple random variables. The candidate PDFs in GGI belong to the Gauss ambiguity set defined as follows:

$$\Omega_{\rm G} = \left\{ f(\Delta w) \middle| \begin{array}{c} f(\Delta w) \in \Omega_{\rm C} \\ f(\Delta w) \text{ is unimodal} \end{array} \right\} \tag{2.110}$$

GGI still comes down to a single SDP [120]

$$\begin{aligned}
&\max_{f(w)\in\Omega_{\rm G}} \quad \Pr\left[w \in \bar{W}_R^D\right] = \max \sum_{i=1}^{k} (\lambda_i - t_{i,0}) \\
&\text{s.t. } z_i \in \mathbb{R}^{N_W},\ Z_i \in \mathbb{S}^{N_W},\ \lambda_i \in \mathbb{R},\ \forall i = 1, \dots, k \\
&\quad t_i \in \mathbb{R}^{l+1},\ l = \lceil \log_2 N_W \rceil,\ \forall i = 1, \dots, k \\
&\quad \begin{pmatrix} Z_i & z_i \\ z_i^T & \lambda_i \end{pmatrix} \succeq 0,\ a_i^T z_i \ge 0,\ t_i \ge 0,\ \forall i = 1, \dots, k \\
&\quad \begin{pmatrix} \dfrac{N_W+2}{N_W}\Theta & 0 \\ 0 & 1 \end{pmatrix} \succeq \sum_{i=1}^{k} \begin{pmatrix} Z_i & z_i \\ z_i^T & \lambda_i \end{pmatrix} \\
&\quad \left\| \begin{matrix} 2\lambda_i b_i \\ t_{i,l} b_i - a_i^T z_i \end{matrix} \right\|_2 \le t_{i,l} b_i + a_i^T z_i,\ \forall i = 1, \dots, k \\
&\quad \left\| \begin{matrix} 2t_{i,j+1} \\ t_{i,j} - \lambda_i \end{matrix} \right\|_2 \le t_{i,j} + \lambda_i, \forall j \in EV, \forall i = 1, \dots, k \\
&\quad \left\| \begin{matrix} 2t_{i,j+1} \\ t_{i,j} - t_{i,l} \end{matrix} \right\|_2 \le t_{i,j} + t_{i,l}, \forall j \in OD, \forall i = 1, \dots, k \\
&\quad EV = \{j \in \{0, \dots, l-1\} : \lceil N_W / 2^j \rceil \text{ is even}\} \\
&\quad OD = \{j \in \{0, \dots, l-1\} : \lceil N_W / 2^j \rceil \text{ is odd }\}
\end{aligned} \tag{2.111}$$

where $\lceil\ \rceil$ stands for the rounding function towards plus infinity.

6. Monte Carlo Simulation Under Given Probability

Even if the exact PDF of renewable generation is known, it's still difficult to compute the probability in an analytical way, because it comes down to a multi-dimensional integration over a complicated polytope. Nevertheless, with the help of dispatchable region, one can perform Monte Carlo simulation to retrieve such a probability, which proceeds as follows: generate N_S renewable generation scenarios from the given PDF; for each scenario, check whether $w \in W_R^D$; let N be the number of scenarios which do not belong to W_R^D, then the probability of infeasible operation is N/N_S. This simulation is inexpensive because testing inclusion $w \in W_R^D$ boils down to validating linear inequalities.

7. Case Study

The IEEE 118-bus system is used to validate the proposed method. System data are provided in [102]. Generation dispatch is determined by solving problem (2.91). In the considered dispatch interval, the total demand is 5000 MW. The regulation cost coefficient d_i^+/d_i^- of each generator is set to be 10% of its production cost coefficient b_i. The ramping limit R_i^+/R_i^- is assumed to be 25% of its maximal output P_i^u. To illustrate the dispatchable region W^D clearly, we consider two wind farms with $C_1^W = C_2^W = 500$ MW and $w_1^e = w_2^e = 250$ MW in this case study. They are connected to the system at bus #70 (Area 1) and bus #100 (Area 3). The impact of the forecast accuracy and the available cost C^R on the probability of operating feasibility is investigated.

According to [106], the root mean-square error of the hourly-ahead wind generation forecast is around 10% of its prediction. In our tests, we change the square root variance σ_j of forecast error from 8% to 12% of w_j^e. Problem (2.102) is called SDP-X method for short, where X is the available order of moments. The optimal value provides an upper bound for the concerning probability of an infeasible operation. Similarly, LP (2.103) is called LPR-X for short, whose optimal value provides a lower bound for the probability of infeasibility in the worst PDF. GCI and GGI models are given in (2.108) and (2.111), respectively. Polyhedral dispatchable regions under different cost C^R are computed by using the method in Sect. 2.5.2, and illustrated in Fig. 2.28. Clearly, W_R^D grows larger with C^R increasing, indicating higher flexibility of operation.

The computation time of SDP method typically varies from 1 to 2 s. As for LPR method, we divide E_B^W into 400 sub-rectangles to retrieve the grid points. The computation time is in the same range. The computation time of GCI and GGI method is around 0.8 s. Detailed results are shown through Figs. 2.29, 2.30 and 2.31. The outcomes offered by SDP method and LPR method with the same X is plotted using the same color. The gap between the upper bound and lower bound shown in Figs. 2.29, 2.30 and 2.31 is acceptable, indicating that both methods provide good approximations for problem (2.96). Clearly, with the available cost C^R increasing, the probability of infeasible operation decreases because W_R^D grows larger. For SDP method, when higher order moments are used, the conservativeness will be reduced. For LPR method, we find if more grid points are used in LP (2.103), the gap $Z_D^S - Z_D^L$ can be further reduced and eventually tends to 0. This result also indicates that the SDP approximation (2.102) is exact. It is worth mentioning that incorporating higher order moments will increase the conditional number and problem size, thus challenging the computation. In view of this, we do not recommend using moments with an order higher than 6. Notice that SDP-6 is only slightly improved compared with SDP-4, the latter may be a suitable compromise in practice.

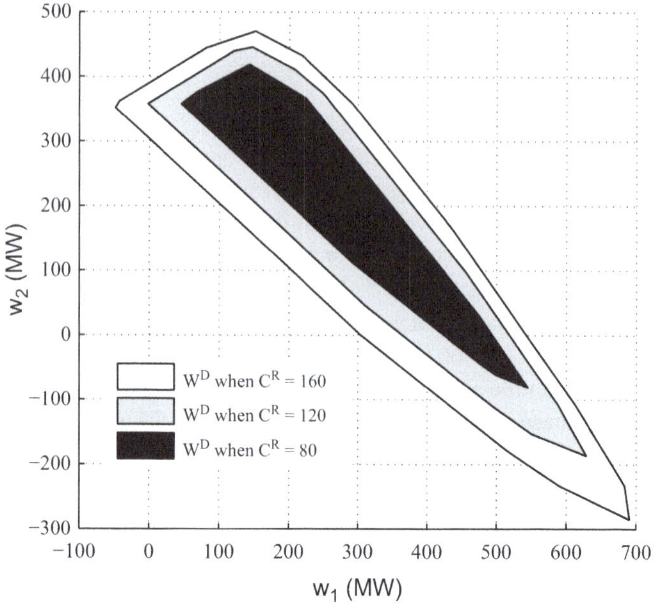

Fig. 2.28 Dispatchable regions under different available costs

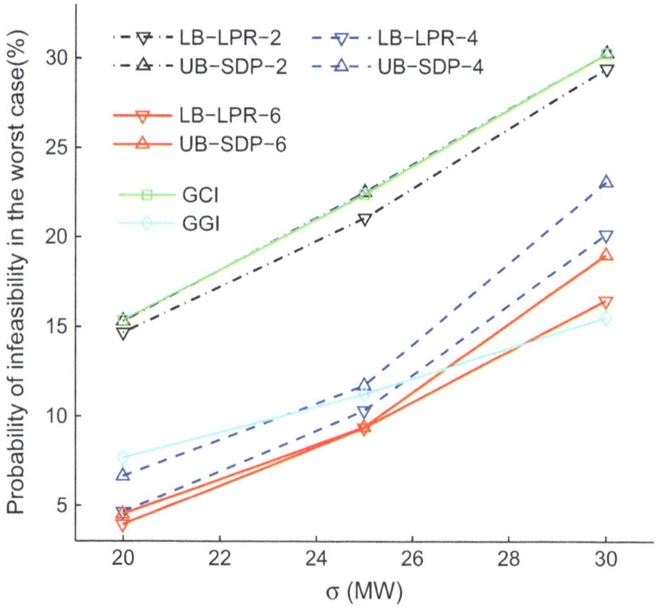

Fig. 2.29 Probability of infeasible operation when $C^R = 80\,\text{MBtu}$

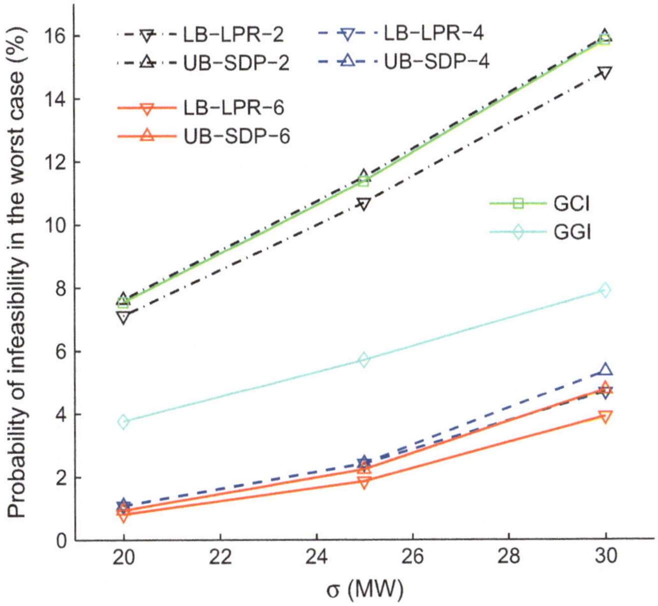

Fig. 2.30 Probability of infeasible operation when $C^R = 120\,\text{MBtu}$

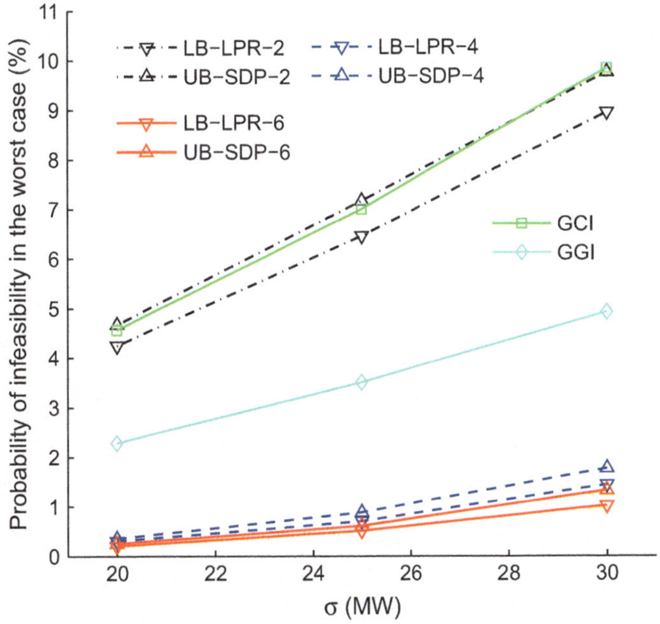

Fig. 2.31 Probability of infeasible operation when $C^R = 160\,\text{MBtu}$

These figures also imply that SDP-2 provides almost the same results as GCI, because they actually yield very similar problem. By considering the unimodality of wind power distribution, GGI remarkably reduces the conservativeness of GCI. It's also interesting to notice that the probability offered by GGI is almost half of that offered by GCI, coinciding with the univariate situation, where the former is 4/9 of the latter. It is also found that SDP-4 and SDP-6 provide a lower probability than GGI in most cases except when C^R is small (W_R^D is small) and the forecast is less accurate (the variance is large). This is because the unimodality in GGI prevents discrete distributions which allocate more chance of data realization around the boundary of dispatchable region.

Please bear in mind that the probabilities offered by all these methods correspond to the worst renewable power distribution. To demonstrate the probability of infeasibility under a specified distribution, say Gaussian distribution, we carry out Monte Carlo simulation. Results are shown in Table 2.22, from which we can see that the infeasible probability under Gauss distribution is smaller than those in Figs. 2.29, 2.30 and 2.31, because Gaussian distribution is only a candidate PDF in set Ω_M^K rather than the worst one, therefore Monte Carlo simulation yields a lower bound of Z_P. The actual infeasible probability may be either lower or higher than the result of Monte Carlo simulation, depending on the actual PDF of renewable power, but must be smaller than the optimal value of problem (2.96).

Finally, we test the computation time of each method with higher dimensional uncertainties. We change the number of wind farms N_W in the system. To maintain a proper penetration level of renewable energy, we keep the total wind power generation at 1000 MW, indicating $w_j^e = 1000/N_W$, and choose $\sigma_j = 0.1w_j^e$ for each wind farm. The computation times of the proposed SDP method, GCI and GGI are listed in Table 2.23.

Table 2.23 demonstrates that the computation time of all methods increases quickly when $N_W = 7$, and SDP-6 fails to return a solution. Other methods can solve the problem in reasonable time and satisfy the requirement of online application. The reason is, the complexity of SDP (2.102) mainly depends on

Table 2.22 Probability of infeasibility under Gaussian distribution

Square root of variance	Probability of infeasible operation (%)		
	$C^R = 80$	$C^R = 120$	$C^R = 160$
$\sigma = 20$ MW	0.81	0.01	0.00
$\sigma = 25$ MW	2.69	0.21	0.00
$\sigma = 30$ MW	6.44	0.82	0.08

Table 2.23 Computation time (s)

	$N_W = 4$	$N_W = 5$	$N_W = 6$	$N_W = 7$
SDP-2	2.42	2.97	6.39	20.2
SDP-4	2.95	4.28	15.4	97.6
SDP-6	8.01	41.6	179	*
GGI	1.89	3.01	11.2	105
GCI	1.64	2.27	7.21	63.5

* fail

N_W, as analyzed before, while the complexity of SDP (2.108) and SDP (2.111) mainly depends on the number of constraints in W_R^D. When the number of wind farms increases, both the dimension of PSD matrix variable in SDP (2.102) and the number of constraints in W_R^D grow quickly. When N_W grows larger, computing W_R^D will become more challenging. Taking the computational efficiency and reliability, solution conservativeness, as well as implementation issues into account, SDP-4 might be the best choice in practical usage.

2.5.4 Dispatchability Maximization

Up to now, we assume the feasible region of re-dispatch is a polytope, which depends on the generation schedule and reserve provision, so does the dispatchable region. It is natural to ask: how can one actively influence and control the dispatchable region in the generation scheduling stage? We have already learnt that the dispatchable region does not have a concise form expressed by generation and reserve scheduling variables, so it is difficult to shape the dispatchable region under the setting in Sect. 2.5.2.

In this section, we consider a new scheme of re-dispatch: the incremental output of generators is a linear function in the renewable power forecast error, which is called the affine policy and widely used in practice. We reveal that the dispatchable region under affine policy based re-dispatch is a polytope whose coefficients are linear functions of the generation and reserve schedule, as well as the gain matrix of affine policy. Based on this polyhedral expression and GGI approach introduced in Sect. 2.5.3, we propose two optimization models for maximizing renewable power dispatchability through joint energy and reserve dispatch. The first one maximizes the so-called radius of dispatchability, which refers to the distance from the current point to the boundaries of dispatchable region, subject to the available budget. The second one minimizes the total production cost subject to a desired reliability level through shaping the dispatchable region. These models make no reference to an uncertainty set or a PDF. For the first model, we develop an efficient SOCP based bisection algorithm to search the optimal solution. For the second one, we develop an SDP based heuristic algorithm to seek an approximate solution, which may not be an optimal one.

1. Dispatchable Region Under the Affine Policy Based Re-Dispatch

We use vector notations for convenience. The renewable power is $w^e + \Delta w$, where w^e is the prediction, and Δw is the forecast error; generator output is $p + \Delta p$, where p is the set point corresponding to w^e, and Δp is the re-dispatch in response to Δw according to affine policy

$$\Delta p_i = \sum_{j=1}^{N_W} \alpha_{ij} \Delta w_j$$

or in a compact form as follows:

$$\Delta p = G\Delta w \tag{2.112}$$

where the gain matrix $G = [\alpha_{ij}], \forall i, j$ reflects how the mismatch between renewable output and its forecast is assigned to eligible generators. Other generators which do not participate in frequency regulation services correspond to a null row of G with all zero elements. Affine policy (2.112) seems restrictive, because Δp is uniquely determined by Δw once G is given, and there is no freedom to adjust Δp in re-dispatch. However, the affine policy turns out to provide reasonable dispatch strategy in power system applications [126–128], and is even proved to have equal optimal value as the fully-adjustable policy in some particular situation [129, 130].

In re-dispatch stage, the generation and reserve schedule $[p^f, r^+, r^-]$ as well as the gain matrix G have already been determined, the generators automatically respond to Δw according to (2.112). Moreover, system operating constraints should be satisfied for any possible Δw in some certain region W_R^D (which will be explained later), requiring

$$1^T(p + G\Delta w) + 1^T(w^e + \Delta w) = 1^T p^d, \ \forall \Delta w \in W_R^D \tag{2.113a}$$

$$-F_L \le \pi_G(p+G\Delta w)+\pi_W(w^e+\Delta w)-\pi_D p^d \le F_L, \ \forall \Delta w \in W_R^D \tag{2.113b}$$

$$-r^- \le G\Delta w \le r^+, \ \forall \Delta w \in W_R^D \tag{2.113c}$$

where (2.113a)–(2.113c) collect the power balancing condition, line flow limits, and reserve capacity under arbitrary renewable generation scenario, respectively. W_R^D is a certain set of the forecast error Δw. When $\Delta w = 0$ and $\Delta p = 0$, (2.113) naturally holds owing to the operating constraints in the nominal scenario. To maintain power balancing for all possible scenarios in W_R^D, it is necessary to require

$$(1^T G + 1^T)\Delta w = 0, \ \forall \Delta w$$

rendering

$$1^T G + 1^T = 0^T \tag{2.114}$$

Due to line flow limits and ramping speed of generators, the region W_R^D cannot be arbitrarily large. The largest one is called the dispatchable region under affine re-dispatch policy.

Definition 2.4 The dispatchable region under affine policy (2.112) is a set W_R^D such that (2.113a)–(2.113c) hold if and only if $\Delta w \in W_R^D$.

Proposition 2.7 *Suppose the gain matrix G satisfies (2.114), the dispatchable region W_R^D under affine policy can be expressed as the following polytope*

$$W_R^D = \left\{ \Delta w \left| \begin{array}{c} (\pi_G G + \pi_W)\Delta w \le \pi_D p^d - \pi_G p - \pi_W w^e + S_L \\ (\pi_G G + \pi_W)\Delta w \ge \pi_D p^d - \pi_G p - \pi_W w^e - S_L \\ -r^- \le G\Delta w \le r^+ \end{array} \right. \right\} \quad (2.115)$$

It is clear that (2.113b) and (2.113c) hold for all $\Delta w \in W_R^D$, while (2.113a) is guaranteed by (2.114). On the contrary, $\forall \Delta w \notin W_R^D$, at least one constraint will be violated. Therefore, $\Delta w \in W_R^D$ is a necessary and sufficient condition for a feasible operation. Unlike the fully-adjustable case, the dispatchable region under affine policy has a concise expression in $[p, r^+, r^-]$ and gain matrix G, which enable us to actively control the shape of W_R^D through jointly scheduling energy and reserve.

Definition 2.5 The radius of dispatchable region (ROD) is quantified by the Euclidean distance from the origin (corresponding to the wind generation forecast w^e, or $\Delta w = 0$) to the boundaries of W_R^D, i.e.

$$R^D = \text{dist}(0, \partial W_R^D)$$

where R^D is the ROD, ∂W_R^D stands for the boundary of W_R^D, the Euclidean distance from a given point (vector) d to a set D is defined as

$$\text{dist}(d, D) = \min_{d_D \in D} \|d - d_D\|_2$$

where $\|d\|_2$ denotes the Euclidean norm of the input vector d.

In this section, all concepts related to dispatchable region maximization refers to maximizing the ROD.

2. Dispatchable Region Maximization with Given Budget

We will discuss how to optimize the generation and reserve schedule $[p^f, r^+, r^-]$ and the gain matrix G, so as to maximize the ROD.

In the nominal scenario, i.e., $\Delta w = 0$, $\Delta p = 0$, i.e., the generation set point p should meet power balancing condition and line flow capacity constraints associated with renewable generation forecast w^e

$$1^T p^f + 1^T w^e = 1^T d \quad (2.116a)$$

$$-F_L \le \pi_G p^f + \pi_W w^e - \pi_D d \le F_L \quad (2.116b)$$

Moreover, reserve provision is accounted for in generation capacity constraints

$$P^L \le p - r^-, \ p + r^+ \le P^U \tag{2.116c}$$

and reserve offer of each unit cannot exceed its ramping speed

$$r^- \le R^- \Delta t, \ r^+ \le R^+ \Delta t \tag{2.116d}$$

Finally, to investigate the impact of economic considerations on the dispatchable region, we incorporate a budget constraint on reserve offer

$$f_R = (c^+)^T r^+ + (c^-)^T r^- \le C^R \tag{2.116e}$$

One can alternatively impose an upper bound on the total production cost as

$$f_G = \sum_{i=1}^{N_G} [a_i (p_i)^2 + b_i p_i] + f_R \le C^P \tag{2.116f}$$

Define feasible set of dispatchable region maximizing energy and reserve dispatch (DM-ERD)

$$\begin{aligned} X &= \{ \, x \mid (2.116a)\text{–}(2.116d) \, \} \\ X(C^R) &= \{ \, x \mid x \in X, (2.116e) \, \} \\ X(C^P) &= \{ \, x \mid x \in X, (2.116f) \, \} \end{aligned} \tag{2.117}$$

where X includes purely operating constraints, $X(C^R)$ and $X(C^P)$ incorporate additional upper bounds on the reserve budget and the total generation cost, respectively.

The compact form of dispatchable region W_R^D is given by

$$W_R^D = \{\Delta w \mid \ H(G)\Delta w \le h(x)\}$$

where

$$H(G) = \begin{bmatrix} \pi_G G + \pi_W \\ -\pi_G G - \pi_W \\ -G \\ G \end{bmatrix}, \quad h(x) = \begin{bmatrix} F_L + \pi_D p^d - \pi_G p - \pi_W w^e \\ F_L - \pi_D p^d + \pi_G p + \pi_W w^e \\ r^- \\ r^+ \end{bmatrix}$$

Now we can see, H is a linear matrix function of the gain matrix G, and h is a linear vector function of the generation and reserve schedule vector $x = [p, r^+, r^-]$.

DM-ERD gives rise to the following problem

$$\max_{x,G} \ \operatorname{dist}(0, \partial W^D) \tag{2.118a}$$

$$\text{s.t. } x \in X(C^R) \tag{2.118b}$$

$$G^T 1 + 1 = 0 \tag{2.118c}$$

In problem (2.118), (2.118b) ensures the operating conditions in (2.116a)–(2.116e); (2.118c) guarantees the power balancing condition in (2.113a) for an arbitrary Δw. According to Definition 2.4 and Proposition 2.7, constraints (2.113b) and (2.113c) hold true for arbitrary $\Delta w \in W_R^D$. Because $0 \in W_R^D$, if the optimal value $R^D = \operatorname{dist}(0, \partial W^D) > 0$, the origin $\Delta w = 0$ is an interior point of W_R^D, and R^D can be regarded as an index of robustness.

Notice that ∂W^D is a non-convex set, problem (2.118) is a non-convex optimization problem. Another difficulty that prevents problem (2.118) from being solved stems from the objective function (2.118a), which is not an explicit function in the decision variables. The following proposition reformulates problem (2.118) as a bi-convex program whose objective and constraints are explicit functions.

Proposition 2.8 *Problem (2.118) is equivalent to the following optimization problem that seeks the largest R*

$$\begin{aligned} \max_{x,G,R} \ & R \\ s.t. \ & R \, \|H_i(G)\|_2 \le h_i(x), \ \forall i \\ & x \in X(C^R), \ G^T 1 + 1 = 0 \end{aligned} \tag{2.119}$$

To see this proposition, we need the following lemma.

Lemma (Cauchy–Schwarz Inequality) *The following inequality holds for vectors a and b with the same dimension*

$$a^T b \le \|a\|_2 \, \|b\|_2$$

Denote by $C(R) = \{\Delta w \mid \|\Delta w\|_2 \le R\}$ the circle whose center is origin and radius is R. The distance function in objective (2.118a) can be verified by set inclusion, i.e., if $\operatorname{dist}(0, \partial W^D) = R$, the following relation holds

$$\begin{aligned} & C(r) \subseteq W_R^D, \ \forall r \le R \\ & C(r) \nsubseteq W_R^D, \ \forall r > R \end{aligned} \tag{2.120}$$

For a given R, $C(R) \subseteq W^D$ means

$$H(G)\Delta w \le h(x), \ \forall \Delta w \in C(R)$$

or equivalently

$$\max_{\|\Delta w\|_2 \le R} H_i(G)\Delta w \le h_i(x), \ \forall i \tag{2.121}$$

where $H_i(G)$ is the i-th row of matrix $H(G)$, $h_i(x)$ is the i-th element of vector $h(x)$. To eliminate the max operator in the left-hand side of inequality (2.121), recall Cauchy-Schwarz inequality, we have

$$\max_{\|\Delta w\|_2 \le R} H_i(G)\Delta w = R \|H_i(G)\|_2$$

As a result, inequality (2.121) is equivalent to

$$R \|H_i(G)\|_2 \le h_i(x), \ \forall i \tag{2.122}$$

Finally, problem (2.118) boils down to problem (2.119) that seeks the largest R such that $C(R) \subseteq W_R^D$ holds true. Compared with the stochastic optimization method which requires the PDF or the robust optimization method which requires the forecast and range of uncertainty, problem (2.119) utilizes the least information, i.e. only the forecast w^e. Nonetheless, the statistical information of renewable generation can be incorporated in the DM-ERD by several means. For example, instead of using the maximal radius of circle $C(R)$, one can alternatively use an ellipsoid $E(Q_V, R) = \{\Delta w \,|\, \|Q_V \Delta w\|_2 \le R\}$ as a measure of dispatchability, where Q_V is a diagonal matrix consists of the variance $\sigma_j, \forall j$ of renewable power forecast error as

$$Q_V = \begin{bmatrix} \sigma_1^{-1} & & \\ & \ddots & \\ & & \sigma_{N_W}^{-1} \end{bmatrix}$$

A simple example of $E(Q_V, R)$ is shown in Fig. 2.32, with

$$\Delta w = \begin{bmatrix} \Delta w_1 \\ \Delta w_2 \end{bmatrix}, \ Q_V = \begin{bmatrix} \sigma_1^{-1} & 0 \\ 0 & \sigma_2^{-1} \end{bmatrix}$$

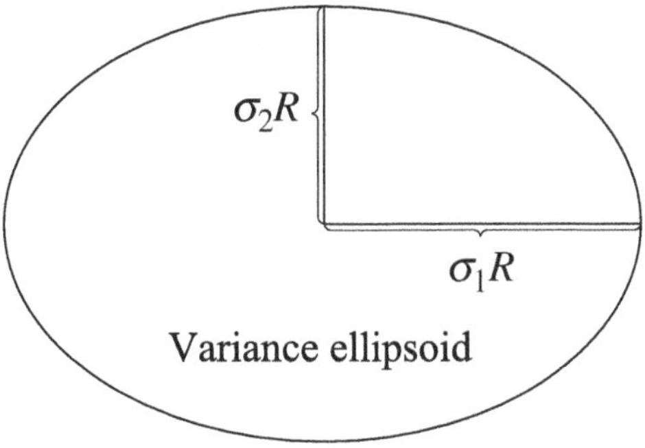

Fig. 2.32 The variance ellipsoid

The motivation is: the larger the variance is, the bigger the security margin should be preserved in that direction. If correlation is taken into account, the off-diagonal elements of Q_V can be non-zero as well, and the corresponding DM-ERD problem boils down to

$$\begin{aligned} \max_{x,G,R} \quad & R \\ \text{s.t.} \quad & R \, \|H_i(G) Q_V\|_2 \leq h_i(x), \ \forall i \\ & x \in X(C^R), \ G^T 1 + 1 = 0 \end{aligned} \tag{2.123}$$

From a geometry perspective, either the circle $C(R)$ or the variance ellipsoid $E(Q_V, R)$ is somehow similar to the uncertainty set in robust optimization. The difference is that the parameter R is not fixed in DM-ERD and should be jointly optimized with the dispatch decisions, reflecting "dispatchability maximization." In addition, in robust optimization, it is not clear whether the system is secure if the real data steps outside the uncertainty set; whereas the dispatchable region characterizes the exact security region in the uncertainty space: whenever Δw exceeds the boundaries of W_R^D, re-dispatch will be infeasible, and emergency measures should be deployed.

Problem (2.119) is still non-convex due to the product of R and $\|H_i(G)\|_2$; When R is fixed, problem (2.119) reduces to a set of SOC constraints. Due to the special property revealed by Eq. (2.120), we develop an SOCP based bisection algorithm to find the optimal R, which is summarized in Algorithm 2.5.

Algorithm 2.5 Bisection for ROD maximization

1: Retrieve the renewable generation forecast w^e. Set $R_l = 0$, $R_u = R_B^U$, where R_B^U is a sufficiently large constant. Choose a small tolerance $\varepsilon > 0$.

2: $R = (R_l + R_u)/2$, solve the following SOCP

$$\begin{aligned} r = \min_{x,G,s} \quad & 1^T s \\ \text{s.t.} \quad & x \in X(C^R),\ G^T 1 + 1 = 0 \\ & R \, \| H_i(G) \|_2 \le b_i(x) + s_i,\ s_i \ge 0,\ \forall i \end{aligned} \tag{2.124}$$

The optimal solution is x^* and G^* and the optimal value is r^*. If $r^* > 0$, $R_u = R$, else if $r^* = 0$, $R_l = R$.

3: If $R_u - R_l \le \varepsilon$, terminate, report the current solution x^*, otherwise, go to step 2.

The convergence of Algorithm 2.5 relies on two facts:

1. Problem (2.119) exhibits a specific structure: fixing R, it renders a set of SOC constraints in variables x and G, which is convex. In problem (2.124), we are actually testing the feasibility of (2.123) under the given value R.
2. Another important feature of problem (2.119) is revealed in Eq. (2.120). As W_R^D is a bounded polytope, suppose the optimal value of problem (2.119) is R^*, then $\forall R > R^*$, $C(R) \nsubseteq W_R^D$, and the optimal value r^* of SOCP (2.124) in step 2 will be strictly positive; $\forall R \le R^*$, $C(R) \subseteq W_R^D$, and the optimal value will be $r^* = 0$. Therefore, we can regard the optimal value r^* of SOCP (2.124) as a univariate monotonically increasing function of R, and problem (2.119) boils down to finding the largest R subject to $r^*(R) = 0$.

Based on the above observations, we can solve problem (2.119) through a line search for the optimal R. The monotonic relation between R and r^* guarantees that Algorithm 2.5 always finds the global optimal solution. The main loop of Algorithm 2.5 is a dichotomy scheme, the length of interval $[R_l, R_u]$ is reduced by half in each iteration. Therefore, the total number of iterations n should satisfy $R_B^U/2^n \le \varepsilon$, or equivalently, $n = \lceil \log_2(R_B^U/\varepsilon) \rceil$, where $\lceil\ \rceil$ stands for rounding towards plus infinity, so n only depends on the parameter R_B^U, ε, and is independent of the system data. Moreover, SOCP (2.124) in step 2 is polynomially solvable. So problem (2.119) always allows efficient computation regardless of the dimension of uncertainty.

3. Cost Minimization Subject to Given Reliability Level

In the DM-ERD problem, the reserve budget or production cost impacts the amount of available resources for mitigating uncertainty. Another important scheme is to minimize the cost subject to certain reliability level, arising the following robust chance-constrained program in variable x and G

$$\begin{aligned} \min_{x,G} \ & f_R \text{ or } f_G \\ \text{s.t. } & x \in X, \ G^T 1 + 1 = 0 \\ & \max_{P \in \Omega_G} \Pr[\Delta w \notin W_R^D(x, G)] \le 1 - \alpha \end{aligned} \tag{2.125}$$

where α is the desired reliability level, P is the PDF of renewable generation. Because P is not known exactly, we consider the worst case in the Gauss ambiguity set Ω_G defined in (2.110). In this way, the chance constraint is robust against all candidate PDFs in Ω. Problem (2.125) is difficult to solve. We outline a heuristic decomposition algorithm that seeks an approximate solution based on maximum dispatchable region and probability evaluation techniques elucidated in Sect. 2.5.3, which is summarized in Algorithm 2.6.

The motivation of this algorithm rests on the following observation: the set W_R^D as well as the ROD will grow larger when the cost parameter C^R/C^P increases, resulting in smaller probability $\Pr[\Delta w \notin W_R^D(x, G)]$. In this regard, we can control the reliability level through simply adjusting the scalar C^R/C^P. Therefore, any one-dimensional search scheme, such as the bisection search, can be used to update C^R/C^P, and the monotonic relation between the cost and reliability level guarantees the convergence of Algorithm 2.6. Finally, this heuristic algorithm only finds an approximate solution of problem (2.125), which may even fail to be a local optimum in rigorous mathematic sense.

Algorithm 2.6 Heuristic method for cost minimization

1: Choose a reliability level α.
2: Solve DM-ERD problem (2.119) using Algorithm 2.5 with the given budget C^P/C^R.
3: Evaluate $\beta^* = \max_{P \in \Omega} \Pr[\Delta w \notin W_R^D(x, G)]$ via any of the three methods presented in Sect. 2.5.3; If $\beta^* \le 1 - \alpha$, decrease C^P/C^R; otherwise, increase C^P/C^R. Then go to step 2. If the desired convergence criterion is reached, terminate and report the optimal solution x, G and optimal value $f_R = C^R$ or $f_G = C^P$.

The maximum reliability problem can be addressed in this way: first solve problem (2.119) without considering budget constraints, and then evaluate the probability β^*. The maximal reliability level is $\alpha^* = 1 - \beta^*$.

We discuss the applicability of the two formulations (2.119) and (2.125) solved by Algorithms 2.5 and 2.6, respectively. The first one is less flexible in dealing with economic considerations due to the objective function. Such consideration is modeled through budget constraints. A proper trade-off between the dispatchability and the cost should be decided manually. Nevertheless, Algorithm 2.5 only requires solving SOCP, which is more tractable than SDP. In summary, when the budget C^P/C^R is clear, or the number of uncertain sources is large, the first formulation and Algorithm 2.5 are recommended to maximize the dispatchable region with limited cost; otherwise, when the reliability level is specified, the second formulation and

Algorithm 2.6 are desired in order to minimize the cost with desired security guarantee. In summary, these methods can provide a meaningful reference for operators with different economic and reliability preferences.

4. Robust Optimization Based Formulation
Recall the definition of objective function f_G/f_R and constraints, the mathematical model of the affine policy robust optimization based energy and reserve dispatch (AR-ERD for short) is formulated as

$$\min_{x,G} \{f_G/f_R \mid x \in X, \text{ (2.113a)–(2.113c)}\} \tag{2.126}$$

where in constraints (2.113), W_R^D is a pre-specified polyhedral uncertainty set

$$W_R^D = \{\Delta w | H \Delta w \leq h\}$$

Despite the similarity of DM-ERD and AR-ERD in the first sight, they are actually different in mathematical substance. In DM-ERD, constraints (2.113a)–(2.113c) are guaranteed under affine policy (2.112) if and only if $\Delta w \in W_R^D$. Moreover, H and h are actively designed to enlarge the dispatchable region. In AR-ERD, constraints (2.113a)–(2.113c) can be guaranteed under affine policy (2.112) as long as $\Delta w \in W$, and the situation is not clear when $\Delta w \notin W$. Moreover, the uncertainty set W is pre-specified, whose parameter is constant.

Next we briefly outline how problem (2.126) can be reformulated as a tractable optimization problem using the method in [131]. Since (2.113a) is guaranteed by (2.114), the remaining (2.113b) and (2.113c) will require the following conditions

$$(\pi_G G + \pi_W)\Delta w \leq 0, \ \forall \Delta w \in W_R^D \tag{2.127a}$$

$$-r^- \leq G\Delta w \leq r^+, \ \forall \Delta w \in W_R^D \tag{2.127b}$$

Because W is a polyhedron, using the duality theory of LP, (2.127a) and (2.127b) are equivalent to the following constraints [131]

$$\Gamma_B h \leq 0, \ \Gamma_B H = \pi_G G + \pi_W, \ \Gamma_B \geq 0 \tag{2.127c}$$

$$\Gamma_U h \leq r^+, \ \Gamma_U H = G, \ \Gamma_U \geq 0 \tag{2.127d}$$

$$\Gamma_L h \leq r^-, \ \Gamma_L H = -G, \ \Gamma_L \geq 0 \tag{2.127e}$$

where Γ_B, Γ_U, Γ_L are matrix variables with compatible dimensions, H and h are given coefficient matrices of W. Therefore, the robust counterpart of AR-ERD (2.126) can be written as

$$\min_{x,G} \{f_G/f_R \mid x \in X,\ (2.114),\ (2.127c)\text{–}(2.127e)\} \tag{2.128}$$

Problem (2.128) is an LP if f_R is optimized, or a convex quadratic program if f_G is optimized. Affine policy robust optimization has been tutored in Appendix C.2.2.

5. Case Study

1. **PJM 5-Bus System**

System topology can be found in Fig. 2.17. It includes four generators. Wind farms W_1 and W_2 are connected to the grid at buses D and A. Deterministic demands at buses B, C, and D are 550 MW, 450 MW, and 350 MW, respectively. Generators and line data are provided in Tables 2.24 and 2.25.

First we neglect the budget constraint and solve problem (2.119) using Algorithm 2.5. The dispatchable region W_R^D with affine policy based re-dispatch is shown in Fig. 2.33. The optimal generation and reserve schedule, the gain coefficients and the network power flow are shown in Fig. 2.34. It's found that a certain security margin is preserved for all transmission lines, preventing them from being congested when the wind generation varies unexpectedly and the generating units automatically respond to the forecast error Δw according to affine policy (2.112).

Next, we investigate the impact of available reserve budget on the dispatchable region by choosing different values of C^R and solving problem (2.119) with $X(C^R)$. Results are shown in Figs. 2.35 and 2.36. Clearly, the dispatchable region grows larger, in terms of its radius, with C^R increasing, which reduces the probability of an infeasible operation. When C^R is large enough, such that constraint (2.116e) never becomes active, W_R^D will eventually become the one shown in Fig. 2.33.

The probability of infeasible operation is computed by GGI and Monte Carlo simulation described in Sect. 2.5.3. Wind power variances are $\sigma_j = 0.1w_j^e, \forall j$ in our tests. In Monte Carlo simulation, we simply assume that the PDF of wind

Table 2.24 Parameters of generators

Unit	p^l (MW)	p^u (MW)	a \$/MW^2	b \$/MW	c^+/c^- \$/MW	Ramp MW/h
G_1	100	200	0.1176	10	1.2	50
G_2	150	300	0.0490	20	1.8	75
G_3	150	400	0.0318	25	2.5	100
G_4	250	600	0.0210	28	3.3	150

Table 2.25 Parameters of transmission lines

Line	AB	AD	AE	BC	CD	DE
x_l (p.u.)	0.0281	0.0304	0.0064	0.0108	0.0297	0.0297
S_l (MW)	608	200	200	100	405	245

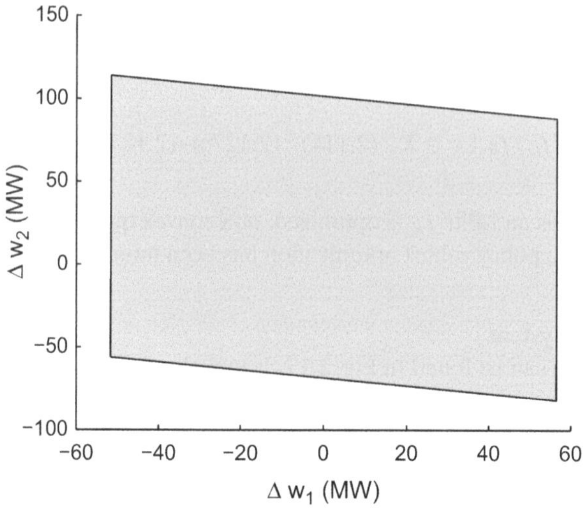

Fig. 2.33 Dispatchable region without budget restriction

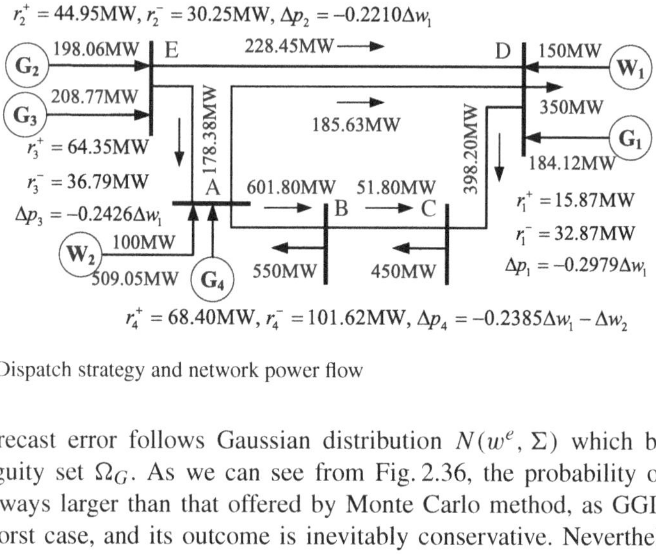

Fig. 2.34 Dispatch strategy and network power flow

power forecast error follows Gaussian distribution $N(w^e, \Sigma)$ which belongs to the ambiguity set Ω_G. As we can see from Fig. 2.36, the probability offered by GGI is always larger than that offered by Monte Carlo method, as GGI accounts for the worst case, and its outcome is inevitably conservative. Nevertheless, such conservatism is helpful in maintaining a higher level of reliability.

We also test AR-ERD, in which the uncertainty set is

$$W = \left\{ \Delta w \left| \begin{array}{c} -w_1^h \le \Delta w_1 \le w_1^h \\ -w_2^h \le \Delta w_2 \le w_2^h \\ \dfrac{|\Delta w_1|}{w_1^h} + \dfrac{|\Delta w_2|}{w_2^h} \le 1.5 \end{array} \right. \right\}$$

The last norm inequality restricts the total forecast error.

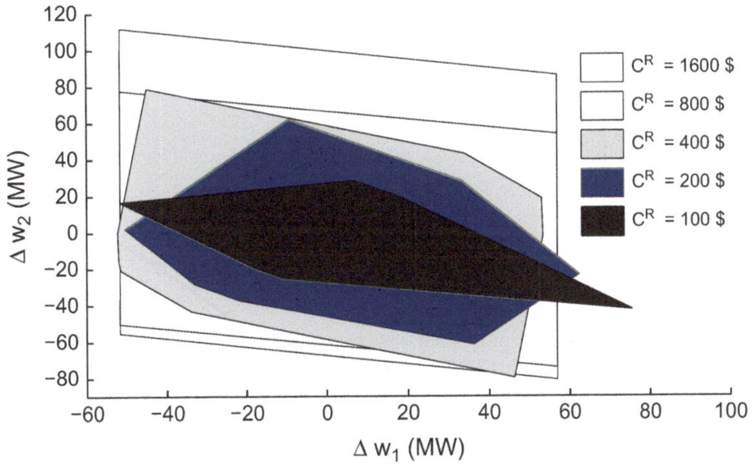

Fig. 2.35 Dispatchable region under different C^R

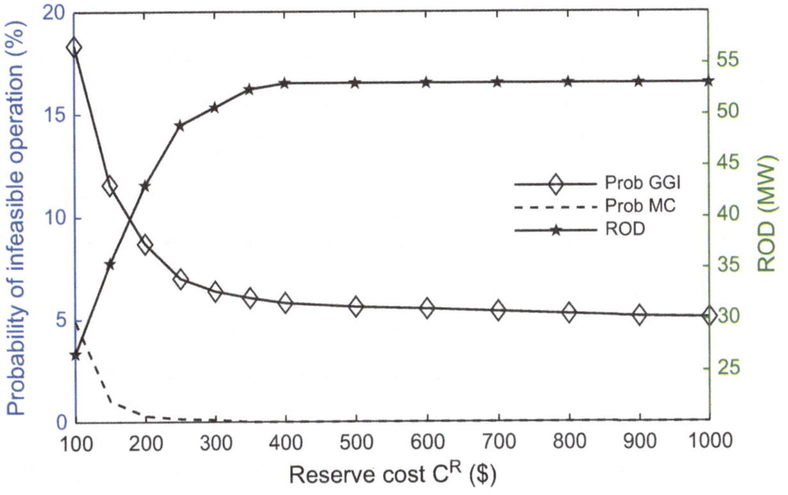

Fig. 2.36 Radius of dispatchable region and probability of infeasible operation

In order to simulate the impact of different level of wind power variability on the reliability level of system operation, we control the size of uncertainty set W through adjusting parameters w_1^h and w_2^h. Specifically, we simply choose $w_1^h = \kappa w_1^e$ and $w_2^h = \kappa w_2^e$, where κ is an adjustable parameter. The generation and reserve schedule $x = [p^f, r^+, r^-]$ and the gain matrix G of affine policy are obtained by solving problem (2.128). The dispatchable region is given by (2.115), and the reliability level can be computed from GGI. If the reliability requirement is not satisfactory, we increase the value of κ to protect the system under a larger uncertainty set.

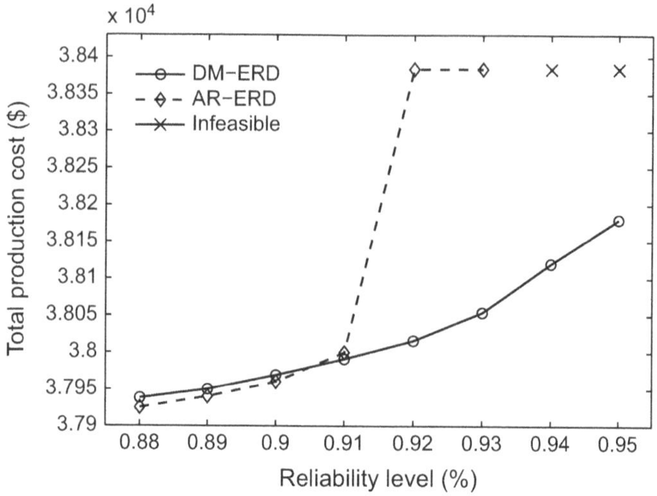

Fig. 2.37 Total cost of DM-ERD and AR-ERD

As for the DM-ERD, we use the second formulation to minimize the total production cost C^P and compare its performance with AR-ERD. Results are shown in Fig. 2.37, from which we can see that when the reliability level $\alpha \leq 91\%$, the optimal values of DM-ERD and AR-ERD are almost the same; when α grows to 92%, the production cost of AR-ERD exhibits a sudden increase; and when $\alpha > 93\%$, AR-ERD becomes infeasible due to the lack of dispatchability no matter how generators are dispatched. This is partly because the manner we define and adjust the uncertainty set (using a single parameter κ to control the size of a polyhedral set) is not flexible enough. In contrast, DM-ERD adaptively controls the shape and size of dispatchable region according to the reliability feedback, and successfully provides a dispatch strategy with the reliability level going up to 95%.

When $w_1^h = 50.0$ MW, $w_2^h = 33.3$ MW, the uncertainty set W is shown in Fig. 2.38. The dispatchable region W_R^D under the optimal dispatch strategy x and gain matrix G offered by AR-ERD is illustrated in the same figure, which clearly demonstrates the inclusive relaxation $W \subset W_R^D$. In fact, this phenomenon also exists in the fully-adjustable robust optimization, examples can be found in Fig. 2.21, in which the uncertainty set is a hypercube, and the dispatchable region is a larger polyhedron.

IEEE 118-Bus System

System data are available in [102]. In the considered dispatch interval, the total demand is 5500 MW. The reserve cost coefficients c_i^+/c_i^- of each generator are assumed to be 10% of its generation cost coefficient b_i. The ramping limits R_i^+/R_i^- of each generator are assumed to be 25% of its maximal output P_i^u. Four wind farms are connected to the system at buses #70 (Area 1), #49, #80 (Area 2), #100 (Area 3). Their predicted output is 200 MW, and variance 400 MW^2. Other parameters without particular mention are the same as those provided online.

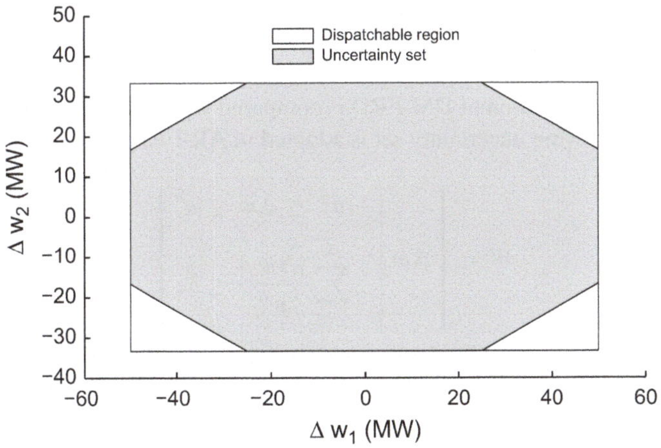

Fig. 2.38 Comparison of uncertainty set and dispatchable region

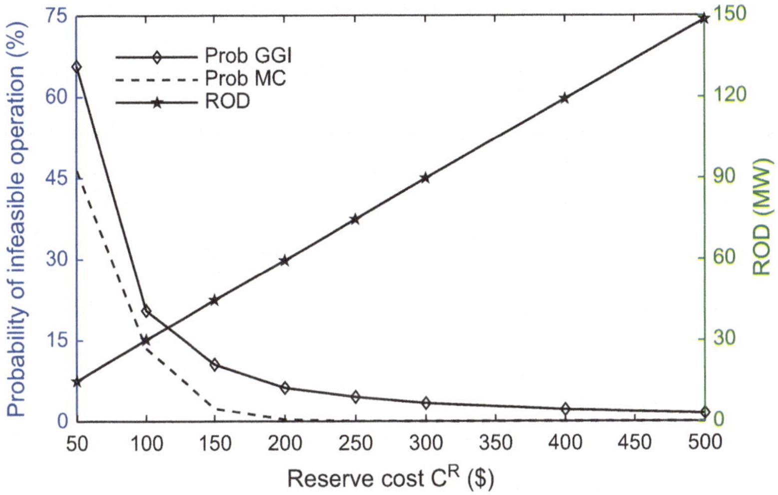

Fig. 2.39 Radius of W^D and probability of infeasible operation

The first formulation of DM-ERD is solved under different C^R using Algorithm 2.5, and results are shown in Fig. 2.39, which clearly demonstrates the radius of dispatchable region grows larger with C^R increasing. In this case, the dispatchable region W_R^D is a polytope in $\mathbb{R}^4$, which cannot be visualized directly. The computation time with different values of C^R varies from 11 s to 13 s, which is fast enough for online applications. Under each dispatch strategy, the probability of infeasible dispatch offered by GGI and Monte Carlo simulation with Gaussian distribution is drawn in Fig. 2.39, which demonstrates that although the GGI method is pessimistic, the gap is acceptable. More importantly, the combination of GGI and dispatchable region offers a comprehensive way to evaluate the operating reliability in the absence of an exact PDF of wind power, and could potentially become an important method to quantify the impact of uncertainty on power system operation.

The second formulation of DM-ERD is compared with AR-ERD on this 118-bus system. The following uncertainty set is adopted in AR-ERD

$$W = \left\{ \Delta w \,\middle|\, \begin{array}{l} -w^h \le \Delta w \le w^h \\ \sum_{j=1}^{4} \frac{|\Delta w_j|}{w_j^h} \le 3 \end{array} \right\}$$

where $w^h = \kappa w^e$, the size of W is controlled by adjusting the value of κ. To validate the optimization results, the total reserve cost C^R is optimized for both methods. The results are shown in Fig. 2.40. In this test, AR-ERD has a higher reserve cost than DM-ERD when $\alpha \le 92.6\%$. When α grows larger, AR-ERD becomes infeasible, while DM-ERD continuously provides valid dispatch strategies although requiring higher reserve costs. Comparing Figs. 2.39 and 2.40, we find the optimal cost provided by Algorithm 2.6 is consistent with that in Fig. 2.39 under the same level of reliability, which validates the applicability of Algorithm 2.6.

Finally, we increase the number of wind farms in this system to demonstrate the performance of the proposed method under higher dimensional uncertainties. When changing N_W in the system, we maintain the total wind power generation is 800 MW, which means $w_j^e = 800/N_W$, and choose $\sigma_j = 0.1 w_j^e$ for each wind farm. Moreover, no wind farm is connected to the same bus in order to avoid degeneracy of the dispatchable region. The computation time of the first formulation is shown

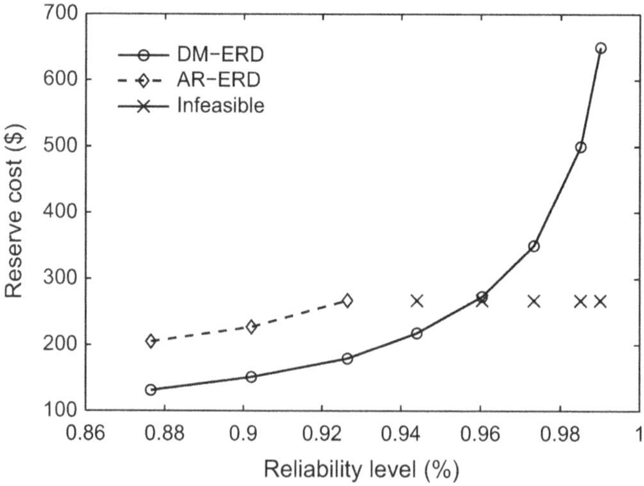

Fig. 2.40 Total cost of DM-ERD and AR-ERD

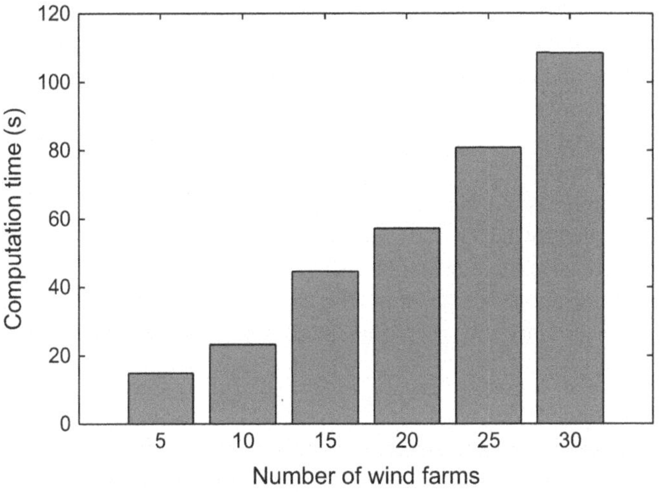

Fig. 2.41 Computational time with different number of wind farms

in Fig. 2.41. We can see that when N_W varies from 5 to 30, the computation time grows moderately and satisfies the requirement of online application. As for the second formulation, Algorithm 2.6 fails to converge in 3 min when $N_W \geq 8$, which may not be efficient enough for online use. Nevertheless, both stochastic and robust optimization encounter computational difficulty while facing a large number of wind farms.

2.6 Robust Unit Commitment

During the past two decades, the electric power industry has been experiencing deregulation and marketization in many countries. Competitive markets have advantages in reducing costs, but they also impose potential risks to system operation, because generators and the network will be operated closer to their security limits, and backup capacity will be cut down due to economic requirements. Meanwhile, the high penetration level of cheap but volatile wind and solar energy adds additional conflict between security and economical considerations. These factors call for reliable solution of the security constrained UC problem in the day-ahead stage in support of system operation and market organization.

Due to the operating characteristics of thermal power plants, power system operation typically involves multiple time scales, and has been explained in Fig. 2.1. Sections 2.4 and 2.5 are devoted to the latter two stages, while assuming the unit commitment (UC) is fixed. However, as the integration of volatile renewable resources is scaling up, the actual condition may be considerably different from the forecasts. As a result, online units may be unable to mitigate the increasing level of uncertainties, whereas uncommitted ones are unavailable to provide backup owing to their slow responses. Under this circumstance, emergency measures such as generator outage and load shedding may be inevitable, which might cause expected contingencies or even cascading failure. In order to offer reliable UC decisions and ensure a feasible real-time operation in the presence of high penetration of renewables, it is important to consider uncertainties in the day-ahead planning stage. Assuming uncertain nodal injection power, this section introduces the concept and formulation of the robust UC. Pioneered by the work in [132–134], there have already been extensive literatures in this line of research. The purpose of this section is to streamline the solution algorithm with concise (compact) mathematical models and different uncertainty sets. Materials in this section come from [132, 135]

2.6.1 Deterministic Formulation of UC

A wide range of system data that will be involved in UC decision-making are summarized below.

Indices

i	Index of generators.
j	Index of renewable power plants.
q	Index of loads.
t	Index of time periods.

Parameters

a_i, b_i	Production cost coefficients of generator i.
F_l	Power flow capacity of transmission line l.
p_i^n	Minimal output of generator i.
p_i^m	Maximal output of generator i.
p_{qt}	Power demand of load q in period t.
R_i^+	Ramp-up limit of generator i.
R_i^-	Ramp-down limit of generator i.
SU_i	Start-up cost coefficient of generator i.
SD_i	Shut-down cost coefficient of generator i.
T_i^{on}	Minimum-on time of generator i.
T_i^{off}	Minimum-off time of generator i.
w_{jt}^f	Generation forecast of renewable plant j in period t
π_{il}	PTDF from generator i to line l.
π_{jl}	PTDF from renewable plant j to line l.
π_{ql}	PTDF from load q to line l.

Variables

u_{it} If generator i is turned on in period t, $u_{it} = 1$; otherwise, $u_{it} = 0$.
v_{it} If generator i is turned off in period t, $v_{it} = 1$; otherwise, $v_{it} = 0$.
z_{it} If generator i is on in period t, $z_{it} = 1$; otherwise $z_{it} = 0$.
p_{it} Output of generator i in period t.

The deterministic UC problem is to determine the most economical commitment and output schedules subject to operating constraints and associated with the forecasted load and renewable generation. Its mathematical formulation is given by the following MILP

$$\begin{aligned} \min \quad & \sum_t \sum_i \left\{ SU_i u_{it} + SD_i v_{it} + a_i p_{it}^2 + b_i p_{it} \right\} \\ \text{s.t.} \quad & \text{Cons-Commitment, Cons-Dispatch} \end{aligned} \tag{2.129}$$

The objective function is the total production cost, including start-up, shut-down, and generation cost. Cons-Commitment for binary variables consists of

$$-z_{i(t-1)} + z_{it} - z_{ik} \le 0,\ \forall i,\ \forall t,\ \forall k \in \{t, t+1, \dots, T_i^{on} + t - 1\} \tag{2.130a}$$

$$z_{i(t-1)} - z_{it} + z_{ik} \le 0,\ \forall i,\ \forall t,\ \forall k \in \{t, t+1, \dots, T_i^{off} + t - 1\} \tag{2.130b}$$

$$-z_{i(t-1)} + z_{it} - u_{it} \le 0,\ \forall i,\ \forall t \tag{2.130c}$$

$$z_{i(t-1)} - z_{it} + v_{it} \le 0,\ \forall i,\ \forall t \tag{2.130d}$$

$$u_{it},\ v_{it},\ z_{it} \in \{0, 1\},\ \forall i,\ \forall t \tag{2.130e}$$

where (2.130a) is the minimum-up time constraint for each generator: if generator i is turned on in period t, indicating $z_{i(t-1)} = 0$ and $z_{it} = 1$, then it must keep working for at least T_i^{on} periods, interpreting $z_{ik} = 1, \forall k \in \{t, t+1, \dots, T_i^{on}+t-1\}$. Constraint (2.130b) for the minimum-down time has a similar interpretation. Binary inequalities (2.130c) and (2.130d) define the logic between on-off status and turn-on/turn-off signals. Particularly, a generator is turned on (off) in period t if and only if $z_{i(t-1)} = 0$ and $z_{it} = 1$ ($z_{i(t-1)} = 1$ and $z_{it} = 0$); when $z_{i(t-1)} = z_{it}$, u_{it} and v_{it} must be 0 because coefficients SU_i and SD_i are positive numbers. The initial status of generator can be assigned with certain values or set as $z_{i0} = z_{iT}, \forall i$ which stipulates equal values on the initial and terminal status in a daily cycle.

In formulation (2.130), the binary variable u_{it}/v_{it} indicating startup/shutdown operations can be eliminated. To this end, in the objective function, replace $SU_i u_{it}$ and $SD_i v_{it}$ with two continuous variables C_{it}^U and C_{it}^D, respectively, and change constraints (2.130c) and (2.130d) into

$$C_{it}^U \geq 0,\ C_{it}^U \geq SU_i(z_{it} - z_{i(t-1)})$$
$$C_{it}^D \geq 0,\ C_{it}^D \geq SD_i(z_{i(t-1)} - z_{it})$$

At the optimal solution, $C_{it}^U = SU_i/C_{it}^D = SD_i$ holds if unit i is turned on/off in period t. Nonetheless, u_{it} and v_{it} could be useful in more meticulous UC models, such as the one in [136].

Cons-Dispatch for continuous variables is comprised of

$$z_{it} p_i^n \leq p_{it} \leq z_{it} p_i^m \tag{2.131a}$$

$$p_{it} - p_{i(t-1)} \leq (2 - z_{i(t-1)} - z_{it}) p_i^n + (1 + z_{i(t-1)} - z_{it}) R_i^+ \tag{2.131b}$$

$$p_{i(t-1)} - p_{it} \leq (2 - z_{i(t-1)} - z_{it}) p_i^n + (1 - z_{i(t-1)} + z_{it}) R_i^- \tag{2.131c}$$

$$\sum_i p_{it} + \sum_j w_{jt}^f = \sum_q p_{qt},\ \forall t \tag{2.131d}$$

$$-F_l \leq \sum_i \pi_{il} p_{it} + \sum_j \pi_{jl} w_{jt}^f - \sum_q \pi_{ql} p_{qt} \leq F_l,\ \forall t \tag{2.131e}$$

where (2.131a) is generation capacity constraint and imposes zero output on uncommitted units; (2.131b) stands for the upward ramping constraint of each generator; in particular, if $z_{i(t-1)} = z_{it} = 0$, the output is forced at 0 in (2.131a), then (2.131b) is naturally met; if $z_{i(t-1)} = z_{it} = 1$, (2.131b) boils down to a traditional ramp limit constraints; if $z_{i(t-1)} = 0$, $z_{it} = 1$, then there must be $p_{i(t-1)} = 0$ and $p_{it} = p_i^n$, i.e., when a generator is turned on, its output in the first-hour must be the minimum; if $z_{i(t-1)} = 1$, $z_{it} = 0$, (2.131b) is redundant since $p_{it} = 0$ and the left-hand side is negative; (2.131c) for the downward ramping capability has a similar interpretation. Equation (2.131d) is the system-wide energy balancing condition in each time period; (2.131e) represents power flow limit for each transmission line.

The quadratic cost function can be approximated by PWL functions in p_{it}, without generality, deterministic UC problem (2.129) can be written as a compact MILP model as

$$\begin{aligned} \min\ & c^T x + d^T p \\ \text{s.t. } & x \in X,\ p \in Y(x, w^f) \end{aligned} \tag{2.132}$$

where vector x includes binary variables u_{it}, v_{it}, z_{it}, $\forall i$, $\forall t$, vector p consists of p_{it}, $\forall i$, $\forall t$, vector w^f is the renewable generation forecast; c and d are corresponding coefficient vectors. Nonetheless, a convex quadratic objective function can be directly handled by commercial solvers and adds no additional difficulty to problem solution. The linear formulation (2.132) is just for the ease of exposition.

Set X represents Cons-Commitment in (2.129) including constraints in (2.130), and

$$Y(x, w^f) = \left\{ p \;\middle|\; By \le b - Ax - Cw^f \right\} \tag{2.133}$$

is the compact form of Cons-Dispatch in (2.129), where matrices A, B, C and vector b correspond to constants in constraints (2.131a)–(2.131e). Deterministic UC problem (2.132) in MILP form can be solved by off-the-shelf solvers, such as CPLEX and GUROBI, or a number of decomposition methods such as Lagrange relaxation and Benders decomposition. A comprehensive survey can be found in [137].

2.6.2 A Heuristic Method to Determine Reserve Level

To reduce the production cost, the number of committed generators are kept as small as possible, and they are usually operated around the rated capacity with a relative lower marginal cost. If renewables produce less electricity, thermal units have to increase their output to maintain power balancing in real-time. This requires the system operator preserve a certain amount of backup in the UC process. This section introduces a heuristic decomposition method to determine the spinning reserve level in the day-ahead UC.

In this section, volatility of renewable power is restricted in a deterministic set, called the uncertainty set. To formalize the discussion, we provide two widely used ones in existing literature.

1. Extreme-Point Description

Let w_{jt}^+ (w_{jt}^-) be the positive (negative) forecast errors of the output power in renewable plant j in period t, and z_{jt}^+ (z_{jt}^-) indicates if the output of renewable plant j reaches its upper (lower) bound in time period t. The set of uncertain nodal injection power in extreme cases can be written as

$$W = \left\{ w \;\middle|\; \begin{array}{l} w_{jt} = w_{jt}^f + z_{jt}^+ w_{jt}^+ - z_{jt}^- w_{jt}^-, \ \forall j, \forall t \\ \sum\limits_j \left(z_{jt}^+ + z_{jt}^- \right) \le \Gamma_{\mathrm{S}}, \ \forall t \\ \sum\limits_t \left(z_{jt}^+ + z_{jt}^- \right) \le \Gamma_{\mathrm{T}}, \ \forall j \\ z_{jt}^+, z_{jt}^- \in \{0, 1\}, \ \forall j, \ \forall t \end{array} \right\} \tag{2.134}$$

The first inequality means that the renewable output is within an interval $[w_{jt}^n, w_{jt}^m]$ with forecasted value w_{jt}^f, where $w_{jt}^n = w_{jt}^f - w_{jt}^-$, $w_{jt}^m = w_{jt}^f + w_{jt}^+$. The values of w^n_{jt}, w^m_{jt} (or equivalently w^+_{jt}, w^-_{jt}) can be obtained based on the historical data or an interval forecast for the renewable generation. Because w_{jt} can run freely in its confidence interval, the worst-case scenario is either $w_{jt} = w^n_{jt}$ or $w_{jt} = w^m_{jt}$, which is apparently over conservative.

As renewable generation centers are distant from each other, their output is unlikely to reach upper or lower bounds simultaneously in a particular time period; meanwhile, the forecast values of a certain plant are also unlikely to reach upper or lower bound in every period. Therefore, we impose spatial and temporal constraints in the second and third inequality to control the total deviation from the forecast, and Γ_S and Γ_T are called the budget of uncertainty. When $\Gamma_S = \Gamma_T = 0$, the uncertainty set W will become a singleton, implying that the prediction is accurate. With the values of Γ_S and Γ_T increasing, set W swells accordingly, which allows a larger total deviation from the forecast, and the system will face a higher level of uncertainty. For example, if $\Gamma_T = 12$ is adopted and the number of planning periods is 24, then w_{jt} can take any value in $[w^n_{jt}, w^m_{jt}]$ in any arbitrary 12 periods, but is assumed to be close to w^f_{jt} in the remaining 12 periods. The system operators can control the robustness of solution by adjusting the values of Γ_S and Γ_T. To mimic real renewable power generation, parameters in W can be selected according to the method in [105].

2. General Polyhedron

For the same consideration, a continuous uncertainty set can be built as

$$W = \left\{ w \;\middle|\; \begin{array}{l} w^n_{jt} \le w_{jt} \le w^m_{jt},\ \forall j, \forall t \\ \sum_j \dfrac{|w_{jt} - w^f_{jt}|}{w^h_{jt}} \le \Gamma_S,\ \forall t \\ \sum_t \dfrac{|w_{jt} - w^f_{jt}|}{w^h_{jt}} \le \Gamma_T,\ \forall j \end{array} \right\} \tag{2.135}$$

where $w^h_{jt} = 0.5(w^m_{jt} - w^n_{jt})$. When the forecast is symmetric, i.e., $w^-_{jt} = w^+_{jt}$, and the budget parameters Γ_S and Γ_T are positive integers, (2.135) is the convex hull of (2.134).

The uncertainty set W defined in (2.135) is a polyhedron. By introducing lifting variables, the absolute value function can be eliminated, and (2.135) can be expressed by a general polytope

$$W = \{w \mid Hw \le h\} \tag{2.136}$$

Please bear in mind that vector w in (2.136) includes the lifting variables. Details can be found in Appendix C.2.3.

For any given w in the uncertainty set W defined by (2.134) or (2.135), Cons-Dispatch becomes

$$Y(x, w) = \{p \mid By \leq b - Ax - Cw\} \tag{2.137}$$

To ensure a secure operation, set $Y(x, w)$ should be always non-empty.

Definition 2.6 A UC decision x is called robust against uncertainty (belongs to set W) if $\forall w \in W$, $Y(x, w) \neq \emptyset$.

Checking whether a given x is robust can boil down to solving a linear max-min problem, which can be further transformed to a bi-convex program (BCP). For the uncertainty set (2.134) consists of extreme points, the BCP is equivalent to an MILP; For the polyhedral one (2.135), the BCP is equivalent to an MPCC, whose global solution can be found via solving an equivalent MILP; Nevertheless, provided with some delicately chosen initial values, a high quality solution can be found by a sequential LP method. As a common problem in robust optimization, details on the linear max-min problem can be found in Appendix C.2.3.

To find a robust UC solution, the most common implementation in power system operating practice is to increase the spinning reserve: more generators are committed online, and each unit preserves a certain margin to its capacity limit. In this way, the total generation will be smoothed by deploying reserve capacity, even if renewable power could be volatile to some extent. To this end, we consider spinning reserve constraints

$$\sum_i u_{it} p_i^m \geq (1 + r) \left(\sum_q p_{qt} - \sum_j w_{jt}^f \right) \tag{2.138a}$$

$$\sum_i u_{it} p_i^n \leq (1 - r) \left(\sum_q p_{qt} - \sum_j w_{jt}^f \right) \tag{2.138b}$$

where r is the system-wide spinning reserve level which should be determined. Equations (2.138a) and (2.138b) guarantee that the discrepancy between load and renewable generation forecast, which is provided by thermal units, should be no greater (less) than the total maximum (minimum) output of online units. However, they are not sufficient to procure a robust UC solution, because ramping and transmission line congestion constraints are neglected. Let

$$X_R^S(r) = \{x \mid x \in X, \text{(2.138a)}, \text{(2.138b)}\} \tag{2.139}$$

be the commitment constraint with fixed spinning reserve level r. By setting

$$r = \max_t \left\{ \frac{\sum_j \max\{w_{jt}^+, w_{jt}^-\}}{\sum_q p_{qt} - \sum_j w_{jt}^f} \right\} \tag{2.140}$$

and solving a deterministic UC problem

$$\begin{aligned} \min \ & c^T x + d^T p \\ \text{s.t. } & x \in X_R^S(r) \\ & p \in Y(x, w^f) \end{aligned} \tag{2.141}$$

the optimal solution x^* is clearly robust, but often appears to be very conservative, because renewable power cannot reach their upper or lower bounds in every periods due to the budget constraints in (2.134) and (2.135). A heuristic algorithm is suggested below to search a smaller r through dichotomy.

Algorithm 2.7 Dichotomy for spinning reserve level

1: Set the lower and upper bounds of spinning reserve level to be $r_{LB} = 0$ and $r_{UB} = r_m$ (determined by (2.141)), and the iteration number $k = 1$. Choose a convergence tolerance $\varepsilon > 0$.
2: Let $r = 0.5(r_{LB} + r_{UB})$, solve problem (2.141), the optimal UC solution is x^*.
3: Check the robustness of x^* (see Appendix C.2.3 for details). If it is robust, update $r_{UB} = r$; otherwise, update $r_{LB} = r$.
4: If $r_{UB} - r_{LB} \le \varepsilon$, terminate and report the robust UC solution x^*; otherwise, update $k = k + 1$ and go to step 2.

Convergence of Algorithm 2.7 is guaranteed by the following observation: when the system has adequate capacity, there exists a threshold value r^*, such that for any $r > r^*$, the UC solution x^* of (2.141) is robust, and for any $r < r^*$, the corresponding x^* is not. Since the length of searching interval $[r_{LB}, r_{UB}]$ is reduced by half in each dichotomy, the total number of iterations spent in Algorithm 2.7 is determined by

$$k = \log_2 \left(\frac{r_m}{\varepsilon} \right) \tag{2.142}$$

The advantage of this approach is that the problem size and complexity remain the same in each iteration, and the number of iteration solely depends on r_m and ε, and is independent of the system. A major deficiency of this approach lies in the fact that the robustness is controlled via a scalar r, and thus less flexible. The UC solution may be sub-optimal. To see this, image a situation in which the output of cheaper generator is restricted by congestion. If a UC solution x^* is not robust, increasing r does not help alleviate congestion and enhance robustness, unless more expensive units are involved. Nonetheless, when the penetration of renewable energy is not very high, or the transmission network is not heavily congested, this approach usually provides satisfactory results. This approach well matches the reserve based generation scheduling paradigm in operation practices.

The proposed method is applied to the IEEE 39-bus system. The daily load curve and predicted renewable generation are shown in Fig. 2.42. The distribution rates (in percentage) of the total load demand among the load buses are given in Table 2.26. Four piecewise linear segments are used to approximate the quadratic cost function. Uncertainty set (2.135) is used, in which $\Gamma_{\mathrm{T}} = \Phi^{-1}(0.95) \times \sqrt{24} = 8.08$ [132].

A wind farm is connected to the grid at bus 29, and the second constraint in (2.135) is redundant (as it is defined for multiple wind farms). The budget Γ_{T} is changed from 0 to 8.08. The UC cost $c^T x$, the dispatch cost $d^T y$, and the total cost are given in Table 2.27. Consistent with common sense, when Γ_{T} grows larger, all kinds of costs increase, and more backup is maintained to tackle higher levels of uncertainties. Because the UC problem is discrete, so the optimal reserve level may not be continuous with respect to Γ_{T}.

Moreover, we fix $\Gamma_{\mathrm{T}} = 8.08$, and divide the renewable plant into smaller identical ones: (1) Three plants which connect to buses 8, 14, and 29, respectively. The budget $\Gamma_{\mathrm{S}} = \Phi^{-1} \times \sqrt{3} \approx 2.85$; (2) Six plants connect to buses 2, 8, 14, 22, 26, and 29, respectively with $\Gamma_{\mathrm{S}} = \Phi^{-1} \times \sqrt{6} \approx 4.03$; (3) Nine plants connect to

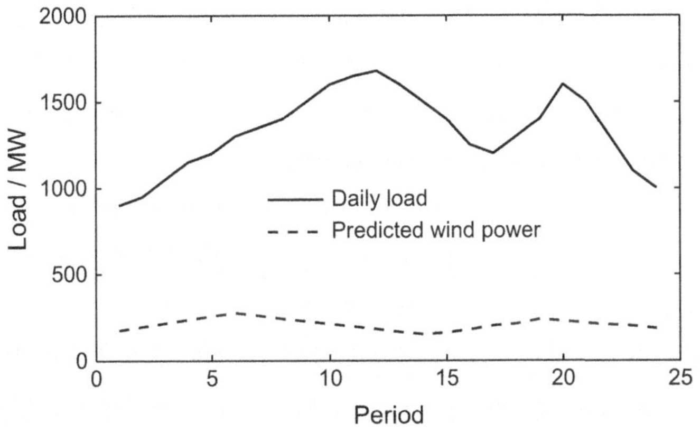

Fig. 2.42 The load curve and predicted renewable generation

Table 2.26 Load distribution rates among load buses

Node	3	4	7	8	12	15	16
Rate	5.24	8.13	3.80	8.49	0.14	5.20	5.35
Node	18	20	21	23	24	25	26
Rate	2.57	11.06	4.46	4.02	5.02	3.64	2.26
Node	27	28	29	31	39		
Rate	4.57	3.35	4.61	0.15	17.95		

Table 2.27 Costs and reserve levels under different values of Γ_T

Γ_T	UC cost ($)	Dispatch cost ($)	Total cost ($)	Reserve level (%)
0	74,760	423,118	497,878	0.00
2.02	77,160	428,043	505,203	1.95
4.04	78,420	430,455	508,875	3.40
6.06	78,420	432,616	511,036	3.40
8.08	78,530	439,664	518,194	4.05

Table 2.28 Costs and reserve levels with different numbers of renewable plants

M	Γ_S	Γ_T	UC cost ($)	Dispatch cost ($)	Total cost ($)	Reserve level (%)
3	2.85	8.08	78,420	424,421	512,841	3.40
6	4.03	8.08	77,990	435,240	513,230	2.58
9	4.92	8.08	76,300	437,753	514,053	1.13

buses 2, 4, 8, 10, 14, 21, 22, 26, and 29 with $\Gamma_S = \Phi^{-1} \times \sqrt{9} \approx 4.92$. Results are shown in Table 2.28.

From Table 2.28 we can see clearly that the more widely the renewable plants disperse, the less the spinning reserve is needed, because the total level of uncertainty is proportional to $1/\sqrt{N}$ where N is the number of renewable plants. This observation implies that if we make better use of geographical regression effects of wind farms, the cost for dealing with uncertainties could be reduced significantly.

2.6.3 *Robust UC with Pumped Storage Hydro*

This section streamlines the application of the ARO in UC problems. To enhance the flexibility in the recourse stage, we consider the dispatch of pumped storage unit, a representative storage-like facility. The robust UC schedule for thermal generators in the day-ahead market protects the system in all possible uncertain parameters in a pre-specified uncertainty set, and minimizes the total cost in the worst-case scenario. In addition to the variables in Sect. 2.6.1, notations of the pumped storage unit are summarized as follows:

Parameters

h_s^0, h_s^e Initial (target) reservoir inventory of pumped-storage unit s in the first (last) period.

η_s^+, η_s^- Charging (discharging) coefficient of pumped-storage unit s.

p_s^{n+}, p_s^{m+} Minimum (maximum) charging power of pumped-storage unit s.

p_s^{n-}, p_s^{m-} Minimum (maximum) discharging power of pumped-storage unit s.

π_{sl} PTDF from pumped-storage unit s to line l.

Variables

z_{st}^+, z_{st}^- $\quad z_{st}^+ = 1/0$ ($z_{st}^- = 1/0$) if pumped storage unit s is working in the charging (discharging) mode or out of service.

p_{st}^+, p_{st}^- $\quad$ Charging (discharging) power of pumped storage unit s in period t.

h_{st} $\quad$ Reservoir inventory of pumped-storage unit s in period t.

The operating constraints of pumped storage units include

$$h_{st} = h_{s(t-1)} + \eta_s^+ p_{st}^+ - \frac{p_{st}^-}{\eta_s^-}, \ \forall s, \ \forall t \tag{2.143a}$$

$$z_{st}^- p_s^{n-} \le p_{st}^- \le z_{st}^- p_s^{m-}, \ \forall s, \ \forall t \tag{2.143b}$$

$$z_{st}^+ p_s^{n+} \le p_{st}^+ \le z_{st}^+ p_s^{m+}, \ \forall s, \ \forall t \tag{2.143c}$$

$$h_{s0} = h_s^0, \ h_{sT} = h_s^e, \ \forall s \tag{2.143d}$$

$$z_{st}^+, z_{st}^- \in \{0, 1\}, \ z_{st}^+ + z_{st}^- \le 1, \ \forall s, \ \forall t \tag{2.143e}$$

where (2.143a) describes reservoir inventory dynamics (measured by MWh); (2.143b) and (2.143c) provide the minimum and maximum power absorbed and generated by the pumped-storage units; (2.143d) defines the initial and terminal water inventory level; Finally, (2.143e) ensures that the pumped-storage units cannot absorb and generate electricity at the same time by introducing binary variables.

The power balancing condition and line flow constraints are modified as

$$\sum_i p_{it} + \sum_j w_{jt}^f + \sum_s (p_{st}^+ - p_{st}^-) = \sum_q p_{qt}, \ \forall t \tag{2.144}$$

$$-S_l \le \sum_i \pi_{il} p_{it} + \sum_j \pi_{jl} w_{jt}^f + \sum_s \pi_{sl}(p_{st}^+ - p_{st}^-) - \sum_q \pi_{ql} p_{qt} \le S_l, \ \forall t \tag{2.145}$$

The compact form of robust UC with pumped storage unit is

$$\min_{x \in X} \left(c^T x + \max_{w \in W} \min_{(p, z^H) \in Y(x,w)} d^T p \right) \tag{2.146}$$

where vector x and w, sets X and W have been defined in Sects. 2.6.1 and 2.6.2; operating pumped-storage hydro incurs little fuel cost, so the commitment cost and dispatch cost in the objective function have the same expressions as (2.129); vector p contains output of conventional units p_{it}, $\forall i$, $\forall t$ and continuous variables of pumped-storage units p_{st}^+, p_{st}^-, h_{st}, $\forall s$, $\forall t$; vector z^H stands for 0–1 variables z_{st}^+, z_{st}^-, $\forall s$, $\forall t$. Feasible region of the second-stage dispatch problem is

$$Y(x, w) = \{(p, z^H) \mid (2.131\text{a})\text{–}(2.131\text{c}), (2.143)\text{–}(2.145)\} \tag{2.147}$$

and involves only linear constraints. Notice that the objective function has a tri-level structure: the inner level $\min_{p \in Y(x,w)} d^T p$ determines the economic dispatch cost under a fixed UC x and renewable scenario w, which is then maximized over the uncertainty set W in the middle level with given x; the optimal UC is calculated in the outer-level by taking into account the worst-case response in the uncertainty set. In this way, the UC solution satisfies Definition 2.6, and the total cost in the worst scenario is minimized.

Problem (2.146) is an ARO with integer recourse decisions. Unlike the formulation in [133] with continuous recourse decisions, non-convexity introduced by discrete variable z^H prevents using of duality theory for the inner economic dispatch problem. A general method for solving such a challenging program is provided in Algorithm C.3 in Appendix C.2.3. However, it is a double-loop algorithm, and an MILP should be solved in both loops. So Algorithm C.3 may not be very efficient. In what follows, we discuss a practical treatment to overcome the difficulty brought by integer variable z^H.

Lower Bounding If $p_s^{n-} = p_s^{n+} = 0$, we can simply drop binary variables z_{st}^+, z_{st}^- and relax constraint (2.143e), and problem (2.146) comes down to an ARO with continuous recourses, which can be solved by Algorithm C.2 in Appendix C.2.3; It is revealed in [138, 139] that under certain conditions, relaxation of complementarity constraints for storage-like facilities provides an exact solution: i.e., $p_{st}^+ p_{st}^- = 0$, $\forall s$, $\forall t$ holds true at the optimal solution.

Upper Bounding If $p_s^{n-} > 0$ and $p_s^{n+} > 0$, or the conditions for exact relaxation are not met, the relaxed problem gives a lower bound of the optimal objective value, and the UC decision is x^*. Then we obtain a statistical bound of the optimal objective value in the following way:

1. Select arbitrary N extreme scenarios $w^1, w^2, \ldots, w^N$ in set W.
2. Solve MILP $\min_{(p,z^H) \in Y(x^*,w^n)} d^T p$, the optimal solution is $y^n, n = 1, 2, \ldots, N$.
3. Calculate a statistical bound as $c^T x^* + \max_n \{d^T y^n\}$.

Extreme scenarios in the first step can be retrieved by computing the vertices polytope W in (2.135) or the convex hull of W in (2.134). Since we don't have to enumerate all of them, the computation time will be acceptable. For some simpler uncertainty sets, such as the following one or its convex hull

$$W = \left\{ w \;\middle|\; \begin{array}{l} w_{jt} = w_{jt}^f + z_{jt}^+ w_{jt}^+ - z_{jt}^- w_{jt}^-, \ \forall j, \forall t \\ \sum_t \left(z_{jt}^+ + z_{jt}^- \right) \le \Gamma_{\mathrm{T}}, \ \forall j \\ z_{jt}^+, z_{jt}^- \in \{0, 1\}, \ \forall j, \ \forall t \end{array} \right\} \tag{2.148}$$

Scenarios can be produced in the following way:

For each renewable plant j in period t, generate a random variable rv subject to uniform distribution over [0,1]; assign

$$\begin{cases} z_{jt}^{+} = 0,\ z_{jt}^{-} = 1 & \text{if rv} \le \Gamma_{\mathrm{T}}/2T \\ z_{jt}^{+} = 1,\ z_{jt}^{-} = 0 & \text{if rv} \ge 1 - \Gamma_{\mathrm{T}}/2T \\ z_{jt}^{+} = 0,\ z_{jt}^{-} = 0 & \text{otherwise} \end{cases}$$

If $\sum_t (z_{jt}^{+} + z_{jt}^{-}) \le \Gamma_{\mathrm{T}}, \forall j$, a scenario is found; otherwise, go back and regenerate rv.

It should be pointed out that the bound obtained in step 3 is not necessarily a valid upper bound for the original problem (2.146), because the true worst-case scenario may be missed from the incomplete samples of extreme points. The computation time in this section corresponds to the Benders decomposition method in [132]. Next, we will present numerical experiments of robust UC on two data sets.

1. Numerical Tests on a 6-Bus System

As shown in Fig. 2.43, the 6-bus system includes four thermal generators at B_1, B_2, B_5, and B_6, a wind farm at B_4, a pumped-storage unit at B_3, and seven transmission lines. The loads are deterministic with mean values 50 MW at B_2, 80 MW at B_3, and 250 MW at B_6. Generator and transmission line data are described in Tables 2.29, 2.30 and 2.31. For the pumped-storage unit, we assume $p_s^{n-} = p_s^{n+} = 0$, $p_s^{m-} = p_s^{m+} = 37.5$ MW, and $\eta_s^{+} = \eta_s^{-} = 0.8$. Wind power uncertainty takes the form in (2.148). We assume that the budget of uncertainty is $\Gamma_{\mathrm{T}} = 6$. That is, we allow six periods in which the wind power output is away from its forecasted value.

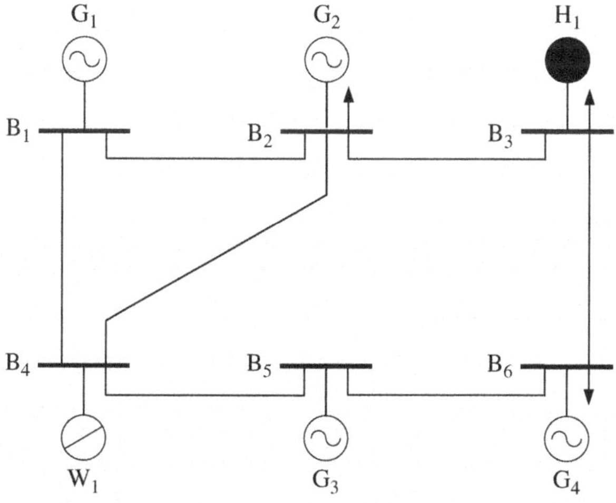

Fig. 2.43 Topology of the 6-bus system

Table 2.29 Generator performance data

Unit	p_i^n (MW)	p_i^m (MW)	T_i^{on} (h)	T_i^{off} (h)	R_i^+/R_i^- (MW/h)
G_1	50	250	8	8	125
G_2	50	200	8	8	100
G_3	30	80	4	4	40
G_4	25	100	5	5	50

Table 2.30 Generator cost data

Unit	a (\$/MWh^2)	b (\$/MWh)	c (\$)	SU_i (\$)
G_1	0.0024	12.33	28.00	1500
G_2	0.0044	13.29	39.00	1500
G_3	0.0459	15.47	74.33	100
G_4	0.0128	17.82	10.15	50

Table 2.31 Transmission line data

Line ID	From	To	Reactance	S_l (MW)
L_1	B_1	B_2	0.0370	200
L_2	B_1	B_4	0.0160	200
L_3	B_2	B_3	0.1015	175
L_4	B_2	B_4	0.1170	175
L_5	B_3	B_6	0.0355	175
L_6	B_4	B_5	0.0370	200
L_7	B_5	B_6	0.1270	200

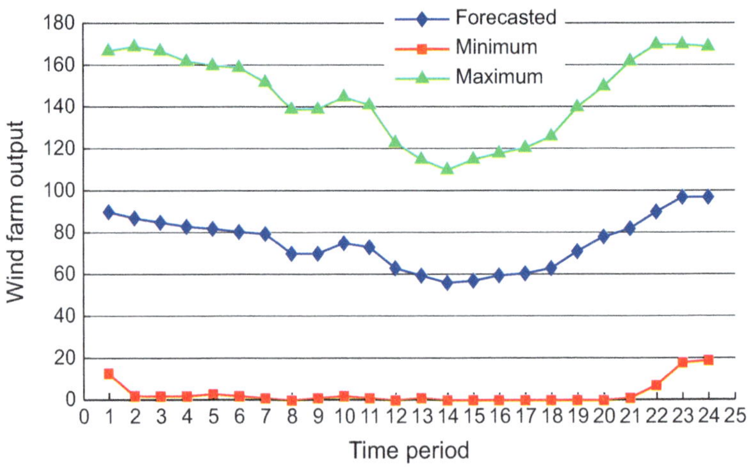

Fig. 2.44 Wind power prediction over the planning horizon

Further, the forecasted value and upper and lower limits of wind power output in each period are shown in Fig. 2.44, which is provided by the statistics from National Renewable Energy Laboratory (NREL).

The robust optimal UC decisions and pumped-storage operations in the worst-case scenario are collected in Table 2.32 with a minimum cost of \$111071.47. Since the wind power output can be considered as negative system load, we can define the net load as the difference of the load and the wind power. If we depict the net load under worst-case scenario and the corresponding water reserve level as in Fig. 2.45. From these results we observe that as time goes, starting from period 10, G_4 is turned on due to net load increment. We also see that the pumped-storage absorbs power when the net load is low (e.g., in periods 1 and 3–6) and generates power when the net load is high (e.g., in periods 8, 9, and 24 to avoid putting G_4 online, and in periods 20 and 21 to avoid putting G_3 online). With this observation, we can conclude that the pumped-storage unit works as a power buffer for the system, by charging/discharging in accordance with the net load information.

Then, we test the case without pumped-storage units. For this case, as compared to the previous one, the robust UC solution turns on G_3 from periods 20 to 23, and extends the on status for G_4 to periods 8, 9, and 24, in addition to the UC schedule for the case with pumped-storage; the corresponding optimal cost is \$111741.36. We

Table 2.32 Robust UC solution of the 6-bus system with a pumped-storage unit

	Thermal generators				Pumped-storage unit	
Period	G_1	G_2	G_3	G_4	p_{st}^- (MW)	p_{st}^+ (MW)
1	1	1	0	0	27.61	0
2	1	1	0	0	0	0
3	1	1	0	0	30.98	0
4	1	1	0	0	37.5	0
5	1	1	0	0	37.5	0
6	1	1	0	0	17.32	0
7	1	1	0	0	0	0
8	1	1	0	0	0	0.93
9	1	1	0	0	0	15.73
10	1	1	0	1	0	0
11	1	1	0	1	0	0
12	1	1	0	1	0	0
13	1	1	0	1	0	0
14	1	1	0	1	0	0
15	1	1	0	1	0	0
16	1	1	0	1	0	0
17	1	1	0	1	0	0
18	1	1	0	1	0	0
19	1	1	0	1	0	0
20	1	1	0	1	0	24.63
21	1	1	0	1	0	31.70
22	1	1	0	1	0	0
23	1	1	0	1	0	0
24	1	1	0	0	0	23.61

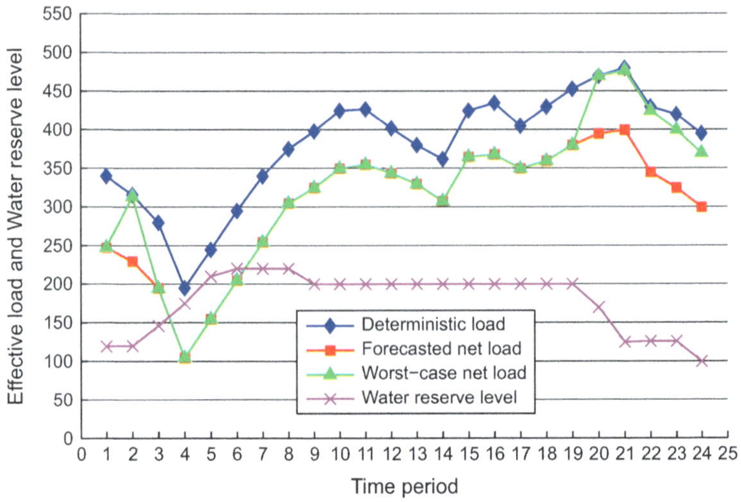

Fig. 2.45 Load and water reserve level curves

can observe that in the absence of pumped-storage units, the system has to commit more thermal generators to ensure desired reliability, which might incur a higher generation cost. In other words, the pumped-storage unit can act as substitutes of thermal generators on increasing the system robustness, and at the same time reduce greenhouse gas emissions.

2. Numerical Tests on the IEEE 118-Bus System

This test system is modified from the IEEE 118-bus system whose data is provided in [102]. In our tests, it has 33 thermal generators, 186 transmission lines, 1 wind farm, and 1 pumped-storage unit. The forecasted wind power output follows the pattern illustrated in Fig. 2.44. Wind power uncertainty takes the form in (2.148).

We first test the system performance under various uncertainty levels and hydro capacities. We allow the budget of uncertainty Γ_T to vary within the interval [2, 10]. We measure the upper bound of electricity the pumped-storage units can absorb or generate per period by its ratio over the average forecasted wind power output. In other words, the pumped-storage unit becomes more capable to absorb or generate power as the ratio increases. We also allow the ratio to vary within the interval [0, 0.5]. The computational results are shown in Tables 2.33 and 2.34.

In Table 2.33, LB and UB are obtained from the lower bounding and upper bounding methods described previously; WV represents the worst-case value of the deterministic UC: we first solve UC problem (2.132) with wind power prediction; Then we randomly generate a set of wind power output scenarios and solve the economic dispatch problem $\min_{p \in Y(x,w)} d^T p$ for each scenario with the obtained UC solution; WV is the sum of the UC cost and the largest dispatch cost in these scenarios. In the experiments, we found that the deterministic UC solutions are

Table 2.33 Computational results for the 118-bus system under various uncertainty levels and hydro capacities-optimal values and worst-case performances

		Γ_T				
Ratio		2	4	6	8	10
0	LB	1,098,380	1,119,588	1,139,856	1,159,481	1,178,612
	UB	1,099,017	1,120,627	1,140,920	1,160,591	1,179,579
	WV	1,188,076	1,278,192	1,346,562	1,396,429	1,419,221
0.1	LB	1,098,179	1,119,652	1,139,800	1,159,433	1,178,730
	UB	1,099,079	1,120,683	1,140,891	1,160,544	1,179,789
	WV	1,181,677	1,263,756	1,318,581	1,350,933	1,373,258
0.2	LB	1,098,066	1,119,464	1,139,909	1,159,593	1,178,482
	UB	1,098,968	1,120,263	1,140,818	1,160,574	1,179,461
	WV	1,186,535	1,252,199	1,308,528	1,348,942	1,371,907
0.3	LB	1,097,770	1,119,336	1,139,997	1,159,372	1,178,678
	UB	1,098,785	1,120,254	1,141,030	1,160,360	1,179,806
	WV	1,176,209	1,255,879	1,306,731	1,350,203	1,373,944
0.4	LB	1,098,360	1,119,681	1,139,928	1,159,409	1,178,648
	UB	1,099,200	1,120,689	1,140,872	1,160,487	1,179,783
	WV	1,165,907	1,231,310	1,277,833	1,311,111	1,334,535
0.5	LB	1,097,871	1,119,577	1,139,796	1,159,487	1,178,644
	UB	1,098,764	1,120,619	1,140,837	1,160,562	1,179,450
	WV	1,160,825	1,224,719	1,270,126	1,303,324	1,326,726

not able to maintain a feasible economic dispatch for sampled scenarios. This observation indicates that it would be risky to underestimate the uncertainty of wind power output and totally rely on the prediction. To compare the performance between these solutions, we introduce a linear penalty function for any unsatisfied demand or transmission capacity/ramp-rate limit violations, and the unit penalty cost is set to be \$7947/MWh. Correspondingly, we calculate the following gaps based on the lower and upper bounds:

1. Opt. Gap = (UB−LB)/LB×100%. It estimates the gap caused by relaxing the storage complementarity constraints.
2. WV. Gap = (WV−LB)/LB×100%. It estimates the difference between the performance of the robust and deterministic UC solutions under the worst-case scenario (the linear penalty function is introduced in calculating WV).

In the experiments, we first observe that statistically our algorithm provides a feasible solution for all instances with $\Gamma_T \in [2, 10]$ and Ratio $\in [0, 0.5]$. Besides, the optimality gaps are less than 0.1%, and the CPU times are less than 1 h in all the instances. This result demonstrates that our algorithm can provide a solution that is very close to the optimal one within a reasonable amount of time. In other words, the proposed method is able to solve large-scale problems, and obtain near-optimal solutions. We observe that WV gap increases with Γ_T increasing and Ratio

Table 2.34 Computational results for the 118-bus system under various uncertainty levels and hydro capacities-optimality gaps, worst-case value gaps, and CPU time

		Γ_T				
Ratio		2	4	6	8	10
0	Opt. Gap	0.06	0.09	0.09	0.10	0.08
	WV. Gap	8.10	14.06	18.02	20.32	20.32
	CPU time	194	353	578	1024	1876
0.1	Opt. Gap	0.08	0.09	0.10	0.10	0.09
	WV. Gap	7.52	12.77	15.57	16.41	16.40
	CPU time	101	309	546	997	542
0.2	Opt. Gap	0.08	0.07	0.08	0.08	0.08
	WV. Gap	7.97	11.78	14.70	16.23	16.32
	CPU time	145	331	908	1852	2053
0.3	Opt. Gap	0.09	0.08	0.09	0.09	0.10
	WV. Gap	7.05	12.11	14.52	16.36	16.46
	CPU time	178	361	736	1116	3101
0.4	Opt. Gap	0.08	0.09	0.08	0.09	0.10
	WV. Gap	6.07	9.87	12.00	12.98	13.12
	CPU time	218	463	1113	1446	1141
0.5	Opt. Gap	0.08	0.09	0.09	0.09	0.07
	WV. Gap	5.65	9.29	11.33	12.30	12.49
	CPU time	178	686	1114	995	3594

decreasing, indicating that the deterministic UC becomes worse with the wind power uncertainty increased and the usage of pumped-storage unit restricted.

Then, we test the system performance under various wind power penetration levels with $\Gamma_T = 6$ and hydro capacity Ratio $= 0.3$. Similar to the hydro capacity, we measure the wind power penetration level by the ratio of the nominal wind farm output over the forecasted value shown in Fig. 2.44. We allow the ratio to run from 0.5 to 1.5 and display in Table 2.35 the lower and upper bounds of the robust optimal value and the corresponding WV gaps under various penetration levels. In addition, we depict the gaps over different penetration levels in Fig. 2.46. From the computational results, we first observe that the proposed method provides tight lower and upper bounds, which make LB a reasonable estimate of the robust optimal value. Second, by comparing LB and WV over various wind penetration levels, we can see that LB decreases slightly with wind power increasingly penetrating into the power system, while WV gap grows significantly. This observation implies that robust optimization methods can generate reliable UC schedules which are not only immune against increasing wind power uncertainty, but also take advantage of the low cost wind energy.

Finally, we test the instances with various transmission line capacities with other parameters fixed (e.g., $\Gamma_T = 6$ and the hydro capacity Ratio $= 0.3$). Again, we measure the transmission line capacity by its ratio over the actual value for the given 118-bus system, and allow the ratio to run from 0.5 to 1.5. We report the

Table 2.35 Computational results for the 118-bus system under various wind power penetration levels-optimality values, gaps, and worst-case performance

Penetration level	LB	UB	WV	Opt. Gap (%)	WV. Gap (%)
0.5	1,219,531	1,220,460	1,241,838	0.08	1.75
0.6	1,202,975	1,203,919	1,231,012	0.08	2.25
0.7	1,186,992	1,187,780	1,232,395	0.07	3.76
0.8	1,171,434	1,172,319	1,242,946	0.08	6.02
0.9	1,155,710	1,156,736	1,259,052	0.09	8.85
1.0	1,139,997	1,141,030	1,306,731	0.09	14.52
1.1	1,125,240	1,126,301	1,490,924	0.09	32.37
1.2	1,111,187	1,112,237	1,614,526	0.09	45.16
1.3	1,098,023	1,098,926	1,812,235	0.08	64.91
1.4	1,085,091	1,086,036	1,919,229	0.09	76.72
1.5	1,072,821	1,073,839	2,412,151	0.09	124.63

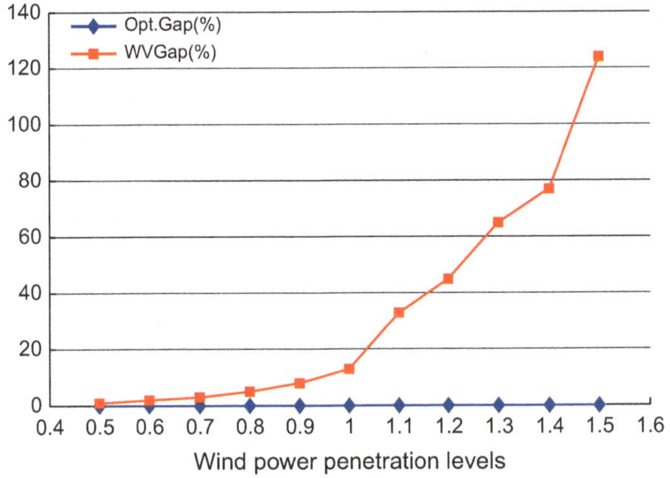

Fig. 2.46 Optimality gaps and worst-case performance under various wind power penetration levels

computational results in Table 2.36 and depict the gaps over different capacities in Fig. 2.47. In Table 2.36, we list the lower and upper bounds of the robust optimal objective value and the corresponding gaps under various transmission capacities. From the computational results, we first notice that the problem becomes infeasible when the transmission line capacity is sufficiently low (e.g., half of the actual value). Second, in view that the bounds we provide are tight, we observe that the robust optimal value decreases slightly with increasing transmission line capacities while WV gap decreases dramatically. This indicates that robust optimization method is insensitive to varying transmission line capacities, while the deterministic UC cost is considerably influenced by the capacity degradation.

Table 2.36 Computational results for the 118-bus system under various transmission line capacities—optimality values, gaps, and worst-case performance

Capacity rate	LB	UB	WV	Opt. Gap (%)	WV. Gap (%)
0.5	$+\infty$	$+\infty$	–	–	–
0.6	1,181,747	1,198,128	2,081,333	1.39	73.72
0.7	1,161,522	1,167,073	1,775,314	0.48	52.12
0.8	1,149,260	1,150,353	1,582,055	0.10	37.53
0.9	1,142,544	1,143,594	1,427,156	0.09	24.80
1.0	1,139,997	1,141,030	1,306,731	0.09	14.52
1.1	1,138,088	1,139,111	1,227,051	0.09	7.72
1.2	1,136,613	1,137,695	1,218,067	0.10	7.06
1.3	1,135,997	1,137,043	1,190,240	0.09	4.68
1.4	1,134,951	1,135,778	1,181,232	0.07	4.00
1.5	1,134,194	1,135,015	1,177,013	0.07	3.70

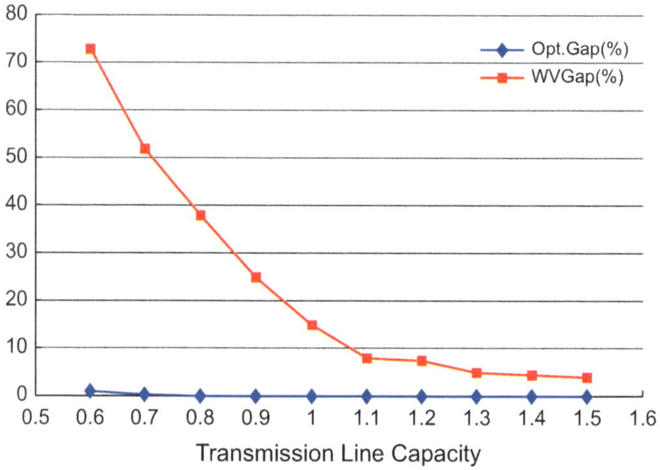

Fig. 2.47 Optimality gaps and worst-case performance under various transmission line capacities

2.7 Summary and Conclusions

This section attempts to cover power system operational issues within the timeframe of 1 day. It introduces applications of convex optimization and robust optimization approaches to tackle the non-convexity of power flow equations as well as the uncertainty arising from the volatility and intermittency of renewable energies. These approaches help procure reliable generation schedules in the most economic manner.

Section 2.2 introduces classical power flow models. Their attributes are summarized and compared in Table 2.37.

Table 2.37 Comparison of different power flow models

		BIM	BFM
Nonlinear model	Convex relaxation	SDP	SOCP
	Variables	Complex bus voltage (rectangular coordinates)	Voltage magnitude line power flow
	Potential usage	Transmission OPF	Distribution OPF
Linearized approximation	Variables and simplifications	Voltage angle voltage magnitude is 1 p.u. no reactive power, lossless	Voltage magnitude line power flow lossless
	Potential usage	Unit commitment energy-reserve dispatch	Distribution market volt/var control

Section 2.3 provides important supplementary materials for the emerging convex relaxation methods of the OPF problem, a fundamental issue in power system steady-state operation. The branch flow model based OPF formulation exhibits a DC structure, and can be solved by a sequential SOCP procedure. It makes no reference to a heuristic initial guess desired by local NLP algorithms, and removes the need of an exactness guarantee required by existing convex relaxation methods, while leveraging the computational superiority of SOCP. Three variants of OPF problems are discussed, covering several classical issues in power generation dispatch. In these regards, the proposed computation framework greatly enhances our ability to solve broader classes of OPF problems, and may become a promising computational tool for power system analysis, due to the elementary role of OPF. For example, it would be useful in distribution market studies, such as electricity pricing and strategic bidding, because calculating distribution locational marginal prices, the Lagrangian dual multipliers associated with nodal active power balancing equalities, relies on an OPF solution. Extensions to the meshed network and SDP relaxation method are also discussed, which might become a viable method for general convex relaxation of QCQPs.

Section 2.4 brings robust optimization into the joint energy and reserve dispatch and proposes two formulations based on different descriptions of the uncertainty. The first one uses an uncertainty set that covers possible realizations of uncertain data and minimizes the total cost in the worst-case scenario; the second one adopts a functional ambiguity set that includes probability distribution functions with the same forecast and variance. In theory, the latter one is expected to be less pessimistic than the first one, because dispersion effect is taken into account. It could become a useful alternative method for power system dispatch problems under uncertain renewable generations, and has the potential to fill the gap between the SO method and the RO method.

Section 2.5 establishes a solid theoretical foundation for the dispatchable region of renewable power generation. It is a security region defined in the uncertainty space, and is shown to be a polytope, which informs how much uncertainty can be dealt with. A constraint generation algorithm is developed to compute the dispatchable region based on solving MILPs. When the affine policy is adopted in real-time dispatch, the dispatchable region has a simple analytical expression. Given the dispatchable region and some statistics data, such as the moments, one can estimate the probability of an infeasible system operation without an exact probability distribution function through generalized probability inequalities, which can be solved via SDPs. Combining these outcomes, a method to optimize the energy and reserve dispatch is proposed based on shaping the dispatchable region. It renders an ARO with a variable uncertainty set, and provides a probability guarantee on the feasibility of real-time operation. The theory of dispatchable region constitutes a promising technique for quantifying the impact of uncertainty and enhancing system security.

Section 2.6 streamlines the mathematical formulation and computation methodology of robust UC problem, which minimizes the total cost under the worst disturbance fluctuation that might happen in real time, while ensuring high utilization of renewable power. A heuristic reserve adjustment approach is suggested to recover a robust UC schedule for power systems which is not heavily congested. Robust UC with pumped storage units is considered, in which the dispatch problem involves integer variables. An algorithm framework that can derive high quality lower and upper bounds of the optimal value is devised.

2.8 Further Reading

Due to the considerable economic benefits, the economic operation of power systems never fails to attract research attentions. More relevant literatures are provided for readers who wish to view a more comprehensive landscape and develop deeper understanding on the mathematical substances.

Convex optimization problems are computationally superior because any local optimizer is also a global one. SDPs and SOCPs are most commonly used convex programs, as they can be solved efficiently via interior point algorithms. However, mathematical models of real-world decision-making problems are often non-convex, which may have many local optimums. A local optimal solution barely shows how good the current solution is, and the global optimal one may significantly improve the objective value. To obtain a valid bound for the objective value with reasonable computation efforts, convex relaxation method is proposed. It enlarges the original feasible region to a convex set by replacing non-convex constraints with convex ones, such that the global optimal solution of the relaxed problem can be readily found. In general, the solution of the relaxed problem is not feasible for the original problem, since the constraint set becomes larger, and its optimal value gives a strict upper/lower bound for the original problem; nevertheless, if the solution of the relaxed problem is proved feasible in the original problem, it also solves the original one, and the relaxation is called exact or tight. Different relaxation methods can lead to different solutions. Readers who are interested in mathematical theory on convex relaxation methods may find extensive information in [65, 140, 141] and references therein for more thorough discussions. Convex relaxation applied to OPF problem can be found in [35–59], state estimation [142], transmission line switching [143–145], expansion planning issues [145–147], unit commitment with AC power flow constraints [148–150], strategic bidding in power markets [151], and generalized network flow problem [152].

To cope with renewable generation uncertainty in hourly-ahead and day-ahead generation scheduling problems, this chapter introduces the application of robust optimization approaches, one of which makes assumptions on the support set of the uncertain data, and the other considers ambiguity in the probability distribution functions. More information on the formulations and algorithms for these robust optimization problems can be found in Appendix C. Their applications in power system optimization problems are briefly introduced as follows.

The first category was firstly used in [153–155] following a static setting: all decisions are made before uncertain data can be observed. This is possible when uncertain data do not appear in an equality constraint (or two opposite inequalities), for example, the electricity price. However, renewable generation impacts the power balancing quality, and there must be some variables which can be adjusted in response to the realization of uncertain data. ARO which follows a dynamic setting was first applied to the unit commitment problem in [132–134], in which unit on-off status is decided with inexact data whereas unit output can be determined when uncertain data are observed. There have been various variations of robust unit commitment: [156–158] use specific uncertainty sets, [158–161] consider multiple stages, in particular, an infeasibility issue caused by the non-causal UC model is reported in [160], and addressed by imposing causality on the affine policy; [162] incorporates a minimax regret criterion; [163] integrates stochastic formulations in the robust formulation and thus can achieve a low expected total cost; nevertheless, it is essentially different from the distributionally robust approach as the probability distribution function is fixed; [164] treats contingencies as the source of uncertainty; [165] comprehensively models continuous and discrete uncertainty/dispatch strategies. From a computation perspective, discrete variables in the dispatch problem are more difficult to handle than those in the uncertainty set.

This kind of robust optimization approach has been tutored in Appendices C.1 and C.2 widely adopted in other decision-making problems in power systems, including generation/transmission expansion planning [166–169], storage devices investment [170], resilient planning against natural disasters [171], active distribution network operation with AC power flow constraints [172–174], economic dispatch [175–177], DCOPF [178, 179], and tie-line scheduling [180, 181], and restoration [182]. In each subject, we only cite a few representative articles. Many others can be found, and the volume of articles in each area is still growing.

The second category takes dispersion effect into consideration. More exactly, the PDF is assumed to be uncertain, and is restricted in a functional ambiguity set which is characterized by the moments and structured property of the underlying data, or the divergence between the PDF and a reference distribution. The model can be built based on min-max expectation or robust chance constraints. More discussions are provided in Appendices C.3 and C.4. Compared with the first category, uncertainty is treated more subtly, therefore, the conservatism can be reduced. However, it is usually challenging to derive a tractable equivalent mathematical program for an arbitrary ambiguity set; the known robust counterparts associated with specific ambiguity sets also differ a lot in their formulations.

This kind of robust optimization approach is relative new in power system studies. The framework is introduced into the stochastic robust unit commitment problem in [183] following a min-max expectation formulation. Computational issues of such unit commitment problems with different ambiguity sets have been studied in [184–187]. Similar idea has been applied to the OPF problem [188–195] and economic dispatch problem [196–198], in which chance constraints and CVaR with distribution uncertainty (described by Chebyshev ambiguity set or Gauss ambiguity set) are discussed. Other applications include reactive power optimization [199] and DG capacity assessment [200], storage device management [201], transmission expansion planning [202], and self-scheduling of combined-cycle units [203]. Distributionally robust optimization may incorporate two kinds of uncertainty sets. One is based on moments, the other is based on divergence. For more details, please see Appendices C.3 and C.4.

The dispatchable region has a close relationship with the uncertainty set in the robust optimization belonging to the first category. If a linear decision rule is adopted in the recourse stage, the dispatchable region maximization problem is a special kind of robust optimization problem with variable uncertainty set [204]. A concept similar to dispatchable region is the do-not-exceed (DNE) limit proposed in [205] and the flexibility measure defined in [206]. Geometrically, these two sets are inner-box approximations of the dispatchable region.

In summary, the mushrooming of robust optimization and convex optimization methods can be attributed to the better management of power systems with large-scale renewable generation by leveraging advanced optimization tools.

References

1. Wood, A.J., Wollenberg, B.F.: Power Generation, Operation, and Control, 2nd edn. Wiley, New York (1996)
2. Zhu, J.: Optimization of Power System Operation, 2nd edn. Wiley, Hoboken (2015)
3. Zimmerman, R.D. http://www.pserc.cornell.edu//matpower/
4. Wu, T., Alaywan, Z., Papalexopoulos, A.D.: Locational marginal price calculations using the distributed-slack power-flow formulation. IEEE Trans. Power Syst. **20**(2), 1188–1190 (2005)
5. Tong, S., Miu, K.N.: A network-based distributed slack bus model for DGs in unbalanced power flow studies. IEEE Trans. Power Syst. **20**(2), 835–842 (2005)
6. Tinney, W.F., Hart, C.E.: Power flow solution by Newton's method. IEEE Trans. Power App. Syst. **86**(11), 1449–1460 (1967)
7. Le Nguyen, H.: Newton-Raphson method in complex form. IEEE Trans. Power Syst. **12**(3), 1355–1359 (1997)
8. Stott, B., Alsaç, O.: Fast decoupled load flow. IEEE Trans. Power Appl. Syst. **93**(3), 859–869 (1974)
9. Wu, F.: Theoretical study of the convergence of the fast decoupled load flow. IEEE Trans. Power Appl. Syst. **96**(1), 268–275 (1977)

10. Van Amerongen, R.A.M.: A general-purpose version of the fast decoupled load flow. IEEE Trans. Power Syst. **4**(2), 760–770 (1989)
11. Monticelli, A.J., Garcia, A., Saavedra, O.R.: Fast decoupled load flow: hypothesis, derivations, and testing. IEEE Trans. Power Syst. **5**(4), 1425–1431 (1990)
12. Stott, B., Jardim, J., Alsaç, O.: DC power flow revisited. IEEE Trans. Power Syst. **24**(3), 1290–1300 (2009)
13. Coffrin, C., Van Hentenryck, P.: A linear-programming approximation of AC power flows. INFORMS J. Comput. **26**(4), 718–734 (2014)
14. Jabr, R.A.: Radial distribution load flow using conic programming. IEEE Trans. Power Syst. **21**(3), 1458–1459 (2006)
15. Esposito, A.G., Ramos, E.R.: Reliable load flow technique for radial distribution networks. IEEE Trans. Power Syst. **14**(3), 1063–1069 (1999)
16. Jabr, R.A.: A conic quadratic format for the load flow equations of meshed networks. IEEE Trans. Power Syst. **22**(4), 2285–2286 (2007)
17. Baran, M.E., Wu, F.F.: Network reconfiguration in distribution systems for loss reduction and load balancing. IEEE Trans. Power Delivery **4**(2), 1401–1407 (1989)
18. Baran, M.E., Wu, F.F.: Optimal capacitor placement on radial distribution systems. IEEE Trans. Power Delivery **4**(1), 725–734 (1989)
19. Baran, M.E., Wu, F.F.: Optimal sizing of capacitors placed on a radial distribution system. IEEE Trans. Power Delivery **4**(1), 735–743 (1989)
20. Eminoglu, U., Hocaoglu, M.H.: Distribution systems forward/backward sweep-based power flow algorithms: a review and comparison study. Electr. Power Compon. Syst. **37**(1), 91–110 (2008)
21. Yeh, H.G., Gayme, D.F., Low, S.H.: Adaptive VAR control for distribution circuits with photovoltaic generators. IEEE Trans. Power Syst. **27**(3), 1656–1663 (2012)
22. Emiroglu, S., Ozdemir, G., Baran, M.E.: Assessment of linear distribution feeder models used in optimization methods. In: Proceedings of IEEE Innovative Smart Grid Technologies Conference (2016)
23. Wang, S., Chen, S., Ge, L., Wu, L.: Distributed generation hosting capacity evaluation for distribution systems considering the robust optimal operation of OLTC and SVC. IEEE Trans. Sustainable Energy **7**(3), 1111–1123 (2016)
24. Quintana-Quiros, P., Yeh, H.G.: Power flow optimization methods for distribution systems. In: Proceedings of IEEE Green Energy and Systems Conference, pp. 18–23 (2015)
25. Carpentier, J.: Contribution to the economic dispatch problem. Bulletin de la Societe Francoise des Electriciens **3**(8), 431–447 (1962)
26. Capitanescu, F.: Critical review of recent advances and further developments needed in AC optimal power flow. Electr. Power Syst. Res. **136**, 57–68 (2016)
27. Wei, H., Sasaki, H., Kubokawa, J., Yokoyama, R.: An interior point nonlinear programming for optimal power flow problems with a novel data structure. IEEE Trans. Power Syst. **13**(3), 870–877 (1998)
28. Wu, Y.C., Debs, A.S., Marsten, R.E.: A direct nonlinear predictor-corrector primal-dual interior point algorithm for optimal power flows. IEEE Trans. Power Syst. **9**(2), 876–883 (1994)
29. Jabr, R.A., Coonick, A.H., Cory, B.J.: A primal-dual interior point method for optimal power flow dispatching. IEEE Trans. Power Syst. **17**(3), 654–662 (2002)
30. Granelli, G.P., Montagna, M.: Security-constrained economic dispatch using dual quadratic programming. Electr. Power Syst. Res. **56**(1), 71–80 (2000)
31. Boggs, P.T., Tolle, J.W.: Sequential quadratic programming for large-scale nonlinear optimization. J. Comput. Appl. Math. **124**(1-2), 123–137 (2000)
32. Boggs, P.T., Tolle, J.W.: Sequential quadratic programming. Acta Numer. **4**, 1–51 (1995)
33. Frank, S., Steponavice, I., Rebennack, S.: Optimal power flow: a bibliographic survey, I: formulations and deterministic methods. Energy Syst. **3**(3), 221–258 (2012)
34. Frank, S., Steponavice, I., Rebennack, S.: Optimal power flow: a bibliographic survey, II: nondeterministic and hybrid methods. Energy Syst. **3**(3), 259–289 (2012)

35. Bai, X., Wei, H., Fujisawa, K., Wang, Y.: Semidefinite programming for optimal power flow problems. Int. J. Electr. Power **30**(6-7), 383–392 (2008)
36. Lavaei, J., Low, S.H.: Zero duality gap in optimal power flow problem. IEEE Trans. Power Syst. **27**(1), 92–107 (2012)
37. Farivar, M., Low, S.H.: Branch flow model: relaxations and convexification. IEEE Trans. Power Syst. **28**(3), 2554–2572 (2013)
38. Wei, W., Wang, J., Li, N., Mei, S.: Optimal power flow of radial networks and its variations: a sequential convex optimization approach. IEEE Trans. Smart Grid **8**(6), 2974–2987 (2017)
39. Wei, W., Wang, J., Wu, L.: Distribution optimal power flow with real-time price elasticity. IEEE Trans. Power Syst. **31**(1), 1097–1098 (2018)
40. Low, S.H.: Convex relaxation of optimal power flow part I: formulations and equivalence. IEEE Trans. Control Network Syst. **1**(1), 15–27 (2014)
41. Low, S.H.: Convex relaxation of optimal power flow part II: exactness. IEEE Trans. Control Netw. Syst. **1**(2), 177–189 (2014)
42. Zhang, B., Tse, D.: Geometry of injection regions of power networks. IEEE Trans. Power Syst. **28**(2), 788–797 (2013)
43. Lavaei, J., Tse, D., Zhang, B.: Geometry of power flows and optimization in distribution networks. IEEE Trans. Power Syst. **29**(2), 572–583 (2014)
44. Sojoudi, S., Lavaei, J.: Exactness of semidefinite relaxations for nonlinear optimization problems with underlying graph structure. SIAM J. Optim. **24**(4), 1746–1778 (2014)
45. Madani, R., Sojoudi, S., Lavaei, J.: Convex relaxation for optimal power flow problem: mesh networks. IEEE Trans. Power Syst. **30**(1), 199–211 (2015)
46. Madani, R., Ashraphijuo, M., Lavaei, J.: Promises of conic relaxation for contingency-constrained optimal power flow problem. IEEE Trans. Power Syst. **31**(2), 1297–1307 (2016)
47. Bose, S., Low, S.H., Teeraratkul, T., Hassibi, B.: Equivalent relaxations of optimal power flow. IEEE Trans. Autom. Control **60**(3), 729–742 (2015)
48. Bose, S., Gayme, D.F., Chandy, K.M., Low, S.H.: Quadratically constrained quadratic programs on acyclic graphs with application to power flow. IEEE Trans. Control Netw. Syst. **2**(3), 278–287 (2015)
49. Molzahn, D.K., Hiskens, I.A.: Convex relaxations of optimal power flow problems: an illustrative example. IEEE Trans. Circuits Syst. Regul. Pap. **63**(5), 650–660 (2016)
50. Lasserre, J.B.: Moments, Positive Polynomials and Their Applications. Imperial College Press, London (2009)
51. Josz, C., Maeght, J., Panciatici, P., Gilbert, J.C.: Application of the moment-SOS approach to global optimization of the OPF problem. IEEE Trans. Power Syst. **30**(1), 463–470 (2015)
52. Molzahn, D.K., Hiskens, I.A.: Sparsity-exploiting moment-based relaxations of the optimal power flow problem. IEEE Trans. Power Syst. **30**(6), 3168–3180 (2015)
53. Ghaddar, B., Marecek, J., Mevissen, M.: Optimal power flow as a polynomial optimization problem. IEEE Trans. Power Syst. **31**(1), 539–546 (2016)
54. Tian, J., Wei, H., Tan, J.: Global optimization for power dispatch problems based on theory of moments. Int. J. Electr. Power **71**, 184–194 (2015)
55. Chen, C., Atamtürk, A., Oren, S.S.: Bound tightening for the alternating current optimal power flow problem. IEEE Trans. Power Syst. **31**(5), 3729–3736 (2016)
56. Phan, D.T.: Lagrangian duality and branch-and-bound algorithms for optimal power flow. Oper. Res. **60**(2), 275–285 (2012)
57. Li, N., Chen, L., Low, S.H.: Exact convex relaxation of OPF for radial networks using branch flow model. In: Proceedings of IEEE Third International Conference on Smart Grid Communications, Tainan, pp. 7–12 (2012)
58. Gan, L., Li, N., Topcu, U., Low, S.H.: Exact convex relaxation of optimal power flow in radial networks. IEEE Trans. Autom. Control **60**(1), 72–87 (2015)

59. Huang, S., Wu, Q., Wang, J., Zhao, H.: A sufficient condition on convex relaxation of AC optimal power flow in distribution networks. IEEE Trans. Power Syst. **32**(2), 1359–1368 (2017)
60. Kocuk, B., Dey, S.S., Sun, X.A.: Inexactness of SDP relaxation and valid inequalities for optimal power flow. IEEE Trans. Power Syst. **31**(1), 642–651 (2016)
61. Kocuk, B., Dey, S.S., Sun, X.A.: Strong SOCP relaxations for the optimal power flow problem. Oper. Res. **64**(6), 1177–1196 (2016)
62. Lipp, T., Boyd, S.: Variations and extension of the convex-concave procedure. Optim. Eng. **17**(2), 263–287 (2016)
63. Han, S.P., Mangasarian, O.L.: Exact penalty functions in nonlinear programming. Math. Program. **17**(1), 251–269 (1979)
64. Pillo, G.D., Grippo, L.: Exact penalty functions in constrained optimization. SIAM J. Control Optim. **27**(6), 1333–1360 (1989)
65. Burer, S., Saxena, A.: The MILP Road to MIQCP. In: Lee J., Leyffer S. (eds.) Mixed Integer Nonlinear Programming. The IMA Volumes in Mathematics and its Applications, vol. 154. Springer, New York (2012)
66. Recht, B., Fazel, M., Parrilo, P.A.: Guaranteed minimum rank solutions to linear matrix equations via nuclear norm minimization. SIAM Rev. **52**(3), 471–501 (2010)
67. Bukhsh, W.A., Grothey, A., McKinnon, K.I., Trodden, P.A.: Local solutions of the optimal power flow problem. IEEE Trans. Power Syst. **28**(4), 4780–4788 (2013)
68. Ajjarapu, V., Christy, C.: The continuation power flow: a tool for steady state voltage stability analysis. IEEE Trans. Power Syst. **7**(1), 416–423 (1992)
69. Chiang, H.D., Flueck, A.J., Shah, K.S., Balu, N.: CPFLOW: a practical tool for tracing power system steady-state stationary behavior due to load and generation variations. IEEE Trans. Power Syst. **10**(2), 623–634 (1995)
70. Li, S.H., Chiang, H.D.: Nonlinear predictors and hybrid corrector for fast continuation power flow. IET Gener. Transm. Distrib. **2**(3), 341–354 (2008)
71. Irisarri, G.D., Wang, X., Tong, J., Mokhtari, S.: Maximum loadability of power systems using interior point nonlinear optimization method. IEEE Trans. Power Syst. **12**(1), 162–172 (1997)
72. Avalos, R.J., Canizares, C.A., Milano, F., Conejo, A.J.: Equivalency of continuation and optimization methods to determine saddle-node and limit-induced bifurcations in power systems. IEEE Trans. Circuits-I **56**(1), 210–223 (2009)
73. Molzahn, D., Lesieutre, B., DeMarco, C.: A sufficient condition for power flow insolvability with applications to voltage stability margins. IEEE Trans. Power Syst. **28**(3), 2592–2601 (2013)
74. Molzahn, D., Hiskens, I., Lesieutre, B.: Calculation of voltage stability margins and certification of power flow insolvability using second-order cone programming. In: Proceedings of 49th Hawaii International Conference on System Sciences, Kauai, Hawaii, pp. 2307–2316 (2016)
75. http://shgl.curent.utk.edu/toolsdemo
76. https://sites.google.com/site/burakkocuk/research
77. https://www.artelys.com/tools/knitro_doc/2_userGuide.html
78. Tawarmalani, M. Sahinidis, N.V.: A polyhedral branch-and-cut approach to global optimization. Math. Program. **103**(2), 225–249 (2005)
79. Marler, R.T., Arora, J.S.: Survey of multi-objective optimization methods for engineering. Struct. Multidisc. Optim. **26**(6), 369–395 (2004)
80. Caramia, M., Dell'Olmo, P.: Multi-objective optimization. In: Multi-Objective Management in Freight Logistics. Springer, Boston, ch. 2, pp. 11–36 (2008)
81. Das, I., Dennis, J.E.: Normal-boundary intersection: a new method for generating the pareto surface in nonlinear multicriteria optimization problems. SIAM J. Optim. **8**(3), 631–657 (1998)
82. Zhou, A., Qu, B., Li, H., Zhao, S., Suganthan, P., Zhang, Q.: Multiobjective evolutionary algorithms: a survey of the state of the art. Swarm Evol. Comput. **1**(1), 32–49 (2011)

83. Roman, C., Rosehart, W.: Evenly distributed Pareto points in multiobjective optimal power flow. IEEE Trans. Power Syst. **21**(2), 1011–1012 (2006)
84. Wächter, A., Biegler, L.T.: On the implementation of a primal-dual interior point filter line search algorithm for large-scale nonlinear programming. Math. Program. **106**(1), 25–57 (2006)
85. Siano, P., Sarno, D.: Assessing the benefits of residential demand response in a real time distribution energy market. Appl. Energy **161**, 533–551 (2016)
86. Liu, Y., Li, J., Wu, L.: Distribution system restructuring: distribution LMP via unbalanced ACOPF. IEEE Trans. Smart Grid **9**(5), 4038–4048 (2018)
87. Papavasiliou, A.: Analysis of distribution locational marginal prices. IEEE Trans. Smart Grid **9**(5), 4872–4882 (2018)
88. Yuan, Z., Hesamzadeh, M.R., Biggar, D.: Distribution locational marginal pricing by convexified ACOPF and hierarchical dispatch. IEEE Trans. Smart Grid **9**(4), 3133–3142 (2018)
89. Alvarado, F.: The stability of power system markets. IEEE Trans. Power Syst. **14**(2), 505–511 (1999)
90. Alvarado, F., Meng, J., DeMarco, C.L., Mota, W.S.: Stability analysis of interconnected power systems coupled with market dynamics. IEEE Trans. Power Syst. **16**(4), 695–701 (2001)
91. Bejestani, A.K., Annaswamy, A.: A dynamic mechanism for wholesale energy market: stability and robustness. IEEE Trans. Smart Grid **5**(6), 2877–2888 (2014)
92. Roozbehani, M., Dahleh, M., Mitter, S.K.: Volatility of power grid under real-time pricing. IEEE Trans. Power Syst. **27**(4), 1926–1940 (2012)
93. Zhao, Z., Wu, L., Song, G.: Convergence of volatile power markets with price-based demand response. IEEE Trans. Power Syst. **29**(5), 2107–2118 (2014)
94. Granas, A., Dugundji, J.: Fixed Point Theory. Springer, New York (2003)
95. Conejo, A.J., Castillo, E., Mínguez, R., Milano, F.: Locational marginal price sensitivities. IEEE Trans. Power Syst. **20**(4), 2026–2033 (2005)
96. Carta, J.A., Ramirez, P., Velazquez, S.: A review of wind speed probability distributions used in wind energy analysis: case studies in the Canary Islands. Renew. Sust. Energy Rev. **13**(5), 933–955 (2009)
97. Delage, E., Ye, Y.: Distributionally robust optimization under moment uncertainty with application to data-driven problems. Oper. Res. **58**(3), 595–612 (2010)
98. Bertsimas, D., Doan, X., Natarajan, K., Teo, C.: Models for minimax stochastic linear optimization problems with risk aversion. Math. Oper. Res. **35**(3), 580–602 (2010)
99. Wei, W., Liu, F., Mei, S., Hou, Y.: Robust energy and reserve dispatch under variable renewable generation. IEEE Trans. Smart Grid **6**(1), 369–380 (2015)
100. Wei, W., Liu, F., Mei, S.: Distributionally robust co-optimization of energy and reserve dispatch. IEEE Trans. Sustainable Energy **7**(1), 289–300 (2016)
101. Wu, L.: A tighter piecewise linear approximation of quadratic cost curves for unit commitment problems. IEEE Trans. Power Syst. **26**(4), 2581–2583 (2011)
102. http://motor.ece.iit.edu/data/JEAS_IEEE118.doc
103. https://docs.mosek.com/modeling-cookbook/index.html
104. https://www.ibm.com/analytics/data-science/prescriptive-analytics/cplex-optimizer
105. Guan, Y., Wang, J.: Uncertainty sets for robust unit commitment. IEEE Trans. Power Syst. **29**(3), 1439–1440 (2014)
106. Foley, A.M., Leahy, P.G., Marvuglia, A., McKeogh, E.J.: Current methods and advances in forecasting of wind power generation. Renew. Energy **37**(1), 1–8 (2012)
107. Wei, W., Liu, F., Mei, S.: Dispatchable region of the variable wind generation. IEEE Trans. Power Syst. **30**(5), 2755–2765 (2015)
108. Wei, W., Liu, F., Mei, S.: Real-time dispatchability of bulk power systems with volatile renewable generations. IEEE Trans. Sustain. Energy **6**(3), 738–747 (2015)
109. Wei, W., Li, N., Wang, J., Mei, S.: Estimating the probability of infeasible real-time dispatch without exact distributions of stochastic wind generations. IEEE Trans. Power Syst. **31**(6), 5022–5032 (2016)

110. Wei, W., Wang, J., Mei, S.: Dispatchability maximization for cooptimized energy and reserve dispatch with explicit reliability guarantee. IEEE Trans. Power Syst. **31**(4), 3276–3288 (2016)
111. Huynh, T., Lassez, C., Lassez, J.: Practical issues on the projection of polyhedral sets. Ann. Math. Artif. Intell. **6**(4), 295–315 (1992)
112. Herceg, M., Kvasnica, M., Jones, C.N., Morari, M.: Multi-Parametric Toolbox 3.0. In: Proceedings of the European Control Conference, Zurich, Switzerland, pp. 502–510 (2013)
113. Ardakani, A., Bouffard, F.: Identification of umbrella constraints in DC-based security-constrained optimal power flow. IEEE Trans. Power Syst. **28**(4), 3924–3934 (2013)
114. Khachiyan, L., Boros, E., Borys, K., Elbassioni, K., Gurvich, V.: Generating all vertices of a polyhedron is hard. Discr. Comput. Geom. **39**(1), 174–190 (2008)
115. Wan, C., Lin, J., Guo, W., Song, Y.: Maximum uncertainty boundary of volatile distributed generation in active distribution network. IEEE Trans. Smart Grid **9**(4), 2930–2942 (2018)
116. Dvijotham, K., Turitsyn, K., Construction of power flow feasibility sets. Tech. Rep. arXiv:1506.07191 (2015)
117. Nguyen, H., Dvijotham, K., Turitsyn, K.: Construction of convex inner approximations of optimal power flow feasibility sets. Tech. Rep. arXiv:1708.06845 (2017)
118. Vandenberghe, L., Boyd, S., Comanor, K.: Generalized Chebyshev bounds via semidefinite programming. SIAM Rev. **49**(1), 52–64 (2007)
119. Bertsimas, D., Popescu, I.: Optimal inequalities in probability theory: a convex optimization approach. SIAM J. Optim. **15**(3), 780–804 (2005)
120. Parys, B.V., Goulart, P.J., Kuhn, D.: Generalized Gauss inequalities via semidefinite programming. Math. Program. **156**(1), 271–302 (2016)
121. Löfberg, J.: Pre- and post-processing sum-of-squares programs in practice. IEEE Trans. Autom. Control **54**(5), 1007–1011 (2009)
122. Constantinescu, E.M., Zavala, V.M., Rocklin, M., Lee, S., Anitescu, M.: A computational framework for uncertainty quantification and stochastic optimization in unit commitment with wind power generation. IEEE Trans. Power Syst. **26**(1), 431–441 (2011)
123. Petra, C.G., Zavala, V.M., Nino-Ruiz, E.D., Anitescu, M.: Economic impacts of wind covariance estimation on power grid operations. Tech. Rep. Argonne National Laboratory, Argonne, IL, ANL/MCS-P5148-0614 (2014)
124. Ma, X., Sun, Y., Fang, H.: Scenario generation of wind power based on statistical uncertainty and variability. IEEE Trans. Sustain. Energy **4**(4), 894–904 (2013)
125. Usaola, J.: Probabilistic load flow in systems with wind generation. IET Gener. Transm. Distrib. **3**(12), 1031–1041 (2009)
126. Jaleeli, N., VanSlyck, L.S., Ewart, D.N., Fink, L.H., Hoffmann, A.G.: Understanding automatic generation control. IEEE Trans. Power Syst. **7**(3), 1106–1122 (1992)
127. Jabr, R.A.: Adjustable robust OPF with renewable energy sources. IEEE Trans. Power Syst. **28**(4), 4742–4751 (2013)
128. Warrington, J., Goulart, P., Mariethoz, S., Morari, M.: Policy-based reserves for power systems. IEEE Trans. Power Syst. **28**(4), 4427–4437 (2013)
129. Bertsimas, D., Iancu, D.A., Parrilo, P.A.: Optimality of affine policies in multistage robust optimization. Math. Oper. Res. **35**(2), 363–394 (2010)
130. Iancu, D.A., Sharma, M., Sviridenko, M.: Supermodularity and affine policies in dynamic robust optimization. Oper. Res. **61**(4), 941–956 (2013)
131. Ben-Tal, A., Goryashko, A., Guslitzer, E., Nemirovski, A.: Adjustable robust solutions of uncertain linear programs. Math. Program. **99**(2), 351–376 (2004)
132. Jiang, R., Wang, J., Guan, Y.: Robust unit commitment with wind power and pumped storage hydro. IEEE Trans. Power Syst. **27**(2), 800–810 (2012)
133. Bertsimas, D., Litvinov, E., Sun, X.A., Zhao, J., Zheng, T.: Adaptive robust optimization for the security constrained unit commitment problem. IEEE Trans. Power Syst. **28**(1), 52–63 (2013)
134. Zhao, L., Zeng, B.: Robust unit commitment problem with demand response and wind energy. In: Proceeedings of IEEE Power and Energy Society General Meeting, San Diego, USA (2012)

135. Wei, W., Liu, F., Mei, S., En, L.: Two-level unit commitment and reserve level adjustment considering large-scale wind power integration. Int. Trans. Electr. Energy Syst. **24**(12), 1726–1746 (2014)
136. Morales-España, G., Latorre, J.M., Ramos, A.: Tight and compact MILP formulation of start-up and shut-down ramping in unit commitment. IEEE Trans. Power Syst. **28**(2), 1288–1296 (2013)
137. Tahanan, M., Van Ackooij, W., Frangioni, A., Lacalandra, F.: Large-scale unit commitment under uncertainty. 4OR **13**(2), 115–171 (2015)
138. Li, Z., Guo, Q., Sun, H., Wang, J.: Sufficient conditions for exact relaxation of complementarity constraints for storage-concerned economic dispatch. IEEE Trans. Power Syst. **31**(2), 1653–1654 (2016)
139. Li, Z., Guo, Q., Sun, H., Wang, J.: Storage-like devices in load leveling: complementarity constraints and a new and exact relaxation method. Appl. Energy 151, 13–22 (2015)
140. Bao, X., Sahinidis, N.V., Tawarmalani, M.: Semidefinite relaxations for quadratically constrained quadratic programming: a review and comparisons. Math. Program. **129**(1), 129–157 (2011)
141. Luo, Z.Q., Ma, W.K., So, A.M.C., Ye, Y., Zhang, S.: Semidefinite relaxation of quadratic optimization problems. IEEE Signal Process. Mag. **27**(3), 20–34 (2010)
142. Zhang, Y., Madani, R., Lavaei, J.: Conic relaxations for power system state estimation with line measurements. IEEE Trans. Control Network Syst. **5**(3), 1193–1205 (2018)
143. Kocuk, B., Dey, S.S., Sun, X.: New formulation and strong MISOCP relaxations for AC optimal transmission switching problem. IEEE Trans. Power Syst. **32**(6), 4161–4170 (2017)
144. Fattahi, S., Lavaei, J., Atamturk, A.: Promises of conic relaxations in optimal transmission switching of power systems. In: IEEE 56th Annual Conference on Decision and Control, pp. 3238–3245 (2017)
145. Hijazi, H., Coffrin, C., Van Hentenryck, P.: Convex quadratic relaxations for mixed-integer nonlinear programs in power systems. Math. Program. Comput. **9**(3), 321–367 (2017)
146. Li, Q., Ayyanar, R., Vittal, V.: Convex optimization for DES planning and operation in radial distribution systems with high penetration of photovoltaic resources. IEEE Trans. Sustainable Energy **7**(3), 985–995 (2016)
147. Taylor, J.A., Hover, F.S.: Conic AC transmission system planning. IEEE Trans. Power Syst. **28**(2), 952–959 (2013)
148. Bai, Y., Zhong, H., Xia, Q., Kang, C., Xie. L.: A decomposition method for network-constrained unit commitment with AC power flow constraints. Energy **88**, 595–603 (2015)
149. Fattahi, S., Ashraphijuo, M., Lavaei, J.,Atamturk, A.: Conic relaxations of the unit commitment problem. Energy **134**, 1079–1095 (2017)
150. Liu, J., Castillo, A., Watson, J.P., Laird, C.: Global solution strategies for the network-constrained unit commitment (NCUC) problem with nonlinear AC transmission models. arXiv:1801.07218 (2018)
151. Ghamkhari, M., Sadeghi-Mobarakeh, A., Mohsenian-Rad, H.: Strategic bidding for producers in nodal electricity markets: a convex relaxation approach. IEEE Trans. Power Syst. **32**(3), 2324–2336 (2017)
152. Sojoudi, S., Fattahi, S., Lavaei, J.: Convexification of generalized network flow problem. Math. Program. **173**, 353–391 (2019)
153. Baringo, L., Conejo, A.J.: Offering strategy via robust optimization. IEEE Trans. Power Syst. **26**(3), 1418–1425 (2011)
154. Hajimiragha, A.H., Canizares, C.A., Fowler, M.W., Moazeni, S., Elkamel, A.: A robust optimization approach for planning the transition to plug-in hybrid electric vehicles. IEEE Trans. Power Syst. **26**(4), 2264–2274 (2011)
155. Chen, Z., Wu, L., Fu, Y.: Real-time price-based demand response management for residential appliances via stochastic optimization and robust optimization. IEEE Trans. Smart Grid **3**(4), 1822–1831 (2012)
156. An, Y., Zeng, B.: Exploring the modeling capacity of two-stage robust optimization: variants of robust unit commitment model. IEEE Trans. Power Syst. **30**(1), 109–122 (2015)

157. Hu, B., Wu, L., Marwali, M.: On the robust solution to SCUC with load and wind uncertainty correlations. IEEE Trans. Power Syst. **29**(6), 2952–2964 (2014)
158. Lorca, A., Sun, X.A.: Multistage robust unit commitment with dynamic uncertainty sets and energy storage. IEEE Trans. Power Syst. **32**(3), 1678–1688 (2017)
159. Zhao, C., Wang, J., Watson, J.P., Guan, Y.: Multi-stage robust unit commitment considering wind and demand response uncertainties. IEEE Trans. Power Syst. **28**(3), 2708–2717 (2013)
160. Lorca, A., Sun, X.A., Litvinov, E., Zheng, T.: Multistage adaptive robust optimization for the unit commitment problem. Oper. Res. **64**(1), 32–51 (2016)
161. Warrington, J., Hohl, C., Goulart, P.J., Morari, M.: Rolling unit commitment and dispatch with multi-stage recourse policies for heterogeneous devices. IEEE Trans. Power Syst. **31**(1), 187–197 (2016)
162. Jiang, R., Wang, J., Zhang, M., Guan, Y.: Two-stage minimax regret robust unit commitment. IEEE Trans. Power Syst. **28**(3), 2271–2282 (2013)
163. Zhao, C., Guan, Y.: Unified stochastic and robust unit commitment. IEEE Trans. Power Syst. **28**(3), 3353–3361 (2013)
164. Wang, Q., Watson, J.P., Guan, Y.: Two-stage robust optimization for N-k contingency-constrained unit commitment. IEEE Trans. Power Syst. **28**(3), 2366–2375 (2013)
165. Hu, B., Wu, L.: Robust SCUC considering continuous/discrete uncertainties and quick-start units: a two-stage robust optimization with mixed-integer recourse. IEEE Trans. Power Syst. **31**(2), 1407–1419 (2016)
166. Dehghan, S., Amjady, N., Kazemi, A.: Two-stage robust generation expansion planning: a mixed integer linear programming model. IEEE Trans. Power Syst. **29**(2), 584–597 (2014)
167. Chen, B., Wang, J., Wang, L., He, Y., Wang, Z.: Robust optimization for transmission expansion planning: minimax cost vs. minimax regret. IEEE Trans. Power Syst. **29**(6), 3069–3077 (2014)
168. Moreira, A., Street, A., Arroyo, J.M.: An adjustable robust optimization approach for contingency-constrained transmission expansion planning. IEEE Trans. Power Syst. **30**(4), 2013–2022 (2015)
169. Jabr, R.A.: Robust transmission network expansion planning with uncertain renewable generation and loads. IEEE Trans. Power Syst. **28**(4), 4558–4567 (2013)
170. Jabr, R.A., Dzafic, I., Pal, B.C.: Robust optimization of storage investment on transmission networks. IEEE Trans. Power Syst. **30**(1), 531–539 (2015)
171. Yuan, W., Wang, J., Qiu, F., Chen, C., Kang, C., Zeng, B.: Robust optimization-based resilient distribution network planning against natural disasters. IEEE Trans. Smart Grid **7**(6), 2817–2826 (2016)
172. Ding, T., Liu, S., Yuan, W., Bie, Z., Zeng, B.: A two-stage robust reactive power optimization considering uncertain wind power integration in active distribution networks. IEEE Trans. Sustainable Energy **7**(1), 301–311 (2016)
173. Ding, T., Li, C., Yang, Y., Jiang, J., Bie, Z., Blaabjerg, F.: A two-stage robust optimization for centralized-optimal dispatch of photovoltaic inverters in active distribution networks. IEEE Trans. Sustainable Energy **8**(2), 744–754 (2017)
174. Gao, H., Liu, J., Wang, L.: Robust coordinated optimization of active and reactive power in active distribution systems. IEEE Trans. Smart Grid **9**(5), 4436–4447 (2018)
175. Li, Z., Wu, W., Zhang, B., Wang, B.: Adjustable robust real-time power dispatch with large-scale wind power integration. IEEE Trans. Sustainable Energy **6**(2), 357–368 (2015)
176. Lorca, A., Sun, X.A.: Adaptive robust optimization with dynamic uncertainty sets for multi-period economic dispatch under significant wind. IEEE Trans. Power Syst. **30**(4), 1702–1713 (2015)
177. Zugno, M., Conejo, A.J.: A robust optimization approach to energy and reserve dispatch in electricity markets. Eur. J. Oper. Res. **247**(2), 659–671 (2015)
178. Jabr, R.A.: Adjustable robust OPF with renewable energy sources. IEEE Trans. Power Syst. **28**(4), 4742–4751 (2013)
179. Jabr, R.A., Karaki, S., Korbane, J.A.: Robust multi-period OPF with storage and renewables. IEEE Trans. Power Syst. **30**(5), 2790–2799 (2015)

180. Li, Z., Wu, W., Zeng, B., Shahidehpour, M., Zhang, B.: Decentralized contingency-constrained tie-line scheduling for multi-area power grids. IEEE Trans. Power Syst. **32**(1), 354–367 (2017)
181. Li, Z., Shahidehpour, M., Wu, W., Zeng, B., Zhang, B., Zheng, W.: Decentralized multiarea robust generation unit and tie-line scheduling under wind power uncertainty. IEEE Trans. Sustainable Energy **6**(4), 1377–1388 (2015)
182. Chen, X., Wu, W., Zhang, B.: Robust restoration method for active distribution networks. IEEE Trans. Power Syst. **31**(5), 4005–4015 (2016)
183. Liu, C., Lee, C., Chen, H., Mehrotra, S.: Stochastic robust mathematical programming model for power system optimization. IEEE Trans. Power Syst. **31**(1), 821–822 (2016)
184. Xiong, P., Jirutitijaroen, P., Singh, C.: A distributionally robust optimization model for unit commitment considering uncertain wind power generation. IEEE Trans. Power Syst. **32**(1), 39–49 (2017)
185. Zhao, C., Guan, Y.: Data-driven stochastic unit commitment for integrating wind generation. IEEE Trans. Power Syst. **31**(4), 2587–2596 (2016)
186. Gourtani, A., Xu, H., Pozo, D., Nguyen, T.: Robust unit commitment with n-1 criteria. Math Method. Oper. Res. **83**(3), 373–408 (2016)
187. Duan, C., Jiang, L., Fang, W., Liu, J.: Data-driven affinely adjustable distributionally robust unit commitment. IEEE Trans. Power Syst. **33**(2), 1385–1398 (2018)
188. Summers, T., Warrington, J., Morari, M., Lygerosa, J.: Stochastic optimal power flow based on conditional value at risk and distributional robustness. Int. J. Electr. Power **72**, 116–125 (2015)
189. Zhang, Y., Shen, S., Mathieu, J.L.: Distributionally robust chance-constrained optimal power flow with uncertain renewables and uncertain reserves provided by loads. IEEE Trans. Power Syst. **32**(2), 1378–1388 (2017)
190. Lubin, M., Dvorkin, Y., Backhaus, S.: A robust approach to chance constrained optimal power flow with renewable generation. IEEE Trans. Power Syst. **31**(5), 3840–3849 (2016)
191. Li, B., Jiang, R., Mathieu, J.L.: Distributionally robust risk-constrained optimal power flow using moment and unimodality information. In: Proceedings of IEEE 55th Conference on Decision and Control, pp. 2425–2430 (2016)
192. Xie, W., Ahmed, S.: Distributionally robust chance constrained optimal power flow with renewables: a conic reformulation. IEEE Trans. Power Syst. **33**(2), 1860–1867 (2018)
193. Guo, Y., Baker, K., Dall'Anese, E., Hu, Z., Summerset, T.: Stochastic optimal power flow based on data-driven distributionally robust optimization. In: IEEE Annual American Control Conference, pp. 3840–3846 (2018)
194. Mieth, R., Dvorkin, Y.: Data-driven distributionally robust optimal power flow for distribution systems. IEEE Contr. Syst. Lett. **2**(3), 363–368 (2018)
195. Duan, C., Fang, W., Jiang, L., Yao, L., Liu, J.: Distributionally robust chance-constrained approximate AC-OPF with Wasserstein metric. IEEE Trans. Power Syst. **33**(5), 4924–4936 (2018)
196. Bian, Q., Xin, H., Wang, Z., Gan, D., Wong, K.P.: Distributionally robust solution to the reserve scheduling problem with partial information of wind power. IEEE Trans. Power Syst. **30**(5), 2822–2823 (2015)
197. Wang, Z., Bian, Q., Xin, H., et al.: A distributionally robust co-ordinated reserve scheduling model considering CVaR-based wind power reserve requirements. IEEE Trans. Sustainable Energy **7**(2), 625–636 (2016)
198. Tong, X., Luo, X., Yang, H., Zhang, L.: A distributionally robust optimization based risk limiting dispatch in power system under moment uncertainty. Int. Trans. Electr. Energy Syst. **27**(8), e2343 (2017)
199. Ding, T., Yang, Q., Yang, Y., Li, C., Bie, Z., Blaabjerg, F.: A data-driven stochastic reactive power optimization considering uncertainties in active distribution networks and decomposition method. IEEE Trans. Smart Grid **9**(5), 4994–5004 (2018)
200. Chen, X., Wu, W., Zhang, B., Lin, C.: Data-driven DG capacity assessment method for active distribution networks. IEEE Trans. Power Syst. **32**(5), 3946–3957 (2017)

201. Darivianakis, G., Eichler, A., Smith, R.S., Lygeros, J.: A data-driven stochastic optimization approach to the seasonal storage energy management. IEEE Control Syst. Lett. **1**(2), 394–399 (2017)
202. Bagheri, A., Wang, J., Zhao, C.: Data-driven stochastic transmission expansion planning. IEEE Trans. Power Syst. **32**(5), 3461–3470 (2017)
203. Pan, K., Guan, Y.: Data-driven risk-averse stochastic self-scheduling for combined-cycle units. IEEE Trans. Ind. Inf. **13**(6), 3058–3069 (2017)
204. Zhang, X., Kamgarpour, M., Georghiou, A., Goulart, P., Lygerosa, J.: Robust optimal control with adjustable uncertainty sets. Automatica **75**, 249–259 (2017)
205. Zhao, J., Zheng, T., Litvinov, E.: Variable resource dispatch through do-not-exceed limit. IEEE Trans. Power Syst. **30**(2), 820–828 (2015)
206. Zhao, J., Zheng, T., Litvinov, E.: A unified framework for defining and measuring flexibility in power system. IEEE Trans. Power Syst. **31**(1), 339–347 (2016)

Chapter 3
Integrated Gas-Electric System

3.1 Introduction

During the past decades, the environmental policy, public awareness of sustainable development, as well as technology breakthroughs in shale gas production [1] have led to the upsurge of natural gas usage in electricity industry [2]. In 2014, 43.5% of global gas production was consumed by gas-fired units for electricity generation [3]. Gas-fired units can be classified into three types:

Steam turbine plants, in which natural gas is burnt to heat water and create steam, and the steam powers a turbine to generate electricity. These plants typically have thermal efficiencies between 30% and 35%.

Combustion turbine plants, in which natural gas is combusted to turn the blades of a turbine (like a jet engine) that drives a generator. The typical sizes of these plant range from 100 to 400 MW with thermal efficiencies between 35% and 40%.

Combined cycle plants, which elaborately integrates the combustion turbine and the steam turbine, have thermal efficiencies between 50% and 60%, because the heat extracted from a gas-fired combustion turbine is utilized to produce steam to power the steam turbine. Its high efficiency also indicates less carbon dioxide emission producing per unit power.

Besides the high efficiency, gas-fired units are also advantageous for their lower capital costs, faster installation, nearly-zero pollutant emissions, and rapid start up/shut down capability, which makes them an appropriate option for base load, intermediate, and peak regulation units, as well as backup resources for volatile renewable generations.

Given large reserves of natural gas resources in shale formations, coal beds, and even methane clathrate formations [4], a subsequent increase in natural gas production is witnessed followed by the price drop [5], which is expected to stabilize the price of global gas market. Meanwhile, the technological advances of gas-fired units would boost their usage in electrical power generation. Experts believe that natural gas will quickly become the largest source of US electricity generation [6].

W. Wei, J. Wang, *Modeling and Optimization of Interdependent Energy Infrastructures*, https://doi.org/10.1007/978-3-030-25958-7_3

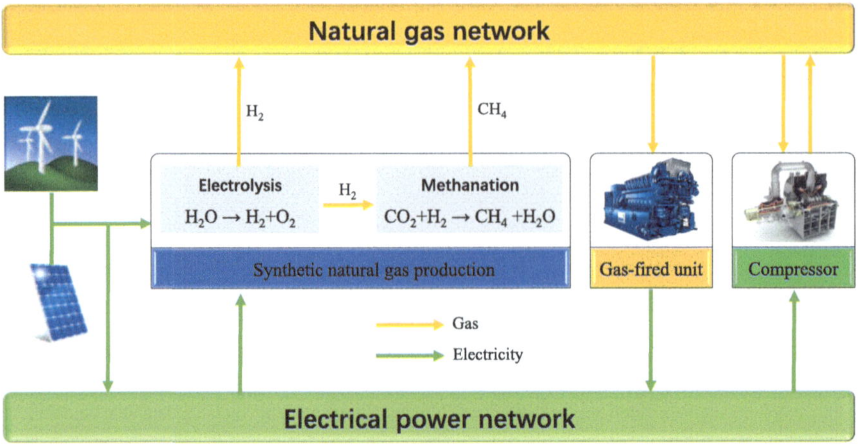

Fig. 3.1 Interdependency between natural gas and power systems

The growing reliance of natural gas in electricity generation has brought the increasing level of interdependence between two capital energy sectors [2, 7, 8]. Understanding the interactions in the gas and electricity supply chain is crucially important for government authorities, system designers, operators, and related stakeholders who are jointly responsible for the overall energy policy and system security. Figure 3.1 illustrates natural gas and electrical power infrastructures coupled by gas-fired units, compressors, and P2G facilities. The bidirectional energy conversion gives occasion to the interdependency.

In the gas-to-power direction, energy conversion is completed in gas-fired units, which play a critical role in the coupled networks. From the power system side, the operator executes UC and OPF to determine cost minimum energy contracts in the day-ahead market and real-time power delivery in the spot market. The price of natural gas directly influences the commitment, dispatch, and profit of the gas-fired unit. From the gas network side, local gas distribution companies historically buy long-term firm contracts without significant intra-day variations [9]. However, gas-fired plants which participate in the power market purchase natural gas according to the short-term and interruptible contracts. The time-varying gas consumption rates of these plants causes pressure fluctuations in gas transmission pipelines, which may have negative impacts on the safety and reliability of gas delivery [10]. Pressure interruptions and the loss of gas supply may also cause outages of multiple gas-fired units and load shedding, compromising the power system security. Although gas storage facilities can offer backup capacity in certain cases, the generation schedule and market decisions would be dramatically affected by the gas availability. With the rapid growing number of gas-fired units, the impact of gas network security on power system operation can no longer be neglected [7], and exerts challenges that require creative incentives to precipitate cross-institutional coordinations.

In the power-to-gas direction, P2G technology allows the conversion of surplus electricity to hydrogen via water electrolysis. A fraction of hydrogen can be injected back into the gas network [11] or used by fuel cell vehicles, and the remainder is further processed by methanation with a suitable carbon source, carbon dioxide for example, to produce synthetic natural gas. Technology and economics of P2G transit are thoroughly reviewed in [12]. The ease of storage of gas resources in tanks or pipelines (the latter is called line pack effect) in conjugation with P2G devices provide a flexible measure to mitigate power system supply-demand imbalance in real time [13].

When gas moves through a pipeline, its pressure would drop due to friction. Hence compressors must be deployed along transmission pipes (usually every 50–100 miles) to boost the gas pressure and maintain constant gas flows. A compressor may be powered by either natural gas or electricity. Optimal placement and operation of compressor stations in gas transmission systems could lead to considerable cost saving and improve the market competition [7]. Given the growing volume and capacity of P2G and compressor stations, the interplay between the natural gas system and the power system in the transmission level will become more prominent, and the electricity price could be influenced by the dispatch strategies of P2G plants and compressor stations. Meanwhile, the gas system operator ought to consider their market power while bidding in the electricity market.

From the spatial point of view, gas and power system interdependence can happen in both transmission and distribution levels. In the transmission level, coupling components include large gas-fired units and P2G facilities. In the distribution level, the two systems are usually connected at the demand side through an energy hub. From the temporal point of view, the transient dynamics in the power system is much faster than that in the gas system, so the steady-state electric power flow model is sufficient for analyzing system interdependence. The gas flow dynamics are described by PDEs, which can be approximated by different means, which will be elaborated later.

This chapter endeavors to establish a holistic modeling framework and analytical tool to address the planning, operation, and marketization issues of the interdependent natural gas and electric power infrastructure. Materials in this chapter originate from authors' publications [14–17]. In particular, the mathematical formulations of the gas flow in pipeline network are introduced in Sect. 3.2. The optimal gas-power flow (OGPF) problem is addressed in Sect. 3.3 via convex optimization and ADMM methods; locational marginal energy prices based bilateral gas-electricity market is studied in Sect. 3.4, and a best-response decomposition algorithm is proposed to compute the market equilibrium based upon convex optimization models of the OGF and the OPF; strategic bidding problems of market participants in the gas-electricity market are presented in Sect. 3.5, in which simplified power flow and gas flow models are adopted to compromise tractability and model accuracy; finally, reinforcing vulnerable components to improve system resilience under malicious attacks is discussed in Sect. 3.6 via robust optimization.

3.2 Mathematical Model of the Natural Gas Network Flow

Natural gas is mined from gas wells and transported to customers at various locations through pipeline networks. A natural gas transmission network is composed of pipes, storage facilities, compressor stations, and auxiliary devices including valves and regulators. Because of the friction between fast moving gas and pipe inner walls, gas pressure drops along the transit direction. Moreover, gas temperature declines over time due to heat dissipativity. Therefore, compressor stations are operated to compensate for energy losses and facilitate gas delivering. Compressors are either powered by electric motors or gas turbines, consuming a significant amount of energy. To manage the gas transmission network in a secure and economic manner, it is imperative to investigate and monitor the gas flow rates and nodal pressures in the pipeline system. Depending on whether the gas flow status changes over time, we distinguish system condition by steady (transient) state if it does not (does) depend on time. In steady (transient) state case, the system model can be described by nonlinear algebraic (partial differential) equations. This section will be mainly devoted to the introduction on steady-state gas flow model elaborated in [18]. We illuminate how the model can be extended to approximate the line pack effect, a relative slow transient process which can be regarded as a sort of storage capability.

3.2.1 Network Components and Topology

Indispensable components for modeling a natural gas system include: gas sources, gas loads, connection points (which are represented by nodes), pipelines and compressor stations (which are represented by branches). To distinguish them, ordinary pipes and compressors are sometimes referred to as passive branches and active branches, respectively. Variable π_i represents the gas pressure at node i. In a steady-state model, the inlet and outlet flows of every passive pipeline are equal. A triplet (i, j, k) where i, j appear in subscript and k appears in superscript labels a variable associated with branch k from node i to node j. Variable f_{ij}^k represents the gas flow rate through branch k. In the line pack model, each pipeline is associated with two gas flow rates: f_{ij}^{kh} at the head node and f_{ij}^{kr} at the tail node, and their difference determines the change in the total amount of gas stored in branch k.

Figure 3.2 shows an example of a natural gas network, which contains two sources at nodes 1 and 2, three loads at nodes 7, 8, and 9, six passive branches, and two active branches (compressor arcs). In order to categorize the topologies of gas pipeline networks, the network topology is reduced in the following manner [19]: first, compressor branches are temporarily removed, and the network is divided into several isolated islands; then the components in each island are merged into an aggregated node; finally, compressor branches are placed back to their places. The reduced network topology is illustrated in the same figure, which has three aggregated nodes and two branches.

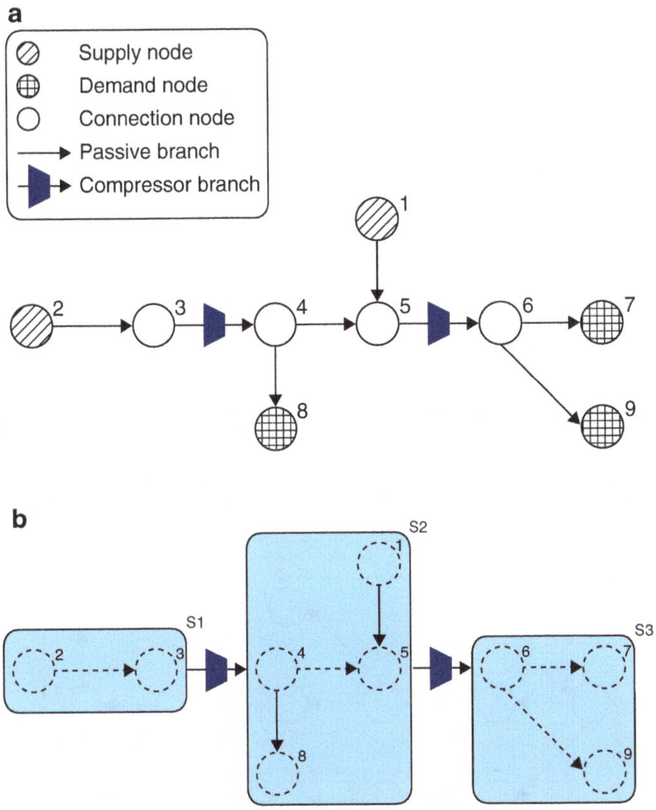

Fig. 3.2 Topology of a simple pipeline network and its reduced network [19]. (**a**) Original network. (**b**) Associated reduced network

Based upon the reduced network topology, gas pipeline networks can be classified into three categories [19]: linear network, if the reduced network is a chain; radial network, if the reduced network is a tree; cyclic network, if the reduced network includes cycles. These three types of topologies have been exhibited through Fig. 3.3, where the original components are plotted by solid nodes and arcs, and the dashed boxes represent aggregated nodes of the reduced network. In Fig. 3.3a, b, although cycles do appear in the original network, they are categorized as non-cyclic network structures, because the network flow can be analyzed through the reduced topology without jeopardizing the physical details of the original system under certain technical assumptions. If the reduced graph has cycles, the flow rates cannot be uniquely determined [20].

The network reduction method stated above offers a rigorous technique to reduce the problem size for large-scale instances in case of need. The discussion in the rest of this section will be restricted to the original network.

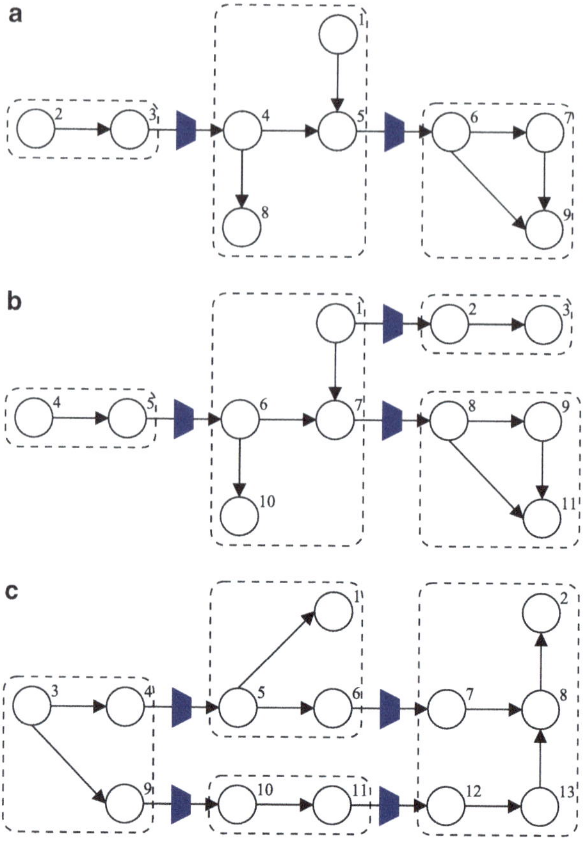

Fig. 3.3 Categories of pipeline network topologies [19]. (**a**) Linear topology. (**b**) Tree topology. (**c**) Cyclic topology

3.2.2 *Matrix Representation of the Network*

Gas pipeline networks usually consist of a large number of components. Matrix notation turns out to be a convenient way to express the network topology. Consider a simple case whose graph representation is shown in Fig. 3.4.

The source node 1 is selected as the reference node, the gas pressure at the reference node is specified. Let $g_i = g_i^s - g_i^d$ be the total gas injection rate at node i, where positive scalar g_i^s (g_i^d) denotes gas supply (demand) at node i. Connection nodes without any load or source have zero injections. They are often used to represent a junction of several arcs. Take the system in Fig. 3.4 for example, the network injection vector is given by

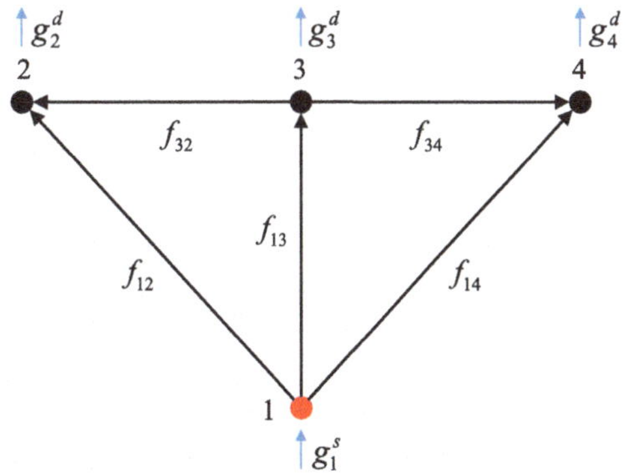

Fig. 3.4 Graphic representation of a gas pipeline network [18]

$$g = \begin{bmatrix} g_1 \\ g_2 \\ g_3 \\ g_4 \end{bmatrix} = \begin{bmatrix} g_1^s \\ -g_2^d \\ -g_3^d \\ -g_4^d \end{bmatrix}$$

To define a matrix representation for the network topology, each branch is assigned with a direction which is assumed to be consistent with the reference direction of gas flow in that branch. The interconnection of network components can be described by the node-arc incidence matrix $A \in \mathbb{M}^{N_B \times N_L}$, where N_B and N_L are the number of nodes and branches, respectively. Its element A_{ij} corresponding to node i and branch j is defined according to the following rule:

$$A_{ij} = \begin{cases} +1, & \text{if branch } j \text{ enters node } i \\ -1, & \text{if branch } j \text{ leaves node } i \\ 0, & \text{if branch } j \text{ does not connect to node } i \end{cases}$$

For example, the node-arc incidence matrix for the system in Fig. 3.4 is

$$A = \begin{bmatrix} -1 & -1 & -1 & 0 & 0 \\ 1 & 0 & 0 & 1 & 0 \\ 0 & 1 & 0 & -1 & -1 \\ 0 & 0 & 1 & 0 & 1 \end{bmatrix}$$

An important fact is that the sum of row vectors of matrix A is always a zero vector, implying that matrix A does not have full rank. In fact, the rank of matrix A is $N_B - 1$. This deficiency can be circumvented by removing the row corresponding to a reference node of the gas system.

3.2.3 Compressor Model

Compressor model describes the relations among the nodal pressures, gas flow rates, and energy consumption in an active branch. Most current compressor stations consist of centrifugal compressor units operated in parallel. In a centrifugal compressor, gas is discharged from an impeller to a diffuser at a very high velocity. The kinetic energy of gas is then converted to pressure. The working details of a compressor can be very complicated and involve the knowledge of thermodynamics. Here we just concern about the static operating status.

A compressor fueled by natural gas insulated from a pipeline is shown in Fig. 3.5. Its inlet (outlet) is node m (n) with gas pressure π_m (π_n). We do not distinguish the pressures at the injection node m and the diffluence node m_c in the compressor station because they are close to each other; otherwise, if they are distant from each other, the pipeline between m and m_c can be isolated and regarded as a passive pipeline. f_{mn} is the gas flow rate at the outlet. τ_{mn} denotes the amount of natural gas consumed by the compressor, so the inflow is given by $f_{mn} + \tau_{mn}$.

A key economic aspect of a compressor is the horsepower, which depends on the pressure boost ratio and the gas flow rate. Such a function for adiabatic compression process can be expressed by [21]

$$H_{mn} = k_c \frac{f_{mn}}{\eta_c} \frac{\alpha}{\alpha - 1} \left[\left(\frac{\pi_n}{\pi_m} \right)^{\frac{\alpha-1}{\alpha}} - 1 \right] \tag{3.1}$$

where k_c is a modification factor which depends on the gas temperature at node m_c, η_c is the compressor efficiency, α is the heat capacity ratio. It is assumed that the natural gas consumption rate can be approximated by a quadratic function

$$\tau_{mn} = a_g + b_g H_{mn} + c_g H_{mn}^2 \tag{3.2}$$

of the horsepower. If the compressor is driven by an electric motor, the power demand can be expressed in the same way, i.e.

$$p_{mn} = a_e + b_e H_{mn} + c_e H_{mn}^2 \tag{3.3}$$

and the fuel branch does not exist, or

$$\tau_{mn} = 0 \tag{3.4}$$

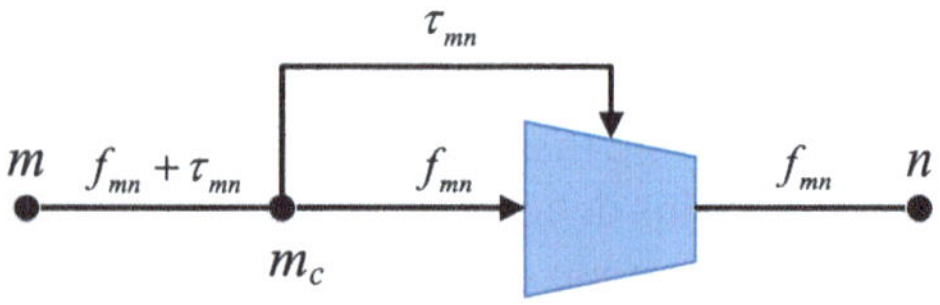

Fig. 3.5 Gas flow in a compressor

The consumption formula (3.1) sometimes can be simplified or approximated. For example, if a compressor is operated with a fixed boost ratio, i.e., π_n/π_m is constant, then H_{mn} would be a linear function in the gas flow rate f_{mn}. Another case is to approximate H_{mn} via a PWL function, if variables vary in fairly large ranges.

To build a matrix network flow model, matrix $T \in \mathbb{M}^{N_B \times N_L}$ is used to specify where gas is withdrawn from the network to power the compressor. Its element T_{ij} corresponding to node i and branch j is defined as follows:

$$T_{ij} = \begin{cases} +1, & \text{if the compressor in branch } j \text{ gets fuel from node } i \\ 0, & \text{otherwise} \end{cases}$$

For example, for the system in Fig. 3.4, if branch 1–3 stands for a compressor station, and the gas is supplied from node 1, the matrix T will be

$$T = \begin{bmatrix} 0 & 1 & 0 & 0 & 0 \\ 0 & 0 & 0 & 0 & 0 \\ 0 & 0 & 0 & 0 & 0 \\ 0 & 0 & 0 & 0 & 0 \end{bmatrix}$$

3.2.4 *Passive Pipeline Model*

The gas flow dynamics in a one-dimensional pipe are given by the isothermal Euler equations, a partial differential equation (PDE) set [21, 22]

$$\frac{\partial \rho}{\partial t} + \frac{\partial(\rho u)}{\partial x} = 0 \tag{3.5a}$$

$$\frac{\partial(\rho u)}{\partial t} + \frac{\partial(\rho u^2)}{\partial x} + \frac{\partial \pi}{\partial x} = -\frac{\rho u|u|}{2D}\lambda \tag{3.5b}$$

where u, π, ρ represent velocity, pressure, and density as functions in position x and time t, respectively; D is the diameter of pipe, λ is the friction factor. Equation (3.5a) interprets the conservation of mass, and (3.5b) states the momentum balance. The right-hand side of (3.5b) represents the friction losses in a pipe. On the left-hand side, the first term $\partial_t(\rho u)$ and second term $\partial_x(\rho u^2)$ can be interpreted as gas inertia and advection, which are usually small compared with the friction loss, and can be neglected. Furthermore, according to the thermodynamic equation of ideal gas, π and ρ has relation

$$\pi = \frac{ZRT}{M_g}\rho \tag{3.5c}$$

where Z is the natural gas compressibility factor; R is the universal gas constant; T is the absolute temperature; M_g is the gas molecular weight. We assume that gas temperature is almost a constant along the pipe, and (3.5c) gives a linear equation. With above notations and simplifications, (3.5) can be arranged into the following equations in terms of the gas pressure π and the mass flow rate $f = \rho u$

$$\frac{\partial \pi}{\partial t} = -\frac{ZRT}{M_g}\frac{\partial f}{\partial x} \tag{3.6a}$$

$$\frac{\partial \pi^2}{\partial x} = 2\pi\frac{\partial \pi}{\partial x} = -\frac{ZRT\lambda}{DM_g} f|f| \tag{3.6b}$$

Since we are discussing steady-state condition, variables do not change over time, $\partial_t = 0$; according to (3.6a), we have $\partial_x f = 0$, so PDE set (3.6) has the following analytical solution

$$f = \text{constant}, \ \pi_0^2 - \pi(x)^2 = axf|f| \tag{3.7}$$

where constant $a = ZRT\lambda/DM_g$.

Now suppose the length of passive pipeline between nodes m and n is L, the gas pressure π_m (π_n) at the head (tail) node and the gas flow f_{mn} must follow relation

$$\pi_m^2 - \pi_n^2 = aLf_{mn}|f_{mn}| \tag{3.8}$$

which is known as the Weymouth equation. Modifications with temperature and gravity calibrations can be found in [18, 21]. Other approximations for the solution of Euler equation (3.5) under different assumptions and simplifications are summarized in [23].

Steady-state condition assumes equal inflow and outflow in every pipeline on a moment-by-moment basis. However, this is rather restrictive in practice. When the inflow and outflow are not balanced, excessive (insufficient) gas will be stored (withdrawn) in (from) the pipeline, which is known as the line pack effect, as a result, the gas pressure changes correspondingly. Harnessing line pack effect could considerably improve the flexibility in gas production, transportation, and short-term operation of the power system under the help of P2G facilities. Exactly modeling pressure transients and propagations entails solving PDE (3.5), which is a challenging task. In view that line pack is a relative slow process, approximation models can be developed based on the finite difference technique, and will be briefly introduced below. More information can be found in [24–27].

Consider pipe k from node m to node n whose volume is V_k and cross-sectional area S_k, recall the thermodynamic equation of ideal gas, then the average pressure $\bar{\pi}_t^k$ and total mass of gas m_t^k stored in the pipe in period t obey the relation

$$\bar{\pi}_t^k V_k = \frac{ZRT}{M_g} m_t^k \tag{3.9}$$

where $\bar{\pi}_t^k$ can be approximated by nodal gas pressures π_{mt}, π_{nt} in period t as

$$\bar{\pi}_t^k = \frac{\pi_{mt} + \pi_{nt}}{2} \tag{3.10}$$

Let f_{mnt}^{kb} and f_{mnt}^{kr} be the inflow and outflow rate at the head node and tail node, the mass dynamics can be written as

$$m_{t+1}^k = m_t^k + (f_{mnt}^{kb} - f_{mnt}^{kr}) S_k \Delta T \tag{3.11}$$

where ΔT is the duration of period t.

Analogously, the Weymouth equation can be expressed via

$$\pi_m^2 - \pi_n^2 = aL \bar{f}_t^k |\bar{f}_t^k| \tag{3.12}$$

where the average gas flow $\bar{f}_t^k$ is approximated by

$$\bar{f}_t^k = \frac{f_{mnt}^{kb} + f_{mnt}^{kr}}{2} \tag{3.13}$$

Equations (3.9)–(3.13) describe the dynamics of nodal gas pressures and branch flows over time. Because the change of variables in a large pipeline network is relatively slow, the reliance of (3.12) on the steady-state solution (3.8) is reasonable and these line-pack equations are quasi steady-state. The accuracy depends on the quality of average approximations (3.10) and (3.13). To improve accuracy, a long pipe can be divided into a series of shorter pipes; In each pipe, the gas temperature is constant, and the arithmetic means (3.10) and (3.13) provide satisfactory approximations for the average pressure $\bar{\pi}_t^k$ and gas flow $\bar{f}_t^k$.

3.2.5 Network Flow Model

Recall the node-arc incidence matrix A and the compressor connection incidence matrix T, the flow balancing equation at each node can be written in a compact form as

$$Af + g - T\tau = 0 \tag{3.14a}$$

For the example given in Fig. 3.4, (3.14a) renders

$$\begin{bmatrix} -1 & -1 & -1 & 0 & 0 \\ 1 & 0 & 0 & 1 & 0 \\ 0 & 1 & 0 & -1 & -1 \\ 0 & 0 & 1 & 0 & 1 \end{bmatrix} \begin{bmatrix} f_{12} \\ f_{13} \\ f_{14} \\ f_{32} \\ f_{34} \end{bmatrix} + \begin{bmatrix} g_1^s \\ -g_2^d \\ -g_3^d \\ -g_4^d \end{bmatrix} - \begin{bmatrix} 0 & 1 & 0 & 0 & 0 \\ 0 & 0 & 0 & 0 & 0 \\ 0 & 0 & 0 & 0 & 0 \\ 0 & 0 & 0 & 0 & 0 \end{bmatrix} \begin{bmatrix} 0 \\ \tau_{13} \\ 0 \\ 0 \\ 0 \end{bmatrix} = 0$$

which can be easily verified.

For each passive pipeline, the nodal gas pressures and pipeline gas flows follow Weymouth equation

$$\pi_i^2 - \pi_j^2 = aLf_{ij}^k |f_{ij}^k|, \ \forall (i, j, k) \in L_P \tag{3.14b}$$

where L_P stands for the set of passive pipelines.

For each (gas-fired) compressor station, the gas consumption rate and inlet/outlet pressure follow

$$\left. \begin{aligned} H_{ij}^k &= k_c \frac{f_{ij}^k}{\eta_c} \frac{\alpha}{\alpha - 1} \left[\left(\frac{\pi_j}{\pi_i} \right)^{\frac{\alpha-1}{\alpha}} - 1 \right] \\ \tau_{ij} &= a_g^k + b_g^k H_{ij}^k + c_g^k H_{ij}^{k2} \end{aligned} \right\}, \ \forall (i, j, k) \in L_C \tag{3.14c}$$

where L_C stands for the set of active pipelines. For electricity powered compressor stations, the power demand is given by (3.3).

Formulation (3.14) is called the natural gas network flow model. In a load flow study, to balance the number of unknown variables and the number of equations, nodes are classified into those with given injections and with given pressures. The former often include sources (with positive injections), loads (with negative injections), and junctions (with zero injections), and the corresponding gas pressures are to be determined; the latter are usually associated with sources and compressor stations, and the corresponding injections are to be determined. Compressors are operated in one of the following modes: constant inflow; constant outlet pressure; constant inlet pressure; constant boost ratio. For more details, please see [18].

This chapter mainly tackles optimization problems in which (3.14) will be treated as constraints. Compressors are operated with constant boost ratio, such that (3.14c) can be simplified; for passive pipelines, non-convexity arises in the Weymouth equation (3.14b), which is a major barrier for developing efficient solution algorithms.

3.3 Optimal Gas-Power Flow

In view of the interaction between the power system and natural gas system, optimal gas-power flow (OGPF) is a fundamental problem for coordinated system operation, just as the role of OPF problem in power system operation. Existing work investigates the OGPF problem with simplified network models: For the power system, many researchers adopt the linear DC power flow model for better computational performance; others may retain the nonlinear AC power flow model for better accuracy while sacrificing tractability. For the natural gas system, the steady-state assumption is widely adopted, i.e., the in-flow and out-flow of each pipeline are equal. Moreover, the nonlinear Weymouth equation is either approximated by MILP based PWL functions, or directly handled by NLP solvers. As the operating conditions are becoming increasingly stressed, more precise models are in great need, and thus call for reliable computational methods. In addition, power systems and gas systems are operated by different institutions. A distributed operating framework that allows operators take charge of their own systems is highly desired.

This section studies the multi-period OGPF problem. The power grid has radial topology and the power flow is described by the BFM discussed in Sect. 2.2.2, whose convex relaxation is proved to give the global optimal solution to the exact AC power flow equations. The optimal gas flow (OGF) incorporates line pack dynamics. The gas flow directions are assumed to be known in advance. The Weymouth equation is represented by a DC function. A sequential convex optimization approach is developed to solve the OGF problem. The ADMM scheme is adopted to solve the whole OGPF problem in a distributed manner. In this way, important computational issues mentioned above have been addressed.

3.3.1 Mathematical Formulation of OGPF

Most symbols and notations used in the mathematical model are defined below for quick reference. Others are defined following their first appearances in case of need.

Sets and Indices

$c \in C$	Gas compressors (Gas active pipelines)
$d_g \in D_g$	Gas loads
$d_p \in D_p$	Electric loads
$g \in G$	Gas-fired units
$i_g \in I_g$	Nodes in the gas system
$l_g \in L_g$	Gas passive pipelines
$i_p \in I_p$	Buses in the power grid
$l_p \in L_p$	Branches in the power grid

$n \in N$	Non-gas units
$t \in T$	Time periods
$w \in W$	Gas retailers

Parameters

p_g^{min}/p_g^{max}	Active power range of gas-fired units
p_n^{min}/p_n^{max}	Active power range of non-gas units
q_g^{min}/q_g^{max}	Reactive power range of gas-fired units
q_n^{min}/q_n^{max}	Reactive power range of non-gas units
p_{d_pt}/q_{d_pt}	Power grid active/reactive load
Q_{wt}^{gas}	Natural gas price
R_{l_g}	Pipeline diameter
$S_{l_p}^{max}$	Square maximum of line current magnitude
r_{l_p}/x_{l_p}	Power grid line resistance/reactance
T_k	Gas temperature
$U_{i_p}^{min}/U_{i_p}^{max}$	Square range of bus voltage magnitude
X_{l_g}	Length of gas pipelines
y_c^{max}	Maximum allowed gas in-flow of compressors
y_{d_gt}	Gas system loads
y_w^u	Supply capacity of gas wells
Z_{l_g}	Compression factor of pipelines
β_g	Gas-electricity conversion factor
χ_c	Electricity consumption factor of compressor
λ	Unit transformation constant
μ	Specific gas constant
ϕ_{l_g}	Weymouth equation coefficient
ρ_0	Gas density in standard condition
$\tau_{i_g}^u/\tau_{i_g}^l$	Gas pressure ranges

Variables

I_{l_pt}	Line current magnitude square
$\dot{m}_{l_gt}$	Average gas mass flow
p_{gt}, p_{nt}	Active power output of generators
q_{gt}, q_{nt}	Reactive power output of generators
P_{l_pt}/Q_{l_pt}	Active/reactive power flow in power grid branches
U_{i_pt}	Nodal voltage magnitude square
u_{i_gt}	Nodal gas pressure
y_{wt}	Purchased gas
$y_{l_gt}^{in}/y_{l_gt}^{out}$	Gas in/out flow of a passive pipeline
y_{ct}^{in}/y_{ct}^{out}	Gas in/out flow of an active pipeline

Assumptions in the OGPF model include:

(1) Gas demand of a gas-fired unit is a linear function in its active power output.
(2) Gas flow dynamics are approximated by the line pack model.

(3) Electricity or gas consumption of compressor is simplified as a linear function, implying that it is operated with a constant boost ratio, and the coefficient c_g/c_e is very small.

The centralized OGPF problem reads as follows:

$$\min \sum_t \sum_n C_n(p_{nt}) + \sum_t \sum_w Q_{wt}^{gas} y_{wt} \tag{3.15a}$$

$$\text{s.t. } p_{\{\cdot\}}^{min} \leq p_{\{\cdot\}t} \leq p_{\{\cdot\}}^{max},\ \{\cdot\} = \{g, n\},\ \forall t \tag{3.15b}$$

$$q_{\{\cdot\}}^{min} \leq q_{\{\cdot\}t} \leq q_{\{\cdot\}}^{max},\ \{\cdot\} = \{g, n\},\ \forall t \tag{3.15c}$$

$$U_{i_p}^{min} \leq U_{i_pt} \leq U_{i_p}^{max},\ \forall i_p, t \tag{3.15d}$$

$$0 \leq I_{l_pt} \leq S_{l_p}^{max},\ \forall l_p, t \tag{3.15e}$$

$$\begin{aligned} &\sum_{\{\cdot\}\in\Psi_{\{\cdot\}}(i_p)} p_{\{\cdot\}t} + \sum_{l\in\Psi_{O_2}(i_p)} (P_{l_pt} - r_{l_p} I_{l_pt}) - \sum_{l\in\Psi_{O_1}(i_p)} P_{l_pt} \\ &- \sum_{d\in\Psi_{d_p}(i_p)} p_{d_pt} - \sum_{c\in\Psi_c(i_p)} \chi_c y_{ct}^{in} = 0,\ \{\cdot\} = \{g, n\},\ \forall i_p, t \end{aligned} \tag{3.15f}$$

$$\begin{aligned} &\sum_{\{\cdot\}\in\Psi_{\{\cdot\}}(i_p)} q_{\{\cdot\}t} + \sum_{l\in\Psi_{O_2}(i_p)} (Q_{l_pt} - x_{l_p} I_{l_pt}) - \sum_{l\in\Psi_{O_1}(i_p)} Q_{l_pt} \\ &- \sum_{d\in\Psi_{d_q}(i_p)} q_{d_qt} = 0,\ \{\cdot\} = \{g, n\},\ \forall i_p, t \end{aligned} \tag{3.15g}$$

$$U_{i_{p2}t} = U_{i_{p1}t} - 2(r_{l_p} P_{l_p} + x_{l_p} Q_{l_p}) + (r_{l_p}^2 + x_{l_p}^2) I_{l_pt},\ \forall (i_{p1}, i_{p2}, l_p), \forall t \tag{3.15h}$$

$$U_{i_{p1}t} I_{l_pt} \geq P_{l_pt}^2 + Q_{l_pt}^2,\ \forall (i_{p1}, i_{p2}, l_p), \forall t \tag{3.15i}$$

$$0 \leq y_{wt} \leq y_w^u,\ \forall w, t \tag{3.15j}$$

$$\tau_{i_g}^l \leq u_{i_gt} \leq \tau_{i_g}^u\ \forall i_g, t \tag{3.15k}$$

$$\begin{aligned} &\sum_{w\in\Theta_w(i_g)} y_{wt} - \sum_{d_g\in\Theta_{d_g}(i_g)} y_{d_gt} - \sum_{\{\cdot\}\in\Theta_{\{\cdot\}_1}(i_g)} y_{\{\cdot\}t}^{out} - \sum_{g\in\Theta_g(i_g)} p_{gt}/\beta_g \\ &+ \sum_{\{\cdot\}\in\Theta_{\{\cdot\}_2}(i_g)} y_{\{\cdot\}t}^{in} = 0,\ \{\cdot\} = \{c, l_g\},\ \forall i_g, t \end{aligned} \tag{3.15l}$$

$$\frac{(y_{l_gt}^{in}+y_{l_gt}^{out})|y_{l_gt}^{in}+y_{l_gt}^{out}|}{4}=\phi_{l_g}[(u_{l_g^1t})^2-(u_{l_g^2t})^2],\ \forall l_g,t \tag{3.15m}$$

$$m_{l_gt}=\frac{\pi}{4}\frac{X_{lg}R_{lg}^2}{\mu T_k Z_{l_g}\rho_0}\frac{u_{l_g^1t}+u_{l_g^2t}}{2},\ \forall l_g,t \tag{3.15n}$$

$$m_{l_gt}=m_{l_g(t-1)}+y_{l_gt}^{in}-y_{l_gt}^{out},\ \forall l_g,t \tag{3.15o}$$

$$u_{i_{gt,c_2}}\le\gamma_c u_{i_{gt,c_1}},\ \forall c,t \tag{3.15p}$$

$$0\le y_{ct}^{in}\le y_c^{max},\ \forall c,t \tag{3.15q}$$

$$y_{ct}^{in}-(1-\alpha_c)y_{ct}^{out}=0\ \forall c,t \tag{3.15r}$$

The objective function (3.15a) is to minimize the total production costs of the coupled system. The first component represents the fuel costs of non-gas units, where $C_n(p_{nt})$ is usually a convex quadratic function. The second component is the gas purchase cost. Gas consumptions include gas system demands and those consumed by gas-fired units. We do not model the payment induced by bilateral energy transaction between the power grid and gas system. Such an issue will be addressed in the next section. Equations (3.15b)–(3.15r) represent operating constraints of coupled systems. For the power system, (3.15b) and (3.15c) enforce active and reactive capacities of generators. Equations (3.15d) and (3.15e) restrict the magnitudes of bus voltage and line current. Equations (3.15f) and (3.15g) are nodal power balancing conditions, where $\Psi_g(i_p)$, $(\Psi_n(i_p), \Psi_{d_p}(i_p), \Psi_c(i_p))$ represent the set of gas-fired units (non-gas units, loads, electrical compressors) connected to bus i_p; $\Psi_{O_1}(i_p)(\Psi_{O_2}(i_p))$ denotes the set of lines whose head (tail) bus is i_p. The last term in (3.15f) is the electric power consumed by electrical compressors. For gas-powered compressors, the electricity consumption factor is $\chi_c = 0$. Equation (3.15h) is the forward voltage drop equation. Equation (3.15i) expresses relations among bus voltages, line currents, and apparent powers. This inequality is active at the optimal solution under mild conditions, as mentioned in Sect. 2.3.1; For the gas system, (3.15j) and (3.15k) define the ranges of gas supply and nodal gas pressures. Equation (3.15l) is the nodal gas balancing condition, where $\Theta_w(i_g)$ $(\Theta_{d_g}(i_g), \Theta_g(i_g))$ represent the set of gas retailer (gas loads, gas-fired units) connected to node i_g; $\Theta_{c_1}(i_g)$ $(\Theta_{c_2}(i_g))$ labels the set of active pipelines whose head (tail) node is i_g; $\Theta_{l_{g_1}}(i_g)$ $(\Theta_{l_{g_2}}(i_g))$ indexes the set of passive pipelines whose head (tail) node is i_g. Equation (3.15m) is the Weymouth equation for passive pipelines, where $u_{l_g^1t}$, $u_{l_g^2t}$ are gas pressures at initial and terminal nodes of l_g, respectively. Equations (3.15n) and (3.15o) model line pack dynamics, which have been discussed in Sect. 3.2.4. Equations (3.15p) and (3.15q) limit the maximum compression ratio and inflow of active pipelines. Equation (3.15r) indicates the gas consumption in active pipelines. For electrical compressors, the fuel consumption coefficient is $\alpha_c = 0$. Equations (3.15p)–(3.15r) constitute the simplified compressor model.

Besides the non-convexity, the sign function in (3.15m) also imposes challenges on solving the OGPF problem. Nevertheless, if we have prior knowledge on the gas flow direction, say, for a radial pipeline network, (3.15m) reduces to

$$(y^{in}_{l_g t} + y^{out}_{l_g t})^2/4 = \phi_{l_g}((u_{l^1_g t})^2 - (u_{l^2_g t})^2),\ u_{l^1_g t} \geq u_{l^2_g t} \geq 0,\ \forall l_g,\ t \tag{3.16}$$

We assume the notations of the head and tail nodes of l_g are consistent with the positive direction of gas flow. Although remains non-convex, (3.16) has a special structure which inspires the use of DC programming method.

3.3.2 Problem Decomposition

In practice, it may not be desirable to solve OGPF in a centralized manner. For one reason, the power system and gas system are operated by different entities. There is no central agency which has full control authority of both infrastructures. For another, operators of both systems may not wish to share their private information. A preferred scheme is that the two systems solve their own problems individually, and exchange information for possible coordination. In such a consideration, the OGPF problem is decomposed into an OPF subproblem which is solved by the power system, and an OGF subproblem, which is solved by the gas system.

1. Power System Subproblem

The power system operator endeavors to minimize the fuel costs of non-gas units subject to operating constraints with fixed electricity demands from the gas system, resulting in

$$\begin{aligned} \min\ & \sum_t \sum_n C_n(p_{nt}) \\ \text{s.t.}\ & \text{(3.15b)–(3.15e), (3.15g)–(3.15i), Couple-PB} \\ & \sum_{\{\cdot\} \in \Psi_{\{\cdot\}}(i_p)} p_{\{\cdot\}t} + \sum_{l \in \Psi_{O_2}(i_p)} (P_{l_p t} - r_{l_p} I_{l_p t}) - \sum_{l \in \Psi_{O_1}(i_p)} p_{l_p t} \\ & - \sum_{d \in \Psi_{d_p}(i_p)} p_{d_p t} = 0,\ \{\cdot\} = \{g, n\},\ \Psi_c(i_p) = \emptyset,\ \forall t \end{aligned} \tag{3.17}$$

where the last constraint represents the active power balancing conditions for power system buses without electrical compressors. The coupling power balancing constraint Couple-PB which involves variables of both systems is given by

$$\begin{aligned} &\sum_{\{\cdot\}\in\Psi_{\{\cdot\}}(i_p)} p_{\{\cdot\}t} + \sum_{l\in\Psi_{O_2}(i_p)} (P_{l_pt} - r_{l_p} I_{l_pt}) - \sum_{l\in\Psi_{O_1}(i_p)} p_{l_pt} - \sum_{d\in\Psi_{d_p}(i_p)} p_{d_pt} \\ &\quad - \sum_{c\in\Psi_c(i_p)} \chi_c y_{ct}^{in} = 0, \ \{\cdot\} = \{g, n\}, \ \Psi_c(i_p) \neq \emptyset, \ \forall t \end{aligned} \tag{3.18}$$

where y_{ct}^{in} is fixed in the OPF subproblem (3.17).

2. Gas System Subproblem

The gas system operator seeks to minimize the gas purchase cost subject to operating constraints with fixed gas demands from the power system, yielding

$$\begin{aligned} \min \ & \sum_t \sum_w Q_{wt}^{gas} y_{wt} \\ \text{s.t.} \ & (3.15\text{j}), (3.15\text{k}), (3.15\text{n})\text{–}(3.15\text{r}), (3.16), \text{Couple-GB} \\ & \sum_{w\in\Theta_w(i_g)} y_{wt} - \sum_{d_g\in\Theta_{d_g}(i_g)} y_{d_gt} - \sum_{\{\cdot\}\in\Theta_{\{\cdot\}_1}(i_g)} y_{\{\cdot\}t}^{out} \\ & \quad + \sum_{\{\cdot\}\in\Theta_{\{\cdot\}_2}(i_g)} y_{\{\cdot\}t}^{in} = 0, \ \{\cdot\} = \{c, l_g\}, \Theta_g(i_g) = \emptyset, \ \forall t \end{aligned} \tag{3.19}$$

where the last constraint represents gas balancing conditions of gas system nodes without connections to gas-fired units. The coupling gas balancing constraint Couple-GB which involves variables of both systems is given by

$$\begin{aligned} &\sum_{w\in\Theta_w(i_g)} y_{wt} - \sum_{d_g\in\Theta_{d_g}(i_g)} y_{d_gt} - \sum_{\{\cdot\}\in\Theta_{\{\cdot\}_1}(i_g)} y_{\{\cdot\}t}^{out} - \sum_{g\in\Theta_g(i_g)} p_{gt}/\beta_g \\ &\quad + \sum_{\{\cdot\}\in\Theta_{\{\cdot\}_2}(i_g)} y_{\{\cdot\}t}^{in} = 0, \ \{\cdot\} = \{c, l_g\}, \ \Theta_g(i_g) \neq \emptyset, \ \forall t \end{aligned} \tag{3.20}$$

where the output of gas-fired unit p_{gt} is fixed in the OGF subproblem (3.19).

3.3.3 Distributed Algorithm

The centralized OGPF problem can be written in a compact form as

$$\begin{aligned} \min \quad & F_p(x) + F_g(z) \\ \text{s.t.} \quad & x \in \Omega_p, \; z \in \Omega_g \\ & Ax + Bz = c \end{aligned} \tag{3.21}$$

In (3.21), x and z are decision variables of OPF and OGF subproblems, respectively. $F_p(\cdot)$ and $F_g(\cdot)$ are corresponding objective functions in (3.17) and (3.19). Ω_p is the feasible region of x which contains constraints in (3.17) except Couple-PB; Ω_g is the feasible region of z which contains constraints in (3.19) except Couple-GB. The last equality represents the coupling constraints given in (3.18) and (3.20). Non-convexity rests in the Weymouth equation (3.16) in Ω_g. In what follows, we first present the distributed algorithm for the OGPF problem, then a method to solve the non-convex OGF subproblem.

1. The ADMM Based Distributed Algorithm

The ADMM method is comprehensively tutored in [28]. This approach is an augmented Lagrangian method with a Gaussian-Seidel type iteration. It is basically suitable for solving block separable convex optimization problems with linear coupled constraints. For the OGPF problem, the power system operator solves problem

$$\min_{x \in \Omega_p} \quad F_p(x) + \xi^T(Ax + Bz^* - c) + \frac{d}{2}\|Ax + Bz^* - c\|^2 \tag{3.22}$$

where ξ is a fixed multiplier vector, d is a constant, and z^* is the current value of gas system decision variable. The gas system operator solves problem

$$\min_{z \in \Omega_g} \quad F_g(z) + \xi^T(Ax^* + Bz - c) + \frac{d}{2}\|Ax^* + Bz - c\|^2 \tag{3.23}$$

The procedures are detailed in Algorithm 3.1. Due to the non-convex gas flow feasible region Ω_g of problem (3.23), there is no rigorous guarantee for the convergence of Algorithm 3.1. Nevertheless, its performance is satisfactory in all our tests. It is also found that parameter d has a notable impact on the convergence rate. Its selection is thoroughly discussed in [29] for broader classes of problems. Equations (3.22) and (3.23) have strongly convex objective functions because of the regularization terms, so each of them has a unique optimal solution. However, the non-monotonicity of objective function may wreck the exactness of SOCP relaxation performed on (3.15i). Nonetheless, we can resort to Algorithms 2.1–2.2 in Sect. 2.3.1 to recover an OPF solution.

2. A Sequential Convex Optimization Approach for the OGF Subproblem
In problem (3.23), all constraints are convex except the Weymouth equation in (3.16), which can be cast as two opposite inequalities

Algorithm 3.1 ADMM for OGPF

1: Initiate x, z, ξ, and iteration index $k = 0$. Set values for parameter d and convergence tolerance ς.
2: Solve problem (3.22) with $z^* = z^k$ and $\xi = \xi^k$; the optimal solutions is x^{k+1}.
3: Solve problem (3.23) with $x^* = x^{k+1}$ and $\xi = \xi^k$; the optimal solution is z^{k+1}.
4: Update the multiplier ξ according to

$$\xi^{k+1} = \xi^k + d(Ax^{k+1} + Bz^{k+1} - c) \tag{3.24}$$

5: If $\|Ax_{k+1} + Bz_{k+1} - c\|_1 \le \varsigma$, terminate; else, $k \leftarrow k+1$, and go to step 2.

$$(y_{l_g t}^{in} + y_{l_g t}^{out})^2/4 + \phi_{l_g}(u_{l_g^2 t})^2 - \phi_{l_g}(u_{l_g^1 t})^2 \le 0 \ \forall l_g, t \tag{3.25}$$

$$\underbrace{\phi_{l_g}(u_{l_g^1 t})^2}_{f_{l_g t}} - \underbrace{((y_{l_g t}^{in} + y_{l_g t}^{out})^2/4 + \phi_{l_g}(u_{l_g^2 t})^2)}_{g_{l_g t}} \le 0, \ \ \forall l_g, t \tag{3.26}$$

where (3.25) defines an SOC whose canonical form is given by

$$\left\| \begin{matrix} (y_{l_g t}^{in} + y_{l_g t}^{out})/2 \\ \sqrt{\phi_{l_g}} u_{l_g^2 t} \end{matrix} \right\|_2 \le \sqrt{\phi_{l_g}} u_{l_g^1 t}, \ u_{l_g^1 t} \ge u_{l_g^2 t} \ge 0, \ \forall l_g, t \tag{3.27}$$

Constraint (3.26) is non-convex, and its left-hand side can be decomposed as the difference of two convex functions $f_{l_g t}(z)$ and $g_{l_g t}(z)$. The latter can be linearized at a given point via first-order Taylor expansion. Motivated by the convex-concave procedure discussed in [30], a sequential convex optimization algorithm for the OGF subproblem is presented in Algorithm 3.2.

Convergence of Algorithm 3.2 is guaranteed by the existence of an exact penalty parameter [30]. Nevertheless, failure of convergence may occur if the OGF problem is infeasible. Long-term solution could be upgrading the gas delivery infrastructure, such as investing on new pipelines and compressor stations. The initial selection of z^0 also influences the computational efficiency. Zero initiation is the default setting for many heuristic algorithms but may not be a good choice for Algorithm 3.2 because the gradient of $g_{l_g t}$ would be zero. To find a good initial point, we can solve problem (3.23) by replacing (3.16) with (3.27) together with

$$(y_{l_g t}^{in} + y_{l_g t}^{out})^2/4 \le \phi_{l_g}(\zeta_{l_g^1 t} - \zeta_{l_g^2 t}), \ \ \forall l_g, t \tag{3.30a}$$

$$(u_{i_g})^2 \le \zeta_{i_g t} \le (\tau_{i_g}^u + \tau_{i_g}^l)(u_{i_g} - \tau_{i_g}^l) + (\tau_{i_g}^l)^2, \ \ \forall i_g, t \tag{3.30b}$$

Algorithm 3.2 Sequential convex optimization for OGF

1: Retrieve x^{k+1}, ξ^k and constant d of Algorithm 3.1. Given an initial value of z^0. Set values for convergence tolerances δ and ϵ, penalty parameters ϱ_0, ϱ_{max}, and $\kappa > 1$. Let initial objective value $f_g^0 = 0$, iteration index $j = 0$.

2: Linearize $g_{l_g t}(z)$ at z^j according to

$$\bar{g}_{l_g t}(z, z^j) = g_{l_g t}(z^j) + \nabla g_{l_g t}(z)|_{z=z^j}(z - z^j) \tag{3.28}$$

and solve the following problem

$$\begin{aligned} f_g^{j+1} = \min_{z, \varsigma_{l_g t}} \ & L_g(z, \xi^k) + \varrho^j \sum_t \sum_{l_g} \varsigma_{l_g t} \\ \text{s.t.} \ & f_{l_g t}(z) - \bar{g}_{l_g t}(z, z^j) \le \varsigma_{l_g t}, \ \forall l_g, t \\ & z \in \Omega_p^C, \ (3.27), \ \varsigma_{l_g t} \ge 0, \ \forall l_g, t \end{aligned} \tag{3.29}$$

where $L_g(z, \xi^k)$ denotes the objective function of OGF subproblem (3.23), and Ω_p^C is obtained from Ω_p by relaxing (3.16). The optimal solutions are z^{j+1}, $\varsigma_{l_g t}^{j+1}$, and the optimal value is f_g^{j+1}.

3: If $|f_g^{j+1} - f_g^j| \le \delta$ and $\varsigma_{l_g t}/|f_{l_g t} - \bar{g}_{l_g t}| \le \epsilon$, $\forall l_g, t$, terminate, otherwise, update $k \leftarrow k+1$, $\varrho^{j+1} = \min(\kappa \varrho^j, \varrho_{max})$, go to step 2.

where ζ represents the square nodal pressure. The geometric interpretation of inequality (3.30b) is that ζ lies between parabola $\zeta = u^2$ and the chord connecting points $(\tau_{i_g}^l, (\tau_{i_g}^l)^2)$ and $(\tau_{i_g}^u, (\tau_{i_g}^u)^2)$. In general, the obtained initial point is not feasible, nevertheless, constraint violation will be gradually eliminated by the growing penalty parameter in (3.29).

3.3.4 Case Studies

1. 13-Bus Power Feeder with 7-Node Gas System

Figure 3.6 depicts the topology of the connected infrastructure. It has 2 gas-fired units, 1 non-gas generator, 2 gas retailers, 2 compressors, 4 passive pipelines, 8 power loads, and 3 gas loads. In this figure, we use P, PL, DG with subscripts to denote the buses, loads, and generators in the power system, and N, C, GR, GL with subscripts to denote nodes, compressors, gas retailers, and gas loads in the gas system. Specially, the fuel of DG_2 and DG_3 is supplied from nodes N_3 and N_1, respectively. Compressors C_1 and C_2 are driven by electricity which comes from buses P_4 and P_5, respectively. The power and gas demands profiles are shown in Fig. 3.7. Detailed system data can be found in [31]. Hereinafter, this system is referred to as the Power13Gas7 system. Parameters of Algorithms 3.1 and 3.2 for this system are listed in Table 3.1.

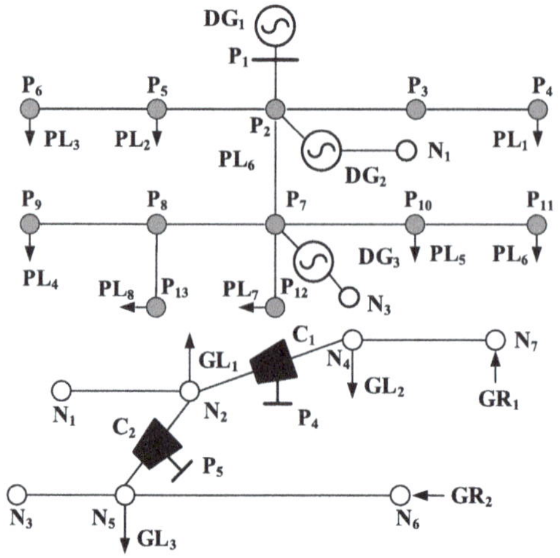

Fig. 3.6 Topology of the Power13Gas7 system

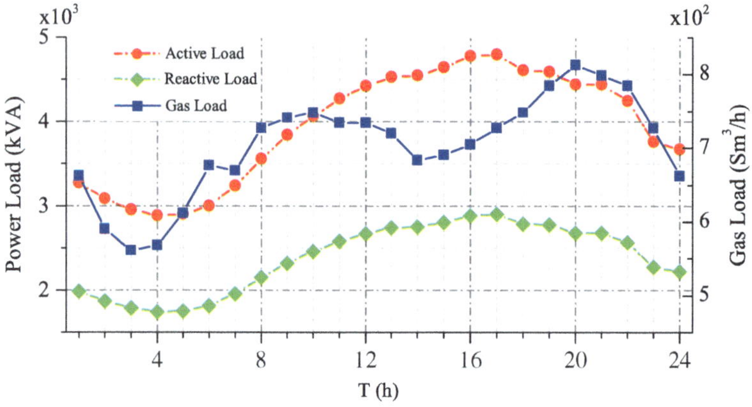

Fig. 3.7 Power and gas daily demands for the Power13Gas7 system

Table 3.1 Parameters of Algorithms 3.1 and 3.2

Algorithm 3.1			Algorithm 3.2					
ς	d	k_{max}	δ	ρ_0	ρ_{max}	ϵ	κ	j^{max}
10^{-3}	100	100	1	0.01	1000	10^{-6}	2	100

In the benchmark case, Algorithm 3.1 converges in three iterations, and the optimal value is \$1.7419 $\times$ 10^4. To quantify the infeasibility, we use MACV and MRCV to represent the maximum absolute constraint violation and maximum relative constraint violation in Algorithm 3.1 and Algorithm 3.2, respectively. The sequences of MACV and MRCV generated in each iteration are plotted in Fig. 3.8. As their values decrease, solutions approach to the feasible region.

Further, dynamics of the total line pack in the gas system is shown in Fig. 3.9. In this test, the gas prices are set nearly proportional to the gas demands. During

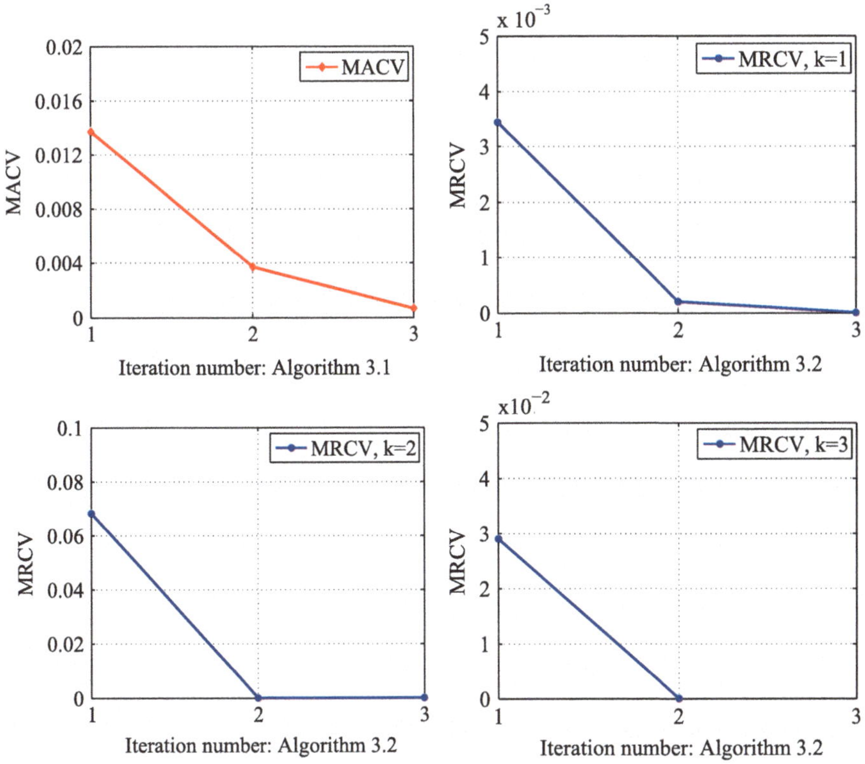

Fig. 3.8 MACV and MRCV sequences generated by Algorithms 3.1 and 3.2

valley load periods, say periods 2–5, extra gas is bought at a low price from the gas retailers and stored in pipelines; during gas peak load periods, say periods 9–11, 17–21, the previously stored gas is extracted to reduce the gas transaction with the retailers because of the high prices. The ratio between stored or extracted gas and the gas demand in each period is also shown in the same figure. The line pack serves about half of gas demands during peaking periods 9–11 and 16–21. In this way, the operation cost of the gas system can be reduced considerably.

The performance of Algorithm 3.2 is compared with three mainstream methods: NLP method, which solves OGF problem in the original NLP format by calling IPOPT solver; the MILP method [27], in which nonlinear functions are approximated by PWL functions with eight segments; as well as the SOCP relaxation method [32] without a feasible solution recovery procedure. Solution quality, constraint violation, and computation time are compared. The time limit is 7200 s. Results are shown in Table 3.2.

It can be observed that the MILP method fails to find the optimal solution in the given time limit. The other three methods successfully offer solutions. The convex relaxation method gives the lowest objective value in the shortest time. However, the solution is infeasible with an MRCV of 1022%, because the relaxation is not exact.

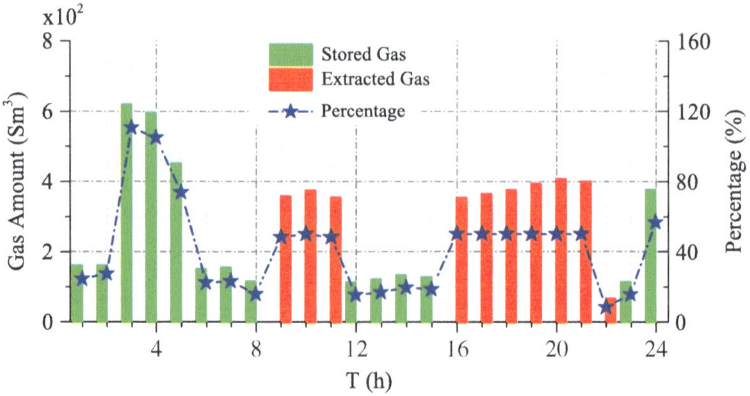

Fig. 3.9 Dynamics of the total line pack

Table 3.2 Comparison with existing methods for gas flow optimization: Power13Gas7 system

Method	Objective ($\times 10^4$\$)	Feasibility		Time (s)
		Coupling	OGF	
IPOPT	1.7639	5.5×10^{-4}	9.1×10^{-7}	82.16
MILP-GUROBI	2.0464	0.12	0.085	7200[a]
Relaxation	1.7022	8.7×10^{-5}	10.22	0.086
Algorithm 3.2	1.7419	7.2×10^{-5}	3.1×10^{-7}	0.574

[a]Means that the computation time exceeds the upper limit 7200s

The objective function value \$17,022 corresponding to the infeasible solution is a lower bound on the true optimal value. IPOPT finds a local optimal solution with an objective value of \$17,639, which is an upper bound on the true optimal value. For Algorithm 3.2, a solution is found in half a second, which is much faster than the NLP method, and the solution quality is also better, because the cost is lower. When the penalty parameters are properly set, the solution offered by Algorithm 3.2 could be very close to the global optimal one. The reason is analogous to the discussions in Sect. 2.3.1.

2. 123-Bus Power Feeder with 20-Node Gas System

In this section, a test system comprised of a modified IEEE 123-bus power feeder and the modified Belgian high-calorific 20-node gas system is used to demonstrate the performance of the proposed algorithms. It has 6 gas-fired units, 4 non-gas generators, 2 gas retailers, 3 compressors, 16 passive pipelines, 85 power loads, and 9 gas loads. Please refer to [31] for the topology, the demand curve as well as other system data. Hereinafter, the test system is referred to as the Power123Gas20 system. Parameters of Algorithms 3.1 and 3.2 are the same with the previous case, except for $d = 1000$ in this case. The ADMM based Algorithm 3.1 converges in 39.7 s after five iterations, and the average number of iterations of Algorithm 3.2 is 5.7. The optimal value is \$549,960.

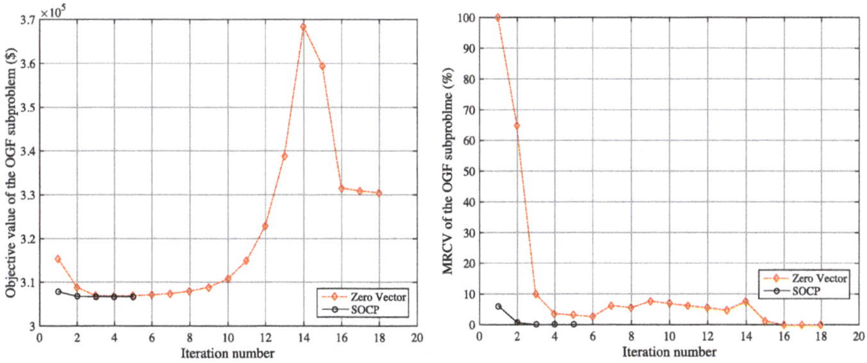

Fig. 3.10 Performance of Algorithm 3.2 with different initial points

To better demonstrate the impacts of initial points on the performance of Algorithm 3.2, we use zero initial point as a comparison, which will be referred to as Case 2, and the SOCP relaxation based initial point selection is termed Case 1. The objective values and the feasibility of the OGF subproblem generated by Algorithm 3.2 in the last iteration of Algorithm 3.1 are presented in Fig. 3.10. In Case 1, the initial point is very close to the optimal one, so constraint violation is quickly eliminated and Algorithm 3.2 converges fast. However, when the zero initial point is used in Case 2, the objective value firstly decreases but exhibits a big rise, and finally converges to a feasible solution which perceives a higher objective value than Case 1. From the feasibility result we can see that although the objective value in the fifth iteration is almost the same as the optimal one, the MRCV is still around 4%, which means that the solution is actually infeasible. When the constraint violation vanishes, Algorithm 3.2 terminates at a feasibility solution, but optimality is not guaranteed in this case. Therefore, a good initial guess has a decisive impact on the performance of Algorithm 3.2, which is similar to the application of convex-concave procedure in OPF problems in Sect. 2.3.1.

As demonstrated in the previous case study, the convex relaxation method may not be exact, while the MILP method may suffer from computation overhead. Therefore, we will focus on the comparison with the NLP method and the proposed method. For Power123Gas20 system, we generate 10 electricity and gas load scenarios, respectively, including the base case scenario, within the ±10% error band with respect to their base values. We test performances of the two computational methods under the 100 load combinations. Results are listed in Table 3.3, which include the objective value and computation time in the base case, the best and the worst cases (the best and worst cases for objective value and solution time may not refer to the same one). The solution time marked with a star superscript means that the solver stops because maximum time limit is reached.

According to Table 3.3, Algorithm 3.2 is significantly faster than IPOPT. Moreover, IPOPT appears to be less robust, because it fails to find an optimal solution in some load scenarios when the number of periods is 8, 12, 20, and 24,

Table 3.3 Comparisons of NLP method and Algorithm 3.2: Power39Gas20 system

	IPOPT						Algorithm 3.2					
	Objective value ($\times 10^5$ \$)			Time (s)			Objective value ($\times 10^5$ \$)			Time (s)		
Periods	Base	Worst	Best	Worst	Average	Best	Base	Worst	Best	Worst	Average	Best
4	0.7881	0.8251	0.7547	772	93.2	59.6	0.7775	0.8166	0.7429	5.88	3.76	2.87
8	2.1754	2.3266	1.7145	7200[a]	5177	1322	1.7237	1.8140	1.6992	16.7	7.01	5.29
12	3.0971	3.5202	2.7010	7200[a]	5451	1962	2.7033	2.9250	2.5927	22.1	9.20	8.25
16	3.6522	3.8665	3.4911	471	203	172	3.6163	3.8209	3.4293	40.7	25.7	22.5
20	4.9028	5.1179	4.4279	7200[a]	6299	2239	4.5393	4.7121	4.3228	32.6	8.22	7.19
24	5.9815	6.2236	5.4631	7200[a]	6501	3078	5.4996	5.6977	5.2101	82.9	40.1	36.4

[a]Means that the computation time exceeds the upper limit 7200s

and the average computation time is also close to the time limit, while Algorithm 3.2 converges in all cases. Meanwhile, the objective values obtained by Algorithm 3.2 are always lower than those offered by IPOPT, which demonstrates its effectiveness and superiority. Intuitively, more time periods implies larger problem sizes, which would result in longer computation times. However, the average computation time of both methods is not strictly increasing with respect to the number of periods. One possible reason might be that the quality of initial values varies in each case.

3.4 Equilibrium of LMP Based Gas-Electricity Markets

The previous section studies distributed computation of OGPF. Although system data are kept private, a central coordinator is responsible for updating the Lagrangian multipliers of the coupled constraints and broadcasting it to both systems. Such an agency may not exist in current industrial practice. Furthermore, one problem remains salient: objective function (3.17) is not the true operating cost of power system, since it does not include the cost associated with gas-fired units, and (3.19) is neither the operating cost of gas system, as it does not involve the cost associated with compressors which are powered by electricity. Bilateral gas and electricity delivery must incur some transaction costs, which depend on the contract or agreement between the two systems. Transaction costs are neglected in the OGPF problem, as they have opposite signs in the cost functions of both systems and lead to zero sums. Given the rapidly growing number of gas-fired units, gas and power systems become more interactive and are operated under more stressed condition. Fixed contract prices may not truly reflect the value of energy resources varying over time, calling for the establishment of market based energy trading among these two energy sectors. In a market environment, transaction costs may not be neglected because they impact the market clearing process.

To encourage gas and power system integration and build a fair and open energy market, we envision a pool-based gas market with a locational marginal pricing scheme which is similar to the current deregulated power markets, because marginal pricing scheme has been well acknowledged for its fairness and ability to carry real-time operating information. We study interdependent markets clearing with marginal prices based bilateral gas and electricity transactions. We suggest a best-response decomposition algorithm to compute the equilibrium of the coupled energy markets, and illustrate geometrical conditions under which an equilibrium appears or no equilibrium exists.

3.4.1 Market Model

The gas and power flow models inherit the ones in Sect. 3.3, so most symbols and notations have the same meanings as what have been clarified there. Others will be defined following their first appearance as required.

1. Gas Market Clearing Model
In analogy to the power market, gas market clearing gives rise to an OGF problem which minimizes the total cost of the gas system

$$\begin{aligned} \min \quad & \sum_t \sum_w Q_{wt}^{gas} y_{wt} + \sum_t \sum_c \xi_{i_p t}^E \chi_c y_{ct}^{in} \\ \text{s.t.} \quad & (3.15\text{j})\text{–}(3.15\text{r}) \end{aligned} \tag{3.31}$$

where $\xi_{i_p t}^E$ is the LMP of electricity (LMEP) at compressor node i_p in period t, which is determined from the electricity market clearing problem (3.32). The fuel demand of gas-fired units p_{gt}/β_g in (3.15l) and LMEP $\xi_{i_p t}^E$ are kept constant in gas market clearing problem (3.31). The first component in the objective function represents gas purchase cost, which is the same as (3.19), and the second one stands for operating cost of electric compressors. Gas flow constraints are consistent with those in Sect. 3.3. Moreover, if gas flow directions can be acquired in advance, we can replace (3.15m) with (3.16), and problem (3.31) can be solved by Algorithm 3.2. Given the optimal solution, the LMP of gas (LMGP) can be easily recovered from the dual variable associating with nodal gas balancing condition (3.15l). To this end, one may solve the KKT optimality condition of problem (3.31) under a fixed OGF solution, which is a set of linear equations in the dual variable.

P2G facilities can be modeled in the way similar to compressors if they are operated by the system operator, a non-profit entity. In such circumstance, they purchase electricity at LMEP and inject gas back to gas pipelines. An additional term representing the output of P2G facilities will occur in the gas balancing condition (3.15l). A preliminary model can be found in [33]. However, under market environment, P2G facilities may be owned by private companies, who pursue maximal profits by offering auxiliary services, participating energy and reserve markets, and selling gas to the gas system. The strategic behaviors and market impacts of emerging P2G economic entities require more dedicated studies. In this section, since bilateral energy transaction is already activated by gas-fired units and electric compressors, we neglect P2G facilities.

2. Electricity Market Clearing Models

Electricity market clearing boils down to an OPF problem which minimizes the total operation cost of the power system

$$\begin{aligned} \min \quad & \sum_t \sum_g \frac{\xi^G_{i_g t} p_{gt}}{\beta_g} + \sum_t \sum_n C_n(p_{nt}) \\ \text{s.t.} \quad & (3.15\text{b})\text{–}(3.15\text{i}) \end{aligned} \tag{3.32}$$

where $\xi^G_{i_g t}$ is the LMGP at node i_g which supplies fuel to a gas-fired unit in period t. It is determined from the gas market clearing problem (3.31). The electricity demand of compressors $\chi_c y^{in}_{ct}$ in (3.15f) and LMGP $\xi^G_{i_g t}$ in the objective function are kept constant in electricity market clearing problem (3.32). The first component in the objective function represents the cost of gas-fired units, and the second one denotes the cost of non-gas-fired units, which is the same as (3.17). Power flow constraints are consistent with those in Sect. 2.3.1. If the convex relaxation performed on (3.15i) is not exact, problem (3.32) can be solved by Algorithm 2.2 in Sect. 2.3.1 without difficulty. Given the OPF solution, the LMEP can be easily recovered from the dual variable associated with nodal power balancing condition (3.15f) via solving a set of equations derived from the KKT optimality condition of (3.32) under a fixed OPF solution.

3. The Bilateral Market Structure

The structure of the coupled natural gas and electricity markets with LMP based bilateral energy transaction is illustrated in Fig. 3.11.

At the electricity side, fuel of gas-fired units is purchased from the gas market at LMGP. The power system operator clears the electricity market with fixed compressor power consumptions submitted by the gas system operator, and determines the optimal schedule of all generation units and LMEP. At the natural gas side, electric power consumed by compressors is purchased from the electricity market at LMEP. The gas system operator clears the gas market with fixed fuel demands submitted by the power system operator, and determines the optimal schedule of

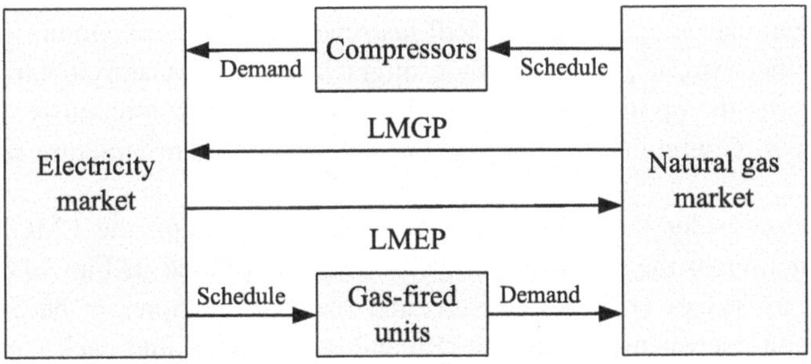

Fig. 3.11 Structure of the coupled natural gas and electricity markets

gas production and LMGP. Several basic problems include: is there an equilibrium in such a market? Under what condition at least one market equilibrium exists? How to find such an equilibrium?

3.4.2 Discussion on Market Equilibria

Let x and y be the dispatch strategies of the power system and gas system, ξ^E and ξ^G the LMEP and LMGP, respectively, two market clearing problems can be simply denoted by

$$[\xi^E, x] = \text{OPF}(\xi^G, y)$$
$$[\xi^G, y] = \text{OGF}(\xi^E, x)$$

Combining them together, the market equilibrium must satisfy

$$[\xi^E, x] = \text{OPF}(\text{OGF}(\xi^E, x))$$
$$[\xi^G, y] = \text{OGF}(\text{OPF}(\xi^G, y))$$

which means that energy prices and dispatch strategies reach a fixed point. Based on this observation, a best-response decomposition algorithm is proposed to identify the market equilibrium, in which OPF and OGF problems are solved interactively through convex optimization.

Because dual variables appear in the objective functions, to the best of our knowledge, there is no rigorous mathematical theory which can guarantee the convergence of Algorithm 3.3. We provide some intuitive discussions to explain under what conditions Algorithm 3.3 is likely to converge or may fail to converge. Our analysis rests on the nodal price-demand curves.

Consider a gas-fired unit g which is connected to bus i_p in the power system. Its fuel is delivered from node i_g in the gas system. The relationship between the gas demand of unit g and the gas price at i_g is illustrated in Fig. 3.12, where gas price is simply denoted by ρ without causing confusion. It can be imaged that if the gas is very cheap, say $\rho \le \rho_1$, unit g will keep working at its maximum generation capacity; otherwise, if $\rho \ge \varrho_2$, unit g might lose all the energy contract. When $\varrho_1 \le \varrho \le \varrho_2$, the optimal generation of g is probably a continuously decreasing function in ϱ. Continuity is indicated by the OPF problem structure revealed in [34, 35].

Now we consider the price response in the gas system: the LMGP at node i_g as a function of the gas demand of g, which is plotted in Fig. 3.12. LMGPs depend on the prices Q^{gas}_{wt} at gas wells and LMEP at compressor buses. Both of them are non-decreasing. When the demand exceeds the line pack capacity, the compressor has to increase its output in order to deliver more gas. As a result, LMGP

Algorithm 3.3 Best-response decomposition

1: Initiate iteration index $j = 0$, LMGP $\xi_{i_g t}^{G,0}$, compressor power demand $y_{ct}^{in,0}$, LMEP $\xi_{i_p t}^{E,0}$, and gas-fired units fuel demands p_{gt}^{0}/β_g. Set the maximum iteration number j^{max} and convergence criterion ϵ.

2: Solve electricity market clearing problem (3.32) with $\xi_{i_g t}^{G,j}$ and $y_{ct}^{in,j}$; record the OPF solution, where the optimal schedule of gas-fired units is p_{gt}^{j+1}; retrieve LMEP ξ^E from the dual variables.

3: Solve gas market clearing problem (3.31) with p_{gt}^{j+1} and ξ^E using Algorithm 3.2, record the OGF solution, where the compressor power demand is $\chi_c y_{ct}^{in,j}$; retrieve LMGP ξ^G from the dual variables.

4: If following criteria are met, terminate; if $j = j^{max}$, quit and report that the algorithm fails to converge; else, update $j = j + 1$, and go to step 2.

$$|p_{gt}^{j+1} - p_{gt}^{j}| \le \epsilon, \ \forall g, t, \ |\xi_{i_p t}^{E,j+1} - \xi_{i_p t}^{E,j}| \le \epsilon, \ \forall i_p, t$$

$$|y_{ct}^{in,j} - y_{ct}^{in,j-1}| \le \epsilon, \ \forall c, t, \ |\xi_{i_g t}^{G,j+1} - \xi_{i_g t}^{G,j}| \le \epsilon, \ \forall i_g, t$$

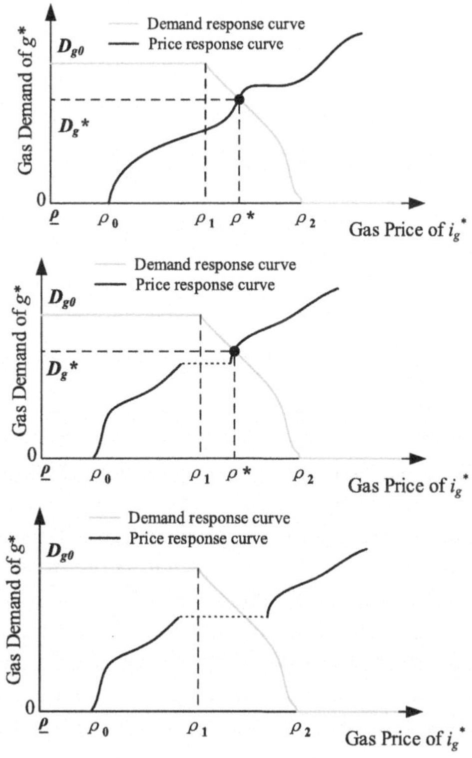

Fig. 3.12 Illustration of gas price and gas demand curves. Top: both curves are continuous; Middle: the price response curve is discontinuous but the two curves have an intersection; Bottom: two curves do not have an intersection

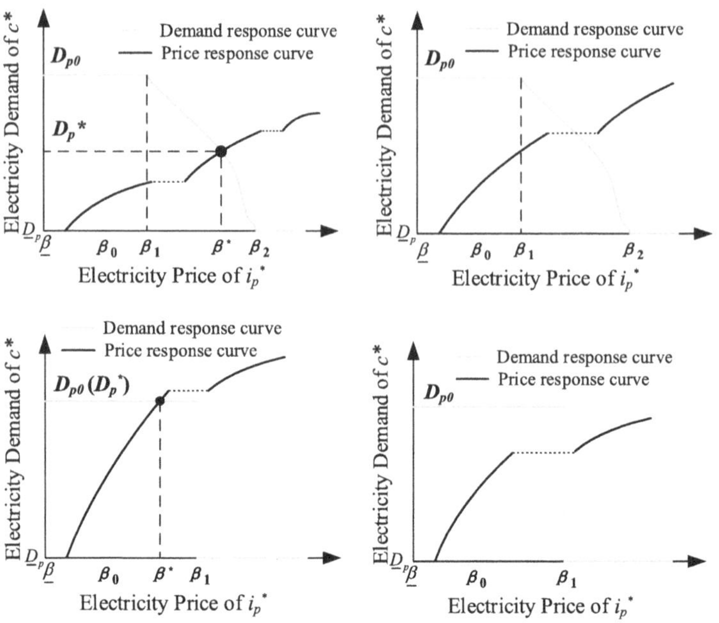

Fig. 3.13 Illustration of electricity price and electricity demand curves. Upper: Demand curve is continuous; Lower: Demand curve is discontinuous; Left: A fixed point exists; Right: No fixed point exists

grows with LMEP rising. However, as LMGP is the dual variable, it is generally discontinuous. Possible outcomes are summarized in Fig. 3.12. If the price response curve is continuous, it should have an intersection with the demand curve, which interprets a fixed point; otherwise, an intersection point may exist or not, depending on system data.

We consider electricity price and demand curves in a similar way. Suppose compressor c is connected to bus i_p in the power system. Its electricity consumption as a function of the LMEP at i_p is portrayed in Fig. 3.13, where electricity price is simply denoted by β without causing confusion. The demand remains unchanged as long as the price $\beta_{i_p^*}$ is either too small or too large. The demand curve can be either continuous or discontinuous, depending on the data of OGF problem. On the other hand, the electricity price, which is determined from the OPF problem, as a function of the demand of compressor c is depicted in the same figure. In general, the price curve would be discontinuous. Take the lossless DCOPF model for example, the LMEP is known to be stepwise constant [36]. Four possible outcomes are summarized in Fig. 3.13.

From the above discussions, it can be concluded that the existence of market equilibrium is system dependent due to the discontinuous nature of the price responsive curves. If no equilibrium exists, Algorithm 3.3 would fail to converge. This may be a weak point of the market mechanism, rather than a limitation of

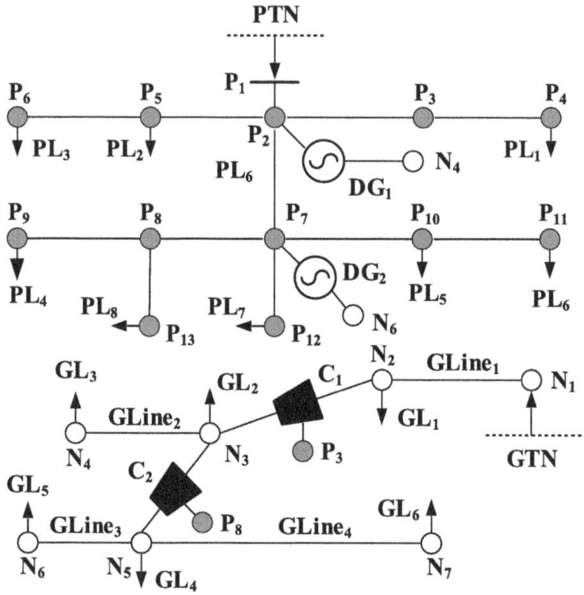

Fig. 3.14 Topology of the Power13Gas7 system

Algorithm 3.3. Enforcing a continuous pricing policy such as the continuous LMP scheme in [36] may remove this pitfall, but computing continuous LMP needs more information than solving a single OPF/OGF problem.

3.4.3 Case Studies

First we present numerical results on a test system composed of a 13-bus power system and a 7-node gas system, whose topology is illustrated in Fig. 3.14, in which P, PL, and DG with subscripts denote electrical buses, power loads, and gas-fired units in the power system, respectively; N, C, GL, and GLine with subscripts denote gas nodes, compressors, gas loads, and passive pipelines in the gas system, respectively. In both systems, electricity and gas can be purchased from a higher level market at time-varying prices and delivered from bus P_1 and node N_1. This is equivalent to a generator (gas well) with a linear cost function and an infinite capacity is connected to P_1 (N_1). The fuel of DG_1 and DG_2 are supplied by nodes N_4 and N_6, respectively. Electric compressors C_1 and C_2 are connected to buses P_3 and P_8, respectively. The daily demand profiles and price forecasts in the higher level electricity market and gas market are shown in Fig. 3.15. Other system data are available in [31]. This system will be referred to as Power13Gas7 later on. Compared with the system in Sect. 3.3 with the same name, the gas system nodes are renumbered and the connection points with the power system are revamped. In

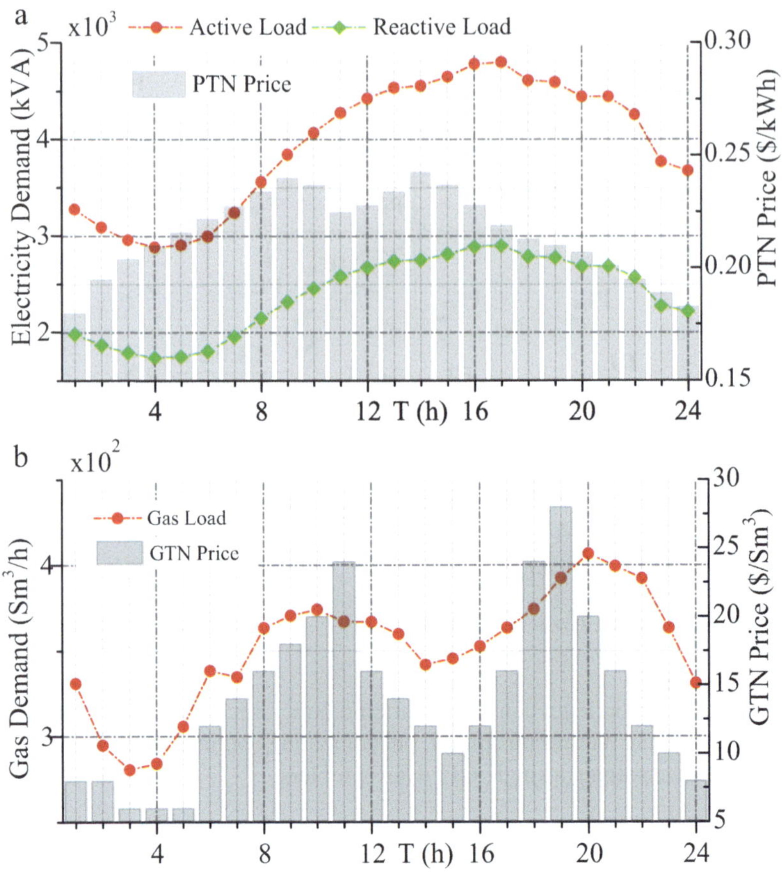

Fig. 3.15 Load profiles and contract energy prices. (**a**) Load profile and contract electricity prices with a higher level power market. (**b**) Load profile and contract gas prices with a higher level gas market

Algorithm 3.2, the penalty parameters are $\varrho_0 = 0.01$, $\varrho_{max} = 1000$, and $\kappa = 2$. The convergence thresholds are $\delta = 1$ and $\epsilon = 10^{-6}$. In Algorithm 3.3, the convergence tolerance is $\epsilon = 10^{-4}$.

With the system configuration, the convex relaxation for power market clearing problem (3.32) is exact; and gas market clearing problem (3.31) is solved by Algorithm 3.2. It converges within three iterations in our tests. Algorithm 3.3 successfully identifies the market equilibrium after four iterations.

Hourly LMEPs and LMGPs are plotted in Fig. 3.16a, b, respectively. The LMEP curve at bus P_1 is identical to the given price curve in Fig. 3.15, and those of other buses grow because the marginal cost of delivering per unit power from the root bus rises with increasing distances to the slack bus owing to the unilateral power flows in the radial network topology and network losses. In the gas system, the node

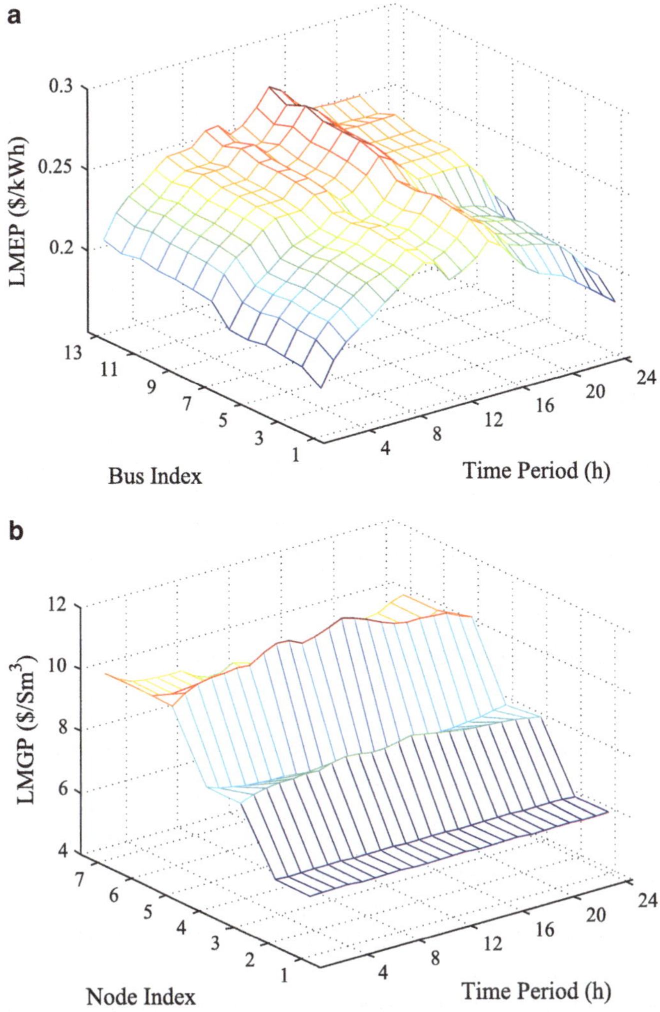

Fig. 3.16 Nodal electricity and gas price curves. (**a**) LMEPs over the scheduling horizon. (**b**) LMGPs over the scheduling horizon

connected with the transmission network is the only source. Since gas delivery may incur electricity consumption, LMGPs are affected by two factors: the gas price in the transmission level market as well as LMEPs at compressor buses. As a result, LMGPs are stepwise constant, as demonstrated in Fig. 3.16b. Moreover, LMGP curves at N_1 and N_2 do not change over time, and price fluctuations at other gas nodes are smaller than the gas price curve in Fig. 3.15b, due to the line pack effect as a sort of gas storage. It allows the system operator to purchase and store extra gas when the gas price is low. In this case, the line pack capacity of $GLine_1$ is adequate to store the gas purchased during periods 3–5, contributing to the constant hourly

Table 3.4 System operating costs at market equilibrium

	Power system ($)	Gas system ($)
With line pack	2.1505×10^4	7.6052×10^4
Without line pack	2.5871×10^4	1.4990×10^5

LMGP curves at nodes N_1 and N_2. For nodes N_3 and N_5, they are downstream nodes of active pipelines C_1 and C_2, respectively, and their LMGP variation patterns are similar to LMEP curves of their electricity supply buses P_3 and P_8, respectively. Because active pipelines do not have line pack effect, and delivery costs for N_3 and N_5 are directly related to LMEPs at P_3 and P_8. For nodes N_4, N_6, and N_7, extra gas purchased in low-price periods can be stored in GLine_2, GLine_3, and GLine_4, respectively, which results in lower LMGPs compared with N_3 and N_5.

The economic impact of the line pack effect is investigated. In a comparative case, the line pack in passive pipelines is neglected, which is implemented by dropping mass flow variables $m_{l_g t}$ and constraints (3.15n)–(3.15o), and imposing equal inflow and outflow for all pipelines in the OGF problem. Operating costs of the power system and the gas system at market equilibrium are listed in Table 3.4. When line pack is not exploited, operating costs of power system and gas system show an increase of 20.3% and 97.1%, respectively, compared with the base case in which line pack is utilized in the optimal way. The economic benefits of line pack effect is clear: it helps reduce energy prices and system operation costs by enabling more flexible gas transactions between both systems. When P2G facilities join the operation, the benefit would be more prominent.

To demonstrate the scalability and efficiency of the proposed methods, they are applied to a larger system, consisting of a modified IEEE 123-bus power feeder and a modified Belgian high-calorific 20-node gas network, which will be referred to as the Power123Gas20 system for short. The system includes 10 gas-fired units, 3 compressors, 16 passive pipelines, 85 power loads, and 9 gas loads. Please refer to [31] for the network topology and system data.

The convergence performances are shown in Fig. 3.17, in which the relative gap is defined in the following way: let x^{j+1} and x^j be a vector of variables in two consecutive iterations, then the relative gap can be expressed as $\|x^{j+1} - x^j\|_\infty / \|x^{j+1}\|_\infty$; A1 and A2 refer to Algorithms 3.2 and 3.3, respectively. Sub-figure (a) shows relative gaps of energy prices produced by the outer-loop Algorithm 3.3; sub-figures (b)–(d) exhibit relative gaps of objective values and slack variables produced by the inner-loop Algorithm 3.2 in each iteration of Algorithm 3.3. Both algorithms converge in three iterations. The average computation time for computing a market equilibrium of the Power123Gas20 system under various load levels is around 10 s on a laptop with Intel(R) Core(TM) 2 Duo 2.2 GHz CPU and 4 GB memory. The high efficiency is largely attributed to the fact that both algorithms leverage the computational superiority of SOCPs.

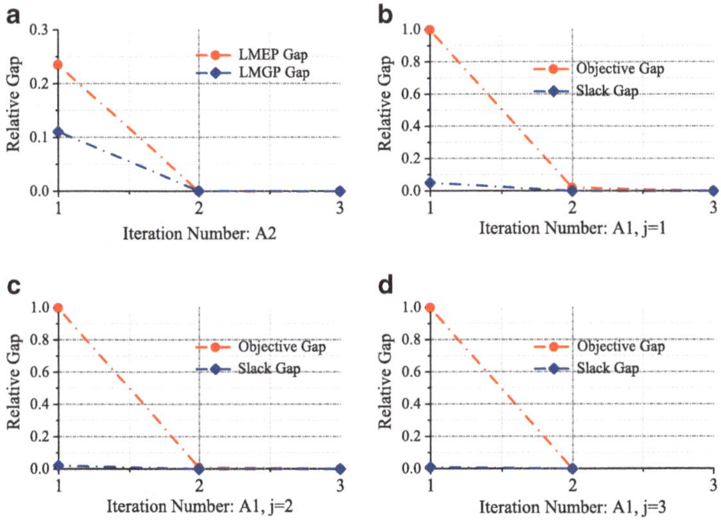

Fig. 3.17 Algorithm performances in the Power123Gas20 system

3.5 Bidding Strategies in Coupled Gas-Electricity Markets

Previous section discusses a market framework for bilateral natural gas and electricity trading. Moderately accurate energy flows are incorporated in market clearing models, and energy producers are assumed to be non-profit. In this section, we consider strategic energy producers who will bid rationally in the markets to maximize their own profits. We propose an equilibrium program with equilibrium constraints (EPEC) model to mimic the gaming configuration, and a special fixed-point algorithm to seek the market equilibrium. For computation tractability issues, simplified energy flow models are used in this section.

3.5.1 Market Settings and Assumptions

1. Pool-Based Market Mechanism

The schematic diagram of the coupled energy markets is shown in Fig. 3.18. At the electricity side, generation companies receive LMGPs from the gas market, and then submit their offering prices to the power market. After the electricity market is cleared based on an OPF problem, LMEPs and natural gas demands of gas-fired units can be obtained. At the gas side, gas producers receive LMEPs from the power market, and then submit their offering prices to the gas market. After the gas market is cleared based on a simplified optimal gas production problem, LMGPs as well as the electricity needs of P2G facilities become clear. It is clear that the power

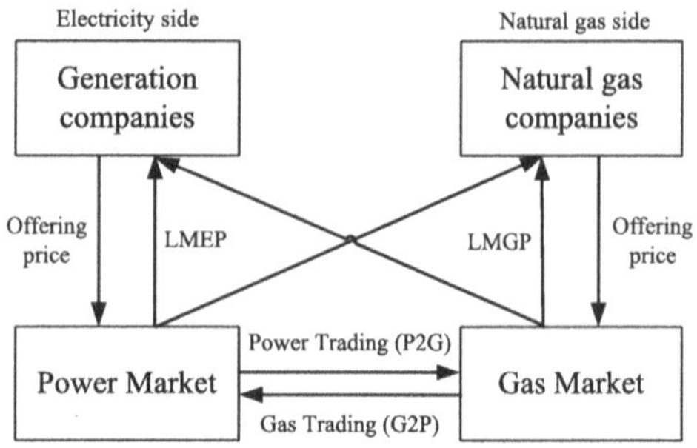

Fig. 3.18 Market structure in the strategic bidding problem

system and the gas system have impacts on each other through bi-directional gas and electricity transactions. Compared with the markets structure illustrated in Fig. 3.11, subjective initiatives of profitable players and their market powers are taken into account. They can rationally forecast the equilibrium in the lower-level market and react to the rivals' strategies, leading to a Nash equilibrium among the upper-level participants.

Above settings assume the same organization in power and natural gas markets. However, existing gas markets are far less competitive. The current natural gas spot market is a daily market and the price is fixed throughout the day [37]. Gas delivery services are categorized as the firm transportation and interruptible transportation, and gas-fired units in power system usually get the latter one [38]. According to [39], gas demand from power generation accounts for 40% of the total gas consumption in 2012, and this proportion will keep increasing in the future. Given the rapidly growing number of gas-fired units and deepening network interdependence, fuel inadequacy is becoming non-negligible. Market reform would be an effective means to provide high quality gas delivery service at minimal social cost.

The main assumptions in this section are summarized as follows:

(1) General settings: the electricity and gas consumptions are charged at LMEPs and LMGPs, respectively. Demands in both markets are non-elastic. Generation companies and gas producers are rational players: they can anticipate the impact of their offers on the market clearing results and make the optimal bidding decisions. The electricity market and the gas market are cleared synchronously.
(2) For the power grid: the lossless DC power flow model is adopted. Unit commitment decisions have been made in a previous stage. Only dispatchable units (changing over time) are considered.
(3) For the gas network: a simplified steady-state model is used, in which the line pack effect is neglected. Gas flows in pipelines are directly controllable and gas nodal pressure variables are omitted, leading to an LP model which has been adopted in [40, 41]. This simplification is reasonable when there are enough regulating devices. Simplified compressor model in [42] is employed. Gas storages and P2G facilities are non-strategic.

3.5.2 EPEC Model for Optimal Bidding

Symbols and notations involved in the model are listed below for quick reference.

Sets and Indices

$b \in B$	Energy blocks of units
$c \in C$	Active pipelines in gas network
$d_g \in D_g$	Gas loads
$d_p \in D_p$	Electricity loads
$e \in E$	Gas storages
$i \in I_s$	Strategic generators
$j \in J_o$	Non-strategic generators
$l_g \in L_g$	Passive pipelines in gas network
$l_p \in L_p$	Power transmission lines
$L_{l_p}(1, n_p)$	Set of transmission lines whose head node is n_p
$L_{l_p}(2, n_p)$	Set of transmission lines whose tail node is n_p
$Line_{l_p}$	Head and tail node sets of line l_p
$m \in M_v$	Gas wells owned by strategic gas producers
$n_g \in N_g$	Gas network nodes
$n_p \in N$	Power grid buses
$o \in O$	Non-strategic electricity producers
$s \in S$	Strategic electricity producers
t	Time periods
$v \in V$	Strategic gas producers
$w \in W$	Non-strategic gas producers
$x \in X_w$	Gas wells owned by non-strategic gas producers
$z \in Z$	P2G facilities
$\Theta(n_p)$	Neighboring nodes of node n_p
Φ	Set of decision variables in the gas market
$\bar{\varphi}^s_{n_g}$	Non-gas-fired units owned by strategic electricity producer s
$\bar{\varphi}^o_{n_g}$	Non-gas-fired units owned by non-strategic electricity producer o
$\varphi^o_{n_g}$	Gas-fired units owned by non-strategic electricity producer o
$\varphi_{\{\cdot\}}(n_g)$	Components connected to node n_g
$\varphi_{\{\cdot\}_1}(n_g)$	Pipelines whose head node is n_g
$\varphi_{\{\cdot\}_2}(n_g)$	Pipelines whose tail node is n_g
$\phi_{\{\cdot\}}(n_p)$	Components connected to bus n_p
$\varphi^{-1}(\cdot)$	Connection node of components in the gas network
$\varphi_1^{-1}(\cdot)$	Head node of a pipeline in the gas network
$\varphi_2^{-1}(\cdot)$	Tail node of a pipeline in the gas network
$\phi^{-1}(\cdot)$	Connection point of components in the power grid

Parameters

$B_{n_{p1}n_{p2}}$	Susceptance of a transmission line
F_{l_p}	Power grid line capacities
$P_{d_p t}$	Electricity load demands

$P_{ib}^{s,max}$ Upper limits of blocks of strategic generators
$P_{jb}^{o,max}$ Upper limits of blocks of non-strategic generators
$P_{i}^{s,min}$ Lower limits of strategic generators
$P_{j}^{o,min}$ Lower limits of non-strategic generators
Q_e Gas storage cost coefficients
$q_{d_g t}$ Gas load demands
q_m^u Upper limits of gas wells owned by strategic gas producers
q_x^u Upper limits of gas wells owned by non-strategic gas producers
q_c^{max} Capacities of active gas pipelines
$q_{l_g}^{max}$ Capacities of passive gas pipelines
$q_{e,max}^{in}$ Maximum gas storage inflow
$q_{e,max}^{out}$ Maximum gas storage outflow
R_i^U Ramping up capacities of strategic generators
R_i^L Ramping down capacities of strategic generators
R_j^U Ramping up capacities of non-strategic generators
R_j^L Ramping down capacities of non-strategic generators
r_e^l, r_e^u Minimum/maximum stored gas
T Number of time periods
α^{max} Maximal offering price of electrical power
δ^{max} Maximal offering price of natural gas
η_{ib}, η_{jb} Efficiencies of units
η_z Efficiencies of P2G facilities
λ_{ib}^s Marginal costs of non-gas-fired strategic generators
λ_{jb}^o Marginal costs of non-gas-fired non-strategic generators
τ Energy conversion constant
ζ_m^v Marginal costs of gas wells owned by strategic gas producers
ζ_x^w Marginal costs of gas wells owned by non-strategic gas producers

Variables

P_{ibt}^s Cleared output of strategic generators
P_{jbt}^o Cleared output of non-strategic generators
P_{zt} Demands of P2G facilities
q_{mt}^v Cleared outputs of gas wells owned by strategic gas producers
q_{xt}^w Cleared outputs of gas wells owned by non-strategic gas producers
q_{ct} Gas flows in active pipelines
$q_{l_g t}$ Gas flows in passive pipelines
q_{et}^{in} Gas storage inflow
q_{et}^{out} Gas storage outflow
r_{et} Stored gas amount
α_{ibt}^s Offering prices of electrical power
$\beta_{n_p t}$ LMEP

δ^{v}_{mt}	Offering prices of natural gas
$\theta_{n_p t}$	Phase angles of power nodes
$\varrho_{n_g t}$	LMGP

1. EPEC Model at the Electricity Side

At the electricity side, every strategic producer (generation company) tries to maximize its own profit (minimize cost) by bidding an offering price to the electricity market. The market is cleared subject to a DCOPF problem, which is modeled in the lower level. Each producer solves an MPEC, constituting the EPEC at the electricity side:

$$\forall s: \min_{\alpha^s_{ibt}, P^s_{ibt}} \sum_t \left(\sum_{(i\in\varphi_i(n_g))b} \tau \varrho_{n_g} \frac{P^s_{ibt}}{\eta_{ib}} + \sum_{(i\in\bar{\varphi}^s_{n_g})b} \lambda^s_{ib} P^s_{ibt} - \sum_{(i\in\phi_i(n_p))b} \beta_{n_p t} P^s_{ibt} \right) \tag{3.33a}$$

$$\text{s.t. } 0 \le \alpha^s_{ibt} \le \alpha^s_{i(b+1)t} \le \alpha^{max}, \ \forall i, \forall b, \forall t \tag{3.33b}$$

$$P^s_{ibt}, \beta_{n_p t} \in \arg\min \sum_t \left(\sum_{o(j\in\varphi^o_{n_g})b} \tau \varrho_{n_g} \frac{P^o_{jbt}}{\eta_{jb}} + \sum_{sib} \alpha^s_{ibt} P^s_{ibt} + \sum_{o(j\in\bar{\varphi}^o_{n_g t})b} \lambda^o_{jb} P^o_{jbt} \right) \tag{3.33c}$$

$$\text{s.t. } 0 \le P^s_{ibt} \le P^{s,max}_{ib} : \beta^{s,min}_{ibt}, \beta^{s,max}_{ibt}, \forall s, \forall i, \forall b, \forall t \tag{3.33d}$$

$$0 \le P^o_{jbt} \le P^{o,max}_{jb} : \beta^{o,min}_{jbt}, \beta^{o,max}_{jbt}, \forall o, \forall j, \forall b, \forall t \tag{3.33e}$$

$$\begin{aligned} &\sum_{s(i\in\phi_i(n_p))b} P^s_{ibt} + \sum_{o(j\in\phi_j(n_p))b} P^o_{jbt} - \sum_{z\in\phi_z(n_p)} P_{zt} - \sum_{d\in\phi_d(n_p)} P_{d_p t} \\ &= \sum_{k\in\Theta(n_p)} B_{n_p k_p}(\theta_{n_p t} - \theta_{k_p t}) : \beta_{n_p t}, \ \forall n_p, \forall t \end{aligned} \tag{3.33f}$$

$$-F_{l_p} \le B_{n_{p1} n_{p2}}(\theta_{n_{p1} t} - \theta_{n_{p2} t}) \le F_{l_p} : \beta^{min}_{l_p t}, \beta^{max}_{l_p t}, \forall l_p, \forall t \tag{3.33g}$$

$$\theta_{n_p t} = 0 : \varepsilon_{1t}, n_p = 1, \forall t \tag{3.33h}$$

$$\sum_b P^s_{ib(t+1)} - \sum_b P^s_{ibt} \le R^U_i : \mu^u_{it}, \ \forall t, \forall i, \forall s \tag{3.33i}$$

$$\sum_b P^s_{ibt} - \sum_b P^s_{ib(t+1)} \le R^L_i : \mu^l_{it}, \ \forall t, \forall i, \forall s \tag{3.33j}$$

$$\sum_b P^o_{jb(t+1)} - \sum_b P^o_{jbt} \leq R^U_j : \mu^u_{jt}, \forall t, \forall j, \forall o \tag{3.33k}$$

$$\sum_b P^s_{ibt} - \sum_b P^s_{ib(t+1)} \leq R^L_j : \mu^l_{jt}, \forall t, \forall j, \forall o \tag{3.33l}$$

$$\sum_b P^s_{ibt} \geq P^{s,min}_i : \nu^{s,min}_{it}, \forall t, \forall i, \forall s \tag{3.33m}$$

$$\sum_b P^o_{jbt} \geq P^{o,min}_j : \nu^{o,min}_{jt}, \forall t, \forall j, \forall o \tag{3.33n}$$

In the above expressions, objective (3.33a) represents the cost minus income of producer s, (3.33b) imposes upper and lower bounds on the offering prices, as well as the monotonicity with respect to energy blocks. Lower-level electricity market clearing problem is shown through (3.33c)–(3.33n), which determines LMEPs $\beta_{n_p t}$ and the energy contracts P^s_{ibt}. Objective (3.33c) expresses the dispatch cost, where the second component is the generation expenses of strategic producers and the remaining two are the costs of gas-fired and non-gas-fired units owned by non-strategic producers. Equations (3.33d) and (3.33e) are the energy block capacity limits. Equation (3.33f) describes the nodal power balancing condition. Equation (3.33g) enforces line power flow limits. Equation (3.33h) defines the phase angle of the reference node. Equations (3.33i)–(3.33l) are the upward and downward ramping limits of generators. Equations (3.33m) and (3.33n) are the minimum output of generators. Dual variables are indicated at the corresponding constraints following a colon. LMEP is the dual variable associated with nodal power balancing condition (3.33f). In each electricity producer's problem, LMGP ϱ_{n_g} in (3.33a) and demand P_{zt} of P2G facility are treated as parameters.

2. EPEC Model of the Natural Gas Side

At the natural gas side, every strategic producer (gas well) tries to maximize its own profit (minimize cost) by bidding an offering price to the natural gas market. The market is cleared subject to an OGF problem, which is modeled in the lower level. Each producer solves an MPEC as follows, constituting the EPEC at the natural gas side:

$$\forall v, \min_{\delta^v_m, q^v_{mt}} \sum_t \left(\sum_m \zeta^v_m q^v_{mt} - \sum_{m \in \varphi_m(n_g)} \varrho_{n_g t} q^v_{mt} \right) \tag{3.34a}$$

$$\text{s.t. } 0 \leq \delta^v_{mt} \leq \delta^{max}, \forall m, \forall t \tag{3.34b}$$

$$q_{mt}^{v}, \varrho_{n_g t} \in \arg\min_{\Phi} \sum_{t} \Big(\sum_{vm} \delta_{mt}^{v} q_{mt}^{v} + \sum_{wx} \zeta_{x}^{w} q_{xt}^{w} + \sum_{z \in \phi_z(n_p)} \beta_{n_p t} P_{zt} + \sum_{e} Q_e q_{et}^{in} \Big) \tag{3.34c}$$

$$\text{s.t. } 0 \le q_{mt}^{v} \le q_{m}^{u} : \rho_{mt}^{min}, \rho_{mt}^{max} \ \forall m, \forall v, \forall t \tag{3.34d}$$

$$0 \le q_{xt}^{w} \le q_{x}^{u} : \rho_{xt}^{min}, \rho_{xt}^{max} \ \forall x, \forall w, \forall t \tag{3.34e}$$

$$-q_{l_g}^{max} \le q_{l_g t} \le q_{l_g}^{max} : \rho_{l_g t}^{min}, \rho_{l_g t}^{max} \ \forall l_g, \forall t \tag{3.34f}$$

$$0 \le q_{ct} \le q_{c}^{max} : \rho_{ct}^{min}, \rho_{ct}^{max} \ \forall c, \forall t \tag{3.34g}$$

$$0 \le P_{zt} : \rho_{zt} \ \forall z, \forall t \tag{3.34h}$$

$$r_e^l \le r_{et} \le r_e^u : \rho_{et}^u, \rho_{et}^l, \ \forall e, \ \forall t \tag{3.34i}$$

$$r_{et} = r_{e,t-1} + q_{et}^{in} - q_{et}^{out} : \rho_{et}, \ \forall e, \ \forall t \tag{3.34j}$$

$$0 \le q_{et}^{in} \le q_{e,max}^{in} : \rho_{et}^{in,min}, \rho_{et}^{in,max}, \ \forall e, \ \forall t \tag{3.34k}$$

$$0 \le q_{et}^{out} \le q_{e,max}^{out} : \rho_{et}^{out,min}, \rho_{et}^{out,max}, \ \forall e, \ \forall t \tag{3.34l}$$

$$\begin{aligned} &\sum_{v(m \in \varphi_m(n_g))} q_{mt}^{v} + \sum_{w(x \in \varphi_x(n_g))} q_{xt}^{w} + \sum_{z \in \varphi_z(n_g)} \tau \eta_z P_{zt} = \\ &\sum_{\{\cdot\} \in \varphi_{\{\cdot\}_1}(n_g)} q_{\{\cdot\}t} - \sum_{\{\cdot\} \in \varphi_{\{\cdot\}_2}(n_g)} q_{\{\cdot\}t} + \sum_{d_g \in \varphi_{d_g}(n_g)} q_{d_g t} \\ &+ \sum_{s(i \in \varphi_i(n_g))b} \tau P_{ibt}^{s} / \eta_{ib} + \sum_{o(j \in \varphi_j(n_g))b} \tau P_{jbt}^{o} / \eta_{jb} + \\ &\sum_{e \in \varphi_e(n_g)} (q_{et}^{out} - q_{et}^{in}) : \varrho_{n_g t} \ \{\cdot\} = \{c, l_g\}, \ \forall n_g, \forall t \end{aligned} \tag{3.34m}$$

In the above expressions, objective (3.34a) represents the negative profit of strategic gas producer v, where the first component is the production cost and the second component is the utility of all its gas wells. Equation (3.34b) restricts the interval of gas offering prices. Lower-level OGF problem is shown through (3.34c)–(3.34m), which determines LMGPs ϱ_{n_g} and the gas contracts q_{mt}^{v}. Objective function (3.34c) defines the total operating cost of the gas system, where the first two terms are the gas production cost of gas wells belonging to strategic and non-strategic gas producers; the third one is the electricity purchase costs of P2G facilities, and the last one is the cost for gas storage. Equations (3.34d) and (3.34e) give the production limits of gas wells. Flow capacities of passive and active pipelines are represented by (3.34f) and (3.34g), respectively. Equation (3.34h) indicates non-negative constraints on the electricity demands of P2G facilities. Equation (3.34i) depicts the capacity limitations of storage facilities. Equation (3.34j) implies the dynamics of storage quantity. Equations (3.34k) and (3.34l) impose the range of inflow and outflow rates of gas storage. Equation (3.34m) is the nodal gas balancing condition. Dual variables are indicated at the corresponding equations following a colon. LMGP is the dual variable associated with gas balancing condition (3.34m). In each gas producer's problem, LMEP $\beta_{n_p t}$ in (3.34c) and power output P^s_{ibt} and P^o_{jbt} of gas-fired units are treated as parameters.

3.5.3 Fixed-Point Algorithm

EPEC belongs to the category of bilevel equilibrium program. A general introduction can be found in Appendix D.5, which helps readers understand the mathematic substance behind the numerous symbols involved in this practical application problem. In this section, the equivalent MILP formulation for each producer's MPEC is developed first, then a fixed-point algorithm is introduced to solve the EPEC model for the bilateral gas and electricity market.

1. Equivalent MILPs for MPECs of Electricity Producers

In the pool based electricity market, the power demands of P2G facilities as well as offering prices of all electricity producers are fixed, so the market clearing problem is an LP, and the following KKT condition is necessary and sufficient for optimality

$$\alpha^s_{ibt} - \beta_{n_p t} + \beta^{s,max}_{ibt} - \beta^{s,min}_{ibt} - \mu^u_{ibt} + \mu^u_{ib(t-1)} + \mu^l_{ibt} - \mu^l_{ib(t-1)} - \nu^{s,min}_{it} = 0$$
$$\forall i, \forall b, \forall s, n_p = \phi_i^{-1}, 2 \le t \le T-1 \tag{3.35a}$$

$$\alpha^s_{ibt} - \beta_{n_p t} + \beta^{s,max}_{ibt} - \beta^{s,min}_{ibt} - \mu^u_{ibt} + \mu^l_{ibt} - \nu^{s,min}_{it} = 0$$
$$\forall i, \forall b, \forall s, n_p = \phi_i^{-1}, t = 1 \tag{3.35b}$$

$$\alpha^s_{ibt} - \beta_{n_p t} + \beta^{s,max}_{ibt} - \beta^{s,min}_{ibt} + \mu^u_{ib(t-1)} - \mu^l_{ib(t-1)} - \nu^{s,min}_{it} = 0$$
$$\forall i, \forall b, \forall s, n_p = \phi_i^{-1}, t = T \tag{3.35c}$$

$$\lambda^o_{jb} - \beta_{n_p t} + \beta^{o,max}_{jbt} - \beta^{o,min}_{jbt} - \mu^u_{jbt} + \mu^u_{jb(t-1)} + \mu^l_{jbt} - \mu^l_{jb(t-1)}$$
$$-\nu^{s,min}_{jt} = 0, \ j \in \varphi^o_{n_g}, \forall b, \forall s, n_p = \phi_j^{-1}, 2 \le t \le T-1 \tag{3.35d}$$

$$\begin{aligned}&\lambda^o_{jb} - \beta_{n_pt} + \beta^{o,max}_{jbt} - \beta^{o,min}_{jbt} - \mu^u_{jbt} + \mu^l_{jbt} - \nu^{s,min}_{jt} = 0\\&j \in \varphi^o_{n_g}, \forall b, \forall s, n_p = \phi^{-1}_j, t = 1\end{aligned} \tag{3.35e}$$

$$\begin{aligned}&\lambda^o_{jb} - \beta_{n_pt} + \beta^{o,max}_{jbt} - \beta^{o,min}_{jbt} + \mu^u_{jb(t-1)} - \mu^l_{jb(t-1)} - \nu^{s,min}_{jt} = 0\\&j \in \varphi^o_{n_g}, \forall b, \forall s, n_p = \phi^{-1}_j, t = T\end{aligned} \tag{3.35f}$$

$$\begin{aligned}&\frac{\tau \varrho_{n_gt}}{\eta_{jb}} - \beta_{n_pt} + \beta^{o,max}_{jbt} - \beta^{o,min}_{jbt} - \mu^u_{jbt} + \mu^u_{jb(t-1)} + \mu^l_{jbt} - \mu^l_{jb(t-1)}\\&-\nu^{s,min}_{jt} = 0,\ j \in \bar{\varphi}^o_{n_g}, \forall b, \forall s, n_p = \phi^{-1}_j, 2 \le t \le T-1\end{aligned} \tag{3.35g}$$

$$\begin{aligned}&\frac{\tau \varrho_{n_gt}}{\eta_{jb}} - \beta_{n_pt} + \beta^{o,max}_{jbt} - \beta^{o,min}_{jbt} - \mu^u_{jbt} + \mu^l_{jbt} - \nu^{s,min}_{jt} = 0\\&j \in \bar{\varphi}^o_{n_g}, \forall b, \forall s, n_p = \phi^{-1}_j, t = 1\end{aligned} \tag{3.35h}$$

$$\begin{aligned}&\frac{\tau \varrho_{n_gt}}{\eta_{jb}} - \beta_{n_pt} + \beta^{o,max}_{jbt} - \beta^{o,min}_{jbt} + \mu^u_{jb(t-1)} - \mu^l_{jb(t-1)} - \nu^{s,min}_{jt} = 0\\&j \in \bar{\varphi}^o_{n_g}, \forall b, \forall s, n_p = \phi^{-1}_j, t = T\end{aligned} \tag{3.35i}$$

$$\begin{aligned}&\sum_{k_p \in \Theta(n_p)} B_{nk}(\beta_{n_pt} - \beta_{k_pt}) - (\varepsilon_{1t})_{n_p=1} + \sum_{n_p \in L_{l_p}(1,n_p)} B_{n_pk_p}(\beta^{max}_{l_pt} - \beta^{min}_{l_pt}) -\\&\sum_{n_p \in L_{l_p}(1,n_p)} B_{n_pk_p}(\beta^{max}_{l_pt} - \beta^{min}_{l_pt}) = 0,\ \forall n_p, \forall t\end{aligned} \tag{3.35j}$$

$$0 \le P^s_{ibt} \perp \beta^{s,min}_{ibt} \ge 0,\ \forall s, \forall i, \forall b, \forall t \tag{3.35k}$$

$$0 \le P^{s,max}_{ib} - P^s_{ibt} \perp \beta^{s,max}_{ibt} \ge 0,\ \forall s, \forall i, \forall b, \forall t \tag{3.35l}$$

$$0 \le P^o_{jbt} \perp \beta^{o,min}_{jbt} \ge 0,\ \forall o, \forall j, \forall b, \forall t \tag{3.35m}$$

$$0 \le P^{o,max}_{jb} - P^o_{jbt} \perp \beta^{o,max}_{jbt} \ge 0,\ \forall o, \forall j, \forall b, \forall t \tag{3.35n}$$

$$\begin{aligned}&0 \le F_{l_p} - B_{n_{p1}n_{p2}}(\theta_{n_{p1}t} - \theta_{n_{p2}t}) \perp \beta^{max}_{l_pt} \ge 0\\&n_{p1}, n_{p2} \in Line_{l_p}, \forall l_p, \forall t\end{aligned} \tag{3.35o}$$

$$\begin{aligned}&0 \le B_{n_{p1}n_{p2}}(\theta_{n_{p1}t} - \theta_{n_{p2}t}) + F_{l_p} \perp \beta^{min}_{l_pt} \ge 0\\&n_{p1}, n_{p2} \in Line_{l_p}, \forall l_p, \forall t\end{aligned} \tag{3.35p}$$

$$0 \leq R_i^U - \sum_b P_{ib(t+1)}^s + \sum_b P_{ibt}^s \perp \mu_{it}^u \geq 0, \forall t < T, \forall i, \forall s \tag{3.35q}$$

$$0 \leq R_i^L - \sum_b P_{ibt}^s + \sum_b P_{ib(t+1)}^s \perp \mu_{it}^l \geq 0, \forall t < T, \forall i, \forall s \tag{3.35r}$$

$$0 \leq R_j^U - \sum_b P_{jb(t+1)}^o + \sum_b P_{jbt}^o \perp \mu_{jt}^u \geq 0, \forall t < T, \forall j, \forall o \tag{3.35s}$$

$$0 \leq R_j^L - \sum_b P_{ibt}^s + \sum_b P_{ib(t+1)}^s \perp \mu_{jt}^l \geq 0, \forall t < T, \forall j, \forall o \tag{3.35t}$$

$$0 \leq \sum_b P_{ibt}^s - P_i^{s,min} \perp \nu_{it}^{s,min} \geq 0, \ \forall t, \forall i, \forall s \tag{3.35u}$$

$$0 \leq \sum_b P_{jbt}^o - P_j^{o,min} \perp \nu_{jt}^{o,min} \geq 0, \ \forall t, \forall j, \forall o \tag{3.35v}$$

where notation $0 \leq a \perp b \geq 0$ stands for complementarity and slackness conditions $a \geq 0, b \geq 0, ab = 0$, which are equivalent to the following linear inequalities with a binary variable z and a large enough constant M

$$0 \leq a \leq Mz, \ 0 \leq b \leq M(1-z), \ z \in \{0, 1\} \tag{3.36}$$

More about this integer formulation trick is elaborated in Appendix B.3.5. Applying the linearization to complementarity conditions in constraints (3.35a)–(3.35v), we obtain the linearized expression of KKT conditions for electricity market clearing (KKTE-MILP for short).

The objective function (3.33a) is nonlinear due to the product term $\beta_{n_p t} P_{ibt}^s$. To find a linear expression for the objective, multiplying both sides of (3.35a)–(3.35c) by P_{ibt}^s and adding them together, yielding the following equation:

$$\sum_{ibt} (\alpha_{ibt}^s + \beta_{ibt}^{s,max} - \beta_{ibt}^{s,min}) P_{ibt}^s - \sum_{(i \in \phi_i(n_p))bt} \beta_{n_p t} P_{ibt}^s +$$
$$\sum_{ib} \left(-\sum_{t=1}^{T-1} (\mu_{it}^u - \mu_{it}^l) P_{ibt}^s + \sum_{t=2}^{T} (\mu_{i(t-1)}^u - \mu_{i(t-1)}^l) P_{ibt}^s \right) - \sum_{ibt} \nu_{it}^{s,min} P_{ibt}^s = 0 \tag{3.37}$$

Notice

$$\sum_{t=2}^{T} (\mu_{i(t-1)}^u - \mu_{i(t-1)}^l) P_{ibt}^s = \sum_{t=1}^{T-1} (\mu_{it}^u - \mu_{it}^l) P_{ib(t+1)}^s$$

Substituting it into (3.37), we have

$$\sum_{ibt}(\alpha_{ibt}^{s} + \beta_{ibt}^{s,max} - \beta_{ibt}^{s^*,min})P_{ibt}^{s} - \sum_{(i\in\phi_i(n_p))bt} \beta_{n_pt}P_{ibt}^{s} - \sum_{ibt} \nu_{it}^{s,min}P_{ibt}^{s}$$

$$-\sum_{i}\sum_{t}^{T-1}\left(\mu_{it}^{u}\left(\sum_{b}P_{ibt}^{s} - \sum_{b}P_{ib(t+1)}^{s}\right) - \mu_{it}^{l}\left(\sum_{b}P_{ibt}^{s} - \sum_{b}P_{ib(t+1)}^{s}\right)\right) = 0 \tag{3.38}$$

From (3.35k)–(3.35l), (3.35q)–(3.35r) and (3.35u), we have

$$\beta_{ibt}^{s,min}P_{ibt}^{s} = 0 \Rightarrow \sum_{ibt}\beta_{ibt}^{s,min}P_{ibt}^{s} = 0 \tag{3.39a}$$

$$\beta_{ibt}^{s,max}P_{ibt}^{s} = \beta_{ibt}^{s,max}P_{ib}^{s,max} \Rightarrow \sum_{ibt}\beta_{ibt}^{s,max}P_{ibt}^{s} = \sum_{ibt}\beta_{ibt}^{s,max}P_{ib}^{s,max} \tag{3.39b}$$

$$\begin{aligned}&\mu_{it}^{u}(\sum_{b}P_{ib(t+1)}^{s} - \sum_{b}P_{ibt}^{s}) = \mu_{it}^{u}R_{i}^{U} \Rightarrow\\ &\sum_{i}\sum_{t}^{T-1}\mu_{it}^{u}(\sum_{b}P_{ib(t+1)}^{s} - \sum_{b}P_{ibt}^{s}) = \sum_{i}\sum_{t}^{T-1}\mu_{it}^{u}R_{i}^{U}\end{aligned} \tag{3.39c}$$

$$\begin{aligned}&\mu_{it}^{l}\left(\sum_{b}P_{ibt}^{s} - \sum_{b}P_{ib(t+1)}^{s}\right) = \mu_{it}^{l}R_{i}^{L}\\ &\sum_{i}\sum_{t}^{T-1}\mu_{it}^{l}\left(\sum_{b}P_{ibt}^{s} - \sum_{b}P_{ib(t+1)}^{s}\right) = \sum_{i}\sum_{t}^{T-1}\mu_{it}^{l}R_{i}^{L}\end{aligned} \tag{3.39d}$$

$$\nu_{it}^{s,min}\sum_{b}P_{ibt}^{s} = \nu_{it}^{s,min}P_{it}^{s,min} \Rightarrow \sum_{ibt}\nu_{it}^{s,min}P_{ibt}^{s} = \sum_{it}\nu_{it}^{s,min}P_{it}^{s,min} \tag{3.39e}$$

Substituting above relations into (3.38) we get

$$\begin{aligned}\sum_{(i\in\phi_i(n_p))bt}\beta_{n_pt}P_{ibt}^{s} &= \sum_{ibt}(\alpha_{ibt}^{s}P_{ibt}^{s} + \beta_{ibt}^{s,max}P_{ib}^{s,max})\\ &+\sum_{i}\sum_{t}^{T-1}(\mu_{it}^{u}R_{i}^{U} + \mu_{it}^{l}R_{i}^{L}) - \sum_{it}\nu_{it}^{s,min}P_{it}^{s,min}\end{aligned} \tag{3.40}$$

From strong duality theorem, we have

$$
\begin{aligned}
&\sum_{ibt}\alpha_{ibt}^{s}P_{ibt}^{s}+\sum_{(-s)ibt}\alpha_{ibt}^{-s}P_{ibt}^{-s}+\sum_{o(j\in\bar{\varphi}_{ng}^{o})bt}\lambda_{jb}^{o}P_{jbt}^{o}+\sum_{o(j\in\varphi_{ng}^{o})bt}\tau\varrho_{n_gt}P_{jbt}^{o}/\eta_{jb}\\
&=-\sum_{oj}\sum_{t=1}^{T-1}(R_j^U\mu_{jt}^u+R_j^L\mu_{jt}^l)-\sum_{ibts}P_{ib}^{s,max}\beta_{ibt}^{s,max}-\sum_{ojbt}P_{jb}^{o,max}\beta_{jbt}^{o,max}\\
&\quad-\sum_{l_pt}F_{l_p}(\beta_{l_pt}^{max}+\beta_{l_pt}^{min})+\sum_{ojt}P_j^{o,min}\nu_{jt}^{o,min}+\beta_{n_pt}\sum_{z\in\phi_z(n_p)t}P_{zt}\\
&\quad+\beta_{n_pt}\sum_{d_p\in\phi_{d_p}(n_p)t}P_{d_pt}+\sum_{sit}P_{it}^{s,min}\nu_{it}^{s,min}-\sum_{si}\sum_{t=1}^{T-1}(R_i^U\mu_{it}^u+R_i^L\mu_{it}^l)
\end{aligned}
\tag{3.41}
$$

Eliminating $\sum_{ibt}\alpha_{ibt}^{s}P_{ibt}^{s}$ in (3.40) using (3.41), we obtain the linear expression of $\sum_{(i\in\phi_i(n_p))bt}\beta_{n_pt}P_{ibt}^{s}$, and the MPEC of electricity producer s can be transformed into an MILP

$$
\begin{aligned}
\min_{\alpha_{ibt}^{s},P_{ibt}^{s}}\ &\sum_{(i\in\varphi_i(n_g))bt}\tau\varrho_{n_gt}P_{ibt}^{s}/\eta_{ib}+\sum_{(i\in\bar{\varphi}_{ng}^{s}))bt}\lambda_{ib}^{s}P_{ibt}^{s}-\sum_{z\in\phi_z(n_p)t}\beta_{n_pt}P_{zt}\\
&+\sum_{l_pt}F_{l_p}(\beta_{l_pt}^{max}+\beta_{l_pt}^{min})+\sum_{-s}\sum_{ibt}P_{ib}^{-s,max}\beta_{ibt}^{-s,max}+\sum_{ojbt}P_{jb}^{o,max}\beta_{jbt}^{o,max}\\
&+\sum_{(-s)ibt}\alpha_{ibt}^{-s}P_{ibt}^{-s}+\sum_{o(j\in\bar{\varphi}_{ng}^{o})bt}\lambda_{jb}^{o}P_{jbt}^{o}-\beta_{n_pt}\sum_{d_p\in\phi_{d_p}(n_p)t}P_{d_pt}\\
&+\sum_{o(j\in\varphi_{ng}^{o})bt}\tau\varrho_{n_gt}\frac{P_{jbt}^{o}}{\eta_{jb}}+\sum_{oj}\sum_{t}^{T-1}(R_j^U\mu_{jt}^u+R_j^L\mu_{jt}^l)-\sum_{ojt}P_j^{o,min}\nu_{jt}^{o,min}\\
&-\sum_{(-s)it}P_{it}^{-s,min}\nu_{it}^{-s,min}+\sum_{(-s)i}\sum_{t=1}^{T-1}(R_i^U\mu_{it}^u+R_i^L\mu_{it}^l)\\
&\text{s.t.}\quad (3.33b),\ \text{KKTE-MILP}(\alpha^{-s})
\end{aligned}
\tag{3.42}
$$

where notation α^{-s} denotes the aggregated bidding strategies of electricity producers except the s-th one. It is treated as a given parameter in problem (3.42).

2. Equivalent MILPs for MPECs of Natural Gas Producers

In the pool based natural gas market, the gas demands of gas-fired units as well as the offering prices of all gas producers are fixed, the market clearing problem is an LP, and the following KKT condition is necessary and sufficient for optimality

$$\delta^v_{mt} + \rho^{max}_{mt} - \rho^{min}_{mt} - \varrho_{n_g t} = 0,\ \forall m, \forall v, n_g = \varphi^{-1}(m), \forall t \tag{3.43a}$$

$$\zeta^w_x + \rho^{max}_{xt} - \rho^{min}_{xt} - \varrho_{n_g t} = 0,\ \forall x, \forall w, n_g = \varphi^{-1}(x), \forall t \tag{3.43b}$$

$$\rho^{max}_{ct} - \rho^{min}_{ct} + \varrho_{n_{g1} t} - \varrho_{n_{g2} t} = 0,\ \forall c, n_{g1} = \varphi_1^{-1}(c), n_{g2} = \varphi_2^{-1}(c), \forall t \tag{3.43c}$$

$$\rho^{max}_{l_g t} - \rho^{min}_{l_g t} + \varrho_{n_{g1} t} - \varrho_{n_{g2} t} = 0,\ \forall l_g, n_{g1} = \varphi_1^{-1}(l_g), n_{g2} = \varphi_2^{-1}(l_g), \forall t \tag{3.43d}$$

$$\beta_{n_p t} - \rho_{zt} - \tau \eta_z \varrho_{n_g t} = 0,\ \forall z, n_p = \phi^{-1}(z), n_g = \varphi^{-1}(z), \forall t \tag{3.43e}$$

$$Q_e + \rho^{in,max}_{et} - \rho^{in,min}_{et} + \rho_{et} - \varrho_{n_g t} = 0,\ \forall e, n_g = \varphi^{-1}(e), \forall t \tag{3.43f}$$

$$\rho^{out,max}_{et} - \rho^{out,min}_{et} - \rho_{et} + \varrho_{n_g t} = 0,\ \forall e, n_g = \varphi^{-1}(e), \forall t \tag{3.43g}$$

$$\rho_{e(t+1)} - \rho_{et} + \rho^u_{et} - \rho^l_{et} = 0, \forall e, t < T \tag{3.43h}$$

$$\rho_{e1} - \rho_{et} + \rho^u_{et} - \rho^l_{et} = 0, \forall e, t = T \tag{3.43i}$$

$$0 \le q^v_{mt} \perp \rho^{min}_{mt} \ge 0,\ \forall m, \forall v, \forall t \tag{3.43j}$$

$$0 \le q^u_m - q^v_{mt} \perp \rho^{max}_{mt} \ge 0,\ \forall m, \forall v, \forall t \tag{3.43k}$$

$$0 \le q^w_{xt} \perp \rho^{min}_{xt} \ge 0,\ \forall x, \forall w, \forall t \tag{3.43l}$$

$$0 \le q^u_x - q^w_{xt} \perp \rho^{max}_{xt} \ge 0,\ \forall x, \forall w, \forall t \tag{3.43m}$$

$$0 \le q_{l_g t} + q^{max}_{l_g} \perp \rho^{min}_{l_g t} \ge 0,\ \forall l_g, \forall t \tag{3.43n}$$

$$0 \le q^{max}_{l_g} - q_{l_g t} \perp \rho^{max}_{l_g t} \ge 0,\ \forall l_g, \forall t \tag{3.43o}$$

$$0 \le q_{ct} \perp \rho^{min}_{ct} \ge 0, \forall c, \forall t \tag{3.43p}$$

$$0 \le q^{max}_c - q_{ct} \perp \rho^{max}_{ct} \ge 0, \forall c, \forall t \tag{3.43q}$$

$$0 \leq P_{zt} \perp \rho_{zt} \geq 0, \forall z, \forall t \tag{3.43r}$$

$$0 \leq r_{et} - r_e^l \perp \rho_{et}^l \geq 0, \ \forall e, \ \forall t \tag{3.43s}$$

$$0 \leq r_e^u r_{et} \perp \rho_{et}^u \geq 0, \ \forall e, \ \forall t \tag{3.43t}$$

$$0 \leq q_{et}^{in} \perp \rho_{et}^{in,min} \geq 0, \ \forall e, \ \forall t \tag{3.43u}$$

$$0 \leq q_{e,max}^{in} - q_{et}^{in} \perp \rho_{et}^{in,max} \geq 0, \ \forall e, \ \forall t \tag{3.43v}$$

$$0 \leq q_{et}^{out} \perp \rho_{et}^{out,min} \geq 0, \ \forall e, \ \forall t \tag{3.43w}$$

$$0 \leq q_{e,max}^{out} - q_{et}^{out} \perp \rho_{et}^{out,max} \geq 0, \ \forall e, \ \forall t \tag{3.43x}$$

By linearizing complementarity constraints in the same way as what has been done in the electricity market, we obtain the linearized expression of KKT conditions of gas market clearing (KKTG-MILP for short). To linearize the objective function (3.34a), where $\varrho_{n_g t} q_{mt}^v$ is nonlinear, multiply both sides of (3.43a) by q_m^v and add them together, resulting in the following equation:

$$\sum_{mt} (\delta_{mt}^v + \rho_{mt}^{max} - \rho_{mt}^{min}) q_{mt}^v - \sum_{(n_g = \varphi^{-1}(m))t} \varrho_{n_g t} q_{mt}^v = 0 \tag{3.44}$$

From (3.43j)–(3.43k), we arrive at

$$\begin{aligned} q_{mt}^v \rho_{mt}^{min} = 0 &\Rightarrow \sum_{mt} q_{mt}^v \rho_{mt}^{min} = 0 \\ q_{mt}^v \rho_{mt}^{max} = q_m^u \rho_{mt}^{max} &\Rightarrow \sum_{mt} q_{mt}^v \rho_{mt}^{max} = \sum_{mt} q_m^u \rho_{mt}^{max} \end{aligned} \tag{3.45}$$

Expanding the first term at the left-hand side of (3.44) and substituting the relations revealed in (3.45), we have

$$\sum_{(n_g = \varphi^{-1}(m))t} \varrho_{n_g t} q_{mt}^v = \sum_{mt} \delta_{mt}^v q_{mt}^v + \sum_{mt} \rho_{mt}^{max} q_m^u \tag{3.46}$$

The first term at the right-hand side of (3.46) is still nonlinear. According to strong duality which declares equal values of primal and dual objective functions, relation

$$
\begin{aligned}
&\sum_{mt}\delta_{mt}^{v}q_{mt}^{v}+\sum_{(-v)mt}\delta_{mt}^{-v}q_{mt}^{-v}+\sum_{(z\in\phi_z(n_p))t}\beta_{n_pt}P_{zt}+\\
&\sum_{wxt}\zeta_{xt}^{w}q_{xt}^{w}+\sum_{e}Q_e q_{et}^{in}=-\sum_{ct}q_c^{max}\rho_{ct}^{max}-\sum_{vmt}\rho_{mt}^{max}q_m^{u}-\\
&\sum_{l_gt}(\rho_{l_gt}^{min}+\rho_{l_gt}^{max})q_{l_g}^{max}-\sum_{wxt}\rho_{xt}^{max}q_x^{u}+(\sum_{d_g\in\varphi_{d_g}(n_g)t}q_{d_gt}\\
&+\sum_{s(i\in\varphi_i(n_g))bt}\tau P_{ibt}^{s}/\eta_{ib}+\sum_{o(j\in\varphi_j(n_g))bt}\tau P_{jbt}^{o}/\eta_{jb})\varrho_{n_gt}-\\
&\sum_{et}(q_{e,max}^{in}\rho_{et}^{in,max}+q_{e,max}^{out}\rho_{et}^{out,max})-\sum_{et}(r_e^{u}\rho_{et}^{u}-r_e^{l}\rho_{et}^{l})
\end{aligned}
\tag{3.47}
$$

holds. From (3.47), nonlinear part $\sum_{mt}\delta_{mt}^{v}q_{mt}^{v}$ in (3.46) can be expressed by linear terms, and the MPEC of natural gas producer v can be transformed into an MILP

$$
\begin{aligned}
\min_{\delta_{mt}^{v},q_{mt}^{v}}\;&\sum_{mt}\zeta_{m}^{v}q_{mt}^{v}+\sum_{-v}\sum_{mt}(\rho_{mt}^{max}q_m^{u}+\delta_{mt}^{-v}q_{mt}^{-v})+\sum_{(z\in\phi_z(n_p))t}\beta_{n_pt}P_{zt}+\sum_{wxt}\zeta_x^{w}q_{xt}^{w}\\
&+\sum_{e}Q_e q_{et}^{in}+\sum_{ct}q_c^{max}\rho_{ct}^{max}+\sum_{l_gt}(\rho_{l_gt}^{min}+\rho_{l_gt}^{max})q_y^{max}+\sum_{wxt}\rho_{xt}^{max}q_x^{u}\\
&-(\sum_{d_g\in\varphi_{d_g}(n_g)t}q_{d_gt}+\sum_{s(i\in\varphi_i(n_g))bt}\tau P_{ibt}^{s}/\eta_{ib}+\sum_{o(j\in\varphi_j(n_g))bt}\tau P_{jbt}^{o}/\eta_{jb})\varrho_{n_gt}\\
&+\sum_{et}(q_{e,max}^{in}\rho_{et}^{in,max}+q_{e,max}^{out}\rho_{et}^{out,max}+r_e^{u}\rho_{et}^{u}-r_e^{l}\rho_{et}^{l})\\
\text{s.t.}\;\;&(3.34\text{b}),\;\text{KKTG-MILP}(\delta^{-v})
\end{aligned}
\tag{3.48}
$$

where δ^{-s} denotes the aggregated bidding strategies of gas producers except the v-th one. It is treated as a given parameter in problem (3.48).

3. The Nested Fixed-Point Algorithm

In view of the interactive structure shown in Fig. 3.18, a natural strategy to seek the market equilibrium includes two loops: the inner loop finds equilibria of the electricity market and the natural gas market, which is summarized in Algorithms 3.4 and 3.5; the outer loop updates energy prices and trading contracts, which is set forth in Algorithm 3.6. For notation brevity, we use $\mathcal{P}_s$ and $\mathcal{G}_v$ to represent the MPEC for each electricity producer s and natural gas producer v in

Algorithm 3.4 Inner loop for the electricity market

1: Get current P_{zt}, $\varrho_{n_g t}$ as input. Set $\alpha_{ibt}^{s,0} = \alpha^{max}, \forall s, i, b, t$. Let $r = 1$, $Flag_p = 0$. Select an allowed number of iteration r^{max} and a convergence tolerance ϵ.
2: For $s = 1, \dots, S$, solve $\mathcal{P}_s$ with the offering prices $\alpha_{ibt}^{-s,0}$ from other electricity producers fixed. The optimal solution is $\alpha_{ibt}^{s,r}$, $P_{ibt}^{s,r}$. Update $\alpha_{ibt}^{s,0} = \alpha_{ibt}^{s,r}$.
3: If $|\alpha_{ibt}^{s,r} - \alpha_{ibt}^{s,r-1}| \leq \epsilon \cdot \max\{\alpha_{ibt}^{s,r}, \alpha_{ibt}^{s,r-1}\}, \forall s$, terminate and report $\alpha_{ibt}^{s,r}$, $P_{ibt}^{s,r}$.
4: If $r = r^{max}$, set $Flag_p = 1$, then quit and report that the algorithm for electricity market fails to converge. Else, update $r \leftarrow r + 1$ and return to step 2.

Algorithm 3.5 Inner loop for the natural gas market

1: Get current P_{ibt}^s, P_{jbt}^o, $\beta_{n_p t}$ as input. Set $\delta_{mt}^{v,0} = \delta^{max}, \forall v, m, t$. Let $r = 1$, $Flag_g = 0$. Select an allowed number of iteration r^{max} and a convergence tolerance ϵ.
2: For $v = 1, \dots, V$, solve $\mathcal{G}_v$ with offering prices $\delta_{mt}^{-v,0}$ from other gas producers fixed. The optimal solution is $\delta_{mt}^{v,r}$, $q_{mt}^{v,r}$. Update $\delta_{mt}^{v,0} = \delta_{mt}^{v,r}$.
3: If $|\delta_{mt}^{v,r} - \delta_{mt}^{v,r-1}| \leq \epsilon \cdot \max\{\delta_{mt}^{v,r}, \delta_{mt}^{v,r-1}\}, \forall v$, terminate and report $\delta_{mt}^{v,r}$, $q_{mt}^{v,r}$.
4: If $r = r^{max}$, set $Flag_g = 1$, then quit and report that the algorithm for gas market fails to converge. Else, update $r \leftarrow r + 1$ and return to step 2.

Algorithm 3.6 Outer loop for the coupled market

1: Set $\varrho_{n_g t}^0 = \delta^{max}$, $P_{zt}^0 = 0$, $P_{ibt}^{s,0} = 0$, $P_{jbt}^{o,0} = 0$, $\beta_{n_p t}^0 = \alpha^{max}$, and $r = 1$. Select a maximal allowed number of iteration r^{max} and a convergence tolerant ϵ.
2: Call Algorithm 3.4 and calculate the equilibrium of the electricity market. Retrieve α_{ibt}^s, $P_{ibt}^{s,r}$, $P_{jbt}^{o,r}$, $\beta_{n_p t}^r$.
3: Call Algorithm 3.5 and calculate the equilibrium of the natural gas market. Retrieve δ_{mt}^v, P_{zt}^r, $\varrho_{n_g t}^r$.
4: If $|P_{zt}^r - P_{zt}^{r-1}| \leq \epsilon \cdot \max\{P_{zt}^r, P_{zt}^{r-1}\}, \forall z$, $|P_{ibt}^{s,r} - P_{ibt}^{s,r-1}| \leq \epsilon \cdot \max\{P_{ibt}^{s,r}, P_{ibt}^{s,r-1}\}, \forall s$, $|P_{jbt}^{o,r} - P_{jbt}^{o,r-1}| \leq \epsilon \cdot \max\{P_{jbt}^{o,r}, P_{jbt}^{o,r-1}\}, \forall o$ hold, then terminate and report α_{ibt}^s, δ_{mt}^v, P_{zt}, q_{mt}^v, P_{ibt}^s, P_{jbt}^o.
5: If $r = r^{max}$, or $Flag_p = 1$, or $Flag_g = 1$, then quit and report the algorithm fails to converge. Else, update $r \leftarrow r + 1$ and return to step 2.

their MILP forms (3.42) and (3.48), respectively. The aggregated problems $\{\mathcal{P}_s\}$, $\forall s$ and $\{\mathcal{G}_v\}$, $\forall v$ represent the EPECs of the electricity and gas markets, respectively.

Due to its complicated nature, it is difficult to prove the convergence of above algorithms rigorously. The NCP based approach for EPEC problems developed in [43] is non-iterative, but the final model is much more challenging to solve than the MILP model of each energy producer. Nevertheless, if an equilibrium does exist, certain heuristic method may be helpful for accelerating the convergence of our algorithms, for example, restricting the change of solutions in two successive iterations if oscillation is observed.

3.5.4 Case Studies

1. 6-Bus Power System with 7-Node Gas System

Topology of the interconnected infrastructure is illustrated in Fig. 3.19a. It has 4 gas-fired units (P_2, P_4, P_6, P_7), 4 non-gas-fired units (P_1, P_3, P_5, P_8), 6 gas wells (G_1-G_6), 1 compressor, 1 gas storage (GS_1), 2 P2G facilities ($P2G_1$, $P2G_2$), 4 power loads and 3 gas loads. System data can be found in [31]. The daily electricity and gas demands in 24 periods are shown in Fig. 3.19b. The total electricity and gas loads in the first period are 525 MW and 5520 Sm^3/h, which will be used in the single-period case. We introduce four strategic electricity producers, who control generators P_1 to P_4, respectively, and the remaining units are owned by the power system operator, a non-profit entity. Likewise, we employ three strategic gas producers, who own gas wells G_1 to G_3, respectively, and the remaining gas wells are managed by the non-profit gas system operator. In Algorithms 3.4–3.6, ϵ is set to 1%, and the maximum number of iterations is 20.

We first consider a single period market, and inter-temporal constraints are removed. We compare the situations in which flow capacity constraints are inactive/

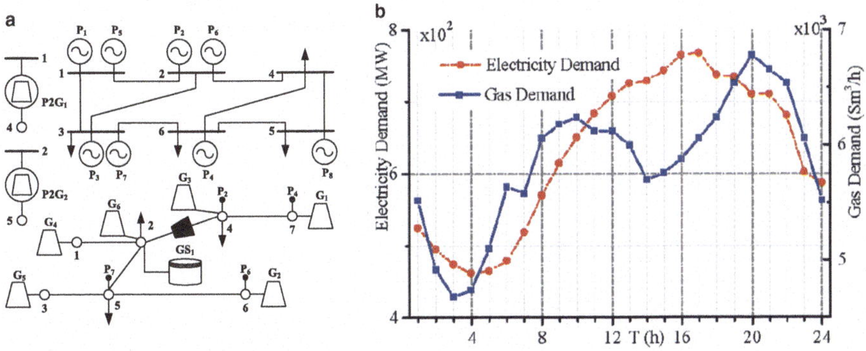

Fig. 3.19 (**a**) System topology. (**b**) Electricity and gas demands

Table 3.5 Results in the congestion-free case

Electricity market			Gas market		
LMEP ($/MWh)	β_1–β_6	13.76	LMGP ($/Sm^3$)	ϱ_1–ϱ_3,ϱ_5–ϱ_6	0.8
				ϱ_4, ϱ_7	2.0
Unit output (MW)	P_1	155	Gas output (Sm^3/h)	G_2	2745
	P_3	124		G_3	1293
	P_4	22.05		G_4	3000
	P_6	68.95	Gas consumption (Sm^3/h)	P_4	372.5
	P_8	155		P_6	1145

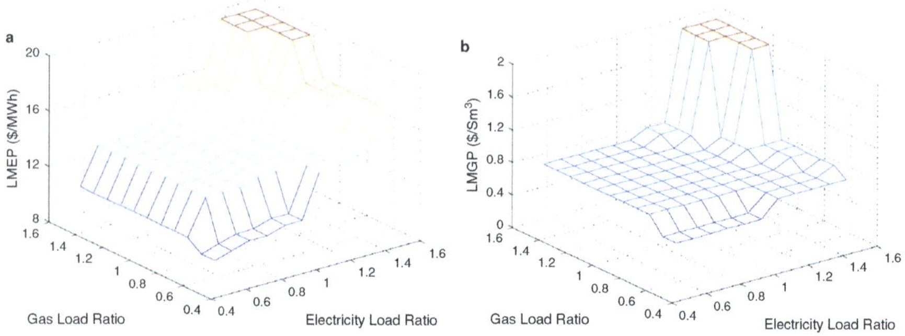

Fig. 3.20 Energy prices under different load ratios. (**a**) LMEP. (**b**) LMGP (Node:1-3, 5-6)

active. In the congestion-free case, Algorithm 3.6 converges in three iterations. Energy prices and production plans are listed in Table 3.5. The output of generators and gas wells which are omitted is 0. In the power grid, all buses share the same LMEP, which is equal to the marginal production cost. In the gas market, the LMGPs could be different at the head and tail nodes of an active pipeline because the operation of compressor incurs additional costs for purchasing electricity from the power grid. In this particular case without congestion, P2G facilities are not dispatched due to their higher marginal costs. Offering prices of electricity and gas producers are consistent with LMEPs and LMGPs, respectively, indicating that under market competition, players will bid according to the marginal costs, which validate the efficiency of the proposed market framework.

We change the electricity and gas load levels by multiplying a ratio with the value of demands. The corresponding LMEPs and LMGPs under different load levels are exhibited in Fig. 3.20. It can be observed that energy prices are nondecreasing with the load ratios, because cheaper resources are always utilized prior to more expensive ones in the absence of congestion, leading to nondecreasing marginal production costs of electricity and natural gas.

The above analysis is performed again with the transmission limits of power and gas flows. Algorithm 3.6 converges in four iterations. Results are summarized in Table 3.6. Generators and the gas wells which are not mentioned have zero output. In the gas market, gas producers could offer higher prices due to congestion,

Table 3.6 Results in the congestion-aware case

Electricity market			Gas market		
LMEP ($/MWh)	β_1	16.69	LMGP ($/Sm^3)	ϱ_1	0.6
	β_2	17.33		ϱ_2, ϱ_5–ϱ_6	0.9
	β_3	16.05		ϱ_3	0.8
	β_4	38.72		ϱ_4,ϱ_7	2
	β_5	39.36	Gas consumption (Sm^3/h)	P_2	214.2
	β_6	40		P_4	342.1
Generation output (MW)	P_1	140.1		P_6	2694
	P_2	12.23	Gas output (Sm^3/h)	G_2	794.4
	P_3	39.83		G_3	1476
	P_4	20.25		G_4	1500
	P_6	157.6		G_5	2000
	P_8	155		G_6	3000

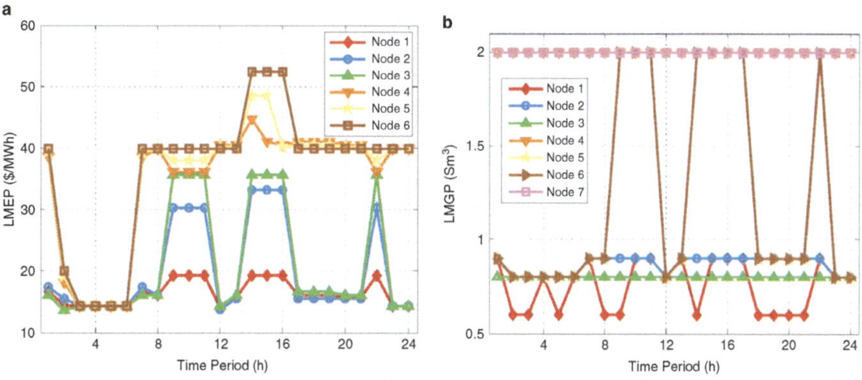

Fig. 3.21 Energy prices in the multi-period case. (**a**) LMEP. (**b**) LMGP

which consequently increases the marginal costs of gas-fired units. Similarly, in the electricity market, electricity producers may offer higher prices because of congestion. Therefore, the offering prices of generator P_4, gas wells G_1 and G_3 reach their upper bounds at the equilibrium. Moreover, gas-fired units produce more electricity so as to mitigate congestion, causing an increase of 1733 Sm^3/h in the gas demand compared with the congestion-free case. The traditional unit P_2 is also put into operation while it is not dispatched in the congestion-free case. P2G facilities are not in service as they are still not cost-effective according to the gas market clearing results. Bidding strategies of producers remain consistent with LMEPs and LMGPs. Compared with congestion-free case, the operating cost of the power system rises from $2835 to $8040. The operating cost of the gas system grows from $6581 to $8868.

In what follows, a multi-period market model with 24 time slots and full transmission capacity constraints are considered. LMEPs and LMGPs are presented in Fig. 3.21, showing strong interdependence across two systems. Gas price peaks

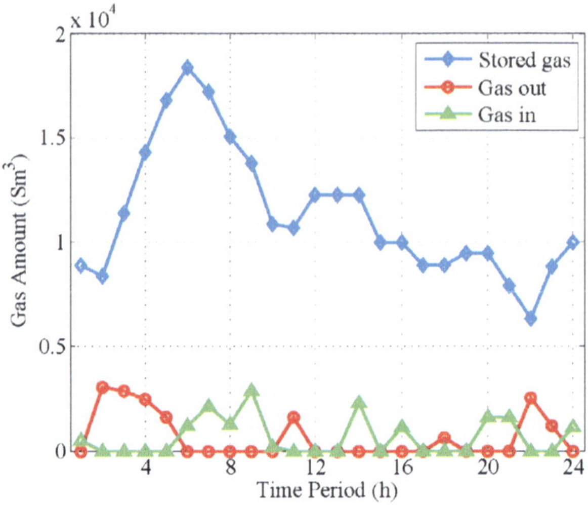

Fig. 3.22 Operating strategy of gas storage

Table 3.7 System operating costs with/without gas storage

	Power system ($)	Gas system ($)
Gas storage	2.5183×10^5	3.0149×10^5
No gas storage	3.2990×10^5	5.0349×10^5

Table 3.8 Ownership of the generators and the gas wells

Natural gas producer	SGP 1	SGP 2	SGP 3	SGP 4	Non-SGP
	G_1, G_2	G_3	G_4	G_5	G_6–G_{10}
Electricity producer	SEP 1	SEP 2	SEP 3	Non-SEP	
	P_1–P_{12}	P_{13}–P_{24}	P_{25}–P_{32}	P_{33}–P_{54}	

occur at nodes 1, 5, and 6 during periods 9 to 11, 14 to 16 and 22; similar trends are found in electricity price curves at buses 1, 2, 3. The amount of gas storage in GS_1 is shown in Fig. 3.22. This facility supplies (stores) gas when LMGP is relatively high (low). The operating costs of both systems with/without gas storage facility are listed in Table 3.7. With the help of gas storage, 31% and 67% cost savings are achieved.

2. 118-Bus Power System with 20-Node Gas System

A larger system consists of a modified IEEE 118-bus power system and a modified Belgian high-calorific 20-node gas system is tested. Topology and parameter of this system can be found in [31]. It includes 30 gas-fired units ($P_1 - P_{30}$), 24 non-gas-fired units, 10 gas wells ($G_1 - G_{10}$), 3 compressors, 4 P2G facilities ($P2G_1 - P2G_4$), 2 gas storage facilities ($GS_1 - GS_2$), 91 power loads, and 9 gas loads. We have 3 (4) strategic producers and 1(1) non-strategic one in the electricity (gas) market. The ownership of generators and gas wells are shown in Table 3.8. Other settings are the same with the previous case.

Table 3.9 Electricity offering prices and electricity market clearing results

Congestion-free case								
Generator	1	2	3	4	5	6	7	8
Offering price ($/MWh)	35.7	40	28.6	28.6	40	0	23.5	0
Cleared quantity (MW)	6	6.5	0	0	64.6	30	0	30
Generator index	9	10	11	12	13	14	15	16
Offering price ($/MWh)	14.3	14.1	20.3	11.6	0	16.8	0	21.9
Cleared quantity (MW)	30	12.3	0	30	30	0	30	0
Generator	17	18	19	20	21	22	23	24
Offering price ($/MWh)	20.4	18.5	17.3	17.9	18.3	40	40	19.2
Cleared quantity (MW)	0	0	0	99	0	0	0	0
Generator	25	26	27	28	29	30	31	32
Offering price ($/MWh)	19.4	40	20	18.8	15.4	14.1	14.3	14.2
Cleared quantity (MW)	72.3	0	41.6	0	18.9	0	0	0
Congestion-aware case								
Generator	1	2	3	4	5	6	7	8
Offering price ($/MWh)	40	40	40	28.9	40	0	23.8	22.8
Cleared quantity (MW)	6.5	6.5	0	4.5	64.5	30	0	0
Generator	9	10	11	12	13	14	15	16
Offering price ($/MWh)	15.8	14.2	20.5	11.6	0	17	22.2	22.3
Cleared quantity (MW)	30	0	16.4	13.3	30	47.7	0	23.4
Generator	17	18	19	20	21	22	23	24
Offering price ($/MWh)	0	18.8	17.9	18.1	18.2	40	40	18.3
Cleared quantity (MW)	30	30	0	54.7	0	0	0	135
Generator	25	26	27	28	29	30	31	32
Offering price ($/MWh)	18.4	40	19.9	18.5	16.6	15.7	15.8	15.8
Cleared quantity (MW)	2.3	0	0	120	0	0	0	0

Offering prices and cleared output of electricity and gas producers in a single-period market with and without congestion constraints are shown in Tables 3.9, 3.10, 3.11. It is interesting to notice that some generators offer free energy to the market and get non-zero quantities, say, P_6, P_8, P_{13}, P_{15} in the congestion-free case and P_6, P_{13}, P_{17} in the congestion-aware case. The reason lies in the fact that generators are paid at LMEPs rather than their bids. They may receive higher incomes if LMEPs are raised by other producers. In the gas system, all producers offer energy at the same price, but only G_2 gets non-zero contracts. System operating costs in both cases are shown in Table 3.11, showing 9.73% and 9.19% increases when congestion is considered.

Finally, the computational efficiency of the proposed algorithm is tested. We change the number of strategic producers and periods in the model. Results are given in Table 3.12, where the tuple (a, b) contains the numbers of players in the electricity market and gas market in the first and second entries. Details of the ownership of units and gas wells can be found in [31]. In all cases, the market equilibrium can be found in reasonable time. Computation time of the multi-period

Table 3.10 Natural gas offering prices and gas market clearing results

Congestion-free case					
Gas well	1	2	3	4	5
Offering price (\$/Sm^3)	0.6	0.6	0.6	0.6	0.6
Cleared quantity (Sm^3/h)	0	3000	0	0	0
Congestion-aware case					
Gas well	1	2	3	4	5
Offering price (\$/Sm^3)	0.8	0.8	0.8	0.8	0.8
Cleared quantity (Sm^3/h)	0	1912	0	0	0

Table 3.11 Operating costs of the electricity and gas markets

	Electricity market (\$)	Gas market (\$)
Congestion-free	1.172×10^5	2.808×10^5
Congestion-aware	1.286×10^5	3.066×10^5

Table 3.12 Computational time with different numbers of producers

Congestion-free					
(a, b)	Time (s)	(a, b)	Time (s)	(a, b)	Time (s)
(3,3)	91	(4,3)	106	(5,3)	43
(3,4)	186	(4,4)	178	(5,4)	185
(3,5)	236	(4,5)	190	(5,5)	247
Congestion-aware					
(a, b)	Time (s)	NSP	Time (s)	(a, b)	Time (s)
(3,3)	37	(4,3)	269	(5,3)	298
(3,4)	35	(4,4)	285	(5,4)	295
(3,5)	43	(4,5)	225	(5,5)	307

Table 3.13 Computational time under different numbers of periods

	Computational time (s)	
Time periods	Congestion-free	Congestion-aware
8	761	669
16	2743	4858
24	6467	–

market model with $(a, b) = (3, 4)$ is listed in Table 3.13. The algorithm fails to converge within 2 h for the 24-period market model.

To improve computational efficiency, it is imperative to reduce the number of binary variables which are induced from linearizing complementarity constraints in KKT conditions. If some constraints in the market clearing problem never become binding, they can be relaxed without impacting the problem solution, and each MILP model will have fewer binary variables. We suggest the following heuristics to pick up redundant constraints: (1) gas-fired units generally have better ramping capabilities than conventional coal-fired units, so ramping constraints of gas-fired units might be omitted. (2) One can run the power transmission line congestion analysis in advance to identify and neglect the redundant transmission capacity constraints [44]. A general method to detect inactive constraints in LP based generation scheduling models is proposed in [45, 46]. (3) Properly chosen

Table 3.14 Computational time of simplified models

	Computational time (s)		
Time periods	Type A	Type B	Type C
8	412	533	331
16	1392	3166	1077
24	2251	–	2019

parameters in the linear expressions of complementarity constraints may give stronger MILP models and reduce computation burden [47].

Applying aforementioned heuristics to the congestion-aware case, the computation times under different level of simplifications are listed in Table 3.14:

Type A: Ramping constraints of gas-fired units are neglected.
Type B: Redundant transmission capacity constraints are neglected.
Type C: Simplifications in Type A and Type B with different big-M parameters in the MILP models for electricity producers and gas producers.

Comparing results in Tables 3.13 and 3.14, we can see that all simplification techniques can reduce the computation time, and neglecting ramping constraints of gas-fired units is more effective in this case, as most of them are indeed redundant. Since the power transmission network is heavily loaded, and only a small fraction of transmission capacity constraints is inactive, Type B is less efficient than Type A. The most simplified Type C enjoys 55% and 79% computational time reducing for market models with 8 and 16 periods, respectively, and makes it possible to solve the 24-period market model.

3.6 System Components Reinforcement

Besides the uncertainty of renewable generation, which would impact power system load balancing and has been well studied in a large body of literature, natural disasters such as hurricanes, floods, and earthquakes, as well as malicious attacks, may also cause catastrophic contingencies in critical infrastructures. These uncertain factors damage system components, and may even trigger a cascading failure. Hence, they should be carefully investigated. In this section, we propose a methodology to identify and protect vulnerable components of interdependent natural gas and electric power systems and guarantee a resilient operation. The proposed formulation renders a tri-level optimization problem. The lower level is a multi-period economic dispatch of the gas-electric system, the middle level distinguishes the most threatening attack on the coupled physical infrastructures, and the upper level provides optimal preventive decisions to reinforce the vulnerable components and improve system survivability against the unfavored attack. A nested column-and-constraint generation (CCG) algorithm is adopted to solve the tri-level model.

3.6.1 Mathematical Formulation

Some prerequisite assumptions and simplifications that facilitate building the model are stated as follows:

For the natural gas pipeline system, we assume: (1) the system is operated in steady state, which means the inflow and outflow of every pipe are instantaneously equal, and line pack dynamics are neglected. In practice, the system will be more resilient under the help of line pack. As robust optimization considers the worst case, our approach offers the most severe condition that might happen, and is inevitably conservative. Line pack can be incorporated without any difficulty in case of need. (2) Compressors are powered by electricity; simplified compressor model is adopted. (3) If a pipeline loses functionality during an attack, the valves close and remove the pipe from the system.

For the electric power system, we assume: (1) the system is operated in steady state, transient dynamics of generators and other components are neglected; (2) the lossless DC power flow model is adopted; (3) start-up and shut-down processes of gas-fired units generators can be instantly completed without delay.

For the rules of the defense and attack, we assume: (1) if a component is not protected, it will necessarily lose functionality when it is attacked; otherwise, it will be fully functional regardless of being attacked or not; (2) only branches can be attacked due to their wide geographical span; gas system nodes and power system buses have already been well protected. Nevertheless, potential attacks on nodes and buses can be modeled in the similar way.

Notations used throughout this section are defined below for quick reference.

1. Sets and Indices

$d_g \in D_G$	Gas loads
$d_p \in D_P$	Power grid loads
$g \in G$	Traditional units
$i_g \in I_G$	Gas network nodes
$i_p \in I_P$	Power grid nodes
$l_c \in L_C$	Compressors stations (active pipelines)
$l_g \in L_G$	Gas passive pipelines
$l_p \in L_P$	Power transmission lines
$l_n \in L_N$	Connection pipelines (pipelines connecting gas-fired units and respective gas network nodes)
$n \in N$	Gas-fired units
$s \in S$	Gas storage tanks
$t \in T$	Time periods
$w \in W$	Gas wells

2. Constant Parameters

F_{l_p}	Power transmission line capacity
$f_g(\cdot)$	Generation cost of traditional units
M_c^f	Mass flow limit of gas compressors

P_g^{min}/P_g^{max}	Output range of traditional units
P_n^{min}/P_n^{max}	Output range of gas-fired units
$p_{d_{pt}}$	Power load demand
Q_{d_p}	Load shedding cost coefficient
Q_w, Q_{d_g}, Q_s	Gas production/shedding/storage cost
$q_{s,max}^{in}$, $q_{s,max}^{out}$	Maximum gas storage in/out range
$q_{d_{gt}}$	Gas load
R_g^+/R_g^-	Ramp up/down limits of traditional units
R_n^+/R_n^-	Ramp up/down limits of gas-fired units
r_s^u, r_s^l	Maximum/minimum gas storage level
Z_{l_g}	Compression factor of the pipeline
β_n	Gas-electric conversion factor
γ_c	Compression factor of the compressor
λ	Unit transformation constant
ϕ_{l_g}	Weymouth equation coefficient
π_{l_p}	Reactance of transmission line
$\tau_{i_g}^u$, $\tau_{i_g}^l$	Gas pressure range
Γ_A	Attack budget
Γ_D	Defense budget

3. Decision Variables

a_{l_p}, a_{l_g}, a_{l_n}, a_{l_c}	Binary variables indicating whether a power transmission line, gas passive pipeline, connection pipeline, active pipeline is attacked
b_{l_p}, b_{l_g}, b_{l_n}, b_{l_c}	Binary variables indicating the operating availability of a power transmission line, gas passive pipeline, connection pipeline, active pipeline
p_{gt}, p_{nt}	Output of generation units
$pf_{l_{pt}}$	Power flow in transmission lines
$q_{l_{gt}}$	Gas flow in passive pipelines
$q_{l_{ct}}$	Gas flow in active pipelines
q_{wt}	Gas well output
q_{st}^{in}, q_{st}^{out}	Gas storage in/out rate
r_{st}	Amount of stored gas
$v_{i_{gt}}$	Gas pressure square
x_{l_p}, x_{l_g}, x_{l_n}, x_{l_c}	Binary variables indicating whether a power transmission line, gas passive pipeline, connection pipeline, active pipeline is defended (reinforced) or not
z_{gt}/z_{nt}	Generator commitment decision
$\Delta p_{d_{pt}}$	Electrical load shedding
$\Delta q_{d_{gt}}$	Natural gas load shedding
$\theta_{i_{pt}}$	Power system phase angle

The proposed formulation involves three levels. In the upper level, the defender (system operator) determines which components should be reinforced; the middle level models a virtual attacker who can sabotage a series of facilities subject to certain budgets; in the lower level, the defender responds to the performed attack by dispatching available resources to minimize the cost and damage, leading to the following tri-level program

$$\min_{\Phi_U} \; F_U \tag{3.49a}$$

$$\text{s.t.} \; \sum_{l_p} x_{l_p} + \sum_{l_g} x_{l_g} + \sum_{l_n} x_{l_n} + \sum_{l_c} x_{l_c} \le \Gamma_D \tag{3.49b}$$

$$x_{l_p},\; x_{l_g},\; x_{l_n},\; x_{l_c} \in \{0, 1\} \tag{3.49c}$$

In the upper-level problem (3.49), the preventive action set is $\Phi_U = \{x_{l_p}, x_{l_g}, x_{l_n}, x_{l_c}\}$; constraint (3.49b) limits the defense budget, where Γ_D is a positive integer; (3.49c) restricts decision variables to be binary. F_U in objective function (3.49a) is the optimal value of the middle-level problem which is given by

$$F_U = \max_{\Phi_M} \; F_M \tag{3.50a}$$

$$\text{s.t.} \; \sum_{l_p} a_{l_p} + \sum_{l_g} a_{l_g} + \sum_{l_n} a_{l_n} + \sum_{l_c} a_{l_c} \le \Gamma_A \tag{3.50b}$$

$$a_{l_p},\; a_{l_g},\; a_{l_n},\; a_{l_c} \in \{0, 1\} \tag{3.50c}$$

In the middle-level problem (3.50), $\Phi_M = \{a_{l_p}, a_{l_g}, a_{l_n}, a_{l_c}\}$ stands for the attack strategy; constraint (3.50b) limits the attack budget, where Γ_A is a positive integer; (3.50c) imposes binary restriction on the attack variable. F_M in objective function (3.50a) is the optimal value of the lower-level problem which is given by

$$F_M = \min_{\Phi_L} \; F_L \tag{3.51a}$$

$$\text{s.t.} \; b_{\{\cdot\}} = 1 - a_{\{\cdot\}} + a_{\{\cdot\}} x_{\{\cdot\}}, \;\; \{\cdot\} = \{l_p, l_g, l_n, l_c\} \tag{3.51b}$$

$$z_{gt} P_g^{min} \le p_{gt} \le z_{gt} P_g^{max} \tag{3.51c}$$

$$z_{nt} b_n P_n^{min} \le p_{nt} \le z_{nt} b_n P_n^{max} \tag{3.51d}$$

$$p_{g,t+1} - p_{gt} \le z_{gt} R_g^+ + (1 - z_{g,t+1}) P_g^{max} \tag{3.51e}$$

$$p_{gt} - p_{g,t+1} \le z_{g,t+1} R_g^- + (1 - z_{gt}) P_g^{max} \tag{3.51f}$$

$$p_{n,t+1} - p_{nt} \le z_{nt} R_n^+ + (1 - z_{n,t+1}) P_n^{max} \tag{3.51g}$$

$$p_{nt} - p_{n,t+1} \le z_{n,t+1} R_n^- + (1 - z_{nt}) P_n^{max} \tag{3.51h}$$

$$0 \le \Delta p_{d_{pt}} \le p_{d_{pt}} \tag{3.51i}$$

$$-\pi \le \theta_{i_{pt}} \le \pi \tag{3.51j}$$

$$\sum_{g \in \Psi_g(i_p)} p_{gt} + \sum_{n \in \Psi_n(i_p)} p_{gt} + \sum_{l \in \Psi_{O_2}(i_p)} pf_{l_{pt}} - \sum_{l \in \Psi_{O_1}(i_p)} pf_{l_{pt}} - \sum_{d \in \Psi_{dp}(i_p)} (p_{d_{pt}} - \Delta p_{d_{pt}}) = 0 \tag{3.51k}$$

$$-F_{l_p} \leq pf_{l_{pt}} \leq F_{l_p} \tag{3.51l}$$

$$\pi_{l_p} pf_{l_{pt}} = b_{l_p}(\theta_{i_{p,l_{p1}}t} - \theta_{i_{p,l_{p2}}t}) \tag{3.51m}$$

$$q_w^l \leq q_{wt} \leq q_w^u \tag{3.51n}$$

$$r_s^l \leq r_{st} = r_{s,t-1} + q_{st}^{in} - q_{st}^{out} \leq r_s^u \tag{3.51o}$$

$$0 \leq q_{st}^{in} \leq q_{s,max}^{in}, \; 0 \leq q_{st}^{out} \leq q_{s,max}^{out} \tag{3.51p}$$

$$(\tau_{i_g}^l)^2 \leq v_{i_g t} \leq (\tau_{i_g}^u)^2 \tag{3.51q}$$

$$\sum_{s \in \Theta_s(i_g)} (q_{st}^{out} - q_{st}^{in}) - \sum_{d_g \in \Theta_{d_g}(i_g)} (q_{d_g t} - \Delta q_{d_g t}) + \sum_{w \in \Theta_w(i_g)} q_{wt} = \sum_{\{\cdot\} \in \Theta_{\{\cdot\}_1}(i_g)} q_{\{\cdot\}t} - \sum_{\{\cdot\} \in \Theta_{\{\cdot\}_2}(i_g)} q_{\{\cdot\}t} + \sum_{n \in \Theta_n(i_g)} p_{nt}/\beta_n, \quad \{\cdot\} = \{l_c, l_g\} \tag{3.51r}$$

$$0 \leq \Delta q_{d_g t} \leq q_{d_g t} \tag{3.51s}$$

$$q_{l_g t}|q_{l_g t}| = \phi_{l_g} b_{l_g}(v_{i_{g,l_{g1}}t} - v_{i_{g,l_{g2}}t}) \tag{3.51t}$$

$$v_{i_{gt,c_2}} \leq \gamma_c^2 b_c v_{i_{gt,c_1}} + (1 - b_c)(\tau_{i_{gt,c_2}}^u)^2 \tag{3.51u}$$

$$0 \leq q_{l_c t} \leq M_c^f b_c \tag{3.51v}$$

With fixed defense and attack strategies, the lower-level problem (3.51) mimics a multi-period economic dispatch problem, which captures the subsequent effects of the attack over time. Commitment of fast unit is allowed, providing more flexible corrective actions. The corrective action set is

$$\Phi_L = \{z_{gt}, z_{nt}, p_{gt}, p_{nt}, \theta_{i_p t}, pf_{l_{pt}}, \Delta p_{d_{pt}}, q_{wt}, q_{st}^{in}, q_{st}^{out}, r_{st}, q_{l_g t}, q_{l_c t}, v_{i_g t}, \Delta q_{d_g t}\}$$

The objective function in (3.51a) represents the total operating cost of both networks, which is expressed as

$$F_L = \sum_{g,t} f_g(p_{gt}) + \sum_{w,t} Q_w q_{wt} + \sum_{s,t} Q_s r_{st} + \sum_{d_p,t} Q_{d_p} \Delta p_{d_{pt}} + \sum_{d_g,t} Q_{d_g} \Delta q_{d_g t}$$

where the first two terms stand for the production cost of power and natural gas, $f_g(\cdot)$ can be approximated by a linear or piecewise linear function; the third term represents gas storage cost, and the last two terms are load shedding costs. The attacker in the middle level strives to seek Γ_A most vulnerable system components such that the minimum of F_L is maximized. In the top level, the operator reinforces potential targets of the attacker in order to mitigate the adverse effect of the severe attack. Certainly, targets of the attacker may change with respect to the use of defensive resources. According to our assumption, reinforced elements are unlikely to be attacked because there is no gain. Equation (3.51b) represents the component availability logic: if $x_{\{\cdot\}} = 1$ for some device, $b_{\{\cdot\}} = 1$ always holds regardless of the value of $a_{\{\cdot\}}$; if $x_{\{\cdot\}} = 0$, then $b_{\{\cdot\}} = 1$ when it is not attacked ($a_{\{\cdot\}} = 0$) and $b_{\{\cdot\}} = 0$ when it is attacked ($a_{\{\cdot\}} = 1$). Equations (3.51c)–(3.51v) constitute the operating constraints of the coupled gas-electric system. For the electricity network, (3.51c) and (3.51d) enforce the generation capacity of traditional units and gas-fired units, respectively. Equations (3.51e)–(3.51h) are the ramping rate limits of traditional units and gas-fired units, respectively. Equation (3.51i) sets the boundary of not-served load. Equation (3.51j) describes the phase angle limits of the power grid. Equation (3.51k) depicts the power balancing condition, where $\Psi_g(i_p)$, $\Psi_n(i_p)$, $\Psi_{d_p}(i_p)$ represent the set of traditional units, gas-fired units, power grid loads connecting to node i_p, respectively, and $\Psi_{O_1}(i_p)$, $\Psi_{O_2}(i_p)$ represent the set of power transmission lines whose initial/terminal node is i_p, respectively. Equation (3.51l) is the active power flow limit in transmission lines. DC power flow equation (3.51m) describes the relation between line flow and phase angle, where $i_{p,l_{p1}}$ and $i_{p,l_{p2}}$ represent the initial and terminal node of l_p. For the gas system, (3.51n) limits the production capacity of gas wells. Equation (3.51o) models capacity and mass dynamics of the gas storage device. Equation (3.51p) imposes charge and discharge rate limits of the gas storage device. Equation (3.51q) defines pressure range of each node. Equation (3.51r) gives the gas balance condition, where $\Theta_s(i_g)$, $\Theta_w(i_g)$, $\Theta_{d_g}(i_g)$ represent the set of gas storage facility/gas well/gas load connecting to node i_g, respectively; $\Theta_n(i_g)$ represents the set of gas-fired units connecting to node i_g through connection line l_n; $\Theta_{c_1}(i_g)$, $\Theta_{c_2}(i_g)$, $\Theta_{l_{g_1}}(i_g)$, $\Theta_{l_{g_2}}(i_g)$ represent the set of active/passive pipelines whose initial/terminal node is i_g, respectively. Equation (3.51s) declares the boundary of gas load curtailment. Weymouth equation (3.51t) characterizes the relation among gas flow and nodal pressures associated with a passive pipeline, where $v_{i_{g,l_{g1}}}$, $v_{i_{g,l_{g2}}}$ are square pressures at the head and tail node of l_g, respectively. Equation (3.51u) describes the pressure relationship between the head node i_{g,c_1} and tail node i_{g,c_2} of an active pipeline. Equation (3.51v) states the gas flow range in an active pipeline. The proposed formulation can be extended in several ways.

1. Modeling Unit Commitment in the Lower Level

Our algorithm allows discrete variables in the lower-level problem, so the start-up costs SU_{gt}, SU_{nt} and shut-down costs SD_{gt}, SD_{nt} of traditional and gas-fired units can be added in the cost function F_L while appending the following constraints in the lower-level problem

$$SU_{gt} \geq 0,\ SU_{gt} \geq C_g^{up}(z_{gt} - z_{g(t-1)}),\ \forall g, t \tag{3.52a}$$

$$SD_{gt} \geq 0,\ SD_{gt} \geq C_g^{dn}(z_{g(t-1)} - z_{gt}),\ \forall g, t \tag{3.52b}$$

$$SU_{gn} \geq 0,\ SU_{gn} \geq C_n^{up}(z_{nt} - z_{n(t-1)}),\ \forall n, t \tag{3.52c}$$

$$SD_{gn} \geq 0,\ SD_{gn} \geq C_n^{dn}(z_{n(t-1)} - z_{nt}),\ \forall n, t \tag{3.52d}$$

where C_g^{up} (C_n^{up}) and C_g^{dn} (C_n^{dn}) are fixed start-up and shut-down costs of traditional units (gas-fired units).

The minimum on and minimum off time constraints of traditional slow-response units can be incorporated as

$$-z_{g(t-1)} + z_{gt} - z_{gk} \leq 0, \forall g, \forall t, \forall k = t, \cdots, t + T_g^{on} - 1 \tag{3.53a}$$

$$z_{g(t-1)} - z_{gt} + z_{gk} \leq 1, \forall g, \forall t, \forall k = t, \cdots, t + T_g^{off} - 1 \tag{3.53b}$$

where T_g^{on} and T_g^{off} are the minimum-on and minimum-off time of unit g.

2. Modeling Monetary Value of Defense/Attack Budgets

In (3.49b) and (3.50b), the defense and attack budgets are counted by the number of elements those are reinforced or sabotaged. To distinguish the efforts or costs spent on protecting/attacking different elements, monetary budgets can be incorporated as

$$\sum_{l_p} \zeta_{l_p}^x x_{l_p} + \sum_{l_g} \zeta_{l_g}^x x_{l_g} + \sum_{l_n} \zeta_{l_n}^x x_{l_n} + \sum_{l_c} \zeta_{l_c}^x x_{l_c} \leq \Gamma_D \tag{3.54a}$$

$$\sum_{l_p} \zeta_{l_p}^a a_{l_p} + \sum_{l_g} \zeta_{l_g}^a a_{l_g} + \sum_{l_n} \zeta_{l_n}^a a_{l_n} + \sum_{l_c} \zeta_{l_c}^a a_{l_c} \leq \Gamma_A \tag{3.54b}$$

where ζ is the cost coefficient parameter; superscripts x, a are adopted to distinguish defense and attack related coefficients, and the subscripts l_p, l_g, l_n, l_c are adopted to distinguish the component-wise coefficients.

3. Modeling of Uncertainty of Renewable Generation

To address the uncertainty of renewable energy generation, one common practice is to treat its volatility as the strategy of the attacker in the middle level, and is constrained in a pre-defined uncertainty set. This technique has been described in Appendix C.2. The solution approach will remain the same. It is also interesting to model the disaster dependent attack strategies, such as the trajectory of a hurricane.

Most constraints in the proposed tri-level formulation are linear. Nonlinear ones include Weymouth equation (3.51t) with gas flow square, which can be approximated by piecewise linear function as follows [27]:

$$\psi(q_{l_g t}) \approx \psi(\Delta_{l_g t 1}) + \sum_{k \in K} (\psi(\Delta_{l_g t, k+1}) - \psi(\Delta_{l_g t k})) \varrho_{l_g t k} \tag{3.55a}$$

$$q_{l_g t} = \Delta_{l_g t 1} + \sum_{k \in K} (\Delta_{l_g t, k+1} - \Delta_{l_g t, k}) \varrho_{l_g t k} \tag{3.55b}$$

$$\varrho_{l_g t, k+1} \le \zeta_{l_g t k} \le \varrho_{l_g t k}, \ \forall k \in K - 1 \tag{3.55c}$$

$$0 \le \varrho_{l_g t k} \le 1, \ \forall k \in K \tag{3.55d}$$

where $\Delta_{l_g t k}$ is the piecewise segment for each passive pipeline and k is the corresponding index; $\varrho_{l_g t k}$ and $\zeta_{l_g t k}$ are the auxiliary continuous and binary variable, respectively; ψ is the value of the nonlinear term. The following fact can be observed: if $\varrho_{l_g t, K_1} = 1$ for some K_1, then $\zeta_{l_g t(K_1-1)} = \varrho_{l_g t(K_1-1)} = 1$ according to (3.55c) and (3.55d); the same holds for $k = 1, \cdots, K_1$. Furthermore, if $\zeta_{l_g t k} = 0$, $\forall k \ge K_1 + 1$, we obtain $\varrho_{l_g t k} = 0$, $\forall k \ge K_1 + 1$ from (3.55c), $q_{l_g t} = \Delta_{l_g t, K_1+1}$ from (3.55b), and $\psi(q_{l_g t}) = \psi(\Delta_{l_g t K_1+1})$ from (3.55a). The approximation error can be controlled by adjusting the size of K. An alternative piecewise linear approximation is the SOS2 formulation discussed in Appendix B.1.1.

For ease of exposition, the tri-level model (3.49)–(3.51) can be written in a compact form as follows:

$$\min_{x \in X} \max_{a \in A} \min_{y, z \in R(x,a)} \ h^T y \tag{3.56}$$

where $X = \{x | Ax \le b\}$ corresponding to (3.49b) and (3.49c) is the feasible set of defense strategies; $A = \{a | Ca \le d\}$ corresponding to (3.50b) and (3.50c) is the feasible set of attack strategies; $R(x, a) = \{y, z | E(x, a)y + F(x, a)z \le g\}$ corresponding to (3.51b)–(3.51v) is the feasible set of the multi-period post-contingency dispatch in the lower level. x, a, z are binary variable vectors, and y is a continuous variable vector. Matrices and vectors A, b, C, d, g, h are constant coefficients. $E(x, a)$ and $F(x, a)$ are variable coefficient matrices but are fixed in the dispatch problem. There is no substantial difficulty if one wishes to consider a more general cost function in the form of $h_1^T x + h_2^T z + h_3^T y$.

3.6.2 A Decomposition Algorithm

A practical method for solving a tri-level min-max-min problem with discrete variables in the lower level is the nested constraint-and-column generation algorithm discussed in Appendix C.2.3, which is employed in this section to tackle problem (3.56). A slight difference is that the lower-level constraint $R(x, a)$ is not linear if x and a are not fixed, therefore, additional linearization should be performed.

1. Inner-Level Algorithm

Given a defense strategy x^*, attack strategy a^*, and a lower-level binary variable z^*, economic dispatch constraint yields the following inequality

$$E(x^*, a^*)y \le g - F(x^*, a^*)z^* \tag{3.57}$$

Then the lower-level problem with $z = z^*$ is a linear program in variable y, whose dual program is given by

$$\begin{aligned} \max_{\mu} \quad & (g - F(x^*, a^*)z^*)^T \mu \\ \text{s.t.} \quad & E(x^*, a^*)^T \mu = h, \mu \le 0 \end{aligned} \tag{3.58}$$

where μ is the dual variable.

The subproblem aims to identify the most severe attack under a given defense strategy which has already been deployed, giving rise to a linear max-min problem

$$\max_{a \in A} \min_{y, z \in R(x,a)} h^T y \tag{3.59}$$

(3.59) is the middle and lower levels of the trilevel problem (3.56). Difficulty arises from the discrete variables in the lower-level dispatch problem, which prevents the use of duality theory to transform (3.59) into a bilinear program, as the most common strategy for solving a linear max-min problem. Exploiting the countability of strategy z, we can conceptually enumerate the choice of z in constraints, so that the discrete variables vanish, and duality theory can be applied. To circumvent the computation overhead caused by the enumeration on z, only critical values are identified and taken into account dynamically on the fly. The procedures for solving (3.59) are described below. More details can be found in Appendix C.2.3 and the references therein.

Algorithm 3.7 Inner CCG

1: Select an arbitrary feasible attack strategy a^*, a convergence tolerant ε_{inner}, and solve the following MILP:

$$\min_{y,z}\{h^T y \mid E(x^*, a^*)y + F(x^*, a^*)z \le g\} \tag{3.60}$$

The optimal solution is y^*, z^*; let $LB = h^T y^*$, $UB = +\infty$, $o = 1$, $z^{1*} = z^*$, $O = \{1\}$, and $Z = \{z^{1*}\}$.

2: Solve the following problem

$$\begin{aligned} &\max_{\theta,a,\mu} \ \theta \\ &\text{s.t. } \theta \le (g - F(x^*, a)z^{r*})^T \mu^r, \ \forall r \in O, \ z^{r*} \in Z \\ &\quad Ca \le d, \ E(x^*, a)^T \mu^r = h, \ \mu^r \le 0, \ \forall r \in O \end{aligned} \tag{3.61}$$

The optimal solution is a^* and θ^*, and update the upper bound $UB = \theta^*$.

3: Solve (3.60) with a^*, the optimal solution is (z^*, y^*); update the lower bound $LB = \max\{LB, h^T y^*\}$.

4: If $UB - LB \le \epsilon_{inner}$, terminate; return the optimal solution a^*, z^*, y^* and optimal value θ^*; otherwise, update $o \leftarrow o + 1$, $O \leftarrow O \cup o + 1$, $z^{o*} = z^*$, $Z = Z \cup z^{o*}$, create new variables μ^o, and append the following constraints

$$\begin{aligned} &\theta \le (g - F(x^*, a)z^{o*})^T \mu^o \\ &E(x^*, a)^T \mu^o = h, \mu^o \le 0 \end{aligned}$$

in problem (3.61), and go to step 2.

Note that there are bilinear terms in problem (3.61) which involve the product of two binary variables or a binary variable and a continuous variable can be further linearized using integer formulation tricks explained in Appendix B.2.2. After performing linearization, problem (3.61) can be solved via commercial MILP solvers. Algorithm 3.7 identifies the worst attack action a^* under fixed defense strategy x^*.

2. Outer-Level Algorithm

The master problem endeavors to identify the optimal defense strategy taking into account the response from the attacker. It starts from a subset of attacker's strategies, and gradually expands the choices of the attacker by identifying more destructive sabotage from the subproblem, until none can be found. The algorithm for the master problem proceeds as follows:

Algorithm 3.8 Outer CCG

1: Set $LB = -\infty$, $UB = +\infty$, $a^{1*} = 0$, $w = 1$, $W = \{w\}$, and select a convergence tolerant ϵ_{outer}.

2: Solve the following problem:

$$\begin{aligned} \min_{x,y,z,\varphi} \quad & \varphi \\ \text{s.t.} \quad & Ax \le b,\ h^T y^f \le \varphi,\ \forall f \in W \\ & E(x, a^{f*}) y^f + F(x, a^{f*}) z^f \le g,\ \forall f \in W \end{aligned} \tag{3.62}$$

The optimal solution is x^*; the optimal value is φ^*; update $LB = \varphi^*$.

3: Given the defense strategy x^*, calculate the worst attack action a^* and corresponding objective value θ^* using Algorithm 3.7; update $UB = \min\{UB, \theta^*\}$.

4: If $UB - LB \le \epsilon_{outer}$, terminate and return the optimal solution x^*, a^*, y^*, z^*; otherwise, update $w = w + 1$, $W = W \cup w$, $a^{w*} = a^*$, create new variables (y^w, z^w), and append the following constraints

$$h^T y^w \le \varphi,\ E(x, a^{w*}) y^w + F(x, a^{w*}) z^w \le g$$

in problem (3.62), and go to step 2.

Similarly, there are bilinear terms in (3.62) which can be further linearized, and problem (3.62) can be solved as an MILP.

Algorithms 3.7 and 3.8 always converge in a few number of iterations. However, the computation effort spent on solving the tri-level model is still high because both loops require solving MILPs. For that cannot be solved in the given time limit, we may accept the current best solution as a compromise.

3.6.3 Case Studies

In this section, we present numerical experiments on two test systems to show the effectiveness of the proposed model and algorithm. Experiments are performed on a laptop with Intel(R) Core(TM) 2 Duo 2.2 GHz CPU and 4 GB memory. Optimization models are coded in MATLAB with YALMIP toolbox. MILP is solved by Gurobi 6.5, and the optimality gap is set as 0.1% without particular mention.

1. Configurations of a 6-Bus Power System with 7-Node Gas System

Figure 3.23 depicts the topology of the connected infrastructure. It possesses 2 gas-fired units, 1 traditional unit, 2 gas wells, 1 compressor, 1 gas storage tank, 3 electricity loads, and 3 gas loads. In this figure, we use PLine, PL, G with subscripts to denote the power transmission lines, power loads, generation units,

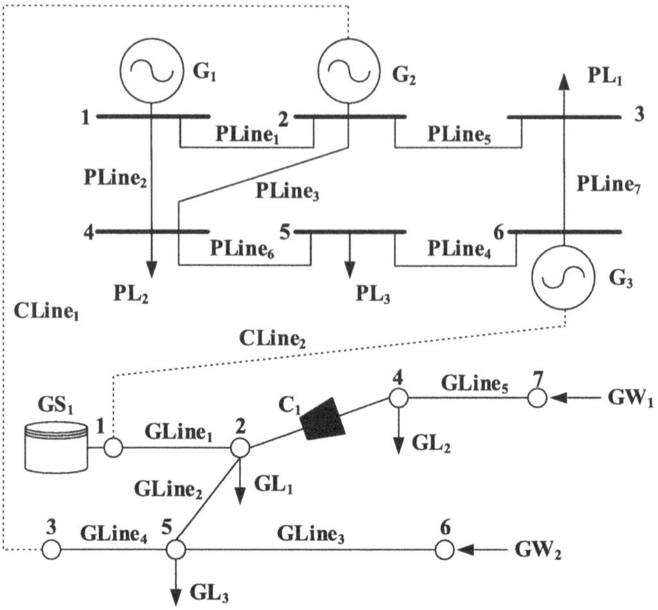

Fig. 3.23 Topology of the 6-bus power system with 7-node gas system

respectively, and C, GW, GS, GLine, GL with subscripts to denote the compressor, gas wells, gas storages, gas passive pipelines, gas loads, respectively. Specially, we use CLine with subscript to denote the connection pipes between power system and gas system. The potential targets of the attacker include 7 power transmission lines, 5 passive pipelines, 1 active pipeline, and 2 connection pipes. System data and unit commitment status can be found in [31]. In the following cases, we consider the problem with 4 periods. The power and gas load profiles are shown in Fig. 3.24, in which four typical time slots defined in Table 3.15 are considered. We use PWL functions with 8 segments to approximate the nonlinear parts in Weymouth equation.

2. Results of the Benchmark Case

In the benchmark case, $\Gamma_D = \Gamma_A = 3$ is used. The corresponding results are summarized in Table 3.16. As the gas system is radial, it is more vulnerable to malicious attacks. From Table 3.16, we can observe that the defense resources are mainly spent on the gas system. Given these reinforcement actions, components in either system may fall a victim to the attack. The average power and gas loss rates in the four time slots are 48.6% and 6.55%, respectively. This difference reveals that the failure of the gas system will have a larger impact on the coupled systems, because a successful attack on the gas system may cause cascading outages of gas-fired units, which further leads to electricity load shedding.

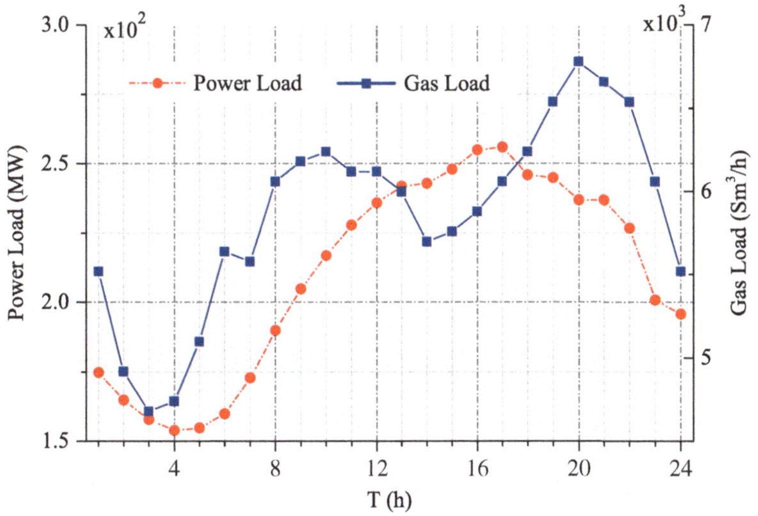

Fig. 3.24 Power and gas demand

Table 3.15 Summary of the selected time slots

Case	Periods	Power	Gas
1	2–5	Low	Low
2	9–12	Mid	Mid
3	14–17	High	Mid
4	19–22	Mid	High

Table 3.16 Results in the benchmark case

Case no.	Reinforced elements	Attacked elements	Cost ($)	Load shedding	
				Power (MWh)	Gas (Sm^3)
1	$GLine_2$	$GLine_1$	6.713×10^5	334	0
	$GLine_3$	$GLine_4$			
	$GLine_5$	$PLine_5$			
2	C_1	$PLine_2$	7.529×10^5	744	0
	$GLine_3$	$PLine_3$			
	$GLine_5$	$PLine_4$			
3	C_1	$GLine_3$	8.615×10^5	0	3900
	$GLine_5$	$CLine_1$			
	$PLine_5$	$PLine_1$			
4	C_1	$GLine_4$	8.047×10^5	530	0
	$GLine_3$	$PLine_1$			
	$GLine_5$	$PLine_2$			

3. Changing the Budgets
We alter the defense and attack budgets in the intervals $[1, L_p+L_g+N+C-1]$ and $[1, L_p+L_g+N+C-A]$, respectively. Plenty of tests are conducted under different

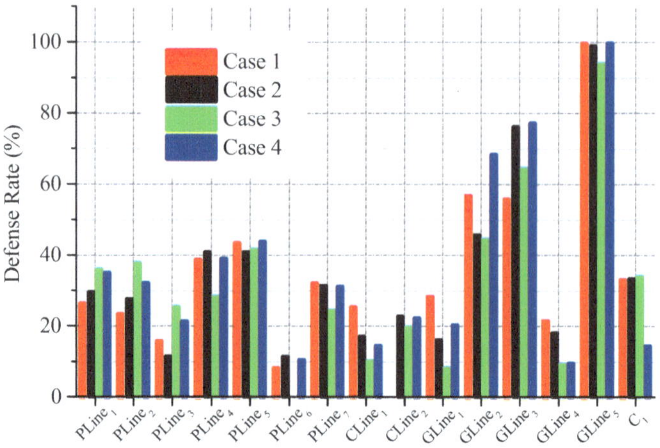

Fig. 3.25 Defense rate of components in the four cases

values of defense and attack budgets, as well as numbers of time slots. We quantify the defense rate of a component as the number of instances in which it is protected divided by the total number of defense and attack budget combinations. Results are shown in Fig. 3.25. Take GLine_5 for example, it is reinforced in 99 out of 105 scenarios of defense and attack budget combinations in Case 3, thus its defense rate is 94.26%. The importance of each component is seen by the relative value of the defense rate in Fig. 3.25, from which we can observe that GLine_5 is always the most vulnerable component, as its defense rate is ranked the highest in all cases. Also, the defense rate of a specific component varies from case to case, which reveals that the defense strategy depends on the load level.

To demonstrate the benefit of proactive protections, the total cost under different combinations of defense and attack budget are shown in Fig. 3.26, from which we can see: with defense budget increasing, the average operation cost under different attack budgets decreases. However, the marginal decrement of total cost also decreases. An appropriate defense budget may be $6 \sim 8$. In such circumstances, the attack has little impact regardless of the attack budget. Moreover, if the attack budget is limited (less than 4 in this test system), the total cost can be well-controlled through a limited defense budget. These observations provide guidance for planning defense resources.

4. Considering Slow-Response Units

We consider G_1 with largest capacity as a slow-response unit. Its minimum on/off time is 2 h. Other system data remain the same. Results are given in Table 3.17 and compared with the results in the benchmark case. The optimal defense strategy, attack strategy as well as the not-served-gas are the same with those in the benchmark case.

From Table 3.17, it can be observed that the electricity load shedding in the case when G_1 is a slow-response unit is much larger than that in the benchmark

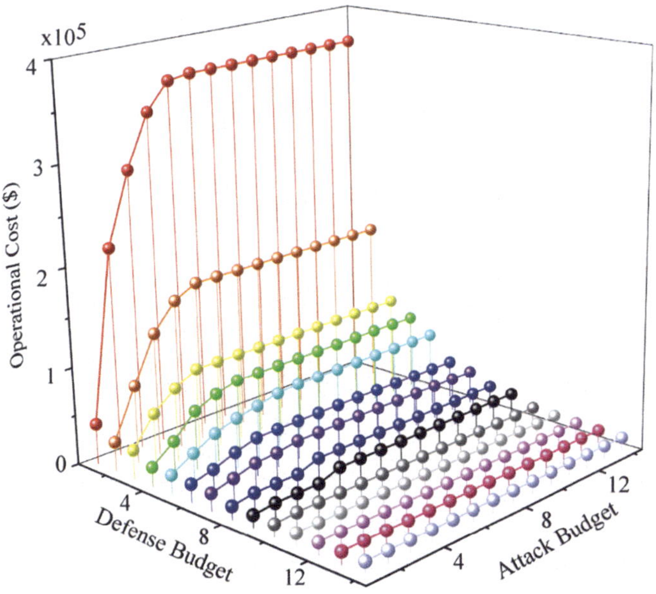

Fig. 3.26 Total cost under different budgets

Table 3.17 Results considering slow-response units

G_1	Slow response		Fast response (benchmark case)	
Case	Cost ($)	Load shedding (MWh)	Cost ($)	Load shedding (MWh)
1	7.507×10^5	464	6.713×10^5	334
2	8.629×10^5	870	7.529×10^5	744
3	1.035×10^6	210	8.615×10^5	0
4	8.512×10^5	875	8.047×10^5	530

case, resulting in dramatic increase in the total cost. Commitment decisions of $G_1 \sim G_3$ in Case 1 are listed in Table 3.18. In Table 3.18, the binary number before/after the slashes is the operating status of generation units before/after the attack, respectively. The major differences are the operating status of G_1 and G_3. In period 3, G_1 is off and G_3 is on after attack in both two cases. However, as the minimum off time of G_1 is 2 h, it remains off in period 4 and cannot be committed until period 5. Therefore, G_3 keeps working to reduce the potential load shedding in period 4. In the fast-response scenario, G_1 can be put back online in period 4 which avoids further load shedding. To sum up, fast-response units can provide important backup in case of contingencies.

5. Efficiency on a 39-Bus Power System with 20-Node Gas Network

The proposed model and algorithm is applied to a larger test system, which consists of the IEEE 39-bus system and a modified Belgian high-calorific 20-node gas network. It includes 3 gas-fired units, 7 traditional units, 2 gas wells, 3 compressors,

Table 3.18 Unit commitment in Case 1

Period	Slow response G_1	Slow response G_2	Slow response G_3	Fast response G_1	Fast response G_2	Fast response G_3
2	1/1	1/0	0/0	1/1	1/0	0/0
3	1/0	1/0	0/1	1/0	1/0	0/1
4	1/0	1/0	0/1	1/1	1/0	0/0
5	1/1	1/0	0/0	1/1	1/0	0/0

Table 3.19 Computation times under different budgets

		Γ_A 1	2	3	4	5
Γ_D	1	65	77	93	111	140
	2	155	172	312	551	673
	3	317	477	699	831	1145
	4	663	927	1401	2022	2712
	5	922	1477	1902	2638	3531

4 gas storage tanks, 19 power loads, and 9 gas loads. Potential targets of the attacker include 46 power transmission lines, 21 passive pipelines, 3 active pipelines, and 3 connection pipes. System topology, data, and unit commitment of this test system is given in [31]. We consider 2 periods economic dispatch and 4-segment PWL approximation for the Weymouth equation. Solver optimality gap is set to be 1%. We select period 2 to 3, 9 to 10, 14 to 15, and 19 to 20, as four target time slots. Table 3.19 summarizes the average computation time under different combinations of defense and attack budgets. The results show that the computation time increases rapidly as the defense and attack budget increase. However, when the defense and attack budgets are limited, the computation time is acceptable.

3.7 Summary and Conclusions

This chapter addresses operation, marketing, and resilience issues of the interdependent natural gas and electric power systems. We develop convex optimization and ADMM based computation approach to tackle the optimal gas-power flow problem, the fundamental issue for coordinated system operation; we envision bilateral energy markets with marginal price based transactions, and devise game theoretical models and methods to analyze the market equilibria and strategic bidding behaviors; we study the vulnerability of the coupled gas and power delivery infrastructures and propose a robust optimization model to identify and protect critical system components so as to enhance system resilience against natural disasters and malicious attacks. These methodologies could potentially become important analytical tools for the planning and operation of natural gas networks and electric power grids with rapidly growing interactions.

3.8 Further Reading

The general concept of infrastructure interdependence is introduced in [48], and the instance for natural gas and electric power systems is firstly studied in [18, 49, 50]. Recent technological breakthroughs in producing natural gas from various formats (which dramatically increases the usable reserve of natural gas) as well as potential economic benefits have precipitated the new upsurge of research in this field. Some active research directions and corresponding representative literatures are reviewed.

There is no doubt that modeling the gas flow in pipeline network is the preliminary task. A comprehensive discussion can be found in [51]. The steady-state gas flow model proposed in [18] is widely adopted in the interdependence studies of gas and power systems. Because natural gas is compressible, the transient process is ubiquitous in gas pipelines. As mentioned in Sect. 3.2.4, the transient behavior is described by a set of PDEs, which are difficult to be incorporated in an optimization model. Except the Weymouth equation, several alternative algebraic approximations are suggested in [23] for different operating conditions. A finite difference scheme is used in [52] to approximate the solution of PDEs based on Weymouth equation with a smaller granularity. An ordinary differential equation based gas flow model is proposed in [53] for optimal control studies. A novel transfer function based model for passive pipelines is devised in [54]. The impact of gas dynamics on generation scheduling problems is investigated in [55]. Transient dynamics with specific thermal model has also been discussed in many literatures, such as [56, 57]. A simulation platform is developed in [58] based on a two-time scale simulation method.

Gas system optimization problems render large-scale NLP problems. Nonlinear constraints originate from compressor energy demand, discrete approximation for the flow dynamic PDEs, and so on. General-purpose NLP solvers is a natural way to solve such kind of problem, for example, the interior point method [59] and SQP method [60]. Due to the non-convexity, a local optimum can be expected at most using an NLP solver. Thanks to the development of MILP algorithms, state-of-the-art MILP solvers are becoming more competent for large-scale problems. Researchers resort to PWL approximations for nonlinear functions [61–65] and seek the global optimum using MILP solvers. Performances of different PWL models are compared in [66]. MILP models may suffer from the exponentially growing worst-case complexity with increasing problem sizes.

Motivated by the success of convex relaxation approach in OPF research, convex optimization is exploited to retrieve either an approximated or exact global solution for gas system optimization problems with polynomial computation complexity. The first attempt is found in [67]. It is revealed that finding a feasible gas flow solution is equivalent to a convex optimization problem whose objective can be interpreted as the energy loss caused by friction. Then the formulation is extended to cost-minimum investment problem. MISOCP relaxation is proposed in [32] for the gas network expansion planning problem under steady-state conditions. The same idea is adopted in [68]. SDP relaxation and the condition for guaranteeing exactness are discussed in [69]. A common downside of relaxation based approaches is that the relaxation may be inexact. In such circumstances, the obtained solution is not feasible in the original problem. The sequential SOCP algorithm developed in Sect. 3.3 overcomes this difficulty by gradually eliminating constraint violation through a penalty approach. Geometric programming is a special convex optimization technique for posynomial optimization problems. Recent study shows that the compressor station siting problem can be formulated as a geometric programming problem [70], which is inspiring since the original problem seems rather complicated.

Coordinated operation of coupled natural gas and power systems has attracted the majority of attentions from the research community. Security-constrained optimal gas and power flow and unit commitment are studied in [62, 63] respectively. The dispatch problem aims to harness more volatile renewable energy and improve the operating condition of power systems by leveraging the flexibility enabled by the gas system and gas storage, which is discussed in [71, 72] using stochastic optimization, in [73–75] using robust optimization, and in [76] using bilevel optimization. The institutional independency calls for distributed operation schemes, which have been addressed in [77, 78] using ADMM. Resilient operation of micro energy grids is tackled in [79].

Expansion planning is another important issue in a long time horizon. Although studies on integrated electricity and natural gas system planning started a decade ago, such as [80, 81], their primary focus is energy adequacy, and operating conditions are usually simplified or neglected. Operating-aware expansion planning of natural gas and electricity transportation infrastructures is investigated in [82] in which reliability index is considered, and [83] where compressor is modeled in detail. Uncertainty issues in planning problems are dealt with in [84, 85] using stochastic optimization and robust optimization, respectively. Energy hub expansion planning is discussed in [86–88].

Sustainable development of natural gas and power systems entails mature energy markets. Current natural gas markets in European countries and the USA are significantly different from power markets in terms of organization and clearing frequency. Mathematic formulations of natural gas markets can be found in [89, 90]. Integrated natural gas and electricity market model is proposed in [91] within the European regulatory framework. In the gas system, pipeline operation is neglected, which greatly simplifies the gas market clearing problem. Another integration approach is envisioned in [92], in which the natural gas and electricity markets are cleared by two independent agents, given the fuel demand of the power system. A mixed complementarity program is proposed to determine the market equilibrium. Compared with the operation and planning issues, research on the integrated energy market has just begun, and many issues should be resolved via policy reform and innovation. Equilibrium analysis would be a central topic in this line of future research.

We refer the interested readers to [93] for a very thorough review on the model and applications of interdependent electric power grid and natural gas network infrastructures.

References

1. Rognerud, E.: The impact of the shale gas revolution. Technical Report. Available at: http://www.wgei.org/wp-content/uploads/2015/11/The-shale-gas-revolution_Eli-W-Rognerud-Aug-2015.pdf
2. Zlotnik, A., Roald, L., Backhaus, S., Chertkov, M., Andersson, G.: Coordinated scheduling for interdependent electric power and natural gas infrastructures. IEEE Trans. Power Syst. **32**(1), 600–610 (2017)
3. International Energy Agency: World Balance. From http://www.iea.org/sankey/#?c=IEATotal&s=Balance. Retrieved 5 July 2017
4. BBC News: China claims breakthrough in mining flammable ice. http://www.bbc.com/news/world-asia-china-39971667
5. Tsai, K., Upchurch, J.: Natural gas prices in 2016 were the lowest in nearly 20 years. https://www.eia.gov/todayinenergy/detail.php?id=29552
6. https://www.eia.gov/todayinenergy/detail.php?id=30832
7. Shahidehpour, M., Fu, Y., Wiedman, T.: Impact of natural gas infrastructure on electric power systems. Proc. IEEE **93**(5), 1042–1056 (2005)
8. Portante, E., Kavicky, J., Craig, B., Talaber, L., Folga, S.: Modeling electric power and natural gas system interdependencies. J. Infrastruct. Syst. **23**(4), 04017035 (2017)
9. Rubio, R., Ojeda-Esteybar, D., Ano, O., Vargas, A.: Integrated natural gas and electricity market: a survey of the state of the art in operation planning and market issues. In Proceedings of the IEEE/PES Transmission and Distribution Conference and Exposition: Latin America, pp. 1–8 (2008)
10. Chertkov, M., Backhaus, S., Lebedev, V.: Cascading of fluctuations in interdependent energy infrastructures: gas-grid coupling. Appl. Energy **160**, 541–551 (2015)
11. Melaina, M.W., Antonia, O., Penev, M.: Blending hydrogen into natural gas pipeline networks: a review of key issues. Technical Report. NREL/TP-5600–51995 (2013)
12. Götz, M., Lefebvre, J., Mörs, F., McDaniel Koch, A., Graf, F., Bajohr, S., Reimert, R., Kolb, T.: Renewable power-to-gas: a technological and economic review. Renew. Energy **85**, 1371–1390 (2016)
13. Clegg, S., Mancarella, P.: Integrated electrical and gas network flexibility assessment in low-carbon multi-energy systems. IEEE Trans. Sustain. Energy **7**(2), 718–731 (2016)
14. Wang, C., Wei, W., Wang, J., et al. Convex optimization based distributed optimal gas-power flow calculation. IEEE Trans. Sustain. Energy **9**(3), 1145–1156 (2018)
15. Wang, C., Wei, W., Wang, J., Wu, L.: Equilibrium of interdependent gas and electricity markets with marginal price based bilateral energy trading. IEEE Trans. Power Syst. **33**(5), 4854–4867 (2018)
16. Wang, C., Wei, W., Wang, J., Liu, F., Mei, S.: Strategic offering and equilibrium in coupled gas and electricity markets. IEEE Trans. Power Syst. **33**(1), 290–306 (2018)
17. Wang, C., Wei, W., Wang, J., Liu, F., Qiu, F., Correa-Posada, C., Mei, S.: Robust defense strategy for gas-electric systems against malicious attacks. IEEE Trans. Power Syst. **32**(4), 2953–2965 (2017)
18. An, S. Natural gas and electricity optimal power flow. Doctoral Dissertation, Oklahoma State University (2004)
19. Ríos-Mercado, R.Z., Borraz-Sánchez, C.: Optimization problems in natural gas transportation systems: a state-of-the-art review. Appl. Energy **147**, 536–555 (2015)
20. Ríos-Mercado, R.Z., Wu, S., Scott, L.R., Boyd, E.A.: A reduction technique for natural gas transmission network optimization problems. Ann. Oper. Res. **117**, 217–234 (2002)
21. Osiadacz, A.: Simulation and Analysis of Gas Networks. Gulf Publishing Company, Houston (1987)
22. Thorley, A., Tiley, C.: Unsteady and transient flow of compressible fluids in pipelines: a review of theoretical and some experimental studies. Int. J. Heat Fluid Flow **8**(1), 3–15 (1987)

23. Herty, M., Mohring, J., Sachers, V.: A new model for gas flow in pipe networks. Math. Methods Appl. Sci. **33**(7), 845–855 (2010)
24. Chaudry, M., Jenkins, N., Strbac, G.: Multi-time period combined gas and electricity network optimisation. Electr. Power Syst. Res. **78**(7), 1265–1279 (2008)
25. Liu, C., Shahidehpour, M., Wang, J.: Coordinated scheduling of electricity and natural gas infrastructures with a transient model for natural gas flow. Chaos: Interdiscipl. J. Nonlinear Sci. **21**(2), 1–12 (2011)
26. Keyaerts, N.: Gas balancing and line-pack flexibility: concepts and methodologies for organizing and regulating gas balancing in liberalized and integrated EU gas markets. Technical Report. D/2012/7515/101 (2012)
27. Correa-Posada, C.M., Sanchez-Martin, P.: Integrated power and natural gas model for energy adequacy in short-term operation. IEEE Trans. Power Syst. **30**(6), 3347–3355 (2015)
28. Boyd, S., Parikh, N., Chu, E., Peleato, B., Eckstein, J.: Distributed optimization and statistical learning via the alternating direction method of multipliers. Found. Trends Mach. Learn. **3**(1), 1–122 (2011)
29. Ghadimi, E., Teixeira, A., Shames, I., Johansson, M.: Optimal parameter selection for the alternating direction method of multipliers (ADMM): quadratic problems. IEEE Trans. Autom. Control **60**(3), 644–658 (2015)
30. Lipp, T., Boyd, S.: Variations and extension of the convex-concave procedure. Optim. Eng. **17**(2), 263–287 (2016)
31. https://sites.google.com/site/chengwang0617/home/data-sheet
32. Borraz-Sánchez, C., Bent, R., Backhaus, S., Hijazi, H., Van Hentenryck, P.: Convex relaxations for gas expansion planning. INFORMS J. Comput. **28**(4), 645–656 (2016)
33. He, C., Wu, L., Liu, T., Shahidehpour, M.: Robust co-optimization scheduling of electricity and natural gas systems via ADMM. IEEE Trans. Sustain. Energy **8**(2), 658–670 (2017)
34. Lavaei, J., Tse, D., Zhang, B.: Geometry of power flows and optimization in distribution networks. IEEE Trans. Power Syst. **29**(2), 572–583 (2014)
35. Zhang, B. Tse, D.: Geometry of injection regions of power networks. IEEE Trans. Power Syst. **28**(2), 788–797 (2013)
36. Li, F.: Continuous locational marginal pricing. IEEE Trans. Power Syst. **22**(4), 1638–1646 (2007)
37. Overview of natural gas: natural gas marketing. Available at http://naturalgas.org/naturalgas/marketing/
38. Natural gas and electricity market coordination issues. PJM Learning Center. Available at https://learn.pjm.com/three-priorities/keeping-the-lights-on/gas-electric-industry/natural-gas-electric-market.aspx
39. International Energy Agency: World energy look 2013. Paris (2013)
40. Gabriel, S., Smeers, Y.: Complementarity Problems in Restructured Natural Gas Markets. Springer, Berlin (2006)
41. Gil, M., Dueñas, P., Reneses, J.: Electricity and natural gas interdependency: comparison of two methodologies for coupling large market models within the European regulatory framework. IEEE Trans. Power Syst. **31**(1), 361–369 (2016)
42. Wolf, D., Smeers, Y.: The gas transmission problem solved by an extension of the simplex algorithm. Manage. Sci. **46**(11), 1454–1465 (2000)
43. Leyffer, S., Munson, T.: Solving multi-leader-common-follower games. Optim. Methods Softw. **25**(4), 601–623 (2010)
44. Ye, H., Li, Z.: Necessary conditions of line congestions in uncertainty accommodation. IEEE Trans. Power Syst. **31**(5), 4165–4166 (2016)
45. Ardakani, A.J., Bouffard, F.: Identification of umbrella constraints in DC-based security-constrained optimal power flow. IEEE Trans. Power Syst. **28**(4), 3924–3934 (2013)
46. Ardakani, A.J., Bouffard, F.: Acceleration of umbrella constraint discovery in generation scheduling problems. IEEE Trans. Power Syst. **30**(4), 2100–2109 (2015)
47. Zhao, L., Zeng, B.: Vulnerability analysis of power grids with line switching. IEEE Trans. Power Syst. **28**(3), 2727–2736 (2013)

48. Rinaldi, S.M., Peerenboom, J.P., Kelly, T.K.: Identifying, understanding, and analyzing critical infrastructure interdependencies. IEEE Control Syst. **21**(6), 11–25 (2001)
49. Shahidehpour, M., Fu, Y., Wiedman, T.: Impact of natural gas infrastructure on electric power systems. Proc. IEEE **93**(5), 1042–1056 (2005)
50. Li, T., Eremia, M., Shahidehpour, M.: Interdependency of natural gas network and power system security. IEEE Trans. Power Syst. **23**(4), 1817–1824 (2008)
51. Osiadacz, A. Simulation and Analysis of Gas Networks. Gulf Publishing Company, Houston (1987)
52. Liu, C., Shahidehpour, M., Wang, J.: Coordinated scheduling of electricity and natural gas infrastructures with a transient model for natural gas flow. Chaos **21**(2), 025102 (2011)
53. Zlotnik, A., Roald, L., Backhaus, S., Chertkov, M., Andersson, G.: Coordinated scheduling for interdependent electric power and natural gas infrastructures. IEEE Trans. Power Syst. **32**(1), 600–610 (2017)
54. Zhou, Y., Gu, C., Wu, H., Song Y.: An equivalent model of gas networks for dynamic analysis of gas-electricity systems. IEEE Trans. Power Syst. **32**(6), 4255–4264 (2017)
55. Yang, J., Zhang, N., Kang, C., Xia, Q.: Effect of natural gas flow dynamics in robust generation scheduling under wind uncertainty. IEEE Trans. Power Syst. **33**(2), 2087–2097 (2018)
56. Chaczykowski, M.: Transient flow in natural gas pipeline-The effect of pipeline thermal model. Appl. Math. Model. **34**(4), 1051–1067 (2010)
57. Abbaspour, M., Chapman, K.: Nonisothermal transient flow in natural gas pipeline. J. Appl. Mech. **75**(3), 031018 (2008)
58. Xu, X., Jia, H., Chiang, H., Wang, D.: Dynamic modeling and interaction of hybrid natural gas and electricity supply system in microgrid. IEEE Trans. Power Syst. **30**(3), 1212–1221 (2015)
59. Steinbach, M.: On PDE solution in transient optimization of gas networks. J. Comput. Appl. Math. **203**(2), 345–361 (2007)
60. Ehrhardt, K., Steinbach, M.: Nonlinear optimization in gas networks. In Modeling, Simulation and Optimization of Complex Processes, pp. 139–148. Springer, Berlin (2005)
61. Martin, A., Möller, M., Moritz, S.: Mixed integer models for the stationary case of gas network optimization. Math. Program. **105**(2–3), 563–582 (2006)
62. Correa-Posada, C., Sánchez-Martin, P.: Security-constrained optimal power and natural-gas flow. IEEE Trans. Power Syst. **29**(4), 1780–1787 (2014)
63. Correa-Posada, C., Sanchez-Martin, P., Lumbreras, S.: Security-constrained model for integrated power and natural-gas system. J. Mod. Power Syst. Clean Energy **5**(3), 326–336 (2017)
64. Correa-Posada, C., Sánchez-Martin, P.: Integrated power and natural gas model for energy adequacy in short-term operation. IEEE Trans. Power Syst. **30**(6), 3347–3355 (2015)
65. Domschke, P., Geißler, B., Kolb, O., Lang, J., Martin, A., Morsi, A.: Combination of nonlinear and linear optimization of transient gas networks. INFORMS J. Comput. **23**(4), 605–617 (2011)
66. Correa-Posada, C., Sánchez-Martin, P.: Gas network optimization: a comparison of piecewise linear models. Technical Report. Available in Optimization Online (2014)
67. Babonneau, F., Nesterov, Y., Vial, J.: Design and operations of gas transmission networks. Oper. Res. **60**(1), 34–47 (2012)
68. Sanchez, C., Bent, R., Backhaus, S., Blumsack, S., Hijazi, H., Van Hentenryck, P.: Convex optimization for joint expansion planning of natural gas and power systems. In Proceedings of the 49th Hawaii International Conference on System Sciences, pp. 2536–2545 (2016)
69. Ojha, A., Kekatos, V., Baldick, R. Solving the natural gas flow problem using semidefinite program relaxation. In Proceedings of the IEEE Power and Energy Society General Meeting, Chicago (2016)
70. Misra, S., Fisher, M., Backhaus, S., Bent, R., Chertkov, M., Pan, F.: Optimal compression in natural gas networks: a geometric programming approach. IEEE Trans. Control Netw. Syst. **2**(1), 47–56 (2015)
71. Alabdulwahab, A., Abusorrah, A., Zhang, X., Shahidehpour, M.: Coordination of interdependent natural gas and electricity infrastructures for firming the variability of wind energy in stochastic day-ahead scheduling. IEEE Trans. Sustain. Energy **6**(2), 606–615 (2015)

72. Zhang, X., Che, L., Shahidehpour, M., Alabdulwahab, A., Abusorrah, A.: Electricity-natural gas operation planning with hourly demand response for deployment of flexible ramp. IEEE Trans. Sustain. Energy **7**(3), 996–1004 (2016)
73. Liu, C., Lee, C., Shahidehpour, M.: Look ahead robust scheduling of wind-thermal system with considering natural gas congestion. IEEE Trans. Power Syst. **30**(1), 544–545 (2015)
74. Martinez-Mares, A., Fuerte-Esquivel, C.: A robust optimization approach for the interdependency analysis of integrated energy systems considering wind power uncertainty. IEEE Trans. Power Syst. **28**(4), 3964–3976 (2013)
75. Bai, L., Li, F., Jiang, T., Jia, H.: Robust scheduling for wind integrated energy systems considering gas pipeline and power transmission $N-1$ contingencies. IEEE Trans. Power Syst. **32**(2), 1582–1584 (2017)
76. Li, G., Zhang, R., Jiang, T., Chen, H., Bai, L., Li, K.: Security-constrained bi-level economic dispatch model for integrated natural gas and electricity systems considering wind power and power-to-gas process. Appl. Energy **194**, 696–704 (2017)
77. He, C., Wu, L., Liu, T., Shahidehpour, M.: Robust co-optimization scheduling of electricity and natural gas systems via ADMM. IEEE Trans. Sustain. Energy **8**(2), 658–670 (2017)
78. Wen, Y., Qu, X., Li, W., Liu, X., Ye, X.: Synergistic operation of electricity and natural gas networks via ADMM. IEEE Trans. Smart Grid **9**(5), 4555–4565 (2017)
79. Manshadi, S., Khodayar, M.: Resilient operation of multiple energy carrier microgrids. IEEE Trans. Smart Grid **6**(5), 2283–2292 (2015)
80. Unsihuay-Vila, C., Marangon-Lima, J., de Souza, A., Perez-Arriaga, I., Balestrassi, P.: A model to long-term, multiarea, multistage, and integrated expansion planning of electricity and natural gas systems. IEEE Trans. Power Syst. **25**(2), 1154–1168 (2010)
81. Unsihuay-Vila, C., Marangon-Lima J., de Souza, A.: Integrated power generation and natural gas expansion planning. In Proceedings IEEE Power Tech, pp. 1404–1409 (2007)
82. Zhang, X., Shahidehpour, M., Alabdulwahab, A., Abusorrah, A.: Security-constrained co-optimization planning of electricity and natural gas transportation infrastructures. IEEE Trans. Power Syst. **30**(6), 2984–2993 (2015)
83. Chaudry, M., Jenkins, N., Qadrdan, M., Wu, J.: Combined gas and electricity network expansion planning. Appl. Energy **113**, 1171–1187 (2014)
84. Ding, T., Hu, Y., Bie, Z.: Multi-stage stochastic programming with nonanticipativity constraints for expansion of combined power and natural gas systems. IEEE Trans. Power Syst. **33**(1), 317–328 (2018)
85. He, C., Wu, L., Liu, T., Bie, Z.: Robust co-optimization planning of interdependent electricity and natural gas systems with a joint $N-1$ and probabilistic reliability criterion. IEEE Trans. Power Syst. **33**(2), 2140–2154 (2018)
86. Zhang, X., Shahidehpour, M., Alabdulwahab, A., Abusorrah, A.: Optimal expansion planning of energy hub with multiple energy infrastructures. IEEE Trans. Smart Grid **6**(5), 2302–2311 (2015)
87. Salimi, M., Ghasemi, H., Adelpour, M., Vaez-ZAdeh, S.: Optimal planning of energy hubs in interconnected energy systems: a case study for natural gas and electricity. IET Gener. Transm. Dis. **9**(8), 695–707 (2015)
88. Zhang, X., Che, L., Shahidehpour, M., Alabdulwahab, A., Abusorrah, A.: Reliability-based optimal planning of electricity and natural gas interconnections for multiple energy hubs. IEEE Trans. Smart Grid **8**(4), 1658–1667 (2017)
89. Egging, R., Gabriel, S., Holz, F., Zhuang, J.: A complementarity model for the European natural gas market. Energy Policy **36**(7), 2385–2414 (2008)
90. Gabriel, S., Zhuang, J., Kiet, S.: A large-scale linear complementarity model of the North American natural gas market. Energy Econ. **27**(4), 639–665 (2005)
91. Gil, M., Dueñas, P., Reneses, J.: Electricity and natural gas interdependency: Comparison of two methodologies for coupling large market models within the European regulatory framework. IEEE Trans. Power Syst. **31**(1), 361–369 (2016)

92. Abrell, J., Weigt, H.: Combining energy networks. Netw. Spat. Econ. **12**(3), 377–401 (2012)
93. He, C., Zhang, X., Liu, T., Wu, L., Shahidehpour, M.: Coordination of interdependent electricity grid and natural gas network-a review. Curr. Sustain. Renew. Energy Rep. **5**(1), 23–36 (2018)

Chapter 4
Heat-Electricity Energy Distribution System

4.1 Introduction

The electricity sector is undergoing a transition to a more sustainable and environmentally friendly one. Given the clean and cost-effective features of renewable energies, the past decade has witnessed dramatic developments of wind and solar power in many countries. However, the volatility of such energy resources makes it challenging to balance electricity demand and supply in power systems with high share of renewables.

Heating and electricity are two major forms of energy demands in urban areas. Combined heat and power production shows clear benefits due to the enhanced usage of waste heat. More importantly, thermal demands can also serve as flexible loads that foster energy storage in form of heat, and therefore facilitate integration of renewable energies. In this regard, combined heat and power generation technologies could become an essential strategy for mitigating greenhouse gas emissions, and play an increasingly important role for realizing sustainable energy production and a low-carbon economy.

Thermal and electrical energies have their unique features. Heat is easy to store, but long distance transmission of thermal energy is not economically efficient. Thermal demands in a city are domestically supplied by a district heating network. Heating devices include gas boilers, CHP units, heat pumps, etc. Electricity is easy for transmission, but large-scale storage of electricity could be expensive. Large wind farms usually locate hundreds kilometers far from load centers. Electrical power is delivered via transmission lines and received by a power distribution network, which supplies urban electricity loads, including those heating devices. In view of this fact, interactions emerge in city-sized energy supply infrastructures: the district heating network (DHN) and the power distribution network (PDN).

W. Wei, J. Wang, *Modeling and Optimization of Interdependent Energy Infrastructures*, https://doi.org/10.1007/978-3-030-25958-7_4

The DHN is composed of heat sources, pipelines, and thermal loads. The thermal energy is produced by heat sources and carried by hot water in a pipeline network. Driven by circulating pump, hot water flows from sources to loads. Thermal energy is extracted from heat exchangers and distributed to consumers. CHP units, heat pumps, electric boilers, and circulation pumps couple electricity and heat networks, since their operations impact energy flows in both physical systems. CHP units produce electricity and heat simultaneously. Electric boilers convert electricity to heat using a resistance heater. Air source heat pumps use a small amount of external power to transfer heat from the surroundings to objects (usually water) with higher temperatures. Using the same amount of electricity, heat pumps are three to four times more effective than simple resistance heaters. Circulation pumps consume electricity to drive water flow in DHN.

Besides traditional CHP units and heat pumps, the cutting-edge concentrating solar power (CSP) generation technology is becoming more attractive in recent years. CSP unit uses high temperature molten salt heat transfer fluid and produces electricity through gas turbines. Its output controllability is much better than PV panels, whose output is scarcely dispatchable. CSP stations are usually built in desert areas, where the sunshine is abundant, and the terrain is flat. If the CSP station is near to a city, it could be a suitable candidate of energy hub which connects the PDN and the DHN. Compressed air energy storage (CAES) consumes excessive energy through compressing air at one time and extracts the energy for use at another time, so as to reduce and shift electricity usage during peak periods and flatten the load over time. An appealing feature of CAES is that the air is much warmer after compression, and much colder after expansion. In this regard, the combination of CSP and CAES essentially constitutes a power-heat-cooling tri-generator.

Operating the DHN and the PDN as an integrated energy distribution system can take advantage of synergies for energy storage and boosting the utilization of intermittent renewable energies. The conceptual architecture of an integrated heat-electricity energy distribution system is provided in Fig. 4.1, showing the energy flow linkages brought by the coupling components, including CHP/CSP units, CAES stations, heat pumps (resistance boilers are omitted because they have the same functionality as heat pumps), and circulation pumps. Dispatching these facilities helps compensate the fluctuations of renewable generations, and strengthen system power balancing capability. However, the wider the adoption of these coupling components, the stronger the interplay between the heat network and power grid. It is becoming urgent to study the operation and market impacts of heating and power system integration.

This chapter addresses several fundamental issues covering the operation, marketization, and planning of the integrated heat-electricity energy distribution systems. Materials come from authors' publications [1–4]. In particular, mathematical model of DHN, including hydraulic and thermal flow models, is introduced in Sect. 4.2; Based on energy trading in the distribution power market, the decentralized operation of interdependent heat-power system is illuminated in Sect. 4.3, which encourages energy transaction based on time-varying LMP thus could improve wind utilization during the night; Furthermore, the marginal cost based energy

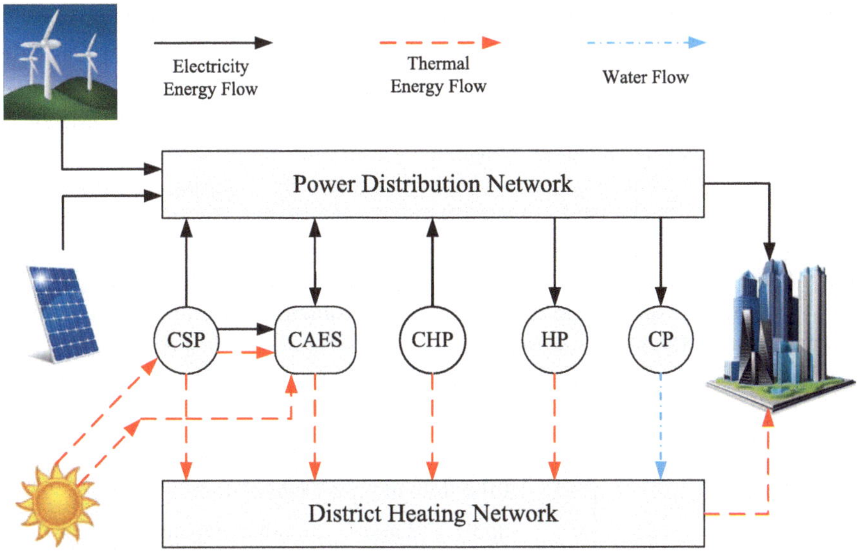

Fig. 4.1 Conceptual architecture of an integrated energy distribution system

pricing policy is applied in the district heating system, and the equilibrium of the spot heating and power markets is discussed in Sect. 4.4, and the impact of elastic demands and strategic producers are thoroughly studied; In the above market structure, heating and electricity markets have direct access to each other.

Under the prevalent trend of energy integration, the energy hub will be a new entity which acts as an interface among different energy systems, with functionalities of energy production, conversion, and storage. Energy hubs can be divided into two categories according to their capacities and clients. Residential-level energy hubs refer to those appearing at the demand side and directly supplying home appliance without considering network constraints, since their capacity is small, and a single hub has tiny effect on the distribution system. Distribution-level energy hubs category those connecting natural gas, power, and heat distribution systems and acting as a prosumer. Because their capacity is comparable to the system load, operation of a distribution-level energy hub could affect network energy flows. The latter one will be studied in this chapter since it impacts the networked infrastructures, which is the theme of this book. Specifically, the strategic behaviors of an energy hub in the heating and electricity distribution markets are analyzed in Sect. 4.5. The capacity planning problem of energy hubs in multi-carrier energy distribution systems is addressed in Sect. 4.6. Investment problem could be either centralized (sponsored or coordinated by the government) or competitive (among investors). This section considers government-leading energy hub capacity planning aiming at tackling renewable and load uncertainties.

4.2 Mathematical Model of the District Heating Network

DHN distributes thermal energy generated in a centralized plant for space or water heating purposes. It is advantageous over localized boilers in terms of efficiency and pollution control. The network usually consists of supply and return pipes that deliver hot water from the sources to the consumers. To describe the steady-state operating status of a DHN, the hydraulic model determines the mass flow rate in each pipeline; the thermal model determines the temperature at each node. Materials of hydraulic and thermal models in this section come from [5–7].

We assume the supply and return pipes are symmetric (have identical topology and parameter). The supply network consists of l loops, m nodes, and n pipes. The node-pipe incidence matrix is defined as

$$A \in \mathbb{M}^{m \times n},\ a_{ij} = \begin{cases} +1, & \text{if the flow in pipe } j \text{ injects node } i; \\ -1, & \text{if the flow in pipe } j \text{ leaves node } i; \\ 0, & \text{pipe } j \text{ and node } i \text{ are not connected;} \end{cases} \tag{4.1}$$

The loop-pipe incidence matrix is defined as

$$B \in \mathbb{M}^{l \times n},\ b_{ij} = \begin{cases} +1, & \text{if the directions of pipe } j \text{ and loop } i \text{ are the same;} \\ -1, & \text{if the directions of pipe } j \text{ and loop } i \text{ are opposite;} \\ 0, & \text{if loop } l \text{ does not include pipe } j; \end{cases} \tag{4.2}$$

To illustrate these concepts, a simple network with 1 loop, 3 nodes, and 3 pipes is depicted in Fig. 4.2. Since the supply and return networks are symmetric, we put emphasis on the supply network, and the return network is plotted with dash lines. The node-pipe incident matrix A and loop-pipe incident matrix B are provided in the same figure.

4.2.1 Hydraulic Model

The hydraulic model relies on three basic laws: the flow continuity condition, the loop pressure equation, and head loss formula. The former two formulate flow conservation and loop pressure drop in a DHN, which are independent of branch characteristics, and similar to the Kirchhoff's current law and the Kirchhoff's voltage law in circuit theory. Graphic representations are reported in [6]. The third one describes pressure drop in each pipe, which is in analogy to the Ohm's law.

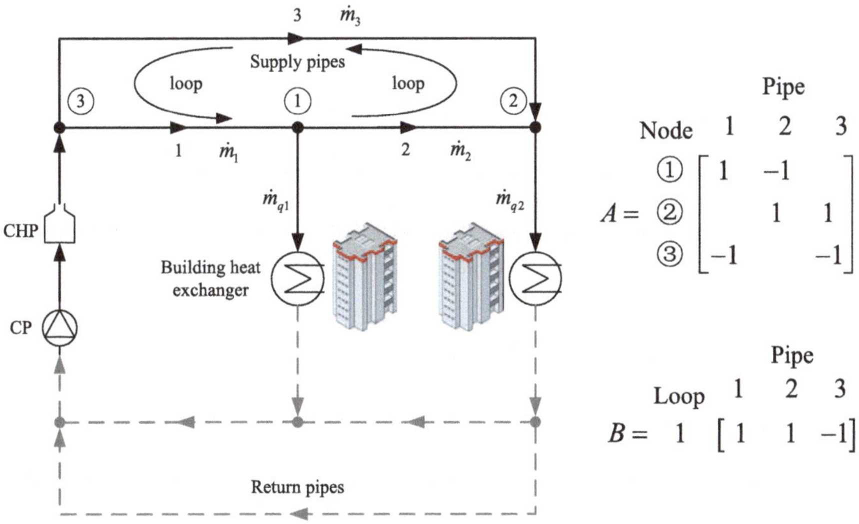

Fig. 4.2 A simple pipeline network and its incident matrices

Flow continuity requires that the sum of inflows should be equal to the sum of outflows at every node, i.e.

$$A\dot{m} = \dot{m}_q \tag{4.3a}$$

where $\dot{m}$ is the vector of mass flows in each pipe; $\dot{m}_q$ is the vector of mass flows between supply side and return side at each node.

Since the network is lossless (for hydraulic conditions), one equation in (4.3a) is redundant, because $1^T A$ is a zero vector, and $\text{rank}(A) = m - 1$. In view of this, we can either eliminate the last row of A and use

$$A = \begin{bmatrix} 1 & -1 & 0 \\ 0 & 1 & 1 \end{bmatrix}$$

or treat the last element $\dot{m}_q$, the mass flow of the heat source, as a variable. We adopt the latter treatment because problems studied in this chapter involve optimization of multiple heat sources.

Head loss refers to the pressure change in pipes due to friction. The loop pressure equation states that the sum of head losses around any closed loop is equal to zero

$$Bh^f = 0 \tag{4.3b}$$

where h^f is the vector of head losses in each pipe. The relation between the head loss and mass flow in each pipe is defined by the head loss formula

$$h_l^f = K_l \dot{m}_l |\dot{m}_l|, \ \forall l \tag{4.3c}$$

where K_l is the resistance coefficient of pipe l, which can be calculated as [7]

$$K_l = \frac{8Lf}{D^5 \rho^2 \pi^2 g} \tag{4.3d}$$

where L is the pipe length; D is the pipe diameter; ρ denotes the density of water; π means the circumference ratio; g stands for the gravitational acceleration. They are all constants. The friction factor f depends on the Reynolds number Re of the fluid, which is calculated from the equation in hydromechanics

$$\mathrm{Re} = \frac{vD}{\mu}$$

where v is the water fluid velocity; μ is the kinematic viscosity of water. The velocity is simply derived from the mass flow rate as

$$v = \frac{4\dot{m}}{\rho \pi D^2}$$

For laminar flow with Re < 2320, an inversely-proportional relationship holds

$$f = \frac{64}{\mathrm{Re}}$$

For turbulent flow with Re > 4000, the friction factor f can be calculated numerically from a transcendental equation

$$\frac{1}{\sqrt{f}} = -2\log_{10}\left(\frac{\varepsilon}{3.7D} + \frac{2.51}{\mathrm{Re}\sqrt{f}}\right)$$

where ε represents the roughness of the pipe.

For 2300 < Re < 4000, f is retrieved from linear interpolation.

For a radial network, the hydraulic model is much simpler. The numbers of nodes and pipes are equal, and A is square. By eliminating one row and one column, the reduced matrix A' is invertible. If the load mass flow rate $\dot{m}_q$ is given, the pipe mass flow rates can be uniquely calculated from (4.3a) without considering the pressure and head losses. The hydraulic conditions in the return network are inverse compared to those in the supply network where the nodal injection is $-\dot{m}_q$, because $A(-\dot{m}) = -\dot{m}_q$ holds. For a meshed network, if all pipes have the same parameter, the hydraulic conditions in supply and return networks are also symmetric like the radial network; otherwise, if the pipes have different friction coefficients, since they are operated under different thermal conditions year after year, the mass flow rates in pipes constituting a cycle might have slight difference at the supply side and the return side. Nevertheless, such a difference can be neglected in practice.

4.2.2 Thermal Model

The thermal model states the distribution of nodal temperatures in the network. Although the mass flow rates in the supply and return pipes are symmetric, the temperatures are not. Supply (return) side of a node has a higher (lower) temperature. We consider a radial network. Related temperatures are elucidated in Fig. 4.3. Variables used throughout the thermal model are defined below for quick reference.

D_L^S	Set of supply pipes
D_L^R	Set of return pipes
D_L	Set of pipes, $D_L = D_L^S \cup D_L^R$
D_N^S	Set of nodes at the supply side
D_N^R	Set of nodes at the return side
D_N	Set of nodes, $D_N = D_N^S \cup D_N^R$
τ_a	Ambient temperature
τ_i^{SN}	Fluid temperature at supply side of node $i \in D_N^S$
τ_i^{RN}	Fluid temperature at return side of node $i \in D_N^R$
l_{ij}^S	Supply pipe $l_{ij}^S \in D_L^S$, its head node is $i \in D_N^S$ and its tail node is $j \in D_N^S$
l_{ij}^R	Return pipe $l_{ij}^R \in D_L^R$, its head node is $i \in D_N^R$ and its tail node is $j \in D_N^R$
$\tau_{l(i,j)}^{S,In}$	Fluid temperature at the head node of supply pipe $l_{ij}^S \in D_L^S$
$\tau_{l(i,j)}^{S,Out}$	Fluid temperature at the tail node of supply pipe $l_{ij}^S \in D_L^S$
$\tau_{l(i,j)}^{R,In}$	Fluid temperature at the head node of return pipe $l_{ij}^R \in D_L^R$
$\tau_{l(i,j)}^{R,Out}$	Fluid temperature at the tail node of return pipe $l_{ij}^R \in D_L^R$
H_N^S	Supply side of heat source node
H_N^R	Return side of heat source node
$P_T^S(i)$	Supply pipes whose tail nodes are $i \in D_N^S$
$P_H^S(i)$	Supply pipes whose head nodes are $i \in D_N^S$

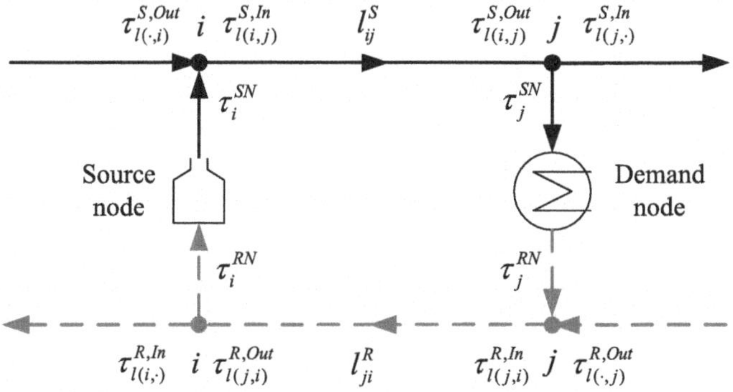

Fig. 4.3 Temperatures in the network

$P_T^R(i)$	Return pipes whose tail nodes are $i \in D_N^R$
$P_H^R(i)$	Return pipes whose head nodes are $i \in D_N^R$
λ	Heat transfer efficiency coefficient
L_l	Length of pipe $l \in D_L$
h_i^e	Heat exchange at node $i \in D_N$
h_i^c/h_i^d	Charge/discharge rate of thermal energy storage unit
η_i^c/η_i^d	Charge/discharge efficiency of thermal energy storage unit
c_f	The specific heat capacity of water
$\dot{m}_{qi}$	Demand mass flow rate at node i from the supply side to the return side
$\dot{m}_l$	Mass flow rate in pipe l
W_i^H	Thermal energy stored at node $i \in D_N$
μ_i^H	Loss rate of thermal energy storage unit

In the pipeline network, each node is associated with three temperature variables. In Fig. 4.3, take node j at the supply side for example, the outlet temperature $\tau_{l(\cdot,j)}^{S,out}$ of pipeline l whose tail is j; the inlet temperature $\tau_{l(j,\cdot)}^{S,in}$ of pipeline l whose head is j; the temperature τ_j^{SN} at the inlet of heat exchanger connecting to the demand. For water divergence (confluence) nodes, these temperatures are equal (different). For source nodes, confluence (divergence) nodes are at the supply (return) side. For demand nodes, confluence (divergence) nodes are at the return (supply) side.

At heat demand nodes, thermal energy is extracted from the carrier fluid, and the fluid temperature drops from τ_i^{SN} to τ_i^{RN}; at heat source nodes, thermal energy is passed to the carrier fluid, and the fluid temperature rises from τ_i^{RN} to τ_i^{SN}. The heat power exchange can be expressed as

$$h_i^e = c_f \dot{m}_{qi}(\tau_i^{SN} - \tau_i^{RN}), \ \forall i \in D_N \tag{4.4a}$$

For a heat source, $\dot{m}_{qi} < 0$, so the energy exchange $h_i^e < 0$, indicating a power injection. Equation (4.4a) is applied to every source/load between a supply node and the corresponding return node. In subsequent sections, we will use slightly different notations in economic operation models with only positive decision variables.

Due to heat dissipativity, the temperature drop along a pipe can be calculated as

$$\begin{aligned} \tau_{l(i,j)}^{S,Out} &= \left[\tau_{l(i,j)}^{S,In} - \tau_a\right] \mathrm{e}^{\frac{\lambda L_l}{c_f \dot{m}_l}} + \tau_a, \ \forall l \in D_L^S \\ \tau_{l(j,i)}^{R,Out} &= \left[\tau_{l(j,i)}^{R,In} - \tau_a\right] \mathrm{e}^{\frac{\lambda L_l}{c_f \dot{m}_l}} + \tau_a, \ \forall l \in D_L^R \end{aligned} \tag{4.4b}$$

Equation (4.4b) indicates that the larger the mass flow rate in a pipe, the smaller the temperature drop along that pipe. However, in order to increase mass flow rates in the whole network, the circulating pump must consume more electricity. It is unrealistic and unnecessary to compensate temperature drop by unlimitedly increasing mass flow rates. Equation (4.4b) is applied to every pipeline in the system.

At the supply side of a heat source, the fluids with different temperatures come across at a confluence node (like node i in Fig. 4.3), and the mixture reaches a new temperature, which is determined from energy conservation law, and then diverges to other pipes with the same temperature, giving the following equations

$$\left(\sum_{l\in P_T^S(j)} \dot{m}_l \tau_{l(\cdot,j)}^{S,Out}\right) = \left(\sum_{l\in P_H^S(j)} \dot{m}_l \tau_{l(j,\cdot)}^{S,In}\right) + \left|\dot{m}_{qj}\right| \tau_j^{SN},\ \forall j \in H_N^S \tag{4.4c}$$

At the supply side of a demand, fluid diverges with the same temperature (like node j in Fig. 4.3), resulting in

$$\tau_{l_1(j,\cdot)}^{S,In} = \tau_j^{SN} = \tau_{l_2(\cdot,j)}^{S,out},\ \forall l_1 \in P_H^S(j),\ \forall l_2 \in P_T^S(j),\ \forall j \in D_N^S - H_N^S \tag{4.4d}$$

Equations (4.4c) and (4.4d) are applied to every node at the supply side, depending on its type (source or demand).

At the return side of a heat demand, the fluids come across at confluence nodes after heat exchange (like node j in Fig. 4.3). Similarly, we have

$$\left(\sum_{l\in P_H^R(j)} \dot{m}_l \tau_{l(j,\cdot)}^{R,In}\right) = \left(\sum_{l\in P_T^R(j)} \dot{m}_l \tau_{l(\cdot,j)}^{R,Out}\right) + \left|\dot{m}_{qj}\right| \tau_j^{RN},\ \forall j \in D_N^R - H_N^R \tag{4.4e}$$

At the return side of a heat source, fluid diverges with the same temperature (like node i in Fig. 4.3), yielding

$$\tau_{l_1(j,\cdot)}^{R,In} = \tau_j^{RN} = \tau_{l_2(\cdot,j)}^{R,out},\ \forall l_1 \in P_H^R(j),\ \forall l_2 \in P_T^R(j),\ \forall j \in H_N^R \tag{4.4f}$$

Equations (4.4e) and (4.4f) are applied to every node at the return side, depending on its type (source or demand).

Thermal energy storage unit plays the role of a prosumer. It can smoothen the heat demand curve over time. So when a thermal energy storage unit is modeled, the operating conditions appear to be time-varying, and all variables and constraints must be duplicated with respect to time periods. For notation brevity, time label is omitted in (4.4a)–(4.4f). Specifically, the equation of storage dynamics is

$$W_{i,t+1}^H = W_{i,t}^H(1-\mu_i^H) + (h_{i,t}^c\eta_i^c - h_{i,t}^d/\eta_i^d)\Delta t,\ \forall i \in D_N, \forall t \tag{4.4g}$$

The heat exchange of thermal energy storage unit is

$$h_{i,t}^e = h_{i,t}^c - h_{i,t}^d,\ \forall i \in D_N, \forall t \tag{4.4h}$$

Because of heat loss, simultaneous charging and discharging is usually not an optimal strategy, i.e., at most one entry of the pair $(h^c_{i,t}, h^d_{i,t})$ could be strictly positive. Nevertheless, strict complementarity can be imposed via auxiliary binary variables.

4.2.3 Operating Characteristics of Heat Sources

DHN and PDN are intertwined through the coupling components, mainly heat sources. To study their impacts on both systems, it is essential to quantify their operating characteristics, which refers to the relations between electrical power output p and heat output h. Some representative components are introduced in this section.

1. CHP Unit

There exist three kinds of CHP: gas turbine based one, internal combustion reciprocating engine based one, and steam turbine based one. The operating characteristics of the former two CHP units are illustrated in the left of Fig. 4.4, and can be expressed by a linear function

$$p = \alpha h \tag{4.5a}$$

where α is a constant. As relation (4.5a) defines a one-to-one correspondence, heat and power outputs depend on each other, and the flexibility in such units is limited.

The third kind of CHP unit can be further categorized as condensing units and back-pressure units, or the combination of both. The remaining heat after the steam turbine can be used for heating purpose (the temperature and pressure of the exhaust steam in back-pressure units are usually higher than those in condensing units). The feasible operating region of such kind of CHP unit, called the extraction-condensing

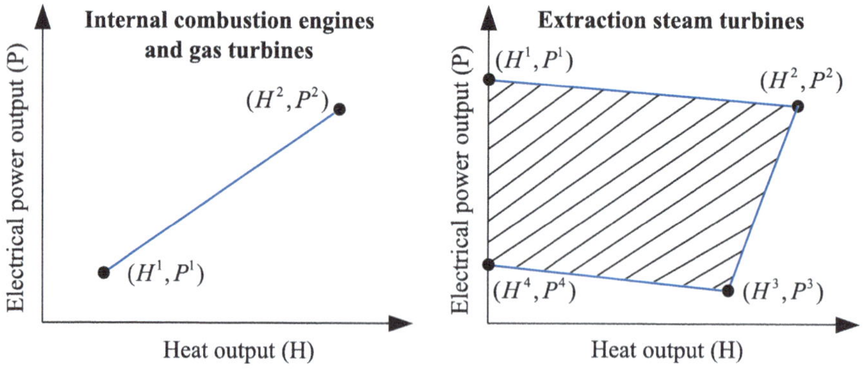

Fig. 4.4 CHP output characteristics

unit, is a polyhedron with several extreme points (H^m, P^m), $m = 1, 2, \cdots$, and the electrical power and heat output can be presented using a convex combination of extreme points as shown in (4.5b):

$$\begin{aligned} & p = \sum_k \alpha^k P^k, \; h = \sum_k \alpha^k H^k, \\ & 0 \le \alpha^k \le 1, \; \forall k, \; \sum_k \alpha^k = 1 \end{aligned} \tag{4.5b}$$

where α^k is the weight coefficient. Compared to (4.5a), either of the heat or power output in (4.5b) can be flexibly controlled in a certain range regardless of the value of the other.

2. Heat Pumps and Electric Boilers

In spontaneous heat transfer, thermal energy flows from warmer objects to colder objects. A heat pump is a device that can transfer thermal energy in the opposite direction, typically absorbing heat from the ambient air or ground with lower temperature, and pumping it into buildings or water with higher temperature, at the cost of a small amount of external power. On the heat-sink side, a volatile evaporating and condensing fluid, also known as a refrigerant, is compressed until it condenses into liquid and releases the heat during condensation. On the heat-source side, the refrigerant enters an evaporator, in which it releases the pressure, boils, and absorbs heat during evaporation. More details on fundamental theories and operating practices of heat pumps can be found in [8, 9]. Recent advances and applications can be found in [10, 11].

The term coefficient of performance (COP) is used to quantify the ratio of useful heat transfer in a heat pump by consuming per unit input energy. The operating characteristics of heat pumps are given by

$$p = \eta^{-1} h \tag{4.6}$$

where η is the COP. The COP values of air-source heat pumps range from 3.2 to 4.5, and those of ground-source heat pumps range from 4.2 to 5.2, according to a recent survey [10]. The operating characteristics of electric boilers can be described through (4.6) as well, in which parameter η stands for the energy conversion efficiency, a constant smaller than 1. The COP of a heat pump may decline with the increase of its load level, while the efficiency of a resistive boiler is usually a constant. Despite the high performance of heat pump, its investment cost is usually much higher than boilers with very simple structure.

3. Circulating Pumps

The circulating pump is installed at heat source nodes to create enough pressure and maintain proper mass flow rates in all supply and return pipes. The electrical power consumed by a circulating pump can be expressed via

$$p = \dot{m}_q \Delta P / \eta \rho \tag{4.7}$$

where $\dot{m}_q$ is the mass flow rate through the pump, η is the efficiency of the pump, ρ is the density of water, and ΔP is the pressure increment offered by the pump, and can be determined via equations in [5, 7].

In a closed pipeline system, the energy offered by the circulating pump equals to the friction losses. When the mass flow rates take moderate values, the circulating pump only consumes a small portion of the total energy consumptions in the DHN, compared with those used by other heating devices.

4.2.4 Optimal Hydraulic-Thermal Flow

In traditional DHN load flow studies, the hydraulic model and thermal model are usually solved independently [6]. Hydraulic conditions are determined from either the Newton-Raphson method [7] for meshed networks or a linear equation set (4.3a) for radial networks. In the thermal model, temperatures at supply sides of source nodes and return sides of demand nodes before mixture are manually specified [5]. Given the mass flow rates obtained from the hydraulic model, thermal constraints (4.4) render a set of linear equations.

In the smart grid era, the distribution market is becoming more competitive; real-time nodal electricity prices encourage consumers to reduce or shift their demands during peak hours, so as to flatten the load profile and alleviate congestions. Spatial difference of electricity price provides the heat producer an opportunity to cut down its expenditure. Moreover, heat losses in pipes are inevitable during network operation. The contrived mass flow rates and nodal temperatures may not be optimal from an economical perspective. This section discusses economic operation of a heating system, which is referred to as an optimal hydraulic-thermal flow (OHTF), in analogy to the OPF in power systems. It is a natural extension of the traditional research on combined heat and power generation by incorporating the network model of heating system. For notation conciseness, we use compact notations hereinafter.

Let vector x^h represent pipe mass flow rate variables in the hydraulic model, x^t denote nodal temperature variables in the thermal model, h be the thermal energy offered by heat sources, p be the electricity consumed by electric heating devices, and $x = [x^h, x^t, h, p]$ be the vector of all decision variables. The hydraulic constraints

$$\text{Cons-H} = \{\ (4.3\text{a}) - (4.3\text{c})\} \tag{4.8}$$

and thermal constraints

$$\text{Cons-T} = \{\ (4.4\text{a}) - (4.4\text{h})\} \tag{4.9}$$

With these notations, the OHTF problem can be stated as

$$\begin{aligned} \min \quad & f(p,h) \\ \text{s.t.} \quad & \text{Cons-H} \\ & \text{Cons-T} \\ & Ap + Bh \le b \\ & x^n \le x \le x^m \end{aligned} \tag{4.10}$$

where the third inequality constraint stands for the operating characteristics of heat sources. The convex combination form (4.5b) can be transformed into a hyperplane form with linear inequalities. A linear equality can be written as a pair of opposite inequalities. x^n and x^m are the lower and upper bounds of decision variables.

The objective function $f(p,h)$ is to minimize the total heat production cost, which can be decomposed into the sum of $f_i(p_i, h_i)$, i.e., the individual costs of heat sources. The fuel cost of CHP unit is a convex quadratic function [12]

$$f_i^{CHP} = a_{0i} + a_{1i}p_i + a_{2i}h_i + a_{3i}p_i^2 + a_{4i}h_i^2 + a_{5i}p_ih_i \tag{4.11}$$

which can be approximated by a bivariate PWL convex function.

Given the electricity price λ_i, the electric boiler cost is

$$f_i^{EB} = \lambda_i p_i = \lambda_i \eta^{-1} h_i \tag{4.12}$$

which is a linear function in its thermal energy output h_i. It should be emphasized that in a market environment, the power demand p_i may influence the electricity price λ_i. This interactive effect will be taken into account in Sects. 4.3 and 4.4, and omitted here, since the power system is exogenous.

For heat pumps, the COP is usually treated as a constant $\eta > 1$ in the current research, and the cost of heat pump shares the same form in (4.12). Nevertheless, experiment results demonstrate that the COP of a heat pump decreases with the compressor frequency increasing [13, 14], indicating a monotonically decreasing relation between the COP and the load level, which can be approximated by a linear function

$$\eta = b - k\frac{h}{c^P} \tag{4.13}$$

where b and k are positive constants derived from experimental data; c^P is the capacity of the heat pump; h/c^P is the output in unit value. When $k = 0$, the COP degenerates into a constant. Let $k^P = k/c^P$, the cost of a heat pump is given by

$$f_i^{HP} = \frac{\lambda_i h_i}{b_i - k_i^P h_i} \tag{4.14}$$

Equation (4.14) is more accurate than a linear function.

Cons-H defined in (4.8) and Cons-T defined in (4.9) are non-convex because of nonlinear terms involving the production of mass flow and temperature variables, as well as the exponential functions. Problem (4.10) can be solved by general purpose NLP solvers. The solution gives optimal mass flow rates, heat and power production levels, which should be deployed. Nodal temperatures are the consequence of dispatch strategies. With growing problem sizes, either due to a more complicated network or the increasing number of scheduling periods, NLP solvers may fail, and it is imperative to develop more tractable convex optimization models and harness the computational superiority of high-performance convex optimization solvers. To this end, the rest of this section will be devoted to a convex approximation model of problem (4.10).

1. Convexifying the Objective Function

In the objective function of (4.10), costs of CHP units and boilers are clearly convex; the cost of heat pump, a fraction in (4.14), seems non-convex, however, as $h_i > 0$ and $b_i - k_i h_i > 0$, its second-order derivative is

$$\frac{\partial^2 f_i^{HP}}{\partial h_i^2} = \frac{2\lambda_i b_i k_i^P}{(b_i - k_i^P h_i)^3}$$

which is always positive over the feasible operating range, indicating the convexity of the fraction in (4.14). To state it in a disciplined convex format, we decompose f_i^{HP} into

$$f_i^{HP} = \frac{\lambda_i b_i}{k_i^P (b_i - k_i^P h_i)} - \frac{\lambda_i}{k_i^P}$$

where the second term is a constant and can be omitted from the objective function; then replace the first term with an auxiliary variable σ_i; finally append the following inequalities in the constraints of (4.10):

$$\left\| \begin{matrix} 2\sqrt{\dfrac{\lambda_i b_i}{k_i^P}} \\ \sigma_i - (b_i - k_i^P h_i) \end{matrix} \right\|_2 \le \sigma_i + (b_i - k_i^P h_i) \tag{4.15}$$

$$\sigma_i \ge 0,\ h_i \ge 0,\ b_i - k_i^P h_i \ge 0 \tag{4.16}$$

where (4.15) is a canonical SOC, and (4.16) is polyhedral; both of them are compatible with SOCP solvers. In fact, (4.15) and (4.16) can boil down to

$$\sigma_i \geq \frac{\lambda_i b_i}{k_i^P (b_i - k_i^P h_i)},\ b_i - k_i^P h_i \geq 0$$

manifesting that we have actually performed an epigraph transformation without loss of accuracy. Inequality $b_i - k_i^P h_i \geq 0$ does not affect heat pump operation, because it is an intrinsic consequence of $\eta > 0$ according to (4.13).

2. Approximating the Constraints

For notation brevity, problem (4.10) is arranged into a compact form

$$\begin{aligned} \min\ & f(p, h) \\ \text{s.t.}\ & C(\dot{m})x^t + Dh = d(\dot{m}) \\ & \text{Cons-OHTF-CVX} \end{aligned} \tag{4.17}$$

where constant matrices and vectors $C(\dot{m})$, D, $d(\dot{m})$ with fixed mass flow rates $\dot{m}$ correspond to coefficients in (4.4a)–(4.4f); all other polyhedral and convex constraints are encapsulated in Cons-OHTF-CVX. Formulation (4.17) shows a special structure: when mass flow rates are fixed, problem (4.17) reduces to a convex optimization problem. This simplification interprets the constant-flow variable-temperature operating mode, one of the typical operating modes of DHN discussed in [15]. Please see more discussions at the end of this section. In what follows, we present a heuristic method to determine near-optimal mass flow rates based on network loss analysis. The rationale rests on the energy conservation law: the total heat production must be equal to the heat demand plus pipeline losses. As the demand is fixed, one natural way to cut down the cost is to reduce the losses.

Heat loss in a pipeline l can be expressed as:

$$\Delta Q_l = c_f \dot{m}_l (\tau_l^{in} - \tau_l^{out}) \tag{4.18}$$

Substituting temperature drop Eq. (4.4b) into (4.18) results in:

$$\Delta Q_l = c_f \dot{m}_l (\tau_l^{in} - \tau_a) \left(1 - \mathrm{e}^{\frac{\lambda L_l}{c_f \dot{m}_l}}\right)$$

where $0 < \lambda L_l / c_f \dot{m}_l \ll 1$, which actually indicates that the inlet temperature and outlet temperature are almost equal [16]. Because $\mathrm{e}^{-x} \approx 1 - x$, we have:

$$\Delta Q_l \approx c_f \dot{m}_l (\tau_l^{in} - T^a) \frac{\lambda L_l}{c_f \dot{m}_l} = \lambda L_l (\tau_l^{in} - T^a) \tag{4.19}$$

From Eq. (4.19) we can see that ΔQ_l is independent of mass flow rate. This means as long as the hydraulic condition makes (4.17) feasible, it has little impact on the total loss. This phenomenon has been corroborated by the experiments on real heating systems [16]. The sole controllable variable that influences pipeline loss ΔQ_l is the inlet temperature. In order to cut down the heat loss and reduce the heat production cost, the overall operating temperature of the DHN should be as low as possible. However, in order to deliver the desired amount of thermal energy, certain temperature differences between the supply side and the return side must be maintained according to (4.4a). In this regard, a two-step hydraulic-thermal decomposition algorithm is suggested to find a near-optimal solution.

Algorithm 4.1 Two-step hydraulic-thermal decomposition

1: Set all nodal temperatures to the lower bounds and calculate heat loss according to equation (4.19).
2: Optimize heat source production; the total output is equal to the sum of heat loads and net loss.
3: Calculate mass flow rates $\dot{m}_{qi}$ in each source/load node according to (4.4a).
4: Calculate mass flow rates $\dot{m}_l$ in pipelines from hydraulic model (4.3).
5: Solve problem (4.17) with obtained mass flow rates, which renders a convex optimization problem. The optimal solution gives the thermal conditions.

If the network is radial, the pipeline mass flow can be uniquely determined. Please be aware that although we set temperatures at lower bounds in Algorithm 4.1, this does not necessarily mean that nodal temperatures will be forced at their lower bounds in the optimal solution. The actual temperature would be slightly higher than their minimum values. When the demand changes, mass flow rates should be reset to guarantee feasibility, resulting in load-dependent hydraulic conditions. It should be pointed out that the water flow in DHN is maintained by a circulating pump, which consumes certain amount of electricity. In normal conditions, the energy consumed by the circulating pump is far less than heat sources, so its cost is neglected.

The proposed method is applied to a 10-node DHN which contains a loop, and its topology is shown in Fig. 4.5a. System data are provided in https://sites.google.com/site/caoyangthu13/data. We randomly generate 50 load scenarios. The total heat demand varies from 3 MW to 6 MW. The exact model (4.10) is solved by Baron [17], a global NLP solver, and the optimal value is Opt-Glo; the convex approximation model (4.17) in which mass flow rates are determined from Algorithm 4.1 is solved by CPLEX, and the optimal value is Opt-Apr. Since certain variables are fixed in (4.17), Opt-Apr $\geq$ Opt-Glo always holds. The optimality gap, defined as (Opt-Apr−Opt-Glo)/Opt-Glo, is shown in Fig. 4.5b. The gap has an order of magnitude of 10^{-4} in all 50 instances, indicating that (4.17) is a high quality approximation for the nonlinear model (4.10) when the mass flow rates are set by Algorithm 4.1.

Mass flow rates of individual pipelines in the minimum, average, and maximum demand scenarios are summarized and compared in Table 4.1 (where Al-4.1 is

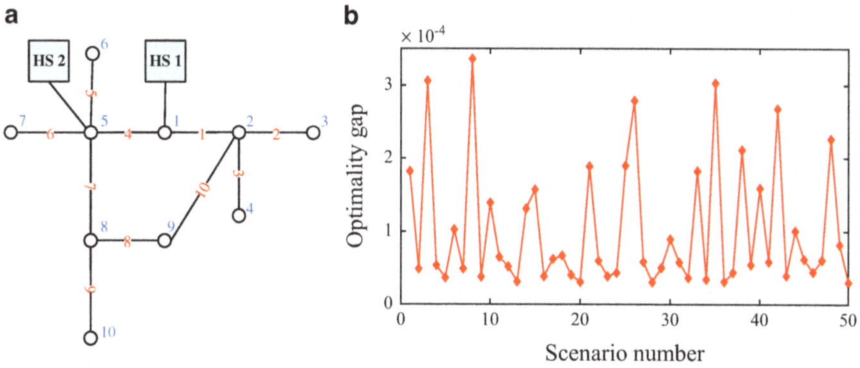

Fig. 4.5 (**a**) Topology of a 10-node meshed DHN. (**b**) Optimality gap

Table 4.1 Calculation results of mass flow rates (kg/s)

	Minimum load			Average load			Maximum load		
Pipe	Al-4.1	Baron	Error	Al-4.1	Baron	Error	Al-4.1	Baron	Error
1	7.39	7.33	0.82%	10.55	10.53	0.21%	13.91	13.90	0.17%
2	3.24	3.20	1.78%	4.61	4.58	0.62%	6.05	6.03	0.36%
3	3.12	3.08	1.45%	4.43	4.41	0.40%	5.82	5.81	0.20%
4	0.65	0.66	3.07%	1.27	1.25	0.91%	1.93	1.95	1.02%
5	3.00	2.98	0.75%	4.26	4.23	0.53%	5.59	5.56	0.45%
6	3.12	3.23	3.48%	4.43	4.50	1.56%	5.82	5.86	0.68%
7	5.23	5.21	0.42%	7.35	7.36	0.04%	9.61	9.60	0.14%
8	2.04	2.02	1.38%	2.83	2.83	0.10%	3.67	3.67	0.11%
9	3.19	3.19	0.18%	4.52	4.53	0.13%	5.94	5.94	0.10%
10	1.02	1.06	3.95%	1.51	1.54	1.57%	2.03	2.05	0.99%

short for Algorithm 4.1). Relative errors on mass flow rates are also very small, and decline with the increase of loads. According to the previous analysis in (4.18) and (4.19), the larger the value of $\dot{m}$, the smaller the value of x, and the more concise the approximation made in (4.19). Experiments conducted in [16] suggest that in most cases, formula (4.19) would be satisfactory if the mass flow rate is greater than 1 kg/s.

The average computation time of Baron and Algorithm 4.1 is 1357 s and 1.14 s, respectively. To test the scalability of the proposed method, we set up several 24-period OHTF instances by incorporating time-varying heat demands across a day. In this test, Baron fails to converge in 10 h, while Algorithm 4.1 finds a solution within 9.8 s in average. The speedup of Algorithm 4.1 mainly benefits from its convex formulation of constraints with fixed mass flow rates, which is based on the property of heat loss revealed in (4.19). We also test IPOPT, a local NLP solver [18]. Mass flow rates provided by Algorithm 4.1 are used to initiate IPOPT, so we expect that the solution would be a global optimum. The average solver time for 24-period

OHTF problems is about 100 s, one order of magnitude longer than Algorithm 4.1. Moreover, the optimums offered by IPOPT and Algorithm 4.1 are almost the same, validating the effectiveness of the hydraulic-thermal decomposition method.

We shed some light on modelling the dynamic behavior of the heating system.

1. The heating system load may have large thermal inertial. The thermal inertial refers to the time delay between heat consumptions and temperature changes. For building space/water heating, the direct request is the reference temperature profile set by the costumer, which can be translated into a thermal energy demand curve by the smart home appliance. For example, if we want to heat a room to 78 °F at 11 a.m., we have to turn on the warmer at 10 a.m., and the heat demand starts at 10, which is done locally. On this account, the thermal inertial effect at the demand side will be considered during the construction of heat demand curve, which can be separated from the OHTF problem.
2. The time frame of thermal transients in pipelines depends on the spatial scale of the DHN and the mass flow velocity. If the network is small, the system can reach a thermal equilibrium in a few minutes, whereas the time granularity of OHTF is 1 h, so temperature transients in pipelines can be neglected in such circumstance. If the DHN scatters in a large area, the transient effect in pipelines could be prominent. The time frame of temperature propagation is shown to be approximately equal to the heat medium circulation time in the DHN [19], and an empirical formula is given to estimate such a time lag. Integrating transient dynamics of water pipelines in the mathematical model of OHTF has been discussed in [5].

Finally, we expound the engineering relevance of the OHTF model. According to [15], DHN can be operated in four different modes: constant-flow constant-temperature mode, constant-flow variable-temperature mode, variable-flow constant-temperature mode, and variable-flow variable-temperature mode. Temperature change depends on the output adjustment of heat sources, which is relatively slow due to the large specific heat capacity of water. Mass flow rate is controlled by the circulating pump; such a mechanical variable can change rapidly. In the single-period model, we fix mass flow rates first for the sake of solving a non-convex OHTF problem through convex optimization, which interprets the constant-flow variable-temperature operating mode; however, from a multi-period perspective, the situation will be somehow different. To see this, in different time periods, the nodal temperature is always set near to the lower bound to reduce losses according to step 1 in Algorithm 4.1; it is the mass flow rates that are changed in response to the variation of demands, according to step 3, and the temperatures have only slight variations. In this regard, a multi-period OHTF actually corresponds to the variable-flow variable-temperature operating mode (but please be aware that the temperature almost keeps constant). The advantage is that the mechanical response is fast enough to follow the time-varying load.

4.3 Market Based Distributed Operation

Most of the existing studies on the operation of integrated heat-power systems rely on a centralized optimization paradigm, in which the heating and power systems are operated by a central authority to leverage the heat storage capacity and ameliorate renewable energy utilization. In current practice, power systems and heating systems are managed by different entities, therefore, distributed decision making is highly desired. In addition, energy trading and market issues in the heat-electricity energy systems have not attracted much attention, the reasons may be: (1) electricity is supplied at fixed prices, so there is not much benefit from improving production plans; (2) there is no clear market mechanism for energy trading between the power system and the heating system; (3) historically, coal-fired or gas-fired boilers are mainstream heating sources. The fuel price is fixed in a fairly long time scale, and the interplay across the two systems is not prominent. However, the proliferation of CHP units and heat pumps greatly promotes the share of loads from heating devices in the total electricity demand, and increases the interaction between the two systems. Recent trends on distribution power system marketization [20] make it possible and superior to trade energy through distribution power markets and operate the system via market-driven approaches. Inspired by the real-time electricity prices, the heating system can save operation costs by strategically scheduling its production plan, which also benefits the power system because the gap between peak and valley demands narrows.

In this section, we envision a distribution power market where electricity consumption is charged at LMP. We set simple rules for bilateral energy transaction. With above market settings, both of power system and heating system can be independently operated by solving an OPF problem and an OHTF problem, given the status of the other network. System-level interaction is reflected by the fact that heat generation schedule changes the electricity demand, which will further impact LMPs; the LMP in turn influences the schedule of heat generation. A best-response algorithm is proposed to identify the optimal operating strategies for both networks at the market equilibrium state. A continuous LMP scheme is adopted to improve convergence performance and enhance market stability.

4.3.1 Basic Settings and Market Equilibrium Model

Heat and electricity transactions obey the following rules:

1. The heating system can purchase heat from CHP units owned by the power system (CHP-PDN) at a fixed contract price ρ_H^P. The quantity is bid by the DHN.
2. The power system can purchase electricity from CHP units owned by the heating system (CHP-DHN) at a fixed contract price ρ_P^H or the LMP.
3. Heat pumps in the heating system participate in the distribution power market and purchase electricity at LMP.

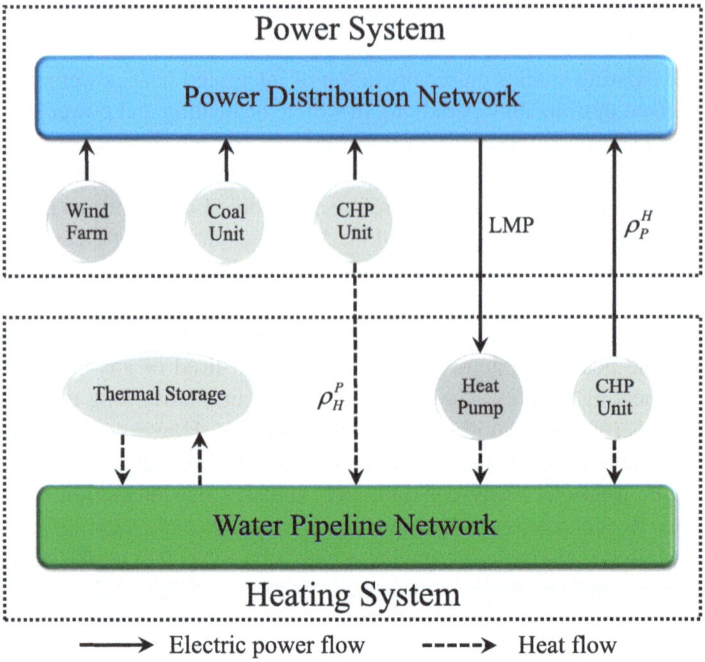

Fig. 4.6 Structure of the distribution market with a district heating system

The market structure is summarized in Fig. 4.6. Symbols and notations used in the objective functions of OPF and OHTF problems have similar interpretation as those in Sect. 4.2.3. At the power system side, the operator collects the heat pump power demand and fixes the heat output of its CHP units according to the requests submitted by the heating system operator, and then clears the distribution power market by solving the following OPF problem:

$$\min \ f_{\text{OPF}} = \sum_t \left(\sum_i C_{i,t}^{GD} + \sum_j C_{j,t}^{PT} + \sum_k C_{k,t}^{HP} + \sum_s C_{s,t}^{WP} \right) \tag{4.20}$$

$$\text{s.t. Cons-BFM, Cons-BND}$$

where expressions of power flow constraints Cons-BFM and variable bounds Cons-BND are given in Sect. 2.3.1. The objective function includes four components: operating cost of CHP-PDN units $C_{i,t}^{GD}$, which is a bivariate convex quadratic function in its electric power output $p_{i,t}^{gD}$ and heat generation $h_{i,t}^{gD}$:

$$C_{i,t}^{GD} = b_i^0 + b_i^1 p_{i,t}^{gD} + b_i^2 h_{i,t}^{gD} + b_i^3 (p_{i,t}^{gD})^2 + b_i^4 (h_{i,t}^{gD})^2 + b_i^5 p_{i,t}^{gD} h_{i,t}^{gD} \tag{4.21}$$

The operation cost of conventional units $C_{j,t}^{PT}$ is modeled by a univariate convex quadratic function in its electric power output $p_{j,t}^{g}$ as

$$C_{j,t}^{PT} = c_j^0 + c_j^1 p_{j,t}^g + c_j^2 (p_{j,t}^g)^2 \tag{4.22}$$

Electricity can be purchased from CHP-DHN units in the heating network, and the payment is:

$$C_{k,t}^{HP} = \rho_P^H p_{k,t}^{gH} \tag{4.23}$$

where ρ_P^H denotes the trading price of electricity; it could be a fixed contract price or the LMP. If the latter is adopted, we fix ρ_P^H at the LMP obtained in the last iteration.

Wind farms provide cheap electricity at a fixed contract price, and the corresponding cost is:

$$C_{s,t}^{WP} = \rho^w p_{s,t}^w \tag{4.24}$$

OPF problem (4.20) can be solved by the method presented in Sect. 2.3.1. The optimal solution gives the best dispatch of generators. LMP can be retrieved from the dual variables associated with the nodal power balancing conditions. This process is called distribution market clearing. After that, the power system operator sends the power demand from CHP-DHN as well as LMP to the heating system.

At the heating system side, the operator fixes the electric power output of CHP-DHN units according to the requests submitted by the power system operator, collects the LMP, and then conducts OHTF problem (4.17) using Algorithm 4.1, where the objective function is given by

$$f_{\mathrm{OHTF}} = \sum_t \left(\sum_i C_{i,t}^{GH} + \sum_j C_{j,t}^{PH} + \sum_k C_{k,t}^{GD} \right) \tag{4.25}$$

where $C_{i,t}^{GH}$ and $C_{j,t}^{PH}$ are the costs of CHP-DHN units and heat pumps, which have been elaborated in Sect. 4.2, and the cost incurred by purchasing heat from CHP-PDN units is:

$$C_{k,t}^{GD} = \rho_H^P h_{k,t}^{gD} \tag{4.26}$$

The optimal solution of OHTF gives the best dispatch of heat sources. Finally, the heating system operator sends the heat requests from CHP-PDN units as well as heat pump electric power demands to the power system.

4.3.2 An Iterative Algorithm

One prominent feature of such a distribution power market is the LMP based energy transaction. As industrial heat pumps possess considerably larger capacities than household appliances, they have the potential to influence LMP, which makes the market more complicated: on the one hand, when the heating system operator performs OHTF and determines heat production plans, LMPs are needed as input; on the other hand, when the power system operator conducts OPF and clears the distribution power market, the demands of heat pumps are needed as input. This interaction constitutes a closed-loop feedback. In accordance with the above setting, a best-response procedure shown in Algorithm 4.2 can be performed.

Algorithm 4.2 Fixed-point algorithm for distributed operation

1: Set a tolerance $\varepsilon > 0$ and a maximum number of iterations N^{max}. All variables are initialized as $\mathbf{x}^0 = 0$; LMPs are initiated as the marginal cost of the largest generator at rated output. The iteration index is set as $n = 0$.

2: Solve OPF problem (4.20) with fixed demands from the DHN and ρ_P^H being the LMP at corresponding bus in last iteration (or the fixed contract price), record the optimal solution in $\mathbf{x}^{n+1}$, and retrieve LMPs; then send LMPs and power demands to the DHN.

3: Solve OHTF problem (4.17) with fixed LMPs and demands from the PDN using Algorithm 4.1, record the optimal solution in $\mathbf{x}^{n+1}$; then send energy demands to the PDN.

4: If $|\mathbf{x}^{(n+1)} - \mathbf{x}^{(n)}| < \varepsilon$, terminate and report $\mathbf{x}^{(n+1)}$ as the final result; else if $n > N^{max}$, quit and report that the algorithm fails to converge; else update $n \leftarrow n + 1$ and go to step 2.

Because the objective function of OHTF involves LMPs, the dual variables of OPF, mathematically provable convergence conditions for Algorithm 4.2 are non-trivial. In our study, we observe that when a heat pump is installed at a downstream bus of some wind farm, the output of this heat pump may fail to converge in the iteration. Consider a heat pump and its corresponding nodal price denoted as LMP. In some iteration, it consumes little electricity and surplus wind generation is curtailed at the upstream bus, then LMP will be approximately the wind power contract price which is at a low level. Hence, the heat pump will increase its output in this iteration, which may eliminate wind power curtailment. As a result, the LMP may exhibit a sudden rise. Then, in the next iteration, the heat pump will in turn decrease its output and the LMP may return to the wind power contract price. In this way, oscillation appears due to the discontinuity of LMP.

A continuous LMP (CLMP) method is suggested in [21] to circumvent the disadvantage of traditional LMP scheme. Basically, it uses a continuous piecewise linear function to approximate the discontinuous LMP-demand curve. CLMP is adopted at electric buses where LMPs are not continuous. The solid line in Fig. 4.7

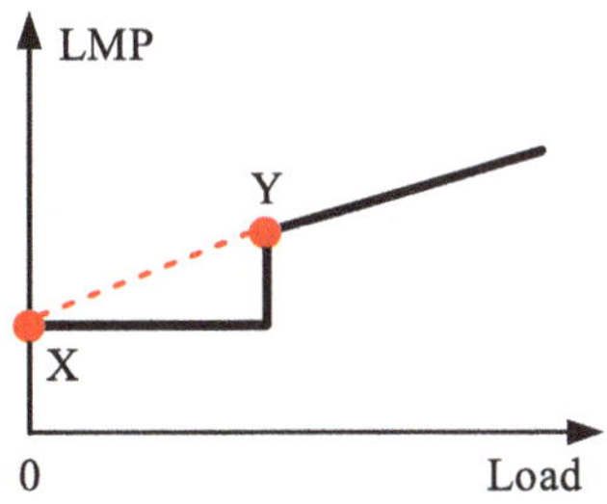

Fig. 4.7 Illustration of CLMP

shows the traditional LMP as a function of the electric load. Y represents the step point. CLMP curve is the dash line between X and Y. In this section, the LMP step is caused by wind power spillage, so a tailored CLMP scheme is suggested as follows.

First, at a heat pump connected bus, set the power consumed by heat pump to zero and solve the OPF problem. The calculated LMP is represented by X; then increase the heat pump power consumption until no wind power is curtailed and solve the OPF problem. The obtained LMP is represented by Y; finally, CLMP curve is represented by the dash line between X and Y.

In such circumstance, at the PDN side, the LMP curves are increasing with the growth of heat pump demands according to OPF results; at the DHN side, the heat pump demands are decreasing with the rising of LMP according to OHTF results. Algorithm 4.2 will terminate at the equilibrium which interprets the intersection of price and demand curves. Such an equilibrium reflects a state where both PDN and DHN have no incentive to alter their operating strategies. Similar discussions can be found in Sect. 3.4.

4.3.3 Case Studies

1. System Configurations

A testing heat-power integrated energy system composed of a 33-bus PDN and a 32-node DHN is used to demonstrate the performance of the proposed method. System topology is shown in Fig. 4.8. Energy production facilities include 3 gas-fired generators, 2 wind farms, 1 CHP-PDN unit, 1 CHP-DHN unit, and 3 heat pumps. Thermal storage units located at nodes 5, 18, 24 in the DHN are not shown in Fig. 4.8. Power and heat load profiles, and forecasted maximum wind generation outputs are shown in Fig. 4.9. We can see that the heat demand has a tendency opposite to the power load curve but consistent with wind generation. Complete system data are provided in [22].

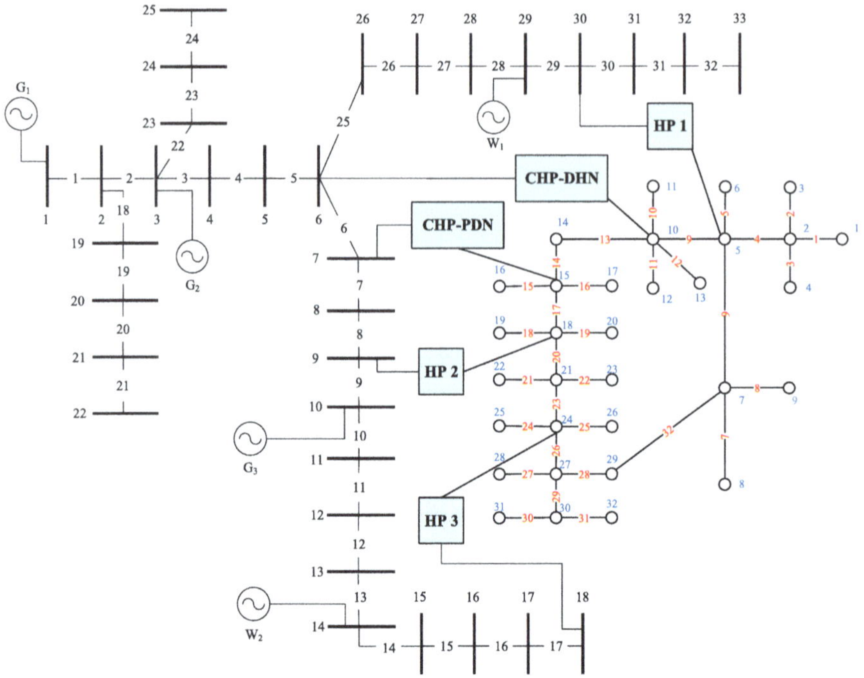

Fig. 4.8 Topology of 32-node DHN

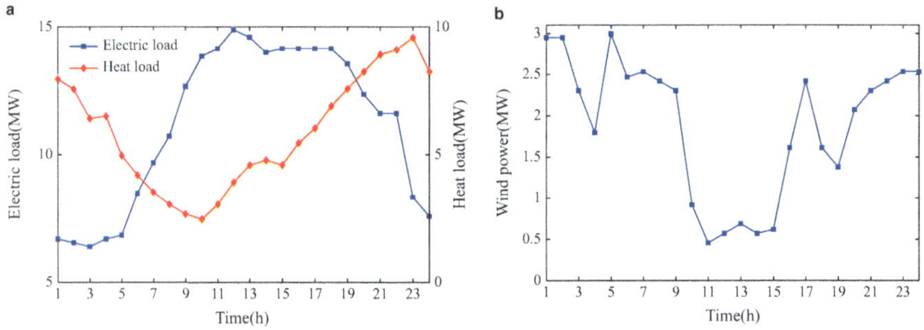

Fig. 4.9 (**a**) Electric and heat loads. (**b**) Wind power forecast

2. Analysis of CLMP Scheme

In the benchmark case, Algorithm 4.2 without CLMP scheme is tested first. Figure 4.10a shows the incremental output of heat pump 1 between two successive iterations in period 5, and Fig. 4.10b gives the LMP at bus 30. Oscillations are observed. The reason has been analyzed before. Then CLMP scheme is applied; results are provided in Fig. 4.10c, d. We can see that Algorithm 4.2 converges successfully after 9 iterations.

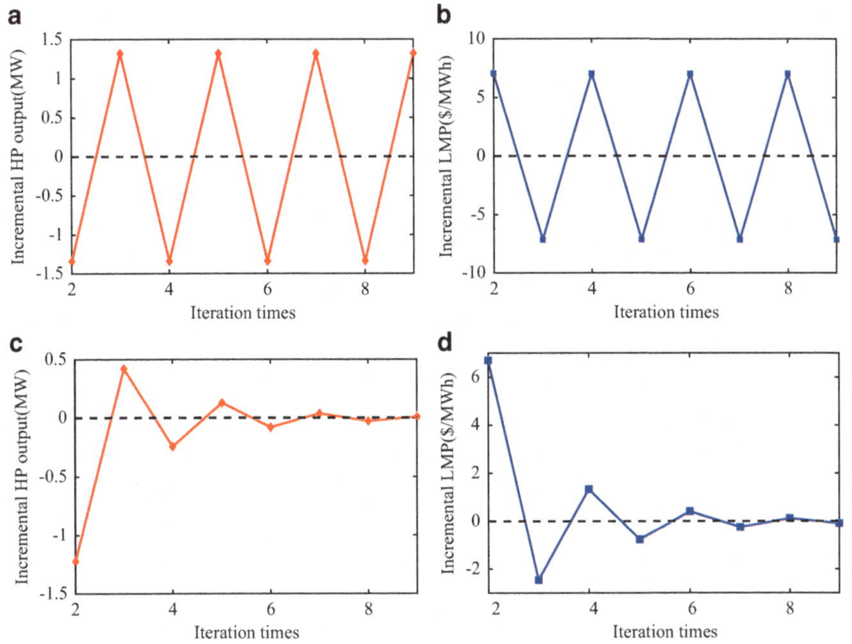

Fig. 4.10 Incremental values between two successive iterations: (**a**) output of heat pump 1 without CLMP method; (**b**) increment of LMP at bus 30 without CLMP method; (**c**) output of heat pump 1 with CLMP method; (**d**) increment of LMP at bus 30 with CLMP method

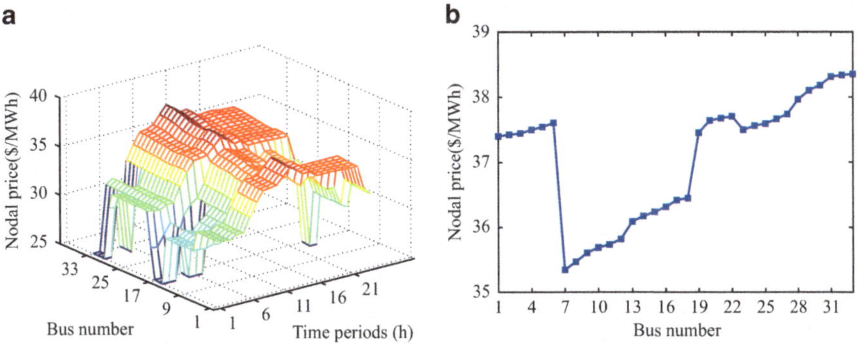

Fig. 4.11 System LMPs. (**a**) Spatial and temporal distribution of LMPs. (**b**) System LMP in period 12

The spatial and temporal variation of LMPs in the PDN is portrayed in Fig. 4.11a. In the temporal horizon, we can see that the LMP curve at each bus follows similar tendency: higher during daytime and lower across night periods because of the difference of load levels. In the spatial horizon, the slice at period 12 is shown in Fig. 4.11b. Clearly, LMP increases along the power flow direction. The LMP at bus

7 is the lowest because there is a CHP unit connecting to the bus, whose marginal cost is low.

3. Comparison of Market-Based Operation and Contract-Based Operation
As a comparison, a pure contract-based situation is considered, in which heat pump power demands are paid at a constant contract price. The network operation is decoupled: the DHN operator solves OHTF first and submits the electricity demands to the PDN, and then the PDN operator solves OPF and determines the electric power generation schedule. The previous market based mode is termed Mark-Oper, and the contract-based one is referred to as Cont-Oper.

The contract price in Cont-Oper mode is set to 34$/MWh, which is approximately the daily average LMP. The dispatched wind power in two modes are compared in Fig. 4.12a. In Cont-Oper mode, wind power curtailment occurs during the night owing to the low electric load and high wind power output. In addition, because the electricity price is flat and heat pumps have less incentive to buy excessive wind power for heat production. For Mark-Oper mode, the LMPs related to heat pumps are shown in Fig. 4.12b. When the wind power is abundant, the LMP is low, encouraging the use of surplus wind power during night which can be observed from Fig. 4.12a and heat pump output curves in Fig. 4.13.

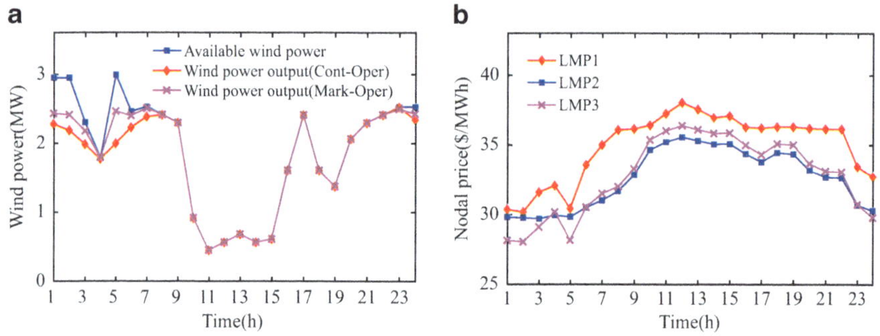

Fig. 4.12 (**a**) Wind power usage in two modes. (**b**) LMPs at heat pump connection nodes

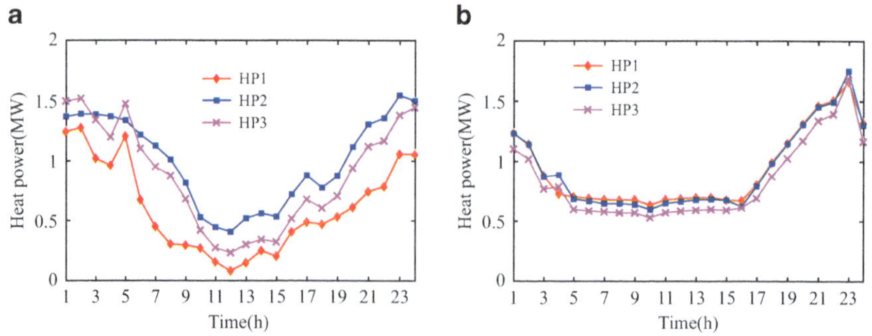

Fig. 4.13 Output of heat pumps in (**a**) Mark-Oper mode. (**b**) Cont-Oper mode

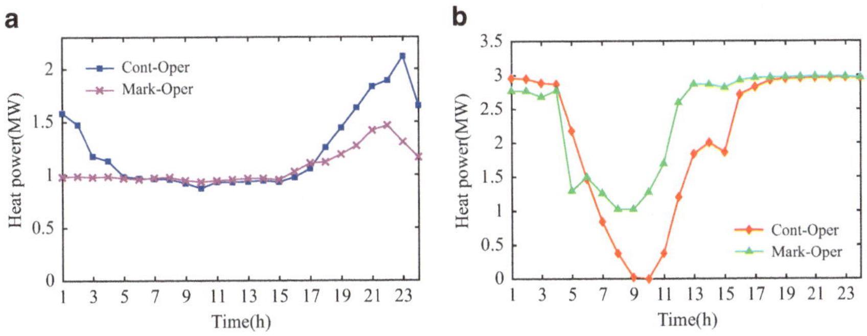

Fig. 4.14 (**a**) Heat produced by CHP-DHN unit. (**b**) Heat produced by CHP-PDN unit

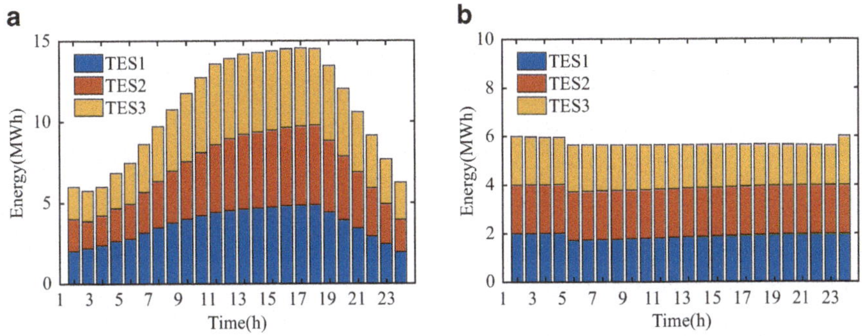

Fig. 4.15 Storage dynamics in (**a**) Mark-Oper mode. (**b**) Cont-Oper mode

The daily operation costs of DHN under Mark-Oper and Cont-Oper modes are \$2428 and \$2698, respectively. Thermal energy produced by CHP units is depicted in Fig. 4.14. The generation cost of CHP units are quadratic functions of heat generation, and the marginal costs increase with the increasing of heat generation. The heat produced by CHP-DHN unit in Mark-Oper mode is always lower than that in Cont-Oper mode, as power to heat conversion enabled by heat pumps enjoys a lower marginal cost, which demonstrates the superiority of the market-driven framework. In addition, when the LMP is high at daytime in Mark-Oper mode, heat pump will consume less electricity compared to what they do in Cont-Oper mode. Therefore, DHN operator tends to purchase more heat from CHP-PDN unit to supply the demand at daytime in Mark-Oper mode.

Storage dynamics in both modes are shown in Fig. 4.15. In Mark-Oper, TSUs are charged in periods 1–11 when electricity is cheap, and begin to discharge at period 17 when electricity price is high and heat load starts to grow; during periods 11–16, the heat demand is low, so the storage levels do not change much. These observations are consistent with the outputs of heat pumps shown in Fig. 4.13. Compared with Cont-Oper scheme, TSUs have deeper charging and discharging levels in Mark-Oper, which help flatten the electricity load profile by shaving peaks

Table 4.2 Results with respect to different contract prices

Contract price ($/MWh)	32	33	34	35	36
Curtailment rate (%)	6.02	6.64	7.23	7.92	8.54
DHN operation cost ($)	2631	2667	2698	2723	2742
PDN operation cost ($)	8896	8874	8853	8836	8822
Total operation cost ($)	11,527	11,541	11,551	11,559	11,564

Table 4.3 Results under different wind power penetration levels

Wind power ratio		0.6	0.8	1.0	1.2	1.4
Mark-Oper	Number of iterations	5	6	9	21	17
	Wind spillage (%)	0	1.02	4.18	7.51	12.83
	DHN operation cost ($)	2594	2523	2428	2352	2282
	PDN operation cost ($)	9103	9014	8967	8912	8872
	Total operation cost ($)	11,697	11,537	11,395	11,264	11,154
Cont-Oper	Wind spillage (%)	0.21	2.52	7.23	12.95	18.97
	DHN operation cost ($)	2698	2698	2698	2698	2698
	PDN operation cost ($)	8997	8914	8853	8803	8764
	Total operation cost ($)	11,695	11,612	11,551	11,501	11,462

and filling valleys. Thus, Mark-Oper scheme driven by time-varying LMP benefits both power grid and heat system.

4. Impact of Contract Price

We investigate the impact of the contract price in Cont-Oper on the curtailment rates, and operation costs. Results are listed in Table 4.2. The wind power spillage grows with the increase in contract price, because heat pumps tend to purchase less electricity due to the rising electricity price. The heating system operation cost increases at the same time, but even when the contract price is 32$/MWh, the economy is still worse than that in Mark-Oper. Power grid operation cost decreases because of the reduction of power consumed by heat pumps.

5. Impact of Wind Penetration Level

We investigate the performance of Algorithm 4.2 under different levels of wind power generation. We multiply the wind power forecast in Fig. 4.9b with a scale ratio varying from 0.6 to 1.4. The numbers of iterations, wind curtailment rates and operation costs are listed in Table 4.3. The curtailment rates grow with the increasing of wind generation. Owing to the increase of wind power, electricity price becomes lower and heating system operation cost decreases in Mark-Oper mode. On the contrary, heating system operation cost remains the same in Cont-Oper mode because of the fixed electricity tariff. Power system operation cost decreases in both schemes because of the increasing use of wind power. In general, Algorithm 4.2 takes more iterations to converge when more wind power is available, because LMPs in more periods become discontinuous. Nevertheless, with the help of CLMP, it still converges successfully.

Table 4.4 Results under different heat pump capacities

Capacity ratio		0.6	0.8	1.0	1.2	1.4
Mark-Oper	Number of iterations	5	8	9	12	14
	Wind spillage (%)	5.49	5.21	4.18	3.23	2.43
	DHN operation cost ($)	2456	2435	2428	2412	2403
	PDN operation cost ($)	8951	8963	8967	8971	8973
	Total operation cost ($)	11,407	11,398	11,395	11,383	11,376
Cont-Oper	Wind spillage (%)	8.45	7.85	7.23	6.79	6.40
	DHN operation cost ($)	2705	2701	2698	2694	2691
	PDN operation cost ($)	8843	8848	8853	8857	8861
	Total operation cost ($)	11,548	11,549	11,551	11,551	11,552

6. Impact of Heat Pump Capacity

Finally, we investigate the impact of heat pump capacity on the performance of Algorithm 4.2. We control their capacities by multiplying c^P with a scale ratio varying from 0.6 to 1.4. Results are listed in Table 4.4. As heat pump capacity increases, wind curtailment rate decreases gradually. Because bigger heat pump capacities may cause larger incremental changes of heat pump power demands and LMPs in two consecutive steps, the iteration number increases consequently. In addition, larger heat pump has a higher efficiency with the same load level according to (4.13). Therefore, heating system operation cost decreases with the increase in heat pump capacity in both Mark-Oper mode and Cont-Oper mode. Power grid operation cost increases because more power is consumed by heat pumps.

4.4 Equilibrium of Interdependent Heat-Power Markets

Section 4.3 discusses a market mechanism which could foster coordination between the power system and the heating system. Indeed, energy markets in the distribution level will provide unique opportunities for energy system integration and promoting energy transactions. Thanks to the development of smart grid technologies, nowadays, end consumers are able to receive and respond to the real-time energy prices, which is known as demand elasticity, a special kind of demand response.

In this section, we generalize the market framework in Sect. 4.3 in two ways. First, we extend the power market paradigm to the heating system, and incorporate a heating market which is cleared according to an OHTF problem. Thermal energy is charged at locational marginal production cost. Second, in the power market and heating market, we consider elastic loads whose demands are continuous functions of energy prices, just as the OPF problem in Sect. 2.3.4. Other settings will be the same as Sect. 4.3. In the presence of elastic demands, the power market and heating market have their individual equilibria, which can be computed iteratively like Algorithm 2.3 in Sect. 2.3.4. In addition, the energy exchange between power system and heating system should be updated in an outer loop, which is similar to Algorithm 4.2 in Sect. 4.3. The detailed flowchart will be explained later.

In contrast to the fixed-point type algorithm, we present closed-form optimality conditions for the market equilibrium with energy trading and elastic demands. To this end, the rotated second-order cone constraints in the power market clearing problem are replaced by their polyhedral approximations [23]. As such, both market clearing problems become linear. Then we concrete the primal-dual optimality condition of the power market clearing problem and the KKT optimality condition of the heating market clearing problem. Performing linearization to a few nonlinear parts, we get the desired concrete formulation, which can be embedded in more dedicated optimization models such as an MPEC.

4.4.1 An LP Model for Power Market Clearing

Recall OPF problem (2.30), which renders an SOCP when the second-order cone relaxation is performed. As the distribution system has radial topology, the relaxation is exact under mild conditions. Without loss of generality, we assume that the objective function is linear, as the commonly used convex quadratic cost can be approximated by a PWL function [24], or the convex combination method introduced in Appendix B.1.1. For a convex function to be minimized, there is no need to append the SOS2 condition. Remaining nonlinearity in the power market clearing problem is in the form of

$$\left\| \begin{matrix} 2P \\ 2Q \\ U - I \end{matrix} \right\| \le U + I \tag{4.27}$$

or

$$\sqrt{(2P)^2 + (2Q)^2 + (U - I)^2} \le (U + I)$$

in arithmetic form. This inequality originates from the relaxation of the branch flow equality, and is equivalent to a pair of SOCs in $\mathbb{R}^3$

$$\sqrt{4P^2 + 4Q^2} \le W \tag{4.28a}$$

$$\sqrt{W^2 + (U - I)^2} \le U + I \tag{4.28b}$$

To see their equivalence, we perform Fourier-Motzkin elimination on (4.28). To do so, squaring both sides of (4.28a) and (4.28b), we have

$$W^2 \geq 4P^2 + 4Q^2$$
$$W^2 \leq (U+I)^2 - (U-I)^2$$

Thus we can eliminate W^2 and get (4.27).

The capacity limits of distribution lines take the form of $\sqrt{P^2+Q^2} \leq S$. Hence we only need to find a polyhedral approximation for the canonical SOC in $\mathbb{R}^3$

$$\sqrt{x_1^2 + x_2^2} \leq x_3 \tag{4.29}$$

The method developed in [23] provides such an approximation. By introducing additional variables $\xi^0, \cdots, \xi^v, \eta^0, \cdots, \eta^v$, the following linear constraints

$$\begin{aligned}
&\xi^0 \geq |x_1| \\
&\eta^0 \geq |x_2| \\
&\xi^j = \cos\left(\frac{\pi}{2^{j+1}}\right)\xi^{j-1} + \sin\left(\frac{\pi}{2^{j+1}}\right)\eta^{j-1},\ j = 1, \cdots, v \\
&\eta^j \geq \left| -\sin\left(\frac{\pi}{2^{j+1}}\right)\xi^{j-1} + \cos\left(\frac{\pi}{2^{j+1}}\right)\eta^{j-1}\right|,\ j = 1, \cdots, v \\
&\xi^v \leq x_3 \\
&\eta^v \leq \tan\left(\frac{\pi}{2^{v+1}}\right)\xi^v
\end{aligned} \tag{4.30}$$

provide a polyhedral approximation for SOC (4.29) in variables (x_1, x_2, x_3). The approximation error of (4.30) is quantified by

$$\sqrt{x_1^2 + x_2^2} \leq [1 + \varepsilon(v)]x_3 \tag{4.31}$$

where the error bound $\varepsilon(v)$ is given by Ben-Tal and Nemirovski [23]

$$\varepsilon(v) = \frac{1}{\cos\left(\frac{\pi}{2^{v+1}}\right)} - 1$$

If we choose the same v for (4.28a) and (4.28b), the approximation error for (4.27) would be

$$\left\| \begin{matrix} 2P \\ 2Q \\ U - I \end{matrix} \right\| \leq \left(\frac{1}{\cos^2\left(\frac{\pi}{2^{v+1}}\right)} - 1 \right)(U + I) \tag{4.32}$$

The geometric interpretation of this approach is explained in [23] as well as in Appendix A.2.3. Applying (4.30) to every SOC constraint in the OPF problem, we obtain the LP model for power market clearing

$$\begin{aligned} \min \ & c_p^T(\alpha)x \\ \text{s.t. } & A_1x = b_1(y) : \lambda^p \\ & A_2x \le b_2(\beta) : \mu^p \end{aligned} \tag{4.33}$$

where λ^p and μ^p following the colon are dual variables of equality and inequality constraints. Electricity prices can be extracted from vector λ^p. Vector y denotes the variable of the heating system and is treated as a constant. In our problem, the coupling constraint in (4.33) is the electrical power demand of heat pumps, so the right-hand side $b_1(y)$ is linear in y. More precisely,

$$b_1(y) = b_0 + \Gamma(My) + \Gamma_p^d(p^d) \tag{4.34}$$

where b_0 is a constant vector reflecting the system constraints without heat pump demands; matrix M selects heat pump demands from vector y, and Γ is a linear operator that maps heat pump demands to corresponding locations in the compact form. Linear operator $\Gamma_p^d(p^d)$ maps elastic power demands to corresponding locations in the compact constraints. $\Gamma(My)$ has the same dimension as vector b_0, so does $\Gamma_p^d(p^d)$. Equation (4.34) is an important relation which will be used later.

In broader classes of applications, there may be private-owned generation companies which bid offering prices α and quantities β in the market. As such, in the market clearing problem, $\alpha_i p_i$ will appear in the objective function, and capacity limit $0 \le p_i \le \beta_i$ should be imposed as constraints. p_i is the variable of market clearing which determines the energy contract with the company, while α_i and β_i are constants in (4.33). Therefore, $c(\alpha)$ and $b_2(\beta)$ are linear in bidding strategies α and β, respectively.

Unlike the DC power flow or the linearized branch flow based market clearing models which neglect network losses, problem (4.33) reserves all details of the nonlinear power flow model, and could provide a promising tool for analyzing the distribution power market. Moreover, elegant theory in linear programming can be applied to derive optimality conditions of (4.33), facilitating the computation of market equilibrium in a more systematic way.

4.4.2 Heating Market Clearing and Thermal Energy Pricing

Recall OHTF problem (4.17) in Sect. 4.2.4, which renders a convex program with linear constraints if the mass flow rate $\dot{m}$ is fixed. In fact, even if $\dot{m}$ should be optimized during market clearing, Algorithm 4.1 can identify the near optimal $\dot{m}$

and then the constraints in problem (4.17) become linear. In this section, we extend the organization of a power market to a heating market, i.e., the market is cleared according to the optimal solution of problem (4.17). Without loss of generality, we assume that the cost function of the heating system is linear, since convex quadratic functions can be linearized. We write out the heating market clearing problem as follows:

$$\begin{aligned} \min \quad & (\lambda^p)^T \Gamma(My) + c_h^T y \\ \text{s.t.} \quad & A_3 y = b_3 : \lambda^h \\ & A_4 y \le b_4 : \mu^h \end{aligned} \tag{4.35}$$

where the first term in the objective function is the operation cost of heat pumps, whose electricity consumption is paid at LMP λ^p, which is treated as a constant in (4.35), and the second one is the cost of other heat sources; λ^h and μ^h following the colon are dual variables of equality and inequality constraints. In the current model, we don't consider strategic players in the heating market, although such entities can be incorporated without any difficulty. The result of heating market clearing will affect the power system demand, which further impacts electricity prices. In view of this interaction, an equilibrium will emerge.

Something more about heat pricing needs clarification. It is [25] that first transplants the LMP scheme to the heating market. If this is adopted, the nodal heat price can be extracted from dual variable λ^h. Nevertheless, in current practice, heat consumption is usually charged at constant price. As a compromise, we suggest a uniform marginal pricing scheme: the heat price is the same across the network, and is equal to the marginal production cost. This can be implemented by adding equality

$$e^T y = \Phi + 1^T h^d : \lambda^{hu} \tag{4.36}$$

in problem (4.35), where vector e has element 1 corresponding to the output of heat sources, and the remaining elements are 0; h^d is the vector of heat demands; Φ is the total network loss. Equation (4.36) interprets energy balancing in the heating system. Its dual variable λ^{hu} is the marginal production cost when the total demand is increased by one unit. Equation (4.36) is merely a definition of Φ, and is redundant if the LMP scheme is adopted. The network loss can be calculated from (4.36) after the market is cleared, or in other words, Eq. (4.36) does not affect the market clearing result. In the uniform marginal pricing scheme, Eq. (4.36) helps retrieve the marginal cost λ^{hu} conveniently. Strictly speaking, the marginal production cost would depend on the location of the incremental load demand, because network losses matter. The dual variable associated with (4.36) simulates an average effect. Researchers also propose non-cost based heat pricing schemes, such as the exergy loss approach proposed in [26]. Existing heat pricing schemes have been thoroughly reviewed in [27].

4.4.3 Modeling Demand Elasticity

We assume that consumers in both systems will adjust their demands according to the energy prices. We consider three sorts of elastic demands.

1. Nodal Elastic Demands in the Power System
We assume that the active power demand p_i^d at bus i can be characterized by a decreasing function in the corresponding nodal electricity price λ_i^p as follows:

$$p_i^d = \Upsilon^p(\lambda_i^p) \tag{4.37}$$

Industrial, commercial, and residential consumers may react differently to the electricity price for gaining profits or saving costs. Function $\Upsilon^p(\lambda_i^p)$ can be calibrated from history energy consumption data and should be continuous. Discrete demand function may jeopardize the existence of market equilibria. If available data are insufficient to calibrate a demand function, five monotonically decreasing functions are proposed in [28] to simulate various elasticity characteristics of end consumers. Specifically, we also consider the following PWL description

$$p_i^d = \begin{cases} p_i^{dm}, & \text{if } \lambda_i^p < \lambda_i^n \\ p_i^{dm} - \dfrac{\lambda_i^p - \lambda_i^n}{\lambda_i^m - \lambda_i^n}(p_i^{dm} - p_i^{dn}), & \text{if } \lambda_i^p \in [\lambda_i^n, \lambda_i^m] \\ p_i^{dn}, & \text{if } \lambda_i^p > \lambda_i^m \end{cases} \tag{4.38}$$

In (4.38), if the electricity price is lower than λ_i^n or higher than λ_i^m, the demand is maintained at a constant p_i^{dm} or p_i^{dn} ($p_i^{dm} > p_i^{dn}$); otherwise, the demand is a linear function in the nodal electricity price. Using MILP formulation tricks explained in Appendix B.1.1, we can easily represent elastic demands in the general form (4.37) or the PWL form (4.38) through linear constraints with integer variables.

2. Nodal Elastic Demands in the Heating System
We assume the nodal heat demand h_i^d at node i can be characterized by a decreasing function in the corresponding nodal heat price λ_i^h as follows:

$$h_i^d = \Upsilon^h(\lambda_i^h) \tag{4.39}$$

Discussions for electricity demands apply to heat demands as well.

3. Industrial Demands
We consider an industrial load such as a chemical plant at bus i produces goods while consuming electricity and heat. It pursues maximum utility according to the following optimization problem

$$\begin{aligned} \max \quad & U(p_i^d, h_i^d) = (p_i^d)^\gamma (h_i^d)^{1-\gamma} \\ \text{s.t.} \quad & \lambda_i^p p_i^d + \lambda_i^h h_i^d \le \hat{I} \end{aligned} \tag{4.40}$$

where the objective is the Cobb-Douglas function [29] which is widely used in microeconomics, and $0 < \gamma < 1$ is a constant; $\hat{I}$ is the available budget for energy consumption. Problem (4.40) is called Cobb-Douglas utility [30].

Because $U(p_i^d, h_i^d)$ is monotonically increasing in its two inputs, maximizing $U(p_i^d, h_i^d)$ is equivalent to maximizing its logarithm $\gamma \log(p_i^d) + (1-\gamma)\log(h_i^d)$. Considering Lagrange function

$$L(p_i^d, h_i^d, \varsigma) = \gamma \log(p_i^d) + (1-\gamma)\log(h_i^d) + \varsigma(\lambda_i^p p_i^d + \lambda_i^h h_i^d - \hat{I})$$

and KKT optimality condition

$$\frac{\partial L\left(p_i^d, h_i^d, \varsigma\right)}{\partial p_i^d} = 0$$
$$\frac{\partial L\left(p_i^d, h_i^d, \varsigma\right)}{\partial h_i^d} = 0$$

we arrive at

$$\frac{\gamma}{p_i^d} + \varsigma\lambda_i^p = 0$$
$$\frac{1-\gamma}{h_i^d} + \varsigma\lambda_i^h = 0$$

Eliminating ς we have

$$\frac{p_i^d\lambda_i^p}{\gamma} = \frac{h_i^d\lambda_i^h}{1-\gamma} \tag{4.41}$$

Substituting (4.41) into the budget constraint, the optimality condition for (4.40) is

$$\begin{aligned} p_i^d\lambda_i^p &= \gamma\hat{I} \\ h_i^d\lambda_i^h &= (1-\gamma)\hat{I} \end{aligned} \tag{4.42}$$

In view of (4.42), Cobb-Douglas utility gives price-dependent demands, which can be expressed via hyperbolic functions.

4.4.4 Market Equilibria

In the underlying market framework, on the one hand, energy markets should be cleared with given nodal demands, and energy prices are determined from dual

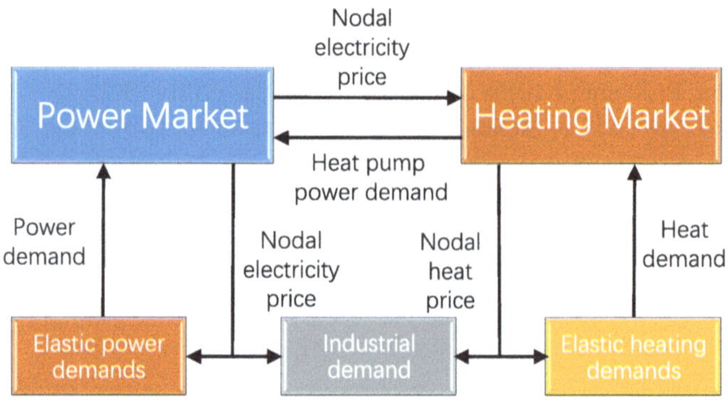

Fig. 4.16 Structure of the considered market

variables associated with nodal energy balancing equalities; On the other hand, elastic loads will adjust their demands in response to the energy prices. The market structure is shown in Fig. 4.16. To model such a market in a holistic way, we have to replace market clearing problem with their optimality conditions. By doing so, minimization operators vanish, and we can consolidate optimality conditions of the two markets into a set of constraints; moreover, dual variables appearing in the optimality conditions interpret the energy prices, which are explicitly modeled. This approach is very useful for studying a marginal pricing based market.

Two kinds of optimality conditions can be applied on the market clearing problem. Their suitability is discussed.

KKT Optimality Condition The advantage of the KKT condition is that it can be applied to both of linear and nonlinear optimization problems. For example, if the objective function of market clearing is a convex quadratic function rather than a linear one, there is no need to perform PWL approximation. The disadvantage is that it may introduce extensive complementarity constraints, depending on the number of inequality constraints. Complementarity constraints are hard to cope with, because they do not satisfy ordinary constraint quantifications. This difficulty is discussed in Appendix D.3. In addition, it is only a necessary condition. For nonlinear programs, the stationary points which satisfy KKT condition may not be an (even local) optimal solution.

Primal-Dual Optimality Condition Primal-dual optimality condition only applies to LPs. It includes primal feasibility constraints, dual feasibility constraints, and an equation which forces equal objective values of the primal problem and dual problem. In contrast to the KKT condition, primal-dual condition does not involve complementarity constraints, and usually has less number of constraints than the KKT condition. For LPs, both optimality conditions are equivalent, which has been discussed in Appendix A.2.1.

Next we examine optimality conditions of the heat-power market equilibrium.

1. Power Market Equilibrium Conditions

We first write out the KKT condition of the power market clearing problem (4.33)

$$\begin{aligned} A_1^T \lambda^p + A_2^T \mu^p &= c_p(\alpha) \\ A_1 x &= b_0 + \Gamma(My) + \Gamma_p^d(p^d) \\ 0 \geq \mu^p \perp A_2 x &- b_2(\beta) \leq 0 \end{aligned} \tag{4.43}$$

Because the last complementarity condition can be linearized by introducing integer variables, (4.43) is compatible with MILP solvers. Since the power market clearing problem has a lot of inequality constraints originating from the polyhedral approximation of SOCs, solving (4.43) may be time consuming. Alternatively, with special treatments on the complementarity constraints, which are discussed in Appendix D.3, (4.43) can be solved by NLP solvers, abandoning a global optimality guarantee.

Then we write out the primal-dual condition of problem (4.33)

$$\begin{aligned} A_1^T \lambda^p + A_2^T \mu^p &= c_p^T(\alpha) \\ A_1 x &= b_0 + \Gamma(My) + \Gamma_p^d(p^d) \\ (\lambda^p)^T [b_0 + \Gamma(My) &+ \Gamma_p^d(p^d)] \\ &+ b_2^T(\beta)\mu^p = c_p^T(\alpha) x \end{aligned} \tag{4.44}$$

Compared with (4.43), the last complementarity condition is replaced with a single equality, which imposes equal values on the primal and dual objective functions, i.e., the strong duality condition, and thus (4.44) has much fewer constraints. This is attractive from a computational perspective. However, there are nonlinear terms in the last equality, and moreover, the feasible region of (4.44) is the set of optimal solutions of (4.33), which is rather restrictive. If an NLP solver is applied to (4.44), it may not be able to find a feasible solution. One way to improve the numeric property of (4.44) is to move the last constraint into the objective function, i.e. to minimize the duality gap

$$c_p^T(\alpha)x - (\lambda^p)^T [b_0 + \Gamma(My) + \Gamma_p^d(p^d)] - b_2^T(\beta)\mu^p$$

Because of weak duality, the duality gap is always non-negative. If the solution gives a zero objective value, it is also a solution of (4.44). However, as the objective is non-convex, the solution may get trapped in a local minimum, which leads to a strictly positive duality gap.

Next we show that if $\Gamma_p^d(p^d) = 0$, i.e., there is no elastic demand in the power system, the only nonlinear part $(\lambda^p)^T \Gamma(My)$ can be linearized through variables in the heating system. Notice the fact that $(\lambda^p)^T \Gamma(My)$ is a part of the heating system cost. Since heating market clearing problem (4.35) is an LP, strong duality holds, resulting in

$$(\lambda^p)^T \Gamma(My) + c_h^T y = b_3^T \lambda^h + b_4^T \mu^h$$

so we obtain a linear expression

$$(\lambda^p)^T \Gamma(My) = b_3^T \lambda^h + b_4^T \mu^h - c_h^T y \tag{4.45}$$

where y, μ^h, and λ^h are primal and dual variables of problem (4.35). Condition (4.45) must be combined with the optimality condition of (4.35) to guarantee that these variables are the true market clearing results.

2. Heating Market Equilibrium Conditions

For the heating market clearing problem (4.35), the majority of OTF constraints are equalities, which will not introduce complementarity constraints in the KKT condition, therefore, the following KKT condition will be adopted

$$\begin{gathered} A_3^T \lambda^h + A_4^T \mu^h = c_h + \Gamma^T(M)\lambda^p \\ A_3 y = b_3 + \Gamma_h^d(h^d) \\ 0 \geq \mu^h \perp A_4 y - b_4 \leq 0 \end{gathered} \tag{4.46}$$

where $\Gamma(M) = \nabla_y \Gamma(My)$ is a constant matrix; linear operator $\Gamma_h^d(h^d)$ maps elastic heating demands to corresponding locations in the compact constraints. Because the last complementarity condition can be linearized by introducing integer variables, (4.46) is compatible with MILP solvers. It should be pointed out that because the strong duality condition has been exploited in linearizing $(\lambda^p)^T \Gamma(My)$ in (4.45), we can no longer use it in the heating market clearing problem.

Now we summarize the concrete equilibrium model of the interdependent heating and power markets:

$$\text{Power Market Equilibrium:} \quad \begin{cases} (4.43) & \text{if } \Gamma_p^d(p^d) \neq 0 \\ (4.44), (4.45) & \text{if } \Gamma_p^d(p^d) = 0 \end{cases} \tag{4.47a}$$

$$\text{Heating Market Equilibrium:} \quad (4.46) \tag{4.47b}$$

$$\text{Demand Elasticity:} \quad (4.37), (4.39), (4.42) \tag{4.47c}$$

In view of the potential large number of complementarity constraints in the power market equilibrium condition, we recommend using the primal-dual condition if $\Gamma_p^d(p^d) = 0$, so nonlinear terms in the strong duality condition (the last equality in (4.44)) vanish. Nevertheless, if the industrial demand in (4.42) is considered, the strong duality condition can be linearized as well. To see this, product terms $p_i^d \lambda_i^p$ and $h_i^d \lambda_i^h$ appearing in (4.47a) and (4.47b) are constant, according to (4.42); furthermore, $(\lambda^p)^T \Gamma(My)$ has a linear expression as in (4.45). The nonlinear

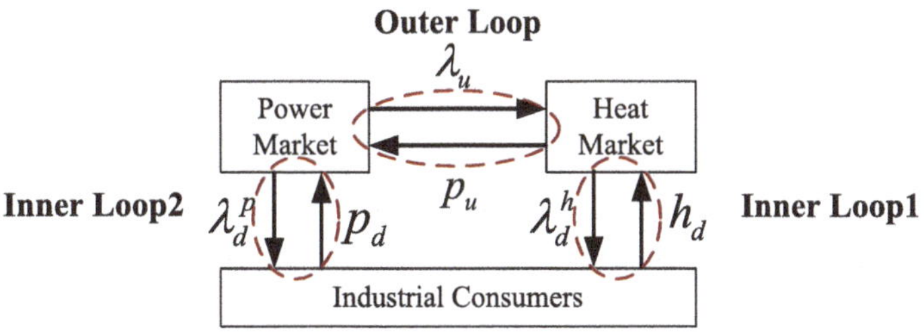

Fig. 4.17 Diagram of the fixed-point algorithm for market equilibrium

relation (4.42) is contained in (4.47c) and can be posed as a hyperbolic function $p_i^d = \gamma \hat{I}/\lambda_i^p$, which can be linearized via the method discussed in Appendix B.1.1. For more general elasticity in form of (4.37), product term $p_i^d \lambda_i^p = \lambda_i^p \Upsilon^p(\lambda_i^p)$ is a univariate nonlinear function, and can be also linearized via the method discussed in Appendix B.1.1. In this way, optimality condition (4.47) can be formulated as an MILP feasibility problem and solved by off-the-shelf solvers. Nevertheless, with some special treatment on the complementarity constraints, (4.47) can be solved by NLP solvers without linearizing the demand functions.

Intuitively, a double-loop fixed-point algorithm can be set up to identify the market equilibrium. In the inner loop, individual equilibria in the heating market and the power market with domestic elastic demands are calculated; in the outer-loop, cross-market energy trading is updated. The relation of each loop and data exchange is illustrated in Fig. 4.17. The flowchart is given in Algorithm 4.3.

Although this double-loop method entails solving only convex optimization problems, it has two apparent shortcomings: One is its efficiency. As analyzed in Sect. 2.3.4, the convergence rate may depend on system parameter, and can be very slow when the demand is sensitive to the price, especially when the inner level will be called repeatedly. The other is its compatibility with more sophisticated analysis. For example, to analyze the optimal action and market power of a strategic provider in the market, the market equilibria serve as the constraints in the decision-making problem of the provider. However, such an iterative algorithm provides only a numeric solution and fails to offer further insights or analytical tools for more sophisticated optimization models. The quality of initial values also affects the performance of Algorithm 4.3.

4.4.5 Case Studies

The proposed model and method is applied on a testing system comprised of the IEEE 33-bus system and a 32-bus heating system. Its topology is shown in Fig. 4.18. Conventional units G_1, G_3–G_5 and CHP unit G_2 connect to the power system at buses 10, 23, 25, 33, and 18, respectively. Two heat pumps at nodes 1 and 13 of the

Algorithm 4.3 Fixed-point algorithm for market equilibrium

1: **Outer Loop**

1.1: Set initial values for variables; choose convergence criterion ε and r_{max}. Let $r = 1$.
1.2: Call **Inner Loop 1** for heating market clearing, and obtain $\lambda_d^{h,r}$, h_d^r, p_u^r.
1.3: Call **Inner Loop 2** for power market clearing, and obtain $\lambda_d^{p,r}$, p_d^r, λ_u^r.
1.4: If $|\lambda_u^r - \lambda_u^{r-1}| \le \varepsilon$ and $|p_u^r - p_u^{r-1}| \le \varepsilon$ hold, then terminate and return results; if $r = r_{max}$, then quit and report that the algorithm fails to converge; else, update $r \leftarrow r+1$ and return to Step 1.2.

2: **Inner Loop 1**

2.1: Get current λ_u^r; set h_d^0 and the convergence criterion ε_1; let $k = 1$.
2.2: Solve heat market clearing problem (4.35), and obtain $\lambda_d^{h,k}$, p_u^k.
2.3: Update consumer demand according to (4.39) and obtain h_d^k.
2.4: If $\|\lambda_d^{h,k} - \lambda_d^{h,k-1}\| \le \varepsilon_1$ and $\|p_h^k - p_h^{k-1}\| \le \varepsilon_1$ hold, then terminate and report $\lambda_d^{h,r} = \lambda_d^{h,k}$, $h_d^r = h_d^k$, $p_u^r = p_u^k$.
2.5: If $k = k_{max}$, then quit and report that the algorithm fails to converge; else, update $k \leftarrow k+1$, and return to step 2.2.

3: **Inner Loop 2**

3.1: Get current p_u^r; set p_d^0 and the convergence criterion ε_2; let $k = 1$.
3.2: Solve power market clearing problem (4.33), and obtain $\lambda_d^{p,k}$, λ_u^k.
3.3: Update consumer's demand according to (4.37) and obtain p_d^k.
3.4: If $\|\lambda_d^{p,k} - \lambda_d^{p,k-1}\| \le \varepsilon_2$ and $\|p_d^k - p_d^{k-1}\| \le \varepsilon_2$ hold, then terminate and report $\lambda_d^{p,r} = \lambda_d^{p,k}$, $p_d^r = p_d^k$, $\lambda_u^r = \lambda_u^k$.
3.5: If $k = k_{max}$ then quit and report that the algorithm fails to converge; else, update $k \leftarrow k+1$, and return to step 3.2.

heating system purchase electricity from buses 2 and 11 in the power system. CHP unit G_2 supplies heat to node 31. A gas-fired boiler connects to the heating system at node 32. We assume G_1 and G_5 are owned by the system operator, whereas G_2–G_4 are managed by a profit-driven company. An industrial demand gets electrical power from bus 3 and thermal energy from node 3. As discussed before, the industrial demand is a particular elastic demand. We don't consider other elastic demands in our tests. Complete system data can be found in [31].

1. Benchmark Case

In market equilibrium model (4.47), we use primal-dual condition for the power market. Nonlinearity in the primal-dual condition (4.44) arises from two factors: one comes from the product of electricity price and elastic demand $(\lambda^p)^T p^d$, which is actually constant according to (4.42). Hyperbolic relations in (4.42) are arranged as univariate functions $p_i^d = \gamma \hat{I}/\lambda_i^p$ and $h_i^d = (1-\gamma)\hat{I}/\lambda_i^h$, and further linearized via PWL functions using SOS2 variables, as introduced in Appendix B.1.1. The other kind of nonlinearity originates from the product of the offering price of energy

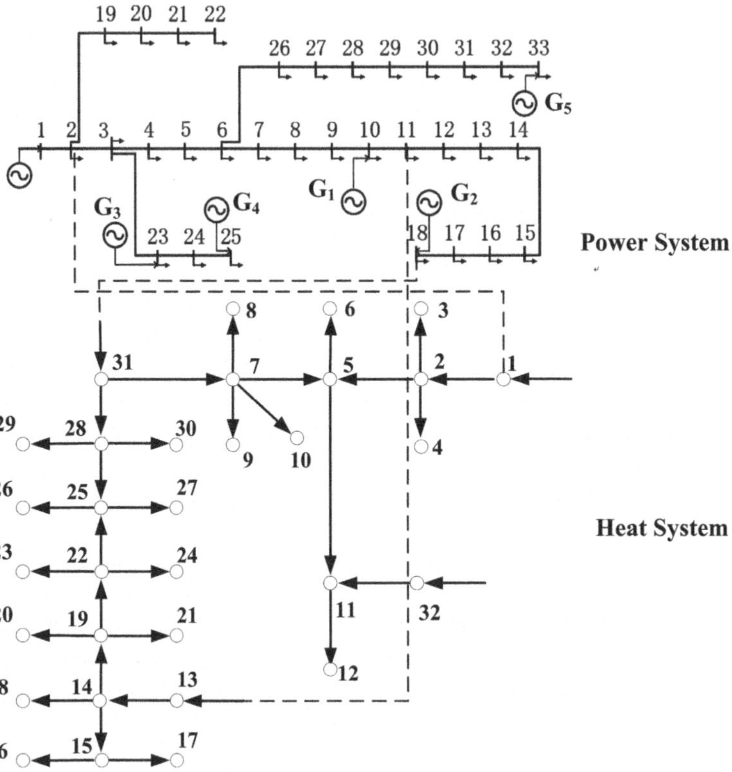

Fig. 4.18 Topology of the testing system

company and its contract power $c_p^T(\alpha)x$. For the energy company, it seeks maximum profit according to the following problem

$$\begin{aligned} \max \quad & \alpha_2^p p_2^g + \alpha_3^p p_3^g + \alpha_4^p p_4^g + \alpha_2^h h_2 \\ \text{s.t.} \quad & \text{offering price limits} \\ & \text{market equilibrium condition} \quad (4.47) \end{aligned} \tag{4.48}$$

where a_i^p, $i = 2, 3, 4$ are the electric power offering prices of units G_2, G_3, and G_4; α_i^h is the heat offering price of unit G_2. The pay-as-bid agreement has been reached between the market and the energy company, so the objective function represents the revenue of the company. One may also consider the cost in the objective function. The company does not bid offering quantities. Unit capacities are known by the market. If the company wishes to bid offering quantities in the market, we can resort to KKT condition for the power market clearing problem, as it avoids non-convex term $b_2^T(\beta)\mu^p$ in the last constraint of (4.44).

Because there is only one strategic player in the market, if the local production levels in the power and heating systems are not adequate, the operators have no choice but purchasing insufficient energy from the company. In such circumstance, the optimal offering price of (4.48) is unbounded: the higher, the better. So we add upper bounds on the offering prices. On the other hand, if the capacity of local generation can meet the demand, the energy company still has a chance to gain profits. If the energy offered by the company is cheaper than the production cost in the system, the operator will opt to purchase certain amount of energy from the company. Later we will investigate the case in which multiple providers exist and competitions occur.

Problem (4.48) is an MPEC, which can be solved by dedicated solvers dealing with complementarity constraints, such as PATH and KNITRO, or general purpose NLP solvers with some special treatments on complementarity constraints, which is discussed in Appendix D.3. If solved successfully, a local optimum of (4.48) can be found. Nevertheless, in this case, there are only a few strategic units; if the offering prices are fixed, term $\alpha_i^p p_i^g$ becomes linear; complementarity constraints in market equilibrium condition (4.47) can be linearized via methods in Appendix B.3.5. We use the pattern search algorithm [32] to find the best bidding strategy of the energy company. It computes objective function values on some sampled points, and updates the sampled points, until certain convergence tolerant is met. The algorithm is derivative-free and has been implemented in MATLAB. For fixed offering prices, the equivalent MILP of (4.47) is solved, which takes approximately 6 s, and the energy contracts and the corresponding profit of the energy company can be evaluated. In theory, the pattern search algorithm terminates at a local optimum. Nonetheless, as the sampling points are scattered and updated in a smart way, it is promising to find the global optimal solution in practice.

We embed the MILP form of (4.47) in MATLAB build-in `patternsearch` function. Results given in Table 4.5 are returned. The optimal costs are \$72785.0 for the power system, and \$23653.7 for the heating system. The profit of the energy company is \$39275.6. From Table 4.5 we can see that CHP unit G_2 and conventional unit G_4 do not bid their price upper bounds. The reason for G_4 is that if it does bid its price upper bound, it will get less energy contract, so bidding highest prices is not always an optimal choice. There are further considerations for the CHP unit, which will be explained later. At the market equilibrium, power demands of heat pumps are 26.29 MW at bus 2 and 21.99 MW at bus 11, respectively. Heat supplies in the heating system include 30.76 MW from the heat boiler and 6.59 MW from the CHP unit.

Table 4.5 Offering prices of the energy company and the energy contracts

Unit	G_2		G_3	G_4
	Power	Heats		
Price limit (\$)	260	210	251	350
Offering price (\$)	252.5	210	251	342.8
Contract (MW)	24.39	6.59	80.0	34.05

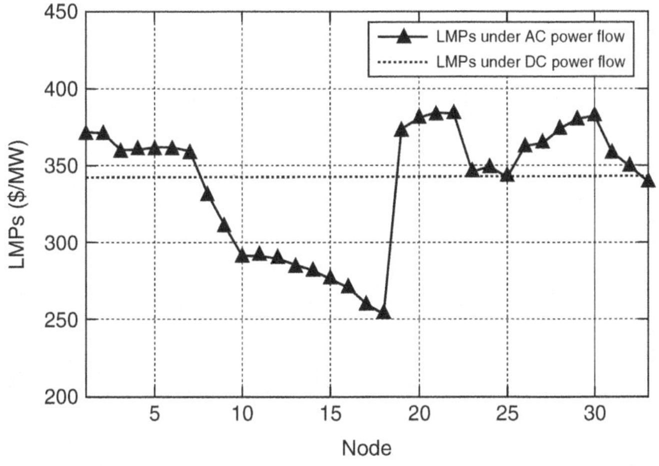

Fig. 4.19 LMP in the PDN

LMPs in the power system are illustrated through Fig. 4.19. We can observe that nodal electricity prices are relatively low at generator buses, and grow gradually with increasing the distance from the power sources. This is because power delivery through distribution lines induces losses. As a result, the further the demand resides, the more losses are incurred for supplying per unit incremental demand there. LMPs calculated from the lossless DC power flow model are plotted in the same figure. Since there is no congestion, the whole network shares the same LMP, which is the marginal production cost. Clearly, two models give significantly different nodal electricity prices, showing the necessity of using AC power flow to clear the distribution power market.

To verify the exactness of the SOCP relaxation and the accuracy of the polyhedral approximation method, the relative errors

$$\varepsilon = \frac{\sqrt{4P^2 + 4Q^2 + (U - I)^2}}{U + I} - 1$$

at all buses are shown in Fig. 4.20. We can observe that the error is less than its theoretical upper bound 1.4×10^{-4} for $\upsilon = 7$, which has two implications: first, the polyhedral approximation achieves desired accuracy; second, inequality (4.27) is active (holds as an equality) at the optimal solution, i.e., the SOCP relaxation is exact.

To validate the accuracy and advantage of the proposed method (LPAC-MILP for short), it is compared with three alternative methods with different compositions of power flow models and solution techniques: (1) LPAC-FP, which consists of the LP approximation of ACOPF market clearing model and Algorithm 4.3; (2) SOCP-FP, which involves the SOCP relaxation of ACOPF without LP approximation and Algorithm 4.3; (3) LPDC-FP, which employs the DCOPF based power market

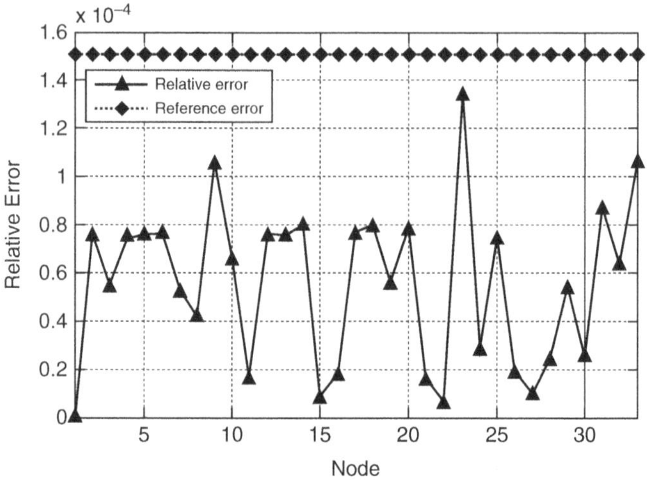

Fig. 4.20 Approximation error at each bus

Table 4.6 Computational times comparison under different initial values (seconds)

Initial value	$0.8\Xi^*$	$0.9\Xi^*$	Ξ^*	$1.1\Xi^*$	$1.2\Xi^*$
LPAC-FP	Fail	22.56	10.29	19.48	Fail
SOCP-FP	Fail	48.17	30.21	48.48	Fail
LPDC-FP	Fail	12.91	11.12	13.37	Fail
LPAC-MILP	5.82				

Table 4.7 Power system cost comparison under different initial values ($)

Initial value	$0.8\Xi^*$	$0.9\Xi^*$	Ξ^*	$1.1\Xi^*$	$1.2\Xi^*$
LPAC-FP	Fail	72784.6	72784.6	72784.6	Fail
SOCP-FP	Fail	72785.0	72784.9	72784.9	Fail
LPDC-FP	Fail	67736.0	67736.0	67736.0	Fail
LPAC-MILP	72785.0				

clearing (please refer to Sect. 2.2.1 for DC power flow model) and Algorithm 4.3. Results with different initial values are given in Tables 4.6 and 4.7, where Ξ^* stands for the market equilibrium in the benchmark case given by LPAC-MILP.

From Table 4.6 we can observe that the performance of Algorithm 4.3 largely depends on the initial value. The closer the initial point leaves from the true equilibrium, the less time is needed; when the initial point departs farther from the true equilibrium, Algorithm 4.3 may fail to converge. LPAC-MILP is the fastest and makes no reference to an initial guess. Besides, when the market equilibrium is embedded into provider's bidding problem, the time difference in once computation will be amplified by the iteration procedure in the upper level. In Table 4.7, when SOCP-FP converges, the equilibrium is exact because no approximation is made in the power flow model; LPAC-MILP and LPAC-FP find the same equilibria (if the latter successfully converges), which is almost the same as the one offered by SOCP-FP, indicating that the LP approximation possesses high accuracy. LPDC-FP reports significantly different costs, because a less accurate power flow model (for the distribution system) is used, which neglects network losses and reactive power, demonstrating the necessity of using AC power flow model to calculate nodal prices in distribution systems. In conclusion, compared with Algorithm 4.3, the proposed MILP based method is advantageous in computational efficiency and robustness, as the true equilibrium is not known in advance.

2. Analysis of the Behavior of CHP Unit

In the integrated heat-power system, a representative entity playing the role of energy company is the so-called energy hub. A key production facility in the energy hub is a CHP unit. It is interesting to study the bidding strategy of the CHP unit. In this test, the offering prices of G_3 and G_4 are fixed at [250, 342]\$/MWh, and G_2 is the only strategic player in problem (4.48).

A particular attribute of CHP unit is that its feasible operating region is a polytope, which means the maximum offering capacities of power and heat are not independent. A viable way to cope with this issue is allowing the CHP unit to bid two offering prices and two maximum offering quantities in the market. Operating feasibility is guaranteed by the bidding strategy. As explained previously, this will introduce additional nonlinear terms in the strong duality condition. Here we use another strategy: a penalty is introduced when the energy contract cannot be delivered due to operating infeasibility, so the CHP unit will bid offering prices in reasonable ranges. We calculate the profits at sampled combinations of power and heat offering prices, which are shown in Fig. 4.21. We can see that the profit is highly non-convex as a function of offering prices.

Because the resource in the heating system is insufficient, to gain more profit, the CHP unit tends to bid a higher price in the heat market. If it rises its offering price in the power market, it loses market share because the power system can dispatch

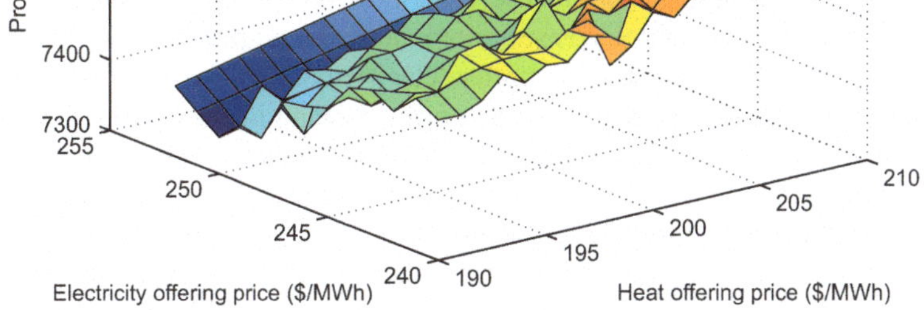

Fig. 4.21 Profit of CHP unit G_2 with different offering prices

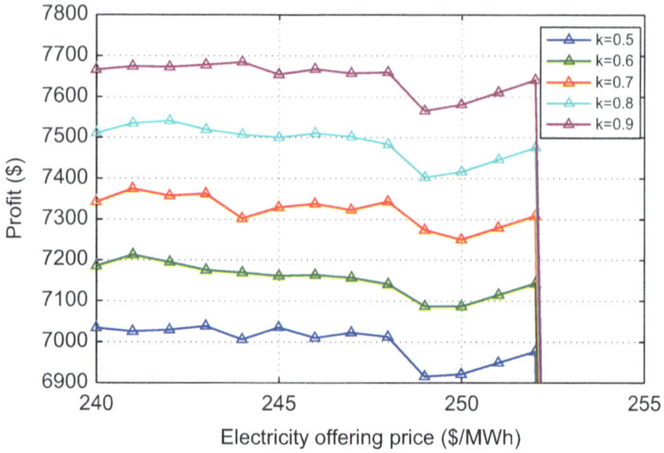

Fig. 4.22 Profit of CHP unit G_2 with a fixed price ratio

conventional units G_1 and G_5. So the best strategy is to reduce the electricity offering price and rising the heat offering price, if they are independently determined. This phenomenon can be observed from Fig. 4.21. Moreover, we consider a constant-ratio bidding scheme: heat offering price $= k \times$ electricity offering price. From Fig. 4.22 we can see that bidding a low price generally gains a higher profit, regardless of the value of k, because more electricity can be sold, which is more expensive than heat.

3. Analysis of the Behavior of Industrial Demand

The demand response problem is described by (4.40) and its optimal solution is given in (4.42). We change the emphasis on electricity and heat, so as to get different consumer types marked by Type A - Type E with $\gamma = 1/4$, $\gamma = 1/3$, $\gamma = 1/2$, $\gamma = 2/3$, and $\gamma = 3/4$, respectively. The budget and energy offering prices are fixed. The energy demands are shown in Fig. 4.23. When γ increases, the desire on electricity rises. So the electric power demand increases and the heat demand decreases. Total costs of the power system and heating system are shown in Fig. 4.24. When the consumer prefers heat (from Type C to Type A), the cost of the heat system goes up quickly as the heat demand increases. The power system cost changes slightly: although the power demand decreases, more electricity is purchased by producing thermal energy. When the consumer prefers electricity (from Type C to Type E), the cost of power system rises greatly as the power demand increases. For the heating system with Type D and Type E loads, the heat demands are almost the same, but the overall electricity prices in the power system are higher, so the cost of heating system is slightly higher with Type E load.

Budget is another important factor that influences the consumer's decision. We multiply the budget with a factor varying from 0.8 to 1.2, and fix $\gamma = 0.5$. Results are shown in Figs. 4.25 and 4.26. As what is expected, when the consumer gets

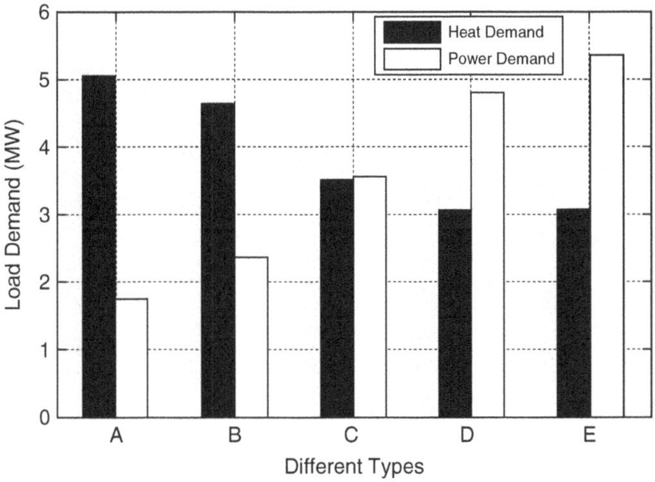

Fig. 4.23 Energy demands of different types of industrial loads

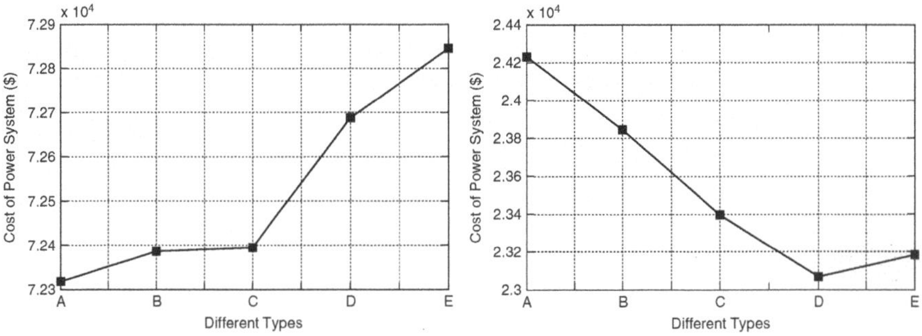

Fig. 4.24 Total costs of the power and heating systems

a larger budget, it will consume more heat and electricity. Moreover, heat prices change in a wider range than the electricity prices, because the growth of electric demands will lead to higher electricity prices, which add extra costs to the heating system other than that induced by the increase of heat demands.

4. Analysis of Mutual Effect Among Consumers

We analyze the mutual impact among different consumers. Two industrial consumers are considered in our tests. Consumer 1 (C1) is connected to node 3 in the DHN and bus 3 in the PDN. We investigate two situations: (1) C2 is connected to node 4 in the DHN and bus 20 in the PDN; (2) C2 is connected to the same node and bus as C1. The impacts of budget and utility function are tested, respectively.

1. Budget. We multiply the budget $\hat{I}$ of C2 with a factor varying between 0.7 and 1.3, while fix $\alpha_1 = \alpha_2 = 1/2$. Results are shown in Figs. 4.27 and 4.28. We

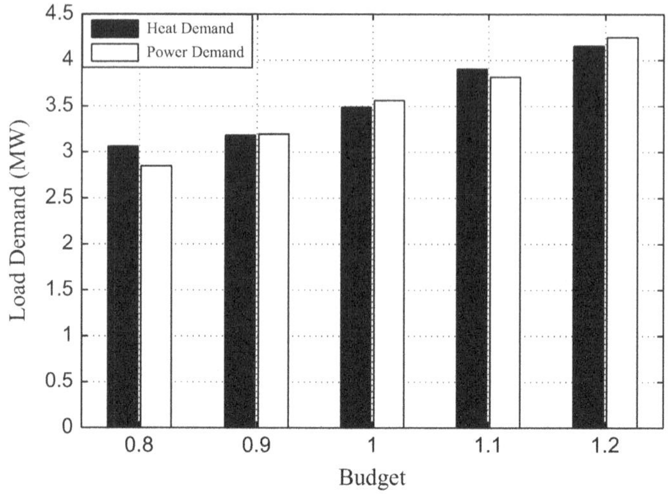

Fig. 4.25 Energy demands of the industrial load with different budgets

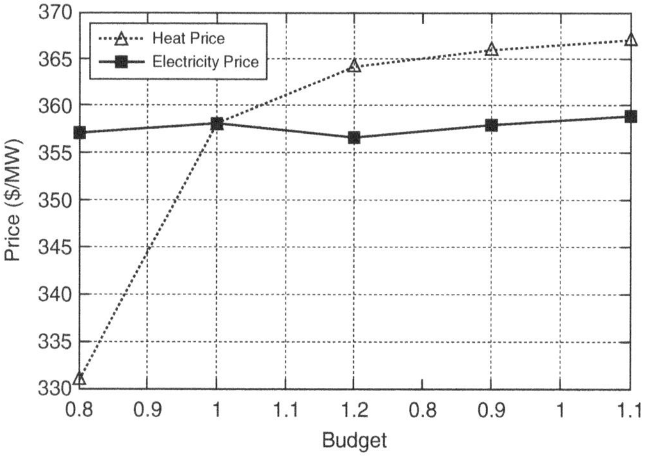

Fig. 4.26 Energy prices at the points where the industrial load connects to the system

can observe that when the budget of C2 increases, the power and heat demand of C2 grow in both cases. The growth of demands leads to the increase of energy prices for C2, especially the price of heat. If the two consumers have no direct connections, the demand variation of C2 has little impact on the energy price for C1, and thus the optimal strategy of C1 almost remains the same. If the two consumers get energy supply from the same place, the heat bought by C1 decreases due to the increase of heat price.

2. Utility function. We change the value of α_1 ($\alpha_1 = 1/3, 2/5, 1/2, 3/5, 2/3$ marked by Type A$'$-Type E$'$) and fix the budget. Results under different C1 types

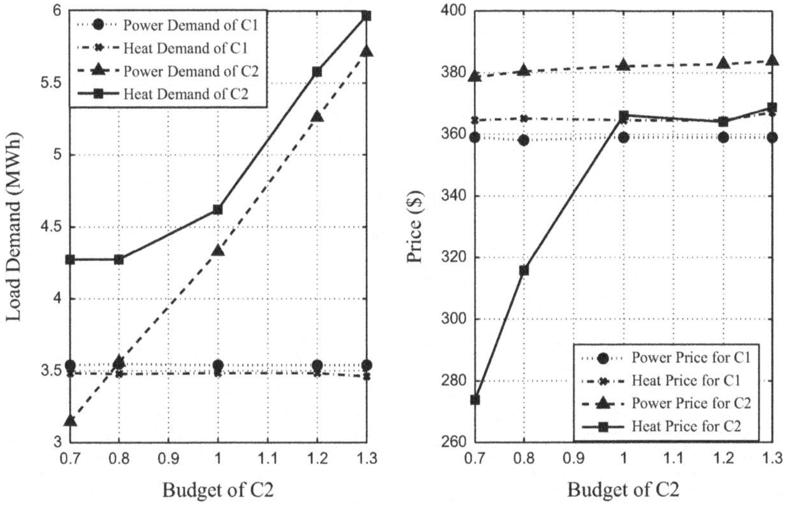

Fig. 4.27 Load demand and price under different budgets when C1 and C2 are connected to different places

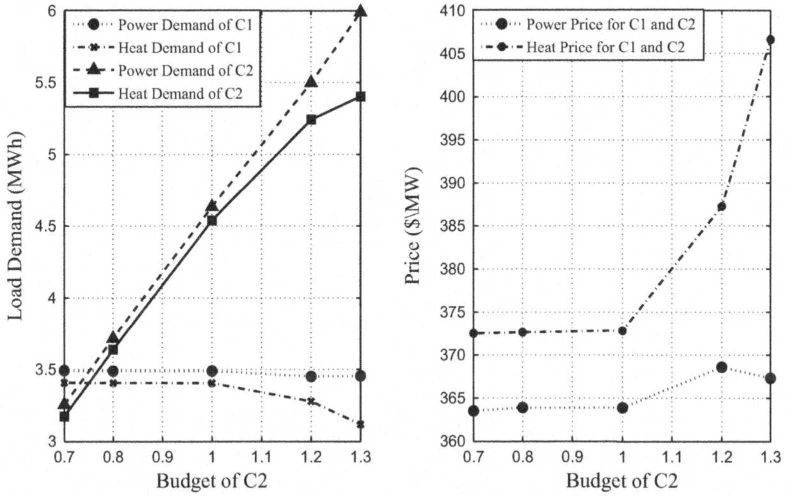

Fig. 4.28 Load demand and price under different budgets when C1 and C2 are connected to the same place

are plotted in Figs. 4.29 and 4.30. We can observe the following facts. With the growth of α_1, the power demand of C1 increases and the heat demand of C1 decreases in both situations, due to the limited budget. In the first situation, when the two consumers are connected to different places, the change in C1's heat demand leads to a decline in heat price for C1 but has little impact on the energy

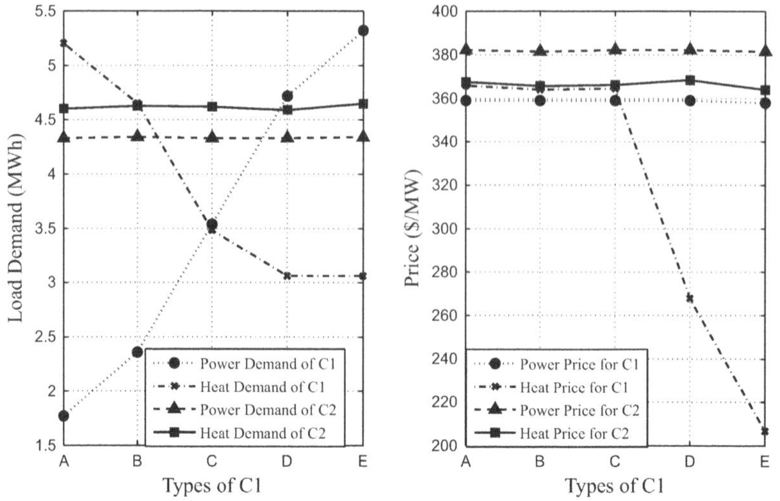

Fig. 4.29 Load demand and price under different budgets when C1 and C2 are connected to different place

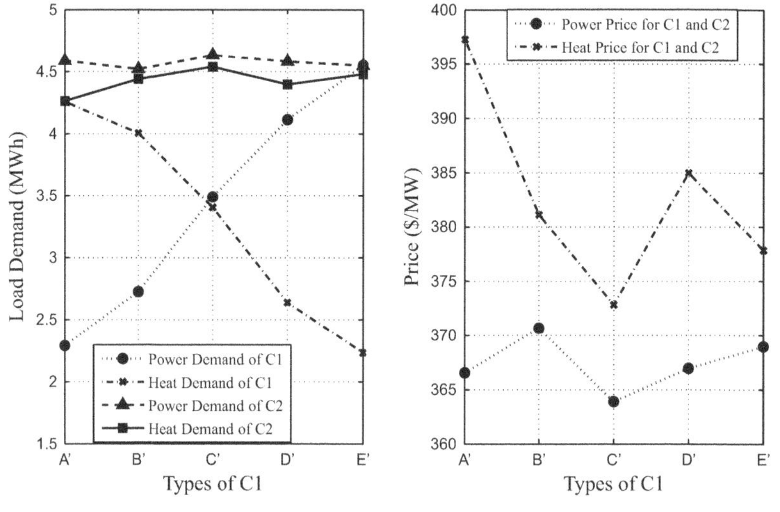

Fig. 4.30 Load demand and price under different budgets when C1 and C2 are connected to the same place

prices for C2. As a result, the power and heat demand of C2 barely change. The second situation is more complicated. In general, if α_1 increases, C1 would consume more power and the electricity price will rise up and the heat price will go down. But for Type-C$'$, the electricity price decreases, because the reduction in heat demand exceeds the increase of power demand, so that the total power demand drops down, so does the electricity price. In such circumstance, both

Table 4.8 Computation time with different number of consumers

Number of consumers	1	2	4	8	16
Computational time (s)	4.52	5.52	9.64	21.55	53.06

power and heat demands of C2 increase slightly. For Type-D′, the growth of power demand of C1 exceeds that caused by the reduction in heat demand, so the electricity price increases again, followed by the heat price. At this time, both power and heat demand of C2 decrease.

As a conclusion, when two consumers are connected at the same place, their mutual influence is prominent; otherwise, the mutual effect is tiny. The situation for multiple consumers is similar. The computational performance in the presence of multiple elastic demands is shown in Table 4.8. We can see that the computational efficiency of the proposed method is satisfactory if there are a moderate number of elastic demands. Because small loads have no access to the distribution power market (actually, they participate in a retail market), the number of elastic demands take part in the distribution-level market will be limited, and the proposed method has no substantial difficulty in coping with a practical system.

5. Analysis of the Market Impact of Competition Among Providers
To investigate the market impact of competition among multiple energy companies, we fix the offering prices of G_2, and assume that G_3 and G_4 are owned by different companies, denoted by EC-1 and EC-2, respectively. A Bertrand game emerges between the two companies. In view of the market equilibrium condition, the problem in fact boils down to an EPEC. Each company solves an MPEC using the pattern search method with the rival's strategy fixed. The best response curves of both companies are drawn in Fig. 4.31, and the intersection interprets the NE of the Bertrand game. We assume that the minimum offering prices of G_3 and G_4 are 250$/MWh and 240$/MWh so as to ensure certain profits of energy companies. Because of the competition, neither company is willing to bid a high price in order to maintain its market share. When more ECs enter the market, a similar phenomenon can be found in general. In view of this, competition can prevent unfairly high prices caused by monopoly, from which consumers can benefit. Nevertheless, in some special cases, when there are some system constraints that restrict the further decrease of the energy expenditure, some providers might possess market power and bid a high price, or possibly none market equilibrium exists. That is case-dependent.

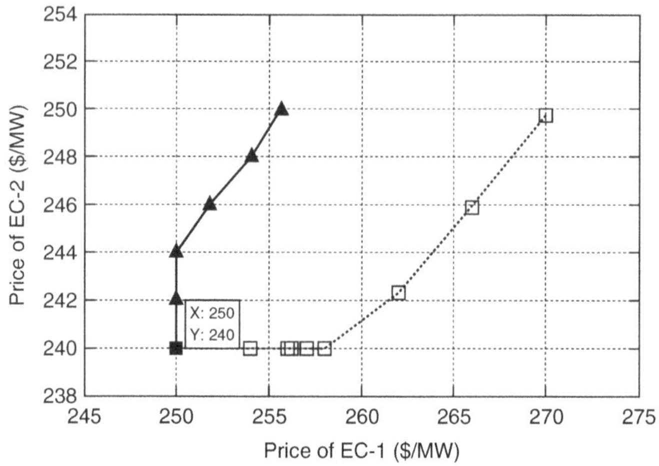

Fig. 4.31 Best response curves of the two energy companies

4.5 Strategic Bidding of Energy Hubs

Section 4.4 gives mathematical characterization on the equilibrium of interdependent heating and power spot markets with energy exchange and responsive demands. The thorough model and computational method can deal with AC power flow constraints and offer accurate LMPs. In the following two situations, the computation is still challenging. First, in a multi-period day-ahead market with inter-temporal constraints, such as those associated with energy storage, special decomposition technique is needed in large-scale instances; Second, if there is a provider who controls many devices in the upper level, the direct search algorithm is usually time-consuming because the search dimension is high.

In this section, we consider a somehow different situation: the PDN and the DHN has no direct linkages; they are connected through a third-party managed energy hub, who owns cogeneration plants, energy conversion devices, such as heat pumps and electric boilers, as well as heat and electricity storage units. Energy exchange is carried out by the energy hub. The energy hub is profit-driven: it buys natural gas and produces electricity and heat, which are then stored or sold to respective markets; it can purchase and sell electricity in the power market like a prosumer. As heat price may not vary significantly intraday, we do not consider arbitrage in the heating market. Because the energy contract between the hub owner and energy markets depends on bidding strategies, to maximize revenue, the energy hub must anticipate the response from the markets, hence its optimal bidding problem has two levels. Furthermore, in order to exploit arbitrage potential enabled by peak and valley prices and storage units, market clearing problems over the entire scheduling horizon, typically 24 h, are incorporated in the lower level, making the problem challenging to solve. Tractable model and method will be developed in this section.

4.5.1 Market Clearing Models

1. Heating Market

We assume that traditional heat sources in the DHN are gas-fired boilers, whose production costs are convex quadratic functions of their thermal output. The energy hub bids a heat offering price ζ^b and a maximum quantity h_m^b it is willing to provide, which are treated as constants in the heating market. In a similar vein, the heating market clearing problem can be formulated as an optimal thermal flow problem, whose compact form is given by

$$\min_{h,\tau} \ \frac{1}{2} h^T Q_H h + c_h^T(\zeta^b) h \tag{4.49a}$$

$$\text{s.t.} \ A_H h + B_H(\dot{m})\tau = b_H \tag{4.49b}$$

$$C_H h + D_H(\dot{m})\tau \le d_H(h_m^b) \tag{4.49c}$$

where vector h collects the thermal output of heat sources; vector τ and $\dot{m}$ denote nodal temperatures and pipeline mass flow rates, respectively; other matrices and vectors are coefficients; symbols in brackets indicate that they could be determined from other problems and will influence the coefficients before them. Particularly, $\dot{m}$ is determined from Algorithm 4.1 in Sect. 4.2.4; ζ^b and h_m^b are given by the energy hub in the upper level. Objective function (4.49a) is to minimize the total cost of DHN, including production expenditure of boilers and the cost paid to the energy hub; (4.49b) describes temperature distributions in the DHN, and (4.49c) represents operating security/capacity bounds. Complete formulations of these constraints have been elaborated in Sect. 4.2.2. Since (4.49) is a convex quadratic program, KKT condition is necessary and sufficient for optimality.

2. Power Market

In the distribution power market, the energy hub bids an electricity offering price ξ^b and a maximum quantity p_m^{gb} it is willing to provide. Meanwhile, the energy hub can purchase electricity from the power market at a price bid χ^b, and the maximum quantity bid is p_m^{db}. The total cost of PDN is

$$C_{\text{PDN}} = \sum_j \left[a_j (p_j^g)^2 + b_j p_j^g \right] + \xi^b p^{gb} - \chi^b p^{db} \tag{4.50}$$

where the first term represents generation cost of local units; a_j and b_j are coefficients in the quadratic function. In this section, the linearized BFM elaborated in Sect. 2.2.2 will be used to depict PDN operating status. Generator at the slack bus with $a_0 = 0$ and $b_0 =$ constant mimics the energy delivery from the transmission network at a fixed price, so the cost paid to the transmission-level market is included in the first term. The second (third) term is paid to (by) the energy hub for purchasing (selling) energy at a rate of p^{gb} (p^{db}). With the bidding strategies of the energy hub fixed, the power market clearing problem can be formulated as an OPF problem, whose compact form is expressed by

$$\min_{p,x} \ \frac{1}{2}p^T Q_P p + c_p^T(\xi^b, \chi^b)p \tag{4.51a}$$

$$\text{s.t.} \ A_P p + B_P x = b_P \tag{4.51b}$$

$$C_P p + D_P x \le d_P(p_m^{gb}, p_m^{db}) \tag{4.51c}$$

where vector p includes the generation dispatch p_j^g and energy transactions p^{gb}, p^{db} with the energy hub; p_m^{gb} and p_m^{db} are the maximal electricity quantities the energy hub is willing to provide or purchase; vector x encapsulates all remaining decision variables, mainly power flow variables. Constraints are classified into equalities (4.51b) and inequalities (4.51c), which involves linearized BFM (2.20a), (2.20b), and (2.20d), as well as security limits. Although network losses are neglected, other factors such as congestions, reactive power, and nodal voltage are taken into account, making it more accurate than the DC power flow model.

Market clearing problems (4.49) and (4.51) are decoupled with respect to time periods. Inter-temporal constraints appear in the energy hub operating conditions in the upper level.

4.5.2 Strategic Bidding Model

1. Energy Hub Model

We consider an energy hub sketched in Fig. 4.32, which links a PDN and a DHN. At the supply side, the electricity comes from the PDN, and natural gas is provided by

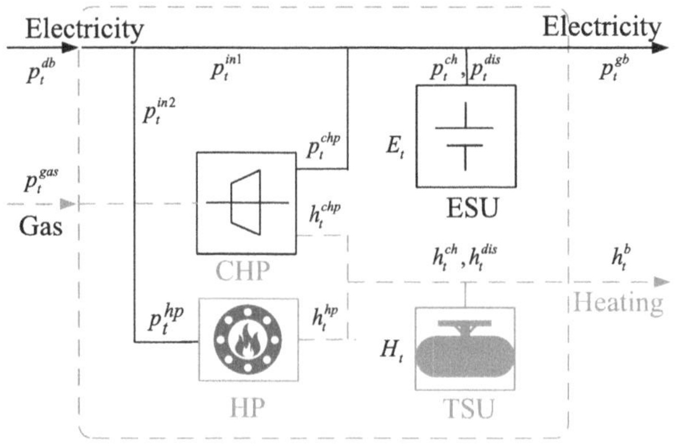

Fig. 4.32 Schematic of the energy hub

a gas company. The gas network constraints are neglected. We can simply impose upper bound on the gas import to mimic security limitation on the gas system side. Different from a residential one, the distribution-level energy hub considered here can sell electricity and heat to the PDN and the DHN, respectively, at its output side. Inside the energy hub, electricity can be used to charge an electricity storage unit (ESU), or consumed by a heat pump which produces heat. Natural gas is burnt by a CHP unit to produce electricity and heat at the same time. Heat can be stored in a thermal storage unit (TSU) if necessary. Energy flows and variables in the energy hub are illustrated in Fig. 4.32. Operating constraints of the energy hub include the following:

$$p_t^{in1} + p_t^{gas}\eta_e^{chp} + p_t^{dis} - p_t^{ch} = p_t^{gb}, \ \forall t \tag{4.52a}$$

$$p_t^{in2}\eta^{hp} + p_t^{gas}\eta_h^{chp} + h_t^{dis} - h_t^{ch} = h_t^{b}, \ \forall t \tag{4.52b}$$

$$p_t^{db} = p_t^{in1} + p_t^{in2}, \ \forall t \tag{4.52c}$$

$$E_{t+1} = E_t + p_t^{ch}\eta_+^{esu} - p_t^{dis}/\eta_-^{esu}, \ \forall t \tag{4.52d}$$

$$H_{t+1} = H_t + h_t^{ch}\eta_+^{tsu} - h_t^{dis}/\eta_-^{tsu}, \ \forall t \tag{4.52e}$$

$$p_{\min}^{gb} \le p_{t,m}^{gb} \le p_{\max}^{gb}, \ \forall t \tag{4.52f}$$

$$p_{\min}^{db} \le p_{t,m}^{db} \le p_{\max}^{db}, \ \forall t \tag{4.52g}$$

$$h_{\min}^{b} \le h_{t,m}^{b} \le h_{\max}^{b}, \ \forall t \tag{4.52h}$$

$$\text{Upper and lower bounds of other variables} \tag{4.52i}$$

where p_t^{gb} and h_t^b are the cleared amount of electric power and thermal energies in the respective markets, and p_t^{db} is the delivered power from the electricity market. These variables are not directly controlled by the hub; instead, they are determined from the market clearing problems. p_t^{gas} is the inflow of gas fuel, E_t and H_t are the stored electricity and heat in the ESU and TSU, respectively. p_t^{ch} and p_t^{dis} are the charging and discharging power of the ESU, respectively. h_t^{ch} and h_t^{dis} are the charging and discharging power of the TSU. Physical meaning of other variables is depicted in Fig. 4.32. $p_{t,m}^{gb}$ and $p_{t,m}^{db}$ are the electricity quantity offer and bid, respectively. $h_{t,m}^b$ is the heat quantity offer. Equations (4.52a) and (4.52b) define the electric and thermal power balance inside the hub; Eq. (4.52c) determines the electric power demands of heat pump and ESU; state-of-charges (SoCs) of the ESU and the TSU are described in (4.52d) and (4.52e), respectively. We assume that the final SoC is identical to the initial one, i.e., $E_T = E_0, H_T = H_0$, to complete a daily cycle; The bounds of $p_{t,m}^{gb}$, $p_{t,m}^{db}$, and $h_{t,m}^b$ are limited in (4.52f)–(4.52h), respectively; the operating security constraints are collected in (4.52i), including the polyhedral operating region of the CHP units (detailed expressions can be found in Sect. 4.2.3), lower and upper bounds of decision variables, as well as charging rate limits. Simultaneous charging and discharging are prevented by introducing binary variables in the charging/discharging rate constraints in (4.52i).

2. Bidding Model of the Energy Hub

Some basic settings and considerations on the market are set forth. We center on the day-ahead electricity and heating markets and use load forecast data. Uncertainty is neglected for model conciseness and can be tackled in the real-time stage, thanks to the energy storage units equipped in the energy hub.

The power grid and the heating system have no direct connection. The energy hub is the intermediary agent between the two systems. The market structure is shown in Fig. 4.33. The energy hub submits its electricity (heat) bidding/offering prices and quantities to the respective markets; and then the operators clear the electricity and heating markets by solving problems (4.49) and (4.51). The energy hub receives the energy contracts and gets paid according to the offering prices following a pay-as-bid agreement.

From the power market side, the energy hub is a prosumer: it can either consume and offer electricity. Different electricity prices in peak and valley hours precipitate arbitrage opportunities; From the heating market side, it is a producer. We don't consider thermal energy import for two reasons: first, heating market is still under study and less mature than the power market; second, the gap between peak and valley prices of heat may not be as large as that of electricity, so arbitrage opportunity in the heating market is not very significant. Nevertheless, there is no difficulty in considering bidirectional heat trading in the mathematical model.

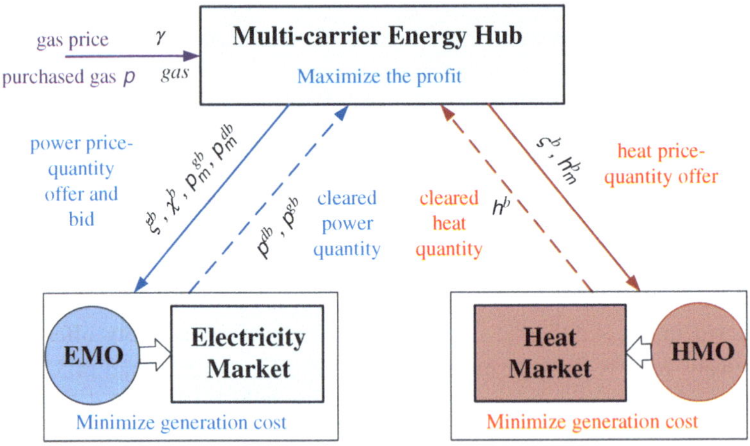

Fig. 4.33 Market structure

In the proposed market, the energy hub seeks maximum profit through the following optimization problem:

$$\max \ (\zeta^b)^T h^b + (\xi^b)^T p^{gb} - (\chi^b)^T p^{db} - (\gamma)^T p^{gas} \tag{4.53a}$$

$$\text{s.t. Hub operating constraints (4.52)} \tag{4.53b}$$

$$\text{heating market clearing (4.49)} \tag{4.53c}$$

$$\text{power market clearing (4.51)} \tag{4.53d}$$

where vectors ζ^b,ξ^b,χ^b,γ represent heat offering prices, electricity offering prices, electricity purchasing prices, and gas prices in the day-ahead market, respectively; energy contract vectors h^b, p^{gb}, p^{db}, p^{gas} stand for heat offering quantities, electricity offering quantities, electricity demands, and gas demands, respectively. Although we could impose upper bounds on the offering prices in (4.53b), bidding the price upper bound in the market is not always a wise decision, because the operators may dispatch other available resources, as such the energy hub would lose market share. As energy contracts are determined from optimization problems (4.49) and (4.51), the strategic bidding model (4.53) is an MPEC.

Since we employ the pay-as-bid scheme in this market framework, and we do not need the accurate LMP, the lossless linearized BFM with reasonable simplification is appropriate in this study. In our current model, uncertain factors are not considered, because we have storage units which can mitigate the negative impact of renewable fluctuations. In other words, system security is not a main concern due to the advent of energy hub. Nonetheless, if the economic impact of these uncertain factors is under investigation, we can use the scenario based stochastic programming approach. Since the problem is totally decoupled with respect to scenario, the profit in each scenario can be calculated in parallel. If the real-time market is considered and a two-stage stochastic model is used to tackle uncertainty factors, the situation would be more complicated, because the day-ahead decision cannot change in the real-time stage. It is difficult to incorporate uncertainty in the bi-level optimization framework using the scenario stochastic approach. One remedy would be restricting the number of bidding strategies of the energy hub so as to reduce the dimension of decision variables in the day-ahead stage. Another is using the proposed deterministic model in the day-ahead market, and adopt a look-ahead bidding (with a few number of periods) in a rolling-horizon fashion for the real-time market, as such uncertainty will be tackled in the real-time stage. The data-driven Stackelberg market framework proposed in [33] could be a promising approach to incorporate uncertainties in such sequential decision-making problems.

3. An MILP Approximation

To solve problem (4.53), the convex quadratic market clearing problems (4.49) and (4.51) are replaced by their respective KKT optimality conditions, then problem (4.53) comes down to an MPCC. The KKT optimality conditions read as follows:

KKT Condition of the Heating Market Clearing Problem

$$A_H h + B_H(\dot{m})\tau = b_H \tag{4.54a}$$

$$C_H h + D_H(\dot{m})\tau \le d_H(h_m^b) \tag{4.54b}$$

$$Q_H h + c_h(\zeta^b) + A_H^T \lambda_h + C_H^T \mu_h = 0 \tag{4.54c}$$

$$B_H(\dot{m})\lambda_h + D_H^T(\dot{m})\mu_h = 0, \ \mu_h \ge 0 \tag{4.54d}$$

$$\mu_h^T \left[C_H h + D_H(\dot{m})\tau - d_H(h_m^b) \right] = 0 \tag{4.54e}$$

where λ_h and μ_h are the vectors of dual variables associated with equality constraint (4.49b) and inequality constraint (4.49c) of the heating market clearing problem (4.49). Coefficient vectors d_H and c_h are linear in h_m^b and ζ^b submitted by the energy hub, which are treated as constants in the heat market clearing problem. Equations (4.54a) and (4.54b) are feasibility constraints of primal variables; Eqs. (4.54c) and (4.54d) are feasibility constraints of dual variables; Eq. (4.54e) represents the complementarity and slackness conditions.

KKT Condition of the Power Market Clearing Problem

$$A_P p + B_P x = b_P \tag{4.55a}$$

$$C_P p + D_P x \le d_P(p_m^{gb}, p_m^{db}) \tag{4.55b}$$

$$Q_P p + c_p(\xi^b, \chi^b) + A_P^T \lambda_p + C_P^T \mu_p = 0 \tag{4.55c}$$

$$B_P^T \lambda_p + D_P^T \mu_p = 0, \ \mu_p \ge 0 \tag{4.55d}$$

$$\mu_p^T \left[C_P p + D_P x - d_P(p_m^{gb}, p_m^{db}) \right] = 0 \tag{4.55e}$$

where λ_p and μ_p are the vectors of dual variables associated with equality constraint (4.51b) and inequality constraint (4.51c) of the power market clearing problem (4.51). Coefficient vectors c_p and d_p are linear in ξ^b, χ^b, p_m^{gb}, and p_m^{db} submitted by the energy hub, which are treated as constants in the electricity market clearing problem. Equations (4.55a) and (4.55b) are feasibility constraints of primal variables; Eqs. (4.55c) and (4.55d) are feasibility constraints of dual variables; Eq. (4.55e) represents the complementarity and slackness conditions.

The MPCC formulation eliminates market clearing optimization problems in the lower level, and is comprised of (4.53a), (4.53b), (4.54), and (4.55). However, it is still nonlinear and non-convex. The difficulty arises from the complementarity and slackness constraints in the KKT conditions which have the form of $x \ge 0$, $y \ge 0$, $x^T y = 0$ as well as production terms $(\zeta^b)^T h^b$, $(\xi^b)^T p^{gb}$, and $(\chi^b)^T p^{db}$

in the objective function (4.53a). For the complementarity constraints, we adopt the linearization method in Appendix B.3.5 to express it as

$$0 \le x \le Mz,\ 0 \le y \le M(1 - z) \tag{4.56}$$

where M is a large enough constant; z is a 0–1 vector with the same dimension as x and y; 1 (0) is the all-one (zero) vector with the same dimension as z.

For the bilinear product terms in the form of xy where x and y are two continuous variables, we employ the binary expansion method in Appendix B.2.3 to linearize them. Particularly, we use 2^K discrete points to approximate possible values of y in its feasible interval $[y^l, y^m]$, which gives rise to

$$y = y^l + \Delta y \sum_{k=1}^{K} 2^{k-1} z_k \tag{4.57}$$

where z_k, $k = 1, \cdots, K$ are binary variables, and the step size Δy is given by

$$\Delta y = \frac{y^m - y^l}{2^K} \tag{4.58}$$

As such, $xy = xy^l + \Delta y \sum_{k=1}^{K} 2^{k-1} x z_k$. Let $v_k = x z_k$, $k = 1, \cdots, K$, the bilinear term xy can be formulated by

$$xy = xy^l + \Delta y \sum_{k=1}^{K} 2^{k-1} v_k \tag{4.59a}$$

$$0 \le x - v_k \le x^m (1 - z_k),\ \forall k \tag{4.59b}$$

$$0 \le v_k \le x^m z_k,\ \forall k \tag{4.59c}$$

If $z_k = 0$, then $v_k = 0$ is forced by (4.59c), and the feasible interval of x is retained in (4.59b). If $z_k = 1$, then $v_k = x$ is forced by (4.59c), and the feasible interval of x is given in (4.59c). In either case, we have the relation $v_k = x z_k$, $k = 1, \cdots, K$, so the right-hand side of (4.59a) provides a linear expression of xy. According to (4.58), the number of binary variables K is a logarithmic function in the given error tolerance.

In our problem, because the energy contracts interpret optimal solutions of market clearing problems, discrete approximation could miss the exact one, which may cause infeasibility of the KKT condition. Therefore, we discrete the bidding strategies $(\zeta^{\mathbf{b}}, \xi^{\mathbf{b}}, \chi^{\mathbf{b}})$ through binary expansion. These variables are physically implementable and do not cause infeasibility issues. Applying (4.59a)–(4.59c) to all product terms in (4.53a), we obtain the linearized objective function (Obj-Lin for short)

$$\begin{aligned}\text{Obj-Lin} =& \sum_t \left[\xi_l p_t^{gb} + \Delta\xi \sum_{k=1}^{K} 2^{k-1} z_{tk}^{gb} \right] - \sum_t \gamma_t p_t^{gas} \\ &+ \sum_t \left[\chi_l h_t^b + \Delta\chi \sum_{k=1}^{K} 2^{k-1} z_{tk}^{hb} \right] - \sum_t \left[\zeta_l p_t^{db} + \Delta\zeta \sum_{k=1}^{K} 2^{k-1} z_{tk}^{db} \right]\end{aligned} \tag{4.60}$$

In summary, the MILP form of the optimal bidding problem can be expressed as

$$\begin{aligned} \max \ & \text{Obj-Lin (4.60)} \\ \text{s.t.} \ & \text{Hub operating constraints (4.52)} \\ & \text{KKT-Lin-Heat} \\ & \text{KKT-Lin-Power} \\ & \text{Cons-BE} \end{aligned} \tag{4.61}$$

where KKT-Lin-Heat and KKT-Lin-Power stand for the linearized KKT conditions (4.54) and (4.55) after performing linear disjunctive formulation (4.56) on (4.54e) and (4.55e); Cons-BE collects all additional constraints in the form of (4.59b)–(4.59c) created during the binary expansion process.

4.5.3 Case Studies

1. Basic Configurations

In this section, numeric experiments are carried out on a testing system comprised of a modified IEEE 33-bus PDN and a 32-node DHN. The energy hub connects to the PDN at Bus 2 and the DHN at Node 31. System topology is shown in Fig. 4.34. 2 gas boilers (GBs) and 2 gas turbines (GTs) produce thermal and electrical energy in the DHN and PDN, respectively. Static var generators with the capacity of 1.0 MVar are placed at Bus 3 and Bus 12 for compensating reactive power and maintaining the voltage profile. Parameters of GTs, GBs, and the energy hub are listed in Tables 4.9 and 4.10. Detailed system data can be found in [34].

In our tests, 128 discrete points ($K = 7$) are used in the binary expansion scheme. For security considerations, the maximum delivered power (p_0^g) from the slack bus is 3 MW (also called the feeder capacity of the PDN), and the maximum gas inflow rate (p^{gas}) delivered to the EH is 1.5 MW. The lower bound, upper bound, and average of heat offering prices (ζ^b) are 12\$/MWh, 30\$/MWh, and 25 \$/MWh,

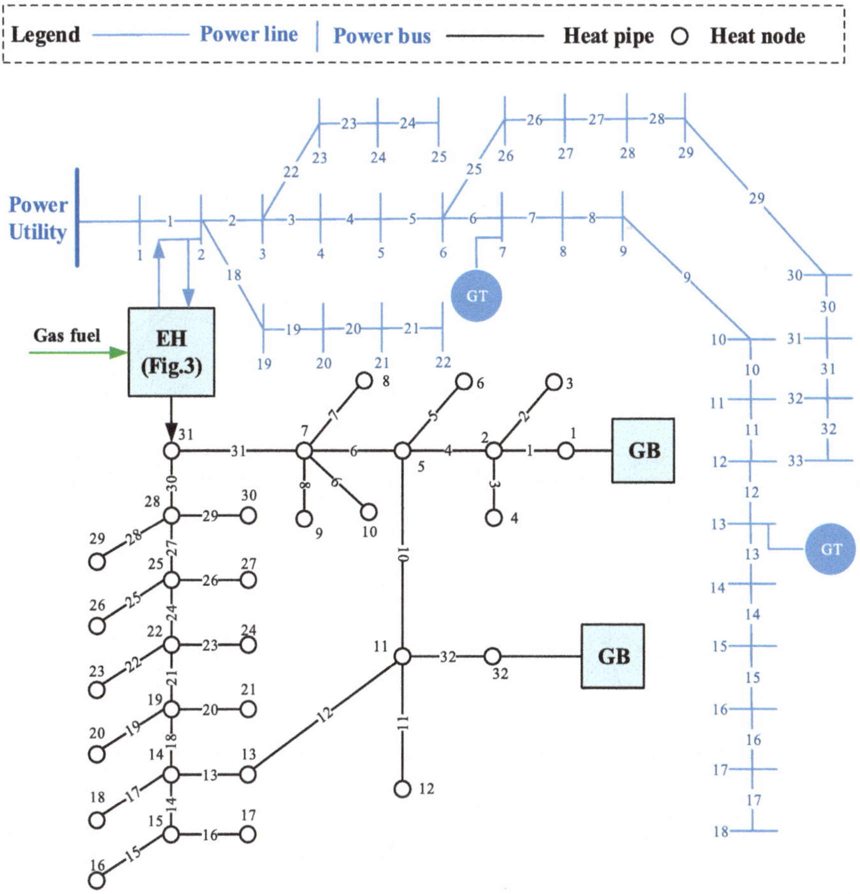

Fig. 4.34 Topology of the testing system

Table 4.9 Generator and heat source data

GT No.	p^g(MW)	q^g(MVar)	a($/MW^2)	b($/MW)
GT 1	[0, 1.5]	[0, 0.5]	0.12	20.0
GT 2	[0, 2.0]	[0, 1.0]	0.09	15.0
GB No.	h^g(MW)	Location	α($/MW^2)	β($/MW)
GB 1	[0, 1.0]	Node 1	0.15	20.0
GB 2	[0, 1.0]	Node 32	0.16	18.0

Table 4.10 Bounds of energy hub decision variables

Variable	Interval	Variable	Interval
p^{ch}	[0, 3.0] MW	E	[0, 10] MWh
p^{dis}	[0, 2.0] MW	H	[0, 10] MWh
h^{ch}	[0, 2.0] MW	p_m^{gb}	[0, 2.0] MW
h^{dis}	[0, 1.5] MW	p_m^{db}	[0, 1.5] MW
p^{gas}	[0, 1.5] MW	h_m^b	[0, 1.5] MW
p_0^g	[0, 3.0] MW		

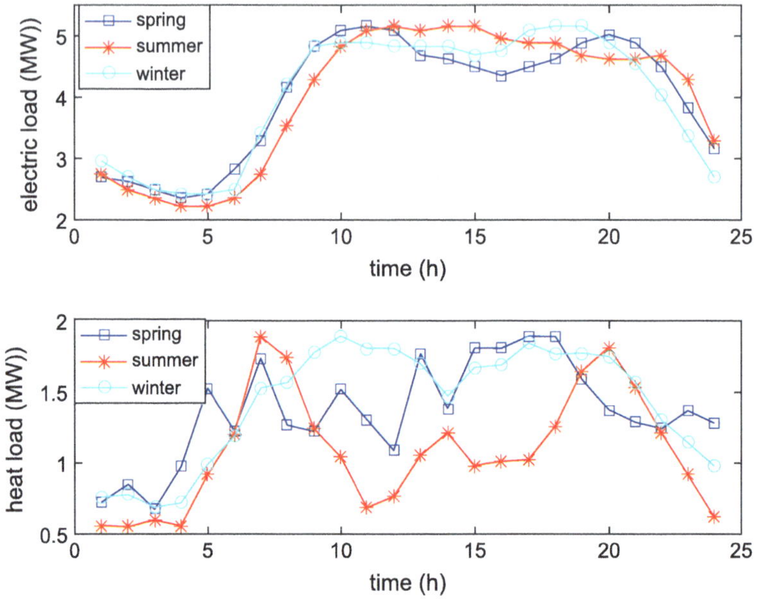

Fig. 4.35 Heat and electricity load profiles

respectively. The electricity price bid (χ^b) should be no less than the retail price at Bus 2 in period t, and the offering price of electricity (ξ^b) should be no greater than the highest daily price multiplies 1.25. The charging and discharging efficiencies of ESU ($\eta_+^{esu}/\eta -^{esu}$) are equal to 98%, the charging and discharging efficiencies of TSU ($\eta_+^{tsu}/\eta -^{tsu}$) are set as 98%, too. The COP of heat pump (η^{hp}) is 3. For the CHP unit, gas to electricity and heat conversion rates ($\eta_e^{chp}/\eta_h^{chp}$) are 0.35 and 0.65, respectively. All the simulations are programmed with YALMIP; MILP models are solved by CPLEX on a laptop with Intel i5-4210M CPU and 16 GB RAM.

We consider following situations in our tests.

1. Different load profiles. We consider heat and electricity demands in spring/fall, winter and summer as shown in Fig. 4.35. The electricity demand in summer is the highest during daytime due to the use of air-conditioners. The heat consumption is the lowest in summer, and highest in winter.

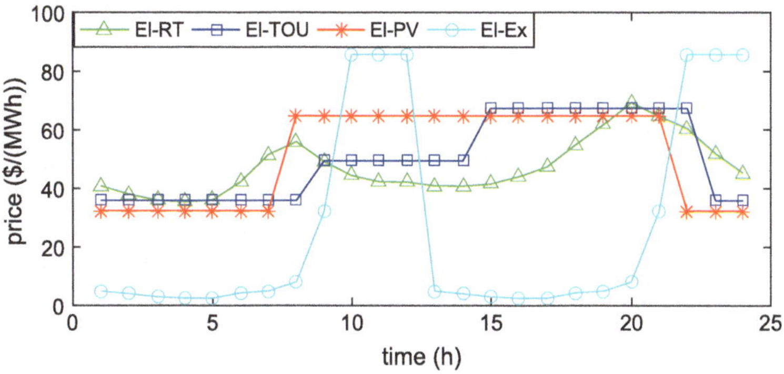

Fig. 4.36 Retail electricity price profiles at Bus 2

2. Different market prices. The electricity retail price at Bus 2 (where the energy hub is connected to) is time-varying. This retail price curve is offered by the power market operator, depending on the pricing policy in PDN, and is independent of the dispatch of energy hub and local generators. We investigate four price curves, the real-time (El-RT) price, the time-of-use (El-TOU) price, the peak-valley (El-PV) price, and an extreme case (El-Ex) price to simulate the different electricity price sequences at Bus 2, as shown in Fig. 4.36.
3. Different gas fuel price. We assume that the price of natural gas fuel is determined by the contract between the energy hub and an external gas company, nevertheless, the values can be either constant or time-varying. We investigate three difference scenarios on gas prices γ: In the benchmark (Gas-BEN) case, $\gamma = 26$\$/MWh and remains unchanged; In the gas extreme (Gas-Ex) case, $\gamma = 40$\$/MWh and keeps constant throughout the day; In the peal-valley scenario (Gas-PV), $\gamma = 30$\$/MWh in periods 7–18 and $\gamma = 20$\$/MWh in the remaining periods.
4. Storage efficiency. Efficiencies of storage units significantly impact the operation of energy hub. For the electricity storage, the round-trip efficiency differs a lot depending on the specific technology. For instance, compressed-air energy storage is around 40–60% [35], pumped storage is approximately 75–85%, and battery storage can reach above 90%. For the TSU, the round-trip efficiency parameters are usually above 98% [36]. In our tests, we decrease the charging and discharging efficiencies ($\eta_+^{esu}/\eta_-^{esu}$) of electricity storage from 98 to 60% (corresponding to decrease the round-trip efficiency from 96 to 36%), and keep the thermal storage efficiency ($\eta_+^{tsu}/\eta_-^{tsu}$) as a constant of 98%.
5. Market power. The energy hub possesses market power and its bidding strategies could influence the clearance of the electricity and heat markets. In normal condition, if the offering price is low, the markets would buy more energies from the hub; otherwise, the hub would gradually lose market share because the system operator could dispatch more local generators or heat sources. Sometimes, due to network congestion or other security considerations, the system operator has

Table 4.11 Correspondence between scenarios and load/price curves

Scenario	Load profile		Price curve	
	Electricity	Heat	Electricity	Natural gas
BEN	Winter	Winter	El-RT	Gas-BEN
El-TOU	Winter	Winter	El-TOU	Gas-BEN
El-PV	Winter	Winter	El-PV	Gas-BEN
El-Ex	Winter	Winter	El-Ex	Gas-BEN
Spring	Spring	Spring	El-RT	Gas-BEN
Summer	Summer	Summer	El-RT	Gas-BEN
Gas-Ex	Winter	Winter	El-RT	Gas-Ex
Gas-PV	Winter	Winter	El-RT	Gas-PV

Table 4.12 Economic performance and computational time for each case

Scenario	Cost ($)		Revenue ($)		Profit ($)	Time (s)
	PDN	Gas	PDN	DHN		
BEN	278.28	598.57	951.66	394.68	469.48	231.2
El-TOU	331.07	489.72	993.02	397.77	569.99	62.03
El-PV	303.23	774.96	1365.0	396.08	682.88	88.97
El-Ex	152.30	519.09	1178.4	396.36	903.39	85.61
Spring	324.41	680.61	1106.4	378.50	479.91	197.5
Summer	301.03	538.06	947.67	309.94	418.53	28.83
Gas-Ex	381.41	134.54	426.12	393.79	303.95	42.77
Gas-PV	309.14	360.00	831.99	393.79	556.64	103.9

no choice but to buy energy from the energy hub. To limit its market power, we consider case MP-RtCap, in which the capacity of the feeder in PDN is changed from 3 MW to 6 MW; case MP-TBPos, in which GT1 and GT2 are moved to Bus 6 and Bus 13, respectively, and their capacities are increased from 1 MW to 2 MW; case MP-GasLim, in which the maximal gas input of energy hub decreases from 1.5 MW to 1 MW.

The load and price curves used in each scenario are summarized in Table 4.11. Other cases are specified following their first appearance.

2. Results

Economic performances in each scenario are summarized in Table 4.12. The computation time is typically a few minutes. Furthermore, we superimpose forecast errors which follow Normal distribution on the load curves shown in Fig. 4.35, and randomly generate 50 demand curves in each scenario. The average computation time over 50 cases in each scenario is listed in the last column of Table 4.12. Because these scenarios are mainly categorized according to the energy price sequence, we conjecture that the price data have more significant impact on the computational efficiency. In the following, offering price and quantity refer to the energy sold by the energy hub to the electricity and heating markets; bidding price and quantity refer to the energy brought by the energy hub from the electricity market.

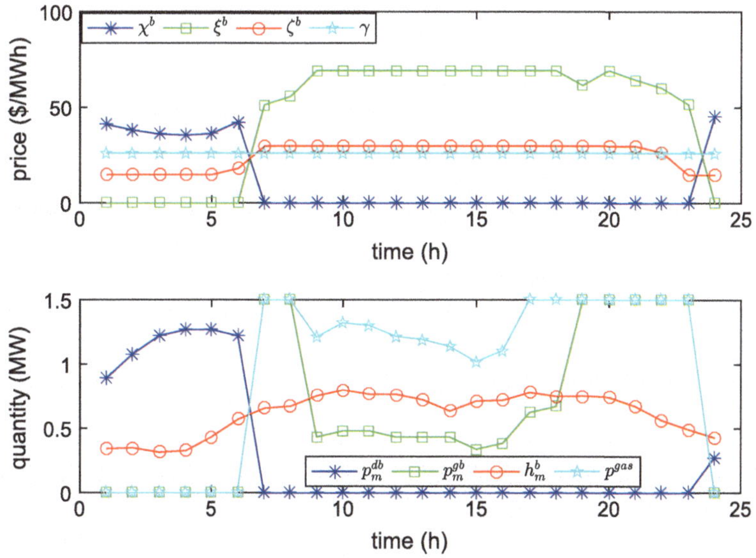

Fig. 4.37 Price and quantity bid/offer in case BEN

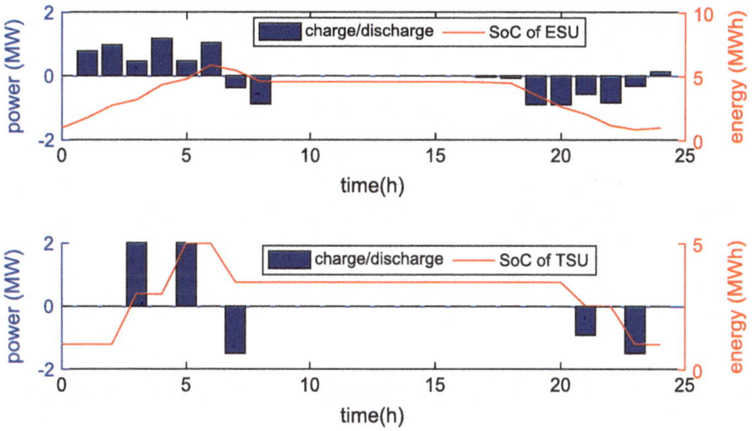

Fig. 4.38 SoCs of the ESU and TSU in the energy hub

1. The benchmark case
   The offering/bidding price and quantity curves are plotted in Fig. 4.37. Scheduling of storage units and their charging dynamics are drawn in Fig. 4.38. The energy hub purchases electricity with a lower price from the PDN in periods 1–6. A fraction of the purchased electricity is stored in the ESU for potential arbitrage during peak hours, e.g., periods 7–8, 19–23. The peak demand of DHN is about 2 MW, gas boilers cannot meet the demands in peak hours of the day. The energy hub constantly maintains a certain level of thermal energy output. In

periods 1–5, as the electricity is cheap, no gas fuel is purchased, and the thermal energy is produced by heat pump. In periods 3 and 5, more heat is produced from the heat pump and used to charge the TSU for future usage.

From period 6, the real-time price begins to rise, and the energy hub switches to consuming gas and producing heat and electricity from the CHP unit. Since the heat pump has a higher efficiency than the CHP unit, the heat offering price ζ^b in periods 1–5 is smaller than those in the remaining periods of the day during which the heat is produced by the CHP unit. Electricity quantity bids and offers (p_m^{gb}, p_m^{db}) and heat quantity offers (h_m^b) are equal to the cleared values. Through the cross-arbitrage among electricity market, gas market, and heating market, the energy hub gains a total profit of \$469.48.

2. Impact of electricity and gas prices
Prices of electricity and gas have pivotal influences on the actions of energy hub. Electricity price bids (χ^b) and offers (ξ^b), and heat price bids (ζ^b) in El-TOU, El-PV, and El-Ex scenarios are shown in Figs. 4.39 and 4.40, respectively. With the given price curve in scenario El-Ex, the energy hub submits the lowest electricity purchasing price (χ^b) during periods 1–6 and highest electricity offering price (ξ^b) during periods 8–24, and gains the highest arbitrage revenue of \$1178.4 and the highest profit of \$903.39. The electricity pricing mechanisms have little impacts on heat offering prices ζ^b as indicated in Fig. 4.40. This is because during periods 7–21, the heat demand is high, so the energy hub possesses strong market power, and the heat offering price quickly reaches the upper bound.

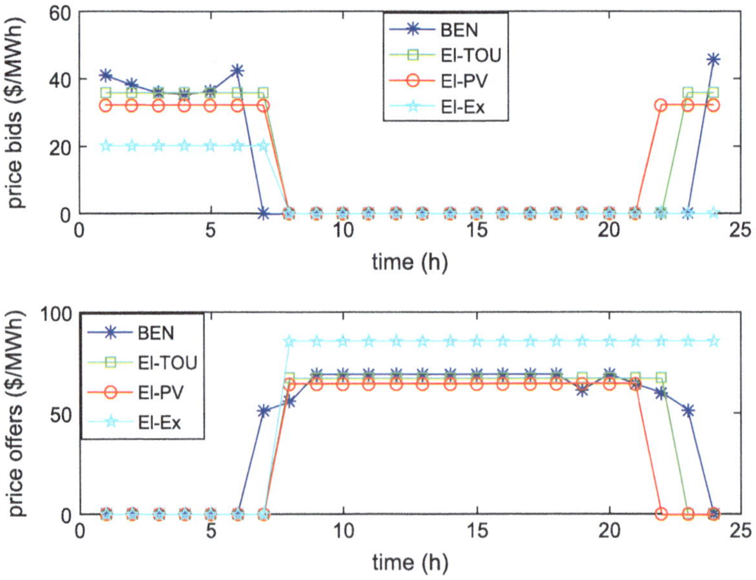

Fig. 4.39 Bidding and offering electricity prices in different retail price scenarios

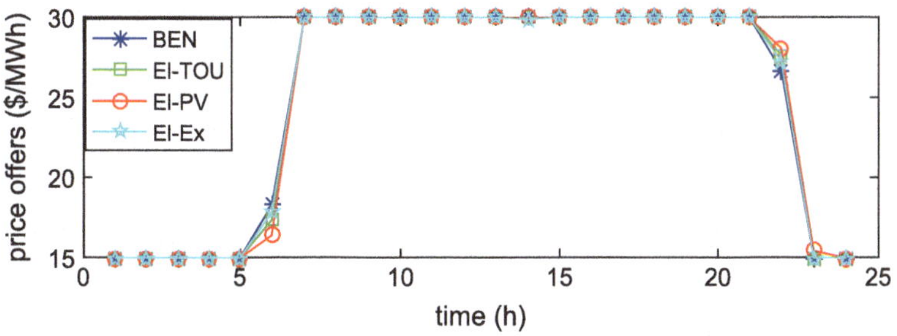

Fig. 4.40 Heat offering price in different retail price scenarios

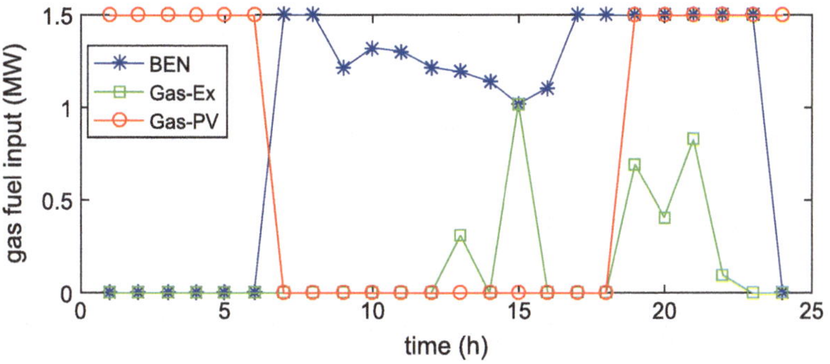

Fig. 4.41 Purchased gas fuel versus gas price curves

The amounts of purchased gas fuel in the BEN scenario, the Gas-PV scenario, and the Gas-Ex scenario are compared in Fig. 4.41. Once the gas price is increased from 26$/MWh to 40$/MWh in case Gas-EX, we can see that the energy hub alters its strategy to purchasing electricity from the PDN instead of purchasing gas in peak-hours as in case BEN. This strategy uses heat pump to produce adequate thermal energy which is reserved in TSU so as to meet heat demands without using expensive natural gas. Consequently, the PDN revenue and the profit of energy hub in the Gas-Ex scenario are the lowest in Table 4.12. With the deepened integration of energy systems, real-time gas market may appear in the future, from which the energy hub can benefit by making full use of the cheap gas during off-peak hours. It is also observed in Fig. 4.41 that the energy hub purchases more natural gas in Gas-PV case than it does in Gas-Ex case as the gas price is lower, and receives more revenue from the electricity market since the production cost of the CHP unit declines. Compared with the BEN scenario, the gas purchasing cost is lower in Gas-PV case because gas is bought during off-peak hours. As a result, the total revenue in Gas-PV case is the highest. Certainly, this conclusion is not universal and depends on actual price data.

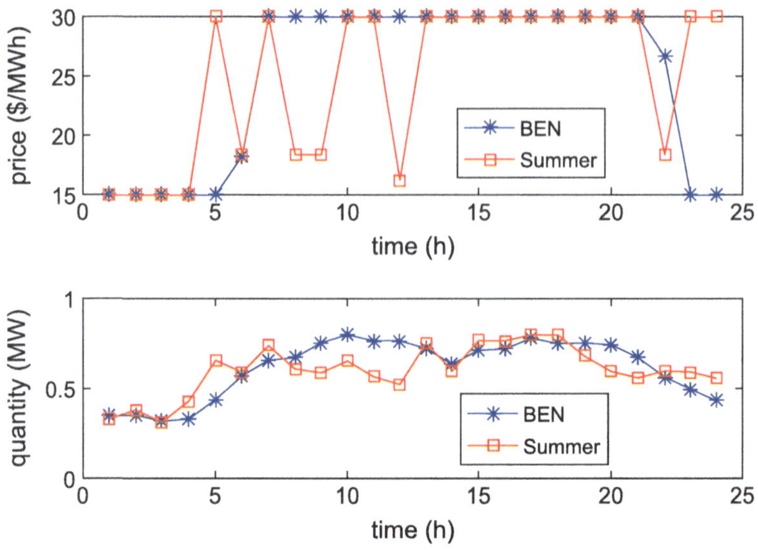

Fig. 4.42 Heat offering price and quantity in BEN and summer

3. Impact of load shape
   The offering prices and quantities of heat in the BEN case (the Winter scenario) and the Summer scenario are compared in Fig. 4.42. Since the total heat demand in Summer is lower than that in Winter, the revenue of selling heat to DHN, as well as the total profit, is smaller than that in the BEN case.
4. Impact of storage efficiency
   In this set of tests, TOU electricity price curve is used. The thermal storage efficiency is equal to 98%, and the electricity storage parameter varies. Results are listed in Table 4.13. It is observed that when $\eta_+^{esu} > 75\%$, the efficiency significantly impacts the total revenue; further decrease of $\eta_+^{esu}/\eta_-^{esu}$ does not have much influence on the total profit, because the energy hub scarcely arbitrages electricity, and the revenue from the PDN mainly comes from selling electricity generated by the CHP unit which burns gas. Purchased electricity from the PDN is stored or converted into heat for supplying demands. From the last three rows of Table 4.13, we can see that more electricity charging cost is necessitated to support the total arbitrage since the electricity charging efficiency is lower. Nevertheless, the minimum profit can be guaranteed by consuming natural gas, demonstrating the advantage of multi-carrier energy integration.
5. Market power mitigation
   Results in the three cases defined previously are illustrated in Table 4.14, Figs. 4.43, 4.44 and 4.45. Through limiting the maximal gas delivery rate in case MP-GasLim, the energy hub consumes less gas during day-time than it does in case BEN (as shown in Fig. 4.43) and imports more electricity in period 24 (as shown in Fig. 4.44). Since the electricity offering price is high, which can be seen

Table 4.13 Profit of energy hub under different ESU efficiency parameters ($)

$\eta_{+}^{esu}/\eta_{-}^{esu}$	Cost		Revenue		Profit
	PDN	Gas	PDN	DHN	
98%	331.07	489.72	993.02	397.77	569.99
95%	330.55	528.97	993.02	397.77	531.27
90%	310.78	594.93	993.02	395.98	483.29
85%	313.13	605.99	946.48	394.68	425.72
80%	315.78	493.53	787.64	397.66	375.98
75%	318.78	493.53	749.08	398.38	335.15
70%	63.41	493.53	474.94	398.39	316.39
65%	67.18	493.53	474.94	398.39	312.62
60%	72.98	493.53	474.94	398.39	306.82

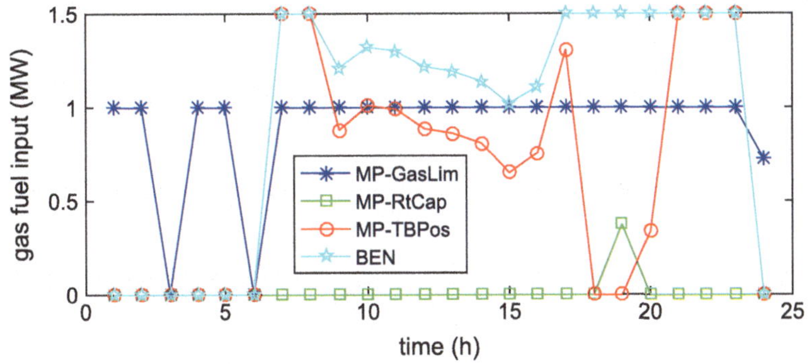

Fig. 4.43 Gas fuel input in scenarios designated for market power mitigation

from Fig. 4.45, the change in the total revenue is tiny ($469.48 v.s. $466.44). The revenue in the heating market remains the same compared to that in case BEN.

In case MP-TBPos, because GT1 in PDN has more capacity to support peak electricity load, the electricity quantity ($p_{t,m}^{gb}$) offered by energy hub during periods 9–17 is lower than that in case BEN, as shown in Fig. 4.44. Since the sold electricity decreases compared to case MP-GasLim, the gas fuel import reduces during peak hours, which can be seen in Fig. 4.43. Meanwhile, more electricity is purchased during off-peak hours (1–6), as indicated in Fig. 4.44, to compensate the decrease of gas contract. Furthermore, since GTs owned by PDN offer more electricity in case MP-TBPos, the electricity offering price (ξ^{b}) in period 18 is lower than that in case BEN. Nevertheless, the amount of electricity sold to the market in this period is still lower than case BEN.

Another way to limit the market power of energy hub is to increase the capacity of the feeder (distribution line connecting to the slack bus). In case MP-RtCap, because more cheaper electricity can be delivered from the slack bus, the energy hub loses certain market share and purchases very little gas as indicated in Fig. 4.43. To maintain total revenue, the energy hub sells electricity with a lower price than other cases in periods 9, 10, 17, and 18, as indicated in Fig. 4.45. Since little gas is

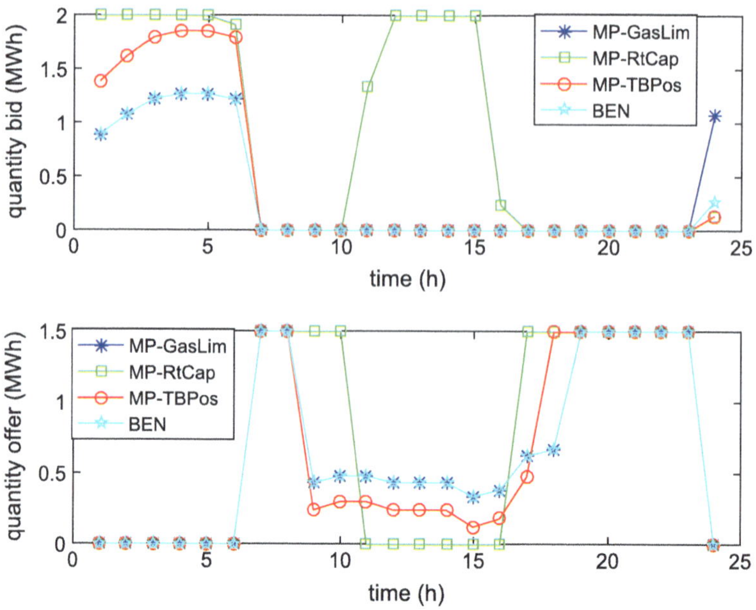

Fig. 4.44 Electricity quantity bid and offer in scenarios designated for market power mitigation

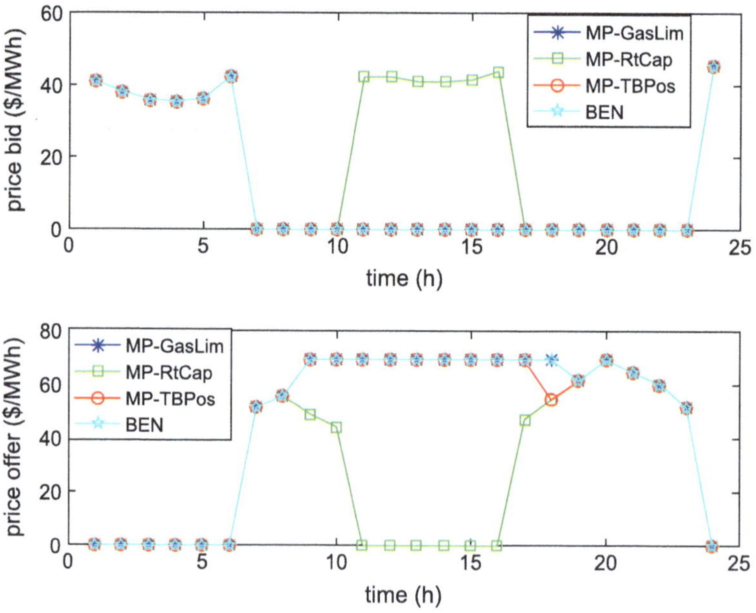

Fig. 4.45 Electricity price bid and offer in scenarios designated for market power mitigation

Table 4.14 Economic performance in scenarios designated for market power mitigation ($)

Scenario	Cost		Revenue		Profit
	PDN	Gas	PDN	DHN	
MP-GasLim	314.96	564.95	951.66	394.68	466.44
MP-TBPos	401.61	416.41	872.68	396.37	451.02
MP-RtCap	863.39	9.77	916.74	393.79	439.94

purchased in this case, heat can only be generated through heat pump by consuming electricity. As a result, more electricity is purchased during periods 1–7 and 11–16 to supply heat demand. Electricity arbitrage is the main source of profit in this case, which can be observed from the price and quantity curves in Figs. 4.44 and 4.45.

4.6 Capacity Planning of Energy Hubs

The integration of power system, heating system and gas system through an energy hub with dual inputs (electricity and natural gas) and dual outputs (electricity and heat) is the most prevalent structure of a multi-carrier energy system at the distribution side. In such infrastructures, energy hub is the interface among three energy systems, which possesses functionalities including energy production, conversion, and storage. The optimal scheduling and management of energy hubs have been well studied in the existing literature under various settings. The capacities of energy production facilities, conversion devices, and storage units in the hub are given parameters in operation related problems, which largely determine how tightly the individual systems are coupled with each other and how flexibly the whole system would behave. The energy hub capacity planning problem is another pivotal problem and has attracted lots of research attention. Nevertheless, existing work focuses on residential-level energy hubs, which directly supply energy to household appliances and have little impact on the distribution systems due to their small sizes.

This section discusses the capacity planning of a distribution-level energy hub whose operation has notable impact on energy flows in the networked energy systems, and proposes a data-driven robust stochastic optimization (DR-SO) model to address the problem with distributional robustness guarantee. Renewable generation and load uncertainties are modeled by sets of ambiguous probability distributions near a reference distribution in the sense of Kullback-Leibler (KL) divergence measure. The objective is to minimize the sum of the construction cost and the expected life-cycle operating cost under the worst-case distribution restricted in an ambiguity set. Network flow is captured in operating constraints in normal conditions; demand supply reliability in extreme conditions is taken into account via robust chance constraints. Through duality theory and sampling average approximation, the proposed model is cast as an equivalent convex program with a nonlinear objective and linear constraints, and is solved by an outer approximation algorithm which entails solving LPs.

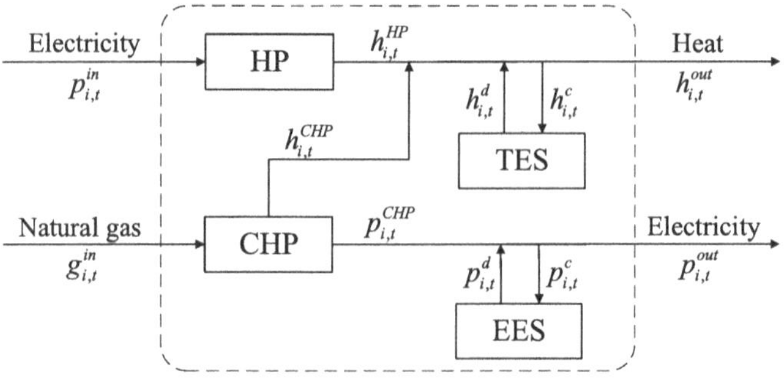

Fig. 4.46 Structure of the energy hub

4.6.1 Deterministic Formulation

1. Hub and System Models

Topology of the energy hub considered in this section is shown in Fig. 4.46. The inputs are electricity and natural gas; the outputs are electricity and heat. Different from residential energy hubs whose output connects to end users, both sides of the hub in Fig. 4.46 connect to energy systems. The energy hub consists of CHP units, heat pumps, as well as ESU and TSU. The operating constraints and energy flows inside the hub can be described as follows:

$$p_{i,t}^{out} = p_{i,t}^{CHP} + p_{i,t}^{d} - p_{i,t}^{c} \tag{4.62a}$$

$$h_{i,t}^{out} = h_{i,t}^{CHP} + h_{i,t}^{HP} + h_{i,t}^{d} - h_{i,t}^{c} \tag{4.62b}$$

$$g_{i,t}^{in} = p_{i,t}^{CHP}/\eta_i^{gp} + h_{i,t}^{CHP}/\eta_i^{gh} \tag{4.62c}$$

$$h_{i,t}^{HP} = \text{COP}_i \cdot p_{i,t}^{in} \tag{4.62d}$$

$$W_{i,t+1}^{E} = W_{i,t}^{E}(1-\mu_i^E) + (p_{i,t}^{c}\eta_{i,c}^{E} - p_{i,t}^{d}/\eta_{i,d}^{E})\Delta t \tag{4.62e}$$

$$W_{i,t+1}^{H} = W_{i,t}^{H}(1-\mu_i^H) + (h_{i,t}^{c}\eta_{i,c}^{H} - h_{i,t}^{d}/\eta_{i,d}^{H})\Delta t \tag{4.62f}$$

where equalities (4.62a) and (4.62b) describe the electric and thermal power balancing conditions; Eqs. (4.62c) and (4.62d) stipulate the input-output relations of CHP unit and heat pump. Constraints (4.62e) and (4.62f) are charging dynamics of ESU and TSU. Complementarity of charging and discharging actions is naturally met under mild conditions owing to dissipativity and thus relaxed in (4.62), because simultaneous charging and discharging is not an optimal strategy. Detailed analysis can be found in [37, 38]. Nevertheless, strict complementarity can be imposed in model (4.62) via binary variables without jeopardizing the solution method discussed later. The connection topology in the energy hub is fixed. The capacity of each component, which restricts the feasible interval of operating variables, is to be optimized.

The PDN is radial in topology. The power flow in PDN can be established recursively from the linearized BFM explained in Sect. 2.2.2 as follows:

$$p_{j,t} + P_{ij,t} = \sum\nolimits_{k \in c(j)} P_{jk,t} + p^d_{j,t} \tag{4.63a}$$

$$q_{j,t} + Q_{ij,t} = \sum\nolimits_{k \in c(j)} Q_{jk,t} + q^d_{j,t} \tag{4.63b}$$

$$V_{j,t} = V_{i,t} - (r_{ij} P_{ij,t} + x_{ij} Q_{ij,t})/V_0 \tag{4.63c}$$

where Eqs. (4.63a) and (4.63b) represent active and reactive power balancing conditions, $p_{j,t}$ denote the total active power injection at bus j, including the active output of local generators $p^{TU}_{j,t}$, wind farms $p^{w}_{j,t}$, and the energy hub p^{out}_t. Equality (4.63c) stands for the forward voltage drop along a distribution line. Model (4.63) neglects network losses, however, reactive power and bus voltage are considered, so it is more appropriate for distribution systems than the DC power flow model which neglects reactive power and assumes constant bus voltage magnitudes. In these regards, the accuracy of the linearized BFM model is satisfactory in a planning oriented problem.

A DHN consists of symmetric supply and return pipelines. At each source (load) node, heat is injected into (withdrawn from) the network via a heat exchanger between the supply side and the return side. The physical DHN model is subject to hydraulic conditions and thermal conditions, as elaborated in Sect. 4.2. In this section, the constant-flow variable temperature operating mode is adopted in which mass flow rates are set to constant values, and the thermal condition can be cast as a compact form

$$A_H h + B_H(\dot{m})\tau = b_H \tag{4.64}$$

where vector h collects the thermal energy output of heat sources; vectors τ and $\dot{m}$ denotes nodal temperatures and pipeline mass flow rates, respectively; other matrices and vectors are coefficients corresponding to the constants in model (4.4a)–(4.4f) in Sect. 4.2.2.

Gas pipeline network is governed by partial differential equations, which has been discussed in Sect. 3.2.4. The steady-state solution renders nonlinear and non-convex algebraic equations. In this section, we don't consider detailed gas network model. Instead, we impose an upper bound on the maximum gas inflow to mimic the capacity of gas supply pipes connecting to the energy hub. This parameter could be time-varying (provided by gas system operator) according to the real-time operating conditions of the gas network. This simplification decomposes hard gas system operation constraints from the planning model, and is acceptable for an investment problem over a long time horizon.

2. Compact Form of the Planning Problem

In the deterministic formulation, renewable output and load demands are known exactly. In the first stage, the capacities of energy hub components (CHP unit, heat pump, ESU and TSU) are determined; In the second stage, operating constraints in three typical days (sampled from spring/autumn, summer, and winter) are considered. The objective function is to minimize the sum of construction cost and life-cycle operation cost. We make the following assumptions in the planning problem:

(a) Energy hub planning is guided by a government agency. The objective is to minimize the investment cost of the hub and the total operation cost of the integrated heat-electricity energy system during the period of 10 years. Because we consider a distribution-level energy hub, whose operation cost is comparable to that of the PDN, the two costs can be added together for minimization.
(b) The connection topology in the energy hub is fixed. The candidate components for investment include CHP unit, heat pump, ESU, and TSU. Electric boilers and different types of battery arrays can be easily included. Nevertheless, for the ease of exposition, we just select one typical facility for each functionality when establishing the model. The operation cost consists of fuel expenditures of local generators in the PDN, the CHP unit in the energy hub, and electricity purchase cost from main grid.
(c) The cost of heat pump is neglected for two reasons: First, it consumes electricity whose cost has already been counted; Second, the cost of heat pump, if exists, is paid by the energy hub to the PDN, i.e., a domestic financial issue in the integrated system, so does not appear in the objective function. If there exist non-electrical heat sources in the DHN, such as coal-fired and gas-fired boilers, their fuel costs should be counted in the objective function.

Conceptually, the deterministic energy hub planning problem can be written as follows:

$$\begin{aligned} \min \quad & f_C + N_d \cdot f_O \\ \text{s.t.} \quad & \text{Cons-PF, Cons-TF} \\ & \text{Cons-EH, Cons-BD} \end{aligned} \tag{4.65}$$

where Cons-PF stands for the linearized BFM in (4.63); Cons-TF is the abbreviation for thermal flow constraints in (4.64); Cons-EH encapsulates energy hub operation conditions in (4.62); Cons-BD collects all lower and upper bound constraints of decision variables. The construction cost is given by

$$f_C = I^{\text{CHP}} C^{\text{CHP}} + I^{\text{HP}} C^{\text{HP}} + I^E C^E + I^T C^T \tag{4.66}$$

which consists of investment costs of CHP unit, heat pumps, ESU and TSU in the energy hub; N_d is the number of services days, which is equal to 3650; The daily operation cost function is defined as

$$fo = \sum_t \left(\sum_i \omega_g g_{i,t}^{\text{CHP}} + \sum_j F(p_{j,t}^{\text{TU}}) + \sum_k \omega_t^e P_t \right) \tag{4.67}$$

where ω_g and ω_t^e denote prices of natural gas and electricity (delivered from the transmission network); $g_{i,t}^{\text{CHP}}$ and P_t represent gas fuel consumed by the CHP unit in the energy hub and the electric power delivered through the distribution line connected to the slack bus. The daily operation cost fo consists of fuel expenditures of CHP units and local generators in PDN represented by the first and the second terms, as well as the expense paid to the main grid in the third term. The convex quadratic function $F(p_j^{\text{TU}})$ can be approximated by a PWL function using well-known methods, such as that in Appendix B.1.1, so we can assume that the objective function is linear without loss of generality. For notation conciseness, one typical day is selected to explain the model; in the implementation, we incorporate three typical days in spring/autumn, summer, and winter with weight coefficients 0.5, 0.25, and 0.25 to calculate the daily operation cost fo.

The bound of energy hub operating variables depends on the capacity planning strategies. Their relations are

$$\left(p_{i,t}^{\text{CHP}} / \eta_i^{gp} + h_{i,t}^{\text{CHP}} / \eta_i^{gh} \right) \leq C^{\text{CHP}} \tag{4.68a}$$

$$p_{i,t}^{\text{CHP}} \geq k_e^l C^{\text{CHP}}, \ h_{i,t}^{\text{CHP}} \leq k_h^m C^{\text{CHP}} \tag{4.68b}$$

$$h_{i,t}^{\text{HP}} \leq C^{\text{HP}} \tag{4.68c}$$

$$W_{i,t}^E \leq C^E, \ 0 \leq p_{i,t}^d \leq k_d^E C^E, \ 0 \leq p_{i,t}^c \leq k_c^E C^E \tag{4.68d}$$

$$W_{i,t}^T \leq C^T, \ 0 \leq h_{i,t}^d \leq k_d^T C^T, \ 0 \leq h_{i,t}^c \leq k_c^T C^T \tag{4.68e}$$

which are included in Cons-BD. Constraints (4.68a)–(4.68b) give the polyhedral operating region of CHP unit, as shown in Fig. 4.47, which consists of three characteristic facets. The left-hand side of (4.68a) interprets the gas fuel input limited by the capacity of CHP, so (4.68a) corresponds to the maximum fuel facet in Fig. 4.47. The thermal output is restricted by a certain value in (4.68b), which is proportional to the capacity and quantified by constant k_h^m; the minimum electric output is also proportional to the capacity and quantified by coefficient k_e^l. Equation (4.68b) corresponds to the maximum heat output and minimum power output facets in Fig. 4.47. (4.68c) is the heat pump capacity constraint. Equation (4.68d) depicts maximum charging/discharging rates (which are proportional to the storage capacity characterized by constants k_d^E and k_c^E) and state-of-charge limits of ESU. We can add an absolute upper bound on the charging/discharging power to mimic

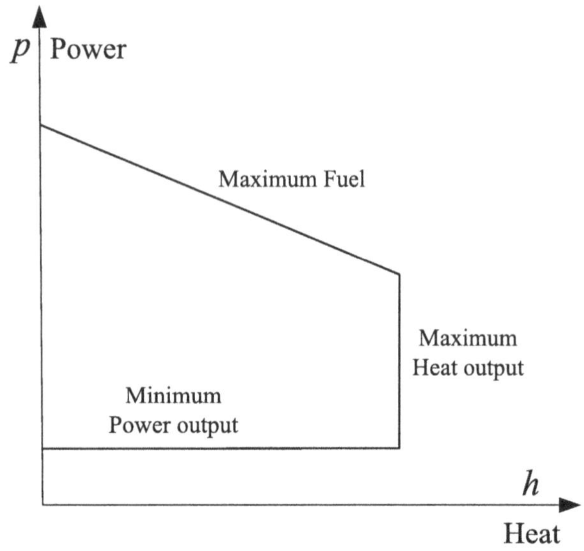

Fig. 4.47 Approximated operating region of the CHP unit

the capacity of the transformer in energy hub. Equation (4.68e) has a similar interpretation for TSU.

Let vector x represent first-stage decision variables, i.e., facility capacities C^{CHP}, C^{HP}, C^{E}, and C^{T}; vector ξ denotes the uncertain parameters including the output of wind farms $p^{w}_{j,t}$ and system loads $p^{d}_{j,t}$ and $h^{L}_{i,t}$; vector y contains second-stage decision variables, including those in the power flow model (4.63) and the thermal flow model (4.64). With notations defined above, the deterministic model for energy hub planning is an LP and can be expressed via a compact matrix form

$$\begin{aligned} \min \quad & c^T x + Q(x, \xi) \\ \text{s.t.}\ & x \in X \end{aligned} \tag{4.69}$$

where $X = \{x | 0 \le x \le x^u\}$ is the feasible set of first-stage decisions; the first term $c^T x$ in the objective function corresponds to the construction cost f_C in (4.66); the second term $Q(x, \xi)$ is the operating cost associated with parameter ξ under given x, which can be expressed via

$$\begin{aligned} Q(x, \xi) = \min_{y} \quad & p^T y \\ \text{s.t.} \quad & Ax + By + C\xi \le d \end{aligned} \tag{4.70}$$

where the objective function corresponds to the operation cost in (4.67). Constraints include Cons-PF, Cons-TF, Cons-EH, and Cons-BD defined in (4.65), except for those in X. The deterministic formulation (4.69) is an LP and can be easily solved.

To explain why the planning problem is posed in a two-stage form, let us see how the uncertainty is dealt with in some traditional methods.

3. Traditional SO and RO Models

To actively consider the uncertain nature of parameter $\boldsymbol{\xi}$, SO and RO models can be set up based on the compact form (4.69). Suppose we have a set of representative samples $\boldsymbol{\xi}^1, \boldsymbol{\xi}^2, \cdots, \boldsymbol{\xi}^n$. For the scenario based SO, if we know the probabilities π_1, $\pi_2, \cdots, \pi_n$ with respect to each sample, then an SO model can be set up as

$$\begin{aligned} \min \;& c^T x + \sum_n \pi_n p^T y^n \\ \text{s.t.}\;& x \in X,\; Ax + By^n + C\xi^n \le d,\; \forall n \end{aligned} \tag{4.71}$$

Otherwise, if no information on the probabilities is available, we can resort to the following RO model

$$\begin{aligned} \min_x \;& c^T x + \max_{\xi^n} \left\{ \min_{y^n} p^T y^n \right\} \\ \text{s.t.}\;& x \in X,\; Ax + By^n + C\xi^n \le d,\; \forall n \end{aligned} \tag{4.72}$$

In RO, parameter ξ is usually assumed to reside in a predetermined uncertainty set. According to the discussion in Appendix C.2.1, the equivalent uncertainty set in formulation (4.72) is the convex hull of sampled scenarios $\boldsymbol{\xi}^1, \cdots, \boldsymbol{\xi}^n$.

From (4.71) and (4.72) we can observe:

(a) The second-stage decision $\boldsymbol{y}^n$ depends on the value $\boldsymbol{\xi}^n$ of uncertain data, while the first-stage decision $\boldsymbol{x}$ does not. This refers to the fact that once the planning strategies are deployed, they cannot change any more; the operating strategies are made when the actual renewable generation and demands are known exactly, at least when they are much more clear than the planning stage.

(b) SO and RO models share the same constraints; they are different in the objective function: the former incorporates an expected cost in the second stage, and the latter considers the worst-case outcome. Clearly, SO requires more information on the uncertainty. In practice, we may not have exact values on the probability distributions. But if we use RO model (4.72), we naturally abandon all the distribution information, and may get over-conservative planning strategies.

SO and RO have their own advantages and drawbacks. For the former one, except for the hard-to-obtain probability distributions, the optimal solution to an SO model could have poor statistical performances if the actual distribution is not identical to the designated one. For the latter one, which is distribution free, as the worst-case scenario rarely happens in reality, the robust strategy could be conservative thus suboptimal in most cases. The above dilemma calls for new methodology that takes full advantage of the above available data, because we can infer useful information from the finite data, such as an empirical distribution, and estimate how far the true one is distant to the reference one.

4.6.2 Data-Driven Robust Stochastic Model

Data-driven robust stochastic optimization (DR-SO) inherits the advantages from both SO and RO approaches. It minimizes the expected cost under the worst-case probability distribution within a family of candidate distributions. The optimal strategy is less conservative than the traditional RO method and is insensitive to the perturbation in probability distributions than the traditional SO models. More details about this approach can be found in Appendix C.4. It is used in this section to determine optimal sizes of components in the energy hub under renewable generation and load variabilities. It differs from SO and RO models mainly in the description of uncertainty, which is the starting point of this section.

1. Ambiguity Set

In the DR-SO model, renewable generation and load uncertainties are modeled by ambiguity PDFs which are close to a reference PDF constructed from available data. The distance between two PDFs is quantified by Kullback-Leibler (KL) divergence measure. The ambiguity set for the family of candidate PDFs can be set up through the following procedures.

(a) Determining a reference distribution P_0. The most widely used empirical distribution for a univariate random variable is the histogram. For example, we have totally M samples to fit into N bins, and there are $M_1, M_2, \cdots, M_N$ samples in each bin. Then the representative scenario in each bin is the expectation of ξ resides in there, and the corresponding probability is $\pi_i = M_i/M$, and the discrete density function of P_0 is $\{\pi_1, \cdots, \pi_N\}$. Otherwise, we may assume that ξ follows some certain distribution, say the Gaussian distribution, and calibrate the parameters in the PDF via curve fitting methods. For a multi-dimensional vector random variable, we can use scenario reduction techniques to create representative samples and determine their respective probabilities. Because we will consider a family of PDFs around P_0, the requirement on an accurate distribution can be relaxed.

(b) Constructing the ambiguity set. We consider all possible probability distributions that are close enough to P_0, or more exactly, all elements in the following set

$$W_\phi^{KL} = \{P(\xi) \mid D_\phi^{KL}(f \| f_0) \le d_{KL}(\beta^*),\ f = \mathrm{d}P/\mathrm{d}\xi\} \tag{4.73}$$

where d_{KL} is the threshold which determines the size of the ambiguity set depending on a confidence level of β^*, such that W_ϕ^{KL} contains the real distribution with a probability no less than β^*. For continuous PDFs, the KL-divergence measure is defined by

$$D_\phi^{KL}(f \parallel f_0) = \int_\Omega f(\xi) \log \frac{f(\xi)}{f_0(\xi)} \mathrm{d}\xi \tag{4.74}$$

For discrete PDFs, the KL-divergence has the form of

$$D_{\phi}^{KL}(f \parallel f_0) = \sum_{n} \pi_n' \log \frac{\pi_n'}{\pi_n} \tag{4.75}$$

In either case, there are infinitely many PDFs contained in W_{ϕ}^{KL} defined by (4.73) when $d_{KL} > 0$. Otherwise, when $d_{KL} = 0$, the ambiguity set W_{ϕ}^{KL} becomes a singleton, and the proposed DR-SO formulation degenerates to a traditional SO model. Using KL-divergence in the ambiguity set gives rise to a convex equivalent program, which greatly facilitates solving the planning problem.

(c) Selecting the confidence level d_{KL}. In practice, the decision marker can specify the value of d_{KL} according to the attitude towards risks. Nevertheless, a recommended value of d_{KL} can be obtained from probability theory. Intuitively, the more historical data we possess, the closer the reference PDF f_0 deviates from the true one, and the smaller d_{KL} should be set. Suppose we have the empirical distribution $P_0 = \{\pi_1, \cdots, \pi_N\}$ from totally M samples. The probability associated with each bin in the true distribution is π_i^r, according to the discussions in [39], random variable $2M \sum_{i=1}^{N} \pi_i^r \log(\pi_i^r/\pi_i)$ follows χ^2 distribution with $N-1$ degrees of freedom. Therefore, the confidence threshold can be calculated from

$$d_{KL}(\beta^*) = \frac{1}{2M} \chi^2_{N-1,\beta^*} \tag{4.76}$$

where χ^2_{N-1,β^*} is the β^* upper quantile of a χ^2 distribution with $N-1$ degrees of freedom.

2. DR-SO Model for Energy Hub Planning

Based on the compact form (4.69) and the ambiguity set W_{ϕ}^{KL} defined in (4.73), the DR-SO model for energy hub planning could be written as follows:

$$\min \ c^T x + \sup\nolimits_{P \in W_{\phi}^{KL}} E_P[Q(x, \xi)] \tag{4.77a}$$

$$\text{s.t. } x \in X \tag{4.77b}$$

$$\inf_{P' \in W_{\phi}'} \Pr\{D_{loss}(\xi) \le 0\} \ge 1 - \alpha \tag{4.77c}$$

where $E_P[Q(x, \xi)]$ denotes the expectation of operation cost value function $Q(x, \xi)$ when the uncertain parameter ξ follows a certain distribution P; D_{loss} means the minimum of maximal unserved load among all buses in PDN and nodes in DHN in the extreme conditions, which is defined as

$$D_{loss}(\xi) = \min \ g \tag{4.78a}$$

$$\text{s.t. } A'x + B'y + C'\xi \le d' \tag{4.78b}$$

$$\sum_{k \in c(j)} P_{jk,t} + p^d_{i,t} - P_{ij,t} - p_{j,t} \le g, \ \forall j \tag{4.78c}$$

$$h^L_{i,t} - c_p \dot{m}^L_{i,t}(\tau^S_{i,t} - \tau^R_{i,t}) \le g, \ \forall i \tag{4.78d}$$

where coefficient matrices A', B', C', and d' in (4.78b) correspond to A, B, C, and d excluding the nodal power balancing conditions, which are replaced by electric load shedding condition (4.78c) and thermal load shedding condition (4.78d), where g interprets the amount of unserved load. The objective function (4.78a) is to seek the minimum value of g, such that the load shedding at any bus (node) will not exceed g. In problem (4.78), ξ appears in items $p^d_{j,t}$, $h^L_{i,t}$, and $p^w_{j,t}$ contained in $p_{j,t}$, and g could take negative values. Excessive generation will be curtailed.

Through constraint (4.77c), the probability that all load can be served in the extreme conditions is maintained greater than a certain level; we call it a robust chance constraint because it considers all PDFs in W^{KL}_{ϕ}; the objective function (4.77a) accounts for the expectation $E_P[Q(x, \xi)]$ in the worst-case distribution. In view of these facts, the planning model (4.77) inherits the advantages from both SO and RO: the exact PDF is not needed, and the optimal strategy is insensitive to the perturbation in the PDF of uncertain parameter ξ. Indeed, due to the storage capabilities of ESU and TSU as well as the flexibility enabled by cogeneration and power-to-heat energy conversion, the system is also robust to the change of parameter ξ. This is simply a physical implication.

In problem (4.77), we set up different ambiguity sets in (4.77a) and (4.77c), because the extreme condition is significantly different from the normal condition, so we use P' and W'_{ϕ} in the last reliability constraint to distinguish them from P and W^{KL}_{ϕ} used in the objective function. It should be pointed out that even if the same ambiguity set is used in (4.77a) and (4.77c), the respective worst-case distributions could be different.

4.6.3 Solution Strategy

In formulation (4.77), the maximum expectation in the objective function (4.77a) and the robust chance constraint (4.77c) over ambiguity set W' prevent it from being solved directly. In this section, we derive tractable reformulations.

1. Reformulation of the Robust Chance Constraint

The robust chance constraint (4.77c) is difficult because it involves a lower bound evaluation over an infinite set. It can be imaged that if (4.77c) is satisfied, then the probability evaluated under the reference distribution P_0 must be greater than $1-\alpha$, and the modifier must depend on the divergence measure and the confidence level

α. If the KL-divergence is used to measure the distance between PDFs, it is proven in [40] that constraint (4.77c) is equivalent to a traditional chance constraint

$$\text{Pr}_0\{D_{loss} \le 0\} \ge 1 - \alpha_+ \tag{4.79}$$

evaluated under the reference distribution P_0, and the modifier α_+ is calculated by

$$\alpha_+ = \max\left\{0, 1 - \inf_{z \in (0,1)} \left\{ \frac{\mathrm{e}^{-d_{KL}} z^{1-\alpha} - 1}{z - 1} \right\}\right\} \tag{4.80}$$

In (4.80), the univariate function $(\mathrm{e}^{-d_{KL}} z^{1-\alpha} - 1)/(z - 1)$ is convex in z over the open interval (0, 1) [40], so its minimum can be easily found from the classical golden section search method [41]. It is also revealed that $\alpha^+ < \alpha$, so (4.77c) is more conservative than a tradition chance constraint evaluated at the reference distribution P_0. Readers who are interested in proofs can find more information in Appendix C.4.1. However, traditional chance constraint (4.79) is still non-convex [42]. A practical way is to find a conservative but convex approximation for (4.79). It is obvious that (4.79) is equivalent to

$$E_{P_0}[\mathbb{I}_+(D_{loss})] = \text{Pr}_0\{D_{loss} > 0\} \le \alpha_+ \tag{4.81}$$

where $E_{P_0}(\cdot)$ denotes the expected function associated with the reference distribution P_0; $\mathbb{I}_+(x)$ is an indicator function, i.e.,

$$\mathbb{I}_+(x) = \begin{cases} 1, & \text{if } x > 0 \\ 0, & \text{otherwise} \end{cases}$$

Now, we just need to find a convex function $\psi(x)$ that over estimates $\mathbb{I}_+(x)$ to ensure the approximation

$$E_{P_0}[\mathbb{I}_+(D_{loss})] \le E_{P_0}[\psi(D_{loss})] \le \alpha_+ \tag{4.82}$$

is conservative. If we require:

(a) $\psi(x)$ is nondecreasing.
(b) $\psi(0) = 1$.

then $\psi(x)$ must over estimate $\mathbb{I}_+(x)$ and (4.79) holds true, since (4.82) is met. In this paper, $\psi(D_{loss})$ is selected as follows,

$$\psi(D_{loss}) = \max\{0, D_{loss}/\beta + 1\} \tag{4.83}$$

where $\beta > 0$ is a constant. Clearly, this choice of $\psi(D_{loss})$ is a convex function as pointwise maximum preserves convexity. In the final problem, parameter β will be optimized as such $\psi(D_{loss})$ can provide a good approximation.

Suppose $\boldsymbol{\xi}^1$, $\boldsymbol{\xi}^2$, $\cdots$, $\boldsymbol{\xi}^K$ are the typical elements in the discrete bins or retrieved from scenario reduction with probabilities $\pi_1, \pi_2, \cdots, \pi_K$. Then inequality $E_{P_0}[\psi(D_{loss})] \le \alpha_+$ comes down to

$$
\begin{aligned}
& D_{loss}(\boldsymbol{\xi}^k) + \beta \le \phi_k, \ \phi_k \ge 0, \quad \forall k \\
& \sum_k \pi_k \phi_k \le \beta\alpha_+, \quad \beta > 0
\end{aligned}
\tag{4.84}
$$

where ϕ_k is an auxiliary variable. The value of $D_{loss}(\boldsymbol{\xi}^k)$ is determined from problem (4.78). In the final equivalent formulation, the min operator in (4.78) is no longer needed. Details can be found in the next subsection. Because we have employed α^+ which is more prudent than α, the reliability requirement in extreme conditions can be satisfied even if probabilities π_1, π_2, $\cdots$, π_K are not entirely accurate. If the reference distribution P_0 is calibrated from known distributions via curve fitting methods, we can easily partition the feasible space of $\boldsymbol{\xi}$ into disjoint bins and create discrete PDF from the continuous one. Equation (4.84) is called a sampling average approximation (SAA) for $E_{P_0}[\psi(D_{loss})] \le \alpha_+$.

It is interesting to note that approximation of chance constraint (4.79) has a close relation with risk theory. In fact, (4.79) is equivalent to a VaR constraint on D_{loss}, which can be interpreted by choosing $\psi(x) = \mathbb{I}_+(x)$, which is non-convex. In a similar vein, the selection of $\psi(x)$ in (4.83) leads to a CVaR approximation for chance constraint (4.79), i.e., the polyhedral set (4.84), which is more conservative but appears to be convex. In-depth discussions can be found in [42]. Risk based formulations for (4.79) are set forth in Appendix C.4.1.

2. Reformulation of the Objective Function

For a given planning strategy x, the worst-case expectation problem reads as follows:

$$
\sup_{P \in W_\phi^{KL}} E_P[Q(x, \xi)] \tag{4.85}
$$

According to [43], the dual problem of (4.85) whose decision variable is a PDF thus has infinite dimensions is

$$
\min_{\lambda \ge 0} \lambda \log \left\{ E_{P_0} \left[e^{Q(x,\xi)/\lambda} \right] \right\} + \lambda d_{KL} \tag{4.86}
$$

whose decision variable is a non-negative scalar λ. For discrete distributions, the expectation can be replaced with a weighted summation, and (4.86) evolves

$$
\min_{\lambda \ge 0} \lambda \log \left\{ \sum_n \pi_n \left[e^{Q(x,\xi^n)/\lambda} \right] \right\} + \lambda d_{KL} \tag{4.87}
$$

In Appendix C.4.1, we have demonstrated that function

$$H(\theta, \lambda) = \lambda \log \left(\sum_n \pi_n e^{\theta_n / \lambda} \right) \tag{4.88}$$

is convex in vector variable $\theta = [\theta_1, \cdots, \theta_n]$ and scalar variable λ. Although θ_n is regarded constant in the dual problem (4.87), however, it will be a decision variable in the final equivalent counterpart to the DR-SO model (4.77). So the convexity refers to both inputs θ and λ.

3. Final Problem and the Outer Approximation Algorithm

Based on the aforementioned discussions, the energy hub planning problem (4.77) can be cast as the following optimization problem with a convex objective and linear constraints

$$\min \; c^T x + \lambda \log \left(\sum_n \pi_n e^{\theta_n / \lambda} \right) + \lambda d_{KL} \tag{4.89a}$$

$$\text{s.t. } x \in X, \; \lambda \geq 0 \tag{4.89b}$$

$$\theta_n = p^T y^n, \; \forall n \tag{4.89c}$$

$$Ax + By^n + C\xi^n \leq d, \; \forall n \tag{4.89d}$$

$$A'x + B'y^k + C'\xi^k \leq d', \; \forall k \tag{4.89e}$$

$$My^k + N\xi^k \leq g^k, \; \forall k \tag{4.89f}$$

$$g^k + \beta \leq \phi_k, \; \phi_k \geq 0, \; \forall k \tag{4.89g}$$

$$\sum_{k=1}^{K} \pi_k \phi_k \leq \beta \alpha_+, \; \beta > 0 \tag{4.89h}$$

In (4.89), scenarios in normal and extreme conditions are generated from three typical days and two extreme days, and independently labeled by n and k, respectively. The optimal second-stage cost $Q(x, \xi^n) = p^T y^n$ is denoted by θ_n in (4.89c); Load shedding is not allowed in normal conditions, as indicated by (4.89d); (4.89e)–(4.89f) quantify the minimum unserved nodal demands, where matrices M and N correspond to the constant coefficients in (4.78c) and (4.78d); (4.89g)–(4.89h) are the reliability constraints for extreme conditions derived in (4.84), where the loss function $D_{loss}(\xi^k) = g^k$.

Because the objective function (4.89a) is shown to be convex, and constraints (4.89b)–(4.89g) are polyhedral, any local algorithm or solver would converge, if succeeds, to the global optimum of (4.89). However, according to our preliminary test, general-purpose NLP solvers, such as IPOPT and NLOPT, fail to solve problem (4.89). To overcome this difficulty, an outer approximation algorithm is developed, which only requires solving LPs. Basic procedures are summarized in Algorithm 4.4. Finite convergence of the outer approximation algorithm for convex programs has been thoroughly established in [44, 45]. According to our experiences, Algorithm 4.4 always converges in no more than 6 iterations for solving problem (4.89).

In Algorithm 4.4, we actually solve the epigraph form of problem (4.89), which minimizes a linear objective function subject to the original constraints and an additional inequality $H(\theta, \lambda) \le \sigma$. This inequality defines a convex region as $H(\theta, \lambda)$ is convex. In step 3, we use linear cuts to approximate its boundary. No feasible points inside the boundary will be removed due to its convexity, and thus the global optimum will be found. From another perspective, in Algorithm 4.4 we are solving a linear relaxation model of the epigraph problem, and generating valid inequalities which successively tighten the relaxation. When Algorithm 4.4 converges, the linear cuts approximate the region $H(\theta, \lambda) \le \sigma$ with high accuracy around the optimal solution, although may not be accurate in the remaining area. An illustration can be found in Appendix C.4.1.

4.6.4 Case Studies

A test system comprised of a modified IEEE 33-bus PDN and a 10-node DHN is used to validate the performance of the proposed model and method. System topology is shown in Fig. 4.48. Two gas-fired units and two wind farms are

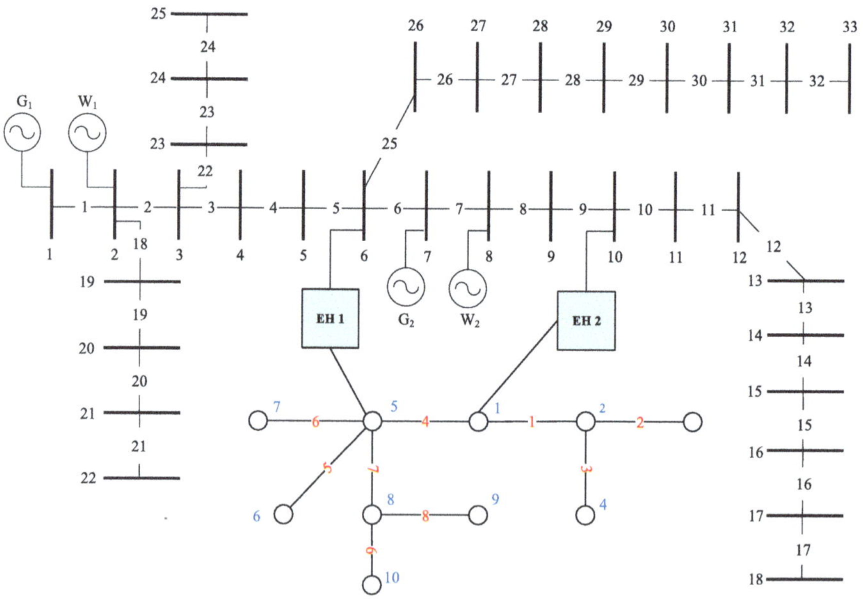

Fig. 4.48 Topology of the test system

connected to the PDN. Two energy hubs are to be invested with a service period of 10 years. Connection points are: buses 6 and 10 at the PDN side, and nodes 1 and 5 at the DHN side. There is no other heat source in the DHN, except energy hubs. Parameters of components in the energy hub are listed in Table 4.15. The coefficients of efficiency η_{gp} and η_{gh} are defined for individual gas-to-power and gas-to-heat conversion, which comply with (4.62c), although could be somehow

Algorithm 4.4 Outer approximation algorithm

1: Choose a convergence tolerance $\varepsilon > 0$; Set iteration index $m = 0$; Initialize θ^0 by solving SO model (4.71) and set $\lambda^0 = 1000$; calculate the value of $H^m = H(\theta^m, \lambda^m)$ as well as the gradient of $H(\theta, \lambda)$ at (θ^m, λ^m)

$$g_H^m = \left[\left(\frac{\partial H}{\partial \theta} \right)^T, \frac{\partial H}{\partial \lambda} \right] \Bigg|_{(\theta^m, \lambda^m)}^T \tag{4.90}$$

where

$$\frac{\partial H}{\partial \theta} = \left(\sum_n \pi_n e^{\theta_n/\lambda} \right)^{-1} \left[\pi_1 e^{\theta_1/\lambda}, \cdots, \pi_N e^{\theta_N/\lambda} \right]^T$$

$$\frac{\partial H}{\partial \lambda} = \log \left(\sum_n \pi_n e^{\theta_n/\lambda} \right) - \lambda^{-1} \left(\sum_n \pi_n e^{\theta_n/\lambda} \right)^{-1} \left(\sum_n \pi_n \theta_n e^{\theta_n/\lambda} \right)$$

2: Solve the following master problem

$$\begin{aligned} \min \ & c^T x + \sigma + \lambda d_{KL} \\ \text{s.t.} \ & (4.89\text{b})-(4.89\text{h}) \\ & H^v + (g_H^v)^T \begin{bmatrix} \theta - \theta^v \\ \lambda - \lambda^v \end{bmatrix} \le \sigma, \\ & v = 0, 1, \cdots, m \end{aligned} \tag{4.91}$$

Update $m \leftarrow m + 1$, and record the optimal solution and the optimal value.

3: If the change of optimal values in two consecutive steps is less than ε, terminate and output the optimal solution; else evaluate H^m and g_H^m according to (4.90), and add a new constraint

$$H^m + (g_H^m)^T \begin{bmatrix} \theta - \theta^m \\ \lambda - \lambda^m \end{bmatrix} \le \sigma \tag{4.92}$$

to the master problem (4.91); go to step 2.

Table 4.15 Component data

	Parameters	Construction cost
CHP	$\eta_{gp} = 0.9$, $\eta_{gh} = 0.9$	1,000,000 ($/MW)
HP	COP = 3	1,500,000 ($/MW)
ESU	$\eta_c^E = 0.98$, $\eta_d^E = 0.98$, $\mu^E = 0.01$	200,000 ($/MWh)
TSU	$\eta_c^H = 0.95$, $\eta_d^H = 0.95$, $\mu^H = 0.01$	150,000 ($/MWh)

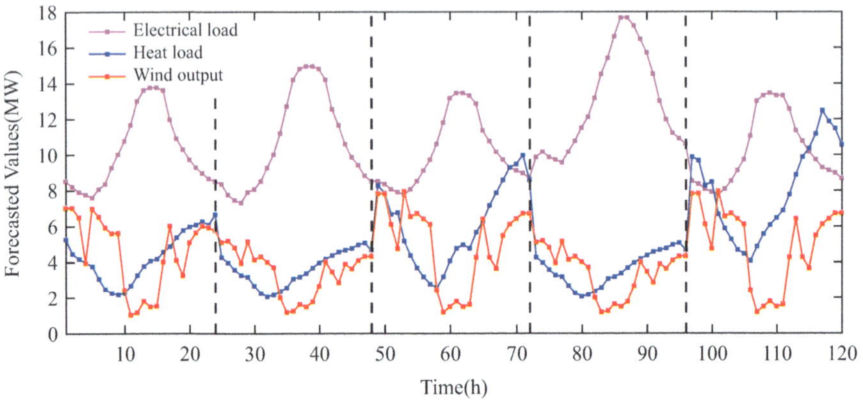

Fig. 4.49 Electrical loads, heat loads, and wind power outputs in different cases. From left to right: three typical days in spring/autumn, summer, and winter; two extreme days in summer and winter

different from the traditional definition of CHP unit efficiency. Remaining data of the system are uploaded in [46].

For the uncertain factors, electric/thermal demands and wind generation profiles in three typical days (spring/autumn, summer, and winter) are employed in normal condition constraint (4.89d); two extremal days in summer and winter are particularly treated in (4.89e)–(4.89f). Generally speaking, the heat demand and wind power output are higher in winter than in summer. Moreover, they have the same tendency of daily variability: higher in the daytime and lower during night, while electric load has the opposite tendency. The predicted data are plotted in Fig. 4.49. We assume that the forecast error obey Normal distribution with zero mean, and the standard deviation is $0.2p^w_{j,t}$ for wind power, and $0.1p^d_{j,t}/0.1h^L_{i,t}$ for system load. To perform SAA, 5000 scenarios are generated for each representative day. Based on the back-forward scenario reduction method [47], 100 scenarios are selected for normal and extreme days. According to (4.76), $d_{KL} = 0.0124$ is used in (4.73) with a confidence level of $\beta^* = 0.95$; In the robust chance constraint (4.77c), the reliability level is maintained no less than 95%, which means $\alpha = 0.05$ and $\alpha_+ = 0.0229$, in light of (4.80).

Table 4.16 Optimal investment strategies

		DR-SP	SP	RO
EH_1	CHP (MW)	5.8732	5.6406	6.1008
	HP (MW)	2.6416	2.6325	2.5780
	ESU (MWh)	3.8214	4.0569	4.0688
	TSU (MWh)	6.8870	6.7553	6.6172
EH_2	CHP (MW)	5.9455	5.7004	6.1711
	HP (MW)	2.6013	2.5917	2.5393
	ESU (MWh)	3.8665	4.0572	4.0786
	TSU (MWh)	6.8611	6.7520	6.5141
Optimum (10^7\$)		8.0282	7.9391	8.6954

Table 4.17 Optimums with different values of d_{KL}

d_{KL} &		β^*		
Optimum (10^7\$)		0.90	0.95	0.99
M	5000	0.0118 & 8.0261	0.0124 & 8.0282	0.0136 & 8.0322
	2000	0.0296 & 8.0800	0.0311 & 8.0844	0.0340 & 8.0930
	1000	0.0592 & 8.1211	0.0622 & 8.1234	0.0679 & 8.1278
	500	0.1185 & 8.1597	0.1243 & 8.1629	0.1358 & 8.1690
	100	0.5925 & 8.3167	0.6217 & 8.3231	0.6790 & 8.3381

1. Benchmark Case

The proposed DR-SO model (4.89) is compared with the traditional SO model (4.71) and RO model (4.72). The planning results offered by the three approaches are shown in Table 4.16. Clearly, in terms of the optimal value, DR-SO is more (less) conservative than the traditional SO (RO), because SO only accounts for the reference distribution P_0, and RO neglects the dispersion effect of uncertain parameters. In other words, RO is equivalent to DR-SO if the ambiguity set assigns 100% probability to the worst-case scenario identified in the RO model. It is also observed that the RO model prefers to invest on CHP unit as it is the most flexible component, and results in the highest cost. Compared to SO, system flexibility in normal conditions and reliability under extreme conditions in DR-SO can be guaranteed at the cost of 1.12% additional expenditure.

Parameter d_{KL} interprets the decision-maker's confidence on the accuracy of reference distribution. The larger d_{KL}, the better the system is protected against distributional uncertainty. In this set of tests, we investigate the impact of data availability and confidence level β^* in W. These two factors lead to different values of d_{KL}, according to (4.76). Results are shown in Table 4.17. The optimal values increase moderately with the increasing of d_{KL}, even if only 100 samples are used for setting up the ambiguity set, the overall cost is still smaller than that in traditional RO. We also observe that the number of data samples has larger impact on the model conservatism than parameter α^*. In all these tests, Algorithm 4.4 successfully converges in 6 iterations and the computation time is less than 900 s, demonstrating the effectiveness of the proposed algorithm for a long term planning problem.

Table 4.18 Failure events with different KL divergence distance

d_p^0		0.005	0.010	0.020	0.030	0.050
DR-SP	Pr_1	2.36%	2.92%	3.74%	4.41%	5.52%
	Pr_2	2.22%	2.75%	3.55%	4.19%	5.27%
SP	Pr_1	4.26%	5.02%	6.15%	7.06%	8.57%
	Pr_2	4.13%	4.88%	5.99%	6.89%	8.37%

2. Sensitivity to PDF Perturbations

A main advantage of the DR-SO model is that its optimal solution is insensitive to PDF perturbations. In this experiment, we generate data sets consisting of 10,000 scenarios whose distribution is different from the reference one. To identify the worst-case distribution, we first test the amount of minimum load shedding in each scenario, and label the ones in which $D_{loss} > 0$. Then we assign larger probabilities to them, and reduce the probabilities of the remaining scenarios, until the KL divergence between the contrived distribution and the reference distribution P_0 reaches a certain value d_p^0. We test the probability of demand loss events under each data set. Results are provided in Table 4.18, where $\mathrm{Pr}_1\ /Pr_2$ denotes the failure probability in the extreme summer/winter day. Under the planning strategies offered by the traditional SO model, the failure probabilities in extreme days exceed the designated value once $d_p^0 > 0.01$. Under the planning strategies offered by the DR-SP model, when $d_p^0 < 0.0124$, the value used in W', the reliability level is greater than 97% even if the real data follows a different distribution, demonstrating the distributional robustness of planning strategies. In fact, the reliability level can be maintained even d_p^0 grows to 0.03. This is because we use convex function $\psi(x)$ to approximate the indicator function $\mathbb{I}_+(x)$, resulting in the conservative inequality (4.82). Further increase of d_p^0 wrecks the requirement on reliability, as shown in the last column of Table 4.18.

3. Impact of Storage Construction Cost

Energy storage units play a central role in the daily operation of energy hub. In the future, the unit capacity costs of both ESU and TSU are expected to drop continuously. In this experiment, we change the investment cost coefficients of storage units by multiplying their values in Table 4.15 with a scalar ζ. Results are given in Table 4.19. It is observed that storage capacity increases rapidly with the deceasing of investment costs. Meanwhile, CHP unit capacity decreases significantly while heat pump capacity grows slightly due to the fast growing size of TSU. The result indicates that the need for energy storage is very likely to witness rapid boosts in the future integrated energy systems.

4. Impact of Wind Penetration Levels

Finally, we investigate the impact of wind penetration levels. In this test, the wind farm output given in Fig. 4.49 is multiplied by a constant. Its capacity refers to the maximum output throughout the day. Results are shown in Table 4.20. We can see that when more wind energy is available to use, the excessive electricity produced during the night can be stored in ESU or converted to heat and stored in TSU, so the capacity of storage units grows significantly. Meanwhile, larger heat

Table 4.19 Investment strategies with different storage costs

ζ		1.0	0.8	0.6	0.4
EH_1	CHP (MW)	5.8732	5.3492	3.3455	3.0995
	HP (MW)	2.6416	2.7203	2.9550	3.0594
	ESU (MWh)	3.8214	5.8367	14.6498	15.7582
	TSU (MWh)	6.8870	8.1579	13.4595	14.2451
EH_2	CHP (MW)	5.9455	5.4157	3.3896	3.1393
	HP (MW)	2.6013	2.6767	2.9060	3.0067
	ESU (MWh)	3.8665	5.8203	14.5244	15.6491
	TSU (MWh)	6.8611	8.1389	13.4434	14.2272
Optimum (10^7\$)		8.0282	7.9438	7.7950	7.5944

Table 4.20 Investment strategies with different wind penetration levels

Wind capacity		6 MW	8 MW	10 MW	12 MW
EH_1	CHP (MW)	6.4103	5.8732	4.9633	3.6916
	HP (MW)	2.5695	2.6416	2.8034	2.9743
	ESU (MWh)	3.0106	3.8214	6.2506	10.7163
	TSU (MWh)	5.5171	6.8870	8.9294	12.1764
EH_2	CHP (MW)	6.4923	5.9455	5.0227	3.7299
	HP (MW)	2.5327	2.6013	2.7678	2.9206
	ESU (MWh)	3.0770	3.8665	6.2979	10.6220
	TSU (MWh)	5.5207	6.8611	8.9499	12.1630
Optimum (10^7\$)		8.8974	8.0282	7.2216	6.4713

pump is built to produce thermal energy from electric power to supply peak heat demand and charge the TSU. Consequently, the capacity of CHP unit decreases because more energy will be produced by wind turbines with zero operation costs, while dispatching CHP unit incurs non-zero gas fuel cost. The above results imply that with the continuous drop of the storage construction cost and the increasing penetration level of renewables, the need for energy storage in the integrated energy system will grow rapidly.

4.7 Summary and Conclusions

The complementary natures and synergetic potentials between thermal and electric energy inspire the combined heat-power generation and utilization. This chapter elucidates a collection of important issues in operation, marketization, and planning of integrated heat-power distribution systems. We devise marginal cost based energy pricing and trading schemes, which streamline the design of multi-carrier energy markets; since low-cost renewable energy contributes to lower electricity prices, such a market framework encourages electrified heating devices to consume excessive renewable power during off-peak hours; we develop mathematical models and computational tools to calculate the equilibrium of the heat-power spot market, and analyze the market impact of strategic producers and elastic demands, with special emphasis on the energy hub, which could be the most influential profit-driven entity in an integrated energy system. We address the capacity planning problem of the energy hub using data-driven robust stochastic optimization, which is capable of tackling inexact probability distributions of renewable power output and load demands; Compared with the prolonged research on economic co-generation, our work considers network energy flow and market models. These methodologies could serve as fundamental analytical tools for the planning and operation of power distribution systems and district heating systems with an increasing level of interdependence.

4.8 Further Reading

Due to the high efficiency of centralized heating production, district heating system becomes popular among regions with long cold winters. Modeling the energy flows in district heating systems consisting of pipelines is the fundamental step for the further study of such an energy distribution infrastructure. The classic steady-state hydraulic-thermal model has been established in [7, 48] based on many references therein. In such a network model, when mass flow rates are fixed, the thermal condition renders a set of linear equations. Researchers find some methods to determine hydraulic condition [49, 50], so that the thermal equations can be easily solved. Different from electricity which travels at the speed of light, temperature propagation in a pipeline is much slower. With the scale of the heating network growing up, the dynamics of temperature propagation matter. The time delays of temperature changes at a source node and a load node vary from a few minutes (residential-scale system) to one or more hours (city-sized system), which are coupled with the time-scale of power system operations. A node method accounting for time delays and energy losses is developed in [5, 51, 52] and validated in [53]. A resistance model is proposed in [16] to calculate the thermal behavior of heating system with parallel pipes. It is found that the heat loss is almost independent of the mass flow rate in the ordinary operating condition, which coincides with the analysis in Sect. 4.2.4. Spatial and temporal thermal transients in district heating systems are thoroughly studied in [54] using a PDE model. Optimization models of CHP units and thermal storage with detailed heat transfer process is discussed in [55–57]. A continuous-time control model for the district heating network is proposed in [58]. Other studies use different models for the heating system in various problems, such as [59–61], to name just a few. It should be pointed out that in thermal engineering, energy may not be the only concern, and other physical quantity should be taken into account, such as exergy and entropy [62]. District heating systems are analyzed in [63–65] from the thermodynamics perspective. In fact, such concepts and methods can be applied to many engineering systems, such as renewable energy system and multi-generation system [66].

As the widespread utilization of electrified heating devices, such as heat pumps, the interdependence between the power system and the heating system becomes more prominent. Co-generation and tri-generation have been extensively studied in the existing literature. However, the network model of the heating system is usually neglected. Some representative work which explicitly considers network models include the following. By concreting the DC power flow model and thermal flow model, optimal generation dispatch and unit commitment problems of the integrated heat-power system are addressed in [67] and [68], respectively, in which temperature propagation dynamics are considered, and the thermal storage capacities of water pipelines are exploited. In [69], the two systems are dispatched using a rolling optimization model. A feasible region method is proposed in [70] to dispatch the integrated heat-power system considering building thermal inertia. The interaction across power and heating system under contingencies is analyzed in [71].

To quantify uncertainty in an integrated heat-power system, an information entropy approach is employed; the dependence between the generation levels and weather conditions is described by a copula-based model in [72]. A scenario generation technique is proposed in [73] based on the optimum quantile method with Wasserstein distance metric. To cope with uncertainty, a probabilistic power flow model is suggested in [74]. A day-ahead heat pump scheduling model is developed in [75], in which load and photovoltaic generation uncertainties are tackled by the distributionally robust optimization approach.

Exploring demand response potentials of heat loads and thermal storage units could greatly enhance system flexibility and facilitate operation [76–78]. A bi-level optimization model is given in [79] for managing a heat-electricity load aggregator. Coupled heat and electricity demand response in microgrids is formulated as a Stackelberg game in [80], while network flow models are neglected. Stochastic optimization method and model predictive control method are used in [81] and [82] for system operation with flexible demands. A geometric approach is developed in [83] to quantify the aggregated flexibility offered by heterogeneous thermostatically controlled loads. More works are introduced in a comprehensive survey [84].

Energy hub suggested in [85, 86] plays important roles in energy conversion and storage and offers unique opportunities for energy system integration. For an energy hub in the integrated heat-power system, it may include CHP units, heat pumps, electricity and thermal storage units. The majority of current research considers combined heat and electricity production of energy hubs in the residential level, which directly supply energy to home appliances. For example, an MILP model is proposed in [87], and a multi-objective optimization model is presented in [88]. In [89], renewable energy is incorporated. To deal with uncertainties resulting from electricity market prices, renewable generation, and heat/electricity demands, probabilistic and robust optimization approaches are employed in [90–92]. Except the residential ones, energy hubs can appear in the distribution level, which connect a power distribution network and a district heating network. Certainly, their capacities would be much larger than the residential ones, and network model should be taken into account during the operation. Such energy hubs are considered in [93, 94]. Considering the topic of this chapter, plenty of research works on energy hubs without thermal energy are not mentioned here.

Most of the above studies rely on a centralized optimization paradigm. In current practice, power systems and heating systems are managed by different entities, therefore, centralized decision making may encounter difficulty in practical implementations. To address this issue, a decentralized algorithm for combined heat and power dispatch is proposed in [95] based on Benders decomposition. In each iteration step, either an optimality cut or a feasibility cut is sent to the power system operator according to the outcome of thermal flow calculation. Physical data of the two systems are kept private. A consensus based distributed algorithm is developed in [96]. An ACOPF subproblem for the power system and a thermal flow optimization problem are solved alternatively. In the thermal flow model, the heating system is approximated by an equivalent circuit, which may not be as accurate as the physical model presented in [7]. In [97], real-time operation of energy hubs is formulated as a potential game, and an online distributed algorithm is proposed to identify the Nash equilibrium. ADMM based distributed optimization approaches are presented in [98] and [99] for energy dispatch of combined heat and power microgrids and integrated gas-heat-electricity systems, respectively. Game theoretical models for demand response management of energy hubs are proposed in [100–102], and distributed algorithms are developed to identify the Nash equilibrium solution.

Further development of the integrated energy system entails new business modes and energy trading initiatives. Energy market is the essential venue for such activities. A preliminary attempt is reported in [103], in which an energy market with multi-lateral natural gas, electricity and heat trading is studied. Unlike the electricity market which has been studied for decades and well developed, the heating market is far less competitive than the electricity market. Due to the measuring issues, how to price heat conveniently is still an open problem. Marginal cost based pricing models for the district heating market are proposed in [104–106]. Nodal energy pricing model for an integrated energy system is established in [107]. Pricing heat and electricity co-generation in local communities is discussed in [108]. It is suggested in [26] to price thermal energy based on exergy losses, and reference [25] provides a locational marginal pricing scheme, which is the same as electricity pricing in power markets. The financial risks under predetermined and marginal pricing policies are analyzed in [109]. The heat pricing issue has been thoroughly reviewed in [27].

As energy distribution infrastructures, the operation of the district heating system and power distribution system share some problems in common. For example, state estimation (including water flows, temperatures, and heat losses) is discussed in [110, 111] using least square method. Reliability is studied in [112], and a repair model of the heating network is developed, which can be used in resilience related studies.

References

1. Cao, Y., Wei, W., Mei, S., Wu, L., Shahidehpour, M., Li, Z.: Distributed operation of interdependent power distribution network and district heating network: a market-driven approach. IEEE Trans. Smart Grid (2019, in press)
2. Chen, Y., Wei, W., Liu, F., Sauma, E., Mei, S.: Energy trading and market equilibrium in integrated heat-power distribution systems. IEEE Trans. Smart Grid. **10**(4), 4080–4094 (2019)
3. Li, R., Wei, W., Mei, S., Hu, Q., Wu, Q.: Participation of an energy hub in electricity and heat distribution markets: an MPEC approach. IEEE Trans. Smart Grid. **10**(4), 3641–3653 (2019)
4. Cao, Y., Wei, W., Mei, S.: Capacity planning of energy hub in multi-carrier energy networks: a data-driven stochastic programming approach. IEEE Trans. Sustain. Energy (2019, in press)
5. Zhao, H.: Analysis, modelling and operational optimization of district heating systems. PhD Thesis, Technical University of Denmark, Lyngby (1995)
6. Hassine, I.B., Ursula E.: Impact of load structure variation and solar thermal energy integration on an existing district heating network. Appl. Therm. Eng. **50**(2), 1437–1446 (2013)
7. Liu, X.Z.: Combined analysis of electricity and heat networks. PhD Thesis, Cardiff University, Cardiff (2013)
8. Sauer, H.J., Howell, R.H.: Heat Pump Systems. John Wiley & Sons, New York (1983)
9. Chiasson, A.D.: Geothermal Heat Pump and Heat Engine Systems: Theory And Practice. John Wiley & Sons, Hoboken (2016)
10. Fischer, D., Hatef, M.: On heat pumps in smart grids: a review. Renew. Sustain. Energy Rev. **70**, 342–357 (2017)
11. Chua, K.J., Chou, S.K., Yang, W.M.: Advances in heat pump systems: a review. Appl. Energy **87**(12), 3611–3624 (2010)
12. Guo, T., Henwood, M.I., van Ooijen, M.: An algorithm for combined heat and power economic dispatch. IEEE Trans. Power Syst. **11**(4), 1778–1784 (1996)
13. Awan, U.: Experimental Analysis of Variable Capacity Heat Pump System Equipped with Vapour Injection and Permanent Magnet Motor. Masters Thesis. KTH, Stockholm (2012)
14. Madani, H., Ahmadi, N., Claesson, J., Lundqvist, P.: Experimental analysis of a variable capacity heat pump system focusing on the compressor and inverter loss behavior. International Refrigeration and Air Conditioning Conference. Paper 1063 (2010)
15. Pirouti, M., Bagdanavicius, A., Ekanayake, J. Wu, J., Jenkins, N.: Energy consumption and economic analyses of a district heating network. Energy **57**, 149–159 (2013)
16. van der Heijde, B., Aertgeerts, A., Helsen, L.: Modelling steady-state thermal behaviour of double thermal network pipes. Int. J. Therm. Sci. **117**, 316–327 (2017)
17. Tawarmalani, M. Sahinidis, N.V.: A polyhedral branch-and-cut approach to global optimization. Math. Program. **103**(2), 225–249 (2005)
18. Wächter, A., Biegler, L.T.: On the implementation of a primal-dual interior point filter line search algorithm for large-scale nonlinear programming. Math. Program. **106**(1), 25–57 (2006)
19. Jie, P., Tian, Z., Yuan, S., Zhu, N.: Modeling the dynamic characteristics of a district heating network. Energy **39**(1), 126–134 (2012)
20. Liu, Y., Li, J., Wu, L.: Distribution system restructuring: Distribution LMP via unbalanced ACOPF. IEEE Trans. Smart Grid **9**(5), 4038–4048 (2018)
21. Li, F.: Continuous locational marginal pricing (CLMP). IEEE Trans. Power Syst. **22**(4), 1638–1646 (2007)
22. https://sites.google.com/site/caoyangthu13/data
23. Ben-Tal, A., Nemirovski, A.: On polyhedral approximations of the second-order cone. Math. Oper. Res. **26**(2), 193–205 (2001)
24. Wu, L.: A tighter piecewise linear approximation of quadratic cost curves for unit commitment problems. IEEE Trans. Power Syst. **26**(4), 2581–2583 (2011)

25. Gebremedhin, A., Moshfegh, B.: Modelling and optimization of district heating and industrial energy system-an approach to a locally deregulated heat market. Int. J. Energy Res. **28**(5), 411–422 (2004)
26. Poredoš, A., Kitanovski, A.: Exergy loss as a basis for the price of thermal energy. Energy Convers. Manag. **43**(16), 2163–2173 (2002)
27. Li, H., Sun, Q., Zhang, Q., Wallin, F.: A review of the pricing mechanisms for district heating systems. Renew. Sustain. Energy Rev. **42**, 56–65 (2015)
28. Zhao, Z., Wu, L., Song, G.: Convergence of volatile power markets with price-based demand response. IEEE Trans. Power Syst. **29**(5), 2107–2118 (2014)
29. Cobb, C.W., Douglas, P.H.: A theory of production. Am. Econ. Rev. **18**(1), 139–165 (1928)
30. https://en.wikipedia.org/wiki/Cobb%E2%80%93Douglas_production_function
31. https://sites.google.com/site/yuechenthu/data-sheet
32. Audet, C., Dennis Jr, J.E.: Analysis of generalized pattern searches. SIAM J. Optimiz. **13**(3), 889–903 (2003)
33. Lu, T., Wang, Z., Wang, J., Ai, Q., Wang, C.: A data-driven Stackelberg market strategy for demand response-enabled distribution systems. IEEE Trans Smart Grid **10**(3), 2345–2357 (2018)
34. Li, R.: Integrated heat-power system dataset. [online] Available at: https://sites.google.com/site/ruilismarterrlc (2018)
35. Drury, E., Denholm, P., Sioshansi, R.: The value of compressed air energy storage in energy and reserve markets. Energy **36**(8), 4959–4973 (2011)
36. Sioshansi, R., Denholm, P.: The value of concentrating solar power and thermal energy storage. IEEE Trans. Sustain. Energy **1**(3), 173–183 (2010)
37. Li, Z., Guo, Q., Sun, H., Wang, J.: Sufficient conditions for exact relaxation of complementarity constraints for storage-concerned economic dispatch. IEEE Trans. Power Syst. **31**(2), 1653–1654 (2016)
38. Li, Z., Guo, Q., Sun, H., Wang, J.: Storage-like devices in load leveling: complementarity constraints and a new and exact relaxation method. Appl. Energy **151**, 13–22 (2015)
39. Pardo, L.: Statistical Inference Based on Divergence Measures, vol. 185. CRC Press, Boca Raton (2006)
40. Jiang, R., Guan, Y.: Data-driven chance constrained stochastic program. Math. Program. **158**(1–2), 291–327 (2016)
41. Antoniou, A., Lu, W.S.: Practical Optimization: Algorithms and Engineering Applications. Springer, New York (2007)
42. Nemirovski, A., Shapiro, A.: Convex approximations of chance constrained programs. SIAM J. Optimiz. **17**(4), 969–996 (2006)
43. Hu, Z., Hong, L.: Kullback-leibler divergence constrained distributionally robust optimization. Optimization Online (2013)
44. Duran, M.A., Grossmann, I.E.: An outer-approximation algorithm for a class of mixed-integer nonlinear programs. Math. Program. **36**(3), 307–339 (1986)
45. Hogan, W.W.: Applications of a general convergence theory for outer approximation algorithms. Math. Program. **5**(1), 151–168 (1973)
46. https://sites.google.com/163.com/caoyang13
47. Dupačová, J., Gröwe-Kuska, N., Römisch, W.: Scenario reduction in stochastic programming. Math. Program. **95**(3), 493–511 (2003)
48. Liu, X.Z., Wu, J., Jenkins, N., Bagdanavicius, A.: Combined analysis of electricity and heat networks. Appl. Energy **162**, 1238–1250 (2016)
49. Chen, Y., Guo, Q., Sun, H., et al.: A water mass method and its application to integrated heat and electricity dispatch considering thermal dynamics. arXiv preprint arXiv:1711.02274 (2017)
50. Wang, H., Wang, H., Zhu, T.: A new hydraulic regulation method on district heating system with distributed variable-speed pumps. Energy Convers. Manag. **147**, 174–189 (2017)
51. Benonysson, A.: Dynamic modelling and operational optimization of district heating systems. Doctoral Dissertation. Tech Univ Denmark, Denmark (1991)

52. Benonysson, A., Bøhm, B., Ravn, H.F.: Operational optimization in a district heating system. Energy. Convers. Manag. **36**(5), 297–314 (1995)
53. Gabrielaitiene, I., Bøhm, B., Sunden, B.: Modelling temperature dynamics of a district heating system in Naestved, Denmark-A case study. Energy. Convers. Manag. **48**(1), 78–86 (2006)
54. Chertkov, M., Novitsky, N.N.: Thermal transients in district heating systems. Working paper, arXiv:1702.07634 (2017)
55. Dai, Y., Chen, L., Min, Y., Chen, Q., Hu, K., Hao, J., Zhang, Y., Xu, F.: Dispatch model of combined heat and power plant considering heat transfer process. IEEE Trans. Sustain. Energy **8**(3), 1225–1236 (2017)
56. Dai, Y., Chen, L., Min, Y., Chen, Q., Hao, J., Hu, K., Xu, F.: Dispatch model for CHP with pipeline and building thermal energy storage considering heat transfer process. IEEE Trans. Sustain. Energy **10**(1), 192–203 (2019)
57. Dai, Y., Chen, L., Min, Y., Mancarella, P., Chen, Q., Hao, J., Hu, K., Xu, F.: A general model for thermal energy storage in combined heat and power dispatch considering heat transfer constraints. IEEE Trans. Sustain. Energy **9**(4), 1518–1528 (2018)
58. Scholten, T., De Persis, C., Tesi, P.: Modeling and control of heat networks with storage: the single-producer multiple-consumer case. IEEE Trans. Contr. Syst. Technol. **25**(2), 414–428 (2017)
59. Chen, Q., Hao, J.H., Zhao, T.: An alternative energy flow model for analysis and optimization of heat transfer systems. Int. J. Heat Mass Tran. **108**, 712–720 (2017)
60. Dorfner, J., Hamacher, T.: Large-scale district heating network optimization. IEEE Trans. Smart Grid **5**(4), 1884–1891 (2014)
61. Jie, P., Zhu, N., Li, D.: Operation optimization of existing district heating systems. Appl. Therm. Eng. **78**, 278–288 (2015)
62. Dincer, I., Cengel, Y.A.: Energy, entropy and exergy concepts and their roles in thermal engineering. Entropy **3**(3), 116–149 (2001)
63. Rosen, M.A., Dincer, I.: Exergy-cost-energy-mass analysis of thermal systems and processes. Energy. Convers. Manag. **44**(10), 1633–1651 (2003)
64. Comakli, K., Yüksel, B., Comakli Ö.: Evaluation of energy and exergy losses in district heating network. Appl. Therm. Eng. **24**(7), 1009–1017 (2004)
65. Ozgener, L., Hepbasli, A., Dincer, I.: Energy and exergy analysis of geothermal district heating systems: an application. Build. Environ. **40**(10), 1309–1322 (2005)
66. Dincer, I., Rosen, M.A.: Exergy. Elesvier, Oxford (2012)
67. Li, Z., Wu, W., Shahidehpour, M., Wang, J., Zhang, B.: Combined heat and power dispatch considering pipeline energy storage of district heating network. IEEE Trans. Sustain. Energy **7**(1), 12–22 (2016)
68. Li, Z., Wu, W., Wang, J, Zhang, B., Zheng, T.: Transmission-constrained unit commitment considering combined electricity and district heating networks. IEEE Trans. Sustain. Energy **7**(2), 480–492 (2016)
69. Wu, C., Gu, W., Jiang, P., Li, Z., Cai, H., Li, B.: Combined economic dispatch considering the time-delay of a district heating network and multi-regional indoor temperature control. IEEE Trans. Sustain. Energy **9**(1), 118–127 (2018)
70. Pan, Z., Guo, Q., Sun, H.: Feasible region method based integrated heat and electricity dispatch considering building thermal inertia. Appl. Energy **192**, 395–407 (2017)
71. Pan, Z., Guo, Q., Sun, H.: Interactions of district electricity and heating systems considering time-scale characteristics based on quasi-steady multi-energy flow. Appl. Energy **167**, 230–243 (2016)
72. Fu, X., Sun, H., Guo, Q., Pan, Z., Xiong, W., Wang, L.: Uncertainty analysis of an integrated energy system based on information theory. Energy **122**, 649–662 (2017)
73. Fu, X., Guo, Q., Sun, H., Pan, Z., Xiong, W., Wang, L.: Typical scenario set generation algorithm for an integrated energy system based on the Wasserstein distance metric. Energy **135**, 153–170 (2017)
74. Fu, X., Sun, H., Guo, Q., Pan, Z., Zhang, X., Zeng, S.: Probabilistic power flow analysis considering the dependence between power and heat. Appl. Energy **191**, 582–592 (2017)

75. Li, Z., Wu, W., Zhang, B., Tai X.: Kullback–Leibler divergence-based distributionally robust optimisation model for heat pump day-ahead operational schedule to improve PV integration. IET Gener. Transm. Distr. **12**(13), 3136–3144 (2018)
76. Klein, K., Herkel, S., Henning, H.M., Felsmann, C.: Load shifting using the heating and cooling system of an office building: quantitative potential evaluation for different flexibility and storage options. Appl. Energy **203**, 917–937 (2017)
77. Nuytten, T., Claessens, B., Paredis, K., Van Bael, J., Six, D.: Flexibility of a combined heat and power system with thermal energy storage for district heating. Appl. Energy **104**, 583–591 (2013)
78. Kiviluoma J, Heinen S, Qazi H, et al.: Harnessing flexibility from hot and cold: Heat storage and hybrid systems can play a major role. IEEE Power Energ. Mag. **15**(1), 25–33 (2017)
79. Shao, C., Ding, Y., Wang, J., Song, Y.: Modeling and integration of flexible demand in heat and electricity integrated energy system. IEEE Trans. Sustain. Energy **9**(1), 361–370 (2018)
80. Liu, N., He, L., Yu, X., Ma, L.: Multi-party energy management for grid-connected microgrids with heat and electricity coupled demand response. IEEE Trans. Ind. Inf. **14**(5), 1887–1897 (2018)
81. Good, N., Mancarella, P.: Flexibility in multi-energy communities with electrical and thermal storage: A stochastic, robust approach for multi-service demand response. IEEE Trans. Smart Grid **10**(1), 503–513 (2017)
82. Verrilli, F., Srinivasan, S., Gambino, G., et al.: Model predictive control-based optimal operations of district heating system with thermal energy storage and flexible loads. IEEE Trans. Autom. Sci. Eng. **14**(2), 547–557 (2017)
83. Zhao, L., Zhang, W., Hao, H., Kalsi, K.: A geometric approach to aggregate flexibility modeling of thermostatically controlled loads. IEEE Trans. Power Syst. **32**(6), 4721–4731 (2017)
84. Wang, J., Zhong, H., Ma, Z., Xia, Q., Kang, C.: Review and prospect of integrated demand response in the multi-energy system. Appl. Energy **202**, 772–782 (2017)
85. Geidl, M., Koeppel, G., Favre-Perrod, P., Klockl, B., Andersson, G., Frohlich, K.: Energy hubs for the future. IEEE Power Energ. Mag. **5**(1), 24–30 (2007)
86. Geidl, M., Andersson, G.: Optimal power flow of multiple energy carriers. IEEE Trans. Power Syst. **22**(1), 145–155 (2007)
87. Rastegar, M., Fotuhi-Firuzabad, M., Lehtonen, M.: Home load management in a residential energy hub. Electr. Power Syst. Res. **119**, 322–328 (2015)
88. Brahman, F., Honarmand, M., Jadid, S.: Optimal electrical and thermal energy management of a residential energy hub, integrating demand response and energy storage system. Energ. Buildings **90**, 65–75 (2015)
89. Rastegar, M., Fotuhi-Firuzabad, M.: Load management in a residential energy hub with renewable distributed energy resources. Energ. Buildings **107**, 234–242 (2015)
90. Rastegar, M., Fotuhi-Firuzabad, M., Zareipour, H, Moeini-Aghtaie, M.: A probabilistic energy management scheme for renewable-based residential energy hubs. IEEE Trans. Smart Grid **8**(5), 2217–2227 (2017)
91. Najafi, A., Falaghi, H., Contreras, J., Ramezani, M.: A stochastic bilevel model for the energy hub manager problem. IEEE Trans. Smart Grid **8**(5), 2394–2404 (2017)
92. Parisio, A., Del Vecchio, C., Vaccaro, A.: A robust optimization approach to energy hub management. Int. J. Electr. Power Energy Syst. **42**(1), 98–104 (2012)
93. Liu, X., Mancarella, P.: Modelling, assessment and Sankey diagrams of integrated electricity-heat-gas networks in multi-vector district energy systems. Appl. Energy **167**, 336–352 (2016)
94. Li, R., Chen, L., Zhao, B., et al.: Economic dispatch of an integrated heat-power energy distribution system with a concentrating solar power energy hub. J. Energy Eng. **143**(5), 04017046 (2017)
95. Lin, C., Wu, W., Zhang, B., Sun, Y.: Decentralized solution for combined heat and power dispatch through Benders decomposition. IEEE Trans. Sustain. Energy **8**(4), 1361–1372 (2017)

96. Li, J., Fang, J., Zeng, Q., Chen, Z.: Optimal operation of the integrated electrical and heating systems to accommodate the intermittent renewable sources. Appl. Energy **167**, 244–254 (2016)
97. Bahrami, S., Toulabi, M., Ranjbar, S., Moeini-Aghtaie, M., Ranjbar, A.: A decentralized energy management framework for energy hubs in dynamic pricing markets. IEEE Trans. Smart Grid **9**(6), 6780–6792 (2018)
98. Liu, N., Wang, J., Wang, L.: Distributed energy management for interconnected operation of combined heat and power-based microgrids with demand response. J. Mod. Power Syst. Clean Energy **5**(3), 478–488 (2017)
99. Zhang, H., Li, Y., Gao, D.W., Zhou, J.: Distributed optimal energy management for energy internet. IEEE Trans. Ind. Inform. **13**(6), 3081–3097 (2017)
100. Sheikhi, A., Rayati, M., Bahrami, S., Ranjbar, A.: Integrated demand side management game in smart energy hubs. IEEE Trans. Smart Grid **6**(2), 675–683 (2015)
101. Sheikhi, A., Bahrami, S., Ranjbar, A.M.: An autonomous demand response program for electricity and natural gas networks in smart energy hubs. Energy **89**, 490–499 (2015)
102. Sheikhi, A., Bahrami, S.: From demand response in smart grid toward integrated demand response in smart energy hub. IEEE Trans. Smart Grid **7**(2), 650–658 (2016)
103. Chen, Y., Wei, W., Liu, F., Mei, S.: A multi-lateral trading model for coupled gas-heat-power energy networks. Appl. Energy **200**, 180–191 (2017)
104. Zhang, J., Ge, B., Xu, H.: An equivalent marginal cost-pricing model for the district heating market. Energ. Policy **63**, 1224–1232 (2013)
105. Difs, K., Trygg, L.: Pricing district heating by marginal cost. Energ. Policy **37**(2), 606–616 (2009)
106. Sjödin, J., Henning, D.: Calculating the marginal costs of a district-heating utility. Appl. Energy **78**(1), 1–18 (2004)
107. Quelhas, A.M., Gil, E., McCalley, J.D.: Nodal prices in an integrated energy system. Int. J. Crit. Infrastruct. **2**(1), 50–69 (2005)
108. Bohman, M., Andersson, R.: Pricing cogenerated electricity and heat in local communities. J. Public Econ. **33**(3), 333–356 (1987)
109. Björkqvist, O., Idefeldt, J., Larsson, A.: Risk assessment of new pricing strategies in the district heating market: a case study at Sundsvall Energi AB. Energ. Policy **38**(5), 2171–2178 (2010)
110. Fang, T., Lahdelma, R.: State estimation of district heating network based on customer measurements. Appl. Therm. Eng. **73**(1), 1211–1221 (2014)
111. Dong, J., Guo, Q., Sun, H., et al.: Research on state estimation for combined heat and power networks. In: Power and Energy Society General Meeting, pp. 1–5 (2016)
112. Shan, X., Wang, P., Lu, W.: The reliability and availability evaluation of repairable district heating networks under changeable external conditions. Appl. Energy **203**, 686–695 (2017)

Chapter 5
Electrified Transportation Network

5.1 Introduction

The excessive consumption of coal and petroleum products not only causes potential shortage of energy resources in the foreseeable future, but also creates various negative impacts on the environment, causing growing concerns on reducing the dependency on fossil fuels and diversifying the energy supply. Transportation and electricity production activities are dominant contributors for green-house gas emissions. Recent technology breakthroughs in electric vehicle (EV) and renewable generation technologies offer new opportunities to overcome the worldwide dilemma of economic growth and environment protection.

At the power system side, renewable energy is paving the way to cleaner electricity production [1, 2]. The rapidly growing penetration level of wind and solar power generation reduces the operating cost and pollution at the same time, which becomes their most attractive features. Meanwhile, the volatility and intermittency can be mitigated or compensated through market trading [3, 4]. As a result, renewables are promising to play a more important role in the energy sector in the near future. At the transportation system side, innovations of battery technologies greatly promote the driving range of EVs. The electrical demand from public charging facilities is growing quickly.

The installation of on-road charging and swapping stations will introduce notable interplay between the transportation network (TN) and the power distribution network (PDN) [5]. On the one hand, traffic congestion patterns and traffic regulation policies in the transportation system will influence routes and charging locations of EVs, which further impact spatial and temporal distributions of the electric load as well as operation of the power grid. On the other hand, the queuing time and electricity prices in fast charging stations (FCSs) also affect travel plans of EVs and consequently the vehicular flows in the TN. Such changes in transportation and energy sectors encourage studies on interdependency and potential synergies between the networked systems.

W. Wei, J. Wang, *Modeling and Optimization of Interdependent Energy Infrastructures*, https://doi.org/10.1007/978-3-030-25958-7_5

In fact, the grid impacts of V2G have been the subject of power system research for more than a decade. Up-to-date surveys on EV charging schemes and grid impacts can be found in [6–9]. In the majority of reviewed researches, a common assumption is that driving patterns of vehicles are known, i.e., arrival rates and timings, as well as amounts of charging requests are specified either in a deterministic or a stochastic manner in advance. This assumption is reasonable for aggregator-level studies, such as the management of a residential parking lot, or a single FCS. However, it may no longer be valid if the system-level interdependency is under investigation.

The distinct feature of urban transportation compared with highway and railway transportation stems from congestion, which translates into time delays. The main target of urban transportation research has been focused on understanding and possibly alleviating congestions. In 1952, Wardrop pointed out the condition which leads to a stable traffic flow pattern in a congested TN [10], called a user equilibrium (UE), such that no traveler has the incentive to change their route unilaterally. Mathematical programming based approaches for computing the UE originated from 1956, when Beckmann showed that the UE condition of Wardrop is equivalent to the optimality condition of a strictly convex traffic assignment problem (TAP) [11]. Introduction to basic traffic theory can be found in [12].

Based on the UE theory, some interdisciplinary research endeavors to study the operation and planning of the coupled transportation and electrical infrastructures while accounting for the system-level interdependency. For example, the FCS planning problem has been addressed in [13]. By solving an OPF problem and a flow capturing problem coordinately, the service area of the FCS is enlarged, and the grid impact in terms of net losses and voltage deviation is minimized. The method is further improved in [14], in which the distribution of vehicle flows in the TN is estimated from the traffic UE pattern, such that the rationality of travelers and the system interaction are better portrayed. A network equilibrium model for the regional coupled transportation system and power distribution system is proposed in [15], which integrates a stochastic UE problem (where stochastic means that the destination of traffic demand follows a certain distribution), a DCOPF problem, and an LMP scheme. The framework is then used to determine the sites of a given number of charging facilities. A similar paradigm is employed in [16] to study electricity price and road toll which will help improve the overall operating condition of the coupled networks. A distributed pricing policy is suggested in [17], in which the TN and the PDN collaborate with each other in order to reach a social optimum operating point while keeping the data of each system private. In particular, the EV routing problem is considered as a shortest path problem on a modified transportation graph, and the economic dispatch problem of the PDN is formulated as a DCOPF problem. In summary, in the traffic UE model, the vehicular flow is non-atomic, i.e., the system impact from the action of a single vehicle is infinitesimal. However, in many of these system-level studies, EVs are not adequately treated. A common assumption is that power demands in FCSs are proportional to road traffic flows, such as [14–16]. Although this assumption may be reasonable in certain cases, it is usually difficult to determine the charging rate coefficient, which has a decisive impact on the interdependency model.

This chapter aims to provide a comprehensive introduction on the planning, operation, and equilibrium analysis of the interdependent urban transportation and power distribution networks. Materials in this chapter come from authors' publications [18–22]. In particular, the mathematical formulations of the TN and static TAPs, including the renowned Backmann model and the relatively new Nesterov model, are introduced in Sect. 5.2; The deterministic optimal traffic-power flow (OTPF) problem is presented in Sect. 5.3; robust operation of the PDN is discussed in Sect. 5.4, in which traffic demand uncertainty is taken into account; coordinated capacity expansion planning of the TN and PDN infrastructures is studied in Sect. 5.5; Vulnerability of the coupled transportation and power system are illuminated in Sect. 5.6, the impact of road capacity degradation is quantified; Finally, by considering the interaction between nodal electricity pricing and response from the traffic flow pattern, the network equilibrium of the coupled infrastructure is addressed in Sect. 5.7, in which EV and other vehicles are treated separately.

5.2 User Equilibrium of Urban Transportation Network

The traffic flow pattern over a TN depends on two interactive mechanisms. On the one hand, system users attempt to travel in the most convenient way. For example, most travelers prefer to choose the route with the minimal travel time. On the other hand, the travel cost of every vehicle may also depend on the other vehicles' choices. To see this, the travel time on each road is an increasing function of the total traffic flow due to the congestion effect. Therefore, traveling on the shortest path may not perceive the shortest travel time, because that path quickly becomes congested if everyone rushes into it. As a result, it may not be easy to determine the flow pattern throughout the network using only geographical information. This section describes how the steady-state traffic flow pattern can be computed by considering congestion and rationality of travelers.

5.2.1 Network and Traffic Flow Model

In what follows we introduce the graph description of the TN and concepts of traffic flows, which do not depend on the particular traffic assignment model. The fundamental theory and mathematical model of the transportation system are well taught in [12].

A TN can be represented by a connected graph $G_T = [T_N, T_A]$, where T_N denotes the set of nodes, representing origins, destinations, and intersections; T_A

denotes the set of links, representing lanes or roadway segments. The node-link incidence matrix $\Lambda \in \mathbb{M}^{|T_N| \times |T_A|}$ depicts network topology

$$\Lambda_{ij} = \begin{cases} +1, & \text{if node } i \text{ is the entrance of link } j; \\ -1, & \text{if node } i \text{ is the exit of link } j; \\ 0, & \text{if there is no connection;} \end{cases}$$

In matrix Γ, each column corresponds to a link and has two non-zero elements: 1 (-1) at the component associated with the entrance (exit) node.

Every vehicle in the TN leaves from an origin r and travels to its destination s. Every node in the TN can be an origin, a destination, or both. The O-D demand matrix $Q \in \mathbb{M}^{|T_N| \times |T_N|}$, and its element Q_{rs} represents the vehicular flow q_{rs} (also called the traffic demand or trip rate) from node r to node s. Q is not symmetric, and the element with a positive value corresponds to an active O-D pair. The set of active O-D pairs is denoted by D_T^{RS}. Later on, "active" is omitted without causing confusion. Each O-D pair $(r, s) \in D_T^{RS}$ is connected by a set of paths, denoted as K^{rs}. A particular path is labeled by an index $k \in K^{rs}$. The path is a chain of connected links. The link-path incidence matrix $\Delta = [\Delta^{rs}], \forall (r, s) \in D_T^{RS}$ depicts the path topology, where $\Delta^{rs} \in \mathbb{M}^{|T_A| \times |K^{rs}|}$ is the sub-matrix associated with O-D pair (r, s), with

$$\delta_{ak}^{rs} = \begin{cases} 1, & \text{if path } k \text{ passes link } a; \\ 0, & \text{otherwise;} \end{cases}$$

We adopt non-atomic vehicular flow which is described by real numbers, because the system impact from the action of a single vehicle is infinitesimal. In accordance with the notations of path and link-path incidence matrix Δ, the link flow x_a can be expressed as a linear combination of path flow f_k^{rs} as follows:

$$x_a = \sum_{rs} \sum_{k} f_k^{rs} \delta_{ak}^{rs}, \ \forall a \in T_A \tag{5.1}$$

or, in a compact form

$$x = \Delta f \tag{5.2}$$

where link flow vector $x = [x_a], \forall a \in T_a$, and path flow vector $f = [f_k^{rs}], \forall k \in K^{rs}, \forall (r, s) \in D_T^{RS}$. Moreover, the path flow should meet the traffic demand, i.e.,

$$\sum_{k \in K^{rs}} f_k^{rs} = q_{rs}, \ \forall r, s \tag{5.3}$$

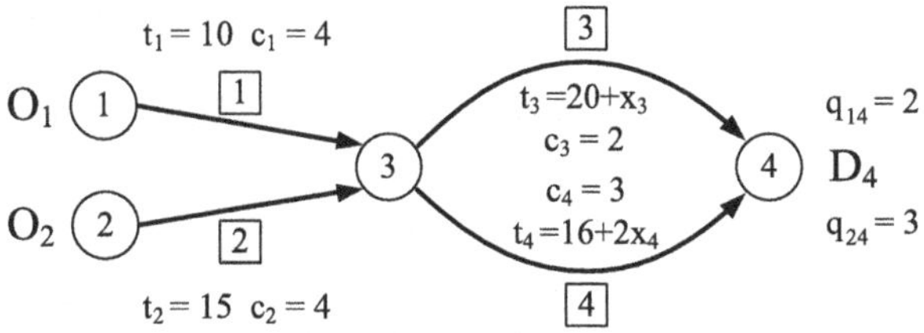

Fig. 5.1 A simple transportation network modified from [12]

or, in a compact form,

$$Ef = q \tag{5.4}$$

where trip rate vector $q = [q_{rs}], \forall(r, s) \in D_T^{RS}$, and E is a matrix consisting of 0 and 1 corresponding to the coefficients in (5.3).

An example modified from [12] is used to explain above concepts and notations more intuitively. A simple TN is shown in Fig. 5.1. The network includes 4 nodes and 4 links. O-D pair O_1-D_4 is connected by paths $\boxed{1} \to \boxed{3}$ and $\boxed{1} \to \boxed{4}$; the trip rate is $q_{14} = 2$. O-D pair O_2-D_4 is connected by paths $\boxed{2} \to \boxed{3}$ and $\boxed{2} \to \boxed{4}$; the trip rate is $q_{24} = 3$. The node-link incidence matrix Γ and link-path incidence matrix Δ can be written as follows:

Node \ Link	1	2	3	4
①	1	0	0	0
②	0	1	0	0
③	−1	−1	1	1
④	0	0	−1	−1

Link \ Path	O_1-D_4 1	O_1-D_4 2	O_2-D_4 1	O_2-D_4 2
1	1	1	0	0
2	0	0	1	1
3	1	0	1	0
4	0	1	0	1

The link flows can be expressed by path flows as

$$x_1 = f_1^{14} + f_2^{14}, \quad x_2 = f_1^{24} + f_2^{24}, \quad x_3 = f_1^{14} + f_1^{24}, \quad x_4 = f_2^{14} + f_2^{24} \tag{5.5}$$

and the flow conservation equations are given by

$$f_1^{14} + f_2^{14} = 2, \quad f_1^{24} + f_2^{24} = 3 \tag{5.6}$$

In transportation engineering, road travel time is the most important metrics to measure congestions. Let t_a denote the travel time on link a; it is a function of the traffic flow. The static TAP aims to find a so-called equilibrium solution

(x, t) that allocates reasonable route choices of drivers. Two representative concepts include:

1. Social Optimum. A traffic flow pattern reaches a social optimum if the total travel time $\sum_a x_a t_a$ is minimized (also referred to as the second Wardrop Principle). This setting requires a central agency which is eligible to decide a travel plan for every driver, while all the drivers are required to behave cooperatively so as to guarantee the most efficient utilization of the entire transportation system. Social optimum is an ideal state for theoretical study, but is unlikely to happen or be implemented in reality.
2. User Equilibrium. Each driver selects his route in order to minimize his own travel time (also referred to as the first Wardrop Principle). In the UE pattern, no traveler has the incentive to change his current route unilaterally. UE captures the selfish behavior of vehicles in urban transportation systems, and has been widely adopted in various research, since it better fits the reality.

Different descriptions of road travel latency t_a lead to completely different TAP models. This chapter introduces the classic Beckmann model and the LP based Nesterov model in the subsequent two sections.

5.2.2 Beckmann Model

In Beckmann model, the link travel time t_a on link a is assumed to be a strictly increasing function that depends only on its own vehicular flow x_a, such as the most widely used function by Bureau of Public Road [23]

$$t_a(x_a) = t_a^0 \left[1 + 0.15 \left(\frac{x_a}{c_a} \right)^4 \right], \quad \forall a \in T_A \tag{5.7}$$

where c_a is the vehicular flow on link a when $t_a = 1.15 t_a^0$. In some literature, it is also called the capacity of link a. However, the capacity limit is not a mandatory constraint, but is penalized by a quickly growing travel time if such a constraint is violated. To model the link travel time with an asymptotic capacity limit, the Davidson function, which is developed in [24] based on the queueing theory, can be employed as follows:

$$t_a(x_a) = t_a^0 \left[1 + J \left(\frac{x_a}{c_a - x_a} \right) \right], \quad \forall a \in T_A \tag{5.8}$$

where c_a represents the capacity limit of link $a \in T_A$, and parameter J controls the shape of Davidson function. Curves of $t_a(x_a)$ with different values of J are plotted in Fig. 5.2.

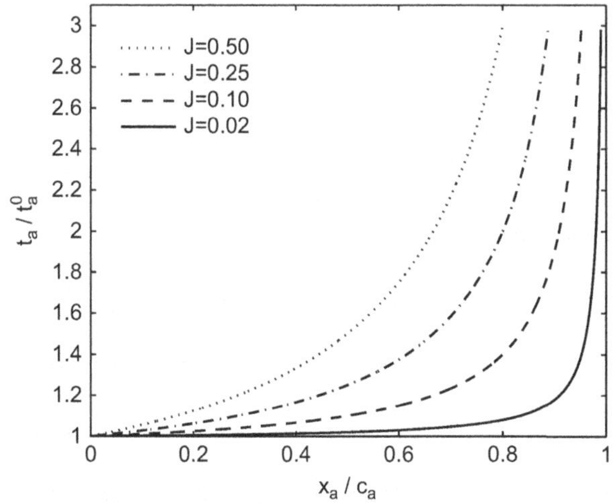

Fig. 5.2 Shape of Davidson function with different values of J

Upon the clarification of link travel time, the total travel time c_k^{rs} on path k between O-D pair (r, s) can be expressed by

$$c_k^{rs} = \sum_{a \in T_A} t_a(x_a)\delta_{ak}^{rs}, \ \forall k \in K^{rs}, \ \forall (r, s) \in D_T^{RS} \tag{5.9}$$

Network flow constraints (5.1) and (5.3) give all possible traffic flow patterns in the TN. To determine a meaningful outcome that may appear in reality, we make some further assumptions on the behavior of the drivers.

Assumption 5.1 (Drivers' Behavior) *Every vehicle travels to fixed destination using the path with minimal travel time.*

Drivers are not allowed to alter their destinations or postpone their travels. It is apparent that if anyone can cut down his travel time by using another path, the flow pattern will be changed. Because of the rationality of minimizing travel time, a stable state emerges when travel expenses on all used paths are equal. The stable traffic flow pattern is also called a UE. The concept can be generalized to consider destination choice, which is called a stochastic UE; to incorporate time effect, which is called a dynamic UE. See further reading at the end of this chapter. The basic UE condition is formally given below.

Proposition 5.1 (Wardrop Principle) *The traffic flow reaches a user equilibrium if travel times on all active paths between any given O-D pair are equal, and no greater than those which would be incurred on any unused path.*

Proposition 5.1 can be mathematically expressed in a logic form:

$$\forall (r,s) \in D_T^{RS},\ \exists u^{rs}: \begin{cases} c_k^{rs} = u^{rs}, & k \in K^{rs} \text{ and } f_k^{rs} > 0 \\ c_k^{rs} > u^{rs}, & k \in K^{rs} \text{ and } f_k^{rs} = 0 \end{cases} \tag{5.10}$$

The logical conditions in (5.10) can be formulated as the following nonlinear complementarity problem

$$0 \le f_k^{rs} \perp c_k^{rs} - u^{rs} \ge 0,\ \forall k \in K^{rs},\ \forall (r,s) \in D_T^{RS} \tag{5.11a}$$

$$c_k^{rs} = \sum_{a \in T_A} t_a(x_a)\delta_{ak}^{rs},\ \forall k \in K^{rs},\ \forall (r,s) \in D_T^{RS} \tag{5.11b}$$

$$\text{Cons-Flow}: \left\{ \begin{array}{l} x_a = \sum_{rs} \sum_{k \in K^{rs}} f_k^{rs}\delta_{ak}^{rs},\ \forall a \in T_A \\ \sum_{k \in K^{rs}} f_k^{rs} = q_{rs},\ \forall (r,s) \in D_T^{RS} \end{array} \right\} \tag{5.11c}$$

where notation $0 \le a \perp b \ge 0$ stands for complementarity and slackness conditions among scalars a and b, i.e., $a \ge 0$, $b \ge 0$, and $ab = 0$. In other words, at most one of a and b can be strictly positive. However, it is generally difficult to compute the UE by directly solving NCP (5.11) via a general purpose NLP solver, because standard constraint qualifications are violated due to the complementarity and slackness condition in (5.11a). A formal discussion can be found in [25]. The MILP reformulation in [18] can be applied if the latency functions $t_a(x_a)$ are represented by PWL functions. Such an NCP formulation is useful when the UE problem is embedded in a bilevel optimization problem, and MILP will provide a global optimal solution for the more challenging bilevel programs in spite of its NP-hard complexity. However, it is not preferred if efficiency becomes more important.

Following the paradigm in [11] and [12], it turns out that constraints in (5.11) constitute the KKT optimality conditions of the following optimization problem, which is called a static TAP

$$\min \sum_{a \in T_A} \int_0^{x_a} t_a(\theta)\mathrm{d}\theta \tag{5.12a}$$

$$\text{s.t. Cons-Flow} \tag{5.12b}$$

$$f_k^{rs} \ge 0,\ \forall k \in K^{rs},\ \forall (r,s) \in D_T^{RS} \tag{5.12c}$$

For the BPR function, the integration can be calculated as

$$\int_0^{x_a} t_a(\theta)\mathrm{d}\theta = t_a^0 \left(x_a + \frac{0.03x_a^5}{c_a^4} \right), \ \forall a \in T_A \tag{5.13}$$

For the Davidson function, we have

$$\int_0^{x_a} t_a(\theta)\mathrm{d}\theta = t_a^0(1-J)x_a + t_a^0 c_a J \ln\left(\frac{c_a}{c_a - x_a} \right), \ \forall a \in T_A \tag{5.14}$$

In either case, $t_a(x_a)$ is differentiable and strictly increasing, hence it is clear that the objective function is strictly convex, because

$$\frac{\mathrm{d}^2}{\mathrm{d}x_a^2} \int_0^{x_a} t_a(\theta)\mathrm{d}\theta = \frac{\mathrm{d}t_a(x_a)}{\mathrm{d}x_a} > 0, \ \forall a \in T_A$$

In this regard, TAP (5.12) is a strictly convex program with linear constraints, and can be globally solved by traditional NLP solvers. Its optimal solution $x_a^*, \forall a$ is unique and determines the traffic flow distributions at the UE pattern. This is a very elegant result because it significantly reduces the computation complexity of UE. If BPR function is used, Cons-Flow is always non-empty, so (5.12) would have a solution. The Davidson function has no definition outside the interval $[0, c_a)$. It actually imposes bound constraints on link traffic flows. TAP (5.12) may become infeasible if the traffic demand is very high. Another explicit assumption in the current formulation of TAP (5.12) is that the sets of all possible paths $K^{rs}, \forall(r, s) \in D_T^{RS}$ are available; however, path enumeration could be difficult and also unnecessary for large-scale TNs, because most paths are redundant (carry zero traffic flow) at the UE. A more practical algorithm will be discussed in Sect. 5.2.4 which identifies active paths on the fly.

The UE of the system shown in Fig. 5.1 is analyzed below. For simplicity, travel time functions on link $\boxed{1}$ and link $\boxed{2}$ are constants, while those corresponding to link $\boxed{3}$ and link $\boxed{4}$ are linear functions, as marked beside each link. Owing to the flow conservation equations, there must be $x_1^* = 2$ and $x_2^* = 3$ for both models. According to Proposition 5.1 the traffic flow on link $\boxed{3}$ and link $\boxed{4}$ must satisfy

$$20 + x_3 = 16 + 2x_4$$

$$x_3 + x_4 = 5$$

which gives $x_3^* = 2$ and $x_4^* = 3$, so the corresponding path travel time is obtained as

$$t_1^{14} = t_1(x_1^*) + t_3(x_3^*) = 32$$
$$t_2^{14} = t_1(x_1^*) + t_4(x_4^*) = 32$$
$$t_1^{24} = t_2(x_2^*) + t_3(x_3^*) = 37$$
$$t_2^{24} = t_2(x_2^*) + t_3(x_4^*) = 37$$

Write out the Beckmann TAP model of this system

$$\begin{aligned} \min \ & 10x_1 + 15x_2 + 20x_3 + 0.5x_3^2 + 16x_4 + x_4^2 \\ \text{s.t.} \ & x_1 = 2, \ x_3 = f_1^{14} + f_1^{24}, \ f_1^{14} + f_2^{14} = 2 \\ & x_2 = 3, \ x_4 = f_2^{14} + f_2^{24}, \ f_1^{24} + f_2^{24} = 3 \end{aligned}$$

Since x_1 and x_2 are fixed, the above problem reduces to

$$\begin{aligned} \min \ & 20x_3 + 0.5x_3^2 + 16x_4 + x_4^2 \\ \text{s.t.} \ & x_3 \geq 0, \ x_4 \geq 0 \, , x_3 + x_4 = 5 \end{aligned}$$

We assume $x_3 > 0$ and $x_4 > 0$ hold at optimum, according to the equal incremental criterion, $20 + x_3 = 16 + 2x_4$ must be satisfied. So the unique optimal solution is $x_3^* = 2$ and $x_4^* = 3$. We observe that in this case, the Wardrop UE principle simply interprets the equal incremental criterion. For path flow variables, we have 4 variables and 4 equations

$$f_1^{14} + f_1^{24} = 2, \ f_1^{14} + f_2^{14} = 2$$
$$f_2^{14} + f_2^{24} = 3, \ f_1^{24} + f_2^{24} = 3$$

However, the linear equations are not independent. The rank of the coefficient matrix is 3, which means that we can derive any one of these equations using the remaining three ones. Let $0 \leq \alpha \leq 2$, then $f_1^{14} = \alpha$, $f_1^{24} = 2 - \alpha$, $f_2^{14} = 2 - \alpha$, $f_2^{24} = 1 + \alpha$ will be the solution of path flows. Non-uniqueness of path flow will not impact the congestion pattern, which only depends on link flows.

5.2.3 *Nesterov Model*

Nesterov and de Palma point out that some assumption on the latency function $t_a(x_a)$ of Beckmann model may not be consistent with the reality [26, 27].

1. A less important criticism relates to the strictly increasing assumption for the latency function $t_a(x_a)$. Clearly, the travel time is a constant when the road is not congested and every vehicle can move at the maximal allowed speed.
2. It is clear that the vehicle flow on link a cannot be arbitrarily large. Ignoring the mandatory road capacity constraint (in BPR function) may lead to infeasible solutions when the traffic flow on a link is greater than its capacity.
3. Most important of all, there is a physical contradiction to the monotonicity of $t_a(x_a)$. According to the definition of traffic flow: flow = speed × density, if the flow x_a is large, neither the speed nor the density can be very small (it is not realistic to compensate for the drop in speed with a dramatic increase in the density, because every vehicle in motion has to maintain a safe distance from the vehicle in front of it). As a result, by assuming a maximal density of vehicles, a further increment of x_a must lead to a rise in the speed, and in turn result in a drop in the travel time.

To mitigate the discrepancy in the road latency function, Nesterov and de Palma abandon making further assumptions on the specific expression of t_a. Instead, they establish a new TAP model based on two weaker assumptions:

Assumption 5.2 (Flow Limit) *Traffic flow on a link cannot exceed its capacity, i.e.,*

$$x_a \leq c_a, \ \forall a \in T_A \tag{5.15}$$

It should be mentioned that the capacity parameter c_a in Nesterov model is not necessarily the same as that in Beckmann model.

Assumption 5.3 (Link Performance) *There is no delay on a link where the traffic flow does not reach its capacity; slowdown only occurs when the traffic flow reaches its capacity, i.e.*

$$\begin{aligned} x_a < c_a &\Rightarrow t_a = t_a^0 \\ x_a = c_a &\Rightarrow t_a \geq t_a^0 \end{aligned} \tag{5.16}$$

We say link a is congested if $t_a > t_a^0$. Congestion occurs in the following manner: when the traffic flow x_a on certain link a reaches its capacity c_a, any additional vehicle will slow down the speed of existing vehicles, as such the travel time t_a increases, while the total flow remains the same (but car density increases). Assumptions 5.2 and 5.3 are weaker than the functional form assumption on the link travel time $t_a(x_a)$ in Beckmann's model. In contrast to $t_a(x_a)$ which is only affected by x_a, road congestions in Nesterove model depend on the utilization of the whole transportation network.

Define vectors $c = [c_a], \forall a \in T_A$, $t^0 = [t_a^0], \forall a \in T_A$, and $t = [t_a], \forall a \in T_A$. Suppose there is a central coordinator who can manage the behavior of vehicles to reach a social optimal pattern. In accordance with its definition in Sect. 5.2.1, the traffic flow vector x and travel time vector t must solve the following NLP

$$\min_{x,f,t} \ x^T t \tag{5.17a}$$

$$\text{s.t. } x - \Delta f = 0 \tag{5.17b}$$

$$Ef = q \tag{5.17c}$$

$$x \le c \tag{5.17d}$$

$$f \ge 0, \ t \ge t^0 \tag{5.17e}$$

where $x^T t = \sum_a x_a t_a$ is the total travel time, constraints (5.17b) and (5.17c) are the compact forms of network vehicular flow constraints (5.1) and (5.3). Inequalities (5.17d) and (5.17e) are the boundary constraints of decision variables. Constraints (5.17b) and (5.17e) naturally suggest $x \ge 0$. Based on this fact, it is clear that at the optimal solution, $t^* = t^0$ regardless of the value of x. Therefore, we arrive at *Nesterov social optimum model*, which is an LP:

$$\min_{x,f} \ x^T t^0 \tag{5.18a}$$

$$\text{s.t. } x - \Delta f = 0 \tag{5.18b}$$

$$Ef = q, \ f \ge 0 \tag{5.18c}$$

$$x \le c : \lambda \tag{5.18d}$$

Problem (5.18) indicates that no congestion happens in the social optimum pattern. Please keep in mind that if the traffic demand exceeds the transit capacity of the transportation network, constraints (5.18b)–(5.18d) will be contradictive, and TAP (5.18) will be infeasible, suggesting that infrastructure upgrading is needed.

Next, we investigate the dual variable $\lambda = [\lambda_a], \forall a \in T_A$ associated with the flow capacity constraint (5.18d), which can be interpreted as the delay that a user would experience when using a congested road, and formulate the Lagrange relaxation of problem (5.18) as

$$\max_{\lambda \ge 0} \min_{f} \ f^T \Delta^T t^0 + \lambda^T (\Delta f - c) \tag{5.19a}$$

$$\text{s.t. } Ef = q, \ f \ge 0 \tag{5.19b}$$

For fixed $\bar{\lambda} \ge 0$, problem (5.19) can be decoupled into the subproblems shown in (5.20) with respect to each O-D pair (r, s), since the remaining constraints are block diagonal. Moreover, the constant term $\bar{\lambda}^T c$ in the objective function can be omitted, then the relaxed social optimum problem for O-D pair $(r, s) \in D_T^{RS}$ can be written as

$$\min_{f^{rs}} \ (t^0 + \bar{\lambda})^T \Delta^{rs} f^{rs} \tag{5.20a}$$

$$\text{s.t.} \ \ E^{rs} f^{rs} = q_{rs}, \ f^{rs} \geq 0 \tag{5.20b}$$

where $f^{rs} = [f_k^{rs}], \forall k \in K^{rs}$, notations Δ^{rs} and E^{rs} represent the sub-matrices in Δ and E corresponds to O-D pair (r, s). Problem (5.20) can be interpreted as follows: when each driver between O-D pair (r, s) makes his own decision given the delay vector $\bar{\lambda}$ without a central coordinator, he must pick up the path with minimal travel time, i.e., $f_K^{rs} = q_{rs}$, where $K = \arg\min_k [t_k^{rs}, \forall k]$, and $f_k^{rs} = 0$, $\forall k \neq K$. The total travel time between O-D pair (r, s) is $t^{rs} = (t^0 + \bar{\lambda})^T \Delta^{rs}$. It is apparent that the social optimum and the UE will provide the same traffic flow pattern in the absence of capacity constraints. Now suppose the optimal solution of LP (5.18) is x^*, and the corresponding dual variable of (5.18d) is λ^*. Because strong duality holds for LPs, which means the duality gap is 0 at the optimal primal-dual pair (x^*, λ^*), and the congestion cannot be further alleviated, then we can arrive at an important conclusion [26, 27]

$$\begin{array}{ll} (x^*, t^0) & \text{is a traffic assignment at social optimum} \\ (x^*, t^0 + \lambda^*) & \text{is a traffic assignment at user equilibrium} \end{array} \tag{5.21}$$

Now we can see that the travel delay caused by congestion is characterized by the Lagrange dual multipliers, which depends on the traffic flow condition in the whole system, rather than a univariate function $t_a(x_a)$ which solely depends on x_a in a particular link. The way of defining travel latency distinguishes Beckmann model and Nesterov model. It is worth mentioning that the UE and the social optimum offered by Nesterov model share the same traffic flow pattern. The two states only differ in the travel times, indicating different vehicle densities.

The simple system in Fig. 5.1 is considered again via Nesterov model. In accordance with TAP (5.18), the link flow x_3 and x_4 must solve the following LP

$$\begin{aligned} \min \ & 20x_3 + 16x_4 \\ s.t. \ & x_3 + x_4 = 5 \\ & x_3 \leq 2 : \ \lambda_3 \\ & x_4 \leq 3 : \ \lambda_4 \end{aligned}$$

The optimal solution is $x_3^* = 2$, $\lambda_3^* = 0$ and $x_4^* = 3$, $\lambda_4^* = 4$, indicating that link $\boxed{4}$ is congested. The link travel times at the UE pattern are given by

$$t_1^{14} = t_1^0 + t_3^0 + \lambda_3^* = 10 + 20 + 0 = 30$$
$$t_2^{14} = t_1^0 + t_4^0 + \lambda_4^* = 10 + 16 + 4 = 30$$
$$t_1^{24} = t_2^0 + t_3^0 + \lambda_3^* = 15 + 20 + 0 = 35$$
$$t_2^{24} = t_2^0 + t_4^0 + \lambda_4^* = 15 + 16 + 4 = 35$$

indicating that no road user can reduce travel time by changing his route unilaterally.

In this particular case, Backmann and Nesterov traffic assignment models offer consistent UE in terms of link traffic flow. Comparative studies on some benchmark transportation networks and real large-scale networks in Switzerland suggest that both traffic assignment models provide similar UE results, especially when capacity constraints are imposed in Backmann model [28]. It is also pointed out in [28] that the travel times of the two models are generally not comparable, as they are established based on different assumptions on link latency.

5.2.4 Approaches Without Path Enumeration

As mentioned above, TAPs in previous two sections render either a convex program with linear constraints or an LP. If all usable paths K^{rs} for every O-D pair are enumerated off-line in advance, TAP can be readily solved by commercial solvers such as IPOPT [29] (for Beckmann model), a primal-dual interior-point algorithm based NLP solver, and CPLEX [30] (for Nesterove model), an MILP solver. However, path enumeration could be exhaustive for large systems, and also introduces a great number of path flow variables, preventing a concise problem setup in solvers. This section discusses two approaches without path enumeration.

1. An O-D Link Flow Based Model

A simple remedy for removing path flow variables is to associate link flow with each O-D pair. To this end, let x_{ij}^{rs} represent traffic flow on link (i, j) contributed by traffic demand between O-D pair (r, s), which is therefore called O-D link flow. Consider flow conservation at node i, which is portrayed in Fig. 5.3, we have

$$\sum_{j \in d(i)} x_{ij}^{rs} - \sum_{k \in u(i)} x_{ki}^{rs} = d_i^{rs} = \begin{cases} q_{rs}, & \text{if } i \text{ is the origin node } r \\ -q_{rs}, & \text{if } i \text{ is the destination node } s \\ 0 & \text{otherwise} \end{cases}$$

where $d(i)/u(i)$ stands for the set of downstream/upstream nodes of node i.

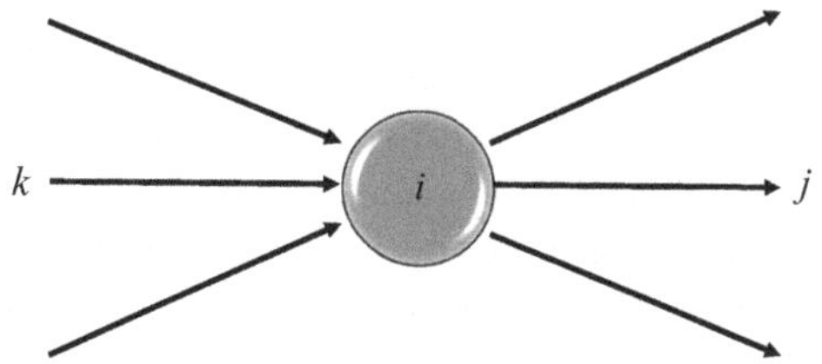

Fig. 5.3 Conservation of link traffic flow

Take the simple system in Fig. 5.1 for example, there is

$$\left\{\begin{array}{l} x_{13}^{14}=q_{14},\ x_{23}^{14}=0 \\ x_{34,1}^{14}+x_{34,2}^{14}-x_{13}^{14}-x_{23}^{14}=0 \\ -x_{34,1}^{14}-x_{34,2}^{14}=-q_{14} \end{array}\right\}, \quad \text{for } \mathrm{O}_1-\mathrm{D}_4$$

$$\left\{\begin{array}{l} x_{13}^{24}=0,\ x_{23}^{24}=q_{24} \\ x_{34,1}^{24}+x_{34,2}^{24}-x_{13}^{24}-x_{23}^{24}=0 \\ -x_{34,1}^{24}-x_{34,2}^{24}=-q_{24} \end{array}\right\}, \quad \text{for } \mathrm{O}_2-\mathrm{D}_4$$

The actual link flow is the summation of O-D link flow. In view of this, the link based TAP model can be formulated as

$$\min \sum_{(i,j)\in T_A} \int_0^{x_{ij}} t_{ij}^a(\theta)\mathrm{d}\theta \tag{5.22a}$$

$$\text{s.t.} \sum_{j\in d(i)} x_{ij}^{rs} - \sum_{k\in u(i)} x_{ki}^{rs} = d_i^{rs},\ \forall i \in T_N, \forall (r,s) \in D_T^{RS} \tag{5.22b}$$

$$x_{ij} = \sum_{(r,s)} x_{ij}^{rs},\ \forall (i,j) \in T_A \tag{5.22c}$$

$$x_{ij}^{rs} \geq 0,\ \forall (i,j) \in T_A,\ \forall (r,s) \in D_T^{RS} \tag{5.22d}$$

where each link a is indexed by the pair of head and tail nodes $(i, j) \in T_A$; the flow conservation constraint (5.22b) is nodal-wise for each O-D pair; link flow is represented by O-D link flow in (5.22c); non-negativity is imposed via (5.22d).

In problem (5.22), the number of variables is equal to $|T_A| \times (|D_T^{RS}|+1)$. Because $|D_T^{RS}|$ grows quickly with respect to the system size, (5.22) is still not scalable for large-scale transportation networks.

2. An Adaptive Path Generation Algorithm

Although there could be a large number of paths connecting an O-D pair of nodes, only a fraction of them will be used actually. To circumvent exhaustive enumeration, the central idea is to identify a subset of paths which are most likely to be active. To do this, we can start with a subset of K^{rs} which includes only a few available paths, and solve a restricted TAP, where "restricted" means that the feasible region of TAP is smaller than the original TAP, as some paths are neglected. At the optimal solution, if the travel time of at least one O-D pair can be reduced by exploring a new path, then the path is added, and K^{rs} will be updated. The iteration proceeds until no one can shorten their travel time by switching to a new route. In such circumstance, the solution to the restricted TAP also solves the original formulation.

This section streamlines the above procedure in a canonical mathematical format. A few new variables in the path generation sub-problem are elaborated. Vector $I^{rs} \in \mathbb{R}^{|T_N|}$ has two non-zero elements, 1 (-1) at the entry corresponding to origin node r (destination node s); $v_p \in \mathbb{B}^{|T_A|}$ is a vector consists of $|T_A|$ binary variables. In view of the definition of node-link incidence matrix Λ, if v_p satisfies $\Lambda v_p = I^{rs}$, the links labeled by 1 consist of a chain of connected roads from r to s, i.e., an available path between (r, s). Taking the system shown in Fig. 5.1 for example again. For paths $\boxed{1} \to \boxed{3}$ and $\boxed{1} \to \boxed{4}$ between O$_1$-D$_4$, it can be verified that

$$I^{14} = \begin{bmatrix} 1 \\ 0 \\ 0 \\ -1 \end{bmatrix}, \quad v_1^{14} = \begin{bmatrix} 1 \\ 0 \\ 1 \\ 0 \end{bmatrix}, \quad v_2^{14} = \begin{bmatrix} 1 \\ 0 \\ 0 \\ 1 \end{bmatrix}, \quad \begin{aligned} \Lambda v_1^{14} &= I^{14} \\ \Lambda v_2^{14} &= I^{14} \end{aligned}$$

Similarly, for paths $\boxed{2} \to \boxed{3}$ and $\boxed{2} \to \boxed{4}$ between O$_2$-D$_4$, we have

$$I^{24} = \begin{bmatrix} 0 \\ 1 \\ 0 \\ -1 \end{bmatrix}, \quad v_1^{24} = \begin{bmatrix} 0 \\ 1 \\ 1 \\ 0 \end{bmatrix}, \quad v_2^{24} = \begin{bmatrix} 0 \\ 1 \\ 0 \\ 1 \end{bmatrix}, \quad \begin{aligned} \Lambda v_1^{24} &= I^{24} \\ \Lambda v_2^{24} &= I^{24} \end{aligned}$$

The link-path incidence matrix $\Delta = \{v_1^{14}, v_2^{14}, v_1^{24}, v_2^{24}\}$, which consists of feasible solutions of $\Lambda v = I^{14}$ and $\Lambda v = I^{24}$.

Proposition 5.2 (Path Enumeration) *The link-path incidence matrix can be expressed as* $\Delta = [\Delta^{rs}], \forall (r, s) \in D_T^{RS}$, *where* $\Delta^{rs} = [v_g^{rs}], \forall v_g^{rs} \in V^{rs}$ *is the submatrix for O-D pair* (r, s), *and set* $V^{rs} = \{v^{rs} \in \mathbb{B}^{|T_A|} \mid \Lambda v^{rs} = I^{rs}\}$ *collects usable paths between* (r, s).

Clearly, the obstacle of path enumeration stems from the construction of V^{rs}, the feasible set of a binary program. In a large network, there could be extensive paths connecting a given O-D pair. In fact, most of them will not be utilized. To formalize the idea mentioned at the beginning of this section, we consider the TAP problem has already been solved using subsets of K^{rs}, $\forall (r, s) \in D_T^{RS}$, and the link flow solution is x^*. The link travel time t_a^* can be calculated via latency function $t_a(x_a)$ or Eq. (5.21) accordingly, and the minimal travel time between (r, s) is u_b^{rs}. To find a possibly better path between (r, s), we solve the following MILP

$$u_c^{rs} = \min_v \sum_a t_a^* v_a \tag{5.23a}$$

$$\text{s.t. } \Lambda v = I^{rs} \tag{5.23b}$$

$$v \in \mathbb{B}^{|T_A|} \tag{5.23c}$$

If $u_c^{rs} < u_b^{rs}$, the optimal solution v indicates a better path, which will be added in K^{rs} as well as in Δ^{rs}. A procedure for solving TAP without path enumeration is shown in Algorithm 5.1. It appears to converge quickly because the number of paths that will be used is very limited.

Algorithm 5.1 Delayed path generation

1: Let $x^* = 0$, compute link travel time t_a^*, $\forall a$ and minimum travel time u_b^{rs} between every O-D pair $(r, s) \in D_T^{RS}$, solve path generation subproblem (5.23) for each O-D pair, and then build link-path incidence matrices Δ.
2: Solve the restricted TAP (5.12) or (5.18) with current Δ, the UE is x^*, update the link travel time t_a^*, $\forall a$, and the minimal travel time u_b^{rs}, $\forall (r, s) \in D_T^{RS}$.
3: solve path generation subproblem (5.23) for each O-D pair, extract the optimal solution v^* and optimal value u_c^{rs}. If $u_c^{rs} \geq u_b^{rs}$, $\forall (r, s)$, terminate and report the current UE solution x^*; otherwise, if $u_c^{rs} < u_b^{rs}$ for some (r, s), $\Delta^{rs} \leftarrow [\Delta^{rs}, v^*]$, update Δ and go to step 2.

Some special tricks are given below.

1. To accelerate convergence of Algorithm 5.1, multiple paths are identified and appended in matrix Δ; some of them may eventually be redundant in the next iteration, since minimal travel times between individual O-D pairs are highly correlated. Nevertheless, this does not harm the usefulness of Algorithm 5.1, as redundant paths only account for a small portion, unlike the situation in the complete path enumeration demonstrated by Proposition 5.2.
2. If we manage to ensure every path included in Δ will be active, then according to Proposition 5.1, the UE in Beckmann model solves the following system of nonlinear equations

$$\sum_{a \in T_A} t_a \delta_{ak}^{rs} = u^{rs}, \ \forall k \in K^{rs}, \ \forall (r, s) \tag{5.24a}$$

$$x_a = \sum_{rs} \sum_{k \in K^{rs}} f_k^{rs} \delta_{ak}^{rs}, \ \forall a \in T_A \tag{5.24b}$$

$$\sum_{k \in K^{rs}} f_k^{rs} = q_{rs}, \ \forall (r, s) \in D_T^{RS} \tag{5.24c}$$

$$t_a = t_a(x_a), \ \forall a \in T_A \tag{5.24d}$$

and nonlinearity only appears in latency functions (5.24d). In view of the convexity of $t_a(x_a)$, (5.24d) can be replaced by an inequality constraint which is convex, and the user equilibrium solves the following convex program

$$\min \left\{ \sum_a t_a \,\middle|\, (5.24a),\ (5.24b),\ (5.24c),\ t_a \geq t_a(x_a),\ \forall a \in T_A \right\} \tag{5.25}$$

As long as there is no redundant path in Δ, the convex relaxation performed on (5.24d) will be exact. In addition, for BPR function, the convex relaxation can be divided into two quadratic inequalities

$$t_a \geq t_a^0(1 + 0.15 y_a^2) \tag{5.26a}$$

$$y_a \geq \left(\frac{x_a}{c_a}\right)^2 \tag{5.26b}$$

and for Davidson function, the convex relaxation can be performed as

$$t_a = t_a^0(1 - J + J z_a) \tag{5.27a}$$

$$y_a z_a \geq c_a,\ y_a \geq 0 \tag{5.27b}$$

$$x_a + y_a = c_a \tag{5.27c}$$

To see the convexity of (5.27b), it is ultimately an SOC

$$\left\| \begin{matrix} 2\sqrt{c_a} \\ y_a - z_a \end{matrix} \right\|_2 \leq y_a + z_a \tag{5.28}$$

3. Formulations in (5.25)–(5.28) may not be very advantageous in the TAP, because it is already convex and easy to solve; furthermore, if an inactive path is included in (5.25), the relaxation performed in (5.26) will be inexact. Nonetheless, these formulations provide convexification potential when the link capacity transits into designing variables. Such kind of problem is known as continuous network design problem [31], which is usually cast as a bilevel program, a challenging mathematical optimization problem [32]. In the upper level, the objective function is defined as the weighted summation of total travel times and link capacity expansion costs; in the lower level, the traffic flow pattern is determined by TAP. In what follows, we discuss how the continuous network

design problem can be solved via convex optimization. The system total travel time $\sum_a x_a t_a$ is nonlinear. Nonetheless, it allows a linear transformation

$$\begin{aligned}\sum_a t_a x_a &= \sum_{rs}\sum_k\sum_a t_a f_k^{rs}\delta_{ak}^{rs} \quad \text{due to (5.1)}\\ &= \sum_{rs}\sum_k c_k^{rs} f_k^{rs} \quad \text{due to (5.9)}\\ &= \sum_{rs} u^{rs} \sum_{k, f_k^{rs}>0} f_k^{rs} \quad \text{due to (5.11a)}\\ &= \sum_{rs} u^{rs} q_{rs} \quad \text{due to (5.3)}\end{aligned} \tag{5.29}$$

The last term is a linear function in u^{rs}. Equation (5.29) is a natural result of Proposition 5.1. For Davidson function based formulation, let $c_a^{sr} = \sqrt{c_a}$, if the expansion costs function $F(c_a^{sr})$ is convex, then the network design problem can be formulated as a convex program

$$\min \left\{ \sum_{rs} u^{rs} q_{rs} \left| \begin{array}{l} \text{(5.24a)–(5.24c), (5.27a), (5.27c)} \\ \left\| \begin{array}{c} 2c_a^{sr} \\ y_a - z_a \end{array} \right\|_2 \le y_a + z_a,\ y_a \ge 0,\ \forall a \in T_A \\ \sum_a F(c_a^{sr}) \le \text{Budget of investment} \end{array} \right. \right\} \tag{5.30}$$

For the BPR function based formulation, let $c_a^{sq} = c_a^2$, if the expansion costs function $F(c_a^{sq})$ is convex, then the network design problem can be formulated as a convex program

$$\min \left\{ \sum_{rs} u^{rs} q_{rs} \left| \begin{array}{l} \text{(5.24a)–(5.24c), (5.26a)} \\ \left\| \begin{array}{c} 2x_a \\ y_a - c_a^{sq} \end{array} \right\|_2 \le y_a + c_a^{sq},\ y_a \ge 0,\ \forall a \in T_A \\ \sum_a F(c_a^{sq}) \le \text{Budget of investment} \end{array} \right. \right\} \tag{5.31}$$

The convexity of $F(c_a^{sr})$ is relatively mild. For example, if the construction cost is a linear function $K_B \Delta c_a = K_B(c_a - c_a^0)$, where c_a^0 is the initial capacity of link a, K_B is a constant, then $F(c_a^{sr}) = K_B[(c_a^{sr})^2 - c_a^0]$. If Δc_a is small compared with c_a, then $c_a^{sq} = (c_a^0 + \Delta c_a)^2 \approx (c_a^0)^2 + 2c_a^0 \Delta c_a$, thus $F(c_a^{sq}) = K_B \Delta c_a \approx K_B[c_a^{sq} - (c_a^0)^2]/2c_a^0$ is linear. Since the continuous network design problem is non-convex, the convex relaxation procedure would be worthwhile.

However, the active paths will depend on the planning strategies, and are more difficult to determine in advance. A remedy is similar to Algorithm 5.1: first solve the convex relaxation model; if the relaxation is inexact, solve the convex TAP with fixed planning strategies and identify inactive paths; if the relaxation is exact, solve the path generation oracle to see if the travel cost can be reduced by exploring a new path.

5.3 Optimal Traffic-Power Flow

EVs have been widely recognized as a promising remedy for the environmental crisis and fuel shortage faced by modern metropolises for a long time. However, the proliferation of EVs in big cities will tighten the relation between the TN and the PDN. Both infrastructures have been separately studied for decades. In the past few years, researchers are beginning to be aware of the interdependency between them brought by the increasing penetration of charging facilities. In such a coupled system, the traffic condition and toll policy in the TN will influence the route choices of EV drivers. The spatial and temporal distribution of vehicle flows in turn impacts the charging load profile, thus affecting the operation of the PDN, which is ignored in most existing research work.

This section envisions an electrified TN enabled by wireless power transfer technology and coupled with a PDN. The cutting-edge wireless power transfer technology has the potential to supply the power consumption of a moving EV at a speed of 75 mph [33, 34]. Wireless power transfer offers a more convenient charging opportunity for EV owners, and eliminates their range anxiety of running out of energy.

We propose an OTPF method for coordinated operation of the interdependent transportation and power distribution systems, which will potentially become a fundamental analytical tool of policy making, planning, and operation of such critical infrastructures in future cities. The mathematical model does not change much if the wireless charging is replaced with FCSs under some natural assumptions.

5.3.1 Mathematical Formulation

In the considered problem, an independent system operator (ISO), a non-profit public entity, is eligible to manage local generation assets and charge congestion tolls on electrified roads with the purpose of minimizing social cost. The route choices of EVs are amenable to the Wardrop UE principle, such that no one can reduce his travel cost by changing route unilaterally. The traffic UE pattern further influences the spatial distribution of the charging loads of the PDN, in which the power flow is described by the BFM presented in Sect. 2.2.2. Major symbols and notations used throughout this section in the TAP and OPF problem are inherited from Sects. 2.2.2 and 5.2.2. There may be slight differences in superscripts and subscripts, without causing confusions. Others are defined following their first appearances.

1. User Equilibrium and Traffic Assignment Problem

The formulations of UE and TAP are similar to those introduced in Sect.5.2, except for the definition of travel cost. In particular, we assume that passing each link will be charged with a congestion toll T_a^C. The travel cost c_k^{rs} of a single vehicle on path k between O-D pair (r, s) is a weighted summation of travel time and congestion toll, which is given by

$$c_k^{rs} = \sum_a (\omega t_a + T_a^C)\delta_{ak}^{rs}, \ \forall k \in K_{rs}, \forall(r,s) \tag{5.32}$$

where ω is the monetary value of travel time. If the battery keeps charging with a constant rate, which is supplied by the wireless charging device, and ω is the charging expense in per unit time, which is proportional to the electricity price, the first term represents the charging cost during completing the trip. The rationality of travelers requires them using the route with minimum travel cost, while considering the congestion pattern over the entire TN. In accordance with discussions in Sect. 5.2, the UE comes down to an NCP

$$0 \le f_k^{rs} \perp c_k^{rs} - u^{rs} \ge 0, \ \forall k \in K^{rs}, \ \forall(r,s) \tag{5.33a}$$

$$c_k^{rs} = \sum_a \left[\omega t_a(x_a) + T_a^C\right]\delta_{ak}^{rs}, \ \forall k \in K_{rs}, \forall(r,s) \tag{5.33b}$$

$$t_a(x_a) = t_a^0 \left[1 + 0.15\left(\frac{x_a}{c_a}\right)^4\right], \ \forall a \in T_A \tag{5.33c}$$

$$\text{Cons-Flow}: \left\{ \begin{aligned} & x_a = \sum_{rs}\sum_{k \in K^{rs}} f_k^{rs}\delta_{ak}^{rs}, \ \forall a \in T_A \\ & \sum_{k \in K^{rs}} f_k^{rs} = q^{rs}, \ \forall(r,s) \in D_T^{RS} \end{aligned} \right\} \tag{5.33d}$$

Following the paradigm in [12], it can be verified that NCP (5.33) constitutes the KKT optimality conditions of the following TAP, a convex optimization problem if the congestion toll T_a^C is specified

$$\begin{aligned} \min \ & F_{TAP} = \omega \sum_a \int_0^{x_a} t_a(\theta)\mathrm{d}\theta + \sum_a T_a^C x_a \\ \text{s.t.} \ & \text{Cons-Flow}, \ f_k^{rs} \ge 0, \ \forall k, \ \forall(r,s) \end{aligned} \tag{5.34}$$

In spite of the technical challenge in solving an NCP, the UE formulation in (5.33) will be useful in this section, because TAP will be embedded in the lower level of OTPF problem. The convexity of TAP (5.34) does not directly facilitate problem resolution. TAP (5.34) will be used in a heuristic decomposition algorithm which gives an approximated OTPF solution.

2. Optimal Power Flow of the Distribution Network

The formulation of OPF problem is similar to that introduced in Sect. 2.3.1, except for the demand. Because EVs are charged in motion, we assume that the charging demand of each electrified road should be a linear function of the vehicle flow, yielding the following charging power equation

$$p_i^d = p_i^{dc} + \eta_a \sum_{a \in C(i)} x_a, \ \forall i \in E_N \tag{5.35}$$

where $C(i)$ is the set of electrified roads served by bus i, p_i^{dc} is the constant traditional power demand at bus i, η_a is the charging rate of unit traffic flow, which is a key factor accounting for the interdependency across the TN and the PDN. This modeling paradigm also applies to other charging modes, such as charging and swapping stations, as long as a proper charging power function $\eta(x_a)$ can be specified.

In accordance with discussions in Sect.2.3.1, the convex relaxation model of OPF problem can boil down to an SOCP

$$\min \quad \sum_i \left[a_i (p_i^g)^2 + b_i p_i^g \right] + \rho \sum_{j \in \pi(0)} P_{0j}^l \tag{5.36a}$$

$$\text{s.t.} \quad P_{ij}^l + p_j^g - r_{ij}^l I_{ij}^l = \sum_{k \in \pi(j)} P_{jk}^l + p_j^d, \ \forall l \tag{5.36b}$$

$$Q_{ij}^l + q_j^g - x_{ij}^l I_{ij}^l = \sum_{k \in \pi(j)} Q_{jk}^l + q_j^d, \ \forall l \tag{5.36c}$$

$$U_j = U_i - 2(r_{ij}^l P_{ij}^l + x_{ij}^l Q_{ij}^l) + (z_{ij}^l)^2 I_{ij}^l, \ \forall l \tag{5.36d}$$

$$I_{ij}^l U_i \geq (P_{ij}^l)^2 + (Q_{ij}^l)^2, \ \forall l \tag{5.36e}$$

$$P_{ij}^l \geq 0, \ Q_{ij}^l \geq 0, \sqrt{(P_{ij}^l)^2 + (Q_{ij}^l)^2} \leq S_l, \ \forall l \tag{5.36f}$$

$$p_i^{gl} \leq p_i^g \leq p_i^{gu}, q_i^{gl} \leq q_i^g \leq q_i^{gu}, U_i^l \leq U_i \leq U_i^m, \ \forall i \tag{5.36g}$$

According to (5.35), the electrical power demand p_j^d in constraint (5.36b) depends on the UE appearing in the TN, which is the optimal solution of TAP (5.34). To avoid an optimization problem in the constraints, we can build the OTPF model by appending optimality condition (5.33) into (5.36), which is explained below.

3. Optimal Traffic-Power Flow
We assume that the coupled system is managed by the non-profit ISO. It is eligible to set congestion tolls on the electrified roads in the TN and dispatch local generators in the PDN. The operator aims to minimize social cost, i.e., the sum of total travel cost on the TN

$$F_T = \sum_a (\omega t_a + T_a^C) x_a \tag{5.37}$$

and the energy service cost of PDN, i.e., the objective function of the OPF problem

$$F_E = \sum_i \left[a_i (p_i^g)^2 + b_i p_i^g \right] + \rho \sum_{j \in \pi(0)} P_{0j}^l \tag{5.38}$$

Although in the current practice, the TN and the PDN are managed by different authorities, new entities may emerge with the progress of transportation electrification in the near future. In this regard, the OTPF can be cast as an OPF problem with UE constraints

$$\begin{aligned} \min \quad & F_E + F_T \\ \text{s.t.} \quad & \text{Cons-PF} : \{(5.36\text{b})\text{–}(5.36\text{g})\} \\ & \text{Cons-UE} : \{(5.33\text{a})\text{–}(5.33\text{d})\} \\ & \text{Cons-Couple} : (5.35) \end{aligned} \tag{5.39}$$

Clearly, OTPF is a non-convex optimization problem, due to the nonlinear complementarity constraint in Cons-UE. We leave the solution method to the next subsection. As a fundamental problem, we highlight the extendability of the proposed formulation.

1. OTPF (5.39) is a single period optimization in its current form. By incorporating a dynamic TAP [35] and a proper interface equation, it is also interesting and useful to extend the OPTF into a dynamic setting, in order to model practical demands, other system regulation measures, and complicated market settings more accurately, such as time-varying travel demands and flexible electricity loads, V2G technology enabled demand response, strategic bidding of energy storage devices, etc.
2. Despite that the congestion toll is a constant for each link in the current model, various pricing schemes, such as a flow-dependent one or a distance dependent one, can be modeled through the cost function c_k^{rs} and incorporated into the proposed model. However, nonlinear functions should be linearized to facilitate computation, which will be discussed later.
3. The travel time is a main concern when the drivers choose their routes. The proposed method is not restricted to the BPR function. Any eligible link latency

function, such as the Davidson function in (5.8), can be adopted in (5.33c), as it can be approximated by a PWL function.
4. Congestion toll is not the only traffic regulation policy. Other effective means, such as traffic signal control, are discussed in [36–38], and can be integrated in the OTPF problem as well.

5.3.2 Solution Algorithm

To find the best generation schedule and congestion tolls, OTPF problem is further reformulated as a mixed-integer convex program, whose global optimal solution is accessible with reasonable computation effort. There are three sources of nonlinearity in problem (5.39).

1. Linearizing the Objective Function
One comes from the objective function F_T, which is represented by link variables x_a, t_a, and T_a^C. According to the UE principle, all paths that carry positive traffic flows between O-D pair r-s must have the same travel cost u^{rs}. Bearing this important feature in mind, the total system travel cost can be written as a linear function in u^{rs} following a manner similar to equation (5.29)

$$\begin{aligned}\sum_a (\omega t_a + T_a^C) x_a &= \sum_{rs}\sum_k\sum_a (\omega t_a + T_a^C) f_k^{rs}\delta_{ak}^{rs} \\ &= \sum_{rs}\sum_k c_k^{rs} f_k^{rs} = \sum_{rs} u^{rs} \sum_{k, f_k^{rs}>0} f_k^{rs} = \sum_{rs} u^{rs} q^{rs}\end{aligned} \tag{5.40}$$

Equation (5.40) is exact and does not exploit any approximation.

2. Linearizing the Latency Function
We perform the PWL approximation on the link latency function $t_a(x_a)$ via special-ordered set of type 2 (SOS2). The set describes a vector of variables with at most two adjacent ones being able to take nonzero values. To this end, the feasible interval of x_a is partitioned into M smaller subintervals by $M+1$ breakpoints x_a^m, $m = 0, 1, \cdots, M$. In each subinterval, $t_a(x_a)$ is approximated by a linear function, which comes down to the following constraints

$$\begin{aligned} & x_a = \sum_m x_a^m \varepsilon_a^m, \ t_a = \sum_m t_a(x_a^m)\varepsilon_a^m, \ \forall a \\ & \sum_m \varepsilon_a^m = 1, \ 0 \le \{\varepsilon_a^m, \forall m\} \in \mathbb{SOS}_2, \ \forall a \end{aligned} \tag{5.41}$$

where $t_a(x_a^m)$ is the value of latency function at breakpoint x_a^m, and $\{\varepsilon_a^m, \forall m\}$ is the set of SOS2 variables, which can be declared using the build-in module of CPLEX and GUROBI solvers. The enhanced formulation with a logarithmic number of binary variables is elaborated in Appendix. SOS2 constraints can be implemented via exploiting integer variables, which is introduced in Appendix B.1. Nevertheless, the dedicated branch-and-bound scheme for SOS2 variables may be more efficient than the traditional branch-and-bound scheme for MILPs. An alternative way to enhance the formulation is the Gray code based approach introduced in Appendix B.1, which only incorporates a logarithmic number of binary variables and constraints.

3. Linearizing Complementarity Constraints
The complementarity condition (5.33a) can be reformulated as the following linear constraints [39] by introducing a binary variable v_k^{rs} for each path

$$\begin{aligned} 0 \le f_k^{rs} &\le M(1 - v_k^{rs}), \ \forall k, r, s \\ 0 \le c_k^{rs} - u_{rs} &\le M v_k^{rs}, \ \forall k, r, s \\ v_k^{rs} &\in \{0, 1\}, \ \forall k, r, s \end{aligned} \tag{5.42}$$

where M is a big enough constant. In practice, we can choose M according to certain heuristic, such as the largest possible travel cost of the corresponding O-D pair. Because the latency function has already been linearized, the path travel cost c_k^{rs} has a linear expression.

4. Final Mixed-Integer Convex Program
Based on the previous discussion, OTPF (5.39) can be cast as the following mixed-integer convex program

$$\begin{aligned} \min \ & F_E + F_T \\ \text{s.t.} \ & \text{Cons-PF} \\ & \text{Cons-UE-Lin} \\ & \text{Cons-Couple} \\ & t_a \ge t_a^0 \left[1 + 0.15 (t_a')^2\right], \ \forall a \\ & t_a' \ge \left(\frac{x_a}{c_a}\right)^2, \ \forall a \end{aligned} \tag{5.43}$$

where t_a' is an auxiliary variable of travel time on link a, F_E and F_T are defined in (5.38) and (5.40), respectively, Cons-PF and Cons-Couple are defined in (5.39), and the linearized UE condition is given by

$$\text{Cons-UE-Lin} : \{(5.33\text{b}),\ (5.33\text{d}),\ (5.41),\ (5.42)\} \tag{5.44}$$

Because the BPR function (5.33c) has already been approximated by (5.41), the last two convex quadratic constraints in (5.43) are redundant and do not change the feasible region. However, they can tighten the bounding problems in the branch-and-bound procedure. Such constraints are known as valid inequalities. According to our experiments, when valid inequalities are imposed, the computation time can be reduced remarkably.

From (5.41) and (5.42) we can see that the number of binary variables in (5.43) depends on the number of O-D pairs, links, and segments of PWL approximation. In this regard, scalability may be a main issue for complex instances with a large TN. We provide three options for such circumstances.

1. Accept the best solution found in a given time limit. We are aware about how good the solution is through the optimality gap provided by the solver. Existing local algorithms do not provide such information.
2. Use the original latency function $t_a(x_a)$ and Cons-UE defined in (5.39). Then OTPF comes down to an NCP, which can be solved by traditional methods in [40–42] while abandoning the global optimality guarantee.
3. As explained before, TAP (5.34) with given link toll T_a^C is a convex optimization problem and is readily solvable, inspiring a heuristic search algorithm.

Algorithm 5.2 Direct searching

1: For a given link toll T_a^C, solve TAP (5.34). The UE is x^*, and the optimal value of F_T^* is evaluated at x^*;
2: Solve the following OPF problem

$$\begin{aligned} \min \quad & F_E \\ \text{s.t.} \quad & \text{Cons-PF} \\ & \text{Cons-Couple} \end{aligned} \tag{5.45}$$

where link flows in Cons-Couple are fixed at x^* obtained in Step 1, and the optimal value is F_E^*.
3: Evaluating the social cost $F_E^* + F_T^*$, find a better choice of link toll T_a^C, until a convergence criterion is reached.

A number of derivative-free searching algorithms are available for the task in Step 3, such as the pattern search method developed in [43] and [44], and direct search algorithms introduced in [45] and the references therein.

In the pattern search procedure, the objective values are evaluated and compared at a set of samples for possible values of T_a^C, then a better T_a^C that decreases the objective value should be found, and local optimality can be guaranteed [46]. Because TAP (5.34) is convex and OPF (5.45) is an SOCP, both of them can be solved very efficiently and reliably. The computation burden stems from the fact that the objective value should be evaluated repeatedly. Nevertheless, this algorithm can be quite useful if only a few roads can be charged, as the dimension of $[T_a^C], \forall a$ is low.

The most important feature of Algorithm 5.2 is that it allows the transportation and electricity authorities to solve their own problems independently, while

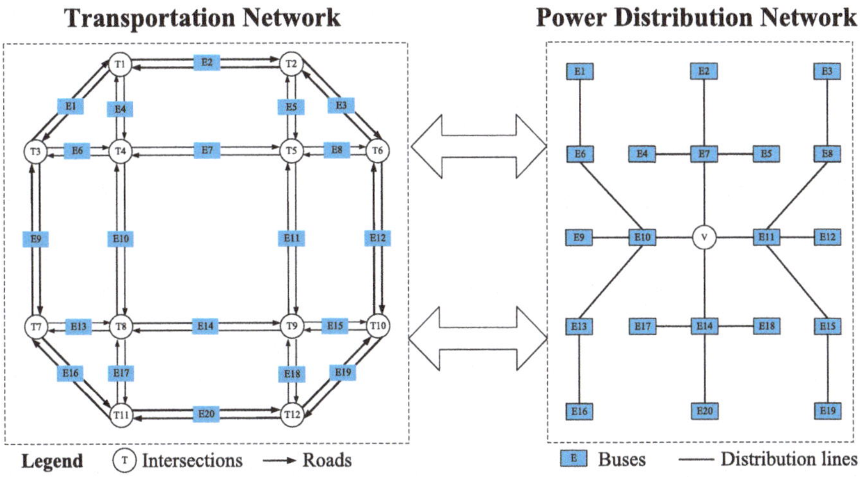

Fig. 5.4 Topology of the coupled transportation and distribution network

preserving their private information. In this regard, the direct searching algorithm also provides a reference paradigm for distributed decision making in future electrified transportation networks.

5.3.3 Case Studies

1. Basic settings

This section exhibits the necessity and benefits of joint management of the transportation and power distribution systems through numerical experiments on a test system. Topologies of the TN and the PDN are shown in Fig. 5.4. The transportation system has an emblematical structure with ring expressways in the outer loop. Each road is served by one electrical bus (represented by a blue block). The physical connection is illustrated in the left part of Fig. 5.4. Parameters of both infrastructures are summarized in Tables 5.1, 5.2, 5.3, and 5.4. We simulate the evening rush hour when the main traffic leaves from the northwest and travels to the east and the south. As for the PDN, we assume that the fixed demand at each electrical bus is $p_i^{dc} = 0.02$ and $q_i^d = 0.01$, the voltage boundary is $U_i^f = 0.8300$ ($V_i^f = 0.9110$) and $U_i^r = 1.10$ ($V_i^r = 1.0488$). The voltage magnitude at the slack bus is $U_0 = 1.04$. The electricity price at the main grid is $\rho = 1500\$/\text{p.u.}$ The monetary value of travel time is $\omega = 10\$/\text{h}$, the charging rate of unit traffic flow is $\eta = 0.01$. All simulations are implemented on a laptop computer with Intel i5-3210M CPU and 4 GB memory which is used throughout this chapter. MISOCP is solved by CPLEX [30].

Table 5.1 Parameters of road segments

Link	c_a (p.u.)	t_a^0 (min)	Bus	Link	c_a (p.u.)	t_a^0 (min)	Bus
T1–T3	18	6.0	E1	T5–T9	13.8	12.5	E11
T1–T2	20	10	E2	T6–T10	20	10.5	E12
T2–T6	17	6.5	E3	T7–T8	8.9	5.8	E13
T1–T4	9.8	5.0	E4	T8–T9	13.2	11	E14
T2–T5	7.9	5.5	E5	T9–T10	9.15	5.9	E15
T3–T4	8.5	6.0	E6	T7–T11	17.5	6.3	E16
T4–T5	13.5	12	E7	T8–T11	9.76	5.7	E17
T5–T6	8.2	6.5	E8	T9–T12	8.97	5.8	E18
T3–T7	19	10.2	E9	T12–T10	18.2	6.1	E19
T4–T8	14	11.5	E10	T11–T12	20	9.8	E20

Table 5.2 O-D pairs and their trip rates (in p.u.)

O-D pair	q^{rs}	O-D pair	q^{rs}	O-D pair	q^{rs}
T1–T6	15	T3–T6	15	T4–T9	5
T1–T10	25	T3–T10	20	T4–T10	10
T1–T12	10	T3–T12	8	T4–T12	15
T1–T11	15	T3–T11	12		

Table 5.3 Parameters of generators

Unit	Node	p_i^f (p.u.)	p_i^r (p.u.)	q_i^f (p.u.)	q_i^r (p.u.)	a_i (\$)	b_i (\$)
G1	E7	0	1.5	−0.2	0.2	228	1382
G2	E10	0	2.0	−0.4	0.4	196	1089
G3	E11	0	2.0	−0.4	0.4	236	834
G4	E14	0	1.6	−0.3	0.3	239	1329

Table 5.4 Parameters of distribution lines (in p.u.)

Line	r	x	Line	r	x	Line	r	x
E0-E7	0.081	0.061	E10-E9	0.135	0.081	E11-E15	0.115	0.082
E0-E10	0.066	0.042	E10-E6	0.107	0.073	E8-E3	0.122	0.083
E0-E11	0.061	0.041	E10-E13	0.111	0.075	E15-E19	0.120	0.080
E0-E14	0.079	0.063	E6-E1	0.127	0.083	E14-E20	0.119	0.077
E7-E2	0.115	0.080	E13-E16	0.119	0.078	E14-E17	0.133	0.090
E7-E4	0.131	0.084	E11-E12	0.132	0.079	E14-E18	0.130	0.088
E7-E5	0.123	0.077	E11-E8	0.105	0.070			

To demonstrate the role of congestion toll for mitigating negative effects caused by traffic overload, we intentionally design the following uncoordinated operating mode: in the first stage, the transportation authority who wishes to minimize the total travel cost F_T solves TAP (5.34) with a given link toll parameter T_a^C, the UE is $x_a, \forall a$; and then in the second stage, the distribution system operator who intends to optimize the energy service cost F_E independently solves OPF (5.45) with the nodal energy demand p_i^d corresponding to the obtained traffic UE pattern. No further iteration is needed. The uncoordinated operating mode will be referred to as UE-OPF(T^C) in the following context, specifically, UE-OPF if $T_a^C = 0, \forall a$. It is worth mentioning that UE-OPF(T^C) exactly constitutes one iteration of Algorithm 5.2. In this regard, OTPF can be interpreted as searching the optimal link toll that gives the best coordination between the transportation and power distribution systems even if they are independently operated.

2. Results

The first task is to reveal that unguided traffic equilibrium may lead to an unsafe operation of the PDN (or equivalently, power flow infeasibility). We first let $T_a^C = 0, \forall a$ and consider non-zero link toll later. When the power flow is infeasible, the SOC relaxation will be inexact, which means some constraints in (5.36e) will be held as strict inequalities. To circumvent this difficulty, we assume that the bus voltage can violate its lower bound at the cost of a penalty, yielding the following objective function

$$F_E = \sum_i \left[a_i (p_i^g)^2 + b_i p_i^g \right] + \rho \sum_{j \in \pi(0)} P_{0j}^l + \kappa \sum_i \Delta U_i$$

where the penalty coefficient is chosen as $\kappa = 50{,}000$\$/p.u.; meanwhile, the voltage boundary constraints in (5.36g) become

$$U_i^f - \Delta U_i \le U_i \le U_i^r, \ \Delta U_i \ge 0, \ \forall i$$

By solving TAP (5.34) with zero tolls, the UE is shown in Fig. 5.5, which illustrates that most travelers prefer to use the ring expressway in the outer loop. Then the nodal power demand p_i^d is calculated through (5.35) in accordance with the obtained UE, and OPF problem (5.45) is then solved. The generation schedule and power flow are given in Table 5.5 and Table 5.6, respectively. Because the electrified ring expressway connects to the terminal buses, heavy traffic flows on these roads pose a threat to the security of the distribution grid. From Table 5.6 we can see that the voltage magnitude at Bus 3 violates the lower bound, which indicates that the PDN might be at the risk of voltage instability.

The proposed OTPF is solved by CPLEX in its mixed-integer convex form (5.43), and the UE is also shown in Fig. 5.5, which illustrates that congestion tolls are charged on three roads. The generation schedule and power flow are given in Tables 5.5 and 5.6 correspondingly. It is interesting to see that the congestion toll is not directly imposed on road T2–T6, which connects to the insecure Bus 3. However, owing to the \$2.603 toll charged on link T6–T10, some vehicles will choose other routes, which helps alleviate traffic overload on link T2–T6. Meanwhile, to prevent voltage violation at other buses, tolls are also imposed on links T1–T3 and T7–T8. In this way, the voltage magnitudes at Bus 3 and Bus 16 reach their lower bounds, but never exceed the bounds. This is the optimal strategy from the perspective of minimum social cost.

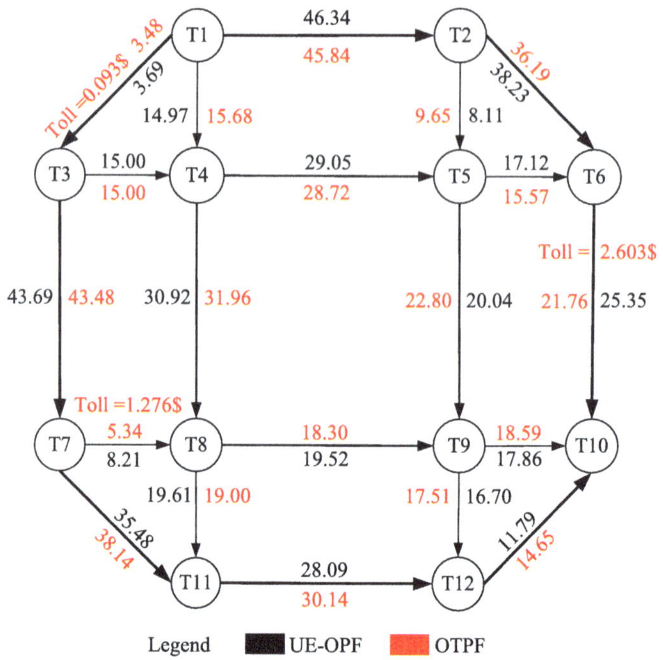

Fig. 5.5 Traffic flow distribution at the user equilibrium

Table 5.5 Optimal generation schedule

Unit	UE-OPF		OTPF	
	p_i^g	q_i^g	p_i^g	q_i^g
G1	0.6065	0.1292	0.6104	0.1294
G2	1.4347	0.4000	1.4947	0.4000
G3	1.5333	0.4000	1.5175	0.4000
G4	0.6048	0.1277	0.6269	0.1644

We further investigate the impact of traffic demand variation on the optimal social cost and traffic congestion level by multiplying the trip rates q^{rs} with a load factor σ and changing σ from 0.85 to 1.05. Results are provided through Figs. 5.6, 5.7, and 5.8. Figure 5.6 demonstrates that in light traffic conditions, the security of distribution network may not be compromised and the congestion toll will not be needed. Hence, UE-OPF and OTPF offer the same strategy with the same social cost. With the growth in traffic demand, voltage boundary constraints cannot be maintained any more. UE-OPF has no choice but to pay for the voltage violation, incurring a higher social cost, while OTPF is able to prevent voltage violation in all cases by charging tolls on the roads with minimal social cost.

The system-level total travel time $T_S = \sum_a x_a t_a$, which is an important measure on the efficiency of the TN, is provided in Fig. 5.7 with different load factors. We can see that T_S grows remarkably when the traffic demand increases, because the

Table 5.6 Optimal power flow

Node	Voltage magnitude (p.u.)		Line	Active power (p.u.)		Reactive power (p.u.)	
	UE-OPF	OTPF		UE-OPF	OTPF	UE-OPF	OTPF
1	0.9873	0.9915	0-7	0.5189	0.5302	−0.0459	−0.0455
2	0.9198	0.9195	0-10	0.1540	0.0988	−0.2766	−0.2762
3	0.9026	0.9110	0-11	0.0000	0.0000	−0.2668	−0.2714
4	0.9579	0.9559	0-14	0.3500	0.3402	−0.0626	−0.0981
5	0.9684	0.9655	7-2	0.5163	0.5107	0.0328	0.0325
6	0.9955	0.9993	7-4	0.1737	0.1811	0.0126	0.0128
7	0.9820	0.9811	7-5	0.1025	0.1183	0.0108	0.0111
8	0.9588	0.9639	10-9	0.4927	0.4898	0.0315	0.0310
9	0.9550	0.9591	10-6	0.2338	0.2315	0.0247	0.0245
10	1.0216	1.0251	10-13	0.5253	0.5258	0.0523	0.0540
11	1.0307	1.0309	6-1	0.0573	0.0552	0.0103	0.0102
12	0.9932	0.9985	13-16	0.3936	0.4229	0.0223	0.0240
13	0.9612	0.9648	11-12	0.2851	0.2461	0.0170	0.0152
14	0.9970	1.0000	11-8	0.6684	0.6231	0.0703	0.0640
15	0.9891	0.9845	11-15	0.3539	0.3948	0.0323	0.0358
16	0.9110	0.9110	8-3	0.4295	0.4062	0.0285	0.0266
17	0.9661	0.9701	15-19	0.1405	0.1703	0.0117	0.0125
18	0.9709	0.9728	14-20	0.3133	0.3357	0.0180	0.0192
19	0.9711	0.9628	14-17	0.2231	0.2168	0.0148	0.0146
20	0.9585	0.9589	14-18	0.1923	0.2009	0.0136	0.0139

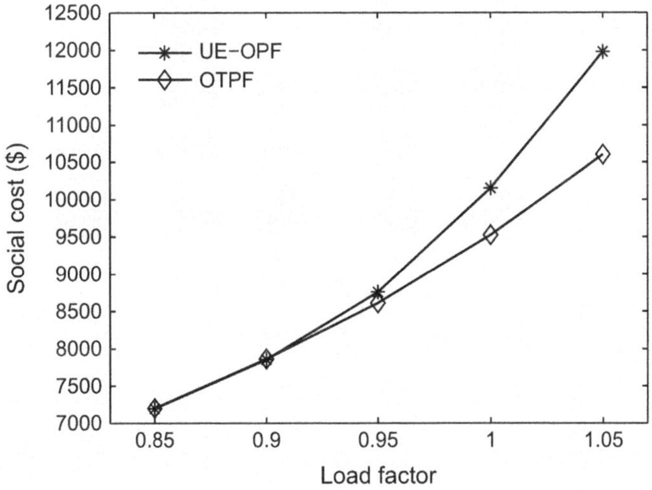

Fig. 5.6 Social costs under different load factors

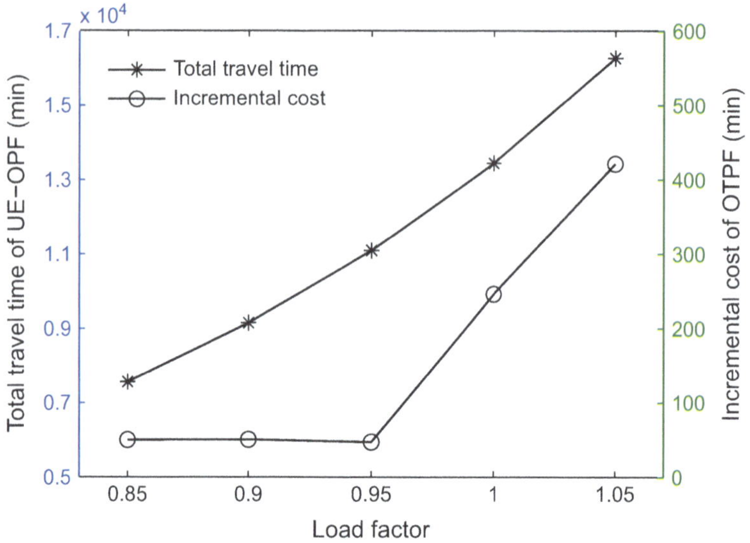

Fig. 5.7 Total travel times under different load factors

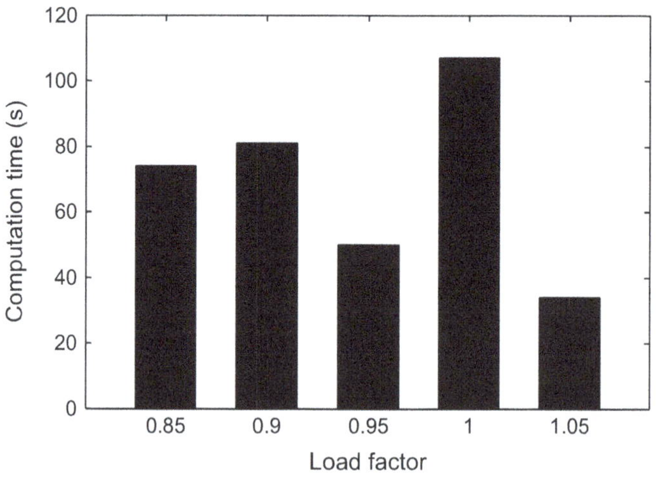

Fig. 5.8 Computation times under different load factors

road latency is amenable to a high-order power function. Moreover, the total travel time offered by OTPF is always higher than that offered by UE-OPF, due to the following two reasons. On the one hand, when the load factor $\sigma \leq 0.95$, the difference is almost a constant (50 min), because in UE-OPF, we directly use the nonlinear latency function $t_a(x_a)$, while in OTPF, we use its PWL approximation. The difference stems from the error of PWL approximation. However, in view that the BPR function (5.33c) itself is also an approximation of the actual situation, this difference appears to be less important for practical usage. On the other hand, when $\sigma \geq 1.0$, the gap increases because the congestion toll drives the traffic flow deviate from the travel cost oriented UE pattern. The benefit is that the power grid is more secure. In view that the total travel time of OTPF only grows slightly (1.86% higher than that of UE-OPF when $\sigma = 1.0$), this is the best choice from a social optimal perspective. It is certified in Fig. 5.8 that the OTPF problem in all the cases can be solved in reasonable time.

Next, Algorithm 5.2 is tested. TAP (5.34) in Step 1 is solved by KNITRO [47], and OPF (5.45) in Step 2 is solved by CPLEX [30]. Derivative-free search in Step 3 is implemented via the build-in `patternsearch` function provided by MATLAB. Please keep in mind that this algorithm is only efficient for instances with a few congestion toll variables. In our implementation, we contrivedly impose zero upper bounds on the roads which are not charged at the global OTPF solution. For example, when the load factor is 1, only links T1–T3, T6–T10, and T7–T8 will be charged, so the dimension of T_a^C is 3. In reality, the chargeable links are pre-specified and we do not need to identify them by exploiting the optimal toll in advance. In our experiment, the computation consumes about half minute in all tests. Results are shown in Fig. 5.9. We can see that `patternsearch` function successfully finds global optimal solutions in all instances, although provable guarantee is non-trivial. The small gap between the optimal values in the two curves stems from the error of PWL approximation for the BPR function. We test the uncoordinated operating mode UE-OPF(T^C). By fixing T^C at the optimal solution offered by model (5.43), UE-OPF(T^C) provides the same solutions as those offered

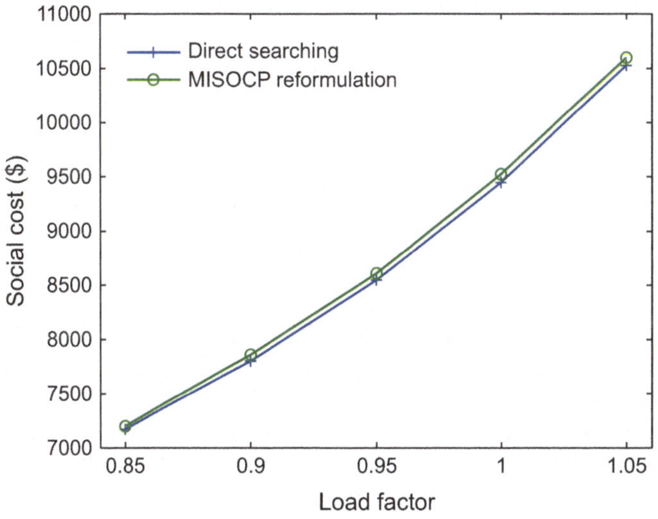

Fig. 5.9 Optimal values offered by MISOCP and direct search

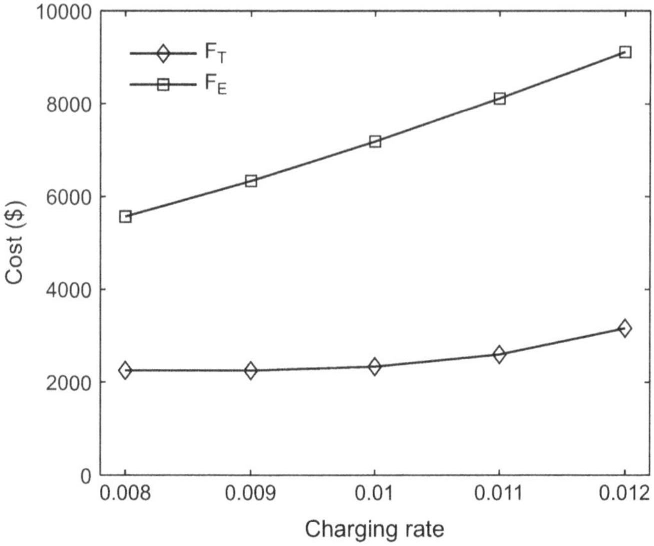

Fig. 5.10 Network costs under different charging rates

by Algorithm 5.2 and shown in Fig. 5.9. These outcomes validate the proposed model and method.

Finally, we study the impact of charging rate parameter η in the interface Eq. (5.35) on network costs and total travel time on the TN. This parameter may depend on several factors, such as the penetration level, the average battery size, and charging power of EVs. In our tests, OTPF (5.43) is solved under different η varying from 0.008 to 0.012. Results are provided through Figs. 5.10, 5.11, and 5.12. The former two suggest that the traffic UE is not influenced by the value of charging rate when $\eta \leq 0.009$, because the power flow constraints can be well maintained without imposing congestion tolls, and the traffic flow reaches a pattern in which the travel time is optimal, but the PDN operating cost F_E increases with the growth of η, because more electricity is needed to meet the charging service. When the charging rate $\eta \geq 0.01$, both F_E and F_T grow monotonically in η, because in order to maintain operating security of the power grid, more roads will be charged with a congestion toll, and the UE will be pulled away from a travel time optimal pattern. However, if η keeps growing, congestion toll will no longer be a good choice for system coordination, in view that the total travel time grows significantly. Long term solution could be upgrading the system infrastructures, such as building additional generators and distribution lines. Figure 5.12 demonstrates that OTPF problem (5.43) with different charging rates can be solved in reasonable time.

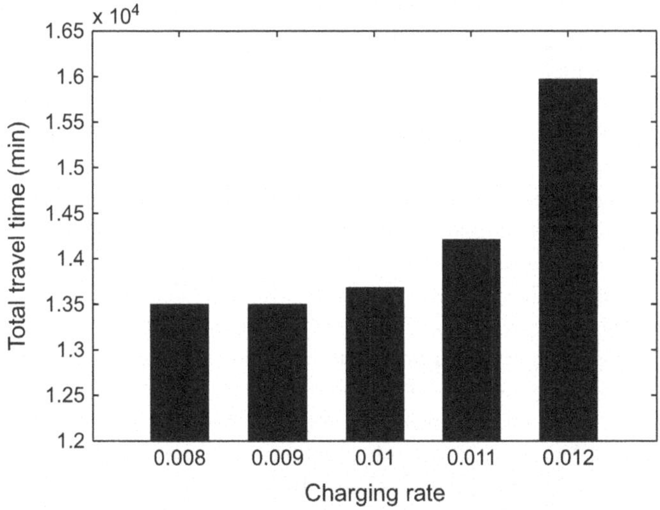

Fig. 5.11 Total travel times under different charging rates

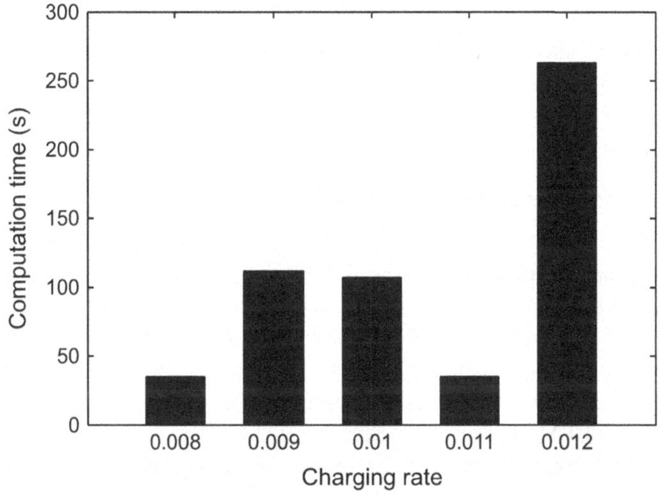

Fig. 5.12 Computation times under different charging rates

5.4 Robust System Operation with Uncertain Traffic Demand

Previous section assumes that the system operator has complete information on the problem data, which may not be exactly known in practice. Uncertainty may originate from the inexact travel time function and inaccurate traffic demand forecast or renewable generation. The first category of uncertainty has been considered in [48, 49]. This section focuses on the second category of uncertainty. We consider interval traffic demands and propose a two-stage robust optimization model for a reliable operation of the PDN which supplies electricity to charging facilities on a TN. It differs from existing robust power system optimization methods in two aspects.

First, the uncertainty set of nodal power demand is not explicitly given, and should be mapped from the set of traffic demand uncertainty through UE conditions. Based on the monotonic relationship between traffic demand and link flow, we reveal the connection between extreme points of traffic demand variation and electrical demand fluctuation. We also discuss how to control the model conservatism by incorporating budget constraints with an adjustable parameter. We derive a polyhedral outer approximation for the electrical demand uncertainty in the presence of the budget constraint. Renewable output uncertainty can be modeled in the traditional way, which boils down to fluctuations in the nodal power injection.

Second, we employ the exact nonlinear BFM and convex relaxation technique to describe the power flow status of the PDN. Therefore, the robust dispatch model gives rise to a min-max-min program with an SOCP lower level. A delayed constraint generation framework is adopted to solve this challenging problem. The master problem is an SOCP. Using duality theory, the scenario generation oracle yields a bi-convex subproblem. We discuss how to solve the non-convex subproblem with different preferences on scalability and optimality.

5.4.1 Mathematical Formulation

Basic settings are clarified. The traffic flow distribution for a given demand vector is amenable to the UE pattern, such that no traveler can reduce his travel cost by changing route unilaterally. Traffic demand of each O-D pair belongs to a certain interval. The electricity demand of charging facilities is a linear function in the road traffic flow. The PDN is operated with tree topology, and the power flow status is described by BFM presented in Sect. 2.2.2. Major symbols and notations are inherited from Sects. 2.2.2 and 5.2. There may be slight differences in superscripts and subscripts, without causing confusions. Specifically, g, p, q, l, u, t, 0, 1, and 2 in the superscript or subscript stand for generator, active power, reactive power, lower bound, upper bound, transportation network, nominal scenario, first-stage, and second-stage related notations, respectively.

For a given traffic demand vector $q^t = [q^t_{rs}], \forall (r,s)$, the UE can be computed from the conventional TAP

$$\begin{aligned} \min \quad & \sum_a \int_0^{x_a} t_a(\theta) \mathrm{d}\theta \\ \text{s.t.} \quad & \text{Cons-TAP } ([q^t_{rs}]) \end{aligned} \tag{5.46}$$

where $t_a(x_a)$ denote the link latency function, and BPR function (5.7) will be used in this study; the feasible set is given by

$$\text{Cons-TAP}\,[q^t_{rs}] = \left\{ \begin{array}{c} x_a = \sum_{rs}\sum_{k} f^{rs}_k \delta^{rs}_{ak},\ \forall a \\ f^{rs}_k \geq 0,\ \forall k, \forall(r,s) \\ \sum_{k\in K^{rs}} f^{rs}_k = q^t_{rs},\ \forall(r,s) \end{array} \right\} \tag{5.47}$$

When the traffic demand vector q^t is uncertain, we assume that the trip rate q^t_{rs} between O-D pair (r,s) varies in the interval $[q^{tl}_{rs}, q^{tu}_{rs}]$, and the uncertainty set of electrical demand p^d will be given by

$$W = \left\{ p^d \left| \begin{array}{c} \exists q^t \in \text{BOX}(q^{tl}, q^{tu}) \\ x_a \text{ solves TAP (5.46)} \\ p^d_i = p^{df}_i + \sum_{a\in C(i)} \eta(x_a),\ \forall i \end{array} \right. \right\} \tag{5.48}$$

where $\text{BOX}(l,u) = \{x | l \leq x \leq u\}$ defines a hypercube. Because the set W in variable p^d is not given in a closed form, we call it an implicit uncertainty set.

The PDN is operated with minimum cost. Because the operator does not have a clear knowledge on the exact electricity demand profile, it will prudently try to minimize the total operation cost in the worst-case situation, leading to a robust dispatch formulation

$$\min_{y\in Y(p^{d0})} \quad F_1 + \max_{p^d\in W} \min_{z\in Z(y,p^d)} F_2 \tag{5.49}$$

where vector $y = [s^1, r^{p+}, r^{p-}, r^{q+}, r^{q-}]$ is the first stage decision variable, r^{p+}_s (r^{p-}_s) and r^{q+}_s (r^{q-}_s) are positive (negative) reserve capacity for active and reactive power; $s^1 = [P^1, Q^1, i^1, U^1, p^{g1}, q^{g1}]$ is the power flow status in the nominal load scenario $p^{d0} = p^{df}_i + \sum_{a\in C(i)} \eta(x^0_a), \forall i$, where x^0_a is the vehicle flow on link a at the UE with $q^t_{rs} = q^{t0}_{rs}, \forall(r,s)$. Vector $z = [s^2, p^{g+}, p^{g-}, q^{g+}, q^{g-}]$ is the second-stage decision variable, p^{g+} (p^{g-}) and q^{g+} (q^{g-}) stand for the amount of upward (downward) regulation power of active and reactive output, and $s^2 = [P^2, Q^2, i^2, U^2, p^{g2}, q^{g2}]$ is the power flow status in the observed load scenario implied by p^d (2 in the superscript is an index rather than a square operator). Function

$$F_1 = \sum_i \left[a^p_i (p^g_i)^2 + b^p_i p^g_i + c^{p+}_i r^{p+}_i + c^{p-}_i r^{p-}_i + c^{q+}_i r^{q+}_i + c^{q-}_i r^{q-}_i \right] \tag{5.50}$$

is the summation of the energy production cost and the reserve cost in the first stage. We assume that reactive power reserve is paid for providing voltage regulation service in the second stage. Function

$$F_2 = \sum_i \left[d_i^{p+} p_i^{g+} + d_i^{p-} p_i^{g-} + d_i^{q+} q_i^{g+} + d_i^{q-} q_i^{g-} \right] + \rho \sum_{j \in \pi(0)} P_{0j}^l \tag{5.51}$$

represents the cost of corrective actions, including generation redispatch and energy trading with the main grid. The objective function of (5.49) is to minimize the total cost in the worst-case traffic demand scenario. Define power flow constraints

$$P_{ij}^l + p_j^g - r_{ij}^l I_{ij}^l = \sum_{k \in \pi(j)} P_{jk}^l + p_j^d, \ \forall j \tag{5.52a}$$

$$Q_{ij}^l + q_j^g - x_{ij}^l I_{ij}^l = \sum_{k \in \pi(j)} Q_{jk}^l + q_j^d, \ \forall j \tag{5.52b}$$

$$U_j = U_i - 2(r_{ij}^l P_{ij}^l + x_{ij}^l Q_{ij}^l) + (z_{ij}^l)^2 i_{ij}^l, \ \forall l \tag{5.52c}$$

$$(I_{ij}^l + U_i)^2 \geq (I_{ij}^l - U_i)^2 + (2P_{ij}^l)^2 + (2Q_{ij}^l)^2, \ \forall l \tag{5.52d}$$

$$I_{ij}^l \geq 0, \ P_{ij}^l \geq 0, \ Q_{ij}^l \geq 0, \ \sqrt{(P_{ij}^l)^2 + (Q_{ij}^l)^2} \leq S_l, \ \forall l \tag{5.52e}$$

$$p_j^{gm} \leq p_j^g \leq p_j^{gu}, q_j^{gm} \leq q_j^g \leq q_j^{gu}, U_j^m \leq U_j \leq U_j^m, \ \forall j \tag{5.52f}$$

and Cons-BPF $(s, p^d) = \{(5.52a)–(5.52f)\}$. The feasible set Y for the first-stage decision can be written as

$$Y(p^{d0}) = \left\{ y \left| \begin{array}{c} p^{g1} + r^{p+} \leq p^r, \ 0 \leq r^{p+} \leq R^+ \Delta t \\ p^{g1} - r^{p-} \geq p^f, \ 0 \leq r^{p-} \leq R^- \Delta t \\ q^{g1} + r^{q+} \leq q^r, \ q^{g1} - r^{q-} \geq q^f \\ p^{g1} \geq 0, \ q^{g1} \geq 0, \ r^{q+} \geq 0, \ r^{q-} \geq 0 \\ \text{Cons-BPF}(s^1, p^{d0}) \end{array} \right. \right\} \tag{5.53}$$

which includes the generation capacity accounting for active/reactive reserve provision, power flow constraints in the nominal scenario, and physical bounds of power flow variables. The feasible set Z for the second-stage decision in scenario p^d with fixed y is given by

$$Z(y, p^d) = \left\{ z \left| \begin{array}{c} p^{g2} = p^{g1} + p^{g+} - p^{g-} \\ 0 \le p^{g+} \le r^{p+}, 0 \le p^{g-} \le r^{p-} \\ q^{g2} = q^{g1} + q^{g+} - q^{g-} \\ 0 \le q^{g+} \le r^{q+}, 0 \le q^{g-} \le r^{q-} \\ \text{Cons-BPF}(s^2, p^d) \end{array} \right. \right\}$$

the former four constraints define the active and reactive power generation in the second stage based on the set points p^{g1} and q^{g1} and regulation power limits restricted by the scheduled reserve capacity r^{p+}, r^{p-}, r^{q+} and r^{q-}. The last one is the power flow constraint with respect to the observed scenario p^d and physical bounds of power flow variables.

For notation conciseness, the second-stage max-min problem can be written as the following compact form

$$\max_{p^d \in W} \min_{z \in Z(y, p^d)} f^T z \tag{5.54}$$

where the matrix form of $Z(y, p^d)$ is given by

$$Z(y, p^d) = \left\{ z \left| \begin{array}{l} Bz \le h_1 - Ay : \mu \\ Cz = h_2 - Dp^d : \lambda \\ S_l z \in \mathbb{L}^4, \ \forall l : \theta^l \end{array} \right. \right\} \tag{5.55}$$

where matrices $A, B, C, D, S_l, \forall l$ and vectors h_1, h_2 correspond to coefficients of constraints in set Z; μ, λ, θ^l following a colon are the dual variables, and $\mathbb{L}^4$ denotes the following second order cone in $\mathbb{R}^4$

$$\mathbb{L}^4 = \left\{ x \in \mathbb{R}^4 \middle| x_4 \ge \sqrt{x_1^2 + x_2^2 + x_3^2} \right\}$$

Some discussions on the modeling framework are provided below. Robust dispatch model (5.49) involves only one period. The temporal sequence of the two stages reflects a joint energy and reserve market: the first-stage deploys set points and reserve capacity of generators in accordance with the forecasted scenario; the second-stage decisions represent recourse actions in response to the observed uncertain data, including the adjustment of active and reactive output. Power flow constraints should be feasible in both stages. If there is no reserve market, the generator can be arbitrarily dispatched after the actual demand is observed, but energy transaction with the main grid should be bit ahead of real-time dispatch. We can choose $F_1 = \rho \sum_{j \in \pi(0)} P^l_{0j}$, $y = s^1$, and $F_2 = \sum_i \left[a^p_i (p^g_i)^2 + b^p_i p^g_i\right]$, $z = s^2$.

The epigraph form of the second-stage cost yields $F_2 = \sum_i \left[\sigma_i + b_i^p p_i^g\right]$ with auxiliary convex quadratic constraints $a_i (p_i^g)^2 \le \sigma_i, \forall i$, which can be represented by SOCs as

$$\left\| \begin{matrix} 2\sqrt{a_i^g b_i^g} p_i^g \\ \sigma_i - b_i^g \end{matrix} \right\|_2 \le \sigma_i + b_i^g, \ \forall i$$

Therefore, the second-stage problem still renders a max-min SOCP similar to (5.54).

5.4.2 Solution Methodology

1. The Box Hull of the Uncertainty Set

The variable in uncertainty set (5.48) is p^d, which depends on the link flow $x_a, \forall a$ at UE pattern, and ultimately affected by the traffic demand q^t. To understand their inherent connection, a set-to-set mapping is illustrated in Fig. 5.13. The traffic demand is restricted in the hypercube BOX(q^{tl}, q^{tu}), which is shown in the left part of the figure. In accordance with the Proposition 5.1, the path flow f_k^{rs} on each used path between O-D pair (r, s) increases if q_{rs}^t grows, so does the vehicle flow x_a on link a, as long as it already carries positive traffic flow. Therefore, the link flow is a monotonic increasing function (or non-decreasing function) in the trip rates. In view of this monotonic feature, the minimal traffic flow vector x^l and the maximal traffic flow vector x^u are given by the following TAPs with extreme traffic demands

$$\begin{aligned} x^l = \arg\min \ & \sum_a \int_0^{x_a} t_a(\theta) \mathrm{d}\theta \\ \text{s.t. } & \text{Cons-TAP } ([q_{rs}^{tl}]) \end{aligned} \tag{5.56a}$$

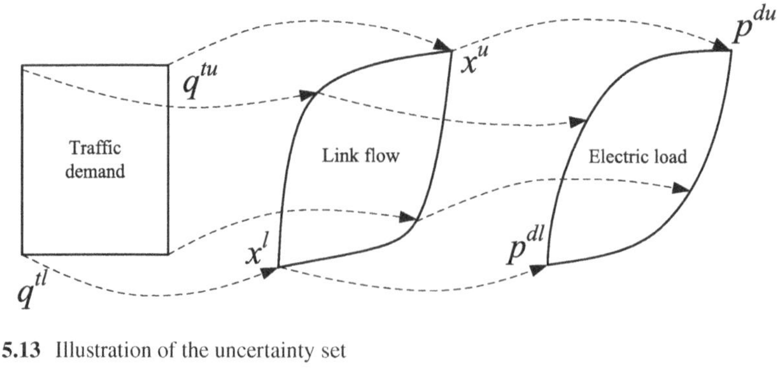

Fig. 5.13 Illustration of the uncertainty set

$$x^u = \arg\min \sum_a \int_0^{x_a} t_a(\theta)\mathrm{d}\theta \qquad (5.56b)$$
$$\text{s.t. Cons-TAP } ([q_{rs}^{tu}])$$

where Cons-TAP with a given traffic demand vector is defined in (5.47). The possible region of the link flow is given by the following implicit set

$$W^{TN} = \left\{ x \,\middle|\, \begin{array}{l} \exists q^t \in \text{BOX}(q^{tl}, q^{tu}) \\ x \text{ solves TAP (5.46)} \end{array} \right\}$$

and is shown in the middle part of Fig. 5.13, where x^l and x^u correspond to q^{tl} and q^{tu}, respectively, as indicated by (5.56a) and (5.56b). Because the charging rate function $\eta(x_a)$ is monotonic (linearity is not a necessity), the lower and upper bounds of active power demand are given as

$$p_i^{dl} = p_i^{df} + \sum_{a \in C(i)} \eta(x_a^l), \ \forall i \qquad (5.56c)$$

$$p_i^{du} = p_i^{df} + \sum_{a \in C(i)} \eta(x_a^u), \ \forall i \qquad (5.56d)$$

where x_a^l and x_a^u are elements of x^l and x^u, respectively.

Finally, the possible region of nodal electrical demand is given by the following implicit set which is equivalent to (5.48)

$$W = \left\{ p^d \,\middle|\, \begin{array}{c} \exists x \in W^{TN} \\ p_i^d = p_i^{df} + \sum_{a \in C(i)} \eta(x_a), \ \forall i \end{array} \right\} \qquad (5.56e)$$

and is shown in the right part of Fig. 5.13. However, it is very difficult to derive a closed form for W. A practical way is to use the box hull of W, i.e., the hypercube $\text{Box}(p^{dl}, p^{du})$. It is important to notice that the corners p^{dl} and p^{du} (correspond to x^l and x^u, respectively) are attainable. In other words, the nodal demands can reach their lower bounds or upper bounds simultaneously, if the traffic demand can reach q^{tl} or q^{tu}. Thus using the box hull will not yield unrealistic results.

2. Reformulation of the Subproblem

To solve the second-stage problem, we can dualize the inner SOCP and convert problem (5.54) into an NLP

$$\max_{p^d \in W, \{\mu, \lambda, \theta^l\} \in D_V} \mu^T(h_1 - Ax) + \lambda^T(h_2 - Dp^d) \qquad (5.57)$$

where the feasible region D_V of dual variables is given by linear constraints and SOCs as follows:

$$D_V = \left\{ \mu, \lambda, \theta^l \left| \begin{array}{c} \mu \le 0,\ \theta^l \in \mathbb{L}^4, \forall l,\ \lambda \text{ free} \\ B^T \mu + C^T \lambda + \sum_l S_l^T \theta = f \end{array} \right. \right\} \tag{5.58}$$

In problem (5.57), the bilinear term $\lambda^T D p^d$ in the objective function is non-convex. Since D and W are separated, when dual variables are optimal, the optimal solution p^d must locate at one of the extreme points of W. In this regard, we can express the extreme points of W through $p^d = p^{dl} + (p^{du} - p^{dl}) \otimes v$, where vector v consists of binary variables, and $\otimes$ stands for the element-wise multiplication. If $v_i = 0/1$, then $p_i^d = p_i^{dl}/p_i^{du}$. The bilinear term can be expanded as $\lambda^T D p^d = \sum_i \sum_j [D_{ij} p_j^{dl} \lambda_i + D_{ij}(p_j^{du} - p_j^{dl}) \lambda_i v_j]$. Let $\kappa_{ij} = \lambda_i v_j$, this equation can be further linearized by using the following big-M constraints

$$L_R = \left\{ \lambda, v, \kappa \left| \begin{array}{c} -M v_j \le \kappa_{ij} \le M v_j, \forall i, j,\ v_j \in \{0, 1\}, \forall j \\ -M(1 - v_j) \le \lambda_i - \kappa_{ij} \le M(1 - v_j), \forall i, j \end{array} \right. \right\} \tag{5.59}$$

where M is a big enough constant.

Finally, the second-stage problem under given y comes down to an MISOCP

$$\begin{aligned} \max \ & \mu^T (h_1 - Ax) + \lambda^T (h_2 - D p^{dl}) \\ & - \sum_i \sum_j \left[D_{ij} (p_j^{du} - p_j^{dl}) \kappa_{ij} \right] \\ \text{s.t.} \ & \{\mu, \lambda, \theta^l\} \in D_V,\ \{\lambda, v, \kappa\} \in L_R \end{aligned} \tag{5.60}$$

It should be pointed out that the value of M will have a notable impact on the computation time. To enhance efficiency, a desired value should be the minimal M that ensures inequality $-M \le \lambda_i \le M$ never becomes binding at the optimal solution of (5.60). However, such a value is generally unclear until we solve the problem. A practical way is the try-and-test heuristic: first use an arbitrary M and solve problem (5.60), then compare the optimal solution λ^* and M, and finally choose a new value of M. Nevertheless, we do not actually need to find the smallest value M^*. Any $M \ge M^*$ which results in a reasonable computation time is acceptable for the task. In this regard, a proper M can be determined from estimating the bound of dual variable λ by certain heuristics.

Another concern might be the scalability. In problem (5.60), the binary variable v takes the place of p^d, whose dimension depends on the number of charging facilities. In this regard, for a large-scale system with hundreds of charging/swapping stations, the computation burden can be prohibitively high.

To alleviate this difficulty, we suggest an efficient heuristic algorithm for NLP (5.57) without a global optimality guarantee, by exploiting the following two features:

1. the constraint sets D_V and W are separated;
2. problem (5.57) exhibits a biconvex structure: fixing p^d, it yields an SOCP; fixing $\{\mu, \lambda, \theta^l\}$, it comes down to an LP. Both problems belong to the category of convex optimization and can be readily solved.

According to the discussion in [50], based on the aforementioned two properties, Algorithm 5.3 will converge to a local optimal solution in a finite number of iterations for a given δ. In fact, Algorithm 5.3 is able to offer a high quality solution provided with multiple initial values of p^d that are elaborately chosen. Algorithm 5.3 is also scalable to large-scale systems as it only requires solving convex programs. Another important advantage of Algorithm 5.3 is that the set W can be replaced by an arbitrary convex outer approximation, rather than the relative restrictive box hull approximation. This algorithm will be used for tackling a polyhedral uncertainty set with budget constraints on traffic demand variations and an adjustable parameter, which controls the model conservatism.

Algorithm 5.3 Mountain climbing

1: Choose a tolerance $\delta > 0$ and a start point $p^{d*} \in W$.
2: Solve the following SOCP with p^{d*}

$$\max_{\{\mu,\lambda,\theta^l\}\in D_V} \quad \mu^T(h_1 - Ax) + \lambda^T(h_2 - Dp^{d*}) \tag{5.61}$$

the optimal value is R_1, and the optimal solution is $\mu^*, \lambda^*, \theta^{l*}$.
3: Solve the following LP with μ^*, λ^*

$$\max_{p^d \in W} (h_1 - Ax)^T\mu^* + (h_2 - Dp^d)^T\lambda^* \tag{5.62}$$

the optimal value is R_2, and the optimal solution is p^{d*}.
4: Terminate if $R_2 - R_1 \leq \delta$, and report the optimal solution p^{d*} and the optimal value R_2, otherwise, go to step 2.

An implicit assumption in problem (5.57) is that the SOCP relaxation in (5.52d) is exact for all possible uncertain parameters in W. For a radial network, this is a relative mild requirement. This dual transformation approach is also adopted in the research on distribution system operation [51–53].

3. The Delayed Constraint Generation Algorithm

To solve the min-max-min problem (5.49), we use the decomposition framework discussed in Appendix C.2.3. It identifies a set of critical scenarios and successively tightens the gap between the upper bound and the lower bound of objective value.

To see the motivation of this algorithm, problem (5.49) can be written as an epigraph form in a scenario enumeration manner

$$\begin{aligned} \min_{y \in Y, z^s, \xi} \quad & F_1 + \xi \\ \text{s.t.} \quad & z^s \in Z(y, p^{ds}), \ \xi \geq f^T z^s \\ & \forall p^{ds} \in \text{Extr}[W] \end{aligned} \tag{5.63}$$

where Extr[W] denotes extreme points of hypercube W. However, the number of extreme points grows exponentially with increasing the dimension of W. Hence, formulation (5.63) is only of theoretical interest. If we only consider a subset of Extr[W] in (5.63), which is called a master problem, its optimum is a lower bound of the optimal value of (5.63), because the feasible region is enlarged. Then we can solve problem (5.57) with fixed y and obtain an upper bound of the optimal value, because y is frozen and the feasible region shrinks. If the lower bound and upper bound are not equal, the component p^d which solves problem (5.57) must be added in the master problem. The delayed constraint generation algorithm identifies necessary extreme points in W, and is summarized in Algorithm 5.4.

Algorithm 5.4 Delayed constraint generation

1: Set $LB = 0$, $UB = +\infty$, $k = 2$, $p^{d1} = p^{dl}$, $p^{d2} = p^{du}$. Choose a convergence tolerance $\varepsilon > 0$.

2: Solve the following master problem (an SOCP)

$$\begin{aligned} \min_{y \in Y, z^s, \xi} \quad & F_1 + \xi \\ \text{s.t.} \quad & z^s \in Z(y, p^{ds}), \ \forall s \leq k \\ & \xi \geq f^T z^s, \ \forall s \leq k \end{aligned} \tag{5.64a}$$

Record the optimal solution (y^*, ξ^*) and the optimal value $F_1^* + \xi^*$.

3: Solve the following subproblem with obtained y^*

$$\max_{p^d \in W} \min_{z \in Z(y^*, p^d)} f^T z \tag{5.64b}$$

The optimal solution is (p^{d*}, z^*), and the optimal value is R^*.

4: Update $LB = F_1^* + \xi^*$, $UB = F_1^* + R^*$. If $UB - LB \leq \varepsilon$, stop and return the optimal solution y^*. Otherwise, update $k = k + 1$, $p^{dk} = p^{d*}$, create new variable z^k and add the following cuts

$$z^k \in Z(y, p^{dk}), \ \xi \geq f^T z^k \tag{5.64c}$$

to the master problem and go to step 2.

Algorithm 5.4 converges in a finite number of iterations that is bounded by 2^{N_W}, where N_W is the dimension of p^d, because there are at most 2^{N_W} elements in

Extr[W]. In practice, it always converges in a few iterations because the subproblem can find out the most critical scenario. We provide further discussions on handling the infeasibility issue.

1. If the subproblem is infeasible, the SOCP relaxation will be inexact, and a feasibility cut $z^k \in Z(y, p^{dk})$ should be added to the master problem after step 3. Because p^{dl} and p^{du} are taken into account in the first iteration, such kind of infeasibility rarely happens in our case study.
2. If problem (5.49) itself is infeasible, this indicates the distribution network is unable to tackle the worst situation with available resources. A long-term solution could be to upgrade the power grid infrastructure. Nevertheless, in a mathematical sense, we can add slack variables to allow voltage violation, at the expense of a penalty cost in the objective function. Since insufficient power can be purchased from the main grid, there is no obstacle for maintaining power balance.

4. Controlling the Conservatism

The presented method relies on the box-hull outer approximation of the uncertainty set. There might be two limitations in practical usage regarding the scalability and conservativeness:

1. The subproblem (5.60) is an MISOCP; large-scale mixed-integer programs are challenging to solve, although it could be a viable way.
2. The traffic demand between each O-D pair can vary, but is unlikely to reach their upper (lower) bound q^{tu} (q^{tl}) simultaneously. Therefore, the original box uncertainty set for the traffic demand may be over conservative, so is the one for the power demand.

We discuss how to control the conservatism by incorporating budget constraints with an adjustable parameter Γ. Intuitively, the upper-right corner p^{du} and the lower-left corner p^{dl} would have a decisive impact on the reserve capacity provision and the operating cost, as they correspond to the maximal and the minimal power demand variations. One possible way is to use a smaller set which excludes the extreme corners q^{tl} and q^{tu}. To do this, we add two inequalities in BOX(q^{tl}, q^{tu}), yielding the following polyhedral description for the traffic demand uncertainty

$$\text{POL-T}(\Gamma) = \left\{ q^t \left| \begin{array}{l} q^t \in \text{BOX}(q^{tl}, q^{tu}) \\ 1^T q^t \le 1^T q^{tu} - \Gamma \\ 1^T q^t \ge 1^T q^{lu} + \Gamma \end{array} \right. \right\}$$

where Γ is an adjustable parameter which reflects the attitude towards risks. It can be interpreted as a measure on the spatial correlation of the traffic demand forecast error. When $\Gamma = 0$, POL-T(Γ) = BOX(q^{tl}, q^{tu}); the conservatism decreases with increasing the value of Γ. However, if $\Gamma > 0$, we are unable to locate extreme points of the nodal power demand in advance, because link flow x_a is no longer a monotonic function when q^t varies on the hyperplane where the budget constraint is active. In this regard, we have to estimate the range of p^d from other means. By restricting $\eta(x_a) = \eta x_a$, we can observe the following fact

$$\sum_i p^d = \sum_i p^{df} + \eta \sum_a x_a = \sum_i p^{df} + \eta \sum_{rs} q^t_{rs}$$

When $q^t \in$ POL-T(Γ), we have

$$1^T q^{tl} + \Gamma \le 1^T q^t \le 1^T q^{tu} - \Gamma$$

then we arrive at the conclusion

$$1^T p^{dl} + \eta\Gamma \le 1^T p^d \le 1^T p^{du} - \eta\Gamma$$

where p^{dl} and p^{du} are defined through (5.56a), (5.56b), (5.56c), and (5.56d). Finally, we can construct the following polyhedral uncertainty set of nodal power demand

$$W = \left\{ p^d \;\middle|\; \begin{array}{l} p^d \in \text{BOX}(p^{dl}, p^{du}) \\ 1^T p^d \le 1^T p^{du} - \eta\Gamma \\ 1^T p^d \ge 1^T p^{lu} + \eta\Gamma \end{array} \right\} \tag{5.65}$$

Please bear in mind that the above formulation relies on the linearity of the interface equation $\eta(x_a) = \eta x_a$. At this time, the uncertainty set W in (5.65) is a polyhedral approximation for the following implicit set:

$$W = \left\{ p^d \;\middle|\; \begin{array}{c} \exists q^t \in \text{POL-T}(\Gamma) \\ x_a \text{ solves TAP (5.46)} \\ p^d_i = p^{df}_i + \sum_{a \in C(i)} \eta x_a, \forall i \end{array} \right\}$$

Because W is no longer a hypercube, the exact MISOCP reformulation is no longer applicable, but Algorithm 5.3 is still valid, and the complexity does not change much, because problem (5.62) remains an LP, except for the two additional linear inequality constraints. Although there is no provable guarantee on the global optimality, a high quality solution can be retrieved by trying multiple initial values. A plausible choice for the initial value is suggested in the case study.

5.4.3 Case Studies

1. Basic Settings

We consider the test system in Fig. 5.4, which has been used in Sect. 5.3. Each road is served by one charging facility. The charging demand equation is $\eta(x_a) = 0.005x_a$. Link parameters are given in Table 5.1. Road segments in the outer loop have larger capacities, as they represent ring expressways which will carry more traffic flows. We have the forecast values of trip rates q_{rs}^{t0} in the evening rush hours, which are shown in Table 5.7, indicating that most vehicles have to travel from the northwest to eastern and southern areas. We assume that actual trip rates are uncertain and can vary in the interval $[q^{tl}, q^{tu}]$, where $q^{tl} = (1-\alpha)q_{rs}^t$, $q^{tu} = (1+\alpha)q_{rs}^t$, and α is defined as the level of uncertainty. We will investigate the impact of $\alpha \in [0, 0.2]$ on the total operation cost. Parameters of generators are provided in Table 5.8. In general, the upward regulation cost d_i^{p+} should be greater than the marginal production cost, because the incremental energy has not been paid in the first stage; the downward regulation cost d_i^{p-} is usually very small, as no additional energy is needed. The remaining cost coefficients of generators are given as follows: $c_i^{p+} = c_i^{p-} = 0.1b_i^p$, $c_i^{q+} = c_i^{q-} = 0.01b_i^p$, $d_i^{q+} = d_i^{q-} = d_i^{p-}$. As for the PDN, distribution line data are listed in Table 5.4; the fixed demand at each bus is $p_i^{df} = 0.10$, $q_i^{df} = 0.01$, the voltage boundary is $U_i^m = 0.8100$, $U_i^r = 1.10$. The voltage magnitude at the slack bus is $U_0 = 1.04$, the electricity price at the slack bus is $\rho = 1500\$/\text{p.u.}$ Simulation environment is the same as that in Sect. 5.3. LPs, SOCPs, and MISOCPs are solved by CPLEX [30].

2. Results

When the level of uncertainty is $\alpha = 0.2$, the minimum and maximum traffic flows (x_a^l, x_a^u) on each link are computed from (5.56a) and (5.56b) and shown in

Table 5.7 O-D pairs and their trip rates

O-D pair	q_{rs}^t (p.u.)	O-D pair	q_{rs}^t (p.u.)	O-D pair	q_{rs}^t (p.u.)
T1–T6	15	T3-T6	13	T4-T9	5
T1-T10	20	T3-T10	17	T4-T10	10
T1-T12	10	T3-T12	8	T4-T12	15
T1-T11	15	T3-T11	12		

Table 5.8 Parameters of generators

Unit	Node	p_i^r (p.u.)	q_i^r (p.u.)	a_i^p (\$)	b_i^p (\$)	d_i^{p+} (\$)	d_i^{p-} (\$)
G1	E7	1.5	0.2	228	1382	1658	120
G2	E10	2.0	0.4	196	1089	1307	150
G3	E11	2.0	0.4	236	834	1001	200
G4	E14	1.6	0.3	239	1329	1595	130

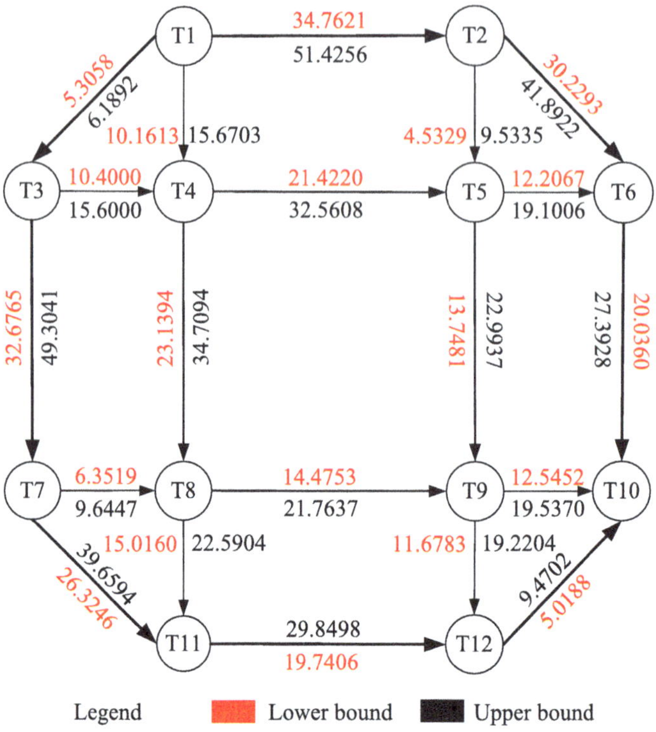

Fig. 5.14 Distribution of traffic flows and their bounds at the UE

Table 5.9 Optimal generation schedule (in p.u.)

Unit	p^g	q^g	r^{p+}	r^{p-}	r^{q+}	r^{q-}
G1	0.5485	0.0931	0	0	0	0
G2	0.8355	0.1595	0.6101	0	0	0
G3	0.6317	0.4348	0.7490	0	0	0.0817
G4	0.5689	0.0955	0	0	0	0

Fig. 5.14, which illustrates that most vehicles prefer to use the ring expressway, imposing higher demands on these charging facilities and potential risks of over low voltage. The optimal generation dispatch and OPF results are elaborated in Tables 5.9 and 5.10. Because the line resistance is not negligible, active power delivery has a remarkable impact on the voltage profile. To tackle possible load variation, upward active power spinning reserve is offered by G2 and G3, as we can see from Table 5.9, for the purpose of maintaining power balance and regulating bus voltage magnitude, while little reactive power reserve is provided in this case. The OPF outcomes in Table 5.10 suggest that the voltage magnitude of bus 3 has already reached its lower bound, and increasing the delivered power in distribution lines will introduce higher voltage drop on Line E0-E11 and Line E8-E3, which may cause voltage violation and stability issue. From Table 5.9 and Table 5.10 we

Table 5.10 Robust optimal power flow (in p.u.)

Node	Voltage	Line	P_{ij}^{l1}	Q_{ij}^{l1}
1	0.937592	0-7	0.335887	−0.02979
2	0.956222	0-10	0.489533	−0.04869
3	0.900000	0-11	0.655952	−0.30596
4	0.972135	0-14	0.241552	−0.03990
5	0.977301	7-2	0.328547	0.019369
6	0.955675	7-4	0.169344	0.013061
7	0.995353	7-5	0.139448	0.011957
8	0.939125	10-9	0.321204	0.019512
9	0.945577	10-6	0.305114	0.029077
10	0.990501	10-13	0.436805	0.041057
11	0.994262	6-1	0.12887	0.011407
12	0.963487	13-16	0.274225	0.016341
13	0.938874	11-12	0.224666	0.014543
14	1.003748	11-8	0.49567	0.045904
15	0.953323	11-15	0.334728	0.031866
16	0.902948	8-3	0.291526	0.018687
17	0.976107	15-19	0.140329	0.011665
18	0.978971	14-20	0.230661	0.014293
19	0.934726	14-17	0.200638	0.014169
20	0.975465	14-18	0.183428	0.013129

can image how the system would respond to uncertainty: If the actual demand is greater than the forecast, generator will deploy upward spinning reserve capacity; otherwise, less electricity will be purchased from the main grid. In this way, the network power flow and voltage profile can be well maintained.

In summary, by implementing the robust dispatch strategy in Table 5.9, all security restrictions can be satisfied by deploying a corrective action in real time dispatch, whatever the actual traffic demand will be. It is worth mentioning that the amount of active power reserve is larger than the demand deviation in the worst-case demand scenario, indicating that even if the actual demand is greater than the forecast, the power delivered from the slack bus has to be reduced in order to maintain voltage constraints of all buses. This example illustrates the role of active power dispatch in voltage regulation. The conclusion is certainly not versatile and depends on the specific system data. For another system whose line reactance is larger than the line resistance, reactive power reserve may play a more important role in voltage regulation.

We further investigate the impact of the level of uncertainty on the total cost and computational efficiency by changing α from 0 to 0.2 with fixed big-M parameter $M = 10{,}000$. Results are provided in Figs. 5.15 and 5.16. Figure 5.15 shows an almost linear growth rate of the total cost with respect to α. On the one hand, this phenomenon is not general and should be parameter dependent; on the other hand, because the worst-case traffic demand usually happens at the upper right corner q^{tu},

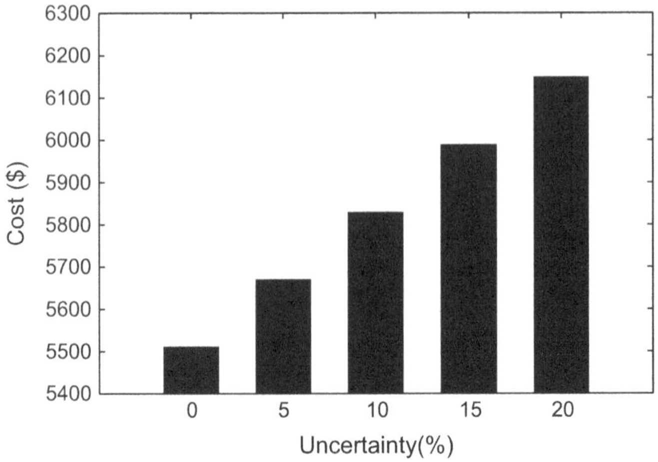

Fig. 5.15 Total costs under different levels of uncertainty

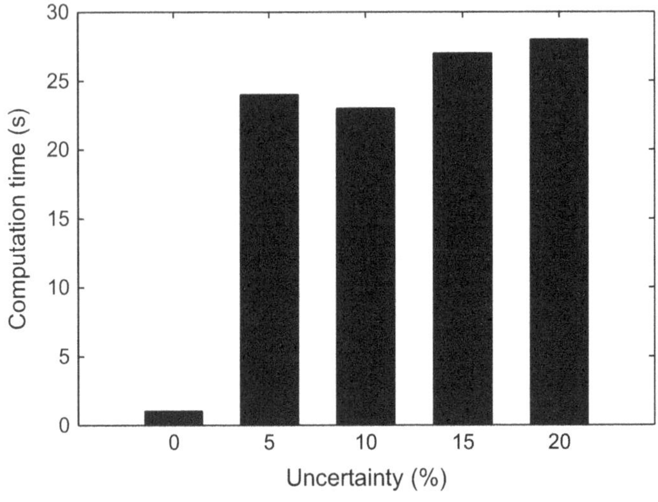

Fig. 5.16 Computation times under different levels of uncertainty

which grows linearly with the level of uncertainty, as a result, the corresponding total electrical demand is given by $\sum_i p_i^d = \sum_i p_i^{df} + \eta \sum_a x_a = \sum_i p_i^{df} + \eta \sum_{rs} q_{rs}^{tu}$, which also grows linearly with respect to α. By observing that the reserve cost, regulating cost, and energy purchasing cost are all linear functions, it is not difficult to understand that the total cost follows a nearly linear growing pattern if the traffic demand only varies in certain range. Figure 5.16 indicates that the level of uncertainty has little impact on the computation time when $\alpha > 0$, because most

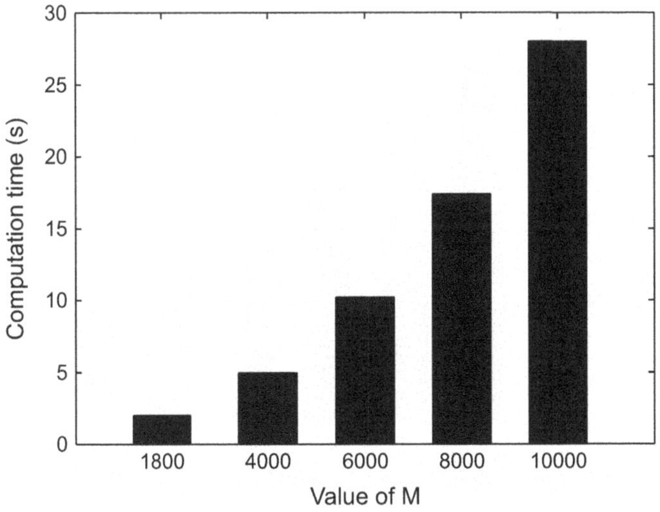

Fig. 5.17 Computation times under different values of M

computation effort will be spent on the MISOCP subproblem whose complexity mainly depends on the number of charging facilities.

Next, we investigate how the value of big-M in (5.59) will affect the computation time. First, we fix $\alpha = 0.2$ and choose $M = 10{,}000$, the optimal solution λ^* satisfies $\max_i\{|\lambda_i^*|\} = 1798$, suggesting that the best choice of M is 1798; then we conduct a series of experiments with different values of M. Results are given in Fig. 5.17. We can see that the computation time can be remarkably reduced if a proper M is used. A general trend is: the smaller, the faster. This is because smaller M will give stronger relaxed problems as well as tighter bounds when the integrality of binary variables is omitted, thus expediting convergence. If M is smaller than 1798, the results will be incorrect, although the computation time can be further reduced.

Finally, we investigate how the value of Γ will impact the total cost and the performance of Algorithm 5.3. We conduct a series of experiments by using different values of Γ from 0 to 50 and the convergence tolerance $\delta = 0.01$. We use two initial values of p^d in step 1 which are given by

$$p^{d1} = p^{du}\left(1 - \frac{\eta\Gamma}{1^T p^{du}}\right)$$

$$p^{d2} = p^{dl}\left(1 + \frac{\eta\Gamma}{1^T p^{dl}}\right)$$

The computations finish in 2 s in all these tests. The total costs are shown in Fig. 5.18, which show a nearly linear decreasing rate. The reason is similar to that of Fig. 5.15. We emphasize that although Algorithm 5.3 is scalable, it only finds a

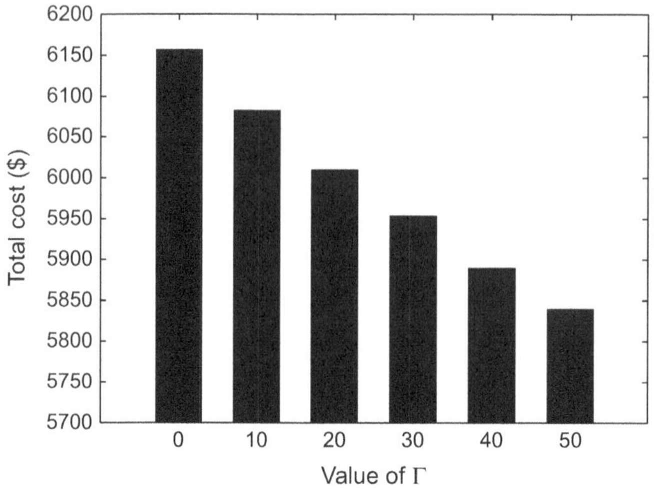

Fig. 5.18 Total costs under different values of Γ

local optimal solution for problem (5.57). When $\Gamma = 0$, we get the same total cost as that in Fig. 5.15.

5.5 Capacity Expansion Planning

The proliferation of EVs has created an emerging trend of transportation electrification, which calls for the installation of brand-new battery charging/swapping infrastructures on the urban transportation system to support EV integration. Coordinated and robust operation of TN and PDN has been studied in Sects. 5.3 and 5.4, respectively. However, the inappropriate deployment of public charging facilities may create negative impacts on the power grid [54], such as congestion and voltage deterioration.

Power system researches have been focusing on the optimal sitting and sizing of charging or swapping stations, while considering the distribution network operating constraints. Road conditions in the TN are usually simplified or ignored, and the driving patterns of EVs are assumed to be exogenously given, either in a deterministic or stochastic manner. Such an assumption is reasonable for instances where the local charging demand profile varies periodically or its probability distribution is known, such as residential parking lots.

The transportation community has been interested in ameliorating the traffic condition through properly placing on-road charging facilities. The problem is often formulated as a bilevel program, in which the lower level describes the UE condition, and the upper level optimizes the planning strategy so as to meet system demands with minimum investment cost. In fact, traffic flow patterns determine spatial and temporal distribution of charging demands, which further impact the power flow of the distribution network. Such interdependency has been well recognized, but operating details of the power grid are usually ignored, which may cause difficulties in distribution system operation.

This section aims to comprehensively address the capacity expansion planning problem of the coupled transportation and power distribution infrastructures. We suggest a holistic model for the networked systems accounting for their interdependency. The UE is depicted by Nesterov TAP, which is more suitable for addressing the planning issue, since it incorporates an explicit bound that the traffic flow on each road cannot exceed, and is also advantageous in computation because of its linearity. We use the linearized BFM introduced in Sect. 2.2.2 to determine the operating status of the PDN. The proposed model comprehensively determines capacity expansion planning strategies of road segments in the TN, on-road charging facilities, generation units, and distribution lines in the PDN, in a coordinated manner. For engineering construction considerations, we assume that the planning variables take discrete values, so the model is essentially a discrete network design problem, which belongs to the category of MPEC because the UE condition appears in the constraints. By using the primal-dual optimality condition transformation and linearization techniques from integer algebra elucidated in Appendix B, we transform the proposed planning model to an equivalent mixed-integer convex program without exploiting approximations. Therefore, a global optimal solution can be found by commercial solvers with affordable computational expense.

5.5.1 Mathematical Formulation

Major symbols and notations in the linearized BFM and Nesterove TAP are inherited from Sects. 2.2.2 and 5.2.3, respectively. There may be slight differences in superscripts and subscripts, without causing confusions. Others are defined in the following for quick reference.

Parameters

n_a^{ce}	Number of existing charging facilities on link a.
Pr_a^r	Cost of building one additional road on link a.
Pr_a^c	Cost of building one charging facility on link a.
Pr_i^g	Cost of building one generator at bus i.
Pr_{ij}^l	Cost of building one additional distribution line between buses i and j.
p_i^{dc}	Traditional power demand at bus i.
p_i^+	Active power generation capacity at bus i.
q_i^+	Reactive power generation capacity at bus i.
V_i^-	Lower bound of voltage magnitude at bus i.
V_i^+	Upper bound of voltage magnitude at bus i.

κ Discount factor, which leverages long-term investment cost and short-term operation cost.
ω Monetary value of vehicle travel time.
η Charging rate of traffic flow.
Δc_a Traffic flow capacity of the expanded link.
Δp_a^c Power capacity of the expanded charging facility.
Δp_i^m Active generation capacity of the expanded unit.
Δq_i^m Reactive generation capacity of the expanded unit.

Variables

n_a^c The number of new charging facilities that should be invested in on link a.
n_i^g The number of new generation units that should be invested in at bus i.
v_{an}^t Additional continuous variable for linearizing the total travel time function.
v_{ijn}^{dl1} Additional continuous variable for linearizing the voltage drop equation.
v_{ijn}^{dl2} Additional continuous variable for linearizing the voltage drop equation.
z_{an}^t Binary variable of road expansion; the number of new roads to be invested in on link a is expressed by a binary expansion $\sum_{n=0}^{N_R} 2^n z_{an}^t$.
z_{ijn}^{dl} Binary variable of distribution line expansion between the head bus i and the tail bus j; the number of new lines to be invested in is given by a binary expansion $\sum_{n=0}^{N_L} 2^n z_{ij}^{dl}$.
π Dual variables of Nesterov TAP.
μ Dual variables of Nesterov TAP.
λ Dual variables of Nesterov TAP.

Basic settings and assumptions for capacity expansion planning are elucidated.

1. New roads, whose number is given by $\sum_{n=0}^{N_R} 2^n z_{an}^t, \forall a$, where $z_{an}^t \in \{0, 1\}$, can be built in parallel with existing links. They have identical capacity Δc_a.
2. New distribution lines, whose number is expressed by $\sum_{n=0}^{N_L} 2^n z_{ijn}^{dl}, \forall (i, j, l)$, where $z_{ijn}^{dl} \in \{0, 1\}$, can be built in parallel with existing branches. For simplicity, we assume that every distribution line between buses i and j has identical parameter. After upgrading, the resistance and reactance are reduced to $1/(1+\sum_{n=0}^{N_L} 2^n z_{ijn}^{dl})$ compared to their original values. Meanwhile, the maximal allowed active and apparent power flow become $(1 + \sum_{n=0}^{N_L} 2^n z_{ijn}^{dl}) P_l^u$ and $(1 + \sum_{n=0}^{N_L} 2^n z_{ijn}^{dl}) S_l$, respectively. From another perspective, the power flow in line l is reduced to $1/(1+\sum_{n=0}^{N_L} 2^n z_{ijn}^{dl})$, while line parameters remain the same, so the voltage drop along the branch can be reduced.
3. New charging facilities with identical capacity Δp_a^c can be built at the electrical bus connecting to a transportation link a. The number is given by a positive integer n_a^c, which should be optimized.
4. New generators with identical parameters can be built at any electrical bus. The number is given by a positive integer variable n_i^g, which should be optimized. We can also specify candidate locations for generators.

We use binary variables to represent the number of lanes and distribution lines because these variables will be multiplied by other continuous variables. These nonlinear terms involving the product of a binary variable and a continuous variable will be linearized by using techniques from integer algebra. More importantly, this expression uses only a logarithmic number of binary variables. For example, if at most 7 additional lanes can be invested in on link a, we use $\log_2(7+1) = 3$ binary variables for link expansion.

Given the desired traffic and electrical demands to be served, the mathematical model of capacity expansion planning is presented as follows:

$$\begin{aligned} \min \;& F_{TN} + F_{PDN} \\ \text{s.t. }& \text{Cons-PDN} \\ & \text{Cons-Couple} \\ & \{x, \lambda\} \text{ is an UE} \end{aligned} \tag{5.66}$$

The first term F_{TN} in the objective function stands for the total cost of TN, which is given by

$$F_{TN} = \sum_{a \in T_A} \left[\omega(t_a + \lambda_a)x_a + \frac{\text{Pr}_a^r}{\kappa} \sum_{n=0}^{N_R} 2^n z_{an}^t \right] \tag{5.67}$$

where $\sum_{a \in T_A}(t_a + \lambda_a)x_a$ is the total vehicle travel time under the UE pattern, ω is the monetary value of time, so this term is the total travel cost of users in the TN; the second term represents the investment cost on new roads, and κ is the discount factor, which leverages long-term investment cost and short-term operation cost. The second term F_{PDN} in the objective function is the total cost of the PDN, which is given by

$$\begin{aligned} F_{PDN} = & \sum_{i \in E_B} \left[a_i (p_i)^2 + b_i p_i \right] + \rho \sum_{j \in \pi(0)} P_{0j}^l \\ & + \sum_{a \in T_A} \frac{\text{Pr}_a^c}{\kappa} n_a^c + \sum_{l \in E_L} \sum_{n=0}^{N_L} \frac{\text{Pr}_{ij}^l}{\kappa} 2^n z_{ijn}^{dl} + \sum_{i \in E_N} \frac{\text{Pr}_i^g}{\kappa} n_i^g \end{aligned} \tag{5.68}$$

where the first term is the generator production cost, which is a convex quadratic function; P_{0j}^l is the active power delivered through distribution line l that connects to the slack bus, $\pi(0)$ is the set of child buses of the slack bus, so that the second term is the purchasing cost paid to the main grid or an upper-level power market; the third to fifth terms represent investments on the new charging facilities, distribution lines, and generation units, respectively, which are discounted into short-term operation cost.

The operation constraint on the PDN is expressed as

$$\text{Cons-PDN} = \text{Cons-LPF} + \text{Cons-BND} \tag{5.69}$$

where Cons-LPF represents linearized branch flow constraints and includes the following constraints

$$P_{ij}^{l} + p_{j}^{g} = \sum_{k \in \pi(j)} P_{jk}^{l} + p_{j}^{d}, \ \forall(i, j, l) \tag{5.70a}$$

$$Q_{ij}^{l} + q_{j}^{g} = \sum_{k \in \pi(j)} Q_{jk}^{l} + q_{j}^{d}, \ \forall(i, j, l) \tag{5.70b}$$

$$V_j = V_i - \frac{r_{ij}^{l} P_{ij}^{l} + x_{ij}^{l} Q_{ij}^{l}}{V_0 \left(1 + \sum\limits_{n=0}^{N_L} 2^n z_{ijn}^{dl}\right)}, \ \forall(i, j, l) \tag{5.70c}$$

See Sect. 2.2.2 for a clear discussion on this power flow model. Cons-BND represents physical limitations of decision variables, which include the following constraints

$$0 \le p_i^{g} \le p_i^{+} + \Delta p_i^{m} n_i^{g} \tag{5.71a}$$

$$-q_i^{+} - \Delta q_i^{m} n_i^{g} \le q_i^{g} \le q_i^{+} + \Delta q_i^{m} n_i^{g}, \forall i \tag{5.71b}$$

$$V_i^{-} \le V_i \le V_i^{+}, \ \forall i, \ Q_{ij}^{l} \ge 0, \forall l \tag{5.71c}$$

$$0 \le P_{ij}^{l} \le \left(1 + \sum_{n=0}^{N_L} 2^n z_{ijn}^{dl}\right) P_l^{u}, \forall l \tag{5.71d}$$

$$\left\| \begin{matrix} P_{ij}^{l} \\ Q_{ij}^{l} \end{matrix} \right\|_2 \le S_l \left(1 + \sum_{n=0}^{N_L} 2^n z_{ijn}^{dl}\right), \forall l \tag{5.71e}$$

Unit active and reactive power output capacities, bus voltage magnitude boundaries, and distribution line power flow limits are taken into account through (5.71a)–(5.71e), respectively. The coupling constraints Cons-Couple include

$$p_i^{d} = p_i^{dc} + \eta \sum_{a \in C(i)} x_a, \ \forall i \tag{5.72a}$$

$$\eta x_a \le \Delta p_a^{c} (n_a^{c} + n_a^{ce}), \ \forall a \tag{5.72b}$$

where the first constraint is the charging demand equation, and the second one requires that the charging demand on each road cannot exceed the service capability that the charging facilities in service can offer.

The traffic flow x and possible delay λ at UE pattern are determined from the following Nesterov TAP

$$\begin{aligned} \min_{x,f} \ & x^T t^0 \\ \text{s.t.} \ & x = \Delta f, \ x \le c : \lambda \\ & Ef = q^t, \ f \ge 0 \end{aligned} \tag{5.73a}$$

where the road capacity is

$$c_a = c_a^0 + \Delta c_a \sum_{n=0}^{N_R} 2^n z_{an}^t, \ \forall a \tag{5.73b}$$

More details about Nesterov TAP can be found in Sect 5.2.3.

Expansion planning problem (5.66) is a bilevel optimization problem because of TAP (5.73a) in the lower level. By replacing Nesterov TAP with its first-order optimality conditions, (5.66) comes down to an MPEC, which is also challenging to solve due to the complementarity and slackness constraints. We endeavor to tackle this problem by developing a mixed-integer convex model and utilizing state-of-the-art solvers.

5.5.2 A Mixed Integer Convex Reformulation

Three kinds of non-convexities prevent an efficient solution of problem (5.66):

1. The optimality condition of Nesterov TAP (5.73a).
2. The product term $\lambda_a x_a$ in F_{TN}.
3. The fractional voltage drop equation (5.70c).

All of them are nonlinear and non-convex. We will derive equivalent linear expressions for them by exploiting the special problem structure.

1. Linearizing the User Equilibrium Constraints

Because Nesterov TAP (5.73a) is an LP, its solution can be characterized by either the KKT optimality condition or the primal-dual optimality condition. We adopt the latter because it involves fewer constraints and a simpler structure without complementarity and slackness conditions. The primal-dual optimality condition of TAP (5.73a) is given as follows:

$$x - \Delta f = 0,\ Ef = q^t,\ x \le c,\ f \ge 0 \tag{5.74a}$$

$$\lambda \ge 0,\ \mu - \lambda = t^0,\ E^T \pi - \Delta^T \mu \le 0 \tag{5.74b}$$

$$\pi^T q^t - \lambda^T c = x^T t^0 \tag{5.74c}$$

where (5.74a) and (5.74b) are the feasible regions of primal and dual variables, respectively, and (5.74c) is the strong duality condition, which enforces equal values on the primal and dual objectives. The feasible solution of (5.74) is also a UE on the TN. In view of equation (5.73b), the nonlinearity only exists in the product term $\lambda^T c = \sum_a \lambda_a (c_a^0 + \Delta c_a \sum_{n=0}^{N_R} 2^n z_{an}^t)$.

We introduce a matrix variable $V_R^t = [v_{an}^t], \forall a, \forall n$, and define the vector of incremental capacity $\Delta c = [\Delta c_a], \forall a$, the vector of initial capacity $c^0 = [c_a^0], \forall a$, and a constant vector $b_R^t = [2^n], \forall n$. If the relationship

$$v_{an}^t = \lambda_a z_{an}^t,\ \forall a,\ \forall n \tag{5.75}$$

holds, then we have a linear expression

$$\lambda^T c = \lambda^T c^0 + (\Delta c)^T V_R^t b_R^t \tag{5.76}$$

Since v_{an}^t represents the product of a binary variable z_{an}^t and a positive continuous variable λ_a, constraint (5.75) can be replaced by the following linear constraints:

$$\begin{gathered} 0 \le v_{an}^t \le M z_{an}^t,\ \forall a, n \\ 0 \le \lambda_a - v_{an}^t \le M(1 - z_{an}^t),\ \forall a, n \end{gathered} \tag{5.77}$$

where M is a big enough constant, which can be interpreted as the maximal possible delay. When $z_{an}^t = 0$, (5.77) enforces $v_{an}^t = 0$ and $0 \le \lambda_a \le M$; otherwise, when $z_{an}^t = 1$, it enforces $v_{an}^t = \lambda_a$ and $0 \le v_{an}^t \le M$. This is equivalent to the original expression in (5.75). The tightest value of M should be $M^* = \min_a\{\lambda_a^*\}$, where $\lambda^* = [\lambda_a], \forall a$ is the optimal dual variable of problem (5.73a) at the solution of (5.66), and is unknown in advance. Any $M \ge M^*$ without causing numeric issue is valid for (5.77). However, a smaller M will yield more efficient computation. A possible value of M would be the maximal delay over transportation links estimated from certain heuristic.

Now we are ready to express the linearized UE constraints as

$$\text{Cons-UE-Lin} = \left\{ \begin{array}{c} x - \Delta f = 0,\ Ef = q^t, f \ge 0 \\ x \le c^0 + \Delta C Z_R^t b_R^t,\ (5.77) \\ \lambda \ge 0, \mu - \lambda = t^0, H^T \pi - \Delta^T \mu \le 0 \\ \pi^T q^r = x^T t^0 + \lambda^T c^0 + (\Delta c)^T V_R^t b_R^t \end{array} \right\} \tag{5.78}$$

where ΔC is a diagonal matrix whose elements are Δc_a, and the binary variable matrix Z_R^t is defined as $Z_R^t = [z_{an}^t]$, $\forall a, n$.

2. Linearizing the Objective Function

The product term $\lambda_a x_a$ in F_{TN} involves two continuous variables. There is no general methodology for linearizing this term without exploiting approximation. We derive an exact linear reformulation based on the special structure of TAP (5.73a).

Bear in mind that x and λ should be the optimal primal and dual solution for LP (5.73a). Recall the KKT optimality condition, complementarity and slackness hold for the inequality constraint and its dual variable, leading to the following equation:

$$\lambda^T (x - c) = 0$$

Hence, we have $\lambda^T x = \lambda^T c$. This equality is easy to understand. Recall assumptions in Nesterov model: if $x_a < c_a$, indicating that road a is not congested, then we must have $t_a = t_a^0$ and $\lambda_a = 0$, thus $\lambda_a x_a = 0$; if road a is congested and $\lambda_a > 0$, we must have $x_a = c_a$ and $\lambda_a x_a = \lambda_a c_a$. In summary, $\lambda^T x = \lambda^T c$ holds at the optimal solution.

Considering equation (5.76), we can arrange F_{TN} as a linear form

$$F_{TN} = \omega(x^T t^0 + \lambda^T c^0 + (\Delta c)^T V_R^t b_R^t) + \sum_{a \in T_A} \sum_{n=0}^{N_R} \frac{\text{Pr}_a^r}{\kappa} 2^n z_{an}^t \tag{5.79}$$

The expressions of F_{TN} in (5.67) and (5.79) are exactly equivalent, as (5.77) has already been considered in (5.78), and no approximation error is involved.

3. Linearizing the Voltage Drop Equality

Multiplying both sides of (5.70c) by $1 + \sum_{n=0}^{N_L} 2^n z_{ijn}^{dl}$ gives the following equation without fractions

$$V_j \left(1 + \sum_{n=0}^{N_L} 2^n z_{ijn}^{dl}\right) = V_i \left(1 + \sum_{n=0}^{N_L} 2^n z_{ijn}^{dl}\right) - \frac{r_{ij}^l P_{ij}^l + x_{ij}^l Q_{ij}^l}{V_0}, \quad \forall (i, j, l)$$

The product terms involving the continuous variable V_i or V_j and the binary variable z_{ijn}^{dl} can be linearized via the method performed in (5.75) and (5.77). By introducing continuous variables $v_{ijn}^{dl1} = V_i z_{ijn}^{dl}$, $v_{ijn}^{dl2} = V_j z_{ijn}^{dl}$, $\forall (i, j, l), \forall n$, (5.70c) can be replaced by the following constraints:

$$\left(V_j + \sum_{n=0}^{N_L} 2^n v_{ijn}^{dl2}\right) = \left(V_i + \sum_{n=0}^{N_L} 2^n v_{ijn}^{dl1}\right) - \frac{r_{ij}^l P_{ij}^l + x_{ij}^l Q_{ij}^l}{V_0}, \quad \forall (i, j, l) \tag{5.80a}$$

$$0 \le v_{ijn}^{dl1} \le V_i^+ z_{ijn}^{dl}, \quad 0 \le V_i - v_{ijn}^{dl1} \le V_i^+ (1 - z_{ijn}^{dl}), \quad \forall (i, j, l) \tag{5.80b}$$

$$0 \le v_{ijn}^{dl2} \le V_j^+ z_{ijn}^{dl}, \quad 0 \le V_j - v_{ijn}^{dl2} \le V_j^+ (1 - z_{ijn}^{dl}), \quad \forall (i, j, l) \tag{5.80c}$$

Now we can express the linearized power flow constraints as

$$\text{Cons-PF-Lin} = \Big\{ (5.70\text{a}),\ (5.70\text{b}),\ (5.71\text{a})\text{–}(5.71\text{e}),\ (5.80) \Big\} \tag{5.81}$$

When the integrality of binary variable is relaxed, Cons-PF-Lin yields mixed-integer linear and SOC constraints.

4. The Final MICP Formulation
On the basis of the previous discussions, the capacity expansion planning problem (5.66) can be cast as the following mixed-integer convex program:

$$\begin{aligned} \min \ & F_{TN} + F_{PDN} \\ \text{s.t.} \ & \text{Cons-PF-Lin} \\ & \text{Cons-UE-Lin} \\ & \text{Cons-Couple} \end{aligned} \tag{5.82}$$

where F_{TN} is defined in (5.79), F_{PDN} is defined in (5.68), Cons-PF-Lin is defined in (5.81), Cons-UE-Lin is defined in (5.78), and Cons-Couple is defined in (5.72). Nonlinearity in (5.82) includes the convex quadratic terms in the objective function F_{PDN} as well as circle constraint in (5.71e). So (5.82) is actually an MISOCP. Nevertheless, both kinds of nonlinearity can be approximated by PWL functions, and (5.82) can be further reduced to an MILP.

The above reformulation relies on the linear models of TAP and OPF. The linearized branch flow model for distribution power flow is only appropriate for radial networks, because radiality is crucial for constructing the voltage drop equation. In current engineering practices, distribution networks are intentionally operated with tree topology. For meshed networks, we can use the DC power flow model discussed in Sect. 2.2.1, which preserves the model linearity, while sacrificing the accuracy on bus voltage. It is a natural assumption that the power grid in the metropolitan area has radial topology. It is also interesting to study the interdependent highway system and power transmission network in a larger spatial scale. The DC power flow model has satisfactory accuracy for a transmission network. Different from an urban transportation system, congestion is not a main concern of a highway system, and every vehicle can travel at the maximum allowed velocity. Hence it is easy to simulate and determine the vehicle behaviors in such systems. In short, the proposed model and method may help government agency make better city planning decisions.

5.5.3 *Case Studies*

1. Basic Settings
The proposed model and method are applied to a test system with the same topology as that shown in Fig. 5.4. System data are provided in Tables 5.11, 5.12, and 5.13 (the values are given in p.u. without particular mention). We simulate the scenario

Table 5.11 Parameters of road segments

Link	c_a	$\Pr_a^r/\kappa$	t_a^0 (min)	n_a^{ce}
T1–T3	18.0	50	6	2
T1–T2	20.0	80	10	4
T2–T6	17.0	50	6.5	4
T1–T4	9.8	35	5	1
T2–T5	7.9	35	5.5	1
T3–T4	8.5	35	6	2
T4–T5	13.5	70	12	3
T5–T6	8.2	40	6.5	2
T3–T7	19.0	80	10.2	4
T4–T8	14.0	75	11.5	3
T5–T9	13.8	75	12.5	2
T6–T10	20.0	80	10.5	3
T7–T8	8.9	35	5.8	1
T8–T9	13.2	70	11	2
T9–T10	9.15	35	5.9	2
T7–T11	17.5	50	6.3	4
T8–T11	9.76	35	5.7	2
T9–T12	8.97	35	5.8	1
T12–T10	18.2	50	6.1	0
T11–T12	20.0	80	9.8	3

Table 5.12 O-D pairs and their trip rates

O-D pair	q_{rs}^t	O-D pair	q_{rs}^t	O-D pair	q_{rs}^t
T1–T6	5	T3–T6	7	T4–T9	6
T1–T10	6	T3–T10	7	T4–T10	7
T1–T12	6	T3–T12	6	T4–T12	5
T1–T11	5	T3–T11	5		

of evening rush hour when the majority of traffic demand leaves from the northwest and travels to the eastern and the southern areas. Details of the O-D pairs and their trip rates are given in Table 5.12. Remaining system parameters are given separately below.

The active and apparent power flow limits of existing distribution lines are $P_l^u = 1.0, \forall l$ and $S_l = 1.2, \forall l$, respectively. The capacity of new road is assumed to be $\Delta c_a = 0.5 c_a, \forall a$. The capacity of every new charging facility and generator is $\Delta p_a^c = 0.1, \forall a$ and $\Delta p_i^m = 1.0$, $\Delta q_i^m = 0.2, \forall i$, respectively. The discounted investment cost for each charging facility and generation unit is $\Pr_a^c/\kappa = \$10$ and $\Pr_a^g/\kappa = \$100$, respectively. The fixed electrical demand at each bus of the distribution network is $p_i^{dc} = 0.02$ and $q_i^d = 0.01$. Most of the active power loads originates from the charging requests of electric vehicles. This corresponds to the situation in which the distribution network only serves on-road charging facilities. This setting highlights the role of charging demand which couples the

Table 5.13 Parameters of distribution lines

Line	r	x	$\mathrm{Pr}_{ij}^{l}/\kappa$ ($)
E0-E7	0.081	0.061	140.8
E0-E10	0.066	0.042	185.2
E0-E11	0.061	0.041	196.1
E0-E14	0.079	0.063	140.8
E7-E2	0.115	0.08	102.6
E7-E4	0.131	0.084	93.0
E7-E5	0.123	0.077	100.0
E10-E9	0.135	0.081	92.6
E10-E6	0.107	0.073	111.1
E10-E13	0.111	0.075	107.5
E6-E1	0.127	0.083	95.2
E13-E16	0.119	0.078	101.5
E11-E12	0.132	0.079	94.8
E11-E8	0.105	0.07	114.3
E11-E15	0.115	0.082	101.5
E8-E3	0.122	0.083	97.6
E15-E19	0.12	0.08	100.0
E14-E20	0.119	0.077	102.0
E14-E17	0.133	0.09	89.7
E14-E18	0.13	0.088	91.7

transportation and electricity infrastructures in this particular test. The growth of traditional power demands can be easily considered with proper forecast on the value of p_i^{dc} in (5.72a). The lower bound and upper bound of bus voltage magnitude are $V_i^f = 0.93$ and $V_i^r = 1.05$, respectively, and the voltage magnitude at the slack bus is $V_0 = 1.04$. The electricity price at the slack bus is $\rho = \$1600$; the monetary value of unit travel time is $\omega = \$0.3333$/min; and the charging rate of unit traffic flow is $\eta = 0.02$. We assume that every existing road and distribution line can be considered for expansion. The charging facilities can be built at the roadside and connected to one electrical bus. The available sites for generators are located at buses E7, E10, E11, and E14. Four identical units, whose parameters are $p_i^+ = 2.0$, $q_i^+ = 0.4$, $a_i = \$30$, and $b_i = \$400$, are in service in the current network. The MISOCP model is solved by CPLEX [30].

2. Results

The capacity expansion planning problem in MISOCP form (5.82) is solved under four traffic demand scenarios. The reference demand is given in Table 5.12. In each scenario, we increase the trip rates of all the O-D pairs according to the same portion. More precisely, parameter $q_{rs}^t, \forall(r,s) \in D_T^{RS}$ is increased by 25%, 50%, 75%, and 100%, respectively. The optimal expansion planning strategies under different demand scenarios are given in Tables 5.14, 5.15, 5.16, and 5.17.

Table 5.14 Road expansion strategies

Link	Demand increase			
	25%	50%	75%	100%
T1–T3	0	0	0	0
T1–T2	0	1	1	1
T2–T6	0	0	1	1
T1–T4	0	0	0	1
T2–T5	0	0	0	0
T3–T4	1	1	1	2
T4–T5	1	2	2	2
T5–T6	1	1	3	3
T3–T7	1	1	2	2
T4–T8	0	1	2	4
T5–T9	0	1	0	0
T6–T10	0	0	0	0
T7–T8	0	1	0	0
T8–T9	0	0	0	2
T9–T10	0	2	0	2
T7–T11	1	0	2	1
T8–T11	0	2	2	3
T9–T12	0	0	0	0
T12–T10	0	0	0	0
T11–T12	0	0	1	1

From Tables 5.14 and 5.15 we can see that more roads and charging facilities are invested in when the traffic demand increases. Table 5.16 demonstrates that one generation unit will be built at bus E10 in the last two scenarios; no generator will be invested in for the other two scenarios with lower traffic demands. Table 5.17 shows that distribution lines E10-E13, E13-E16, E11-E8, and E8-E3 are candidates for expansion in this problem, while others remain unchanged. As a result, the growing trends of expansion costs corresponding to different components are plotted in Fig. 5.19, showing the dominant status of road investments over others. To highlight the impact of EV charging requests on both networks, we neglect the traditional demand. If the traditional power demands and their growth are considered, the investment on the distribution network will have a larger share of the total cost.

These results can be explained by exploring the system topology and the equilibrium traffic flow pattern shown in Fig. 5.20. We can observe that links T4–T8, T7–T11, T2–T6, and T5–T6 are carrying heavy traffic flows, leading to higher charging demands at buses E10, E16, E3, and E8. Moreover, E3 and E16 are the terminal buses. Delivering power to these buses introduces higher voltage drops along distribution lines. In this regard, to enhance the voltage magnitude at these buses, distribution lines E10-E13, E13-E16, E11-E8, and E8-E3 should be given higher priorities for upgrading. In the first scenario, a new distribution line is put into service. It is also interesting to notice that although the traffic demand is increased by 50% in the second scenario which is higher than the growth in the first scenario,

Table 5.15 Charging facility expansion strategies

Link	Demand increase			
	25%	50%	75%	100%
T1–T3	1	0	0	0
T1–T2	0	1	2	2
T2–T6	0	1	1	2
T1–T4	0	1	1	2
T2–T5	0	0	0	0
T3–T4	0	1	1	1
T4–T5	1	3	3	2
T5–T6	1	1	3	3
T3–T7	2	2	3	4
T4–T8	0	2	3	6
T5–T9	0	2	0	0
T6–T10	1	1	2	2
T7–T8	0	1	0	1
T8–T9	0	1	1	4
T9–T10	0	2	0	2
T7–T11	1	0	2	2
T8–T11	0	2	2	3
T9–T12	0	0	0	0
T12–T10	0	0	0	0
T11–T12	0	0	2	2

Table 5.16 Generator expansion strategies

Link	Demand increase			
	25%	50%	75%	100%
E7	0	0	0	0
E10	0	0	1	1
E11	0	0	0	0
E14	0	0	0	0

Table 5.17 Distribution line expansion strategies

Link	Demand increase			
	25%	50%	75%	100%
E10-E13	0	0	1	1
E13-E16	1	0	0	0
E11-E8	0	0	0	1
E8-E3	0	0	1	0

no generator or distribution lines will be built, from a system-level perspective, because additional road segments will be constructed to ameliorate the power flow conditions in the PDN. This apparently shows the importance of considering the interdependency and exploiting coordination in system planning.

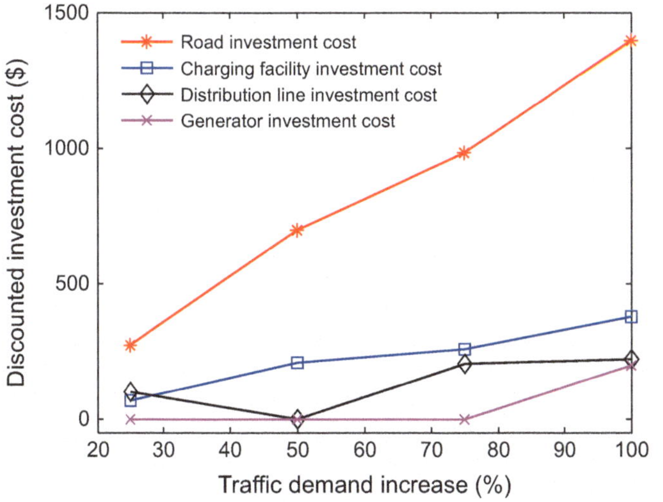

Fig. 5.19 Investment costs for different load scenarios

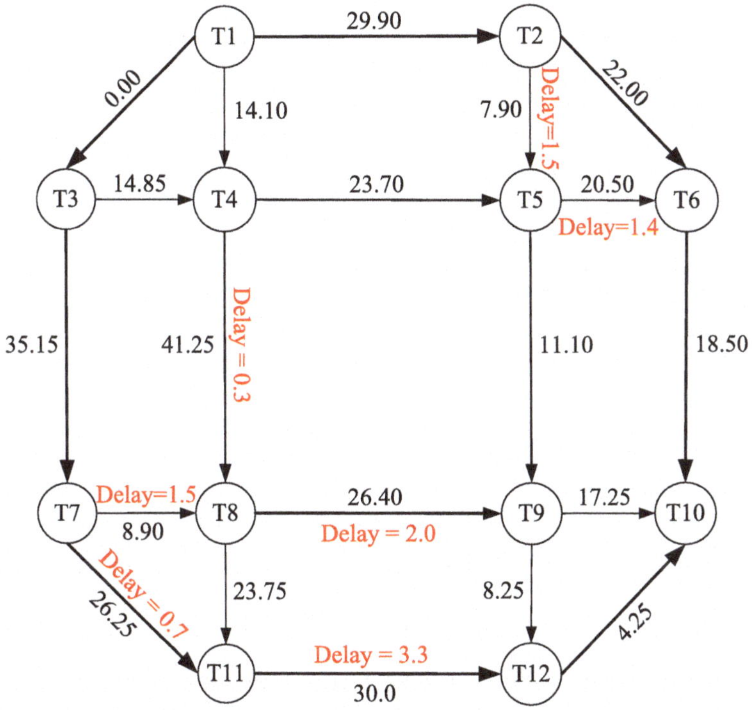

Fig. 5.20 Distribution of traffic flows and delay at the UE

Table 5.18 Other performance indices

Quantities	Demand increase			
	25%	50%	75%	100%
Composite total cost (p.u.)	3482.9	4536.9	5814.3	7030.0
Average travel time (min)	27.3	26.8	26.98	27.42
Computation time (s)	2.34	15.9	126.1	101.5

A few more observations about upgrading the generators and distribution lines are explained. The ultimate goal of capacity expansion is to meet the charging demand without violating operating constraints. There are two options for acquiring electricity: producing it from local generators, or purchasing it from the main grid. If the marginal cost of energy production is lower than the price in the main grid, and the generators can be arbitrarily placed where they are needed, then there is no need to build additional distribution lines. On the other hand, when producing electricity is expensive and more energy is delivered from the slack bus, or the available sites for generators are restricted, additional distribution lines may need reinforcement to circumvent potential voltage or power flow violation.

The composite total cost (defined as $F_{TN} + F_{PDN}$), the average vehicle travel time of TN (defined as $\sum_a x_a(t_a^0 + \lambda_a)/\sum_{rs} q_{rs}^r$), and the computation times in different scenarios are provided in Table 5.18, showing that when the traffic demand grows, the total cost increases, but the average travel time is maintained at about 27 min per vehicle, which indicates that the road expansion planning strategy is reasonable. The congestion pattern shown in Fig. 5.20 associating with the highest traffic demand scenario demonstrates that the delay on each congested road is moderate and acceptable. The computation time does not necessarily (but is very likely to) increase as the traffic demand grows. However, this will not become a critical limitation for practical usage. On the one hand, because the time scale of planning problems is long, it is acceptable and desired to spend a few hours or even longer time to seek a better solution which may cut down the expense notably in the long run. On the other hand, urban infrastructures should be upgraded every few years; in this regard, the traffic demand is unlikely to grow remarkably (usually less than 100%). A multi-period model can also be used to tackle long term planning problems. Moreover, solving mixed-integer programs for realistic large-scale instances is still very challenging. In such circumstances, one can accept the best solution found in a pre-specified time limit.

5.6 Vulnerability of Electrified Transportation System

The vulnerability of a network is usually measured by the change in system performance if some of its components are degraded or removed. In the electrified transportation network, the car accident is the most commonly seen contingency, which deteriorates the road capacity temporarily. As a result, traffic flow will be redistributed according to the UE condition stated in Proposition 5.1. In fact, the vulnerability of the TN has been studied for a long time [55–60]. A comprehensive survey can be found in [61]. This section strives to fill the gap between researches on the interdependence analysis of coupled transportation-electricity infrastructures and the vulnerability analysis of the transportation system alone, and will be committed to evaluating the impact of road capacity degradation on the performances of both systems.

Specifically, we propose rigorous mathematical formulations for identifying the most critical links whose deterioration will bring about the most severe impact on either the total vehicle travel time (TNV for short) or the operating cost of the distribution network (PDNV for short). In the proposed models, a virtual attacker seeks a combination of target links that will maximize the network cost subject to his budget. The traffic flow pattern and power flow distribution are determined from UE and OPF, respectively, such that no single vehicle can reduce its travel time by altering its route unilaterally, and the power grid is operated in the most economic manner. Owing to the presence of UE conditions, TNV yields an MPCC, which is reformulated as an MILP via methods in [39, 62–64]. Owing to the reactions from the generators after a contingency, PDNV can boil down to a max-min MPCC. By using the duality theory of LP, it is reformulated as a single-level MPCC, which can be solved by traditional NLP methods [40–42] and commercial solvers. A direct searching algorithm is suggested for the instance with a moderate dimension of attack strategy. It solves the TAP and the OPF problem alternately, both of which render convex optimization and can be readily solved.

5.6.1 User Equilibrium and Optimal Power Flow Models

Dedicated UE and OPF models used in this section are briefly reviewed here. Major symbols and notations used throughout this section in OPF and TAP formulations are inherited from Sects. 2.2.2 and 5.2, respectively. There may be slight differences in superscripts and subscripts, without causing confusions. Others are defined following their first appearances.

1. User Equilibrium and Traffic Assignment Formulations

Theoretical foundation and mathematical formulation of UE and TAP with constant road capacity have been explained in Sect. 5.2. In this section, the road capacity loss Δc will be treated as the decision variable of a virtual attacker. The UE pattern $x(\Delta c)$ under a given Δc is determined from Beckmann model as follows:

$$\begin{aligned} \min \quad & \sum_{a \in T_A} \int_0^{x_a} t_a(\theta, \Delta c_a) \mathrm{d}\theta \\ \text{s.t.} \quad & \text{Cons-TN} \end{aligned} \tag{5.83a}$$

where Cons-TN is given by

$$\text{Cons-TN} = \left\{ \begin{array}{c} x_a = \sum_{rs} \sum_{k} f_k^{rs} \delta_{ak}^{rs}, \ \forall a \in T_A \\ \sum_{k \in K_{rs}} f_k^{rs} = q_{rs}^t, \ \forall r, s \\ f_k^{rs} \geq 0, \ \forall k \in K_{rs}, r, s \end{array} \right\} \tag{5.83b}$$

In view of road capacity loss, the travel time function on link a is given by

$$t_a(x_a, \Delta c_a) = t_a^0 \left[1 + 0.15 \left(\frac{x_a}{c_a - \Delta c_a} \right)^4 \right] \tag{5.83c}$$

For a given Δc, TAP (5.83a) is a convex optimization problem. Its KKT optimality condition can be expressed via

$$\text{TAP-KKT} = \left\{ \begin{array}{c} x_a = \sum_{rs} \sum_{k} f_k^{rs} \delta_{ak}^{rs}, \ \forall a \in T_A \\ \sum_{k \in K_{rs}} f_k^{rs} = q_{rs}^t, \ \forall r, s, \ (5.83c) \\ 0 \leq f_k^{rs} \perp t_k^{rs} - u_{rs} \geq 0, \ \forall k \in K_{rs}, r, s \\ t_k^{rs} = \sum_{a \in T_A} t_a(x_a, \Delta c_a) \delta_{ak}^{rs}, \ \forall k \in K_{rs}, r, s \end{array} \right\} \tag{5.84}$$

Because the link flow x_a is not upper bounded, TAP (5.83a) is always feasible and TAP-KKT must have a stationary point. TAP-KKT will be used in TNV and PDNV to model the change in traffic flow patterns in response to a given attack strategy Δc.

2. Optimal Power Flow Formulation

Theoretical foundation and mathematical formulation of OPF with constant demands have been discussed in Sect 2.3.1. In this section, slight voltage violation is allowed and penalized to quantify the impact of road capacity loss on the operation of PDN. The linearized BFM introduced in Sect. 2.2.2 will be used in the model of vulnerability analysis, which is summarized as follows:

$$\text{Cons-LBF} = \left\{ \begin{array}{c} P_{ij}^l + p_j^g = \sum_{k \in \pi(j)} P_{jk}^l + p_j^d, \ \forall j \in E_N \\ Q_{ij}^l + q_j^g = \sum_{k \in \pi(j)} Q_{jk}^l + q_j^d, \ \forall j \in E_N \\ V_j = V_i - \dfrac{r_{ij}^l P_{ij}^l + x_{ij}^l Q_{ij}^l}{V_0}, \ \forall j \in E_N \\ p_j^d = p_j^{df} + \sum_{a \in C(j)} \gamma_a x_a, \ \forall j \in E_N \end{array} \right\} \tag{5.85}$$

where nodal active power demand p_j^d consists of a fixed part p_j^{df} and the EV charging request, which is proportional to the link flow x_a at the UE status according to a coefficient γ_a, $C(i)$ represents charging facilities on link a that are served by bus i.

The economic generation dispatch can be calculated from an OPF problem:

$$\begin{aligned} \min \ & \sum_i f_i^g + \rho \sum_{j \in \pi(0)} P_{0j}^l + \sum_i f_i^V \\ \text{s.t.} \ & \text{Cons-LBP, Cons-BND} \end{aligned} \tag{5.86}$$

where $f_i^g = a_i (p_i^g)^2 + b_i p_i^g$ is the production cost of generator i; the second term is the payment to the main grid, where $\sum_{j \in \pi(0)} P_{0j}^l$ denotes the total power delivered from the slack bus; ΔV_j^+ and ΔV_j^- are slack variables in voltage boundary constraints, and $f_i^V = \pi_V(\Delta V_i^+ + \Delta V_i^-)$ is the penalty cost for possible voltage violation; π_V is the penalty coefficient. Security boundary constraints include

$$\text{Cons-BND} = \left\{ \begin{array}{c} V_i^f - \Delta V_i^- \le V_i \le V_i^r + \Delta V_i^+, \forall i \in E_N \\ 0 \le \Delta V_i^+ \le 0.1, 0 \le \Delta V_i^- \le 0.1, \forall i \in E_N \\ p_i^l \le p_i^g \le p_i^m,\ q_i^l \le q_i^g \le q_i^m, \forall i \in E_N \\ P_{ij}^l \ge 0,\ Q_{ij}^l \ge 0,\ \forall l \in E_L \\ \sqrt{(P_{ij}^l)^2 + (Q_{ij}^l)^2} \le S_l, \forall l \in E_L \end{array} \right\} \tag{5.87}$$

The quadratic cost function f_i^g can be approximated by the following convex combination method: We first select $K+1$ breakpoints $p_{ik}^g, k = 0, 1, \cdots, K$ on the interval $[p_i^f, p_i^r]$, and let $f_{ik}^g = a_i (p_{ik}^g)^2 + b_i p_{ik}^g$ be the corresponding function values. Then we can add the following linear constraints to Cons-PDN:

$$p_i^g = \sum_{k=0}^K \lambda_{ik} p_{ik}^g,\ f_i^g = \sum_{k=0}^K \lambda_{ik} f_{ik}^g,\ \forall i$$

$$\lambda_{ik} \ge 0,\ \forall i, \forall k,\ \sum_{k=0}^K \lambda_{ik} = 1,\ \forall i$$

with λ_{ik} being the additional decision variable. Since f_i^g to be minimized is convex, the weights $\{\lambda_{ik}\}, \forall k$ automatically constitute an $\mathbb{SOS}_2$, in which at most two adjacent elements can take strictly positive values. Other PWL approximation methods can be found in [65]. The circular constraint in Cons-BND can be approximated by 8 linear inequalities [66]. In this regard, the economic power dispatch can be written as a compact LP:

$$\begin{aligned} F_E(p^d) = \min \ & d^T y \\ \text{s.t. } & A_1 y + B p^d = h_1 \\ & A_2 y = h_2 \\ & A_3 y \le h_3 \end{aligned} \tag{5.88}$$

where $y = [P, Q, V, p^g, q^g, \lambda]$ is the decision variable, active power demand p^d is determined from the UE in the TN, and coefficient matrices and vectors correspond to constants in Cons-LBP and Cons-BND.

5.6.2 Vulnerability of the Total Vehicle Travel Time

1. Mathematical Model

This section aims to quantify how link deterioration will impact the total vehicle travel time. In the proposed model, the vulnerability is simulated by a two-stage process: a virtual attacker seeks a combination of target links subject to a given budget; then the traffic flow in the disrupted TN forms a new UE in response to the attack. In order to identify the critical combination of vulnerable links, the attacker always tries to seek a strategy that would bring the largest total travel time to the system, leading to the following problem:

$$\begin{aligned} \max \ & \sum_{a \in T_A} x_a t_a^c \\ \text{s.t. } & \Delta c \in C_B \\ & \text{TAP-KKT} \end{aligned} \tag{5.89}$$

where TAP-KKT in problem represents the UE condition after an attack, and the allowable set of Δc is given by

$$C_B = \left\{ \Delta c \middle| 0 \le \Delta c_a \le \Delta c_a^m, \forall a, \sum_a \Delta c_a \le \Gamma \right\} \tag{5.90}$$

where parameter Δc_a^m describes how a usual contingency would impact road capacity. Because most trivial accidents are unable to block the entire road, thus Δc_a^m is usually strictly smaller than c_a. TNV problem (5.89) is an MPCC which is challenging to solve. To find the most severe attack and corresponding vulnerable links, we present an MILP approximation, whose global optimal solution can be found by off-the-shelf solvers.

2. MILP Reformulation

MPCC (5.89) includes three kinds of nonlinearities. According to the discussions in Sect. 5.3, the total vehicle travel time can be represented by

$$\begin{aligned}\sum_a t_a^c x_a &= \sum_{rs}\sum_k\sum_a t_a^c f_k^{rs}\delta_{ak}^{rs} = \sum_{rs}\sum_k t_k^{rs} f_k^{rs} \\ &= \sum_{rs} u_{rs} \sum_{k,f_k^{rs}>0} f_k^{rs} = \sum_{rs} u_{rs} q_{rs}^t\end{aligned} \tag{5.91}$$

which is linear in u_{rs}. The complementarity and slackness conditions in TAP-KKT are equivalent to the following linear constraints:

$$\begin{aligned}&0 \le f_k^{rs} \le M(1-\xi_k^{rs}),\ \forall k,r,s \\ &0 \le t_k^{rs} - u_{rs} \le M\xi_k^{rs},\ \forall k,r,s\end{aligned} \tag{5.92}$$

where ξ_k^{rs} is a binary variable that enforces complementarity, and M is a sufficiently large constant.

Combining the methods in [62–64], we develop a PWL approximation for the bivariate nonlinear latency function $t_a^c(x_a, \Delta c_a)$ using a logarithmic number of binary variables and constraints with respect to the number of breakpoints.

The typical shape of $t_a^c(x_a, \Delta c_a)$ is plotted in Fig. 5.21. The entire feasible region is partitioned into $M \times N$ disjoint sub-rectangles by $M + N + 2$ breakpoints x_a^n,

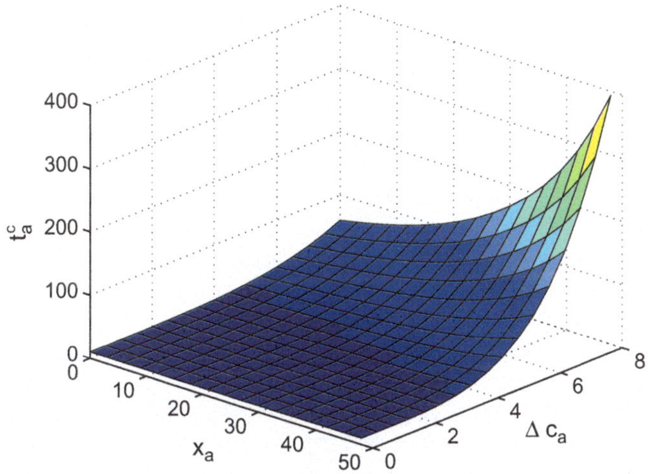

Fig. 5.21 Shape of latency function $t_a^c(x_a, \Delta c_a)$

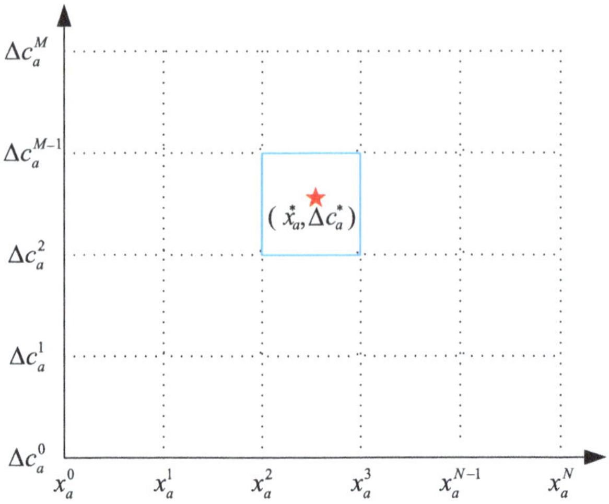

Fig. 5.22 Breakpoints and active rectangle for PWL approximation

$n = 0, 1, \cdots, N$ and Δc_a^m, $m = 0, 1, \cdots, M$, as illustrated in Fig. 5.22, and the corresponding function values are t_a^{mn}. By introducing a weight parameter λ_a^{mn} associated with each grid point that satisfies

$$\begin{aligned} &\lambda_a^{mn} \geq 0, \ \forall m, n, a \\ &\sum_{m=0}^{M} \sum_{n=0}^{N} \lambda_a^{mn} = 1, \forall a \end{aligned} \tag{5.93a}$$

we can present any point $(x_a, \Delta c_a)$ in the feasible region by a convex combination of the extreme points of the sub-rectangle it resides in, giving rise to:

$$\begin{aligned} x_a &= \sum_{m=0}^{M} \sum_{n=0}^{N} \lambda_a^{mn} x_a^n \\ \Delta c_a &= \sum_{m=0}^{M} \sum_{n=0}^{N} \lambda_a^{mn} \Delta c_a^m \end{aligned} \tag{5.93b}$$

and its function value

$$t_a^c = \sum_{m=0}^{M} \sum_{n=0}^{N} \lambda_a^{mn} t_a^{mn} \tag{5.93c}$$

is also a convex combination of the function values at the corner points.

As we can see from Fig. 5.22, in a valid representation, if $(x_a^*, \Delta c_a^*)$ belongs to a sub-rectangle, only the weight parameter associated with the four corner points can be non-negative, while others should be forced at 0. In fact, at most three of the four corner points can be assigned with uniquely determined non-negative weights.

Detecting the active sub-rectangle requires additional constraints on the weight parameter λ_a^{mn}. In this study, we employ the method in [63, 64] which only involves a logarithmic number of binary variables and constraints with respect to the number of break points. Let λ_a^n and λ_a^m be the aggregated weights for x_a and Δc_a, respectively, i.e.,

$$\lambda_a^n = \sum_{m=0}^{M} \lambda_a^{mn}, \quad \lambda_a^m = \sum_{n=0}^{N} \lambda_a^{mn} \tag{5.93d}$$

and introduce the following constraints:

$$\text{For } x_a : \left\{ \begin{array}{c} \sum_{n \in L_k^x} \lambda_a^n \le z_{ak}^x \\ \sum_{n \in R_k^x} \lambda_a^n \le 1 - z_{ak}^x \\ z_{ak}^x \in \{0, 1\} \end{array} \right\}, \ \forall a, \ \forall k \in K_a^x \tag{5.93e}$$

$$\text{For } \Delta c_a : \left\{ \begin{array}{c} \sum_{m \in L_k^c} \lambda_a^m \le z_{ak}^c \\ \sum_{m \in R_k^c} \lambda_a^m \le 1 - z_{ak}^c \\ z_{ak}^c \in \{0, 1\} \end{array} \right\}, \ \forall a, \ \forall k \in K_a^c \tag{5.93f}$$

where L_k^x, L_k^c and R_k^x, R_k^c are index sets of weights λ_a^n and λ_a^m, K_a^x and K_a^c are index sets corresponding to the number of binary variables. The dichotomy sequences $\{L_k^x, R_k^x\}_{k \in K_a^x}$ and $\{L_k^c, R_k^c\}_{k \in K_a^c}$ constitute a branching scheme on the indices of weights, such that constraints (5.93e) and (5.93f) guarantee that at most two adjacent elements of λ_a^n and λ_a^m can take strictly positive values, so as to detect the active sub-rectangle. The required number of binary variables z_{ak}^x and z_{ak}^c for each a is $\lceil \log M \rceil + \lceil \log N \rceil$, where $\lceil \cdot \rceil$ stands for rounding up towards plus infinity. More details about this procedure can be found in Appendix B.1.2.

In summary, we have the following linearized KKT condition for TAP (5.83a):

$$\text{TAP-KKT-LIN} = \left\{ \begin{array}{c} x_a = \sum_{rs} \sum_{k} f_k^{rs} \delta_{ak}^{rs}, \ \forall a \\ \sum_{k \in K_{rs}} f_k^{rs} = q_{rs}^t, \ \forall (r, s) \\ (5.92), \ (5.93a)\text{–}(5.93f) \\ t_k^{rs} = \sum_{a \in T_A} t_a(x_a, \Delta c_a) \delta_{ak}^{rs}, \ \forall k, \forall (r, s) \end{array} \right\} \tag{5.94}$$

and MPCC (5.89) can be approximated by the following MILP

$$\begin{aligned} \max \ & \sum_{rs} u_{rs} q_{rs}^t \\ \text{s.t.} \ & \Delta c \in C_B \\ & \text{TAP-KKT-LIN} \end{aligned} \tag{5.95}$$

Another option is to solve MPCC (5.89) via its equivalent NLP formulation [40–42], while abandoning a provable global optimality guarantee.

3. Case Studies

We consider the test system in Fig. 5.4, which has been used in Sect. 5.3. Each road is served by one charging facility. Link parameters are given in Table 5.1. O-D pairs and their trip rates q_{rs}^t are shown in Table 5.19. MILP (5.95) is solved to find out the most threatening attack and corresponding vulnerable link combination in the TN under different attack budgets Γ. The feasible interval of each link flow variable x_a (link capacity loss Δc_a) is divided into 2^3 (2^2) segments, indicating that 9 (5) breakpoints and 3 (2) binary variables will be used in the PWL approximation for latency function. An upper bound on the link capacity loss $\Delta c_a^m = 0.25 c_a$ is used in our tests. MILP (5.95) is solved by CPLEX [30]. Results are provided in Table 5.20 and Figs. 5.23 and 5.24.

From Table 5.20 we can see that links T1–T2, T4–T5, T3–T7, and T4–T8 will become the targets regardless of the attack budget. This is easy to understand because the major traffic flow leaves from nodes T1, T3, and T4. With increasing the value of Γ, more adjacent roads around these source nodes will be disrupted. Hence,

Table 5.19 O-D pairs and their trip rates (in p.u.)

O-D pair	q_{rs}^t	O-D pair	q_{rs}^t	O-D pair	q_{rs}^t
T1–T6	12	T3–T6	10	T4–T9	6
T1–T10	13	T3–T10	12	T4–T10	10
T1–T12	8	T3–T12	8	T4–T12	9
T1–T11	7	T3–T11	10		

Table 5.20 Vulnerable links under different values of Γ in TNV

Budget	Attacked link	Capacity loss (p.u.)
5	T1–T2	4.4064
	T4–T5	0.0705
	T3–T7	0.3251
	T4–T8	0.1980
10	T1–T2	4.8201
	T4–T5	0.1528
	T3–T7	4.6324
	T4–T8	0.3947
15	T1–T2	4.5118
	T4–T5	3.1975
	T3–T7	3.8134
	T4—-T8	3.4772
20	T1–T2	5.0000
	T2–T6	3.5169
	T4–T5	3.2331
	T3–T7	4.7500
	T4–T8	3.5000
25	T1–T2	5.0000
	T2–T6	4.2500
	T4–T5	3.3750
	T5–T6	1.8433
	T3–T7	4.7500
	T4–T8	3.5000
	T7–T11	2.2817
30	T1–T2	5.0000
	T2–T6	4.2500
	T4–T5	3.3750
	T5–T6	2.0500
	T3–T7	4.7500
	T4–T8	3.5000
	T5–T9	0.3581
	T9–T10	2.2875
	T7–T11	4.3750
	T8–T11	0.0544
35	T1–T2	5.0000
	T2–T6	4.2500
	T4–T5	3.3750
	T5–T6	1.9395
	T3–T7	4.7500
	T4–T8	3.5000
	T5–T9	2.8819
	T8–T9	2.9297
	T7–T11	4.2992
	T8–T11	2.0747

(continued)

Table 5.20 (continued)

40	T1–T2	5.0000
	T2–T6	4.2500
	T4–T5	3.3750
	T5–T6	2.0500
	T3–T7	4.7500
	T4–T8	3.5000
	T5–T9	2.8606
	T8–T9	3.0157
	T9–T10	2.1810
	T7–T11	4.3750
	T11–T12	4.6427

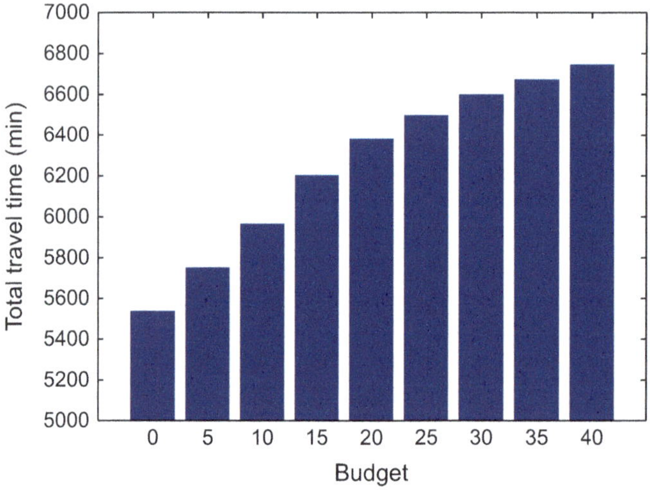

Fig. 5.23 Total travel times under different values of Γ

effective measures are desired to reinforce these critical links, which are close to the source nodes in order to enhance system resilience. Since the traffic demand is time varying, it is necessary to check the critical links under various traffic demand scenarios. Figure 5.23 shows the growing trend of total travel time with increasing the attack budget Γ. Certainly, the growth rate is case-dependent and is also related to the upper bound of link capacity loss Δc_a^m. Figure 5.24 demonstrates that all instances can be solved in reasonable times, because the formulation uses only a few binary variables and constraints. When we apply the original PWL method in [62] to the latency function with the same number of breakpoints, the computation times are much longer in all cases, and in some instances the problem cannot be successfully solved because the size of the branch-and-bound tree in computation goes beyond the memory space of MATLAB.

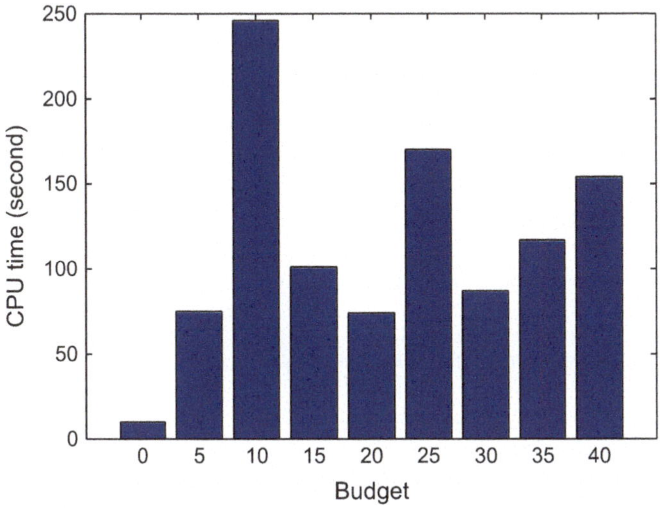

Fig. 5.24 Computation times under different values of Γ

5.6.3 Vulnerability of the Operating Cost

1. Mathematical Model

As mentioned before, the traffic flow pattern will influence the spatial distribution of electrical demands, thus affecting economic generation dispatch of the PDN. Moreover, failure to meet operating security constraints, such as voltage violation, may incur the risk of instability and will be penalized. This section aims to reveal how link deterioration will impact the operation cost of PDN. A framework similar to the TNV problem is adopted, which means the attacker will seek a strategy that would bring the largest cost, leading to a max-min formulation:

$$\begin{aligned} &\max_{\Delta c \in C_B} \min_{y \in Y(p^d)} d^T y \\ &\text{s.t. TAP-KKT} \end{aligned} \tag{5.96}$$

where the electrical demand p^d of PDN depends on link flow x_a at UE pattern, and the UE is determined from TAP-KKT; the feasible set of y is given in (5.88) as

$$Y(p^d) = \left\{ y \;\middle|\; \begin{aligned} &A_1 y = h_1 - Bp^d \\ &A_2 y = h_2 \\ &A_3 y \le h_3 \end{aligned} \right\} \tag{5.97}$$

Unlike TNV problem (5.89), PDNV problem (5.96) is no longer a traditional MPCC because of reactions from the economic generation dispatch, which is modeled by the inner minimization problem.

2. NLP Reformulation and a Direct Search Algorithm

To solve max-min problem (5.96), we reformulate it as a traditional MPCC based on the duality theory. Since the inner minimization problem is an LP, we can write out its dual LP as follows:

$$\max_{\{\mu_1,\mu_2,\mu_3\}\in D} \mu_1^T(h_1 - Bp^d) + \mu_2^T h_2 + \mu_3^T h_3 \tag{5.98}$$

where the feasible region D of dual variables μ_1, μ_2, μ_3 is given by

$$D = \left\{ \mu_1, \mu_2, \mu_3 \;\middle|\; \begin{array}{c} \mu_1 \text{ free},\ \mu_2 \text{ free},\ \mu_3 \le 0 \\ A_1^T \mu_1 + A_2^T \mu_2 + A_3^T \mu_3 = d \end{array} \right\} \tag{5.99}$$

Strong duality holds for LPs: for a given p^d, dual problem (5.98) and its primal form (5.88) share the same optimal value. In this regard, we can replace the inner minimization problem in (5.96) with its dual form (5.98), which is a maximization problem. In this way, we are able to reduce the PDNV problem to a traditional MPCC:

$$\begin{aligned} \max\ & \mu_1^T(h_1 - Bp^d) + \mu_2^T h_2 + \mu_3^T h_3 \\ \text{s.t.}\ & \Delta c \in C_B,\ \{\mu_1, \mu_2, \mu_3\} \in D \\ & \text{TAP-KKT} \end{aligned} \tag{5.100}$$

Besides the complementarity constraints in TAP-KKT and the nonlinear latency function $t_a^c(x_a, \Delta c_a)$, the bilinear term $\mu_1^T Bp^d$ in the objective function is also non-convex. Although a number of methods are available to approximate it using MILP, such as the binary expansion approach discussed in Appendix B.2.3, in view of the potential computational burden, we abandon an equivalent MILP reformulation and the global optimal solution. Instead, we can solve problem (5.100) as an NLP using the methods in [40–42]. The general purpose NLP solver KNITRO [47] can directly handle complementarity constraints. A high-quality solution can be recovered by trying multiple initial values of Δc. Although it is quite challenging to solve problem (5.100) directly, its solution can be found in a more robust way.

Given the following two observations:

1. Given an attack strategy Δc, the UE solves TAP (5.83a), which is a strictly convex optimization problem, and the electrical demand $p^d(\Delta c)$ is readily available. There is no need to solve TAP-KKT with nonlinear functions and complementarity constraints.

2. Given the electrical demand $p^d(\Delta c)$ at the UE, LP (5.88) can be readily solved. Therefore, the objective value $d^T y(\Delta c)$ for a given attack strategy can be evaluated with reasonable computation efforts.

These observations motivate a direct search procedure shown in Algorithm 5.5. A variety of derivative-free searching algorithms are available for the task in step 4, such as those discussed in [43–46] and references therein. MATLAB global optimization toolbox provides build-in pattern search functionality. Since (5.83a) and (5.88) should be solved repeatedly, Algorithm 5.5 may be less efficient but more reliable. When the dimension of Δc is small, i.e., only a few roads can be attacked, Algorithm 5.5 will be very useful.

Algorithm 5.5 Direct searching for PDNV

1: For a given attack strategy Δc, solve TAP (5.83a), the UE is x^*.
2: Update the electrical demand p^d at x^*.
3: Solve economic dispatch problem (5.88) with the obtained p^d in step 2.
4: Evaluate the operation cost $d^T y(\Delta c)$; if a convergence criterion is not met, find a better choice of attack strategy Δc, and go to step 1.

If the nonlinear BFM and SOCP relaxation is used for the economic generation dispatch, its dual will be an SOCP, and the PDNV problem still comes down to an MPCC. However, if some charging stations are overloaded after a contingency, the SOCP relaxation may become inexact due to the extreme operating condition. In Algorithm 5.5, the nonlinear BFM based dispatch model can be used in step 3, and solved by Algorithm 2.1 in Sect. 2.3.1, no matter whether the convex relaxation is exact or not.

3. Case Studies

Topology of the coupled networks is shown in Fig. 5.4. Transportation system is the same as the one used in Sect. 5.3. Generator and distribution line data are provided in Tables 5.3 and 5.4. Specifically, the charging demand equation is $p_j^d = p_j^{df} + 0.01x_a^j$. The fixed demand at each bus is $p_i^{df} = 0.02$ and $q_i^{df} = 0.01$; the voltage boundary is $V_i^l = 0.95$ and $V_i^m = 1.05$. The voltage magnitude at the slack bus is $V_0 = 1.02$; the electricity price is $\rho = 1500\$$; the voltage penalty coefficient is $\pi_V = 10,000\$$. The PDNV problem in form of MPCC (5.100) is solved by KNITRO [47]. Results are provided in Table 5.21 and Figs. 5.25 and 5.26.

From Table 5.21 we can see that the critical links are different from those in Table 5.20. As analyzed before, the attacker tries to produce a traffic pattern that will cause voltage violation in the PDNV problem. At this time, Bus 3 and Bus 16 become the vulnerable buses of the distribution network, because they are far away from the slack bus, so their voltage magnitudes are usually low. Meanwhile, Link T2–T6 and Link T7–T11 are parts of the ring expressway which are preferred by most travelers. This preference will worsen the voltage situations at these two buses. The values of voltage violations ΔV_3^- and ΔV_{16}^- under different values of Γ and corresponding operating costs are shown in Fig. 5.25 and Fig. 5.26, respectively,

Table 5.21 Vulnerable links under different values of Γ in PDNV

Budget	Attacked link	Capacity loss (p.u.)
5	T1–T3	0.5000
	T1–T2	4.5000
10	T1–T3	0.2500
	T1–T2	4.7500
	T1–T4	1.0000
	T12–T10	4.0000
15	T1–T2	5.0000
	T1–T4	2.0000
	T3–T4	2.0000
	T4–T5	2.0000
	T12–T10	4.0000
20	T1–T2	5.0000
	T1–T4	2.0000
	T2–T5	1.0000
	T4–T5	2.0000
	T5–T6	2.0000
	T4–T8	2.0000
	T5–T9	2.0000
	T12–T10	4.0000
25	T1–T2	5.0000
	T1–T4	2.0000
	T3–T4	2.0000
	T4–T5	2.0000
	T5–T6	2.0000
	T4–T8	2.0000
	T5–T9	2.0000
	T7–T8	2.0000
	T8–T9	2.0000
	T12–T10	4.0000
30	T1–T2	5.0000
	T1–T4	2.0000
	T2–T5	1.0000
	T3–T4	1.0000
	T4–T5	3.0000
	T5–T6	2.0000
	T4–T8	2.0000
	T5–T9	2.0000
	T7–T8	2.0000
	T9–T10	2.0000
	T8–T11	2.0000
	T9–T12	2.0000
	T12–T10	4.0000

(continued)

Table 5.21 (continued)

Budget	Attacked link	Capacity loss (p.u.)
35	T1–T2	5.0000
	T1–T4	2.0000
	T2–T5	1.0000
	T3–T4	2.0000
	T4–T5	3.0000
	T5–T6	2.0000
	T4–T8	3.0000
	T5–T9	3.0000
	T7–T8	2.0000
	T8–T9	2.0000
	T9–T10	2.0000
	T8–T11	2.0000
	T9–T12	2.0000
	T12–T10	4.0000
40	T1–T2	5.0000
	T1–T4	2.4060
	T2–T5	1.8750
	T3–T4	2.0940
	T4–T5	3.3750
	T5–T6	2.0000
	T4–T8	3.5000
	T5–T9	3.3750
	T7–T8	2.1250
	T8–T9	3.2500
	T9–T10	2.2500
	T8–T11	2.2500
	T9–T12	2.0000
	T12–T10	4.5000

which corroborate the previous analysis. One possible way to circumvent this situation is to build generators at Bus 3 and Bus 16, so as to reduce the power flow going through distribution lines, thereby decreasing the line voltage drop and enhancing the voltage magnitude.

5.7 Network Equilibrium of Electrified Transportation Systems

In the previous sections of this chapter, an assumption in common is that power demands of charging facilities are proportional to road traffic flows, and EVs and GVs are not explicitly distinguished. Although this assumption may be reasonable

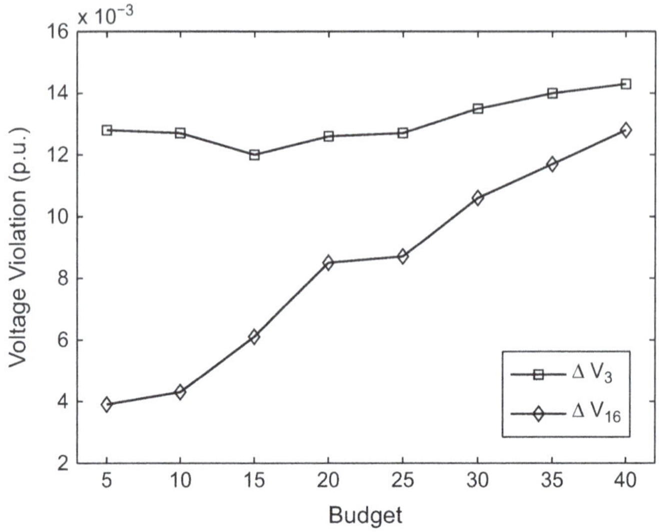

Fig. 5.25 Voltage violations at Bus-3 and Bus-16 under different values of Γ

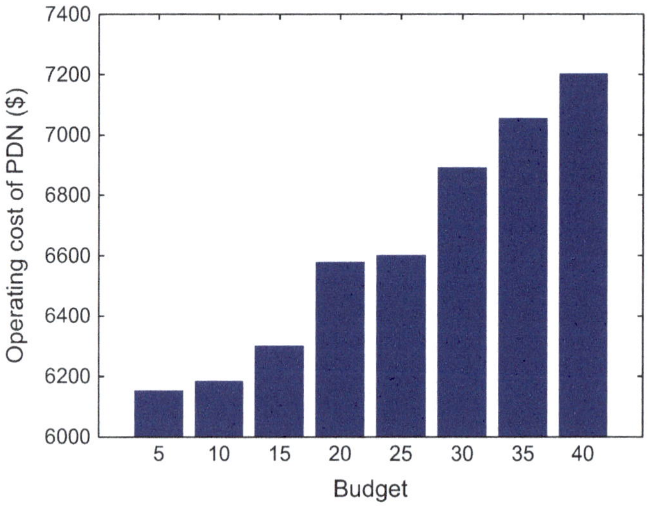

Fig. 5.26 Operating costs under different values of Γ

in certain system-level studies, it is still desired to reveal how the behavior of EVs will impact the charging rate coefficient, which has a decisive impact on the interdependency model.

This section will develop a new modeling framework for analyzing and computing the interdependency between the TN and the PDN brought by the progress of transportation electrification. The focus will be put on the interaction among traffic, energy, and cash flows in the coupled networks, instead of shedding more light on either side. By jointly considering route choices of GVs and EVs, nodal prices of electricity, and the interface at FCSs, we reveal that the coupled network will reach an equilibrium state with three implications. First, in the TN, every vehicle travels on the route with the minimal expense, yielding a UE pattern. Unilateral departure from the UE will increase its travel expense. Second, in the PDN, the electricity is produced in the most economic manner. Third, consumers pay for their energy consumptions at the marginal production cost. In these regards, a social optimum can be achieved. We conduct fundamental studies on the network equilibrium, including its fixed-point formulation and geometric intuitions on its existence, stability, and uniqueness conditions. We develop a tractable computation method to find the network equilibrium via solving convex optimization problems.

5.7.1 Mixed User Equilibrium Model

A dedicated traffic assignment model for the UE state is discussed. It quantifies route choices of GVs and EVs under the individual rationality of minimum travel expense, with combined consideration of route selections, charging opportunities, electricity prices, and capacities of charging facilities. The mixed UE model is formulated as a convex TAP over an extended TN. Some basic considerations and assumptions are summarized below to facilitate building the mathematical model.

1. All EVs in need of recharging service have the same battery size. Although this assumption seems rather restrictive, it could be relaxed by incorporating several aggregated classes of batteries categorized by their energy demands. A more detailed description on the heterogeneous battery information is difficult and may be unnecessary in the system-level study, because this issue can be well addressed by the single vehicle routing problem. This assumption will be further elaborated in Item 3 below.
2. When traveling to their destinations, a GV seeks the route which has minimal travel time, and an EV looks for the route which minimizes its travel expense composed of the travel time and charging cost. EVs which do not need battery recharge are considered in the same way as GVs, as they have the same route selection criterion.
3. We envision that in the near future, the average maximal driving range of EV can reach 300 km (actual driving ranges of current electric vehicle models vary from 85 km to 586 km according to the survey in [67]), while a single trip in a city is typically less than 40 km. Moreover, drivers may consider to recharge their cars when the remaining range is no greater than 50 km. In this regard, we assume: (a) An EV is able to reach any FCS without running out of energy. (b) In an FCS, the battery will be fully recharged. (c) According to the driving range data, the energy consumption before an EV arrives at an FCS is small compared to its battery capacity. In view of the settings considered in Item 1

above, we further assume: d) All EVs receive the same amount of energy E_B when recharged. (e) The time for recharging a vehicle is proportional to E_B, thus the same for all vehicles. Assumptions (b–e) can be relaxed by incorporating multiple vehicle groups. The above settings make the recharging action of an EV similar to the refueling behavior of a GV. These assumptions will reasonably simplify the model without significant deviations from the reality. Refueling of GVs in gas stations can be modeled through the same way but neglected in this chapter, because it has no direct impact on the PDN.

4. The distribution electricity market will be cleared via an ACOPF model. The charging service is paid at the LMP, which will be discussed later. The queueing time increases with the vehicle flow passing the FCS growing. The total time spent for recharging is approximated by the Davidson function [24] developed via queueing theory, which imposes an upper bound on the vehicle flow, reflecting the capacity of an FCS.

1. Transportation Network Model

The TN model is similar to that in Sect. 5.2, except for the FCSs, which are not explicitly considered previously. In order to simulate the choice of an EV to stop for charging or skip and keep driving, an FCS is represented by a two-node two-link module shown in Fig. 5.27. It consists of an entrance node I, an exit node E, a charging link I→FCS→E, and a bypass link I→E with zero travel time. In this way, an FCS is described as nodes and links compatible with a traditional TN model. It also provides a plausible way to evaluate the charging demand, which is proportional to the EV flow in the charging link. The extended TN in this section includes three kinds of links:

1. Regular link without an FCS (denoted by $a \in T_A^R$). The latency function $t_a(x_a)$ is characterized by the BPR function

$$t_a(x_a) = t_a^0 \left[1 + 0.15 \left(\frac{x_a}{c_a} \right)^4 \right], \ \forall a \in T_A^R \tag{5.101}$$

where t_a^0 is the free travel time, i.e., the length of link a divided by the speed limit, and c_a is the traffic flow on link a when $t_a = 1.15 t_a^0$.

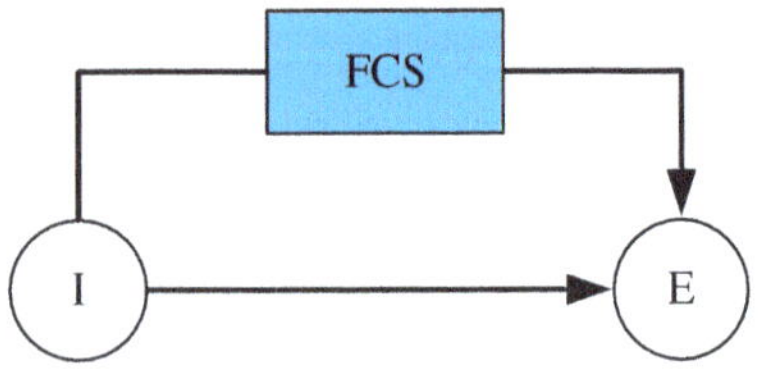

Fig. 5.27 Representation of a fast charging station

2. Charging link with an FCS (denoted by $a \in T_A^C$), i.e., I→FCS→E in Fig. 5.27. The time spent on link $a \in T_A^C$ consists of two parts: charging time and queueing time. The former is a constant t_a^0 which is proportional to the total charging energy E_B; the latter should be slightly increasing if $x_a < c_a$, and asymptotic to the capacity limit. We employ Davidson function to model such a consideration

$$t_a(x_a) = t_a^0 \left[1 + J \left(\frac{x_a}{c_a - x_a} \right) \right], \ \forall a \in T_A^C \tag{5.102}$$

where c_a represents the capacity of charging station on link $a \in T_A^C$, and parameter J controls the shape of the function. Curves of $t_a(x_a)$ with different values of J are plotted in Fig. 5.2. $J = 0.05$ will be used in this section.
3. Bypass link (denoted by $a \in T_A^B$), i.e., I→E in Fig. 5.27. As the length of the bypass link is very short, the travel time on a bypass link can be neglected, i.e.

$$t_a(x_a) = 0, \ \forall a \in T_A^B \tag{5.103}$$

If the FCS sits at the node in the TN, the node is replaced with the composite structure shown in Fig. 5.27; if the FCS sits somewhere beside a road segment, when it is separated from the link, the original branch is divided into two new links: one enters node I, and the other leaves node E.

2. Network Flow and Travel Expense Model

Major symbols and notations are inherited from Sect 5.2. There may be slight differences in superscripts and subscripts, without causing confusions. Specifically, e and g in the superscript or subscript stand for EV and GV related symbols, respectively. The definition of path is generalized to distinguish two kinds of vehicles.

Definition 5.1 A feasible path labeled by $k \in K_g^{rs}$ ($k \in K_e^{rs}$) for GVs (EVs) between O-D pair (r, s) is a chain consisting of connected links from the origin r to the destination s, if it does not (does) visit one FCS.

The feasible path depends on the network topology, and is independent of the congestion pattern and travel expenses. All feasible paths can be calculated and enumerated off-line. However, most of them are redundant and will not be used. Inspired by Algorithm 5.1 in Sect. 5.2.4, a special path generation oracle will be discussed later to circumvent exhaustive path enumeration in large-scale applications. For now we consider that all feasible paths as well as the link-path incidence matrix Δ_G (Δ_E) are available, without jeopardizing the soundness of discussion.

Let f_{kg}^{rs} (f_{ke}^{rs}) represent the GV (EV) flow on path k between O-D pair (r, s). With the notation of Δ_G (Δ_E), the network flow constraints can be written as follows:

$$x_a = \sum_{rs} \sum_{k \in K_g^{rs}} f_{kg}^{rs} \delta_{akg}^{rs} + \sum_{rs} \sum_{k \in K_e^{rs}} f_{ke}^{rs} \delta_{ake}^{rs}, \ \forall a \in T_A \tag{5.104a}$$

$$\left\{ f_{kg}^{rs} \geq 0, \ \forall k \in K_g^{rs}, \ \sum_{k \in K_g^{rs}} f_{kg}^{rs} = q_g^{rs} \right\}, \ \forall (r, s) \tag{5.104b}$$

$$\left\{ f_{ke}^{rs} \geq 0, \ \forall k \in K_e^{rs}, \ \sum_{k \in K_e^{rs}} f_{ke}^{rs} = q_e^{rs} \right\}, \ \forall (r, s) \tag{5.104c}$$

where equality (5.104a) defines link flow in terms of path flow, while (5.104b) and (5.104c) are flow conservation conditions.

Travel time t_{kg}^{rs} and travel expense c_{kg}^{rs} of GVs on path k between O-D pair (r, s) are given as

$$t_{kg}^{rs} = \sum_{a \in T_A} t_a(x_a) \delta_{akg}^{rs}, \ c_{kg}^{rs} = \omega t_{kg}^{rs}, \ \forall k \in K_g^{rs}, \ \forall (r, s) \tag{5.105a}$$

where ω is the monetary value of travel time.

Similarly, travel time t_{ke}^{rs} and travel expense c_{ke}^{rs} of EVs on path k between O-D pair (r, s) can be expressed by

$$t_{ke}^{rs} = \sum_{a \in T_A} t_a(x_a) \delta_{ake}^{rs}, \ \forall k \in K_e^{rs}, \ \forall (r, s) \tag{5.105b}$$

$$c_{ke}^{rs} = \omega t_{ke}^{rs} + \lambda_a^i E_B, \ \forall k \in K_e^{rs}, \ \forall (r, s) \tag{5.105c}$$

where $\lambda_a^i, a \in T_A^C$ is the nodal electricity price at the FCS; it is a constant in the mixed UE model; i is index of the electric bus in the distribution network to which the FCS is connected; E_B is the received energy, which is a constant.

3. The Mixed User Equilibrium Condition

Network flow constraints (5.104) give all possible traffic flow patterns in the TN. It is apparent that if anyone can cut down his travel expense by switching to another path, the flow pattern will be changed. Because of the rationality of minimizing travel expense, a stable state emerges when travel expenses on all used paths are equal. The stable traffic flow distribution of GVs and EVs is called a mixed user equilibrium. In the mixed UE model, the paths for GVs and EVs are different.

By applying Proposition 5.1, we know that the traffic flow reaches a UE if travel expenses of GVs (EVs) on all active paths between any given O-D pair are equal, and no greater than those which would be incurred on any unused path. As an extension that includes two kinds of vehicles, each O-D pair is associated with two minimal travel expenses: u_g^{rs} for gasoline ones, and u_e^{rs} for electric ones. Bearing this in mind, the mixed UE condition can be revamped in a logic form:

$$\forall (r,s) \in D_T^{RS},\ \exists u_g^{rs},\ \text{such that for } k \in K_g^{rs}: \begin{cases} \text{if } f_{kg}^{rs} > 0,\ \text{then } c_{kg}^{rs} = u_g^{rs} \\ \text{if } f_{kg}^{rs} = 0,\ \text{then } c_{kg}^{rs} \geq u_g^{rs} \end{cases} \tag{5.106a}$$

$$\forall (r,s) \in D_T^{RS},\ \exists u_e^{rs},\ \text{such that for } k \in K_e^{rs}: \begin{cases} \text{if } f_{ke}^{rs} > 0,\ \text{then } c_{ke}^{rs} = u_e^{rs} \\ \text{if } f_{ke}^{rs} = 0,\ \text{then } c_{ke}^{rs} \geq u_e^{rs} \end{cases} \tag{5.106b}$$

or in an NCP form

$$0 \leq f_{kg}^{rs} \perp c_{kg}^{rs} - u_g^{rs} \geq 0,\ \forall k \in K_g^{rs},\ \forall (r,s) \tag{5.107a}$$

$$0 \leq f_{ke}^{rs} \perp c_{ke}^{rs} - u_e^{rs} \geq 0,\ \forall k \in K_e^{rs},\ \forall (r,s) \tag{5.107b}$$

$$\text{Cons-Flow} = \{\ (5.104a),\ (5.104b),\ (5.104c)\ \} \tag{5.107c}$$

$$\text{Cons-Cost} = \{\ (5.105a),\ (5.105b),\ (5.105c)\ \} \tag{5.107d}$$

The MILP reformulation in Sect. 5.3 can be applied if the latency functions $t_a(x_a)$ in c_{kg}^{rs} and c_{ke}^{rs} are represented by PWL functions. Such a formulation is useful when the mixed UE problem is embedded in a bilevel optimization problem.

Following the paradigm in Sect. 5.2.2, it turns out that constraints (5.107) constitute the KKT conditions of the following TAP

$$\begin{aligned} &\min\ F_{TAP} \\ &\text{s.t. Cons-Flow} \end{aligned} \tag{5.108}$$

where the objective function F_{TAP} is

$$\sum_{a \in T_A^R} \int_0^{x_a} \omega t_a(\theta) \mathrm{d}\theta + \sum_{a \in T_A^C} \left(\int_0^{x_a} \omega t_a(\theta) \mathrm{d}\theta + \lambda_a^n E_B x_a \right)$$

As demonstrated in Sect. 5.2, the integration terms are convex in x_a, and TAP (5.108) with linear constraints can be globally solved by general purpose NLP solvers. The existence and uniqueness of the mixed UE are discussed below.

Assumption 5.4 *The transportation network is strongly connected, and the total charging demand is no greater than the total capacity of FCSs in the TN.*

Strong connection means that there is at least one feasible path to visit each FCS, regardless of the O-D pair. This assumption is mild for real TNs. In addition, the capacity inequality ensures that TAP (5.108) is feasible. In view of the correspondence of the optimal solution of TAP (5.108) and the stationary point of NCP (5.107), we can confirm the existence of mixed UE.

Proposition 5.3 *The mixed UE exists if Assumption 5.4 is met.*

Uniqueness is guaranteed by the strict convexity of F_{TAP} in link flow variables. It should be mentioned that although F_{TAP} is not strictly convex in bypass link flows x_a, $\forall a \in T_A^B$, as they do not appear in the objective function, these variables can be eliminated by transformation $x_a = x_{r(a)} - x_{c(a)}$, where $x_{r(a)}$ is the total traffic flow in the regular link $r(a)$ connected to the entrance node of the bypass link a, and $x_{c(a)}$ is the EV flow in the FCS corresponding to the bypass link a. In this regard, TAP (5.108) is in fact a strictly convex optimization problem with linear constraints.

Proposition 5.4 *The mixed UE is uniquely determined under Assumption 5.4.*

Multiple solutions to path variables may exist, but this is not an issue because only link flow impacts system operation.

4. A Path Generation Algorithm

Despite the convexity of TAP (5.108), enumerating all possible paths could be exhaustive when the TN becomes larger. This issue has been discussed in Sect. 5.2.4. The central idea is to identify a subset of paths which are likely to be used. In this section, we extend results in Sect. 5.2.4 in order to include EVs. Some symbols are introduced. Vector $I^{rs} \in \mathbb{R}^{|T_N|}$ denotes a vector with two non-zero elements: 1 and -1 at the entries corresponding to nodes r and s; $d \in \mathbb{B}^{|T_A|}$ is the charging availability vector, whose element is 1 for $a \in T_A^C$, and 0 otherwise.

Proposition 5.5 *All possible paths for gasoline vehicles and electric vehicles between O-D pair (r, s) can be expressed by the following sets*

$$V_g^{rs} = \{v_g | \Lambda v_g = I^{rs},\ d^T v_g = 0,\ v_g \in \mathbb{B}^{|T_A|}\} \tag{5.109a}$$

$$V_e^{rs} = \{v_e | \Lambda v_e = I^{rs},\ d^T v_e = 1,\ v_e \in \mathbb{B}^{|T_A|}\} \tag{5.109b}$$

In view of the definition of node-link matrix Λ, $v_g \in V_g^{rs}$ ($v_e \in V_e^{rs}$) represents a chain of connected links from r to s which does not (does visit one FCS). Let $\Delta_G^{rs} = [v_g^i]$, $\forall v_g^i \in V_G^{rs}$ and $\Delta_E^{rs} = [v_e^i]$, $\forall v_e^i \in V_E^{rs}$, the link-path incidence matrices are given by $\Delta_G = [\Delta_G^{rs}]$, $\forall (r, s) \in D_T^{RS}$ and $\Delta_E = [\Delta_E^{rs}]$, $\forall (r, s) \in D_T^{RS}$.

To obviate path enumeration, we first solve TAP (5.108) with a subset of V_g^{rs} (V_e^{rs}), then check if the travel expense can be further reduced. To do so, suppose the current restricted UE is x_a^*, $\forall a$, and then corresponding link travel expenses for GVs and EVs can be calculated as

$$p_g^a = \omega t_a(x_a^*),\ p_e^a = \omega t_a(x_a^*) + f_e^a,\ \forall a \in T_A \tag{5.110}$$

where the charging cost is $f_e^a = \lambda_a^i E_B, \forall a \in T_A^C$, and $f_e^a = 0, \forall a \in T_A^R \cup T_A^B$. The minimal travel expenses u_g^{rs} and u_e^{rs} between O-D pair (r, s) can be readily evaluated through (5.105) and (5.106) with the given traffic flow $x_a^*, \forall a$.

To find a possibly better path between (r, s) for GVs and EVs, we solve the following two MILPs

$$u_{gc}^{rs} = \min_{v_g \in \mathbb{B}^{|T_A|}} \left\{ \sum_a p_g^a v_g^a \middle| \Lambda v_g = I^{rs},\ d^T v_g = 0 \right\} \tag{5.111a}$$

$$u_{ec}^{rs} = \min_{v_e \in \mathbb{B}^{|T_A|}} \left\{ \sum_a p_e^a v_e^a \middle| \Lambda v_e = I^{rs},\ d^T v_e = 1 \right\} \tag{5.111b}$$

If $u_{gc}^{rs} < u_g^{rs}$ ($u_{ec}^{rs} < u_e^{rs}$), the solution v_g (v_e) indicates a better path for GVs (EVs). A procedure for solving TAP (5.108) without a complete path enumeration is given in Algorithm 5.6. It converges in a finite number of iterations which is bounded by $\max\{\max_{rs}[N_{ga}^{rs}], \max_{rs}[N_{ea}^{rs}]\}$, where N_{ga}^{rs} and N_{ea}^{rs} are the numbers of active paths for GVs and EVs between O-D pair (r, s) at the UE.

Algorithm 5.6 TAP with path generation

1: Let $x_a = 0, \forall a$, compute p_a^g and p_a^e, solve problems (5.111a) and (5.111b) for each O-D pair, and then build link-path incidence matrices Δ_G and Δ_E.
2: Solve the restricted TAP (5.108) with current Δ_G and Δ_E, and the UE is $x_a^*, \forall a$.
3: With the obtained UE solution, update travel expenses p_a^g and p_a^e on each link; find minimal travel expenses u_g^{rs} and u_e^{rs} between (r, s); solve (5.111a) and (5.111b) for each O-D pair; the optimal solutions are v_g^{rs} and v_e^{rs}.
4: If $u_{gc}^{rs} \geq u_g^{rs}$, $u_{ec}^{rs} \geq u_e^{rs}$, $\forall (r, s)$, terminate and report the current UE solution; otherwise, if $u_{gc}^{rs} < u_g^{rs}$ for some (r, s), $\Delta_G^{rs} \leftarrow [\Delta_G^{rs}, v_g^{rs}]$; if $u_{ec}^{rs} < u_e^{rs}$ for some (r, s), $\Delta_E^{rs} \leftarrow [\Delta_E^{rs}, v_e^{rs}]$; update Δ_G and Δ_E, and go to step 2.

5. An Illustrative Example

A small example is given in this section to illustrate related concepts. The topology of the TN is shown in Fig. 5.28, which includes 2 nodes (T1, T2), 3 links, 1 O-D pair, and 2 FCSs (C1, C2). The extended network topology is given in Fig. 5.29, which is made up of 6 nodes and 9 links, including 5 regular links, 2 charging links, and 2 bypass links. System data are provided in the same figure. The based value is 100 vehicles/h. The total traffic demand is 100, 10% of which is made up of EVs, while half of them need recharging. This setting leads to $q_g^{rs} = 95$ and $q_e^{rs} = 5$. The energy demand of each EV is $E_B = 50$ kWh. The fixed charging time is 20 min. In Davidson function (5.102), $J = 0.05$ is adopted.

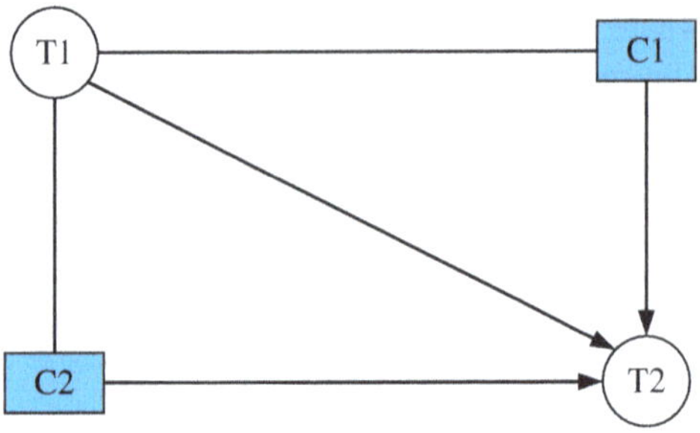

Fig. 5.28 Topology of the transportation network

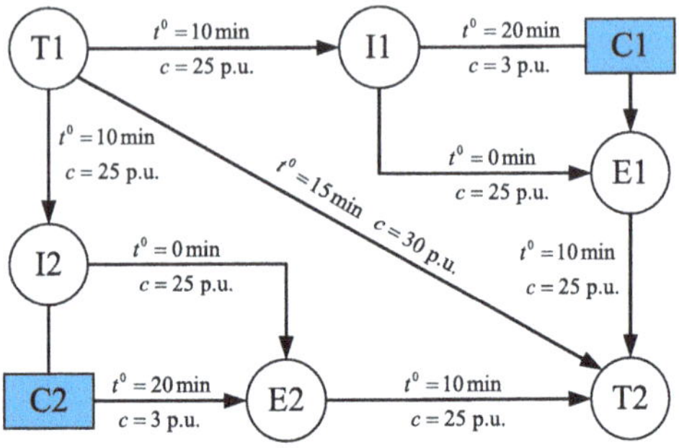

Fig. 5.29 Topology and data of the extended transportation network

The node-link incidence matrix can be written as

$$\Lambda = \begin{bmatrix} 1 & 0 & 0 & 0 & 1 & 0 & 0 & 0 & 1 \\ -1 & 1 & 1 & 0 & 0 & 0 & 0 & 0 & 0 \\ 0 & -1 & -1 & 1 & 0 & 0 & 0 & 0 & 0 \\ 0 & 0 & 0 & 0 & -1 & 1 & 1 & 0 & 0 \\ 0 & 0 & 0 & 0 & 0 & -1 & -1 & 1 & 0 \\ 0 & 0 & 0 & -1 & 0 & 0 & 0 & -1 & -1 \end{bmatrix} \begin{matrix} \text{T1} \\ \text{I1} \\ \text{E1} \\ \text{I2} \\ \text{E2} \\ \text{T2} \end{matrix}$$

where each row (column) of Λ corresponds to a particular node (link). For example, the first column corresponds to link T1-I1, and the last column corresponds to link T1–T2. Node correspondence is shown following the matrix. Please be aware that C1 is not a node, and the two links corresponding to the second and third columns of Λ between I1 and E1 share the same entrance and exit. So do the two links associated with the sixth and seventh columns of Λ.

All possible paths in this simple network can be enumerated. The link-path incidence matrices Δ_G and Δ_E are given by

$$\Delta_G = \begin{bmatrix} 1 & 0 & 0 \\ 0 & 0 & 0 \\ 1 & 0 & 0 \\ 1 & 0 & 0 \\ 0 & 1 & 0 \\ 0 & 0 & 0 \\ 0 & 1 & 0 \\ 0 & 1 & 0 \\ 0 & 0 & 1 \end{bmatrix} \quad \Delta_E = \begin{bmatrix} 1 & 0 \\ 1 & 0 \\ 0 & 0 \\ 1 & 0 \\ 0 & 1 \\ 0 & 1 \\ 0 & 0 \\ 0 & 1 \\ 0 & 0 \end{bmatrix} \quad \begin{matrix} \text{T1-I1} \\ \text{I1-C1-E1} \\ \text{I1-E1} \\ \text{E1-T2} \\ \text{T1-I2} \\ \text{I2-C2-E2} \\ \text{I2-E2} \\ \text{E2-T2} \\ \text{T1-T2} \end{matrix} \tag{5.112}$$

where each row (column) of Δ_G (Δ_E) is associated with a particular link (path). For example, the first column of Δ_E corresponds to the path T1→I1→C1→E1→T2. Link correspondence is shown following the matrix.

We consider two scenarios to show the impact of electricity price on the UE. In the first scenario, we set the electricity price as $\lambda = 100$\$/MWh in both charging stations. TAP (5.108) is solved by BARON, and the UE is exhibited in the left of Fig. 5.30. It is clear that the traffic flow is symmetrically distributed since the network itself (including electricity price) is symmetric. The aggregated charging power in C1 and C2 can be calculated as $p_1^c = x_a E_B = 12.5$MW$= p_2^c$. Travel times and expenses on available paths are shown in Table 5.22. We can see that travel expenses of GVs (EVs) are equal, which is consistent with condition (5.106), as there is only one O-D pair. The travel expense of EV is higher than that of GV due to the waiting time and cost for recharging. We do not include the refueling cost of the latter. Moreover, with the progress in advanced battery technology, the driving range of EV may catch up with that of GV in the near future. The average cost of EV may be lower owing to different prices of electricity and gasoline.

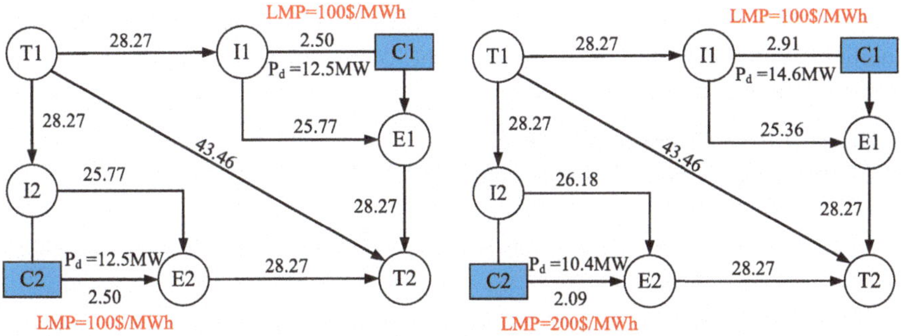

Fig. 5.30 User equilibrium patterns with different electricity prices

Table 5.22 Path travel times and expenses

		Scenario 1		Scenario 2	
		Time (min)	Expense ($)	Time (min)	Expense ($)
GV-Path	T1-I1-E1-T2	24.91	4.1511	24.91	4.1511
	T1-I2-E2-T2	24.91	4.1511	24.91	4.1511
	T1-T2	24.91	4.1511	24.91	4.1511
EV-Path	T1-I1-C1-E1-T2	49.91	13.318	77.20	17.867
	T1-I1-C1-E1-T2	49.91	13.318	47.20	17.867

To show the impact of electricity price on the route choice of electric vehicles, the price in C2 is doubled in the second scenario. The UE is portrayed in the right of Fig. 5.30, from which we can see that more EV drivers prefer to charge their cars in C1 where the electricity is cheaper. Power demands in C1 and C2 are $p_1^c =$ 14.6 MW and $p_2^c = 10.4$ MW, respectively. Moreover, GVs respond to the changes in EV traffic flows, and the total traffic flow (not include that in the FCSs) is still symmetric. Certainly, prices of electricity will be affected by the charging demands, which will be discussed at the power system side.

5.7.2 Distribution System and Network Equilibrium

Economic operation of the PDN and electricity market clearing are formulated as an ACOPF problem, which accounts for network losses and obviates the inaccuracy introduced by the DC power flow or linearized BFM. Convex relaxation is performed on branch flow equations. The optimal generation dispatch can be computed from an SOCP, and LMPs of electricity can be extracted from Lagrangian dual multipliers associated with nodal power balancing equalities. The process is based on convex optimization. The network equilibrium is formulated as a fixed-point problem. Intuitions on the existence, stability, and uniqueness of the network equilibrium are discussed. A best-response algorithm is suggested to identify the equilibria of the coupled network, in which the TAP and the OPF problem are solved iteratively until a fixed point is reached.

1. Optimal Power Flow and LMP

The OPF problem is similar to that in Sect. 2.3.1

$$\min \quad \sum_{i \in E_N} \left[a_i p_{gi}^2 + b_i p_{gi} \right] + \rho \sum_{j \in \pi(0)} P_{0j}^l \tag{5.113a}$$

$$\text{s.t.} \quad \text{Cons-BPF, Cons-BND, Cons-CP} \tag{5.113b}$$

where

$$
\text{Cons-BPF} = \left\{ \begin{array}{c}
P^l_{ij} + p^g_j - r^l_{ij} I^l_{ij} = \sum\limits_{k \in \pi(j)} P^l_{jk} + p^d_j : \lambda^j_a, \ \forall l \in E_L \\
Q^l_{ij} + q^g_j - x^l_{ij} I^l_{ij} = \sum\limits_{k \in \pi(j)} Q^l_{jk} + q^d_j, \ \forall l \in E_L \\
U_j = U_i - 2(r^l_{ij} P^l_{ij} + x^l_{ij} Q^l_{ij}) + (z^l_{ij})^2 I^l_{ij}, \ \forall l \in E_L \\
I^l_{ij} U_i \ge (P^l_{ij})^2 + (Q^l_{ij})^2, \ \forall l \in E_L
\end{array} \right\}
$$

represents convexified power flow equations, Lagrangian dual multiplier λ^j_a following the colon after active power balancing equality is the LMP of electricity at the FCS connected to bus j, which can be easily retrieved when the primal optimal solution is known, and

$$
\text{Cons-BND} = \left\{ \begin{array}{c}
p^{gn}_j \le p^g_j \le p^{gm}_j, \ \forall j \in E_N \\
q^{gn}_j \le q^g_j \le q^{gm}_j, \ \forall j \in E_N \\
U^n_j \le U_j \le U^m_j, \ \forall j \in E_N \\
\sqrt{(P^l_{ij})^2 + (Q^l_{ij})^2} \le S_l, \ \forall l \in E_L \\
P^l_{ij} - r^l_{ij} I^l_{ij} \ge 0, \forall l \in E_L \\
Q^l_{ij} - x^l_{ij} I^l_{ij} \ge 0, \forall l \in E_L
\end{array} \right\}
$$

gives secure operating boundaries of system components, where the last two inequalities prevent reverse active and reactive power flows in distribution lines. Nonetheless, they are not mandatory and can be relaxed according to actual operation situations. The coupling constraint

$$
\text{Cons-CP} = \{p^d_j = p^{d0}_j + p^{dc}_j, \ p^{dc}_j = x^j_a E_B, \ \forall j \in E_N\} \tag{5.114}
$$

depicts the nodal demand in the PDN. It consists of a fixed part p^{d0}_j and the charging request p^{dc}_j from FCS a connecting to bus j, which is a linear function in the EV flow passing through the FCS. This equation validates the linear charging demand function used in the previous sections, and reveals that the coefficient depends on the received energy of each EV. For bus j that does not serve an FCS, $p^{dc}_j = 0$.

2. Network Equilibrium

The interdependency between vehicle routing and generation dispatch is shown in Fig. 5.31: vehicles plan their routes according to road congestion levels and electricity prices in FCSs; the charging demand determines the load profile of

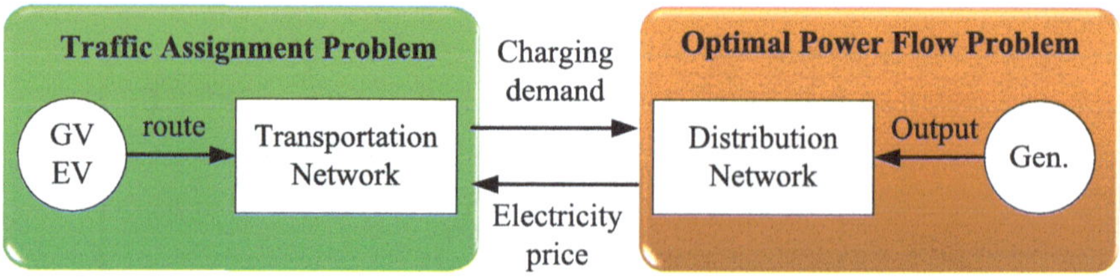

Fig. 5.31 Interdependency between vehicle routing and generation dispatch

the PDN, and further impacts the LMPs, which in turn influence the traffic flow distribution. Under this interplay, an equilibrium state will emerge in the coupled networks. The equilibrium encapsulates the UE over the TN where no vehicle is able to reduce its travel expense by changing route unilaterally, as well as the OPF state in the PDN, such that electricity is produced and delivered in the most economic manner; meanwhile, consumers pay for their energy consumptions at marginal production costs.

Let $\mathbf{x}^\mathbf{T}$ be the vector of variables in TAP (5.108), $\mathbf{x}^\mathbf{E}$ be the vector of primal and dual variables in OPF problem (5.113), and vector $\mathbf{x} = [\mathbf{x}^\mathbf{T}; \mathbf{x}^\mathbf{E}]$. The mapping $\mathcal{S}_{TAP}(\mathbf{x})$ stands for the UE under given LMPs, and $\mathcal{S}_{OPF}(\mathbf{x})$ represents the OPF solution and LMP under the given UE. Denote mapping $\mathcal{S}_{NTE}(\mathbf{x}) = [\mathcal{S}_{TAP}(\mathbf{x}); \mathcal{S}_{OPF}(\mathbf{x})]$, then we can define the network equilibrium as a fixed-point problem.

Definition 5.2 The network equilibrium of the coupled transportation and power distribution systems is the solution of the following fixed-point problem

$$\mathbf{x} = \mathcal{S}_{NTE}(\mathbf{x}) \tag{5.115}$$

According to the relation between an equilibrium problem and a variational inequality, one possible way to express the network equilibrium in a mathematical programming formulation is the KKT optimality condition based approach. Let OPF-KKT($\mathbf{x}$) be the KKT condition of OPF problem (5.113), in which $\mathbf{x}^\mathbf{T}$ (appears in the charging demand function) is parameter and $\mathbf{x}^\mathbf{E}$ is the decision variable; let TAP-KKT($\mathbf{x}$) be the UE condition (5.107), in which $\mathbf{x}^\mathbf{E}$ (appears in the travel expense) is parameter and $\mathbf{x}^\mathbf{T}$ is the decision variable. Then the network equilibrium boils down to the stationary point of the following NCP

$$\left\{ \begin{array}{c} \text{TAP-KKT}(\mathbf{x}) \\ \text{OPF-KKT}(\mathbf{x}) \end{array} \right\} \tag{5.116}$$

Please bear in mind that TAP-KKT can be formulated as linear constraints with integer variables via the method in Sect. 5.3; the SOC constraint in Con-BPF can be approximated by polyhedral approximation for $\mathbb{L}^3$ recursively [68] (this technique has been used in Sect. 4.4), therefore OPF problem (5.113) can be solved as an LP, and its KKT condition renders an LCP. According to the linearization tricks for LCPs explained in Appendix D.4.2, problem (5.116) can finally come down to an MILP without an objective function. Due to the presence of numerous complementarity and slackness conditions, solving (5.116) as an MILP would be extremely challenging, if not impossible.

Before proceeding to the resolution method, we reveal some basic properties of the network equilibrium. To facilitate understanding, some notations are clarified here. In TAP (5.108), $x^L = [x_a], \forall a$ denotes the link flow vector; $f_G^P = [f_{kg}^{rs}], \forall(r,s), \forall k \in K_G^{rs}$ and $f_E^P = [f_{ke}^{rs}], \forall(r,s), \forall k \in K_E^{rs}$ stand for path flow vectors of GVs and EVs, respectively; vector $f^P = [f_G^P; f_E^P]$.

The feasible region of TAP (5.108) is defined as

$$\mathcal{F}^{TAP} = \{(x^L, f^P) | \text{ Cons-Flow (5.107c)}\}$$

Since all constraints are linear and all variables are bounded, $\mathcal{F}^{TAP}$ is a bounded polytope. The feasible region of link flow is the projection of $\mathcal{F}^{TAP}$ on the x^L-space:

$$\mathcal{F}^{TL} = \{x^L | \exists f^P : (x^L, f^P) \in \mathcal{F}^{TAP}\} \tag{5.117}$$

$\mathcal{F}^{TL}$ is also a bounded polytope [69] (see Appendix B.2), thus convex and compact.

Define the charging load injection vector $p^{dc} = [p_i^{dc}], \forall i$, where $p_i^{dc} = x_a^i E_B$ for bus i connecting to a charging link $a \in T_A^C$, and x_a^i is the EV flow on that link; $p_i^{dc} = 0$ for bus i which does not supply electricity to an FCS. The load injection space yields

$$P_D^C = \{p^{dc} \mid p^{dc} = [p_i^{dc}], \forall i, \forall x^L \in \mathcal{F}^{TL}\} \tag{5.118}$$

2. Properties of the Network Equilibrium

(a) Existence

We demonstrate the existence of a network equilibrium by using Brouwer fixed-point theorem. To construct a continuous self-mapping, we make further assumptions on the UE and the OPF problem.

Assumption 5.5 *The user equilibrium is continuous with respect to the electricity price at each fast charging stations.*

This assumption generally holds true. From Proposition 5.3, a UE exists if Assumption 5.4 is met, which is independent of the electricity price. Moreover, from Proposition 5.1 we can image that path flows should be continuous in the electricity price, because all travel expense functions in (5.105) are continuous over $\mathcal{F}^{TL}$ and linearly depend on the LMP λ_a^i, and a small change in the charging cost will lead to a small change in the path flow. This continuity property can also be understood from the fact that TAP is strictly convex in x^L, and electricity prices are coefficients of the linear term of its objective function. Finally, according to (5.104a), the link flow x^L is a linear function of path flow f^P, thus also continuous in the electricity price.

Assumption 5.6 *OPF problem (5.113) is feasible for any given* $p^{dc} \in P_D^C$*; electricity prices at charging station buses are continuous with respect to* p^{dc} *on* P_D^C.

The feasibility requirement can be satisfied as long as generation capacities, line flow limits, and bus voltage intervals are properly assigned, otherwise, upgrading the distribution system could resolve insufficient power supply or alleviate line congestion. However, the continuity of LMP may not be guaranteed. For example, the DCOPF based LMP is stepwise constant [70]. In general, LMP should be piecewise continuous. Nevertheless, continuity can be expected in the following situations:

1. The set P_D^C is small enough. This is true when p_j^{dc} only accounts for a small part of the total demand p_j^d.
2. A continuous pricing policy is enforced, such as the continuous LMP scheme described in [70].
3. Congestion and voltage boundary constraints are inactive. Theoretical analysis and illustrative examples are given in [71] based on a uniform pricing model without power flow constraints.

Proposition 5.6 *A network equilibrium exists if Assumptions 5.5–5.6 are met.*

To see this, Assumption 5.5 implies that $\mathcal{S}_{TAP}(\mathbf{x})$ is a continuous mapping. The mapping $x \mapsto p_d^c$ from link flow to load injection shown in Cons-CP and (5.118) is linear, thus continuous. Combined with Assumption 5.6, $\mathcal{S}_{OPF}(\mathbf{x})$ is a continuous mapping. Consider the following mapping: give a link flow pattern $x^L \in \mathcal{F}^{TL}$, first solve OPF problem (5.113) and retrieve the LMP, then solve TAP (5.108) with the obtained LMP, and the outcome is again a link flow pattern $x^L \in \mathcal{F}^{TL}$, which can be expressed as a composite self-mapping

$$\mathcal{M} : \mathcal{F}^{TL} \to \mathcal{F}^{TL} = \mathcal{S}_{TAP}\left(\mathcal{S}_{OPF}(\mathbf{x})\right)$$

which is continuous according to the above analysis. Since polytope $\mathcal{F}^{TL}$ is convex and compact, according to Brouwer's fixed point theorem (Corollary 6.6 in [72]), $\mathcal{S}_{TAP}\left(\mathcal{S}_{OPF}(\cdot)\right)$ has a fixed point x^L, which gives the traffic flow pattern at the network equilibrium. OPF variables at the network equilibrium can be calculated through $\mathcal{S}_{OPF}(\mathbf{x})$.

(b) Stability

The network equilibrium, if exists, may be stable or not, depending on problem parameter. An illustrative explanation is provided in Fig. 5.32. Consider an FCS connecting to bus i. First consider the OPF problem. When the charging demand grows, more electricity is produced and network losses may also increase, leading to a rise in the electricity price. In this regard, the offering curve in Fig. 5.32

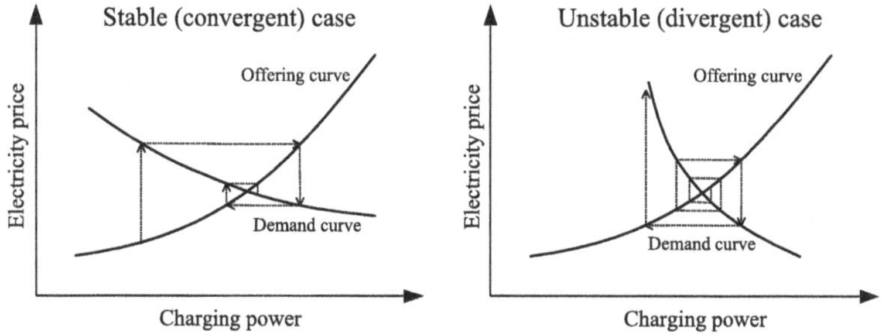

Fig. 5.32 Stable and unstable network equilibrium

is increasing. Then consider the TAP. When the electricity price rises, the path travel expense becomes higher; according to the minimum expense rationality, some drivers may switch to other places for receiving recharging services. Therefore, the demand curve in Fig. 5.32 is decreasing. The stability of the equilibrium depends on slopes of two curves at the intersection point. The slope of LMP curve is given by $\partial\lambda_a^i/\partial p_i^{dc}$, which is positive; the slope of the demand curve can be expressed via

$$\left|\frac{\partial p_i^{dc}}{\partial\lambda_a^i}\right| = \left|\frac{\partial p_i^{dc}}{\partial x_a}\frac{\partial x_a}{\partial\lambda_a^i}\right| = \left|\frac{\partial x_a}{\partial\lambda_a^i}\right| E_B$$

Proposition 5.7 *The network equilibrium is locally stable, if*

$$\frac{\partial\lambda_a^i}{\partial p_i^{dc}}\left|\frac{\partial x_a}{\partial\lambda_a^i}\right| E_B < 1, \ \forall a \in T_A^C \tag{5.119}$$

holds at the equilibrium.

Stability of the network equilibrium can be understood in the following way. First, a UE in the TN is stable according to its definition. Second, condition (5.119) indicates that $\mathcal{S}_{TAP}(\mathcal{S}_{OPF}(\cdot))$ is locally contractive. The Banach contraction mapping theory [73] ensures the stability of equilibrium. The local contractivity also implies that if multiple network equilibria exist, they are isolated: there is no other equilibria in a small enough neighborhood of an equilibrium. Condition (5.119) is difficult to check a priori, unless the equilibrium is known. For theoretical interests, it can be used to verify the stability of an equilibrium offered by NCP model (5.116).

We notify that (5.119) is only an intuitive condition, because it implicitly assumes that the demand at bus j mainly affects the local LMP and has little influence on other LMPs, which needs further justification. A rigorous condition will require that the spectral radius of the self mapping is less than 1, involving the sensitivity between nodal charging demand and all the LMPs in the PDN.

An important implication of the network equilibrium is that under the rationality of minimum travel expense, EV drivers attempt to circumvent expensive charging services, which prevents demands and LMPs in those heavily loaded FCSs from rising higher. This adaptivity of UE creates system-wide responsive demands which are helpful to the operation of PDN. If the price sensitivity $\partial \lambda_a^i / \partial p_i^{dc}$ and demand sensitivity $\partial x_a / \partial \lambda_a^i$ are relatively small, the demand response dynamics will be stable, and the system state will converge to the network equilibrium.

(c) Uniqueness

Because electricity prices appearing in TAP (5.108) are dual variables of OPF problem (5.113), provable uniqueness guarantee from existing theories is non-trivial. Motivated by the stability condition (5.119), we conjecture that monotonic and bounded price sensitivity might lead to a unique network equilibrium, i.e.,

$$\left\{ \frac{\partial \lambda_a^i}{\partial p_i^{dc}} \left| \frac{\partial x_a}{\partial \lambda_a^i} \right| < E_B^{-1}, \forall a \in T_A^C \right\}, \forall p^{dc} \in P_D^C \tag{5.120}$$

Under condition (5.120), $\mathcal{S}_{TAP}(\mathcal{S}_{OPF}(\cdot))$ is a contractive mapping, according to contraction mapping theory [73], the UE solution will be unique. The OPF solution under fixed p^{dc} is likely to be unique due to the geometric property of power flow in radial networks revealed in [74]. However, condition (5.120) would be restrictive in general and cannot be checked conveniently. In fact, the LMP may be discontinuous, thus the partial derivative and the equilibria may not even exist.

3. A Best-Response Algorithm

In view of the interaction illustrated in Fig. 5.31, the network equilibrium can be computed from a best-response procedure shown in Algorithm 5.7. The advantage relies on the convexity of TAP (5.108) and OPF problem (5.113), both of which can be solved reliably and efficiently.

Algorithm 5.7 Best response iteration

1: Choose a convergence tolerance $\varepsilon > 0$, an allowed number of iterations K_{it}^M, and an initial LMP vector $\lambda \in \mathbb{R}^{|E_N|}$. The current iteration number is $k = 1$. Let the initial traffic flow $x^L = \mathbf{0}^{|T_A|}$.
2: Solve TAP (5.108) with fixed electricity price λ_a^i at each FCS; the optimal solution (or UE) is x^{L*}; update $\Delta^{TN} = \|x^{L*} - x^L\|$, and $x^L = x^{L*}$.
3: Solve OPF problem (5.113) with the obtained UE x^L; retrieve LMP vector λ^* at the OPF solution; update $\Delta^{PDN} = \|\lambda^* - \lambda\|$, and $\lambda = \lambda^*$.
4: If $\Delta^{TN} + \Delta^{PDN} \le \varepsilon$, terminate, and return the UE, OPF solution, and LMPs as the network equilibrium; else if $k = K_{it}^M$, terminate and report that the algorithm fails to converge.
5: Update $k = k + 1$, and go to step 2.

In step 3, if the convex relaxation of OPF problem (5.113) is not exact, Algorithm 2.1 in Sect. 2.3.1 should be performed. Algorithm 5.7 converges provided that the initial value is close enough to the equilibrium solution, if there is any; otherwise, it may fail to converge, even an equilibrium exists. According to our experiences, an equilibrium exists in most cases as long as cost functions of generators are almost linear. The reason is explained as follows. The stability criterion (5.119) implies that the network equilibrium is more likely to be stable, if E_B is smaller, and the offering curve is flatter. Because the marginal cost of generation unit is $2a_i p_i^g + b_i$, the sensitivity $\partial \lambda_a^i / \partial p_i^{dc}$ is bounded by $2a_m$, where $a_m = \max_i\{a_i\}, \forall i$. In these regards, smaller a_m and E_B will contribute to a more stable equilibrium. Practical treatments which facilitate convergence are discussed.

1. Initialization. When the penetration level of EV is low, p_j^{dc} only accounts for a small portion of the load in PDN, we can initialize the LMP vector λ by solving OPF problem (5.113) with $p^{dc} = \mathbf{0}^{|E_B|}$. Otherwise, we assume that each FCS serves the same amount of EVs (no matter whether this pattern is a UE) and calculate p^{dc} accordingly, and then solve OPF problem (5.113) and recover an initial value of λ.
2. System data. Properly widening the bounds of bus voltages and line flow limits or relaxing certain constraints in Cons-BND without jeopardizing a secure operation would generally contribute to more smooth LMPs and help Algorithm 5.7 converge.

5.7.3 Case Studies

1. Basic settings

This section exhibits experimental results on a test system. The topology of the TN is shown in Fig. 5.33, which consists of four classes of roads and eight FCSs in the outer and inner ring expressway loops. Road data are given in Table 5.23. The base value is 100 vehicle per hour. In the Davidson function for a charging link $a \in T_A^C$, $t_a^0 = 20$ min and $c_a = 15$ are used; for the bypass link, $t_a^0 = 0$, and its capacity c_a is equal to the road segment it locates at, i.e., bypass link capacities of C1, C3, C6, C8 are 100, and those of C2, C4, C5, C7 are 80. O-D pairs and their traffic demands are listed in Table 5.24, indicating that the main traffic goes from the northwest to the eastern and southern areas. The monetary value of travel time is $\omega = 10\$/\text{h}$.

The distribution network topology is shown in Fig. 5.34. The base value of power is 100MVA. Distribution line data are provided beside each line. Apparent power flow capacity is $S_l = 1.5$ for all lines. The fixed part of electrical demand at each bus is $p_i^{d0} = 0.1$ and $q_i^{d0} = 0.1$. The lower and upper bounds of voltage magnitude at bus j are $\sqrt{U_i^n} = 0.88$ and $\sqrt{U_i^m} = 1.05$, respectively. Voltage magnitude of the slack bus is $\sqrt{U_0} = 1.04$. Generators G1-G8 share the same parameter: $p_i^{gn} = 0$, $p_i^{gm} = 1.5$, $q_i^{gn} = -1.0$, $q_i^{gm} = 1.0$, $a_i = 0.3\$/\text{MW}^2\text{h}$, $b_i = 150\$/\text{MWh}$. The contract price with the main grid is $\rho = 140\$/\text{MWh}$. The energy demand of each EV is $E_B = 50\,\text{kWh}$.

2. Results

Algorithm 5.7 is applied to compute the network equilibrium. TAP (5.108) is solved by IPOPT [29], and the OPF problem (5.113) in SOCP form is solved by MOSEK [75]. The convergence tolerance is set as $\varepsilon = 0.5$. The maximal number of iterations

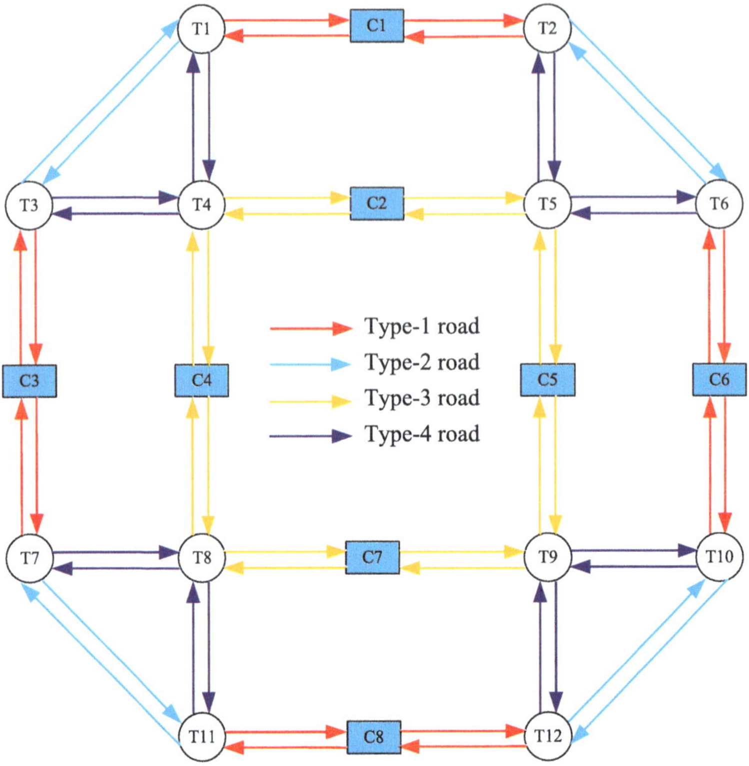

Fig. 5.33 Topology of the ring expressway transportation system

Table 5.23 Parameter of roadway segments

Road	Type-1	Type-2	Type-3	Type-4	Charging	Bypass
c_a (p.u.)	100	100	80	60	15	–
t_a^0 (min)	5	8	5	7	20	0

Table 5.24 O-D pairs and their trip rates (in p.u.)

O-D pair	q_g^{rs}	q_e^{rs}	O-D pair	q_g^{rs}	q_e^{rs}
T1–T6	30	10	T3–T6	30	10
T1–T10	60	5	T3–T10	50	5
T1–T11	40	10	T3–T11	40	10
T1–T12	40	5	T3–T12	50	5
T4–T9	50	10	T4–T10	40	10
T4–T12	40	10			

is $K_{it}^M = 30$. The initial LMP is given by the OPF result with zero charging demands. Under the above settings, Algorithm 5.7 terminates in 4 iterations. Results are shown in Figs. 5.35 and 5.36 and Tables 5.25, 5.26, and 5.27. If Algorithm 5.7 is not interrupted when the convergence tolerant is met, the values of $\Delta^{TN} + \Delta^{PDN}$

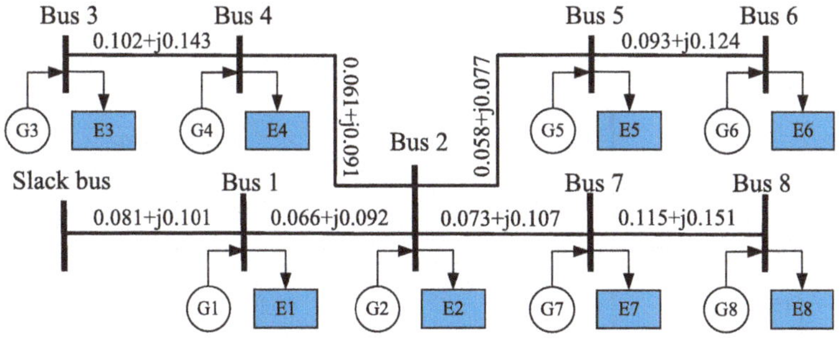

Fig. 5.34 Topology of the distribution network

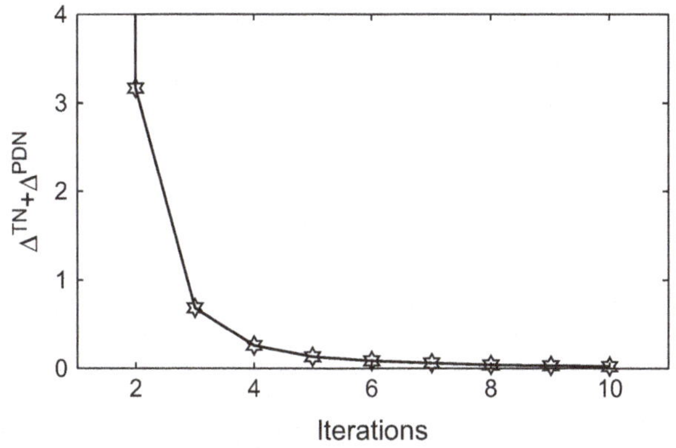

Fig. 5.35 Convergence performance of Algorithm 5.7

generated in the first ten iterations are shown in Fig. 5.35. It is equal to 700.26 in the first iteration, and monotonically decreases with a value less than 1 after three iterations.

The UE in the TN is depicted in Fig. 5.36, which illustrates that most vehicles travel on ring expressways in the inner and outer loops, because they have larger capacities and fewer traffic signals (reflected by a smaller t_a^0). Trip information for each O-D pair is listed in Table 5.25. Similar to the simple test system in the previous subsection, EVs spend more time and cost than GVs do due to the waiting time and payment for recharging. If the refueling cost is taken into account, the average travel expense of GVs would increase due to the higher price of fuel. Nevertheless, the average waiting time for refueling a GV is typically shorter than charging an EV. LMPs, active power demands at FCSs, and optimal generation dispatch strategies are given in Table 5.26. Two facts can be observed: (1) Since reverse power flow is banned, the bus voltage decreases and LMP increases from the root bus to the leaf bus; (2) EV flows in charging stations C3, C6, and C8 are smaller than those in the remaining ones, because they connect to leaf buses, where LMPs are higher. Meanwhile, the rationality of minimum travel expense constitutes a spontaneous

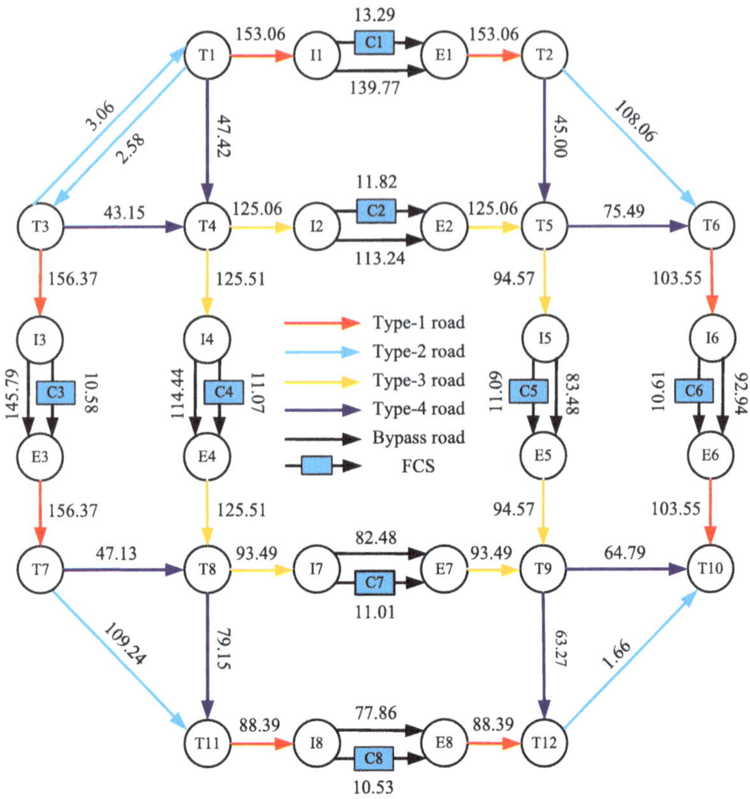

Fig. 5.36 Traffic flows at the user equilibrium pattern

Table 5.25 Summary of trip information

O-D pair	Travel time (min)		Travel expense ($)	
	GV	EV	GV	EV
1–6	27.87	55.66	4.6448	17.6728
1–10	39.59	62.01	6.5989	19.6268
1–11	36.68	59.07	6.1129	19.1409
1–12	46.79	69.63	7.7986	20.8265
3–6	35.87	59.58	5.9781	19.0060
3–10	47.59	69.95	7.9322	20.9601
3–11	28.68	51.07	4.7796	17.8076
3–12	39.59	61.95	6.5989	19.6268
4–9	31.89	54.64	5.3142	18.3421
4–10	40.31	62.73	6.7188	19.7467
4–12	40.18	62.54	6.6973	19.7252

Table 5.26 Summary of LMPs, charging demands, and generation dispatch

Node	LMP ($/MWh)	p_i^{dc} (MW)	p_{gi} (MW)	q_{gi} (MVar)
1	167.92	66.47	29.87	12.63
2	181.52	59.08	52.53	10.39
3	185.91	52.91	59.84	9.98
4	184.49	55.37	57.48	10.03
5	184.44	55.44	57.41	10.03
6	185.84	53.04	59.73	09.98
7	184.70	55.04	57.83	10.03
8	186.04	52.65	60.07	9.98

Table 5.27 OPF at the network equilibrium

Node	V_i (p.u.)	Line	P_{ij}^l (MW)	Q_{ij}^l (MVar)
1	0.9520	0–1	105.89	10.57
2	0.9154	1–2	50.81	2.63
3	0.9045	2–4	11.05	0.136
4	0.9080	4–3	3.08	0.033
5	0.9081	2–5	11.45	0.124
6	0.9047	5–6	3.33	0.035
7	0.9075	2–7	9.88	0.127
8	0.9042	7–8	2.59	0.029

Table 5.28 LMPs and convergence rates under different values of E_B

E_B (kWh)		20	30	40	60
LMP	1	155.64	159.57	163.67	172.32
	2	163.98	169.72	175.56	187.59
	3	167.22	173.48	179.71	192.12
	4	165.98	172.09	178.27	190.79
	5	165.93	172.05	178.22	190.77
	6	167.13	173.39	179.63	192.12
	7	166.17	172.31	178.48	190.92
	8	167.42	173.68	179.87	192.05
Iterations		3	3	3	6

demand response, preventing demands in C3, C6, C8 from rising higher. OPF results in the PDN are provided in Table 5.27. From Tables 5.26 and 5.27 we can see that all security constraints are maintained. The energy consumption in C1 is mainly offered by the main grid, and LMP in C1 is the lowest due to the cheaper electricity price at the slack bus.

Finally, we investigate the impact of battery energy demand E_B on LMPs. Network equilibria under different values of E_B are enumerated in Table 5.28. Clearly, LMPs rise when E_B grows larger. Meanwhile, the gap between the highest and the lowest LMPs is also widened. The number of iterations in each scenario is also given in Table 5.28 (4 for $E_B = 50$ kWh, which can be observed from Fig. 5.35). In general, when E_B grows, more iterations are needed before Algorithm 5.7 converges, because the interdependency between the two networks appears to be stronger.

5.8 Summary and Conclusions

The proliferation of electric vehicles is paving the way to greener urban transportation with less dependence on fossil fuels. The transportation system is experiencing a transition for electrification. In this chapter, we bring new vision that the charging behavior of electric vehicles will knot the traffic flow in the transportation network and power flow in the power grid. We show that neglecting this interplay and managing two systems independently may lead to insecure operating condition and unstable flow dynamics. Hence, we propose analytical tools and methods to address a variety of issues ranging from planning, operation, to market analysis, while acknowledging the interdependence created by on-road charging facilities, with the purpose of moving both infrastructures towards better operating conditions.

The optimal traffic-power flow in Sect. 5.3 studies the coordinated operation of the transportation network and the power grid. The essential technique is the MPEC modeling paradigm of the OTPF problem and its MISOCP reformulation. The joint operation achieves a socially optimal traffic flow and power flow pattern, and offers an opportunity to mitigate potential adverse effects on the security of the electricity network.

The robust dispatch method in Sect. 5.4 tackles the operation of distribution network with uncertain traffic demands. The modeling trick stems from the characterization of nodal demand space affected by uncertain traffic demand. The ACOPF based robust dispatch model and its solution method provide a versatile methodology to operate radial distribution networks with nodal power injection uncertainty, such as those induced by renewable generations and load forecast errors.

The capacity expansion planning problem investigated in Sect. 5.5 discusses a comprehensive model for the optimal deployment of roads, charging facilities, generators, and distribution lines simultaneously. Technical features are twofold. One originates from the Nesterov TAP for the transportation network, which does not depend on the particular road latency function and defines the delay through Lagrangian dual multipliers. It is suitable for the planning problem because it incorporates explicit capacity limits on road traffic flows. The other rests on the exact mixed-integer program formulation for the capacity expansion planning model without exploiting approximation. This approach is able to find the global optimal solution for moderately sized systems with reasonable computation effort.

The vulnerability assessment method studied in Sect. 5.6 provides the system operator with important information and insights on the priority of vulnerable links in a transportation system that need to be reinforced in order to enhance system resilience against unexpected contingencies. More interesting research direction is the modeling of traffic flow dynamics after a contingency. It is also important to study the impact of power grid failure on the traffic user equilibrium.

The mixed user equilibrium model developed in Sect. 5.7 treats gasoline vehicles and electric vehicles separately, and the network equilibrium framework encapsulates vehicle routing and charging, electricity pricing, and economic energy management with system interdependency in a holistic model. The basic properties of the network equilibrium, although not very rigorous, provide important implications and reference on the market design and system operation; the best-response algorithm leverages computation superiorities of convex optimization.

5.9 Further Reading

The transportation system and power system have been the subject of engineering research for several decades. Extensive publications can be found. Some representative ones are summarized.

For readers who are interested in a more extensive coverage of transportation system modeling, we refer to [12] for discussions of basic user equilibrium problem and its extensions which consider variable traffic demands, destination choices, link interactions, and travel pattern selections. Link capacity constrained user equilibrium models and methods are in [76–78]. Recent user equilibrium research highlights the specific route choice behavior of electric vehicles due to their limited driving ranges. For instance, a formulation with length constraints on electric vehicle paths is investigated in [79], and is further generalized in [80] to include trip chains. A basic mixed user equilibrium model with gasoline and electric vehicles is developed in [81], and two efficient algorithms are suggested to solve the proposed model. A comprehensive user equilibrium model which jointly incorporates destination, route, and parking choices of gasoline and electric vehicle subject to the driving range limit is presented in [82]. More dedicated user equilibrium models with battery electric vehicles are proposed in [83]. Under different assumptions on the dependency of battery energy consumption and recharging time on traffic flows, three models are suggested in [83]; two of which are formulated as convex TAPs, and the third one yields an NCP. The trip chain of electric vehicle is incorporated in the user equilibrium model studied in [84]. Limited driving range and recharging needs of electric vehicle are main considerations. A user equilibrium model with charging-while-driving lanes are portrayed in [85]. A multi-class user equilibrium model with both conventional vehicles and autonomous vehicles in the presence of dedicated lanes is formulated in [86]. Impact of the travel time display on route choices of drivers under recurrent congestion is investigated through a user equilibrium method [87]. These works emphasize traffic flow distribution and neglect the operation of power systems.

Taxi market is becoming a hot topic in recent years, owing to the presence of various e-hailing platforms, which provide convenient services to passengers. Spatial equilibrium of taxi supply and demand under smart phone enabled e-hailing application is modeled in [88], and the taxi market equilibrium is characterized in [89]. Equilibrium with credit trading schemes is analyzed in [90]. Equilibrium of a taxi market with both street-hailing and e-hailing services is studied in [91]. Reservation and cancellation behaviors from smart phone users are considered, and the optimal penalty/compensation strategies are suggested.

A pioneering work that addresses the interdependence of regional coupled transportation and power distribution network is [15], which encapsulates a user equilibrium with destination choice, a DCOPF problem, and marginal electricity pricing scheme in an integrated model. The similar paradigm is employed in [92, 93] to study electricity price and road toll which will help improve the over-all operation condition of the coupled networks. A distributed pricing policy is suggested in [17], in which the transportation system and the power system collaborate with each other in order to reach a social optimum operating point while keeping the data of each system private. Both traffic assignment and power flow optimizations are taken into account in these works, which open a new direction of interdisciplinary research. Collaborative transportation and charging scheduling of self-driving electric vehicles under real-time electricity price is considered in [94], where vehicles seek individual maximum total transportation and charging utilities. The problem is formulated as a cake cutting game whose equilibrium is shown to be unique. Network models are neglected in this work.

A straightforward application of user equilibrium model is the planning of public charging facilities. Along this line, optimal deployment of public charging stations is discussed in [15], based on the proposed network equilibrium model. A multi-objective model for charging station planning is suggested in [13, 14]. In the former one, the shortest route of electric vehicles which is independent of congestion is pre-determined. Charging stations are located so as to capture the electric vehicle flow as much as possible. In the latter one, by jointly optimizing a TAP and an OPF problem, the resulting Pareto optimal strategy minimizes the investment cost and energy losses, while maximizes the captured traffic flow simultaneously. The limited driving range and interaction between the transportation and power systems are also considered in [95], but congestion is ignored in the capacitated-flow refueling location model. Planning of charging facilities has long been the topic of the power system community. Only those shed light on both transportation and electricity infrastructures as well as their interdependency are mentioned here.

Thanks to the proliferation of e-hailing platforms on smart phones, ridesharing (or carpooling) becomes a promising mean to take full advantage of unused car capacity and help alleviate traffic congestion. Analyzing the interactions between the level of road congestion and choices of commuters is fundamental for an enterprise to pricing its service and for government authorities to regulate the ridesharing market as well as the transportation system. The main factors a traveler will consider to decide whether to take part in ridesharing include the price, congestion level, and inconvenience perceived during such an activity. These tradeoffs would be balanced in an equilibrium which determines the percentages of different types of travelers, the ridesharing prices, and road congestion levels. The user equilibrium is modeled by a mixed complementarity problem in [96] and modified Beckmann models in

[97, 98]. Another complementarity model is set forth in [99] based on the one in [96], which makes different assumptions on ridesharing revenues. We believe that ridesharing will continue to be a hot topic in the next a few years, and will precipitate cross-discipline researches.

The static user equilibrium model introduced in Sect. 5.2 assumes all motorists leave the origin at the same time and complete their trips instantly. In fact, not all vehicles begin to travel simultaneously. To simulate postponed travel and spatial-temporal distributions of traffic flows more precisely, dynamic traffic assignment models should be used. Indeed, as the power system load and electricity price are time-varying, dynamic traffic assignment is more appropriate for studying system interdependence and will be an important direction in the future. Unlike the static user equilibrium which has a streamlined and well-acknowledged formulation, dynamic traffic assignment theory is less mature than the static one. There is no universal model; existing ones rely on specific assumptions and require more sophisticated mathematical tools. A review on some earlier dynamic traffic assignment models can be found in [35], and an up-to-date survey in [100].

Continuous-time spatial-temporal traffic flow can be described by a kinematic wave equation, or namely, the classic Lighthill-Whitham-Richards (LWR) model [101, 102]. Cell transmission model proposed in [103, 104] is a discrete-time approximation of the LWR model. It simulates traffic flow and density along a corridor. Discrete-time and continuous-time dynamic user equilibrium models based on the cell transmission model are developed in [105, 106], respectively. The discrete-time model is cast as complementarity constraints in [107]. Other discrete time formulations are discussed in [108–111]. In authors' opinion, discrete-time models are more likely to be used in the system interdependence research, because mainstream power system dispatch and power market models appear to be discrete-time optimization problems.

Continuous-time dynamic traffic assignment models have been proposed in form of optimal control [112], differential games [113–115], differential variational inequality [116], differential complementarity problem [117], and partial-differential complementarity problem [118]. Extension to elastic demands can be found in [119]. Combining dynamic user equilibrium and continuous-time power system operation models [120, 121] constitutes a more plausible way to study transient behaviors and stability of network flow evolution. However, the computational tractability may be a main issue for practical usage.

We refer the interested readers to [122] for a very thorough review on the model and applications of interdependent power grid and transportation network infrastructures. Prospective topics in this young and active research field are also envisioned.

References

1. Panwar, N.L., Kaushik, S.C., Kothari, S.: Role of renewable energy sources in environmental protection: a review. Renew. Sustain. Energy Rev. **15**(3), 1513–1524 (2011)
2. Dincer, I.: Renewable energy and sustainable development: a crucial review. Renew. Sustain. Energy Rev. **4**(2), 157–175 (2000)
3. Morales, J.M., Conejo, A.J., Madsen, H., Pinson, P., Zugno, M.: Integrating Renewables in Electricity Markets: Operational Problems. Springer, Berlin (2013)
4. Navid, N., Rosenwald, G.: Market solutions for managing ramp flexibility with high penetration of renewable resource. IEEE Trans. Sustain. Energy **3**(4), 784–790 (2012)
5. Galus, M., Waraich, R., Noembrini, F., Steurs, K., Georges, G., Boulouchos, K., Axhausen, K., Andersson, G.: Integrating power systems, transport systems and vehicle technology for electric mobility impact assessment and efficient control. IEEE Trans. Smart Grid **3**(2), 934–949 (2012)
6. Shuai, W., Maille, P., Pelov, A.: Charging electric vehicles in the smart city: a survey of economy-driven approaches. IEEE Trans. Intell. Transp. **17**(8), 2089–2106 (2016)
7. Kong, P.Y., Karagiannidis, G.K.: Charging schemes for plug-in hybrid electric vehicles in smart grid: a survey. IEEE Access **4**, 6846–6875 (2016)
8. Richardson, D.B.: Electric vehicles and the electric grid: a review of modeling approaches, Impacts, and renewable energy integration. Renew. Sustain. Energy Rev. **19**, 247–254 (2013)
9. Yilmaz. M., Krein. P.T.: Review of the impact of vehicle-to-grid technologies on distribution systems and utility interfaces. IEEE Trans. Power Electron. **28**(12), 5673–5689 (2013)
10. Wardrop, J.G.: Some theoretical aspects of road traffic research. Proc. I. Civil Eng. **1**(3), 325–362 (1952)
11. Beckmann, M., McGuire, C., Winsten, C.: Studies in the Economics of Transportation. Yale University Press, New Haven (1956)
12. Sheffi, Y.: Urban Transportation Network: Equilibrium Analysis with Mathematical Programming Methods. Prentice-Hall, Upper Saddle River (1985)
13. Wang, G., Xu, Z., Wen, F., Wong, K.P.: Traffic-constrained multi-objective planning of electric-vehicle charging stations. IEEE Trans. Power Del. **28**(4), 2363–2372 (2013)
14. Yao, W., Zhao, J., Wen, F., Dong, Z., Xue, Y., Xu, Y., Meng, K.: A multi-objective collaborative planning strategy for integrated power distribution and electric vehicle charging systems. IEEE Trans. Power Syst. **29**(4), 1811–1821 (2014)
15. He, F., Wu, D., Yin, Y., Guan, Y.: Optimal deployment of public charging stations for plug-in hybrid electric vehicles. Transp. Res. B Methodol. **47**, 87–101 (2013)
16. He, F., Yin, Y., Wang, J., Yang, Y.: Sustainability SI: optimal prices of electricity at public charging stations for plug-in electric vehicles. Netw. Spat. Econ. **16**(1), 131–154 (2016)
17. Alizadeh, M., Wai, H., Chowdhury, M., Goldsmith, A., Scaglione, A., Javidi, T.: Optimal pricing to manage electric vehicles in coupled power and transportation networks. IEEE Trans. Control Netw. Syst. **4**(4), 863–875 (2017)
18. Wei, W., Mei, S., Wu, L., Shahidehpour, M. Fang, Y.: Optimal traffic-power flow in urban electrified transportation networks. IEEE Trans. Smart Grid **8**(1), 84–95 (2017)
19. Wei, W., Mei, S., Wu, L., Wang, J., Fang. Y.: Robust operation of distribution networks coupled with urban transportation infrastructures. IEEE Trans. Power Syst. **32**(3), 2118–2130 (2017)
20. Wei, W., Wu, L., Wang, J., Mei, S.: Expansion planning of urban electrified transportation networks: a mixed integer convex programming approach. IEEE Trans. Transp. Electrific. **3**(1), 210–224 (2017)
21. Wei, W., Wang, J., Wu, L.: Quantifying the impact of road capacity loss on urban electrified transportation networks: an optimization based approach. Int. J. Transport. Sci. Technol. **5**(4): 268–288 (2016)
22. Wei, W., Wu, L., Wang, J., Mei, S.: Network equilibrium of coupled transportation and power distribution systems. IEEE Trans. Smart Grid **9**(6) 6764–6779 (2018)

23. Bureau of Public Roads: traffic assignment manual. U.S. Department of Commerce, Washington, DC, 1964
24. Davidson, K.B.: A flow-travel time relationship for use in transportation planning. In: Proceedings of the Conference on Australian Road Research Board, Melbourne, Australia, 183–194 (1966)
25. Ralph, D.: Mathematical programs with complementarity constraints in traffic and telecommunications networks. Philos Trans. R. Soc. A **366**, 1973–1987 (2008)
26. Nesterov, Y.: Stable traffic equilibria: properties and applications. Optim. Eng. **1**(1), 29–50 (2000)
27. Nesterov, Y., de Palma, A.: Stationary dynamic solutions in congested transportation networks: summary and perspectives. Netw. Spat. Econ. **3**(3), 371–395 (2003)
28. Chudak, F., Eleuterio, V., Nesterov Y.: Static traffic assignment problem: a comparison between Beckmann (1956) and Nesterov and de Palma (1998) models. In: Proceedings of Swiss Transport Research Conference, pp. 1–23. Monte-Verità, Ascona (2007)
29. Wächter, A., Biegler, L.T.: On the implementation of an interior-point filter line-search algorithm for large-scale nonlinear programming. Math. Program. **106**(1), 25–57 (2006)
30. https://www.ibm.com/analytics/data-science/prescriptive-analytics/cplex-optimizer
31. Chiou, S.W.: Bilevel programming for the continuous transport network design problem. Transp. Res. B Methodol. **39**(4), 361–383 (2005)
32. Dempe, S., Kalashnikov, V., Pérez-Valdés, G.A., Kalashnykova, N.: Bilevel Programming Problems: Theory, Algorithms and Applications to Energy Networks. Springer, Berlin (2015)
33. Madawala, U., Thrimawithana, D.J.: A bidirectional inductive power interface for electric vehicles in V2G systems. IEEE Trans. Ind. Electron. **58**(10), 4789–4796 (2011)
34. EDL: Energy dynamics laboratory achieves historic milestone for electric vehicle, available at: http://www.prnewswire.com/news-releases/energydynamics-laboratory-achieves-historic-milestone-for-electric-vehicles-125005864.html 2011
35. Peeta, S., Ziliaskopoulos, A.K.: Foundations of dynamic traffic assignment: the past, the present and the future. Netw. Spat. Econ. **1**(3), 233–265 (2001)
36. Chiou, S.W.: Optimization of area traffic control for equilibrium network flows. Transport. Sci. **33**(3), 279–289 (1999)
37. Yang, H., Yagar, S.: Traffic assignment and signal control in saturated road networks. Transp. Res. A Policy Pract. **29**(2), 125–139 (1995)
38. Yang, H., Yagar, S.: Traffic assignment and traffic control in general freeway-arterial corridor systems. Transp. Res. B Methodol. **28**(6), 463–486 (1994)
39. Fortuny-Amat, J., McCarl, B.: A representation and economic interpretation of a two-level programming problem. J. Oper. Res. Soc. **32**(9), 783–792 (1981)
40. Fletcher, R., Leyffer, S.: Solving mathematical programs with complementarity constraints as nonlinear programs. Optim. Methods Softw. **19**(1), 15–40 (2004)
41. Fletcher, R, Leyffer, S., Ralph, D., Scholtes, S.: Local convergence of SQP methods for mathematical programs with equilibrium constraints. SIAM J. Optim. **17**(1), 259–286 (2006)
42. Ralph, D., Wright, S.J.: Some properties of regularization and penalization schemes for MPECs. Optim. Methods Softw. **19**(5), 527–556 (2004)
43. Custódio, A.L., Vicente, L.N.: Using sampling and simplex derivatives in pattern search methods. SIAM J. Optim. **18**(2), 537–555 (2007)
44. Custódio, A.L., Rocha, H., Vicente, L.N.: Incorporating minimum Frobenius norm models in direct search. Comput. Optim. Appl. **46**(2), 265–278 (2010)
45. Kolda, T.G., Lewis, R.M., Torczon, V.: Optimization by direct search: new perspectives on some classical and modern methods. SIAM Rev. **45**(3), 385–482 (2003)
46. Audet, C., Dennis, J.E.: Analysis of generalized pattern searches. SIAM J. Optim. **13**(3), 889–903 (2003)
47. Byrd, R.H., Nocedal, J., Waltz, R.A.: Knitro: an integrated package for nonlinear optimization. In: Pillo, G. D., Roma, M. (eds.) Large-Scale Nonlinear Optimization, chap. 4, pp. 35–59. Springer, Berlin (2006)
48. Ordóñez, F., Stier-Moses, N.E.: Wardrop equilibria with risk-averse users. Transp. Sci. **44**(1), 63–86 (2010)

49. Ito, Y.: Robust Wardrop Equilibria in the Traffic Assignment Problem with Uncertain Data. Master Dissertation, Kyoto University (2011)
50. Korski, J., Pfeuffer, F., Klamroth, K.: Biconvex sets and optimization with biconvex functions: a survey and extensions. Math. Methods Oper. Res. **66**(3), 373–407 (2007)
51. Ding, T., Li, C., Yang, Y., Jiang, J., Bie, Z., Blaabjerg, F.: A two-stage robust optimization for centralized-optimal dispatch of photovoltaic inverters in active distribution networks. IEEE Trans. Sustain. Energy **8**(2), 744–754 (2017)
52. Ding, T., Yang, Q., Yang, Y., Li, C., Bie, Z., Blaabjerg, F.: A data-driven stochastic reactive power optimization considering uncertainties in active distribution networks and decomposition method. IEEE Trans. Smart Grid **9**(5), 4994–5004 (2018)
53. Ding, T., Liu, S., Yuan, W., Bie, Z., Zeng, B.: A two-stage robust reactive power optimization considering uncertain wind power integration in active distribution networks. IEEE Trans. Sustain. Energy **7**(1), 301–311 (2016)
54. Clement-Nyns, K., Haesen. E., Driesen, J.: The impact of vehicle-to-grid on the distribution grid. Electr. Power Syst. Res. **81**(1), 185–192 (2011)
55. Chen, A., Yang, C., Kongsomsaksakul, S., Lee, M.: Network-based accessibility measures for vulnerability analysis of degradable transportation networks. Netw. Spat. Econ. **7**(3), 241–256 (2007)
56. Scott, D.M., Novak, D.C., Aultman-Hall, L., Guo, F.: Network robustness index: a new method for identifying critical links and evaluating the performance of transportation networks. J. Transp. Geogr. **14**(3), 215–227 (2006)
57. Nagurney A., Qiang, Q.: A relative total cost index for the evaluation of transportation network robustness in the presence of degradable links and alternative travel behavior. Int. Trans. Oper. Res. **16**(1), 49–67 (2009)
58. Al-Deek H., Emam, E.B.: New methodology for estimating reliability in transportation networks with degraded link capacities. J. Intell. Transp. Syst. **10**(3), 117–129 (2006)
59. Chiou, S.W.: Vulnerability analysis of a signal-controlled road network for equilibrium flow. In: Advanced Business Analytics, pp. 109–142. Springer, Cham (2014)
60. Wang, D.Z.W., Liu, H., Szeto, W.Y., Chow, A.H.F.: Identification of critical combination of vulnerable links in transportation networks: a global optimization approach. Transport. A: Transp. Sci. **12**(4), 346–365 (2016).
61. Erath, A.L.: Vulnerability assessment of road transport infrastructure. Ph.D. Dissertation, ETH Zürich, Switzerland (2011). http://e-collection.library.ethz.ch/view/eth:5164
62. Misener, R., Floudas, C.A.: Piecewise-linear approximations of multidimensional functions. J. Optim. Theory Appl. **145**(1), 120–147 (2010)
63. Vielma, J.P., Nemhauser, G.L.: Modeling disjunctive constraints with a logarithmic number of binary variables and constraints. Math. Program. **128**(1), 49–72 (2011)
64. Vielma, J.P., Ahmed, S., Nemhauser, G.: Mixed-integer models for nonseparable piecewise-linear optimization: unifying framework and extensions. Oper. Res. **58**(2), 303–315 (2010)
65. Wu, L.: A tighter piecewise linear approximation of quadratic cost curves for unit commitment problems. IEEE Trans. Power Syst. **26**(4), 2581–2583 (2011)
66. Chen, X., Wu, W., Zhang, B.: Robust restoration method for active distribution networks. IEEE Trans. Power Syst. **31**(5), 4005–4015 (2016)
67. Grunditz, E., Thiringer, T. Performance analysis of current BEVs based on a comprehensive review of specifications. IEEE Trans. Transp. Electrific. **2**(3), 270–289 (2016)
68. Ben-Tal, A., Nemirovski, A.: On polyhedral approximations of the second-order cone. Math. Oper. Res. **26**(2), 193–205 (2001)
69. Ben-Tal, A., Nemirovski, A.: Lectures on modern convex optimization. http://www2.isye.gatech.edu/nemirovs/LectModConvOpt.pdf
70. Li, F.: Continuous locational marginal pricing. IEEE Trans. Power Syst. **22**(4), 1638–1646 (2007)
71. Zhao, Z., Wu, L., Song, G.: Convergence of volatile power markets with price-based demand response. IEEE Trans. Power Syst. **29**(5), 2107–2118 (2014)
72. Border, K.C.: Fixed Point Theorems with Applications to Economics and Game Theory. Cambridge University Press, Cambridge (1989)

73. Granas, A., Dugundji, J.: Fixed Point Theory. Springer, New York (2003)
74. Lavaei, J., Tse, D., Zhang, B.: Geometry of power flows and optimization in distribution networks. IEEE Trans. Power Syst. **29**(2), 572–583 (2014)
75. https://docs.mosek.com/modeling-cookbook/index.html
76. Larsson, T., Patriksson, M.: An augmented Lagrangean dual algorithm for link capacity side constrained traffic assignment problems. Transp. Res. B Methodol. **29**(6), 433–455 (1995)
77. Larsson, T., Patriksson, M.: Side constrained traffic equilibrium models—analysis, computation and applications. Transp. Res. B Methodol. **33**(4), 233–264 (1999)
78. Nie, Y., Zhang, H.M., Lee, D.H.: Models and algorithms for the traffic assignment problem with link capacity constraints. Transp. Res. B Methodol. **38**(4), 285–312 (2004)
79. Jiang, N., Xie, C., Waller, T.: Path-constrained traffic assignment: model and algorithm. Transp. Res. Rec. J. Transp. Res. Board **2283**, 25–33 (2012)
80. Wang, T., Xie, C., Xie, J., Waller, T.: Path-constrained traffic assignment: a trip chain analysis under range anxiety. Transp. Res. C Emerg. Technol. **68**, 447–461 (2016)
81. Jiang, N., Xie, C.: Computing and analyzing mixed equilibrium network flows with gasoline and electric vehicles. Comput. Aided Civil Infrastruct. Eng. **29**(8), 626–641 (2014)
82. Jiang, N., Xie, C., Duthie, J.C., Waller, T.: A network equilibrium analysis on destination, route and parking choices with mixed gasoline and electric vehicular flows. Euro. J. Transp. Logist. **3**(1), 55–92 (2014)
83. He, F., Yin, Y., Lawphongpanich, S.: Network equilibrium models with battery electric vehicles. Transp. Res. B Methodol. **67**, 306–319 (2014)
84. He, F., Yin, Y.: Deploying public charging stations on urban road networks. Transp. Res. C Emerg. Technol. **60**, 227–240 (2015)
85. Chen, Z., He, F. and Yin, Y.: Optimal deployment of charging lanes in transportation networks. Transp. Res. B Methodol. **91**, 344–365 (2016)
86. Chen, Z., He, F., Zhang L., Yin Y.: Optimal deployment of autonomous vehicle lanes with endogenous market penetration. Transp. Res. C Emerg. Technol. **72**, 143–156 (2016)
87. Li, M., Lin, X., He, F., Jiang, H.: Optimal locations and travel time display for variable message signs. Transp. Res. C Emerg. Technol. **69**, 418–435 (2016)
88. He, F., Shen, Z.M.: Modeling taxi services with smartphone-based e-hailing applications. Transp. Res. C Emerg. Technol. **58**, 93–106 (2015)
89. Wang, X., He, F., Yang, H., Gao, H.O.: Pricing strategies for a taxi-hailing platform. Transport. Res. E **93**, 212–231 (2016)
90. He, F., Yin, Y., Nima, S., Nie, Y.: Tradable credit schemes on networks with mixed equilibrium behaviors. Transp. Res. B Methodol. **57**, 47–65 (2013)
91. He, F., Wang, X., Lin, X, et al.: Pricing and penalty/compensation strategies of a taxi-hailing platform. Transp. Res. C Emerg. Technol. **86**, 263–279 (2018)
92. He, F., Yin, Y., Wang, J., Yang, Y.: Optimal prices of electricity at public charging stations for plug-in electric vehicles. Netw. Spat. Econ. **16**(1), 131–154 (2016)
93. He, F., Yin, Y., Zhou, J.: Integrated pricing of roads and electricity enabled by wireless power transfer. Transp. Res. C Emerg. Technol. **34**, 1–15 (2013)
94. Zhu, M., Liu, X.Y., Wang, X.: Joint transportation and charging scheduling in public vehicle systems: a game theoretic approach. IEEE Trans. Intell. Transport. Syst. **19**(8): 2407–2419 (2018)
95. Zhang, H., Moura, S., Hu, Z., Song, Y.: PEV fast-charging station siting and sizing on coupled transportation and power networks. IEEE Trans. Smart Grid **9**(4), 2595–2605 (2018)
96. Xu, H., Pang, J.S., Ordóñez, F., Dessouky, M.: Complementarity models for traffic equilibrium with ridesharing. Transp. Res. B Methodol. **81**, 161–182 (2015)
97. Xu, H., Ordóñez, F., Dessouky, M.: A traffic assignment model for a ridesharing transportation market. J. Adv. Transp. **49**(7), 793–816 (2015)
98. Bahat, O., Bekhor, S.: Incorporating ridesharing in the static traffic assignment model. Netw. Spat. Econ. **16**(4), 1125–1149 (2016)
99. Di, X., Ma, R., Liu, H.X., Ban, X.: A link-node reformulation of ridesharing user equilibrium with network design. Transp. Res. B Methodol. **112**, 230–255 (2018)

100. Wang, Y., Szeto, W.Y., Han, K., Friesz, T.L.: Dynamic traffic assignment: a review of the methodological advances for environmentally sustainable road transportation applications. Transp. Res. B Methodol. **111**, 370–394 (2018)
101. Lighthill, M., Whitham, G.: On kinematic waves. II. A theory of traffic flow on long crowded roads. Proc. R. Soc. Lond. A Math. Phys. Sci. **229**(1178), 317–345 (1955)
102. Richards, P.I.: Shockwaves on the highway. Oper. Res. **4**(1), 42–51 (1956)
103. Daganzo, C.F.: The cell transmission model: a dynamic representation of highway traffic consistent with the hydrodynamic theory. Transp. Res. B Methodol. **28**(4), 269–287 (1994)
104. Daganzo, C.F.: The cell transmission model part II: network traffic. Transp. Res. B Methodol. **29**(2), 79–93 (1995)
105. Ukkusuri, S.V., Han, L, Doan, K.: Dynamic user equilibrium with a path based cell transmission model for general traffic networks. Transp. Res. B Methodol. **46**(10), 1657–1684 (2012)
106. Friesz T.L., Han, K., Neto, P.A., Meimand, A., Yao, T.: Dynamic user equilibrium based on a hydrodynamic model. Transp. Res. B Methodol. **47**(1), 102–126 (2013)
107. Han, L., Ukkusuri, S., Doan, K.: Complementarity formulations for the cell transmission model based dynamic user equilibrium with departure time choice, elastic demand and user heterogeneity. Transp. Res. B Methodol. **45**(10), 1749–1767 (2011)
108. Nahapetyan, A., Lawphongpanich, S.: Discrete-time dynamic traffic assignment models with periodic planning horizon: system optimum. J. Global Optim. **38**(1), 41–60 (2007)
109. Zheng, Q.P., Arulselvan, A.: Discrete time dynamic traffic assignment models and solution algorithm for managed lanes. J. Global Optim. **51**(1), 47–68 (2011)
110. Nakayama, S., Connors, R.: A quasi-dynamic assignment model that guarantees unique network equilibrium. Transport. A: Transp. Sci. **10**(7), 669–692 (2014)
111. Waller, S.T., Fajardo, D., Duell, M., Dixit, V.: Linear programming formulation for strategic dynamic traffic assignment. Netw. Spat. Econ. **13**(4), 427–443 (2013)
112. Friesz, T.L., Luque, J., Tobin, R.L., Wie, B.W.: Dynamic network traffic assignment considered as a continuous time optimal control problem. Oper. Res. **37**(6), 893–901 (1989)
113. Wie, B.W.: A differential game approach to the dynamic mixed behavior traffic network equilibrium problem. Eur. J. Oper. Res. **83**(1), 117–136 (1995)
114. Friesz, T.L. Dynamic Optimization and Differential Games. Springer, New York (2010)
115. Wagner, N. The dynamic user equilibrium on a transport network: mathematical properties and economic applications. Tech. Rep. pastel-00939008 (2014)
116. Friesz, T.L., Han, K.: The mathematical foundations of dynamic user equilibrium. Transp. Res. B Methodol. **126**, 309–328 (2019)
117. Liu, H.X., Ban, X.J., Pang, J.S.: Modeling and solving continuous-time instantaneous dynamic user equilibria: a differential complementarity systems approach. Transp. Res. B Methodol. **46**(3), 389–408 (2012)
118. Wang, D.Z.W., Du, B.: Continuum modelling of spatial and dynamic equilibrium in a travel corridor with heterogeneous commuters-a partial differential complementarity system approach. Transp. Res. B Methodol. **85**, 1–18 (2016)
119. Han, K., Friesz, T.L., Szeto, W.Y., Liu, H.: Elastic demand dynamic network user equilibrium: Formulation, existence and computation. Transp. Res. B Methodol. **81**, 183–209 (2015)
120. Parvania, M., Scaglione, A.: Unit commitment with continuous-time generation and ramping trajectory models. IEEE Trans. Power Syst. **31**(4), 3169–3178 (2016)
121. Parvania, M., Khatami, R.: Continuous-time marginal pricing of electricity. IEEE Trans. Power Syst. **32**(3), 1960–1969 (2017)
122. Wei, W., Wu, D., Wu, Q., Shafie-khah, M., Catalão, J.: Transportation system and power system interdependence: a comprehensive review on models and applications. J. Mod. Power Syst. Clean Energy **7**(3), 433–448 (2019)

Appendix

Tutorials on Advanced Optimization Methods

W. Wei, J. Wang, *Modeling and Optimization of Interdependent Energy Infrastructures*, https://doi.org/10.1007/978-3-030-25958-7

Appendix A
Basics of Linear and Conic Programs

The great watershed in optimization isn't between linearity and nonlinearity, but convexity and non-convexity.

−Ralph Tyrrell Rockafellar

The mathematical programming theory has been thoroughly developed in width and depth since its birth in the 1940s, when George Dantzig invented simplex algorithm for linear programming. The most influential findings in the field of optimization theory can be summarized as [1]:

1. Recognition of the fact that under mild conditions, a convex optimization program is computationally tractable: the computational effort under a given accuracy grows moderately with the problem size even in the worst case. In contrast, a non-convex program is generally computationally intractable: the computational effort of the best known methods grows prohibitively fast with respect to the problem size, and it is reasonable to believe that this is an intrinsic feature of such problems rather than a limitation of existing optimization techniques.
2. The discovery of interior-point methods, which was originally developed in the 1980s to solve LPs and could be generalized to solve convex optimization problems as well. Moreover, between these two extremes (LPs and general convex programs), there are many important and useful convex programs. Although nonlinear, they still possess nice structured properties, which can be utilized to develop more dedicated algorithms. These polynomial-time interior-point algorithms turn out to be considerably more efficient than those exploiting only the convex property.

The superiority of formulating a problem as a convex optimization problem is apparent. The most appealing advantage is that the problem can be solved reliably and efficiently. It is also convenient to build the associated dual problem, which gives insights on sensitivity information and may help develop distributed algorithm for solving the problem. Convex optimization has been applied in a number of energy system operational issues, and well acknowledged for its computational

W. Wei, J. Wang, *Modeling and Optimization of Interdependent Energy Infrastructures*, https://doi.org/10.1007/978-3-030-25958-7

superiority. We believe that it is imperative for researchers and engineers to develop certain understanding on this important topic.

As we have already learnt in the previous chapters, many optimization problems in energy system engineering can be formulated as or converted to convex programs. The goal of this chapter is to help readers develop necessary background knowledge and skills to apply several well-structured convex optimization models, including LPs, SOCPs, and SDPs, i.e., to formulate or transform their problems as these specific convex programs. Certainly, convex transformation (or convexification) may be rather tricky and require special knowledge and skills. Nevertheless, the attempt often turns out to be worthwhile. We also pay special attention to nonconvex QCQPs, which can model various decision-making problems in engineering, such as optimal power flow in Chap. 2 and optimal gas flow in Chap. 3. We discuss convex relaxation technique based on SDP,which is shown to be very useful to get a high-quality objective lower bound. We also present MILP formulations for some special QPs; because of the special problem structure, these MILP models can tackle practically sized problems in reasonable time.

Most materials regarding convex sets and functions come from [2] and its solution manual [3]; extensions of duality theory from linear programming to conic programming follow from [1]. We consolidate necessary contents in a convenient way to make this book self-contained and easy to understand.

A.1 Basic Notations

A.1.1 Convex Sets

A set $C \in \mathbb{R}^n$ is convex if the line segment connecting any two points in C is contained in C, i.e., for any $x_1, x_2 \in C$, we have $\theta x_1 + (1-\theta)x_2 \in C, \forall \theta \in [0,1]$. Roughly speaking, standing at anywhere in a convex set, you can see every other point in the set. Figure A.1 illustrates a simple convex set and a non-convex set in $\mathbb{R}^2$.

The convex combination of k points $x_1, \cdots, x_k$ is defined as $\theta_1 x_1 + \cdots + \theta_k x_k$, where $\theta_1, \cdots, \theta_k \geq 0$, and $\theta_1 + \cdots + \theta_k = 1$. A convex combination of points can be regarded as a weighted average of the points, with θ_i the weight of x_i in the mixture.

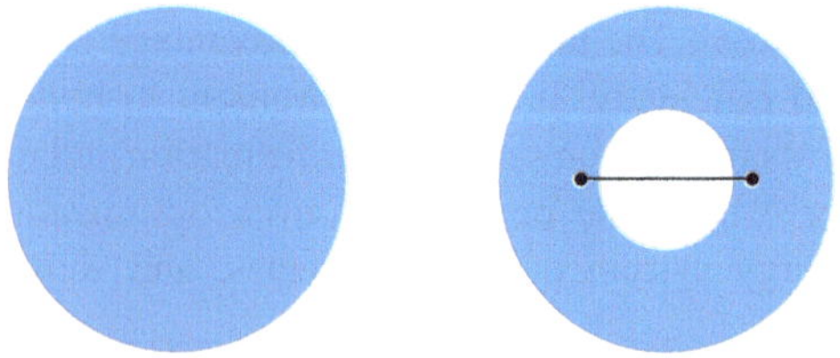

Fig. A.1 Left: the circle is convex; right: the ring is non-convex

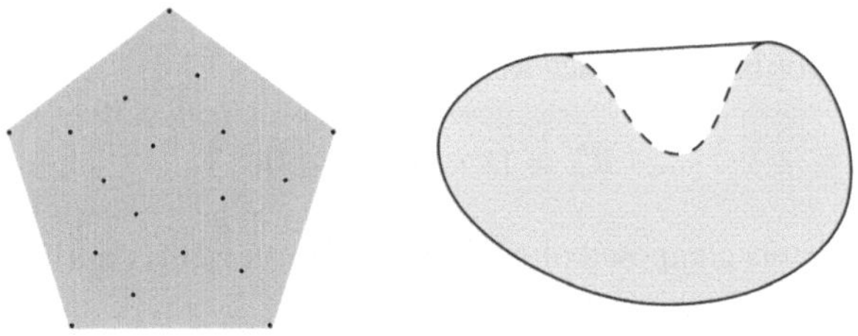

Fig. A.2 Left: The convex hull of eighteen points. Right: The convex hull of a kidney shaped set

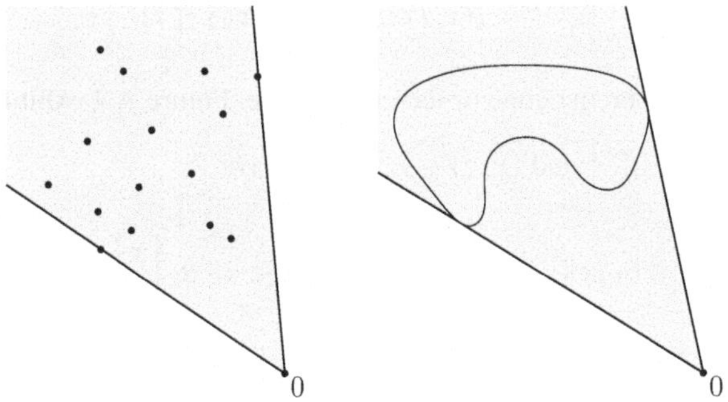

Fig. A.3 The conic hulls of the two sets [2]

The convex hull of set C, denoted conv(C), is the smallest convex set that contains C. Particularly, if C has finite elements, then

$$\text{conv}(C) = \{\theta_1 x_1 + \cdots + \theta_k x_k \mid x_i \in C,\ \theta_i \geq 0,\ i = 1,\ \cdots, k,\ \theta_1 + \cdots + \theta_k = 1\}$$

Figure A.2 illustrates the convex hulls of two sets in $\mathbb{R}^2$.

Some useful convex sets are briefly introduced.

1. Cones

A set C is called a cone, or nonnegative homogeneous, if for any $x \in C$, we have $\theta x \in C$, $\forall \theta \geq 0$. A set C is a convex cone if it is convex and a cone: for any $x_1, x_2 \in C$ and $\theta_1, \theta_2 \geq 0$, we have $\theta_1 x_2 + \theta_2 x_2 \in C$.

The conic combination (or nonnegative linear combination) of k points $x_1, \cdots, x_k$ is defined as $\theta_1 x_1 + \cdots + \theta_k x_k$, where $\theta_1, \cdots, \theta_k \geq 0$. If a set of finite points $\{x_i\}$, $i = 1, 2 \cdots$ resides in a convex cone C, then every conic combination of $\{x_i\}$ remains in C. Conversely, a set C is a convex cone if and only if it contains all conic combinations of its elements. The conic hull of set C is the smallest convex cone that contains C. Figure A.3 illustrates the conic hulls of two sets in $\mathbb{R}^2$.

Some widely used cones are introduced.

a. The Nonnegative Orthan
The nonnegative orthant is defined as

$$\mathbb{R}^n_+ = \{x \in \mathbb{R}^n \mid x \geq 0\} \tag{A.1}$$

It is the set of vectors composed of non-negative entries. It is clearly a convex cone.

b. Second-Order Cone
The unit second-order cone is defined as

$$\mathbb{L}^{n+1}_C = \{(x,t) \in \mathbb{R}^{n+1} \mid \|x\|_2 \leq t\} \tag{A.2}$$

It is also called the Lorentz cone or ice-cream cone. Figure A.4 exhibits $\mathbb{L}^3_C$.

For any $(x,t) \in \mathbb{L}^{n+1}_C$ and $(y,z) \in \mathbb{L}^{n+1}_C$, we have

$$\|\theta_1 x + \theta_2 y\|_2 \leq \theta_1 \|x\|_2 + \theta_2 \|y\|_2 \leq \theta_1 t + \theta_2 z \Rightarrow \theta_1 \begin{bmatrix} x \\ t \end{bmatrix} + \theta_2 \begin{bmatrix} y \\ z \end{bmatrix} \in \mathbb{L}^{n+1}_C$$

which means that the unit second-order cone is a convex cone.

Sometimes, it is convenient to use the following inequality to represent a second-order cone in optimization problems

$$\|Ax + b\|_2 \leq c^T x + d \tag{A.3}$$

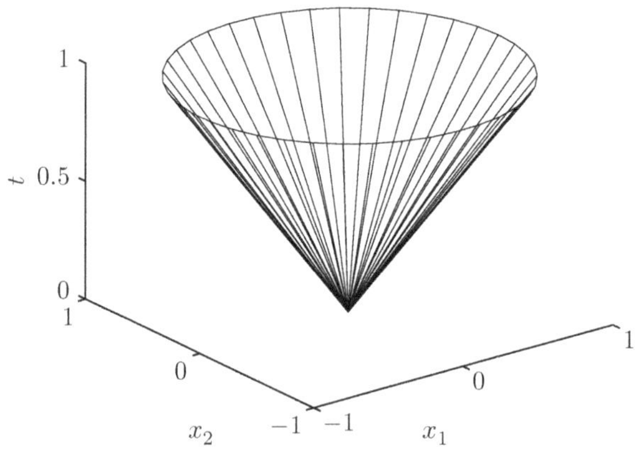

Fig. A.4 $\mathbb{L}^3_C = \left\{(x_1, x_2, t) \,\middle|\, \sqrt{x_1^2 + x_2^2} \leq t\right\}$ in $\mathbb{R}^3$ [2]

where $A \in \mathbb{R}^{m \times n}$, $b \in \mathbb{R}^m$, $c \in \mathbb{R}^n$, $d \in \mathbb{R}$. It is the inverse image of the unit second-order cone under the affine mapping $f(x) = (Ax + b, c^T x + d)$, and hence is convex. Second-order cones in forms of (A.2) and (A.3) are interchangeable.

$$\|Ax + b\|_2 \le c^T x + d \Leftrightarrow \begin{bmatrix} A \\ c^T \end{bmatrix} x + \begin{bmatrix} b \\ d \end{bmatrix} \in \mathbb{L}_C^{m+1}$$

and hence is convex.

c. Positive Semidefinite Cone

The set of symmetric $m \times m$ matrices is denoted by

$$\mathbb{S}^m = \{X \in \mathbb{R}^{m \times m} \mid X = X^T\}$$

which is a vector space with dimension $m(m+1)/2$.

The set of symmetric positive semidefinite matrices is denoted by

$$\mathbb{S}_+^m = \{X \in \mathbb{S}^m \mid X \succeq 0\}$$

The set of symmetric positive definite matrices is denoted by

$$\mathbb{S}_{++}^m = \{X \in \mathbb{S}^m \mid X \succ 0\}$$

Clearly, $\mathbb{S}_+^m$ is a convex cone: if $A, B \in \mathbb{S}_+^m$, then for any $x \in \mathbb{R}^m$ and positive scalars $\theta_1, \theta_2 \ge 0$, we have

$$x^T(\theta_1 A + \theta_2 B)x = \theta_1 x^T A x + \theta_2 x^T B x \ge 0$$

implying $\theta_1 A + \theta_2 B \in \mathbb{S}_+^m$.

A positive semidefinite cone in $\mathbb{R}^2$ can be expressed via three variables x, y, z as

$$\begin{bmatrix} x & y \\ y & z \end{bmatrix} \succeq 0 \Leftrightarrow x \ge 0,\ xz \ge y^2$$

which is plotted in Fig. A.5. In fact, $\mathbb{L}_C^3$ and $\mathbb{S}_+^2$ are equivalent to each other. To see this, the hyperbolic inequality $xz \ge y^2$ with $x \ge 0, z \ge 0$ defines the same feasible region in $\mathbb{R}^3$ as the following second-order cone

$$\left\| \begin{matrix} 2y \\ x - z \end{matrix} \right\|_2 \le x + z,\ x \ge 0,\ z \ge 0$$

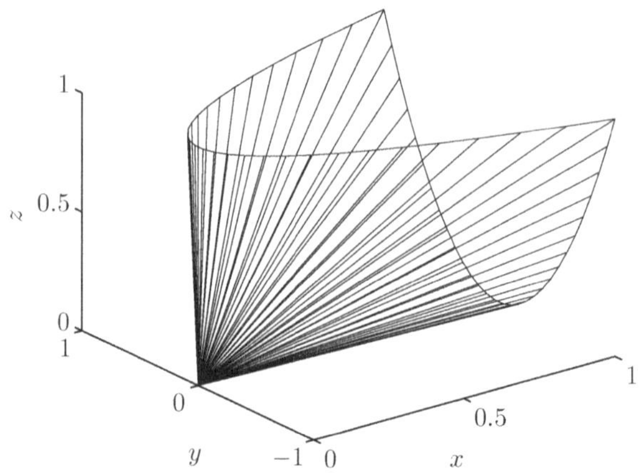

Fig. A.5 Positive semidefinite cone in $\mathbb{S}^2$ (or in $\mathbb{R}^3$) [2]

In higher-order dimensions, every second-order cone can be written as an LMI via Schur complement as

$$\|Ax+b\|_2 \le c^T x + d \Rightarrow \begin{bmatrix} \left(c^T x + d\right) I & Ax+b \\ (Ax+b)^T & c^T x + d \end{bmatrix} \succeq 0 \tag{A.4}$$

In this sense of representability, positive semidefinite cones are more general than second-order cones. However, the transformation in (A.4) may not be superior from the computational perspective, because SOCPs are more tractable than SDPs.

d. Copositive Cone

A copositive cone $\mathbb{C}_+^n$ consists of symmetric matrices whose quadratic form is nonnegative over the nonnegative orthant $\mathbb{R}_+^n$:

$$\mathbb{C}_+^n = \{A \mid A \in \mathbb{S}^n,\ x^T A x \ge 0,\ \forall x \in \mathbb{R}_+^n\} \tag{A.5}$$

The copositive cone $\mathbb{C}_+^n$ is closed, pointed, and convex [4]. Clearly, $\mathbb{S}_+^n \subseteq \mathbb{C}_+^n$, and every entry-wise nonnegative symmetric matrix A belongs to $\mathbb{C}_+^n$. Actually, $\mathbb{C}_+^n$ is significantly larger than the positive semidefinite cone and the nonnegative symmetric matrix cone.

2. Polyhedra

A polyhedron is defined as the solution set of a finite number of linear inequalities:

$$P = \{x \mid Ax \le b\} \tag{A.6}$$

Equation (A.6) is also called a hyperplane representation for a polyhedron. It is easy to show that polyhedra are convex sets. Sometimes, a polyhedron is also called a polytope. The two concepts are often used interchangeably in this book. Because of physical bounds of decision variables, the polyhedral feasible regions in practical energy system optimization problems are usually bounded, which means that there is no extreme ray.

Polyhedra can be expressed via the convex combination as well. The convex hull of a finite number of points

$$\text{conv}\{v_1, \cdots, v_k\} = \{\theta_1 v_1 + \cdots + \theta_k v_k \mid \theta \geq 0,\ 1^T \theta = 1\} \tag{A.7}$$

defines a polyhedron. Equation (A.7) is called a convex hull representation. If the polyhedron is unbounded, a generalization of this convex hull representation is

$$\{\theta_1 v_1 + \cdots + \theta_k v_k \mid \theta \geq 0,\ \theta_1 + \cdots + \theta_m = 1, m \leq k\} \tag{A.8}$$

which considers nonnegative linear combinations of v_i, but only the first m coefficients whose summation is 1 are bounded, and the remaining ones can take arbitrarily large values. In view of this, the convex hull of points $v_1, \cdots, v_m$ plus the conic hull of points $v_{m+1}, \cdots, v_k$ is a polyhedron. The reverse is also correct: any polyhedron can be represented by convex hull and conic hull.

How to represent a polyhedron depends on what information is available: if its boundaries are expressed via linear inequalities, the hyperplane representation is straightforward; if its extreme points and extreme rays are known in advance, the convex-conic hull representation is more convenient. With the growth in dimension, it is becoming more difficult to switch (derive one from the other) between the hyperplane representation and the hull representation.

A.1.2 Generalized Inequalities

A cone $K \subseteq \mathbb{R}^n$ is called a proper cone if it satisfies:

1. K is convex and closed.
2. K is solid, i.e., it has non-empty interior.
3. K is pointed, i.e., $x \in K, -x \in K \Rightarrow x = 0$.

A proper cone K can be used to define a generalized inequality, a partial ordering on $\mathbb{R}^n$, as follows

$$x \preceq_K y \Longleftrightarrow y - x \in K \tag{A.9}$$

We denote $x \succeq_K y$ for $y \preceq_K x$. Similarly, a strict partial ordering can be defined by

$$x \prec_K y \Longleftrightarrow y - x \in \text{int}(K) \tag{A.10}$$

where $\text{int}(K)$ stands for the interior of K, and write $x \succ_K y$ for $y \prec_K x$.

The nonnegative orthant $\mathbb{R}^n_+$ is a proper cone. When $K = \mathbb{R}^n_+$, the partial ordering $\preceq_K$ comes down to the element-wise comparison between vectors: for $x, y \in \mathbb{R}^n$, $x \preceq_{\mathbb{R}^n_+} y$ means $x_i \le y_i$, $i = 1, \cdots n$, or the traditional notation $x \le y$.

The positive semidefinite cone $\mathbb{S}^n_+$ is a proper cone in $\mathbb{S}^n$. When $K = \mathbb{S}^n_+$, the partial ordering $\preceq_K$ comes down to a linear matrix inequality between symmetric matrices: for $X, Y \in \mathbb{S}^n$, $X \preceq_{\mathbb{S}^n_+} Y$ means $Y - X$ is positive semidefinite. Because it arises so frequently, we can drop the subscript $\mathbb{S}^n_+$ when we write a linear matrix inequality $Y \succeq X$ or $X \preceq Y$. It is understood that such a generalized inequality corresponds to the positive semidefinite cone without particular mention.

A generalized inequality is equivalent to linear constraints with $K = \mathbb{R}^n_+$; for other cones, such as the second-order cone $\mathbb{L}^{n+1}_C$ or the positive semidefinite cone $\mathbb{S}^n_+$, the feasible region is nonlinear but remains convex.

A.1.3 Dual Cones and Dual Generalized Inequalities

Let K be a cone in $\mathbb{R}^n$. Its dual is defined as the following set

$$K^* = \{y \mid x^T y \ge 0, \ \forall x \in K\} \tag{A.11}$$

Because K^* is the intersection of homogeneous half spaces (half spaces passing through the origin). It is a closed convex cone.

The interior of K^* is given by

$$\text{int}(K^*) = \{y \mid x^T y > 0, \ \forall x \in K, \ x \ne 0\} \tag{A.12}$$

To see this, if $y^T x > 0$, $\forall x \in K$, then $(y+u)^T x > 0$, $\forall x \in K$ holds for all u that is sufficiently small; hence, $y \in \text{int}(K^*)$. Conversely, if $y \in K^*$ and $\exists x \in K : y^T x = 0$, $x \ne 0$, then $(y - tx)^T x < 0$, $\forall t > 0$, indicating $y \notin \text{int}(K^*)$.

If $\text{int}(K) \ne \emptyset$, then K^* is pointed. If this is not true, suppose $\exists y \ne 0$: $y \in K^*$, $-y \in K^*$, i.e., $y^T x \ge 0$, $\forall x \in K$ and $-y^T x \ge 0$, $\forall x \in K$, so we have $x^T y = 0$, $\forall x \in K$, which is in contradiction with $\text{int}(K) \ne \emptyset$.

In conclusion, K^* is a proper cone, if the original cone K is so; K^* is closed and convex, regardless of the original cone K. Figure A.6 shows a cone K (the region between L_2 and L_3) and its dual cone K^* (the region between L_1 and L_4) in $\mathbb{R}^2$.

In light of the definition of K^*, a non-zero vector y is the normal of a homogeneous half space which contains K if and only if $y \in K^*$. The intersection of all such half spaces containing K constitutes the cone K (if K is closed), in view of this

$$K = \bigcap_{y \in K^*} \left\{x \mid y^T x \ge 0\right\} = \{x \mid y^T x \ge 0, \ \forall y \in K^*\} = K^{**} \tag{A.13}$$

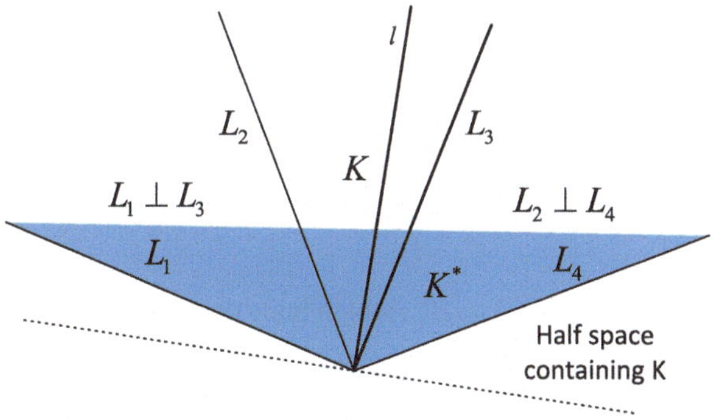

Fig. A.6 Illustration of a cone and its dual cone in $\mathbb{R}^2$

This fact can be also understood in $\mathbb{R}^2$ from Fig. A.6. The extreme cases for the normal vector y such that the corresponding half space contains K are L_1 and L_4, and the intersection of these half spaces for all $y \in K^*$ turns out to be the original cone K.

Next, we investigate the dual cones of three special proper cones, i.e., $\mathbb{R}^n_+$, $\mathbb{L}^{n+1}_C$, and $\mathbb{S}^n_+$, respectively.

1. The Nonnegative Orthant
By observing the fact

$$x^T y \geq 0, \ \forall x \geq 0 \Longleftrightarrow y \geq 0$$

we naturally have $(\mathbb{R}^n_+)^* = \mathbb{R}^n_+$; in other words, the nonnegative orthant is self-dual.

2. The Second-Order Cone
Now, we show that the second-order cone is also self-dual: $(\mathbb{L}^{n+1}_C)^* = \mathbb{L}^{n+1}_C$. To this end, we need to demonstrate

$$x^T u + tv \geq 0, \ \forall (x, t) \in L^{n+1}_C \Longleftrightarrow \|u\|_2 \leq v$$

$\Rightarrow$: Suppose the right-hand condition is false, and $\exists (u, v) : \|u\|_2 > v$, by recalling Cauchy-Schwarz inequality $|a^T b| \leq \|a\|_2 \|b\|_2$, we have

$$\min_x \left\{ x^T u \;\middle|\; \text{s.t. } \|x\|_2 \leq t \right\} = -t\|u\|_2$$

In such circumstance, $x^T u + tv = t(v - \|u\|_2) < 0, \forall t > 0$, which is in contradiction with the left-hand condition.

$\Leftarrow$: Again, according to Cauchy-Schwarz inequality, we have

$$x^T u + tv \geq -\|x\|_2 \|u\|_2 + tv \geq -\|x\|_2 \|u\|_2 + \|x\|_2 v = \|x\|_2 (v - \|u\|_2) \geq 0$$

3. The Positive Semidefinite Cone
We investigate the dual cone of $\mathbb{S}_+^n$. The inner product of $X, Y \in \mathbb{S}^n$ is defined by the element-wise summation

$$\langle X, Y\rangle = \sum_{i=1}^{n}\sum_{j=1}^{n} X_{ij}Y_{ij} = \operatorname{tr}\left(XY^T\right)$$

We establish this fact: $(\mathbb{S}_+^n)^* = \mathbb{S}_+^n$, which boils down to

$$\operatorname{tr}\left(XY^T\right) \geq 0,\ \forall X \succeq 0 \Longleftrightarrow Y \succeq 0$$

$\Rightarrow$: Suppose $Y \notin \mathbb{S}_+^n$, then $\exists q \in \mathbb{R}^n$ such that

$$q^T Y q = \operatorname{tr}\left(qq^T Y^T\right) < 0$$

which is in contradiction with the left-hand condition because $X = qq^T \in \mathbb{S}_+^n$.

$\Leftarrow$: Now suppose $X, Y \in \mathbb{S}_+^n$. X can be expressed via its eigenvalues $\lambda_i \geq 0$ and eigenvectors q_i as $X = \sum_{i=1}^n \lambda_i q_i q_i^T$, then we arrive at

$$\operatorname{tr}(XY^T) = \operatorname{tr}\left(Y\sum_{i=1}^{n}\lambda_i q_i q_i^T\right) = \sum_{i=1}^{n}\lambda_i q_i^T Y q_i \geq 0$$

In summary, it follows that the positive semidefinite cone is self-dual.

4. The Completely Positive Cone
Following the same concept of matrix inner product, it is shown that $(\mathbb{C}_+^n)^*$ is the cone of the so-called completely positive matrices and can be expressed as [5]

$$(\mathbb{C}_+^n)^* = \operatorname{conv}\left\{xx^T \mid x \in \mathbb{R}_+^n\right\} \tag{A.14}$$

In contrast to previous three cones, the copositive cone $\mathbb{C}_+^n$ is not self-dual.

When the dual cone K^* is proper, it induces a generalized inequality $\preceq_{K^*}$, which is called the dual generalized inequality of the one induced by cone K (if K is proper). According to the definition of dual cone, an important fact relating a generalized inequality and its dual is

1. $x \preceq_K y$ if and only if $\lambda^T x \leq \lambda^T y$, $\forall \lambda \in K^*$.
2. $x \prec_K y$ if and only if $\lambda^T x < \lambda^T y$, $\forall \lambda \in K^*, \lambda \neq 0$.

When $K = K^{**}$, the dual generalized inequality of $\preceq_{K^*}$ is $\preceq_K$, and the above property holds if the positions of K and K^* are swapped.

A.1.4 Convex Function and Epigraph

A function $f : \mathbb{R}^n \to \mathbb{R}$ is convex if its feasible region X is a convex set, and for all $x_1, x_2 \in X$, the following condition holds:

$$f(\theta x_1 + (1-\theta)x_2) \leq \theta f(x_1) + (1-\theta) f(x_2), \ \forall \theta \in [0, 1] \tag{A.15}$$

The geometrical interpretation of inequality (A.15) is that the chord connecting points $(x_1, f(x_1))$ and $(x_2, f(x_2))$ always lies above the curve of f between x_1 and x_2 (see Fig. A.7). Function f is strictly convex if strict inequality holds in (A.15) when $x_1 \neq x_2$ and $0 < \theta < 1$. Function f is called (strictly) concave if $-f$ is (strictly) convex. An affine function is both convex and concave.

The graph of a function $f : \mathbb{R}^n \to \mathbb{R}$ is defined as

$$\text{graph } f = \{(x, f(x)) \mid x \in X\} \tag{A.16}$$

which is a subset of $\mathbb{R}^{n+1}$.

The epigraph of a function $f : \mathbb{R}^n \to \mathbb{R}$ is defined as

$$\text{epi } f = \{(x, t) \mid x \in X, \ f(x) \leq t\} \tag{A.17}$$

which is a subset of $\mathbb{R}^{n+1}$. These definitions are illustrated through Fig. A.7.

Epigraph bridges the concepts of convex sets and convex functions: A function is convex if and only if its epigraph is a convex set. Epigraph is frequently used in formulating optimization problems. A nonlinear objective function can be replaced by a linear objective and an additional constraint in epigraph form. In this sense, we can assume that any optimization problem has a linear objective function. Nonetheless, this does not facilitate solving the problem, as non-convexity moves to the constraints, if the objective function is not convex. Nonetheless, the solution to an optimization problem with a linear objective can always be found at the boundary of the convex hull of its feasible region, implying that if we can characterize the

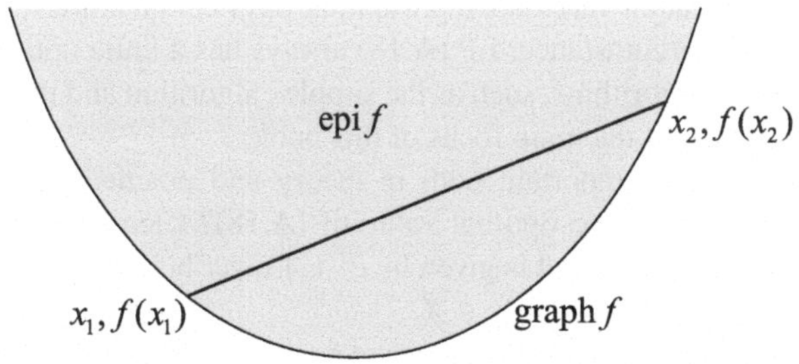

Fig. A.7 Illustration of the graph of a convex function $f(x)$ (the solid line) and its epigraph (the shaded area) in $\mathbb{R}^2$

convex hull, a problem in epigraph form admits an exact convex hull relaxation. However, in general, it is difficult to express convex hull in an analytical form.

Analyzing convex functions is a well-developed field. Broadening the knowledge in convex analysis could be mathematically demanding, especially for readers who are primarily interested in applications. We will not pursue in sophisticated theories in depth any more. Readers are referred to the literature suggested at the end of this chapter for further information.

A.2 From Linear to Conic Program

Linear programming is one of the most mature and tractable mathematical programming problems. In this section, we first investigate and explain the motivation of linear programming duality theory, then provide a unified model for conic programming problems. LPs, SOCPs, and SDPs are special cases of conic programs associated with generalized inequalities $\preceq_K$ where $K = \mathbb{R}^n$, $\mathbb{L}_C^{n+1}$, and $\mathbb{S}_+^n$, respectively. Our aim is to help readers who are not familiar with conic programs build their decision-making problems in these formats with structured convexity, and write out their dual problems more conveniently. The presentation logic is consistent with [1], and most of the presented materials in this section also come from [1].

A.2.1 Linear Program and its Duality Theory

A linear program is an optimization program with the form

$$\min\{c^T x \mid Ax \geq b\} \tag{A.18}$$

where x is the vector of decision variables, A, b, c are constant coefficient matrices with compatible dimensions. We assume LP (A.18) is feasible, i.e., its feasible set $X = \{x \mid Ax \geq b\}$ is a non-empty polyhedron; moreover, because of the limited ranges of decision variables representing physical quantities, we assume X is bounded. In such circumstance, LP (A.18) always has a finite optimum. LPs can be solved by mature algorithms, such as the simplex algorithm and the interior-point algorithm, which are not the main focus of this book.

A question which is important both in theory and practice is: how to find a systematic way to bound the optimal value of (A.18)? Clearly, if x is a feasible solution, an instant upper bound is given by $c^T x$. Lower bounding is to find a value a, such that $c^T x \geq a$ holds for all $x \in X$.

A trivial answer is to solve the problem and retrieve its optimal value, which is the tightest lower bound. However, there may be a smarter way to retrieve a valid lower bound with much cheaper computational expense. To outline the basic motivation, let us consider the following example

$$\min \left\{ \sum_{i=1}^{6} x_i \;\middle|\; \begin{array}{l} 2x_1 + 1x_2 + 3x_3 + 8x_4 + 5x_5 + 3x_6 \geq 5 \\ 6x_1 + 2x_2 + 6x_3 + 1x_4 + 1x_5 + 4x_6 \geq 2 \\ 2x_1 + 7x_2 + 1x_3 + 1x_4 + 4x_5 + 3x_6 \geq 1 \end{array} \right\} \tag{A.19}$$

Although LP (A.19) is merely a toy case for modern solvers and computers, one may guess it is still a little bit complicated for mental arithmetic. In fact, we can claim the optimal value is 0.8 at a glance without any sophisticated calculation: summing up the three constraints yields an inequality

$$10(x_1 + x_2 + x_3 + x_4 + x_5 + x_6) \geq 8 \tag{A.20}$$

which immediately gives the optimal value is 0.8. To understand why such a value is indeed the optimum, by adding the constraints together and dividing both sides by 10, inequality (A.20) implies that the objective function must get a value which is greater than or equal to 0.8 at any feasible point; moreover, to demonstrate that 0.8 is attainable, we can find a point x^* which activates the three constraints simultaneously, so (A.20) becomes an equality. LP duality is merely a formal generalization of this simple trick.

Multiplying each constraint in $Ax \geq b$ with a non-negative weight λ_i, and adding all constraints together, we will see

$$\lambda^T Ax \geq \lambda^T b$$

If we choose λ elaborately such that $\lambda^T A = c^T$, then $\lambda^T b$ will be a valid lower bound of the optimal value of (A.18). To improve the lower bound estimation, one may optimize the weighting vector λ, giving rise to the following problem

$$\max_{\lambda} \{\lambda^T b \mid A^T \lambda = c, \ \lambda \geq 0\} \tag{A.21}$$

where λ is the vector of decision variables or dual variables, and the feasible region $D = \{\lambda \mid A^T \lambda = c, \ \lambda \geq 0\}$ is a polyhedron. Clearly, (A.21) is also an LP, and is called the dual problem of LP (A.18). Correspondingly, (A.18) is called the primal problem. From the above construction, we immediately conclude $c^T x \geq \lambda^T b$.

Proposition A.1 (Weak Duality) *The optimal value of (A.21) is less than or equal to the optimal value of (A.18).*

In fact, the optimal bound offered by (A.21) is tight.

Proposition A.2 (Strong Duality) *Optimal values of (A.21) and (A.18) are equal.*

To see this, an explanation is given in [1]. If a real number a is the optimal value of the primal LP (A.18), the system of linear inequalities

$$S^P : \begin{cases} -c^T x > -a : \lambda_0 \\ Ax \geq b : \lambda \end{cases}$$

must have an empty solution set, indicating that at least one of the following two systems does have a solution (called separation property later)

$$S_1^D : \begin{cases} -\lambda_0 c + A^T \lambda = 0 \\ -\lambda_0 a + b^T \lambda \geq 0 \\ \lambda_0 > 0, \ \ \lambda \geq 0 \end{cases}$$

$$S_2^D : \begin{cases} -\lambda_0 c + A^T \lambda = 0 \\ -\lambda_0 a + b^T \lambda > 0 \\ \lambda_0 \geq 0, \ \ \lambda \geq 0 \end{cases}$$

We can show that S^P has no solutions if and only if S_1^D has a solution.

S_1^D has a solution $\Rightarrow$ S^P has no solution is clear. Otherwise, suppose that S^P has a solution x, because λ_0 is strictly positive, the weighted summation of inequalities in S^P leads to

$$0 = 0^T x = \left(-\lambda_0 c + A^T \lambda\right)^T x = -\lambda_0 c^T x + \lambda^T A x > -\lambda_0 a + \lambda^T b$$

which is in contradiction with the second inequality in S_1^D.

S^P has no solution $\Rightarrow$ S_1^D has a solution. Suppose S_1^D has no solution, S_2^D must have a solution owing to the separation property (Theorem 1.2.1 in [1]). Moreover, if $\lambda_0 > 0$, the solution of system S_2^D also solves system S_1^D, so there must be $\lambda_0 = 0$. As a result, the solution of S_2^D is independent of the values of a and c. Let $c = 0$ and $a = 0$, the solution λ of S_2^D satisfies $A^T \lambda = 0$, $b^T \lambda > 0$. Therefore, for any x with a compatible dimension, $\lambda^T (Ax - b) = \lambda^T A x - \lambda^T b < 0$ holds. In addition, because $\lambda \geq 0$, we can conclude that $Ax \geq b$ has no solution, a contradiction to the assumption that (A.18) is feasible.

Now, consider the solution of S_1^D. Without loss of generality, we can assume $\lambda_0 = 1$; otherwise, if $\lambda_0 \neq 1$, $(1, \lambda/\lambda_0)$ also solves S_1^D. In view of this, in normalized condition ($\lambda_0 = 1$), S_1^D comes down to

$$S_3^D : \begin{cases} A^T \lambda = c \\ b^T \lambda \geq a \\ \lambda \geq 0 \end{cases}$$

Now we can see the strong duality: Let a^* be the optimal solution of (A.18). For any $a < a^*$, S^P has no solution, so S_1^D has a solution $(1, \lambda^*)$. According to S_3^D, the optimal value of (A.21) is no smaller than a, i.e., $a \le b^T\lambda^* \le a^*$. When a tends to a^*, we can conclude that the primal and dual optimal values are equal. Since the primal problem always has a finite optimum (as we assumed before), so does the dual problem, as they share the same optimal value. Nevertheless, even if the primal feasible region is bounded, the dual feasible set D may be unbounded, and the dual problem is always bounded above. Please refer to [6–8] for more information on duality theory in linear programming.

Proposition A.3 (Primal-Dual Optimality Condition) *If LP (A.18) is feasible and X is bounded, then any feasible solution to the following system*

$$\begin{aligned} & Ax \ge b \\ & A^T\lambda = c, \ \lambda \ge 0 \\ & c^T x = b^T \lambda \end{aligned} \tag{A.22}$$

solves the original primal-dual pair of LPs: x^* is the optimal solution of (A.18), and λ^* is the optimal solution of (A.21).

Equation (A.22) is also called the primal-dual optimality condition of LPs. It consists of linear inequalities and equalities, and there is no objective function to be optimized.

Substituting $c = A^T\lambda$ into the last equation of (A.22) gives $\lambda^T Ax = \lambda^T b$, i.e.

$$\lambda^T(b - Ax) = 0$$

Since $\lambda \ge 0$ and $Ax \ge b$, above equation is equivalent to

$$\lambda_i(b - Ax)_i = 0$$

where notation $(b - Ax)_i$ and λ_i stand for the i-th components of vectors $b - Ax$ and λ, respectively. This condition means that at most one of λ_i and $(b - Ax)_i$ can take a strictly positive value. In other words, if the i-th inequality constraint is inactive, then its dual multiplier λ_i must be 0; otherwise, if $\lambda_i > 0$, then the corresponding inequality constraint must be binding. This phenomenon is called the complementarity and slackness condition.

Applying KKT optimality condition for general nonlinear programs to LP (A.18) we have:

Proposition A.4 (KKT Optimality Condition) *If LP (A.18) is feasible and X is bounded, the following system*

$$\begin{aligned} & 0 \le \lambda \perp Ax - b \ge 0 \\ & A^T\lambda = c \end{aligned} \tag{A.23}$$

has a solution (x^*, λ^*) *(may not be unique), where* $a \perp b$ *means* $a^T b = 0$, x^* *solves (A.18) and* λ^* *solves (A.21).*

The question that which one of (A.22) and (A.23) is better can be subtle and has very different practical consequences. At the first look, the former one seems more tractable because (A.22) is a linear system while (A.23) contains complementarity and slackness conditions. However, the actual situation in practice is more complicated. For example, to solve a bilevel program with an LP lower level, the LP is often replaced by its optimality condition. In a bilevel optimization structure, some of the coefficients A, b, and c are optimized by the upper-level agent, say, the coefficient vector c representing the price is controlled by the upper level decision maker, while A and b are constants. If we use (A.22), the term $c^T x$ in the single-level equivalence becomes non-convex, although c is a constant in the lower level, preventing a global optimal solution from being found easily. In contrast to this, if we use (A.23) and linearize the complementarity and slackness condition via auxiliary integer variables, the single-level equivalent problem can be formulated as an MILP, whose global optimal solution can be procured with reasonable computation effort.

The dual problem of LPs which maximize its objective can be derived in the same way. Consider the LP

$$\max\{c^T x \mid Ax \le b\} \tag{A.24}$$

For this problem, we need an upper bound on the objective function. To this end, associating a non-negative dual vector λ with the constraint, and adding the weighted inequalities together, we have

$$\lambda^T Ax \le \lambda^T b$$

If we intentionally choose λ such that $\lambda^T A = c^T$, then $\lambda^T b$ will be a valid upper bound of the optimal value of (A.24). The dual problem

$$\min_{\lambda} \left\{\lambda^T b \mid A^T \lambda = c,\ \lambda \ge 0\right\} \tag{A.25}$$

optimizes the weighting vector λ to offer the tightest upper bound.

Constraints in the form of equality and $\ge$ inequality can be considered using the same paradigm. Bearing in mind that we are seeking an upper bound, so we need a certification for $c^T x \le a$, so the dual variables for equalities have no signs and those for $\ge$ inequalities should be negative.

Sometimes it is useful to define the dual cone of a polyhedron, despite that a bounded polyhedron is not a cone. Recall its definition, the dual cone of a polyhedron P can be defined as

$$P^* = \left\{y \mid x^T y \ge 0,\ \forall x \in P\right\} \tag{A.26}$$

where $P = \{x \mid Ax \geq b\}$. As we have demonstrated in Sect. A.1.3, the dual cone is always closed and convex; however, for a general set, its dual cone does not have an analytical expression.

For polyhedral sets, the condition in (A.26) holds if and only if the minimal value of $x^T y$ over P is non-negative. For a given vector y, let us investigate the minimum of $x^T y$ through an LP

$$\min_x \left\{ y^T x \mid Ax \geq b \right\}$$

It is known from Proposition A.1 that $y^T x \geq b^T \lambda$, $\forall \lambda \in D_P$, where $D_P = \{\lambda \mid A^T \lambda = y, \lambda \geq 0\}$. Moreover, if $\exists \lambda \in D_P$ such that $b^T \lambda < 0$, Proposition A.2 certifies the existence of $x \in P$ such that $y^T x = b^T \lambda < 0$. In conclusion, the dual cone of polyhedron P can be cast as

$$P^* = \left\{ y \mid \exists \lambda : b^T \lambda \geq 0,\ A^T \lambda = y,\ \lambda \geq 0 \right\} \tag{A.27}$$

which is also a polyhedron. It can be observed from (A.27) that all constraints in P^* are homogeneous, so P^* is indeed a polyhedral cone.

A.2.2 General Conic Linear Program

Linear programs cover vast topics in engineering optimization problems. Its duality program provides informative quantifications and valuable insights of the problem at hand, which help develop efficient algorithms for itself and facilitate building tractable reformulations for more complicated mathematical programming models, such as robust optimization, multi-level optimization, and equilibrium problems. The algorithms of LPs, which are perfectly developed by now, can solve quite large instances (with up to hundreds of thousands of variables and constraints). Nevertheless, there are practical problems which cannot be modeled by LPs. To cope with these essentially nonlinear cases, one needs to explore new models and computational methods beyond the reach of LPs.

The broadest class of optimization problems which the LP can be compared with is the class of convex optimization problems. Convexity marks whether a problem can be solved efficiently, and any local optimizer of a convex program must be a global optimizer. Efficiency is quantified by the number of arithmetic operations required to solve the problem. Suppose that all we know about the problem is its convexity: its objective and constraints are convex functions in decision variables $x \in \mathbb{R}^n$, and their values along with their derivatives at any given point can be evaluated within M arithmetic operations. The best known complexity for finding an ϵ-solution turns out to be [1]

$$O(1)n\left(n^3 + M\right)\ln\left(\frac{1}{\epsilon}\right)$$

Although this bound grows polynomially with n, the computation time may be still unacceptable for a large n like $n = 1000$, which is in contrast to LPs which are solvable with $n = 100{,}000$. The reason is: linearity is much stronger than convexity; the structure of an affine function $a^T x + b$ solely depends on its constant coefficients a and b; function values and derivatives are never evaluated in a state-of-the-art LP solver. There are many classes of convex programs which are essentially nonlinear, but still possess nice analytical structure, which can be used to develop more dedicated algorithms. These algorithms may perform much more efficiently than those exploiting only convexity. In what follows, we consider such a class of convex program, i.e., the conic program, which is a simple extension of LP. Its general form and mathematical model are briefly introduced, while the details about interior-point algorithms are beyond the scope of this book, which can be found in [1, 2].

1. Mathematical Model

When we consider to add some nonlinear factors in LP (A.18), the most common way is to replace a linear function $a^T x$ with a nonlinear but convex function $f(x)$. As what has been explained, this may not be advantageous from a computational perspective. In contrast to this, we sustain all functions to be linear, but inject non-linearity in the comparative operators $\geq$ or $\leq$. Recall the definition of generalized inequalities $\succeq_K$ with cone K, we consider the following problem in this section

$$\min_x \left\{ c^T x \mid Ax \succeq_K b \right\} \tag{A.28}$$

which is called a conic programming problem. An LP is a special case of the conic program with $K = \mathbb{R}^n_+$. With this generalization, we are able to formulate a much wider spectrum of optimization problems which cannot be modeled as LPs, while enjoy nice properties of structured convexity.

2. Conic Duality

Aside from developing high-performance algorithms, the most important and elegant theoretical result in the area of LP is its duality theorem. In view of their similarities in mathematical appearances, how can the LP duality theorem be extended to conic programs? Similarly, the motivation of duality is the desire of a systematic way to certify a lower bound on the optimal value of conic program (A.28). Let us try the same trick: multiplying the dual vector λ on both sides of $Ax \succeq_K b$, and adding them together, we obtain $\lambda^T Ax$ and $b^T \lambda$; moreover, if we are lucky to get $A^T \lambda = c$, we guess $b^T \lambda$ can serve as a lower bound of the optimum of (A.28) under some condition. The condition can be translated into: what is the admissible region of λ, such that the inequality $\lambda^T Ax \geq b^T \lambda$ is a consequence of $Ax \succeq_K b$? A nice answer has been given at the end of Sect. A.1.3. Let us explain the problem from some simple cases.

Particularly, when $K = \mathbb{R}^n_+$, the admissible region of λ is also $\mathbb{R}^n_+$, because we have already known the fact that the dual variable of $\geq$ inequalities in an LP which minimizes its objective should be non-negative. However, $\mathbb{R}^n_+$ is no longer a feasible region of λ for conic programs with generalized inequality $\succeq_K$ if $K \neq \mathbb{R}^n_+$. To see this, consider $\mathbb{L}^3_C$ and the corresponding generalized inequality

$$\begin{bmatrix} x \\ y \\ z \end{bmatrix} \succeq_{\mathbb{L}^3_C} \begin{bmatrix} 0 \\ 0 \\ 0 \end{bmatrix} \Longleftrightarrow z \geq \sqrt{x^2 + y^2}$$

$(x, y, z) = (-1, -1, 1.5)$ is a feasible solution. However, the weighted summation of both sides with $\lambda = [1, 1, 1]^T$ gives a false inequality $-0.5 \geq 0$.

To find the feasible region of λ, consider the condition

$$\forall a \succeq_K 0 \Rightarrow \lambda^T a \geq 0 \tag{A.29}$$

If (A.29) is true, we have the following logical inferences

$$\begin{aligned} & Ax \succeq_K b \\ \Leftrightarrow \quad & Ax - b \succeq_K 0 \\ \Rightarrow \quad & \lambda^T (Ax - b) \geq 0 \\ \Leftrightarrow \quad & \lambda^T Ax \geq \lambda^T b \end{aligned}$$

Conversely, if λ is an admissible vector for certifying

$$\forall (a, b : a \succeq_K b) \Rightarrow \lambda^T a \geq \lambda^T b$$

then, (A.29) is clearly true by letting $b = 0$. Therefore, the admissible set of λ for generalized inequality $\succeq_K$ with cone K can be written as

$$K^* = \left\{ \lambda \mid \lambda^T a \geq 0, \ \forall a \in K \right\} \tag{A.30}$$

which contains vectors whose inner products with all vectors belonging to K are nonnegative. Recall the definition in (A.11), we can observe that the set K^* is actually the dual cone of cone K.

Now we are ready to set up the dual problem of conic program (A.28). As in the case of LP duality, we try to recover the objective function from the linear combination of constraints by choosing a proper dual variable λ, i.e., $\lambda^T Ax = c^T x$, in addition, $\lambda \in K^*$ ensures $\lambda^T Ax \geq \lambda^T b$, implying that $\lambda^T b$ is a valid lower bound of the objective function. The best bound one can expect is the optimum of the problem

$$\max_{\lambda} \left\{ b^T \lambda \mid A^T \lambda = c, \ \lambda \succeq_{K^*} 0 \right\} \tag{A.31}$$

which is also a conic program, and called the dual problem of conic program (A.28). From the above construction, we have already known that $c^T x \geq b^T \lambda$ is satisfied for all feasible x and λ, which is the weak duality of conic programs.

In fact, the primal-dual pair of conic programs has following properties:

Proposition A.5 (Conic Duality Theorem [1]) *The following conclusions hold true for conic program (A.28) and its dual (A.31).*

(1) Conic duality is symmetric: the dual problem is still a conic one, and the primal and dual problems are dual to each other.
(2) Weak duality holds: the duality gap $c^T x - b^T \lambda$ is nonnegative over the primal and dual feasible sets.
(2) If either of the primal problem or the dual problem is strictly feasible and has a finite optimum, then the other is solvable, and the duality gap is zero: $c^T x^* = b^T \lambda^*$ for some x^* and λ^*.
(3) If either of the primal problem or the dual problem is strictly feasible and has a finite optimum, then a pair of primal-dual feasible solutions (x, λ) solves the respective problems if and only if

$$\begin{aligned} Ax &\succeq_K b \\ A^T \lambda &= c \\ \lambda &\succeq_{K^*} 0 \\ c^T x &= b^T \lambda \end{aligned} \tag{A.32}$$

or

$$\begin{aligned} 0 \preceq_{K^*} \lambda \perp Ax - b \succeq_K 0 \\ A^T \lambda = c \end{aligned} \tag{A.33}$$

where (A.32) is called the primal-dual optimality condition, and (A.33) is called the KKT optimality condition.

The proof can be found in [1] and is omitted here. To highlight the role of strict feasibility in Proposition A.5, consider the following example

$$\min_x \left\{ x_2 \;\middle|\; \begin{bmatrix} x_1 \\ x_2 \\ x_1 \end{bmatrix} \succeq_{\mathbb{L}^3_C} 0 \right\}$$

The feasible region is

$$\sqrt{x_1^2 + x_2^2} \leq x_1 \Leftrightarrow x_2 = 0,\ x_1 \geq 0$$

So its optimal value is 0. As explained before, second-order cones are self-dual: $(\mathbb{L}_C^3)^* = \mathbb{L}_C^3$, it is easy to see the dual problem is

$$\max_{\lambda} \left\{ 0 \;\middle|\; \lambda_1 + \lambda_3 = 0,\ \lambda_2 = 1,\ \lambda \succeq_{\mathbb{L}_C^3} 0 \right\}$$

The feasible region is

$$\left\{ \lambda \;\middle|\; \sqrt{\lambda_1^2 + \lambda_2^2} \le \lambda_3,\ \lambda_3 \ge 0,\ \lambda_2 = 1,\ \lambda_1 = -\lambda_3 \right\}$$

which is empty, because $\sqrt{(-\lambda_3)^2 + 1} > \lambda_3$.

This example demonstrates that the existence of a strictly feasible point is indispensable for conic duality. But this condition is not necessary in LP duality, which means strong duality holds in conic programming with stronger assumptions.

Several classes of conic programs with particular cones are of special interest. The cones in these problems are self-dual, so we can set up the dual program directly, which allows to explore deeply into the original problem, or convert it into equivalent formulations which are more computationally friendly. The structure of these relatively simple cones also helps develop efficient algorithms for corresponding conic programs. In what follows, we will investigate two extremely important classes of conic programs.

A.2.3 Second-Order Cone Program

1. Mathematical Models of the Primal and Dual Problems

Second-order cone program is a special class of conic problem with $K = \mathbb{L}_C^{n+1}$. It minimizes a linear function over the intersection of a polytope and the Cartesian product of second-order cones, and can be formulated as

$$\min_{x} \left\{ c^T x \;\middle|\; Ax - b \succeq_K 0 \right\} \tag{A.34}$$

where $x \in \mathbb{R}^n$, and $K = \mathbb{L}_C^{m_1} \times \cdots \times \mathbb{L}_C^{m_k} \times \mathbb{R}_+^{m_p}$, in other words, the conic constraints in (A.34) can be expressed as k second-order cones $A_i x - b_i \succeq_{\mathbb{L}^{m_i}} 0$, $i = 1, \cdots, k$ plus one polyhedron $A_p x - b_p \ge 0$ with the following matrix partition

$$[A; b] = \begin{bmatrix} [A_1; b_1] \\ \vdots \\ [A_k; b_k] \\ [A_p; b_p] \end{bmatrix}$$

Recall the definition of second-order cone, we further partition the sub-matrices A_i, b_i into

$$[A_i; b_i] = \begin{bmatrix} D_i & d_i \\ p_i^T & q_i \end{bmatrix}, \; i = 1, \cdots, k$$

where $D_i \in \mathbb{R}^{(m_i-1)\times n}$, $p_i \in \mathbb{R}^n$, $d_i \in \mathbb{R}^{m_i-1}$, $q_i \in \mathbb{R}$. Then we can write (A.34) as

$$\begin{aligned} \min_x \;\; & c^T x \\ \text{s.t.} \;\; & A_p x \geq b_p \\ & \|D_i x - d_i\|_2 \leq p_i^T x - q_i, \; i = 1, \cdots, k \end{aligned} \tag{A.35}$$

Equation (A.35) is often more convenient for model builders.

It is easy to see that the cone K in (A.34) is self-dual, as both second-order cone and non-negative orthant are self-dual. In this regard, the dual problem of SOCP (A.34) can be expressed as

$$\max_\lambda \left\{ b^T \lambda \;\middle|\; A^T \lambda = c, \; \lambda \succeq_K 0 \right\} \tag{A.36}$$

Partitioning the dual vector as

$$\lambda = \begin{bmatrix} \lambda_1 \\ \vdots \\ \lambda_k \\ \lambda_p \end{bmatrix}, \; \lambda_i \in \mathbb{L}_C^{m_i}, \; i = 1, \cdots, k, \; \lambda_p \geq 0$$

We can write the dual problem as

$$\begin{aligned} \max_\lambda \;\; & \sum_{i=1}^k b_i^T \lambda_i + b_p^T \lambda_p \\ \text{s.t.} \;\; & \sum_{i=1}^k A_i^T \lambda_i + A_p^T \lambda_p = c \\ & \lambda_i \in \mathbb{L}_C^{m_i}, \; i = 1, \cdots, k \\ & \lambda_p \geq 0 \end{aligned} \tag{A.37}$$

We further partition λ_i according to the norm representation in (A.35)

$$\lambda_i = \begin{bmatrix} \mu_i \\ \nu_i \end{bmatrix}, \; \mu_i \in \mathbb{R}^{m_i-1}, \; \nu_i \in \mathbb{R}$$

all second-order cone constraints are associated with dual variables as

$$\begin{bmatrix} D_i x \\ p_i^T x \end{bmatrix} - \begin{bmatrix} d_i \\ q_i \end{bmatrix} \in \mathbb{L}_C^{m_i} : \begin{bmatrix} \mu_i \\ \nu_i \end{bmatrix}, \ i = 1, \cdots, k$$

so the admissible region of dual variables (μ_i, ν_i) is

$$\begin{bmatrix} \mu_i \\ \nu_i \end{bmatrix} \in (\mathbb{L}_C^{m_i})^* \Rightarrow \|\mu_i\|_2 \leq \nu_i$$

Finally, we arrive at the dual form of (A.35)

$$\begin{aligned} \max_{\lambda} \ & \sum_{i=1}^{k} \left(\mu_i^T d_i + \nu_i q_i \right) + b_p^T \lambda_p \\ \text{s.t.} \ & \sum_{i=1}^{k} \left(D_i^T \mu_i + \nu_i p_i \right) + A_p^T \lambda_p = c \\ & \|\mu_i\|_2 \leq \nu_i, \ i = 1, \cdots, k \\ & \lambda_p \geq 0 \end{aligned} \tag{A.38}$$

Equations (A.35) and (A.38) are more convenient than (A.34) and (A.37), respectively, because norm constraints can be recognized by most commercial solvers, whereas generalized inequalities $\succeq_K$ and constraints with the form $\in \mathbb{L}_C^{m_i}$ are supported only in some dedicated packages. Strict feasibility can be expressed in a more straightforward manner via norm constraints: the primal problem is strictly feasible if $\exists x : \|D_i x - d_i\|_2 < p_i^T x - q_i, \ i = 1, \cdots, k, \ A_p x > b_p$; the dual problem is strictly feasible if $\|\mu_i\|_2 < \nu_i, \ i = 1, \cdots, k, \ \lambda_p > 0$. In view of this, (A.35) and (A.38) are treated as the standard forms of an SOCP and its dual by practitioners whose primary interests are applications.

2. What Can Be Expressed via SOCPs?

Mathematical programs raised in engineering applications may not always appear in standard convex forms, and convexity may be hidden in seemingly non-convex expressions. Therefore, an important step is to recognize the potential existence of a convex form that is equivalent to the original formulation. This task can be rather tricky. We introduce some frequently used functions and constraints that can be represented by second-order cone constraints.

a. Convex Quadratic Constraints

A convex quadratic constraint has the form

$$x^T P x + q^T x + r \leq 0 \tag{A.39}$$

where $P \in \mathbb{S}^n_+$, $q \in \mathbb{R}^n$, $r \in \mathbb{R}$ are constant coefficients. Let $t = q^T x + r$, we have

$$t = \frac{(t+1)^2}{4} - \frac{(t-1)^2}{4}$$

Performing the Cholesky factorization $P = D^T D$, (A.39) can be represented by

$$\|Dx\|_2^2 + \frac{(t+1)^2}{4} \le \frac{(t-1)^2}{4}$$

So (A.39) is equivalent to the following second-order cone constraint

$$\left\| \begin{matrix} 2Dx \\ q^T x + r + 1 \end{matrix} \right\|_2 \le q^T x + r - 1 \tag{A.40}$$

However, not every second-order cone constraint can be expressed via a convex quadratic constraint. By squaring $\|Dx - d\|_2 \le p^T x - q$ we get an equivalent quadratic inequality

$$x^T (D^T D - pp^T) x + 2(qp^T - d^T D)x + d^T d - q^2 \le 0 \tag{A.41}$$

with $p^T x - q \ge 0$. The matrix $M = D^T D - pp^T$ is not always positive semidefinite. Indeed, $M \succeq 0$ if and only if $\exists u, \|u\|_2 \le 1 : p = D^T u$. On this account, SOCPs are more general than convex QCQPs.

b. Hyperbolic Constraints

Hyperbolic constraints are frequently encountered in engineering optimization problems. They are non-convex in their original forms but can be represented by a second-order cone constraint. A hyperbolic constraint has the form

$$x^T x \le yz, \ y > 0, \ z > 0 \tag{A.42}$$

where $x \in \mathbb{R}^n$, $y, z \in \mathbb{R}_{++}$. Noticing the fact that $4yz = (y+z)^2 - (y-z)^2$, (A.42) is equivalent to the following second-order cone constraint

$$\left\| \begin{matrix} 2x \\ y - z \end{matrix} \right\|_2 \le y + z, \ y > 0, \ z > 0 \tag{A.43}$$

However, a hyperbolic constraint cannot be expressed via a convex quadratic constraint, because the compact quadratic form of (A.42) is

$$\begin{bmatrix} x \\ y \\ z \end{bmatrix}^T P \begin{bmatrix} x \\ y \\ z \end{bmatrix} \le 0, \ P = \begin{bmatrix} 2I & 0 & 0 \\ 0 & 0 & -1 \\ 0 & -1 & 0 \end{bmatrix}$$

where the matric P is indefinite.

Many instances can be regarded as special cases of hyperbolic constraints, such as the upper branch of hyperbola

$$\{(x, y) \mid xy \geq 1,\ x > 0\}$$

and the epigraph of a fractional-quadratic function $g(x, s) = x^T x/s, s > 0$

$$\left\{(x, s, t) \,\middle|\, t \geq \frac{x^T x}{s},\ s > 0\right\}$$

c. Composition of Second-Order Cone Representable Functions
A function is called second-order cone representable if its epigraph can be represented by second-order cone constraints. Second-order cone representable functions are closed under composition [9]. Suppose two univariate convex functions $f_1(x)$ and $f_2(x)$ are second-order cone representable, and $f_1(x)$ is monotonically increasing, the composition $g(x) = f_1(f_2(x))$ is also second-order cone representable, because its epigraph $\{(x, t) \mid g(x) \leq t\}$ can be expressed by

$$\{(x, t) \mid \exists s : f_1(s) \leq t,\ f_2(x) \leq s\}$$

where $f_1(s) \leq t$ and $f_2(x) \leq s$ essentially come down to second-order cone constraints.

d. Maximizing the Production of Concave Functions
Suppose two functions $f_1(x)$ and $f_2(x)$ are concave with $f_1(x) \geq 0,\ f_2(x) \geq 0$, and $-f_1(x)$ and $-f_2(x)$ are second-order cone representable [which means $f_1(x) \geq t_1$ and $f_2(x) \geq t_2$ are (equivalent to) second-order cone constraints]. Consider the maximum of their production

$$\max_x \{f_1(x) f_2(x) \mid x \in X,\ f_1(x) \geq 0,\ f_2(x) \geq 0\} \tag{A.44}$$

where the feasible region X is the intersection of a polyhedron and second-order cones. It is not instantly clear whether problem (A.44) is a convex optimization problem or not. This formulation frequently arises in engineering applications, such as the Nash Bargaining problem and multi-objective optimization problems. An example can be found in Sect. 2.3.3.

By introducing auxiliary variables t, t_1, t_2, it is immediately seen that problem (A.44) is equivalent to the following SOCP

$$\begin{aligned} \max_{x,t,t_1,t_2} \ & t \\ \text{s.t. } & x \in X \\ & t_1 \geq 0,\ t_2 \geq 0,\ t_1 t_2 \geq t^2 \\ & f_1(x) \geq t_1,\ f_2(x) \geq t_2 \end{aligned} \tag{A.45}$$

At the optimal solution, $f_1(x) f_2(x) = t^2$.

3. Polyhedral Approximation of Second-Order Cones
Although SOCPs can be solved very efficiently, the state-of-the-art in numerical computing of SOCPs is still incomparable to that in LPs. The salient computational superiority of LPs inspires a question: can we approximate an SOCP by an LP without dramatically increasing the problem size? There have been other reasons to explore LP approximations for SOCPs. For example, to solve a bilevel program with an SOCP lower level, the SOCP should be replaced by its optimality conditions. However, the primal-dual optimality condition (A.32) may introduce bilinear terms, while the second-order cone complementarity constraints in KKT optimality condition (A.33) cannot be linearized easily. If the SOCP can be approximated by an LP, then the KKT optimality condition can be linearized and the original bilevel program can be reformulated as an MILP. Clearly, if we only work in original variables, the number of additional constraints would quickly grow unacceptable with the increasing problem dimension and required accuracy. In this section, we introduce the technique developed in [10], which lifts the problem into higher dimensions with moderate numbers of auxiliary variables and constraints.

We start with the basic question: find a polyhedral ϵ-approximation Π for $\mathbb{L}_C^3$ such that:

1. If $x \in \mathbb{L}_C^3$, then $\exists u : (x, u) \in \Pi$.
2. If $(x, u) \in \Pi$ for some u, then $\sqrt{x_1^2 + x_2^2} \le (1+\epsilon)x_3$.

Geometrically, the polyhedral cone Π includes a system of homogeneous linear equalities and inequalities in variables x, u; its projection on x-space is an ϵ-outer approximation of $\mathbb{L}_C^3$, and the error bound is quantified by ϵx_3. The answer to this question is given in [10]. It is shown that Π can be expressed by

$$\begin{aligned}
&\text{(a)} \quad \begin{cases} \xi^0 \ge |x_1| \\ \eta^0 \ge |x_2| \end{cases} \\
&\text{(b)} \quad \begin{cases} \xi^j = \cos\left(\dfrac{\pi}{2^{j+1}}\right)\xi^{j-1} + \sin\left(\dfrac{\pi}{2^{j+1}}\right)\eta^{j-1} \\ \eta^j \ge \left| -\sin\left(\dfrac{\pi}{2^{j+1}}\right)\xi^{j-1} + \cos\left(\dfrac{\pi}{2^{j+1}}\right)\eta^{j-1} \right| \end{cases}, \quad j = 1, \cdots, v \\
&\text{(c)} \quad \begin{cases} \xi^v \le x_3 \\ \eta^v \le \tan\left(\dfrac{\pi}{2^{v+1}}\right)\xi^v \end{cases}
\end{aligned} \tag{A.46}$$

Formulation (A.46) can be understood from a geometric point of view:

1. Given $x \in \mathbb{L}_C^3$, set $\xi^0 = |x_1|$, $\eta^0 = |x_2|$, which satisfies (a) in (A.46), and point $P^0 = (\xi^0, \eta^0)$ belongs to the first quadrant. Let

$$\begin{cases} \xi^j = \cos\left(\frac{\pi}{2^{j+1}}\right)\xi^{j-1} + \sin\left(\frac{\pi}{2^{j+1}}\right)\eta^{j-1} \\ \eta^j = \left|-\sin\left(\frac{\pi}{2^{j+1}}\right)\xi^{j-1} + \cos\left(\frac{\pi}{2^{j+1}}\right)\eta^{j-1}\right| \end{cases}$$

which ensures (b). Point $P^j = (\xi^j, \eta^j)$ is obtained from P^{j-1} according to the following operation: rotate P^{j-1} by angle $\phi_j = \pi/2^{j+1}$ clockwise and get a mediate point Q^{j-1}; if Q^{j-1} resides in the upper half-plane, $P^j = Q^{j-1}$; otherwise P^j is the reflection of Q^{j-1} with respect to the x-axis. By this construction, it is clear that all vectors from the origin to P^j have the same Euclidean norm, i.e., $\|[x_1, x_2]\|_2$. Moreover, as P^0 belongs to the first quadrant, the angle of Q^0 must satisfy $-\pi/4 \le \arg(Q^0) \le \pi/4$, and $0 \le \arg(P^1) \le \pi/4$. With the procedure going on, we have $|\arg(Q^j)| \le \pi/2^{j+1}$, and $0 \le \arg(P^{j+1}) \le \pi/2^{j+1}$, for $j = 1, \cdots, v$. In the last step, $\xi^v \le \|P^v\|_2 = \|[x_1, x_2]\|_2 \le x_3$ and $0 \le \arg(P^v) \le \pi/2^{v+1}$ hold, ensuring condition (c). In this manner, a point in $\mathbb{L}_C^3$ has been extended to a solution of (A.46).
2. Given $(x, u) \in \Pi$, where $u = \{\xi^j, \eta^j\}$, $j = 1, \cdots, v$. Define $P^j = [\xi^j, \eta^j]$, and it directly follows from (a) and (b) that all P^j belongs to the first quadrant, and $\|P^0\|_2 \ge \sqrt{x_1^2 + x_2^2}$. Moreover, recall the construction of Q^j in previous analysis, it is seen $\|P^j\|_2 = \|Q^j\|_2$; the absolute value of the vertical coordinate of P^{j+1} is no less than that of Q^j; therefore, $\|P^{j+1}\|_2 \ge \|Q^j\|_2 = \|P^j\|_2$. At last

$$\|P^v\|_2 \le \frac{x_3}{\cos\left(\frac{\pi}{2^{v+1}}\right)}$$

so we arrive at $\sqrt{x_1^2 + x_2^2} \le (1+\epsilon)x_3$, where

$$\epsilon = \frac{1}{\cos\left(\frac{\pi}{2^{v+1}}\right)} - 1 \tag{A.47}$$

In this way, a solution of (A.46) has been approximately extended to $\mathbb{L}_C^3$.

Now, let us consider the general case: approximating

$$\mathbb{L}_C^{n+1} = \left\{(y, t) \,\middle|\, \sqrt{y_1^2 + \cdots + y_n^2} \le t\right\}$$

via a polyhedral cone. Without loss of generality, we assume $n = 2^K$. To make use of the outcome in (A.46), y is split into 2^{K-1} pairs $(y_1, y_2), \cdots, (y_{n-1}, y_n)$, which are called variables of generation 0. A successor variable is associated with each pair, which is called variable of generation 1, and is further divided into 2^{K-2} pairs and associated with variable of generation 2, and so on. After $K - 1$ steps of dichotomy, we complete variable splitting with two variables of generation $K - 1$. The only variable of generation K is t. For notation convenience, let y_i^l be i-th variable of generation l, the original vector $y = [y_1^0, \cdots, y_n^0]$, and $t = y_1^K$. The "parents" of y_i^l are variables $y_{2i-1}^{l-1}, y_{2i}^{l-1}$. The total number of variables in the "tower" is $2n - 1$.

Using the tower of variables $y^l, \forall l$, the system of constraints

$$\sqrt{(y_{2i-1}^{l-1})^2 + (y_{2i}^{l-1})^2} \le y_i^l,\ i = 1, \cdots, 2^{K-l},\ l = 1, \cdots, K \tag{A.48}$$

gives the same feasible region on y as L_C^{n+1}, and each second-order cone in $\mathbb{L}_C^3$ in (A.48) can be approximated by a polyhedral cone given in (A.46). To see the first claim, a simple example is given in Sect. 4.4.

The size of this polyhedral approximation is unveiled in [10]:

1. The dimension of the lifted variable is $p \le n + O(1) \sum_{l=1}^{K} 2^{K-l} v_l$.
2. The number of constraints is $q \le O(1) \sum_{l=1}^{K} 2^{K-l} v_l$.

The quality of the approximation is [10]

$$\beta = \prod_{l=1}^{K} \frac{1}{\cos\left(\frac{\pi}{2^{v_l+1}}\right)} - 1$$

Given a desired tolerance ϵ, choose

$$v_l = \left\lfloor O(1) l \ln \frac{2}{\epsilon} \right\rfloor$$

with a proper constant $O(1)$, we can guarantee the following bounds:

$$\begin{aligned} &\beta \le \epsilon \\ &p \le O(1) n \ln \frac{2}{\epsilon} \\ &q \le O(1) n \ln \frac{2}{\epsilon} \end{aligned}$$

which implies that the required numbers of variables and constraints grow linearly in the dimension of the target second-order cone.

A.2.4 Semidefinite Program

1. Notation Clarification
In this section, variables appear in the form of symmetric matrices, some notations should be clarified first.

The Frobenius inner product of two matrices $A, B \in \mathbb{M}^n$ is defined by

$$\langle A, B \rangle = \text{tr}\left(AB^T\right) = \sum_{i=1}^{n}\sum_{j=1}^{n} A_{ij}B_{ij} \tag{A.49}$$

The Euclidean norm of a matrix $X \in \mathbb{M}^n$ can be defined through the Frobenius inner product as follows:

$$\|X\|_2 = \sqrt{\langle X, X \rangle} = \sqrt{\text{tr}(X^T X)} \tag{A.50}$$

Equipped with the Frobenius inner product, the dual cone of a given cone $K \subset \mathbb{S}^n$ is defined by

$$K^* = \{Y \in \mathbb{S}^n \mid \langle Y, X \rangle \geq 0, \ \forall X \in K\} \tag{A.51}$$

Among the cones in $\mathbb{S}^n$, this section talks about the positive semidefinite cone $\mathbb{S}^n_+$. As what has been demonstrated in Sect. A.1.3, S^n_+ is self-dual, i.e., $(\mathbb{S}^n_+)^* = \mathbb{S}^n_+$. The interior of cone $\mathbb{S}^n_+$ consists of all $n \times n$ matrices that are positive definite, and is denoted by $\mathbb{S}^n_{++}$.

2. Primal and Dual Formulations of SDPs
When $K = \mathbb{S}^n_+$, conic program (A.28) boils down to an SDP

$$\min_x \left\{c^T x \mid Ax - b \in \mathbb{S}^n_+\right\}$$

which minimizes a linear objective over the intersection of affine plane $y = Ax - b$ and the positive semidefinite cone $\mathbb{S}^n_+$. However, the notation in such a form is a little confusing: $Ax - b$ is a vector, which is not dimensionally compatible with the cone $\mathbb{S}^n_+$. In fact, we have met a similar difficulty at the very beginning: the vector inner product does not apply to matrices, which is consequently replaced with the Frobenius inner product. There are two prevalent ways to resolve the confliction in dimension, leading to different formulations which will be discussed.

a. Formulation Based on Vector Decision Variables
In this formulation, b is replaced with a matrix $B \in \mathbb{S}^n$, and Ax is replaced with a linear mapping $\mathcal{A}x : \mathbb{R}^n \to \mathbb{S}^n$. In this way, $\mathcal{A}x - B$ becomes an element of $\mathbb{S}^n$. A simple way to specify the linear mapping $\mathcal{A}x$ is

$$\mathcal{A}x = \sum_{j=1}^{n} x_j A_j, \ x = [x_1, \cdots, x_n]^T, \ A_1, \cdots, A_n \in \mathbb{S}^n$$

With all these input matrices, an SDP can be written as

$$\min_x \left\{ c^T x \mid x_1 A_1 + \cdots + x_n A_n - B \succeq 0 \right\} \tag{A.52}$$

where the cone $\mathbb{S}_+^n$ is omitted in the operator $\succeq$ without causing confusion. The constraint in (A.52) is an LMI. This formulation is general enough to capture the situation in which multiple LMIs exist, because

$$\mathcal{A}_i x - B_i \succeq 0, \ i = 1, \cdots, k \Leftrightarrow \mathcal{A}x - B \succeq 0$$

with $\mathcal{A}x = \text{Diag}(\mathcal{A}_1 x, \cdots, \mathcal{A}_k x)$ and $B = \text{Diag}(B_1, \cdots, B_k)$.

The general form of conic duality can be specified in the case when the cone $K = \mathbb{S}_+^n$. Associating a matrix dual variable Λ with the LMI constraint, and recalling the fact that $(\mathbb{S}_+^n)^* = \mathbb{S}_+^n$, the dual program of SDP (A.52) reads:

$$\max_\Lambda \{\langle B, \Lambda \rangle \mid \langle A_i, \Lambda \rangle = c_i, \ i = 1, \cdots, n, \ \Lambda \succeq 0\} \tag{A.53}$$

which remains an SDP.

Apply conic duality theorem given in Proposition A.5 to SDPs (A.52) and (A.53).

1. Suppose $A_1, \cdots, A_n$ are linearly independent, i.e., no nontrivial linear combination of $A_1, \cdots, A_n$ gives an all zero matrix.
2. The primal SDP (A.52) is strict feasible, i.e., $\exists x : x_1 A_1 + \cdots + x_n A_n \succ B$, and is solvable (the minimum is attainable)
3. The dual SDP (A.53) is strict feasible, i.e., $\exists \Lambda \succ 0 : \langle A_i, \Lambda \rangle = c_i, i = 1, \cdots, n$, and is solvable (the maximum is attainable).

The optimal values of (A.52) and (A.53) are equal, and the complementarity and slackness condition

$$\langle \Lambda, x_1 A_1 + \cdots + x_n A_n - B \rangle = 0 \tag{A.54}$$

is necessary and sufficient for a pair of primal and dual feasible solutions (x, Λ) to be optimal for their respective problems. For a pair of positive semidefinite matrices, it can be shown that

$$\langle XY \rangle = 0 \Leftrightarrow XY = YX = 0$$

indicating that the eigenvalues of these two matrices in some certain basis are "complementary": for every common eigenvector, at most one of the two eigenvalues of X and Y can be strictly positive.

b. Formulation Based on Matrix Decision Variables

This formulation directly incorporates a matrix decision variable $X \in \mathbb{S}^n_+$, and imposes other restrictions on X through linear equations. In the objective function, the vector inner product $c^T x$ is replaced by a Frobenius inner product $\langle C, X \rangle$. In this way, an SDP can be written as

$$\begin{aligned} \min_{X} \ & \langle C, X \rangle \\ \text{s.t.} \ & \langle A_i, X \rangle = b_i : \lambda_i, \ i = 1, \cdots, m \\ & X \succeq 0 : \Lambda \end{aligned} \tag{A.55}$$

By introducing dual variables (following the colon) for individual constraints, the dual program of (A.55) can be constructed as

$$\begin{aligned} \max_{\lambda, \Lambda} \ & b^T \lambda + \langle 0, \Lambda \rangle \\ \text{s.t.} \ & \Lambda + \lambda_1 A_1 + \cdots + \lambda_n A_n = C \\ & \Lambda \succeq 0 \end{aligned}$$

Eliminating Λ, we obtain

$$\begin{aligned} \max_{\lambda} \ & b^T \lambda \\ \text{s.t.} \ & C - \lambda_1 A_1 - \cdots - \lambda_n A_n \succeq 0 \end{aligned} \tag{A.56}$$

It is observed that (A.55) and (A.56) are in the same form compared with (A.53) and (A.52), respectively, except for the signs of some coefficients.

SDP handles positive semidefinite matrices, so it is especially powerful in eigenvalue related problems, such as Lyapunov stability analysis and controller design, which are the main field of control theorists. Moreover, every SOCP can be formulated as an SDP because

$$\|y\|_2 \le t \Leftrightarrow \begin{bmatrix} tI & y \\ y^T & t \end{bmatrix} \succeq 0$$

Nevertheless, solving SOCPs via SDP may not be a good idea. Interior-point algorithms for SOCPs have much better worst-case complexity than those for SDPs. In fact, SDPs are extremely popular in the convex relaxation technique for non-convex quadratic optimization problems, owing to its ability to offer a nearly global optimal solution in many practical applications, such as the OPF problem in power systems. The SDP based convex relaxation method for non-convex QCQPs will be discussed in the next section. Here we talk about some special cases involving homogeneous quadratic functions or at most two non-homogeneous quadratic functions.

3. Homogeneous Quadratic Programs

Consider the following quadratic program

$$\begin{aligned} \min\ & x^T B x \\ \text{s.t.}\ & x^T A_i x \geq 0,\ i = 1, \cdots, m \end{aligned} \tag{A.57}$$

where $A_1, \cdots, A_m, B \in \mathbb{S}^n$ are constant coefficients. Suppose that problem (A.57) is feasible. Due to its homogeneity, the optimal value is clear: $-\infty$ or 0, depending on whether there is a feasible solution x such that $x^T Bx < 0$ or not. But it is unclear which situation takes place, i.e., to judge $x^T Bx \geq 0$ over the intersection of homogeneous inequalities $x^T A_i x \geq 0$, $i = 1, \cdots, m$, or whether the implication

$$x^T A_i x \geq 0,\ i = 1, \cdots, m \Rightarrow x^T Bx \geq 0 \tag{A.58}$$

holds.

Proposition A.6 *If there exist $\lambda_i \geq 0$, $i = 1, 2, \cdots$ such that $B \succeq \sum_i \lambda_i A_i$, then the indication in (A.58) is true.*

To see this, $B \succeq \sum_i \lambda_i A_i \Leftrightarrow x^T(B - \sum_i \lambda_i A_i)x \geq 0 \Leftrightarrow x^T Bx \geq \sum_i \lambda_i x^T A_i x$; therefore, $x^T Bx$ is a direct consequence of $x^T A_i x \geq 0$, $i = 1, \cdots, m$, as the right-hand side of the last inequality is non-negative. Proposition A.6 provides a sufficient condition for (A.58), and necessity is generally not guaranteed. Nevertheless, if $m = 1$, the condition is both necessary and sufficient.

Proposition A.7 (S-Lemma) *Let $A, B \in \mathbb{S}^n$ and a homogeneous quadratic inequality*

$$(a) \quad x^T Ax \geq 0$$

is strictly feasible. Then the homogeneous quadratic inequality

$$(b) \quad x^T Bx \geq 0$$

is a consequence of (a) if and only if $\exists \lambda \geq 0 : B \succeq \lambda A$.

Proposition A.7 is called the S-Lemma or S-Procedure. It can be proved by many means. The most instructive one, in our tastes, is based on the semidefinite relaxation, which can be found in [1].

4. Non-homogeneous Quadratic Programs with a Single Constraint
Consider the following quadratic program

$$\begin{aligned} \min \quad & f_0(x) = x^T A_0 x + 2b_0^T x + c_0 \\ \text{s.t.} \quad & f_1(x) = x^T A_1 x + 2b_1^T x + c_1 \le 0 \end{aligned} \tag{A.59}$$

Let f^* denote the optimal solution, so $f_0(x) - f^* \ge 0$ is a consequence of $-f_1(x) \ge 0$. A sufficient condition for this implication is $\exists \lambda \ge 0 : f_0(x) - f^* + \lambda f_1(x) \ge 0$. The left-hand side is a quadratic function with matrix form

$$\begin{bmatrix} x \\ 1 \end{bmatrix}^T \begin{bmatrix} A_0 + \lambda A_1 & b_0 + \lambda b_1 \\ (b_0 + \lambda b_1)^T & c_0 + \lambda c_1 - f^* \end{bmatrix} \begin{bmatrix} x \\ 1 \end{bmatrix}$$

Its non-negativeness is equivalent to

$$\begin{bmatrix} A_0 + \lambda A_1 & b_0 + \lambda b_1 \\ (b_0 + \lambda b_1)^T & c_0 + \lambda c_1 - f^* \end{bmatrix} \succeq 0 \tag{A.60}$$

Similar to the homogeneous case, this condition is also sufficient. In view of this, the optimal value f^* of (A.59) solves the following SDP

$$\begin{aligned} \min_{\lambda, f} \quad & f \\ \text{s.t.} \quad & \begin{bmatrix} A_0 + \lambda A_1 & b_0 + \lambda b_1 \\ (b_0 + \lambda b_1)^T & c_0 + \lambda c_1 - f \end{bmatrix} \succeq 0 \end{aligned} \tag{A.61}$$

This conclusion is known as the non-homogeneous S-Lemma:

Proposition A.8 (Non-homogeneous S-Lemma) *Let $A_i \in \mathbb{S}^n$, $b_i \in \mathbb{R}^n$, and $c_i \in \mathbb{R}$, $i = 0, 1$, if $\exists x : x^T A_1 x + 2b_1^T x + c_1 < 0$, the implication*

$$x^T A_1 x + 2b_1^T x + c_1 \le 0 \Rightarrow x^T A_0 x + 2b_0^T x + c_0 \le 0$$

holds if and only if

$$\exists \lambda \ge 0 : \begin{bmatrix} A_0 & b_0 \\ b_0^T & c_0 \end{bmatrix} \preceq \lambda \begin{bmatrix} A_1 & b_1 \\ b_1^T & c_1 \end{bmatrix} \tag{A.62}$$

Because the implication can boil down to the maximum of quadratic function $x^T A_0 x + 2b_0^T x + c_0$ being non-positive over set $\{x | x^T A_1 x + 2b_1^T x + c_1 \le 0\}$, which is a special case of (A.59) by letting $f_0(x) = -x^T A_0 x - 2b_0^T x - c_0$, Proposition A.8 is a particular case of (A.61) with the optimum $f^* = 0$.

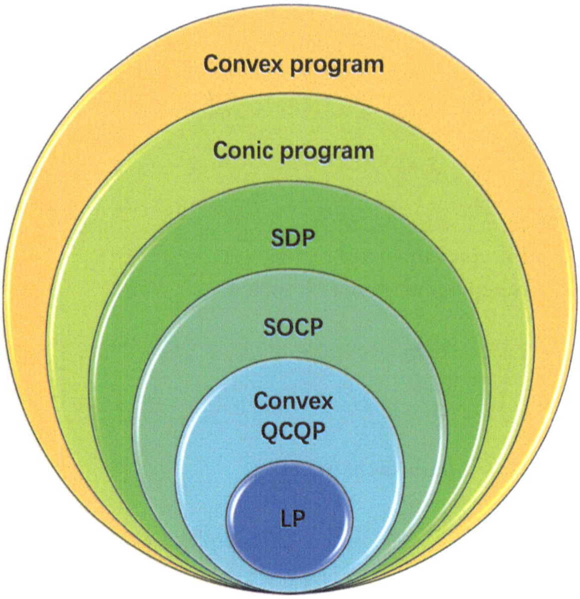

Fig. A.8 Relations of the discussed convex programs

A formal proof based on semidefinite relaxation is given in [1]. Since a quadratic inequality describes an ellipsoid, Proposition A.8 can be used to test whether an ellipsoid is contained in another one.

As a short conclusion, we summarize the relation of discussed convex programs in Fig. A.8.

A.3 Convex Relaxation Methods for Non-convex QCQPs

One of the most prevalent and promising applications of SDP is to build tractable approximations of computationally intractable optimization problems. One of the most quintessential appliances is the convex relaxation of quadratically constrained quadratic programs (QCQPs), which cover vast engineering optimization problems. QCQPs are generally non-convex and could have more than one locally optimal solution, and each of them may yield significant different objective values. However, gradient based algorithms can only find a local solution which largely depends on the initial point. One primary interest is to identify the global optimal solution or determine a high-quality bound for the optimum, which can be used to quantify the optimality gap of a given local optimal solution. The SDP relaxation technique for solving non-convex QCQPs is briefly reviewed in this section.

A.3.1 SDP Relaxation and Valid Inequalities

A standard fact of quadratic expression is

$$x^T Qx = \langle Q, xx^T \rangle \tag{A.63}$$

where $\langle \cdot \rangle$ stands for the Frobenius inner product.

Following the logic in [11], we focus our attention on QCQPs in the following form

$$\min \ \{x^T Cx + c^T x \mid x \in F\} \tag{A.64}$$

where

$$F = \left\{ x \in \mathbb{R}^n \ \middle| \ x^T A_k x + a_k^T x \le b_k, \ k = 1, \cdots, m, \ l \le x \le u \right\} \tag{A.65}$$

All coefficient matrices and vectors have compatible dimensions. If $A_k = 0$ in all constraints, then the feasible set F is a polyhedron, and (A.64) reduces to a quadratic program (QP); If $A_k \succeq 0$, $k = 1, \cdots, m$ and $C \succeq 0$, (A.64) is a convex QCQP, which is easy to solve. Without loss of generality, we assume A_k, $k = 1, \cdots, m$ and C are indefinite, F is a non-convex set, and the objective is a non-convex function. In fact, a number of hard optimization problems can be cast as non-convex QCQP (A.64). For example, a polynomial optimization problem can be reduced to a QCQP by introducing a tower of condensing variables, e.g., $x_1x_2x_3x_4$ could be replaced by quadratic term $x_{12}x_{34}$ with $x_{12} = x_1x_2$ and $x_{34} = x_3x_4$. Moreover, a binary constraint $x \in \{0, 1\}$ is equivalent to quadratic equality $x(x-1) = 1$ where x is continuous.

A common idea to linearize non-convex terms $x^T A_k x$ is to define new variables $X_{ij} = x_i x_j$, $i = 1, \cdots, n$, $j = 1, \cdots, n$. In this way, $x^T A_k x = \sum_i \sum_j A_{ij} x_i x_j = \sum_{ij} A_{ij} X_{ij}$, and the last term is linear. Recall (A.63), this fact can be written in a compact form

$$x^T A_k x = \langle A_k, X \rangle, \ X = xx^T$$

With this transformation, QCQP (A.64) becomes

$$\min \ \{\langle C, X \rangle + c^T x \mid (x, X) \in \hat{F}\} \tag{A.66}$$

where

$$\hat{F} = \left\{ (x, X) \in \mathbb{R}^n \times \mathbb{S}^n \ \middle| \ \begin{array}{c} \langle A_k, X \rangle + a_k^T x \le b_k, \ k = 1, \cdots, m \\ l \le x \le u, \ X = xx^T \end{array} \right\} \tag{A.67}$$

In problem (A.66), non-convexity is concentrated in the relation between the lifting variable X and the original variable x, whereas all other constraints are linear. Moreover, if we replace $\hat{F}$ with its convex hull conv($\hat{F}$), the optimal solution of (A.66) will not change, because its objective function is linear. However, conv($\hat{F}$) does not have a closed-form expression. Convex relaxation approaches can be interpreted as attempting to approximate conv($\hat{F}$) through structured convex constraints which can be recognized by existing solvers.

We define the following linear relaxation

$$\hat{L} = \left\{ (x, X) \in \mathbb{R}^n \times \mathbb{S}^n \;\middle|\; \begin{array}{c} \langle A_k, X\rangle + a_k^T x \le b_k \\ k = 1, \cdots, m \\ l \le x \le u \end{array} \right\} \tag{A.68}$$

which contains only linear constraints. Now, let us consider the lifting constraint

$$X = xx^T \tag{A.69}$$

which is called a rank-1 constraint. However, a rank constraint is non-convex and cannot be accepted by most solvers. Notice the fact that if (A.69) holds, then

$$\begin{bmatrix} 1 & x^T \\ x & X \end{bmatrix} = \begin{bmatrix} 1 & x^T \\ x & xx^T \end{bmatrix} = \begin{bmatrix} 1 \\ x \end{bmatrix} \begin{bmatrix} 1 \\ x \end{bmatrix}^T \succeq 0$$

Define an LMI constraint

$$\text{LMI} = \left\{ (x, X) \;\middle|\; Y = \begin{bmatrix} 1 & x^T \\ x & X \end{bmatrix} \succeq 0 \right\} \tag{A.70}$$

The positive semi-definiteness condition is true over conv($\hat{F}$).

The basic SDP relaxation of (A.66) replaces the rank-1 constraint in $\hat{F}$ with a weaker but convex constraint (A.70), giving rise to the following SDP

$$\begin{aligned} \min \;& \langle C, X\rangle + c^T x \\ \text{s.t.} \;& (x, X) \in \hat{L} \cap \text{LMI} \end{aligned} \tag{A.71}$$

Clearly, the LMI constraint enlarges the feasible region defined by (A.69), so the optimal solution to (A.71) may not be feasible in the original QCQP, and the optimal value is a strict lower bound. In this situation, the SDP relaxation is inexact. Conversely, if matrix Y is indeed rank-1 at the optimal solution, then the SDP relaxation is exact and x solves the original QCQP (A.64).

The basic SDP relaxation model (A.71) can be further improved by enforcing additional linkages between x and X, which are called valid inequalities. Suppose linear inequalities $\alpha^T x \le \alpha_0$ and $\beta^T x \le \beta_0$ are chosen from $\hat{L}$, then the quadratic inequality

$$\left(\alpha_0 - \alpha^T x\right)\left(\beta_0 - \beta^T x\right) = \alpha_0\beta_0 - \alpha_0\beta^T x - \beta_0\alpha^T x + x^T\alpha\beta^T x \ge 0$$

holds for all $x \in \hat{L}$. The last quadratic term can be linearized via the lifting variable X, resulting in the following linear inequality

$$\alpha_0\beta_0 - \alpha_0\beta^T x - \beta_0\alpha^T x + \langle \beta\alpha^T, X\rangle \ge 0 \tag{A.72}$$

Any linear inequality in $\hat{L}$ (possibly the same) can be used to construct valid inequalities. Because additional constraints are imposed on X, the relaxation could be tightened, and the feasible region shrinks but may still be larger than conv($\hat{F}$).

If we construct valid inequality (A.72) from side constraint $l \le x \le u$, we get

$$\left.\begin{aligned} (x_i - l_i)(x_j - l_j) \ge 0 \\ (x_i - l_i)(u_j - x_j) \ge 0 \\ (u_i - x_i)(x_j - l_j) \ge 0 \\ (u_i - x_i)(u_j - x_j) \ge 0 \end{aligned}\right\}, \quad \forall i, j = 1, \cdots, n, \ i \le j \tag{A.73}$$

Expanding these quadratic inequalities, the coefficients of quadratic terms $x_i x_j$ are equal to 1, and we obtain simple bounds on X_{ij}

$$\begin{aligned} x_i l_j + x_j l_i - l_i l_j \le X_{ij} \\ x_i u_j + x_j l_i - l_i u_j \ge X_{ij} \\ u_i x_j - u_i l_j + x_i l_j \ge X_{ij} \\ u_i x_j + x_i u_j - u_i u_j \le X_{ij} \end{aligned}$$

or in a compact matrix form [11]

$$\text{RLT} = \left\{ (x, X) \;\middle|\; \begin{aligned} lx^T + xl^T - ll^T \le X \\ ux^T + xu^T - uu^T \le X \\ xu^T + lx^T - lu^T \ge X \end{aligned} \right\} \tag{A.74}$$

Equation (A.74) is known as the reformulation-linearization technique after the term appeared in [12]. These constraints have been extensively studied since it was proposed in [13], due to the simple structure and satisfactory performance in various applications. The improved SDP relaxation with valid inequalities can be written as

$$\begin{aligned} \min \quad & \langle C, X \rangle + c^T x \\ \text{s.t.} \quad & (x, X) \in \hat{L} \cap \text{LMI} \cap \text{RLT} \end{aligned} \tag{A.75}$$

From the construction of $\hat{L}$, LMI, and RLT, it is directly concluded that

$$\text{conv}(\hat{F}) \subseteq \hat{L} \cap \text{LMI} \cap \text{RLT} \tag{A.76}$$

The inclusion becomes tight only in some very special situations, such as those encountered in the homogeneous and non-homogeneous S-Lemma. Nevertheless, what we really need is the equivalence between the optimal solution of the relaxed problem (A.75) and that of the original problem (A.64): if the optimal matrix variable of (A.75) allows a rank-1 decomposition

$$\begin{bmatrix} 1 & x^T \\ x & X \end{bmatrix} = \begin{bmatrix} 1 \\ x \end{bmatrix} \begin{bmatrix} 1 \\ x \end{bmatrix}^T$$

which indicates X has a rank-1 decomposition $X = xx^T$, then x is optimal in (A.64), and the SDP relaxation is said to be exact, although $\text{conv}(\hat{F})$ may be a strict subset of $\hat{L} \cap \text{LMI} \cap \text{RLT}$.

A.3.2 Successively Tightening the Relaxation

If the matrix X has a rank higher than 1, the corresponding optimal solution x in (A.75) may be infeasible in (A.64). The rank-1 constraint on X can be exactly described by a pair of LMIs $X \succeq xx^T$ and $X \preceq xx^T$. The former one is redundant to (A.70) indicated by the Schur complement theorem; the latter one is non-convex, which is simply neglected in the SDP relaxation.

1. A Dynamical Valid Inequality Generation Approach

An approach is proposed in [11] to generate valid inequalities dynamically by harnessing the constraint violations in $X \preceq xx^T$. The motivation comes from the fact that

$$X - xx^T \preceq 0 \Leftrightarrow \langle X, v_i v_i^T \rangle \le \left(v_i^T x\right)^2, \ i = 1, \cdots, n$$

where $\{v_1, \cdots, v_n\}$ is a set of orthogonal basis of $\mathbb{R}^n$. To see this, any vector $h \in \mathbb{R}^n$ can be expressed as the linear combination of the orthogonal basis as $h = \sum_{i=1}^n \lambda_i v_i$, therefore, $h^T(X - xx^T)h = \langle X, hh^T \rangle - (h^T x)^2 = \sum_{i=1}^n \lambda_i^2 [\langle X, v_i v_i^T \rangle - (v_i^T x)^2] \le 0$. In view of this,

$$\hat{F} = \hat{L} \cap \text{LMI} \cap \text{NSD}$$

where

$$\begin{aligned}\text{NSD} &= \left\{(x, X) \mid X - xx^T \preceq 0\right\} \\ &= \left\{(x, X) \mid \left\langle X, v_i v_i^T \right\rangle \le \left(v_i^T x\right)^2, i = 1, \cdots, n\right\}\end{aligned}$$

If $\{v_1, \cdots, v_n\}$ is the standard orthogonal basis,

$$\text{NSD} = \{(x, X) \mid X_{ii} \le x_i^2,\ i = 1, \cdots, n\} \tag{A.77}$$

It is proposed in [11] to construct NSD as

$$\text{NSD} = \left\{(x, X) \mid \langle X, \eta_i \eta_i^T \rangle \le (\eta_i^T x)^2,\ i = 1, \cdots, n\right\} \tag{A.78}$$

where $\{\eta_1, \cdots, \eta_n\}$ are the eigenvectors of matrix $X - xx^T$, because they exclude infeasible points with respect to $X - xx^T \preceq 0$ most effectively.

Non-convex constraints in (A.77) and (A.78) can be handled by a special disjunctive programming derived in [14] and the convex-concave procedure investigated in [15]. The former one is an exact approach which requires binary variables to formulate disjunctive constraints; the latter is a heuristic approach which only solves convex optimization problems. We do not further detail these techniques here.

2. A Rank Penalty Method [16]

In view of the rank-1 exactness condition, another way to tighten SDP relaxation is to work on the rank of the optimal solution. A successive rank penalty approach is proposed in [16]. We consider problem (A.66) as a rank-constrained SDP

$$\min \ \{\langle \Omega, Y \rangle \mid Y \in \hat{L} \cap \text{LMI} \cap \text{RLT},\ \text{rank}(Y) = 1\} \tag{A.79}$$

where

$$\Omega = \begin{bmatrix} 0 & 0.5c^T \\ 0.5c & C \end{bmatrix}, \quad Y = \begin{bmatrix} 1 & x^T \\ x & X \end{bmatrix}$$

constraints $\hat{L}$, LMI, and RLT (rearranged for variable Y) are defined in (A.68), (A.70), and (A.74), respectively. The last constraint in (A.79) ensures that Y has a rank-1 decomposition such that $X = xx^T$. Actually, LMI and RLT are redundant to the rank-1 constraint, but will give a high quality convex relaxation when the rank constraint is relaxed.

To treat the rank-1 constraint in a soft manner, we introduce a dummy variable Z, and penalize the matrix rank in the objective function, giving rising to the following problem

$$\begin{aligned} \min_{Y} \ & \left\{ \langle \Omega, Y \rangle + \min_{Z} \frac{\rho}{2} \|Y - Z\|_2^2 \right\} \\ \text{s.t.} \ & Y \in \hat{L} \cap \text{LMI} \cap \text{RLT}, \ \text{rank}(Z) = 1 \end{aligned} \tag{A.80}$$

If the penalty parameter ρ is sufficiently large, the penalty term will be zero at the optimal solution, so $Y = Z$ and $\text{rank}(Z) = 1$. One advantage of this treatment is that the constraints on Y and Z are decoupled, and the inner rank minimization problem has a closed-form solution.

To see this, if $\text{rank}(Y) = k > 1$, the singular value decomposition of Y has the form $Y = U \Sigma V^T$, where

$$\Sigma = \text{diag}(S, 0), \ S = \text{diag}(\sigma_1, \cdots, \sigma_k), \ \sigma_1 \geq \cdots \geq \sigma_k > 0$$

U and V are orthogonal matrices. Let matrix D have the same dimension as Y, $D_{11} = \sigma_1$, and $D_{ij} = 0, \forall (i, j) \neq (1, 1)$, we have

$$\begin{aligned} & \min_{Z} \left\{ \frac{\rho}{2} \|Y - Z\|_2^2 \,\middle|\, \text{rank}(Z) = 1 \right\} \\ & = \min_{Z} \left\{ \frac{\rho}{2} \|U(Y - Z)V^T\|_2^2 \,\middle|\, \text{rank}(Z) = 1 \right\} \\ & = \min_{Z} \left\{ \frac{\rho}{2} \|\Sigma - UZV^T\|_2^2 \,\middle|\, \text{rank}(Z) = 1 \right\} \\ & = \frac{\rho}{2} \|\Sigma - D\|_2^2 = \frac{\rho}{2} \sum_{i=2}^{k} \sigma_i^2 \\ & = \frac{\rho}{2} \|Y\|_2^2 - \frac{\rho}{2} \sigma_1^2(Y) \end{aligned}$$

To represent the latter term via a convex function, let matrix Θ have the same dimension as Y, $\Theta_{11} = 1$, and $\Theta_{ij} = 0, \forall (i, j) \neq (1, 1)$, we have

$$\text{tr}(Y^T U \Theta U^T Y) = \text{tr}(V \Sigma U^T U \Theta U^T U \Sigma V^T) = \text{tr}(V \Sigma \Theta \Sigma V^T) = \text{tr}(\Sigma \Theta \Sigma) = \sigma_1^2(Y)$$

Define two functions $f(Y) = \langle \Omega, Y \rangle + \frac{\rho}{2} \|Y\|_2^2$ and $g(Y) = \text{tr}(Y^T U \Theta U^T Y)$. Because $\|Y\|_2$ is convex in Y (Example 3.11, [2]), so is $\|Y\|_2^2$ (composition rule, page 84, [2]); clearly, $f(Y)$ is a convex function in Y, as it is the sum of a linear function and a convex function. For the latter one, the Hessian matrix of $g(Y)$ is

$$\nabla_Y^2 g(Y) = U \Theta U^T = U \Theta^T \Theta U^T = (\Theta U^T)^T \Theta U^T \succeq 0$$

so $g(Y)$ is also convex in Y. Substituting above results into problem (A.80), the rank constrained SDP (A.79) boils down to

$$\min_{Y} \left\{ \langle \Omega, Y \rangle + \frac{\rho}{2} \|Y\|_2^2 - \frac{\rho}{2} \mathrm{tr}(Y^T U \Theta U^T Y) \;\middle|\; Y \in \hat{L} \cap \mathrm{LMI} \cap \mathrm{RLT} \right\} \tag{A.81}$$

The objective function is a DC function, and the feasible region is convex, so (A.81) is a DC program. One can employ the convex-concave procedure discussed in [15] to solve this problem. The flowchart is summarized in Algorithm A.1.

For the convergence of Algorithm A.1, we have the following properties.

Proposition A.9 ([16]) *The optimal value sequence $F(Y^i)$ generated by Algorithm A.1 is monotonically decreasing.*

Denote by $F(Y) = f(Y) - \frac{\rho}{2} g(Y)$ the objective function of (A.81) in the DC form, and $H(Y, Y^i) = f(Y) - \frac{\rho}{2} g_L(Y, Y^i)$ the convexified objective function in (A.83) by linearizing the concave term in $F(Y)$. Two basic facts help explain this proposition:

Algorithm A.1 Sequential SDP

1: Choose an initial penalty parameter ρ^0, a penalty growth rate $\tau > 0$, and solve the following SDP relaxation model

$$\min \; \{ \langle \Omega, Y \rangle \mid Y \in \hat{L} \cap \mathrm{LMI} \cap \mathrm{RLT} \}$$

The optimal solution is Y^*.

2: Construct the linear approximation of $g(Y)$ as

$$g_L(Y, Y^*) = g(Y^*) + \langle \nabla g(Y^*), Y - Y^* \rangle \tag{A.82}$$

Solve the following SDP

$$\min_{Y} \left\{ f(Y) - \frac{\rho}{2} g_L(Y, Y^*) \;\middle|\; Y \in \hat{L} \cap \mathrm{LMI} \cap \mathrm{RLT} \right\} \tag{A.83}$$

The optimal solution is Y^*.

3: If $\mathrm{rank}(Y^*) = 1$, terminate and report the optimal solution Y^*; otherwise, update $\rho \leftarrow (1 + \tau)\rho$, and go to step 2.

---

1. $g_L(Y^*, Y^*) = g(Y^*)$, $\forall Y^*$ which directly follows from the definition in (A.82).
2. For any given Y^*, $g(Y) \ge g_L(Y, Y^*)$, $\forall Y$, because the graph of a convex function must lie over its tangent plane at any fixed point.

First we can asset inequality $H(Y^{i+1}, Y^i) \le H(Y^i, Y^i)$, because $H(Y, Y^i)$ is optimized in problem (A.83). The optimum $H(Y^{i+1}, Y^i)$ deserves a value no greater than that at any feasible point. Furthermore, with the definition of $H(Y^i, Y^i)$, we have

$$H\left(Y^{i+1}, Y^i\right) \le H\left(Y^i, Y^i\right) = f\left(Y^i\right) - \frac{\rho}{2} g_L\left(Y^i, Y^i\right) = f\left(Y^i\right) - \frac{\rho}{2} g\left(Y^i\right)$$
$$= F\left(Y^i\right)$$

On the other hand,

$$H\left(Y^{i+1}, Y^i\right) = f\left(Y^{i+1}\right) - \frac{\rho}{2} g_L\left(Y^{i+1}, Y^i\right) \ge f\left(Y^{i+1}\right) - \frac{\rho}{2} g\left(Y^{i+1}\right)$$
$$= F\left(Y^{i+1}\right)$$

Consequently, we arrive at the monotonic property

$$F(Y^{i+1}) \le F(Y^i)$$

Proposition A.10 ([16]) *The solution sequence Y^i generated by Algorithm A.1 approaches to the optimal solution of problem (A.79) when $\rho \to \infty$.*

It is easy to understand that whenever ρ is sufficiently large, the penalty term will tend to 0, and the rank-1 constraint in (A.79) is met. A formal proof can be found in [16]. A few more remarks are given below.

1. The convex-concave procedure in [17] is a local algorithm under mild conditions and needs a manually supplied initial point. Algorithm A.1, however, is elaborately initiated at the solution offered by the SDP relaxation model, which usually appears to be close to the global optimal one for many engineering optimization problems. Therefore, Algorithm A.1 generally performs well and will identify the global optimal solution, although a provable guarantee is non-trivial.
2. In practical applications, Algorithm A.1 could converge without the penalty parameter approaching infinity, because when some constraint quantification holds, there exists an exact penalty parameter ρ^*, such that the optimal solution leads to a zero penalty term for any $\rho \ge \rho^*$ [18, 19], and Algorithm A.1 converges in a finite number of steps. If the exact penalty parameter does not exist, Algorithm A.1 may fail to converge. In such circumstance, one can impose an upper bound on ρ, and use an alternative convergence criterion: the change of the objective value $F(Y)$ in two consecutive steps is less than a given threshold value. As a result, Algorithm A.1 will be able to find an approximate solution of problem (A.79), and the rank-1 constraint may not be enforced.
3. From the numeric computation perspective, a very large ρ may cause ill-conditioned problem and lead to numerical instability, so it is useful to gradually increase ρ from a small value. Another reason for the moderate growth of ρ is that it does not cause dramatic change of optimal solutions in two successive iterations. As a result, $g_L(Y, Y^*)$ can provide relatively accurate approximation for $g(Y)$ in every iteration.

4. The penalty term $\rho_i p(Y^i)/2 = \rho_i \left[\|Y^i\|_2^2 - \text{tr}(Y^{iT} U \Theta U^T Y^i)\right]/2$ gives an upper bound on the optimality gap induced by rank relaxation. To see this, let ρ^* and Y^* be the exact penalty parameter and corresponding optimal solution of (A.83), i.e., $p(Y^*) = 0$; ρ_i and Y^i be the penalty parameter and optimal solution in i-th iteration. According to Proposition A.9, we have $\langle \Omega, Y^* \rangle \le \langle \Omega, Y^i \rangle + \rho_i p(Y^i)/2$; moreover, since the rank-1 constraint is relaxed before Algorithm A.1 could converge, $\langle \Omega, Y^i \rangle \le \langle \Omega, Y^* \rangle$ holds. Therefore, $\langle \Omega, Y^i \rangle$ and $\langle \Omega, Y^i \rangle + \rho_i p(Y^i)/2$ are lower and upper bounds for the optimal value of problem (A.79). In this regard, $\rho_i p(Y^i)/2$ is an estimation on the optimality gap.

A.3.3 Completely Positive Program Relaxation

Inspired by the convex hull expression in (A.14), researchers have shown that most non-convex QCQPs can be modeled as linear programs over the intersection of a completely positive cone and a polyhedron [20–22]. For example, consider minimizing a quadratic function over a standard simplex

$$\begin{aligned} \min \ & x^T Q x \\ \text{s.t.} \ & e^T x = 1 \\ & x \ge 0 \end{aligned} \tag{A.84}$$

where $Q \in \mathbb{S}^n$, and e denotes the all-one vector with n entries. Following the paradigm similar to (A.66), let $X = xx^T$, and then we can construct a valid inequality

$$1 = x^T e e^T x = x^T E x = \langle E, X \rangle$$

where $E = ee^T$ is the all-one matrix. According to (A.14), $\text{conv}\{xx^T | x \in \mathbb{R}^n_+\}$ is given by $(\mathbb{C}^n_+)^*$. Therefore, problem (A.84) transforms to

$$\begin{aligned} \min \ & \langle Q, X \rangle \\ \text{s.t.} \ & \langle E, X \rangle = 1 \\ & X \in (\mathbb{C}^n_+)^* \end{aligned} \tag{A.85}$$

Problem (A.85) is a convex relaxation of (A.84). Because the objective is linear, the optimal solution must be located at one extremal point of the convex hull of the feasible region. In view of the representation in (A.14), the extremal points are exactly rank-1, so the convex relaxation (A.85) is always exact.

Much more general results are demonstrated in [21] that every quadratic program with linear and binary constraints can be rewritten as a completely positive program.

More precisely, a mixed-integer quadratic program

$$\begin{aligned} \min\ & x^T Qx + 2c^T x \\ \text{s.t.}\ & a_i^T x = b_i,\ i = 1, \cdots, m \\ & x \geq 0,\ x_j \in \{0, 1\},\ j \in B \end{aligned} \tag{A.86}$$

and the following completely positive program

$$\begin{aligned} \min\ & \langle Q, X\rangle + 2c^T x \\ \text{s.t.}\ & a_i^T x = b_i,\ i = 1, \cdots, m \\ & \langle a_i a_i^T, X\rangle = b_i^2,\ i = 1, \cdots, m \\ & x_j = X_{jj},\ j \in B \\ & X \in (C_+^n)^* \end{aligned} \tag{A.87}$$

have the same optimal solution, as long as problem (A.86) satisfies: $a_i^T x = b_i, \forall i$ and $x \geq 0$ implies $x_j \leq 1, \forall j \in B$. Actually, this is a relatively mild condition [21]. Complementarity constraints can be handled in the similar way. Whether problems with general quadratic constraints can be restated as completely positive programs in the similar way remains an open question.

The NP-hardness of problem (A.86) makes (A.87) NP-hard itself. The complexity has been encapsulated into the last cone constraint. The relaxation model is still interesting due to its convexity. Furthermore, it can be approximated via a sequence of SDPs with growing sizes [23] given an arbitrarily small error bound.

A.3.4 MILP Approximation

SDP relaxation technique introduces a squared matrix variable that contains $n(n+1)/2$ independent variables. Although exploiting the sparse pattern of X via graphic theory is helpful to expedite problem solution, the computational burden is still high especially when the initial relaxation is inexact and a sequence of SDPs should be solved. Inspired by difference-of-convex programming an alternative choice is to express the non-convexity of QCQP by univariate concave functions, and approximate these concave functions via PWL functions compatible with mixed-integer programming solvers. This approach has been expounded in [24].

Consider nonconvex QCQP

$$\begin{aligned} \min\ & x^T A_0 x + a_0^T x \\ \text{s.t.}\ & x^T A_k x + a_k^T x \leq b_k,\ k = 1, \cdots, m \end{aligned} \tag{A.88}$$

We can always find $\delta_0, \delta_1, \cdots, \delta_m \in \mathbb{R}^+$, such that $A_k + \delta_k I \succeq 0$, $k = 0, \cdots, m$. For example, δ_k can take the absolute value of the most negative eigenvalue of A_k, and $\delta_k = 0$ if $A_k \succeq 0$. Then, problem (A.88) can be cast as

$$\begin{aligned} \min\ & x^T(A_0 + \delta_0 I)x + a_0^T x - \delta_0 1^T y \\ \text{s.t.}\ & x^T(A_k + \delta_k I)x + a_k^T x - \delta_k 1^T y \le b_k,\ k = 1, \cdots, m \\ & y_i = x_i^2, i = 1, \cdots, n \end{aligned} \tag{A.89}$$

Problem (A.89) is actually a difference-of-convex program; however, the nonconvex terms are consolidated in much simpler parabolic equalities, which can be linearized via the SOS2 based PWL approximation technique discussed in Appendix B.1. Except for the last n quadratic equalities, remaining constraints and objective function of problem (A.89) are all convex, so the linearized problem gives rise to a mixed-integer convex quadratic program.

Alternatively, we can first perform convex relaxation by replacing $y_i = x_i^2$ with $y_i \ge x_i^2, i = 1, \cdots, n$; if strict inequality holds at the optimal solution, a disjunctive cut is generated to remove this point from the feasible region. However, the initial convex relaxation can be very weak ($y = +\infty$ is usually an optimal solution). Predefined disjunctive cuts can be added [24].

Finally, nonconvex QCQP is a hard optimization problem. Developing an efficient algorithm should leverage specific problem structure. For example, SDP relaxation is suitable for OPF problems; MILP approximation can be used for small and dense problems. Unlike SDP relaxation works on a squared matrix variable, the number of auxiliary variables in (A.89) and its mixed-integer convex quadratic program approximation is moderate. Therefore, this approach is promising to tackle practical problems whose coefficient matrices are usually sparse. Furthermore, no particular assumption is needed to guarantee the exactness of relaxation, so this method is general enough to tackle a wide spectrum of engineering optimization problems.

A.4 MILP Formulation of Nonconvex QPs

In a non-convex QCQP, if the constraints are all linear, it is called a nonconvex QP. There is no doubt that convex relaxation methods presented in the previous section can be applied to nonconvex QPs. However, the relaxation is generally inexact. In this section, we introduce exact MILP formulations to globally solve such a nonconvex optimization problem; unlike the mixed-integer programming approximation method in Sect. A.3.4, in which approximation error is inevitable, by using duality theory, the MILP models will be completely equivalent to the original QP. Thanks to the advent of powerful MILP solvers, this method is becoming increasingly competitive compared to existing global solution methods and is attracting more attention from the research community.

A.4.1 Nonconvex QPs Over Polyhedra

The presented approach is devised in [25]. A nonconvex QP with linear constraints has the form of

$$\begin{aligned} \min \ & \frac{1}{2}x^T Qx + c^T x \\ \text{s.t.} \ & Ax \leq b \end{aligned} \tag{A.90}$$

where Q is a symmetric, but indefinite matrix; A, b, c are constant coefficients with compatible dimensions. We assume that finite lower and upper limits of the decision variable x have been included, and thus the feasible region is a bounded polyhedron. The KKT conditions of (A.90) can be written as:

$$\begin{aligned} & c + Qx + A^T \xi = 0 \\ & 0 \leq \xi \perp b - Ax \geq 0 \end{aligned} \tag{A.91}$$

If there is a multiplier ξ so that the pair (x, ξ) of primal and dual variables satisfies KKT condition (A.91), then x is said to be a KKT point or a stationary point. The complementarity and slackness condition in (A.91) gives $b^T \xi = x^T A^T \xi$. For any primal-dual pair (x, ξ) that satisfies (A.91), the following relations hold

$$\begin{aligned} \frac{1}{2}x^T Qx + c^T x &= \frac{1}{2}c^T x + \frac{1}{2}x^T (c + Qx) \\ &= \frac{1}{2}c^T x - \frac{1}{2}x^T A^T \xi = \frac{1}{2}\left(c^T x - b^T \xi\right) \end{aligned} \tag{A.92}$$

As such, the non-convex quadratic objective function is equivalently stated as a linear function in the primal and dual variables without loss of accuracy. Thus, if problem (A.90) has an optimal solution, then the solution can be retrieved by solving an LPCC

$$\begin{aligned} \min \ & \frac{1}{2}\left(c^T x - b^T \xi\right) \\ \text{s.t.} \ & c + Qx + A^T \xi = 0 \\ & 0 \leq \xi \perp b - Ax \geq 0 \end{aligned} \tag{A.93}$$

which is equivalent to the following MILP

$$\begin{aligned}
\min \quad & \frac{1}{2} c^T x - b^T \xi \\
\text{s.t.} \quad & c + Qx + A^T \xi = 0 \\
& 0 \le \xi \le M(1 - z) \\
& 0 \le b - Ax \le Mz \\
& z \text{ binary}
\end{aligned} \tag{A.94}$$

where M is a sufficiently large constant; z is a vector of binary variables. Regardless of the value of z_i, at most one of ξ_i and $(b - Ax)_i$ can take a strictly positive value. For more rigorous discussions on this method, please see [25], in which an unbounded feasible region is considered. More tricks in MILP reformulation technique can be found in the next chapter.

It should be pointed out that the set of optimal solutions of (A.90) is a subset of stationary points described by (A.91), because (A.91) is only a necessary condition for optimality but not sufficient. Nevertheless, as we assumed that the feasible region is a bounded polytope (thus compact), QP (A.90) must have a finite optimum, then according to [26], the optimal value is equal to the minimum of objective function values perceived at stationary points. Therefore, MILP (A.94) provides an exact solution to (A.90).

Finally, we shed some light on the selection of M, since it has notable impact on the computational efficiency of (A.94). An LP based bound preprocessing method is thoroughly discussed in [27], which is used in a finite branch-and-bound method for solving LPCC (A.93). Here we briefly introduce the bounding method.

For the primal variable x which represents physical quantities or measures, its bounds depend on practical situations and security considerations, and we assume that the bound is $0 \le x \le U$. The bound can be tightened by solving

$$\min(\max) \, \{x_j \mid Ax \le b, \; 0 \le x \le U\} \tag{A.95}$$

In (A.95), we can incorporate individual bounds for the components of vector x, which never wrecks the optimal solution and can be supplemented in (A.94).

For the dual variables, we consider (A.90) again with explicit bounds on primal variable x

$$\begin{aligned}
\min \quad & \frac{1}{2} x^T Q x + c^T x \\
\text{s.t.} \quad & Ax \le b : \xi \\
& 0 \le x \le U : \lambda, \rho
\end{aligned}$$

where ξ, λ, ρ following the colon are dual variables. Its KKT condition reads

$$c + Qx + A^T\xi - \lambda + \rho = 0 \tag{A.96a}$$

$$0 \le \xi \perp b - Ax \ge 0 \tag{A.96b}$$

$$0 \le x \perp \lambda \ge 0 \tag{A.96c}$$

$$0 \le U - x \perp \rho \ge 0 \tag{A.96d}$$

Multiplying both sides of (A.96a) by a feasible solution x^T

$$c^T x + x^T Qx + x^T A^T \xi - x^T \lambda + x^T \rho = 0 \tag{A.96e}$$

Substituting $\xi^T Ax = \xi^T b$, $x^T \lambda = 0$, and $x^T \rho = \rho^T U$ concluded from (A.96b)–(A.96d) into (A.96e) outcomes

$$c^T x + x^T Qx + b^T \xi + U^T \rho = 0 \tag{A.96f}$$

The upper bounds (lower bounds are 0) on the dual variables required for MILP (A.94) can be computed from the following LP:

$$\max \ \lambda_j \tag{A.97a}$$

$$\text{s.t. } c + Qx + A^T\xi - \lambda + \rho = 0 \tag{A.97b}$$

$$\text{tr}(Q^T X) + c^T x + b^T \xi + U^T \rho = 0 \tag{A.97c}$$

$$\text{Cons-RLT} = \{(x, X) \mid (\text{A.74})\} \tag{A.97d}$$

$$0 \le x \le U, \ Ax \le b, \lambda, \xi, \rho \ge 0 \tag{A.97e}$$

In (A.97c), quadratic equality (A.96f) is linearized by letting $X = xx^T$, and (A.97d) is a linear relaxation for above rank-1 condition, as explained in Sect. A.3.1.

By exploiting the relaxation revealed in (A.97c), it has been proved that problem (A.97) always has a finite optimum, because the recession cone of the set comprised of the primal and dual variables as well as their associated valid inequalities is empty, see the proof of Proposition 3.1 in [27]. This is a pivotal theoretical guarantee. Other bounding techniques which only utilize KKT conditions hardly ensure a finite optimum.

A.4.2 Standard Nonconvex QPs

The presented approach is devised in [28]. A standard nonconvex QP entails minimizing a nonconvex quadratic function over a unit probability simplex

$$\begin{aligned} v(Q) = \min\ & x^T Qx \\ \text{s.t. } & x \in \Delta_n \end{aligned} \tag{A.98}$$

where Q is a symmetric matrices, and unit simplex

$$\Delta_n = \{x \in \mathbb{R}^n_+ \mid e^T x = 1\}$$

where e is all-one vector. A nonhomogeneous objective can always be transformed to a quadratic form given the simplex constraint Δ_n:

$$x^T Qx + 2c^T x = x^T (Q + ec^T + ce^T)x, \ \forall x \in \Delta_n$$

Standard nonconvex QPs have wide applications in portfolio optimization, quadratic resource allocation, graphic theory, and so on. In addition, for a given symmetric matrix Q, a necessary and sufficient condition for Q being copositive is $v(Q) \geq 0$. Copositive programming is a young and active research field, and can help the research in convex relaxation. A fundamental problem is copositivity test, which entails solving (A.98) globally.

Problem (A.98) is a special case of nonconvex QP (A.90), so the methods in previous subsection also work for (A.98). The core trick is to select a big-M parameter in linearizing complementarity and slackness conditions. Due to its specific structure, the valid big-M parameter for problem (A.98) can be chosen in a much more convenient way. To see this, the KKT condition of (A.98) reads as

$$Qx - \lambda e - \mu = 0 \tag{A.99a}$$

$$e^T x = 1 \tag{A.99b}$$

$$x \geq 0 \tag{A.99c}$$

$$\mu \geq 0 \tag{A.99d}$$

$$x_j \mu_j = 0, \ j = 1, \cdots, n \tag{A.99e}$$

where λ and μ are dual variables associated with equality constraint $e^T x = 1$ and inequality constraint $x \geq 0$. Because the feasible region is polyhedral, constraint quantification always holds, and any optimal solution of (A.98) must solve KKT system (A.99).

Multiplying both sides of (A.99a) by x results in $x^T Qx = \lambda x^T e - x^T \mu$; substituting (A.99b) and (A.99e) into the right-hand side concludes $x^T Qx = \lambda$.

Provided with eligible big-M parameter, problem (A.98) is (exactly) equivalent to the following MILP

$$\begin{aligned} \min \ & \lambda \\ \text{s.t. } & Qx - \lambda e - \mu = 0 \\ & e^T x = 1, \ 0 \le x \le y \\ & 0 \le \mu_j \le M_j(1 - y_j), \ j = 1, \cdots, n \end{aligned} \tag{A.100}$$

where $y \in \{0, 1\}^n$, and M_j is the big-M parameter. It is the upper bound of dual variable μ_j. To estimate such a bound, according to (A.99a)

$$\mu_j = e_j^T Qx - \lambda, \ j = 1, \cdots, n$$

where e_j is the j-th column of $n \times n$ identity matrix. For the first term,

$$x^T Qe_j \le \max_{i \in \{1,\cdots,n\}} Q_{ij}, \ j = 1, \cdots, n$$

As for the second term, we know $\lambda \ge v(Q)$, so any known lower bound of $v(Q)$ can be used to obtain an upper bound of M_j. One possible lower bound of $v(Q)$ is suggested in [28] as

$$l(Q) = \min_{1 \le i,j \le n} Q_{ij} + \frac{1}{\sum_{k=1}^n \left(Q_{kk} - \min_{1 \le i,j \le n} Q_{ij} \right)^{-1}}$$

If the minimal element of Q locates on the main diagonal, the second term vanishes and $l(Q) = \min_{1 \le i,j \le n} Q_{ij}$.

In summary, a valid choice of M_j would be

$$M_j = \max_{i \in \{1,\cdots,n\}} Q_{ij} - l(Q), \ j = 1, \cdots, n \tag{A.101}$$

It is found in [28] that if we relax (A.99a) as an inequality and solve the following MILP

$$\begin{aligned} \min \ & \lambda \\ \text{s.t. } & Qx - \lambda e - \mu \le 0 \\ & e^T x = 1, \ 0 \le x \le y \\ & 0 \le \mu_j \le M_j(1 - y_j), \ j = 1, \cdots, n \end{aligned} \tag{A.102}$$

which is a relaxed version of (A.100), the optimal solution will not change. However, in some instances, solving (A.102) is significantly faster than solving (A.100). More thorough theoretical analysis can be found in [28].

A.5 Further Reading

Decades of wonderful research has resulted in elegant theoretical developments and sophisticated computational software, which have brought convex optimization to an unprecedented dominating stage where it serves as the baseline and reference model for optimization problems in almost every discipline. Only problems which can be formulated as convex programs are regarded as theoretically solvable. We suggest following materials for readers who want to build a solid mathematical background or know more about applications in the field of convex optimization.

1. Convex analysis and convex optimization. Convex analysis is a classic topic in mathematics, and focuses on basic concepts and topological properties of convex sets and convex functions. We recommend monographs [29–31]. The last one sheds more light on optimization related topics, including DC programming, polynomial programming, and equilibrium constrained programming, which are originally non-convex. The most popular textbooks on convex optimization include [1, 2]. They contain important materials that everyone who wants to apply this technique should know.
2. Special convex optimization problems. The most mature convex optimization problems are LPs, SOCPs, and SDPs. We recommend [6–8] for the basic knowledge of duality theory, simplex algorithm, interior-point algorithm, and applications of LPs. The modeling abilities of SOCPs and SDPs have been well discussed in [1, 2]. A geometric program is a type of optimization problem whose objective and constraints are characterized by special monomials and posynomial functions. Through a logarithmic variable transformation, a geometric program can be mechanically converted to a convex optimization problem. Geometric programming is relatively restrictive in structure, and it may not be apparent to see whether a given problem can be expressed by a geometric program. We recommend a tutorial paper [32] and references therein on this topic. Copositive program is a relatively young field in operational research. It is a special class of conic programming which is more general than SDP. Basic information on copositive/completely positive programs is introduced in [33–35]. They are particularly useful in combinatorial and quadratic optimization. Though very similar to SDPs in appearances, copositive programs are NP-hard. Algorithms and applications of copositive and completely positive programs have continued to be highly active research fields [36–38].
3. General convex optimization problems. Beside above mature convex optimization models that can be specified without high level of expertise, recognizing the convexity of a general mathematical programming problem may be rather tricky.

A deep understanding on convex analysis is unavoidable. Furthermore, to solve the problem using off-the-shelf solvers, a user must find a way to transform the problem into one of the standard forms (if a general purpose NLP solver fails to solve it). The so-called disciplined convex programming method is proposed in [39] to lower this expertise barrier. The method consists of a set of rules and conventions that one must follow when setting up the problem such that the convexity is naturally sustained. This methodology has been implemented in cvx toolbox under Matlab environment.

4. Convex relaxation methods. One major application of convex optimization is to derive tractable approximations for non-convex programs, so as to facilitate problem resolution in terms of computational efficiency and robustness. A general QCQP is a quintessential non-convex optimization problem. Among various convex relaxation approaches, the SDP relaxation is shown to be able to offer high quality solutions for many QCQPs raised in signal process [40, 41] and power system energy management [42, 43]. Decades of excellent studies on SDP relaxation methods for QCQPs are comprehensively reviewed in [11, 44, 45]. Some recent advances are reported in [46–52]. The rank of the matrix variable has a decisive impact on the exactness (or tightness) of the SDP relaxation. Low rank SDP method are attracting increasing attentions from researchers, and many approaches are proposed to recover a low-rank solution. More information can be found in [53–57] and references therein.
5. Sum-of-squares (SOS) programming is originally devised in [58] to decompose a polynomial $f(x)$ as the square of another polynomial $g(x)$ (if there exists), such that $f(x) = [g(x)]^2$ must be non-negative. Non-negativity of a polynomial over a semi-algebraic set can be certified in a similar way via Positivstellensatz refutations. This can be done by solving a structured SDP [58], and implemented in a Matlab based toolbox [59]. Based on these outcomes, a promising methodology is quickly developed for polynomial programs, which cover a broader class of optimization problems than QCQPs. It is proved that the global solution of a polynomial program can be found by solving a hierarchy of SDPs under mild conditions. This is very inspiring since polynomial programs are generally non-convex while SDPs are convex. We recommend [60] for a very detailed discussion on this approach, and [61–64] for some recent advances. However, users should be aware that this approach may be unpractical because the size of the relaxed SDP quickly becomes unacceptable after a few steps. Nonetheless, the elegant theory still marks a milestone in the research field.

Appendix B
Formulation Tricks in Integer Programming

There is no problem in all mathematics that cannot be solved by direct counting. But with the present implements of mathematics many operations can be performed in a few minutes which without mathematical methods would take a lifetime.

–Ernst Mach

As stated in Appendix A, generally speaking, convex optimization problems can be solved efficiently. However, the majority of optimization problems encountered in practical engineering are non-convex, and gradient based NLP solvers terminate at a local optimum, which may be far away from the global one. In fact, any nonlinear function can be approximated by a PWL function with adjustable errors by controlling the granularity of partitions. A PWL function can be expressed via a logic form or incorporating integer variables. Thanks to the latest progress in branch-and-cut algorithms and the development of state-of-the-art MILP solvers, a large-scale MILP can often be solved globally within reasonable computational efforts [65], although the MILP itself is proved to be NP-hard. In view of this fact, PWL/MILP approximation serves as a viable option to tackle real-world non-convex optimization problems, especially those with special structures.

This chapter introduces PWL approximation methods for nonlinear functions and linear representations of special non-convex constraints via integer programming techniques. When the majority of a problem at hand is linear or convex, while non-convexity arises from nonlinear functions with only one or two variables, linear complementarity constraints, logical inferences and so on, it is worth trying the methods in this chapter, in view of the fact that MILP solvers are becoming increasingly efficient to retrieve a solution with a pre-defined optimality gap.

W. Wei, J. Wang, *Modeling and Optimization of Interdependent Energy Infrastructures*, https://doi.org/10.1007/978-3-030-25958-7

B.1 Piecewise Linear Approximation of Nonlinear Functions

B.1.1 Univariate Continuous Function

Considering a nonlinear continuous function $f(x)$ in a single variable x, we can evaluate the function values $f(x_0)$, $f(x_1)$, $\cdots$, $f(x_n)$ at given breakpoints x_0, x_1, $\cdots$, x_k, and replace $f(x)$ with the following PWL function

$$f(x) = \begin{cases} m_1 x + c_1, & x \in [x_0, x_1] \\ m_2 x + c_2, & x \in [x_1, x_2] \\ \vdots & \vdots \\ m_k x + c_k, & x \in [x_{k-1}, x_k] \end{cases} \tag{B.1}$$

As an illustrative example, two curves of the original nonlinear function and its PWL approximation are portrayed in part (a), Fig. B.1. The PWL function

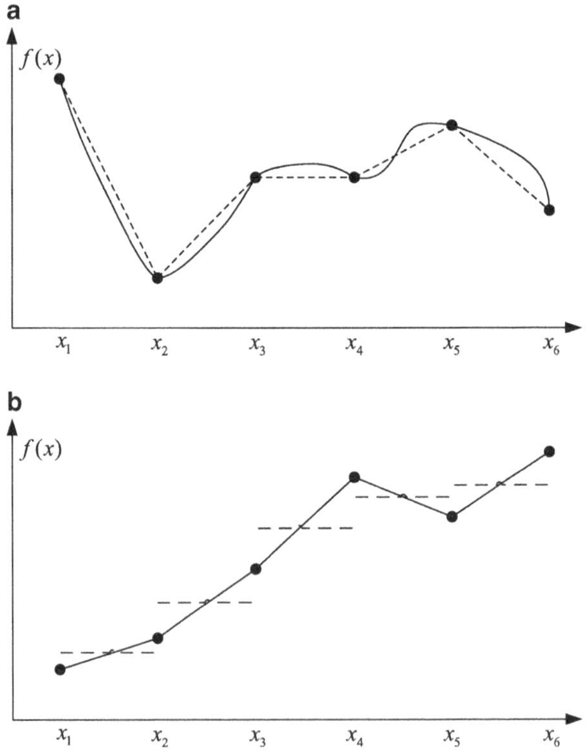

Fig. B.1 Piecewise linear and piecewise constant approximations. (**a**) Piecewise linear approximation (dash line). (**b**) Piecewise constant approximation (dash line)

in (B.1) is a finite union of line segments, but still non-convex. Moreover, the logic representation in (B.1) is not compatible with commercial solvers. Given the fact that any point on a line segment can be expressed as a convex combination of two terminal points, (B.1) can be written as

$$\begin{aligned} & x = \sum_i \lambda_i x_i \\ & y = \sum_i \lambda_i f(x_i) \\ & \lambda \geq 0, \ \sum_i \lambda_i = 1 \\ & \lambda \in \mathbb{SOS}_2 \end{aligned} \tag{B.2}$$

where $\mathbb{SOS}_2$ stands for the special ordered set of type 2, describing a vector of variables with at most two adjacent ones being able to take nonzero values. The $\mathbb{SOS}_2$ constraint on λ can be declared via the build-in module of commercial solvers such as CPLEX or GUROBI. Please note that if $f(x)$ is convex and to be minimized, then the last $\mathbb{SOS}_2$ requirement is naturally met (thus can be relaxed), because the epigraph of $f(x)$ is a convex region. Otherwise, relaxing the last $\mathbb{SOS}_2$ constraint in (B.2) gives rise to the convex hull of the sampled points $(x_0, y_0), \cdots, (x_k, y_k)$. In general, the relaxation is inexact.

Branch-and-bound algorithms which directly working on SOS variables exhibit good performance [66], but it is desirable to explore equivalent MILP formulations to leverage the superiority of state-of-the-art solvers. To this end, we first provide an explicit form using additional integer variables.

$$\begin{aligned} & \lambda_0 \leq z_1 \\ & \lambda_1 \leq z_1 + z_2 \\ & \lambda_2 \leq z_2 + z_3 \\ & \quad \vdots \\ & \lambda_{k-1} \leq z_{k-1} + z_k \\ & \lambda_k \leq z_k \\ & z_i \in \{0, 1\}, \ \forall i, \ \sum\nolimits_{i=1}^k z_i = 1 \\ & \lambda_i \geq 0, \ \forall i, \ \sum\nolimits_{i=0}^k \lambda_i = 1 \end{aligned} \tag{B.3}$$

Formulation (B.3) illustrates how integer variables can be used to enforce $\mathbb{SOS}_2$ requirements on the weighting coefficients. This formulation does not involve any manually supplied parameter, and often gives stronger bounds when the integrality of binary variables is relaxed.

Sometimes, it is more convenient to use a piecewise constant approximation, especially when the original function $f(x)$ is not continuous. An example is exhibited in part (b), Fig. B.1. In this approach, the feasible interval of x is partitioned into $S-1$ segments (associated with binary variables θ_s, $s = 1, \cdots, S-1$) by S breakpoints $x_1, \cdots, x_S$ (associated with S continuous weight variables λ_s, $s = 1, \cdots, S$); In the s-th interval between x_s and x_{s+1}, the function value $f(x)$ is approximated by the arithmetic mean $f_s = 0.5[f(x_s) + f(x_{s+1})]$, $s = 1, \cdots, S-1$, which is a constant as illustrated in Fig. B.1. With an appropriate number of partitions, an arbitrary function $f(x)$ can be approximated by a piecewise constant function as follows:

$$x = \sum_{s=1}^{S} \lambda_s x_s, \ y = \sum_{s=1}^{S-1} \theta_s f_s \tag{B.4a}$$

$$\lambda_1 \le \theta_1, \ \lambda_S \le \theta_{S-1} \tag{B.4b}$$

$$\lambda_s \le \theta_{s-1} + \theta_s, \ s = 2, \cdots, S-1 \tag{B.4c}$$

$$\lambda_s \ge 0, \ s = 1, \cdots, S, \ \sum\nolimits_{s=1}^{S} \lambda_s = 1 \tag{B.4d}$$

$$\theta_s \in \{0, 1\}, s = 1, \cdots, S-1, \ \sum\nolimits_{s=1}^{S-1} \theta_s = 1 \tag{B.4e}$$

In (B.4), binary variable $\theta_s = 1$ indicates interval s is activated, and constraint (B.4e) ensures that only one interval will be activated; Furthermore, constraints (B.4b)–(B.4d) enforce weight coefficients α_s, $s = 1, \cdots, S$ to be $\mathbb{SOS}_2$; Finally, constraint (B.4a) expresses y and x via the linear combination of sampled values. The advantage of piecewise constant formulation (B.4) lies in the binary expression of function value y, such that the product of y and another continuous variable can be easily linearized via integer programming technique, which can be seen in Sect. B.2.3.

Clearly, the required number of binary variables introduced in formulation (B.3) is k, which grows linearly with respect to the number of breakpoints, and the final MILP model may suffer from computational overheads due to the presence of a large number of binary variables when more breakpoints are involved for improving accuracy. In what follows, we present a useful formulation that only engages a logarithmic number of binary variables and constraints. This technique is proposed in [67–69]. Consider the following constraints:

$$\sum_{i \in L_n} \lambda_i \leq z_n, \ \forall n \in N$$

$$\sum_{i \in R_n} \lambda_i \leq 1 - z_n, \ \forall n \in N \tag{B.5}$$

$$z_n \in \{0, 1\}, \ \forall n \in N$$

$$\lambda \geq 0, \ \sum\nolimits_{i=0}^{k} \lambda_i = 1$$

where L_n and R_n are index sets of weights λ_i, N is an index set corresponding to the number of binary variables. The dichotomy sequences $\{L_n, R_n\}_{n \in N}$ constitute a branching scheme on the indices of weights, such that constraint (B.5) guarantees that at most two adjacent elements of λ can take strictly positive values, so as to meet the $\mathbb{SOS}_2$ requirement. The required number of binary variables z_n is $\lceil \log_2 k \rceil$, which is significantly smaller than that involved in formulation (B.3).

Next, we demonstrate how to design the sets L_n and R_n based on the concept of Gray codes. For notation brevity, we restrict the discussion to the instances with 2 and 3 binary variables (which are shown in Fig. B.2), indicating 5 and 9 breakpoints (or 4 and 8 intervals) in consequence.

As shown in Fig. B.2, Gray codes G_1–G_8 form a binary system where any two adjacent numbers only differ in one bit. For example, G_4 and G_5 differ in the first bit, and G_5 and G_6 differ in the second bit. Such Gray codes are used to describe which two adjacent weights are activated. In general, sets R_n and L_n are constructed as follows: the index $v \in L_n$ if the binary values of the n-th bit of two successive

λ^0					R_1	R_2	R_3
	G_1	0	0	0			
λ^1					R_1	R_2	
	G_2	0	0	1			
λ^2					R_1		L_3
	G_3	0	1	1			
λ^3					R_1	L_2	
	G_4	0	1	0			
λ^4						L_2	R_3
	G_5	1	1	0			
λ^5					L_1		R_3
	G_6	1	0	0			
λ^6					L_1	R_2	
	G_7	1	0	1			
λ^7					L_1		L_3
	G_8	1	1	1			
λ^8					L_1	L_2	L_3

λ^0				R_1	R_2
	G_1	0	0		
λ^1				R_1	
	G_2	0	1		
λ^2					L_2
	G_3	1	1		
λ^3				L_1	
	G_4	1	0		
λ^4				L_1	R_2

Fig. B.2 Gray codes and sets L_n, R_n for two and three binary variables codes G_n and G_{n+1} are equal to 1, or $v \in R_n$ if they are equal to 0. This principle can be formally defined in a mathematical way as

$$L_n = \left\{ v \left| \begin{array}{l} (G_v^n = 1 \text{ and } G_{v+1}^n = 1) \\ \cup\, (v = 0 \text{ and } G_1^n = 1) \\ \cup\, (v = k \text{ and } G_k^n = 1) \end{array} \right. \right\} \tag{B.6}$$

$$R_n = \left\{ v \left| \begin{array}{l} (G_v^n = 0 \text{ and } G_{v+1}^n = 0) \\ \cup\, (v = 0 \text{ and } G_1^n = 0) \\ \cup\, (v = k \text{ and } G_k^n = 0) \end{array} \right. \right\} \tag{B.7}$$

where G_v^n stands for the n-th bit of code G_v.

For example, sets R_1, R_2, R_3 and L_1, L_2, L_3 for Gray codes G_1–G_8 are shown in Fig. B.2. In such a way, we can establish the rule that only two adjacent weights can be activated via (B.5). To see this, consider that if $\lambda_i > 0$ for $i = 4, 5$ and $\lambda_i = 0$ for other indices, we let $z_1 = 1$, $z_2 = 1$, $z_3 = 0$, which leads to the following constraint set:

$$\begin{cases} \lambda_0 + \lambda_1 + \lambda_2 + \lambda_3 \le 1 - z_1 = 0 \\ \lambda_5 + \lambda_6 + \lambda_7 + \lambda_8 \le z_1 = 1 \\ \lambda_0 + \lambda_1 + \lambda_6 \le 1 - z_2 = 0 \\ \lambda_3 + \lambda_4 + \lambda_8 \le z_2 = 1 \\ \lambda_0 + \lambda_4 + \lambda_5 \le 1 - z_3 = 1 \\ \lambda_2 + \lambda_7 + \lambda_8 \le z_3 = 0 \\ \lambda_i \ge 0, \forall i, \sum_{i=0}^{8} \lambda_i = 1 \end{cases}$$

Thus we can conclude that

$$\lambda_4 + \lambda_5 = 1, \lambda_4 \ge 0, \lambda_5 \ge 0,$$
$$\lambda_0 = \lambda_1 = \lambda_2 = \lambda_3 = \lambda_6 = \lambda_7 = \lambda_8 = 0$$

This mechanism can be interpreted as follows: $z_1 = 1$ enforces $\lambda_i = 0, i = 0, 1, 2, 3$ through set R_1; $z_2 = 1$ further enforces $\lambda_6 = 0$ through set R_2; finally, $z_3 = 0$ enforces $\lambda_7 = \lambda_8 = 0$ through set L_3. Then the remaining weights λ_4 and λ_5 constitute the positive coefficients. In this regard, only $\log_2 8 = 3$ binary variables and $2\log_2 8 = 6$ additional constraints are involved. Compared with formulation (B.3), the gray code can be regarded as extra branching operation enabled by problem structure, so the number of binary variables in expression (B.5) is greatly reduced in the case with a large value of k.

As a special case, consider the following problem

$$\min \left\{ \sum_i f_i(x_i) \,\middle|\, x \in X \right\} \tag{B.8}$$

where $f_i(x_i), i = 1, 2, \cdots$ are convex univariate functions, and X is a polytope. This problem is convex but nonlinear. The DCOPF problem, a fundamental issue in power market clearing, is given in this form, in which $f_i(x_i)$ is a convex quadratic function. Although any local NLP algorithm can find the global optimal solution of (B.8), there are still reasons to seek approximated LP formulations. One is that problem (B.8) may be embedded in another optimization problem and serve as its constraint. This is a pervasive modeling paradigm to study the strategic behaviors and market powers of energy providers, where the electricity market is cleared according to a DCOPF, and delivered energy of generation companies and nodal electricity prices are extracted from the optimal primal variables and dual variables associating with power balancing constraints, respectively. An LP representation allows to exploit the elegant LP duality theory for further analysis, and helps characterize optimal solution through primal-dual or KKT optimality conditions. To this end, we can opt to solve the following LP

$$\begin{aligned}
\min_{x,y,\lambda} \ & \sum_i y_i \\
\text{s.t.} \ & y_i = \sum_k \lambda_{ik} f_i(x_{ik}), \ \forall i \\
& x_i = \sum_k \lambda_{ik} x_{ik}, \ \forall i, \ x \in X \\
& \lambda \geq 0, \ \sum_k \lambda_{ik} = 1, \ \forall i
\end{aligned} \tag{B.9}$$

where $x_{ik}, k = 1, 2, \cdots$ are break points (constants) for variable x_i, and the associated weights are λ_{ik}. Because $f_i(x_i)$ are convex functions, the $\mathbb{SOS}_2$ requirement on the weight variable λ is naturally met, so it is relaxed from the constraints.

B.1.2 Bivariate Continuous Nonlinear Function

Consider a continuous nonlinear function $f(x, y)$ in two variables x and y. The entire feasible region is partitioned into $M \times N$ disjoint sub-rectangles by $M+N+2$ breakpoints x_n, $n = 0, 1, \cdots, N$ and y_m, $m = 0, 1, \cdots, M$, as illustrated in Fig. B.3, and the corresponding function values are $f_{mn} = f(x_m, y_n)$. By introducing a planar weighting coefficient matrix $\{\lambda_{mn}\}$ for each grid point that satisfies

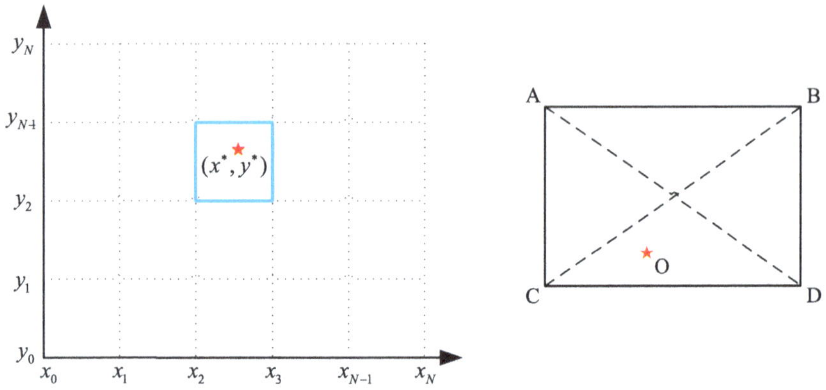

Fig. B.3 Breakpoints and active rectangle for PWL approximation

$$\begin{aligned} &\lambda_{mn} \geq 0, \ \forall m, \forall n \\ &\sum_{m=0}^{M} \sum_{n=0}^{N} \lambda_{mn} = 1 \end{aligned} \tag{B.10a}$$

we can present any point (x, y) in the feasible region by a convex combination of the extreme points of the sub-rectangle it resides in:

$$\begin{aligned} x &= \sum_{m=0}^{M} \sum_{n=0}^{N} \lambda_{mn} x_n = \sum_{n=0}^{N} \left(\sum_{m=0}^{M} \lambda_{mn} \right) x_n \\ y &= \sum_{m=0}^{M} \sum_{n=0}^{N} \lambda_{mn} y_m = \sum_{m=0}^{M} \left(\sum_{n=0}^{N} \lambda_{mn} \right) y_m \end{aligned} \tag{B.10b}$$

and its function value

$$f(x, y) = \sum_{m=0}^{M} \sum_{n=0}^{N} \lambda_{mn} f_{mn} \tag{B.10c}$$

is also a convex combination of the function values at the corner points.

As we can see from Fig. B.3, in a valid representation, if (x^*, y^*) belongs to a sub-rectangle, only the weight parameter associated with the four corner points can be non-negative, while others should be forced at 0. In such a pattern, the sum of columns/rows of matrix $\Lambda = [\lambda_{mn}], \forall m, n$, which remains a vector, should constitute an $\mathbb{SOS}_2$, and Λ is called a planar $\mathbb{SOS}_2$, which can be implemented via two $\mathbb{SOS}_1$ constraints on the marginal weight vectors. In fact, at most three of the four corner points can be associated with uniquely determined non-negative weights. Consider point O and the active rectangle ABCD shown in Fig. B.3. The location of O can be expressed by a linear combination of the coordinates of corner points x_A, x_B, x_C, x_D associating with non-negative weights $\lambda_A, \lambda_B, \lambda_C, \lambda_D$ as:

$$x_O = \lambda_A x_A + \lambda_B x_B + \lambda_C x_C + \lambda_D x_D \tag{B.11a}$$

In the first case

$$\lambda_A^1,\ \lambda_B^1,\ \lambda_C^1,\ \lambda_D^1 \geq 0,\ \lambda_A^1 + \lambda_B^1 + \lambda_C^1 + \lambda_D^1 = 1 \tag{B.11b}$$

In the second case

$$\lambda_A^2, \lambda_C^2, \lambda_D^2 \geq 0,\ \lambda_B^2 = 0,\ \lambda_A^2 + \lambda_C^2 + \lambda_D^2 = 1 \tag{B.11c}$$

In the third case

$$\lambda_B^3, \lambda_C^3, \lambda_D^3 \geq 0,\ \lambda_A^3 = 0,\ \lambda_B^3 + \lambda_C^3 + \lambda_D^3 = 1 \tag{B.11d}$$

We use superscripts 1, 2, 3 to distinguish values of weights in different representations. According to Caratheodory theorem, the non-negative weights are uniquely determined in (B.11f) and (B.11g), and in the former (latter) case, we say ΔACD (ΔBCD) is activated or selected. Denote function values in these three cases by

$$f_1(x_O) = \lambda_A^1 f(x_A) + \lambda_B^1 f(x_B) + \lambda_C^1 f(x_C) + \lambda_D^1 f(x_D) \tag{B.11e}$$

$$f_2(x_O) = \lambda_A^2 f(x_A) + \lambda_C^2 f(x_C) + \lambda_D^2 f(x_D) \tag{B.11f}$$

$$f_3(x_O) = \lambda_B^3 f(x_B) + \lambda_C^3 f(x_C) + \lambda_D^3 f(x_D) \tag{B.11g}$$

Suppose $f(x_A) < f(x_B)$, the plane defined by points B, C, D lies above that defined by points A, C, D, hence $f_2(x_O) < f_1(x_O) < f_3(x_O)$. If a smaller (larger) function value is in favor, then ΔACD (ΔBCD) will be activated at the optimal solution. Please bear in mind that as long as A, B, C, D are not in the same plane, $f_1(x_O)$ will be strictly less (greater) than $f_3(x_O)$ ($f_2(x_O)$). Therefore, (B.11e) will not become binding at the optimal solution, and the weights for active corners are uniquely determined. If rectangle ABCD is small enough, such a discrepancy can be neglected. Nonetheless, non-uniqueness of the corner weights has little injury on its application, because the optimal solution x_O and optimal value will be consistent with the original problem. The weights do not correspond to physical strategies that need to be deployed, and the linearization method can be considered as a black box to the decision maker, who provides function values at x_A, x_B, x_C, x_D and receives a unique solution x_O.

Detecting the active sub-rectangle that (x^*, y^*) resides in requires additional constraints on the weight parameter λ_{mn}. The aforementioned integer formulation is used to impose planar $\mathbb{SOS}_2$ constraints. Let λ^n and λ^m be the aggregated weights for x and y, respectively, i.e.,

$$\begin{aligned} \lambda^n &= \sum_{m=0}^{M} \lambda_{mn}, \ \forall n \\ \lambda^m &= \sum_{n=0}^{N} \lambda_{mn}, \ \forall m \end{aligned} \tag{B.12}$$

which are also called the marginal weight vectors, and introduce the following constraints:

$$\text{For } x: \left\{ \begin{aligned} \sum_{n \in L_k^1} \lambda^n &\le z_k^1 \\ \sum_{n \in R_k^1} \lambda^n &\le 1 - z_k^1 \\ z_k^1 &\in \{0, 1\} \end{aligned} \right\}, \ \forall k \in K_1 \tag{B.13}$$

$$\text{For } y: \left\{ \begin{aligned} \sum_{m \in L_k^2} \lambda^m &\le z_k^2 \\ \sum_{m \in R_k^2} \lambda^m &\le 1 - z_k^2 \\ z_k^2 &\in \{0, 1\} \end{aligned} \right\}, \ \forall k \in K_2 \tag{B.14}$$

where L_k^1, L_k^2 and R_k^2, R_k^2 are index sets of weights λ^n and λ^m, K_1 and K_2 are index sets of binary variables. The dichotomy sequences $\{L_k^1, R_k^1\}_{k \in K_1}$ and $\{L_k^2, R_k^2\}_{k \in K_2}$ constitute a branching scheme on the indices of weights, such that constraints (B.13) and (B.14) would guarantee that at most two adjacent elements of λ^n and λ^m can take strictly positive values, so as to detect the active sub-rectangle. In this approach, the required number of binary variables is $\lceil \log_2 M \rceil + \lceil \log_2 N \rceil$. The construction of these index sets has been explained in the univariate case.

Likewise, the discussions for problems (B.8) and (B.9) can be easily extended if the objective function is the sum of bi-variate convex functions, implying that the planar $\mathbb{SOS}_2$ condition is naturally met.

B.1.3 Approximation Error

This section answers a basic question: For a given function, how many intervals (break points) are needed to achieve certain error bound ε? For the ease of understanding, we restrict our attention to univariate function, including the quadratic function $f(x) = ax^2$, and more generally, the continuous function $f(x)$ that is three times continuously differentiable. Let $\phi_f(x)$ be the PWL approximation for function $f(x)$ on $X = \{x | x_l \le x \le x_m\}$, the absolute maximum approximation error is defined by $\Delta = \max_{x \in X} |f(x) - \phi_f(x)|$.

First let us consider the quadratic function $f(x) = ax^2$, which has been thoroughly studied in [70]. The analysis is briefly introduced here. Choose an arbitrary interval $[x_{i-1}, x_i] \subset X$, the PWL approximation can be parameterized in a single variable $t \in [0, 1]$ as

$$x(t) = x_{i-1} + t(x_i - x_{i-1})$$
$$\phi_f(x(t)) = ax_{i-1}^2 + at(x_i^2 - x_{i-1}^2)$$

Clearly, $f(x(0)) = \phi_f(x(0)) = x_{i-1}^2$, $f(x(1)) = \phi_f(x(1)) = x_i^2$, and $\phi_f(x(t)) > f(x(t)), \forall t \in (0, 1)$. The maximal approximation error in the interval must be found at a critical point which satisfies

$$\begin{aligned}
&\frac{d}{dt}\left(\phi_f(x(t)) - f(x(t))\right) \\
&= \frac{d}{dt} a \left[x_{i-1}^2 + t(x_i^2 - x_{i-1}^2) - (x_{i-1} + t(x_i - x_{i-1}))^2\right] \\
&= \frac{d}{dt} a(x_i - x_{i-1})^2(t - t^2) \\
&= a(x_i - x_{i-1})^2(1 - 2t) \\
&= 0 \Rightarrow t = \frac{1}{2}
\end{aligned}$$

implying that $x(1/2)$ is always a critical point where the approximation error reaches maximum, regardless of the partition of intervals, and the error is given by

$$\begin{aligned}
\Delta &= \phi\left(x\left(\frac{1}{2}\right)\right) - f\left(x\left(\frac{1}{2}\right)\right) \\
&= a\left[x_{i-1}^2 + \frac{1}{2}(x_i^2 - x_{i-1}^2) - \frac{1}{4}(x_{i-1} + x_i)^2\right] \\
&= \frac{a}{4}(x_i - x_{i-1})^2
\end{aligned}$$

which is quadratic in the length of the interval and independent of its location. In this regard, the intervals must be evenly distributed with equal length in order to get the best performance. If X is divided into n intervals, the absolute maximum approximation error is

$$\Delta = \frac{a}{4n^2}(x_m - x_l)^2$$

Therefore, for a given tolerance ϵ, the number of intervals should satisfy

$$n \geq \sqrt{\frac{a}{\varepsilon}} \frac{x_m - x_l}{2}$$

For quadratic function $f(x) = ax^2$, coefficient a determines its second-order derivative. For more general situations, above discussion implies that the number of intervals needed to perform a PWL approximation for function $f(x)$ may depend on its second-order derivative. This problem has been thoroughly studied in [71]. The conclusion is: for a three times continuously differentiable $f(x)$ in the interval $[x_l, x_m]$, the optimal number of segments $s(\varepsilon)$ under given error tolerance ε can be selected as

$$s(\varepsilon) \propto \frac{c}{\sqrt{\varepsilon}}, \quad \varepsilon \to 0^+$$

where

$$c = \frac{1}{4} \int_{x_l}^{x_m} \sqrt{|f''(x)|}$$

The conclusion still holds if $\sqrt{|f''(x)|}$ has integrable singularities at the endpoints.

B.2 Linear Formulation of Product Terms

Product of two variables, or a bilinear term, naturally arises in optimization models from various disciplines. For one example, in economic studies, if the price c and the quantity q of a commodity are variables, then the cost cq would be a bilinear term. For another, in circuit analysis, if both of the voltage v and the current i are variables, then the electric power vi would be a bilinear term. Bilinear terms are non-convex. Throughout history, linearizing bilinear terms using linear constraints and integer variables is a frequently used technique in optimization community. This section presents several techniques for the central question of product linearization: how to enforce constraint $z = xy$, depending on the types of x and y.

B.2.1 Product of Two Binary Variables

If $x \in \mathbb{B}$ and $y \in \mathbb{B}$, then $z = xy$ is equivalent to the following linear inequalities

$$\begin{gathered} 0 \le z \le y \\ 0 \le x - z \le 1 - y \\ x \in \mathbb{B},\ y \in \mathbb{B},\ z \in \mathbb{B} \end{gathered} \tag{B.15}$$

It can be verified that if $x = 1$, $y = 1$, then $z = 1$ is achieved; if $x = 0$ or $y = 0$, then $z = 0$ is enforced, regardless of the value of y or x. This is equivalent to the requirement $z = xy$.

If $x \in \mathbb{Z}^+$ belongs to interval $[x^L, x^U]$, and $y \in \mathbb{Z}^+$ belongs to interval $[y^L, y^U]$, given the following binary expansion

$$\begin{aligned} x &= x^L + \sum_{k=1}^{K_1} 2^{k-1} u_k \\ y &= y^L + \sum_{k=1}^{K_2} 2^{k-1} v_k \end{aligned} \tag{B.16}$$

where $K_1 = \lceil \log_2(x^U - x^L) \rceil$, $K_2 = \lceil \log_2(y^U - y^L) \rceil$. To develop a vector expression, define vectors $b^1 = [2^0, 2^1, \cdots, 2^{K_1-1}]$, $u = [u_1, u_2, \cdots, u_{K_1}]^T$, $b^2 = [2^0, 2^1, \cdots, 2^{K_2}]$, $v = [v_1, v_2, \cdots, v_{K_2}]^T$, and matrices $B = (b^1)^T b^2$, $z = uv^T$, then

$$xy = x^L y^L + x^L b^2 v + y^L b^1 u + \langle B, z \rangle$$

where product matrix $z = uv^T$, and $\langle B, z \rangle = \sum_i \sum_j B_{ij} z_{ij}$. Relation among u, v, and z can be linearized via Eq. (B.15) element-wise. Its compact form is given by

$$\begin{gathered} \mathbf{0}^{K_1 \times K_2} \le z \le \mathbf{1}^{K_1 \times 1} v^T \\ \mathbf{0}^{K_1 \times K_2} \le u \mathbf{1}^{1 \times K_2} - z \le \mathbf{1}^{K_1 \times K_2} - \mathbf{1}^{K_1 \times 1} v^T \\ u \in \mathbb{B}^{K_1 \times 1},\ v \in \mathbb{B}^{K_2 \times 1},\ z \in \mathbb{B}^{K_1 \times K_2} \end{gathered} \tag{B.17}$$

B.2.2 Product of Integer and Continuous Variables

We consider the binary-continuous case. If $x \in \mathbb{R}$ belongs to interval $[x^L, x^U]$, and $y \in \mathbb{B}$, then $z = xy$ is equivalent to the following linear inequalities

$$\begin{aligned} x^L y \le z \le x^U y \\ x^L(1-y) \le x - z \le x^U(1-y) \\ x \in \mathbb{R},\ y \in \mathbb{B},\ z \in \mathbb{R} \end{aligned} \tag{B.18}$$

It can be verified that if $y = 0$, then z is enforced to be 0 and $x^L \le x \le x^U$ is naturally met; if $y = 1$, then $z = x$ and $x^L \le z \le x^U$ must be satisfied, indicating the same relationship on x, y, and z. As for the integer-continuous case, the integer variable can be represented as (B.16) using binary variables, yielding a linear combination of binary-continuous products.

It should be mentioned that the upper bound x^U and the lower bound x^L are crucial for creating linearization inequalities. If explicit bounds are not available at hand, one can incorporate a constant M that is big enough. The value of M will have a notable impact on the computation time. To enhance efficiency, a desired value should be the minimal M that ensures that inequality $-M \le x \le M$ never becomes binding at optimum, as it leads to the strongest bound if integrality of binary variables is neglected, expediting the converge of the branch-and-bound procedure. However, such a value is generally unclear before we solve the problem. Nevertheless, we do not actually need to find the smallest value M^*. Any $M \ge M^*$ produces the same optimal solution and is valid for linearization. Please bear in mind that an over-large M not only deteriorates the computation time, but also cause numeric instability due to a large conditional number. So a proper tradeoff must be made between efficiency and accuracy. A proper M can be determined from estimating the bound of x from certain heuristics, which is problem-dependent.

B.2.3 Product of Two Continuous Variables

If $x \in \mathbb{R}$ belongs to interval $[x^L, x^U]$, and $y \in \mathbb{R}$ belongs to interval $[y^L, y^U]$, there are three options for linearizing their product xy. The first one considers $z = xy$ as a bivariate function $f(x, y)$, and applies the planar $\mathbb{SOS}_2$ method in Sect. B.1.2. The second one discretizes y, for example, as follows:

$$\begin{aligned} y = y^L + \sum_{k=1}^{K} 2^{k-1} u_k \Delta y \\ \Delta y = \frac{y^U - y^L}{2^K},\ u_k \in \mathbb{B},\ \forall k \end{aligned} \tag{B.19}$$

and

$$xy = xy^L + \sum_{k=1}^{K} 2^{k-1} v_k \Delta y \tag{B.20}$$

where $v_k = u_k x$ can be linearized through Eq. (B.18) as

$$\begin{aligned} & x^L u_k \le v_k \le x^U u_k, \ \forall k \\ & x^L(1-u_k) \le x - v_k \le x^U(1-u_k), \ \forall k \\ & x \in \mathbb{R}, \ u_k \in \mathbb{B}, \ \forall k, \ v_k \in \mathbb{R}, \ \forall k \end{aligned} \tag{B.21}$$

In practical problems, bilinear terms often appear as the inner production of two vectors. For convenience, we present the compact linearization of $x^T y$ via binary expansion. Let y be the candidate vector variable to be discretized; perform (B.19) on each element of y

$$y_j = y_j^L + \sum_{k=1}^{K} 2^{k-1} u_{jk} \Delta y_j, \ \forall j$$

and thus

$$x_j y_j = x_j y_j^L + \sum_{k=1}^{K} 2^{k-1} v_{jk} \Delta y_j, \ \forall j, \ v_{jk} = u_{jk} x_j, \ \forall j, \forall k$$

Relation $v_{jk} = u_{jk} x_j$ can be expressed via linear constraints

$$x_j^L u_{jk} \le v_{jk} \le x_j^U u_{jk}, \ x_j^L(1-u_{jk}) \le x_j - v_{jk} \le x_j^U(1-u_{jk}), \ \forall j, \forall k$$

Denote by V and U are matrix variables consisting of v_{jk} and u_{jk}, respectively; 1_K stands for all-one column vector with a dimension of K; Δ_Y is a diagonal matrix with Δy_j being non-zero entries; vector $\zeta = [2^0, 2^1, \cdots, 2^{K-1}]$. Combining all the above element-wise expressions together, we have the linear formulation of $x^T y$ in a compact matrix form

$$x^T y = x^T y^L + \zeta V^T \Delta y$$

in conjunction with

$$\begin{aligned} & y = y^L + \Delta_Y U \zeta^T \\ & (x^L \cdot 1_K^T) \otimes U \le V \le (x^U \cdot 1_K^T) \otimes U \\ & (x^L \cdot 1_K^T) \otimes (1-U) \le x \cdot 1_K^T - V \le (x^U \cdot 1_K^T) \otimes (1-U) \\ & x \in \mathbb{R}^J, \ y \in \mathbb{R}^J, \ U \in \mathbb{B}^{J \times K}, \ V \in \mathbb{R}^{J \times K} \end{aligned} \tag{B.22}$$

where $\otimes$ represents element-wise product of two matrices with the same dimension.

One possible drawback of this formulation is that the discretized variable is no longer continuous. The approximation accuracy can be improved by increasing the number of breakpoints without introducing too many binary variables, whose number is given by $\lceil \log_2(y^U - y^L)/\Delta y \rceil$. Furthermore, the variable to be discretized must have clear upper and lower bounds. This is not restrictive because decision variables of engineering problems are subject to physical operating limitations, such as the maximum and minimum output of a generator. Nevertheless, if x, for example, is unbounded in formulation, but the problem has a finite optimum, we can replace x^U (x^L) in (B.21) with a large enough big-M parameter $M(-M)$, so that the true optimal solution remains feasible. It should be pointed out that the value of M may influence the computational efficiency of the equivalent MILP, as mentioned previously. The optimal choice of M in general cases remains an open problem, but there could be heuristic methods for specific instances. For example, if x stands for the marginal production cost, which is a dual variable whose bounds are unclear, one can alternatively determine a suitable bound from historical data or price forecast.

An alternative formulation for the second option deals with product term $xf(y)$, and xy is a special case when $f(y) = y$. By performing the piecewise constant approximation (B.4) on function $f(y)$, the product becomes $xy = \sum_{s=1}^{S-1} x\theta_s y_s$, where y_s is constant, x is continuous, and θ_s is binary. The products $x\theta_s$, $s = 1, \cdots, S-1$ can be readily linearized via the method in Sect. B.2.2. In this approach, the continuity of x and y are retained. However, the number of binary variables in the piecewise constant approximation for $f(y)$ grows linearly in the number of samples on y.

The third one converts the product into a separable form, and then performs PWL approximation for univariate nonlinear functions. To see this, consider a bilinear term xy. Introduce two continuous variables u and v defined as follows:

$$\begin{aligned} u &= \frac{1}{2}(x + y) \\ v &= \frac{1}{2}(x - y) \end{aligned} \tag{B.23}$$

Now we have

$$xy = u^2 - v^2 \tag{B.24}$$

In (B.24), u^2 and v^2 are univariate nonlinear functions, and can be approximated by the PWL method presented in Sect. B.1.1. Furthermore, if $x_l \le x \le x_u$, $y_l \le y \le y_u$, then the lower and upper bounds of u and v are given by

$$\begin{aligned} &\frac{1}{2}(x_l + y_l) \le u \le \frac{1}{2}(x_u + y_u) \\ &\frac{1}{2}(x_l - y_u) \le v \le \frac{1}{2}(x_u - y_l) \end{aligned}$$

Formulation (B.24) has a connotative advantage. If xy appears in the objective function which is to be minimized and is not involved in constraints, we only need to approximate v^2 because u^2 is convex and $-v^2$ is concave. The minimum amount of binary variables in this method is a logarithmic function in the number of break points, as explained in Sect. B.1.1.

The bilinear term xy can be replaced by a single variable z in the following situation: (1) if the lower bounds x_l and y_l are nonnegative; (2) either x or y is not referenced anywhere else except in xy. For instance, y is such a variable, then the bilinear term xy can be replaced by variable z and constraint $xy_l \le z \le xy_u$. Once the problem is solved, y can be recovered by $y = z/x$ if $x > 0$, and the inequality constraint on z guarantees $y \in [y_l, y_u]$; otherwise if $x = 0$, then y is undetermined and has no impact on the optimum.

B.2.4 Monomial of Binary Variables

Previous cases discuss linearizing the product of two variables. Now we consider a binary monomial with n variables

$$z = x_1 x_2 \cdots x_n, \; x_i \in \{0, 1\}, \; i = 1, 2, \cdots, n \tag{B.25}$$

Clearly, this monomial takes a binary value. Since the product of two binary can be expressed by a single one in light of (B.15), the monomial can be linearized recursively. Nevertheless, by making full use of the binary property of z, a smarter and concise way to represent (B.25) is given by

$$z \in \{0, 1\} \tag{B.26}$$

$$z \le \frac{x_1 + x_2 + \cdots + x_n}{n} \tag{B.27}$$

$$z \ge \frac{x_1 + x_2 + \cdots + x_n - n + 1}{n} \tag{B.28}$$

If at least one of x_i is equal to 0, because $\sum_{i=1}^{n} x_i - n + 1 \le 0$, (B.28) becomes redundant; moreover, $\sum_{i=1}^{n} x_i / n \le 1 - 1/n$, which removes $z = 1$ from the feasible region, so z will take a value of 0; otherwise, if all x_i are equal to 1, $\sum_{i=1}^{n} x_i / n = 1$, and the right-hand side of (B.28) is $1/n$, which removes $z = 0$ from the feasible region. Hence z is forced to be 1. In conclusion, linear constraints (B.26)–(B.28) have the same effect as (B.25).

In view of the above transformation technique, a binary polynomial program can always be reformulated as a binary linear program. Moreover, if a single continuous variable appears in the monomial, the problem can be reformulated as an MILP.

B.2.5 Product of Functions in Integer Variables

First, let us consider $z = f_1(x_1)f_2(x_2)$, where decision variables are positive integers, i.e., $x_i \in \{d_{i,1}, d_{i,2}, \cdots, d_{i,r_i}\}, i = 1, 2$. Without particular tricks, f_1 and f_2 can be expressed as

$$f_1 = \sum_{j=1}^{r_1} f_1(d_{1,j})u_{1,j},\ u_{1,j} \in \{0, 1\},\ \sum_{j=1}^{r_1} u_{1,j} = 1$$

$$f_2 = \sum_{j=1}^{r_2} f_2(d_{2,j})u_{2,j},\ u_{2,j} \in \{0, 1\},\ \sum_{j=1}^{r_2} u_{2,j} = 1$$

and the product of two binary variables can be linearized via (B.15). The above formulation introduces a lot of intermediary binary variables, and is not propitious to represent a product with more functions recursively.

Reference [72] suggests another choice

$$z = \sum_{i=1}^{r_2} f_2(d_{2,j})\sigma_{2,j},\ \sum_{i=1}^{r_2} \sigma_{2,j} = f_1(x_1),\ \sigma_{2,j} = f_1(x_1)u_{2,j} \tag{B.29}$$

where $u_{2,j} \in \{0, 1\}$ and $\sum_{j=1}^{r_2} u_{2,j} = 1$. Although $f_1(x_1)u_{2,j}$ remains nonlinear because of decision variable x_1, (B.29) can be used to linearize a product with more than two nonlinear functions.

To see this, denote by $z_1 = f_1(x_1)$, $z_i = z_{i-1}f_i(x_i)$, $i = 2, \cdots, n$; integer variable $x_i \in \{d_{i,1}, d_{i,2}, \cdots, d_{i,r_i}\}$, $f_i(x_i) > 0$, $i = 1, \cdots, n$. By using (B.29), $z_i, i = 1, \cdots, n$ have the following expressions [72]

$$\begin{aligned}
& z_1 = \sum_{j=1}^{r_1} f_1(d_{1,j})u_{1,j} \\
& z_2 = \sum_{j=1}^{r_2} f_2(d_{2,j})\sigma_{2,j},\ \sum_{i=1}^{r_2} \sigma_{2,j} = z_1,\ \cdots \\
& z_n = \sum_{j=1}^{r_n} f_n(d_{n,j})\sigma_{n,j},\ \sum_{i=1}^{r_n} \sigma_{n,j} = z_{n-1} \\
& \left.\begin{array}{l} 0 \le z_{i-1} - \sigma_{i,j} \le \bar{z}_{i-1}(1 - u_{i,j}) \\ 0 \le \sigma_{i,j} \le \bar{z}_{i-1}u_{i,j},\ u_{i,j} \in \{0, 1\} \end{array}\right\},\ \begin{array}{l} j = 1, \cdots, r_i, \\ i = 2, \cdots, n \end{array} \\
& x_i = \sum_{j=1}^{r_i} d_{i,j}u_{i,j},\ \sum_{j=1}^{r_i} u_{i,j} = 1,\ i = 1, 2, \cdots, n
\end{aligned} \tag{B.30}$$

In (B.30), the number of binary variables is $\sum_{i=1}^{n} r_i$, and grows linearly in the dimension of x and the interval length of each x_i. To reduce the number of auxiliary binary variable $u_{i,j}$, the dichotomy procedure in Sect. B.1.1 for SOS2 can be applied, which is discussed in [72].

B.2.6 Log-Sum Functions

We consider log-sum function $\log(x_1 + x_2 + \cdots + x_n)$, which arises from solving a signomial geometric programming problem. The basic element in such a problem has a form of

$$c_k \prod_{j=1}^{l} y_j^{a_{jk}} \tag{B.31}$$

where $y_j > 0$, c_k is a constant, and $a_{jk} \in \mathbb{R}$. Non-integer value of a_{jk} makes signomial geometric programming problem even harder than polynomial programs. Under some variable transformation, the non-convexity of a signomial geometric program can be concentrated in some log-sum functions [73]. In view of the form in (B.31), we discuss log-sum function in Sect. B.2.

We aim to represent function $\log(x_1 + x_2 + \cdots + x_n)$ in terms of $\log x_1$, $\log x_2$, $\cdots$, $\log x_n$. Following the method in [73], define a univariate function $F(X) = \log(1+e^X)$ and let $X_i = \log x_i$, $\Gamma_i = \log(x_1+\cdots+x_i)$, $i = 1, \cdots, n$. The relation between X_i and Γ_i can be revealed. Because

$$\begin{aligned} F(X_{i+1} - \Gamma_i) &= \log\left(1 + e^{\log x_{i+1} - \log(x_1+\cdots+x_i)}\right) \\ &= \log\left(1 + \frac{x_{i+1}}{x_1 + \cdots + x_i}\right) = \Gamma_{i+1} - \Gamma_i \end{aligned}$$

By stipulating $W_i = X_{i+1} - \Gamma_i$, we have the following recursive equations

$$\begin{aligned} \Gamma_{i+1} &= \Gamma_i + F(W_i),\ i = 1, \cdots, n-1 \\ W_i &= X_{i+1} - \Gamma_i,\ i = 1, \cdots, n-1 \end{aligned} \tag{B.32}$$

Function $F(W_i)$ can be linearized using the method in Sect. B.1.1. Based on this technique, an outer-approximation approach is proposed in [73] to solve signomial geometric programming problem via MILP.

B.3 Other Frequently Used Formulations

B.3.1 Minimum Values

Let x_1, x_2, $\cdots$, x_n be continuous variables with known lower bound x_i^L and upper bound x_i^U, and $L = \min\{x_1^L, x_2^L, \cdots, x_n^L\}$, then their minimum $y = \min\{x_1, x_2, \cdots, x_n\}$ can be expressed via linear constraints

$$\begin{aligned} & x_i^L \le x_i \le x_i^U, \ \forall i \\ & y \le x_i, \forall i \\ & x_i - (x_i^U - L)(1 - z_i) \le y, \forall i \\ & z_i \in \mathbb{B}, \ \forall i, \ \sum_i z_i = 1 \end{aligned} \tag{B.33}$$

The second inequality guarantees $y \le \min\{x_1, x_2, \cdots, x_n\}$; in addition, if $z_i = 1$, then $y \ge x_i$, hence y achieves the minimal value of $\{x_i\}$. According to the definition of L, $x_i - y \le x_i^U - L$, $\forall i$ holds, thus the third inequality is inactive for the remaining $n - 1$ variables with $z_i = 0$.

B.3.2 Maximum Values

Let x_1, x_2, $\cdots$, x_n be continuous variables with known lower bound x_i^L and upper bound x_i^U, and $U = \max\{x_1^U, x_2^U, \cdots, x_n^U\}$, then their maximum $y = \max\{x_1, x_2, \cdots, x_n\}$ can be expressed via linear constraints

$$\begin{aligned} & x_i^L \le x_i \le x_i^U, \ \forall i \\ & y \ge x_i, \forall i \\ & x_i + (U - x_i^L)(1 - z_i) \ge y, \forall i \\ & z_i \in \mathbb{B}, \ \forall i, \ \sum_i z_i = 1 \end{aligned} \tag{B.34}$$

The second inequality guarantees $y \ge \max\{x_1, x_2, \cdots, x_n\}$; in addition, if $z_i = 1$, then $y \le x_i$, hence y achieves the maximal value of $\{x_i\}$. According to the definition of U, $y - x_i \le U - x_i^L$, $\forall i$ holds, thus the third inequality is inactive for the remaining $n - 1$ variables with $z_i = 0$.

B.3.3 Absolute Values

Suppose $x \in \mathbb{R}$ and $|x| \le U$, the absolute value function $y = |x|$, which is nonlinear, can be expressed via PWL function as

$$\begin{aligned} &0 \le y - x \le 2Uz, \ U(1-z) \ge x \\ &0 \le y + x \le 2U(1-z), \ -Uz \le x \\ -&U \le x \le U, \ z \in \mathbb{B} \end{aligned} \tag{B.35}$$

When $x > 0$, the first line yields $z = 0$ and $y = x$, while the second line is inactive. When $x < 0$, the second line yields $z = 1$ and $y = -x$, while the first line is inactive. When $x = 0$, either $z = 0$ or $z = 1$ gives $y = 0$. In conclusion, (B.35) has the same effect as $y = |x|$.

B.3.4 Linear Fractional of Binary Variables

A linear fractional of binary variables takes the form of

$$\frac{a_0 + \sum_{i=1}^n a_i x_i}{b_0 + \sum_{i=1}^n b_i x_i} \tag{B.36}$$

We assume $b_0 + \sum_{i=1}^n b_i x_i \ne 0$ for all $x \in \{0, 1\}^n$. Define a new continuous variable

$$y = \frac{1}{b_0 + \sum_{i=1}^n b_i x_i} \tag{B.37}$$

The lower bound and upper bound of y can be easily computed. Then the linear fractional shown in (B.36) can be replaced with a linear expression

$$a_0 y + \sum_{i=1}^n a_i z_i \tag{B.38}$$

with constraints

$$b_0 y + \sum_{i=1}^n b_i z_i = 1 \tag{B.39}$$

$$z_i = x_i y, \ \forall i \tag{B.40}$$

where (B.40) describes a product of a binary variable and a continuous variable, which can be linearized through Eq. (B.18).

B.3.5 Disjunctive Inequalities

Let $\{P^i\}, i = 1, 2, \cdots, m$ be a finite set of bounded polyhedra. Disjunctive inequalities usually arise when the solution space is characterized by the union $S = \cup_{i=1}^m P^i$ of these polyhedra. Unlike intersection operator which preserves convexity, disjunctive inequalities form a non-convex region. It can be represented by MILP model using binary variables. We introduce three emblematic methods.

1. Big-M Formulation

The hyperplane representations of polyhedra are given by $P^i = \{x \in \mathbb{R}^n | A^i x \le b^i\}, i = 1, 2, \cdots, m$. By introducing binary variables $z_i, i = 1, 2, \cdots, m$, an MILP formulation for S can be written as

$$\begin{aligned} & A^i x \le b^i + M^i(1 - z_i), \ \forall i \\ & z_i \in \mathbb{B}, \ \forall i, \ \sum_{i=1}^m z_i = 1 \end{aligned} \tag{B.41}$$

where M^i is a vector such that when $z_i = 0$, $A^i x \le b^i + M^i$ holds. To show the impact of the value of M on the tightness of formulations (B.41) when integrality constraints $z_i \in \mathbb{B}, \forall i$ are relaxed as $z_i \in [0, 1], \forall i$, we contrivedly construct 4 polyhedra in $\mathbb{R}^2$, which are depicted in Fig. B.4. The continuous relaxations of (B.41) with different values of M are illustrated in the same graph, showing that the smaller the value of M, the tighter the relaxation of (B.41).

From a computational perspective, the element in M should be as small as possible, because a huge constant without any insights about problem data will feature a bad conditional number. Furthermore, the continuous relaxation of MILP model will be very weak, resulting in poor objective value bounds and excessive branch-and-bound computation. The goal of big-M parameter selection is to create a model whose continuous relaxation is close to the convex hull of the original constraint, i.e. the smallest convex set that contains the original feasible region. A possible selection of the big-M parameter is

$$\begin{aligned} M_l^i &= \left(\max_{j \ne i} M_l^{ij} \right) - b_l^i \\ M_l^{ij} &= \max_x \left\{ [A^i x]_l : A^j x \le b^j \right\} \end{aligned} \tag{B.42}$$

where subscript l stands for the l-th element of a vector or l-th row of a matrix. As polyhedron P^i are bounded, all bound parameters in (B.42) are well defined.

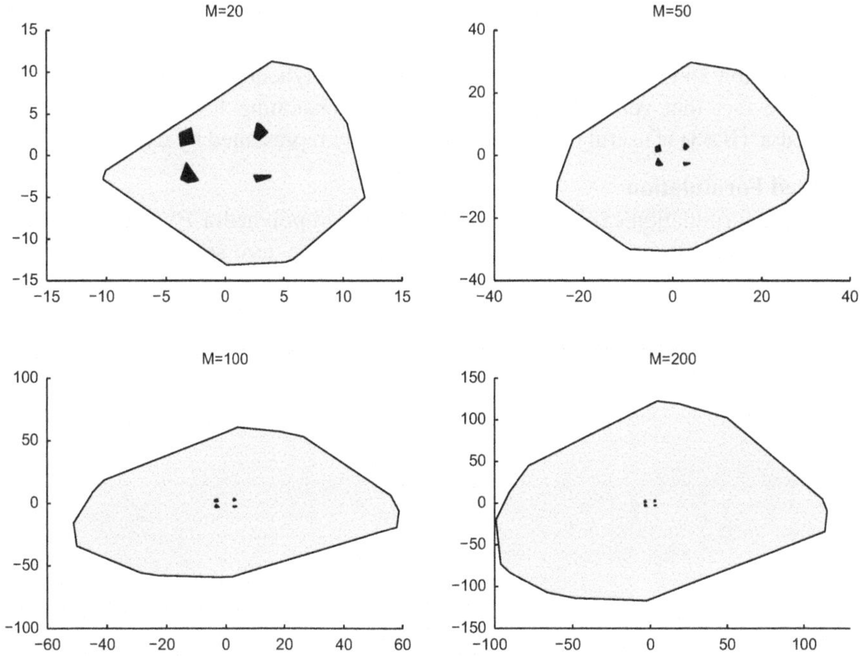

Fig. B.4 Big-M formulation and their relaxed regions

However, even the tightest big-M parameter will yield a relaxed solution space that is generally larger than the convex hull of the original feasible set. In many applications, good variable bounds can be estimated from certain heuristic methods which explore specific problem data and structure.

2. Convex Hull Formulation

Let $\text{vert}(P^i) = \{v_l^i\}, l = 1, 2, \cdots, L^i$ denote sets of vertices of polyhedra $\{P^i\}, i = 1, 2, \cdots, m$, where L^i is the number of vertices of P^i. The set of extreme rays is empty since P^i is bounded. By introducing binary variables $z_i, i = 1, 2, \cdots, m$, an MILP formulation for S is given by

$$\begin{aligned} & \sum_{i=1}^{m} \sum_{l=1}^{L^i} \lambda_l^i v_l^i = x \\ & \sum_{l=1}^{L^i} \lambda_l^i = z_i, \ \forall i \\ & \lambda_l^i \geq 0, \ \forall i, \ \forall l \\ z_i \in \mathbb{B}, \ \forall i, \ & \sum_{i=1}^{m} z_i = 1 \end{aligned} \tag{B.43}$$

Formulation (B.43) does not rely on manually supplied parameter. Instead, it requires enumerating all extreme points of polyhedra P^i. Although the vertex representation and hyperplane representation of a polyhedron are interchangeable, given the fact that vertex enumeration is time consuming for high-dimensional polyhedra, (B.43) is useful only if P^i are originally represented by extreme points.

3. Lifted Formulation

A smarter formulation exploits the fact that bounded polyhedra P^i share the same recession cone $\{0\}$, i.e., equation $A^i x = 0$ has no non-zero solutions. Otherwise, suppose $A^i x^* = 0$, $x^* \neq 0$, and $y \in P^i$, then $y + \lambda x^* \in P^i$, $\forall \lambda > 0$, because $A^i(y + \lambda x^*) = A^i y \le b^i$. As a result, P^i is unbounded. Bearing this in mind, an MILP formulation for S is given by

$$\begin{aligned} & A^i x^i \le b^i z^i, \ \forall i \\ & \sum_{i=1}^m x^i = x \\ & z_i \in \mathbb{B}, \ \forall i \\ & \sum_{i=1}^m z_i = 1 \end{aligned} \tag{B.44}$$

Formulation (B.44) is also parameter-free. Since it incorporates additional continuous variable for each polytope, we call it a lifted formulation. It is easy to see that the feasible region of x is the union of P^i: if $z_i = 0$, $x^i = 0$ as analyzed before; otherwise, if $z_i = 1$, $x = x^i \in P^i$.

4. Complementarity and Slackness Condition

Complementarity and slackness condition naturally arises in the KKT optimality condition of a mathematical programming problem, an equilibrium problem, a hierarchical optimization problem, and so on. It is a quintessential law to characterize the logic condition under which a rational decision-making progress must obey. Here we pay attention to the linear case and equivalent MILP formulation, because nonlinear cases give rise to MINLPs, which are challenging to solve and not superior from the computational point of view.

A linear complementarity and slackness condition can be written as

$$0 \le y \perp Ax - b \ge 0 \tag{B.45}$$

where vectors x and y are decision variables; A and b are constant coefficients with compatible dimensions; notation $\perp$ stands for the orthogonality of two vectors. In fact, (B.45) encompasses the following nonlinear constraints in traditional form

$$y \ge 0, \ Ax - b \ge 0, \ y^T(Ax - b) = 0 \tag{B.46}$$

In view of the non-negativeness of y and $Ax - b$, the orthogonality condition is equivalent to the element-wise logic form $y_i = 0$ or $a_i x - b_i = 0$, $\forall i$, where a_i is the i-th row of A; in other words, at most one of y_i and $a_i x - b_i$ can take a strictly positive value, implying that the feasible region is either the slice $y_i = 0$ or the slice $a_i x - b_i = 0$. Therefore, (B.45) can be regarded as a special case of the disjunctive constraints.

In practical application, (B.45) usually serves as constraints in an optimization problem. For example, in a sequential decision making or a linear bilevel program, the KKT condition of the lower-level LP appears in the form of (B.45), which is the constraint of the upper-level optimization problem. The main computation challenge arises from the orthogonality condition, which is nonlinear and non-convex, and violates the linear independent constraint qualification, see Appendix D.3 for an example. Nonetheless, in view of the switching logic between y_i and $a_i x - b_i$, we can introduce a binary variable z_i to select which slice is active [74]

$$\begin{aligned} &0 \le a_i x - b_i \le M z_i, \ \forall i \\ &0 \le y_i \le M(1 - z_i), \ \forall i \end{aligned} \tag{B.47}$$

where M is a large enough constant. According to (B.47), if $z_i = 0$, then $(Ax-b)_i = 0$ must hold, and the second inequality is redundant; otherwise, if $z_i = 1$, then we have $y_i = 0$, and the first inequality becomes redundant. (B.47) can be written in a compact form as

$$\begin{aligned} &0 \le Ax - b \le Mz \\ &0 \le y \le M(1 - z) \end{aligned} \tag{B.48}$$

It is worth mentioning that the big-M parameter M has a notable impact on the feasible region of the relaxed problem as well as the computational efficiency of the MILP model, as illustrated in Fig. B.4. One should make sure that (B.48) would not remove the optimal solution from the feasible set. If both x and y have clear bounds, then M can be easily estimated; otherwise, we may prudently employ a large M, at the cost of sacrificing the computational efficiency.

Furthermore, if we are aiming to solve (B.45) without an objective function and other constraints, such a problem is called a linear complementarity problem (under some proper transformation), for which we can build parameter-free MILP models. More details can be found in Appendix D.4.2.

B.3.6 Logical Conditions

Logical conditions are associated with indicator constraints with a statement like "if event A then event B." An event can be described in many ways. For example, a binary variable $a = 1$ can stand for event A happens, and otherwise $a = 0$; a

Table B.1 Linear form of some typical logic conditions

If A then B	$b \geq a$
Not B	$1 - b$
If A then not B	$a + b \leq 1$
If not A then B	$a + b \geq 1$
A if and only if B	$a = b$
If A then B and C	$b + c \geq 2a$
If A then B or C	$b + c \geq a$
If B or C then A	$2a \geq b + c$
If B and C then A	$a \geq b + c - 1$
If M or more of N events then A	$(N - M + 1)a \geq b + c + \cdots - M + 1$

point x belongs to a set X can denote a system is under secure operating condition, and otherwise $x \notin X$. In view of this, the disjunctive constraints discussed above is a special case of logical condition. In this section, we expatiate on how some usual logical conditions can be expressed via linear constraints. Let A, B, C, $\cdots$ associated with binary variables a, b, c, $\cdots$ represent events. The main results for linearizing typical logical conditions are summarized in Table B.1 [75].

Logical AND is formulated as a function of two binary inputs. Specifically, $c = a$ AND b can be expressed as $c = \min\{a, b\}$ or $c = ab$. The former one can be linearized via (B.33) and the letter one through (B.15), and both of them render

$$c \leq a, \ c \leq b, \ c \geq a + b - 1, \ c \geq 0 \tag{B.49}$$

For the case with multiple binary inputs, i.e., $c = \min\{c_1, \cdots, c_n\}$, or $c = \prod_{i=1}^{n} c_i$, (B.49) can be generalized as

$$c \leq c_i, \ \forall i, \ c \geq \sum\nolimits_{i=1}^{n} c_i - n + 1, \ c \geq 0 \tag{B.50}$$

Logical OR is formulated as a function of two binary inputs, i.e., $c = \max\{a, b\}$, which can be linearized via (B.34), yielding

$$c \geq a, \ c \geq b, \ c \leq a + b, \ c \leq 1 \tag{B.51}$$

For the case with multiple binary inputs, i.e., $c = \max\{c_1, \cdots, c_n\}$, (B.51) can be generalized as

$$c \geq c_i, \ \forall i, \ c \leq \sum\nolimits_{i=1}^{n} c_i, \ c \leq 1 \tag{B.52}$$

B.4 Further Reading

Throughout the half-century long research and development, MILP has become an indispensable and unprecedentedly powerful modeling tool in mathematics and engineering, thanks to the advent of efficient solvers that encapsulate many state-of-the-art techniques [76]. This chapter aims to provide an overview on formulation tricks that transform complicated conditions into MILPs, so as to take full advantages of off-the-shelf solvers. The paradigm is able to deal with a fairly broad class of hard optimization problems.

Readers who are interested in the strength of MILP model may find in-depth discussions in [77] and references therein. For those interested in the PWL approximation of nonlinear functions, we refer to [78–80] and references therein, for various models and methods. The most promising one may be the convex combination model with a logarithmic number of binary variables, whose implementation has been thoroughly discussed in [67–69]. For those who are interested in the polyhedral study of single-term bilinear sets and MILP based methods for bilinear programs may find extensive information in [81, 82] and references therein. For those who need more knowledge about mathematical program with disjunctive constraints, in which constraint activity is controlled by logical conditions, we recommend [83]; specifically, the choice of big-M parameter is discussed in [84]. For those who wish to learn more about integer programming techniques, we refer to [85] for the formulation of union of polyhedra, [86] for the representability of MILP, and [87, 88] for more general mixed-integer conic programming as well as its duality theory. To the best of our knowledge, dissertation [89] launches the most comprehensive and in-depth study on MILP approximation of non-convex optimization problems. The state-of-the-art MILP formulations which balance problem size, strength, and branching behavior are developed and compared, including those mentioned above. The discussions in [89] offer insights on designing efficient MILP models that perform extremely well in practice, despite their theoretically non-polynomial complexity in the worst case.

Appendix C
Basics of Robust Optimization

To be uncertain is to be uncomfortable, but to be certain is to be ridiculous.

Real-world decision-making models often involve unknown data. Reasons for data uncertainty could come from inexact measurements or forecast errors. For example, in power system operation, the wind power generation and system loads are barely known exactly at the time when the generation schedule should be made; in inventory management, market price and demand volatility is the main source of financial risks. In fact, optimal solutions to mathematical programming problems can be highly sensitive to parameter perturbations [90]. The optimal solution to the nominal problem may be highly suboptimal or even infeasible in reality due to parameter inaccuracy. Consequently, there is a great need of a systematic methodology that is capable of quantifying the impact of data inexactness on the solution quality, and is able to produce robust solutions that are insensitive to data uncertainty.

Optimization under uncertainty has been a focus of the operational research community for a long time. Two approaches are prevalent to deal with uncertain data in optimization, namely stochastic optimization (SO) and robust optimization (RO). They differ in the ways of modeling uncertainty. The former one assumes that the true probability distribution of uncertain data is known or can be estimated from available information, and minimizes the expected cost in its objective function. SO provides strategies that are optimal in the sense of statistics. However, the probability distribution itself may be inexact owing to the lack of enough data, and the performance of the optimal solution could be sensitive to the probability distribution chosen in the SO model. The latter one considers uncertain data resides in a pre-defined uncertainty set, and minimizes the cost in the worst-case scenario in its objective function. Constraint violation is not allowed for all possible data realizations in the uncertainty set. RO is popular because it relies on simple data and distribution-free. From the computational perspective, it is equivalent to convex optimization problems for a variety of uncertainty sets and problem types; for the intractable cases, it can be solved via systematic iteration algorithms. For more technical details about RO, we refer to [90–93], survey articles

W. Wei, J. Wang, *Modeling and Optimization of Interdependent Energy Infrastructures*, https://doi.org/10.1007/978-3-030-25958-7

[94, 95], and many references therein. Recently, distributionally robust optimization (DRO), an emerging methodology that inherits the advantages of SO and RO, has attracted wide attention. In DRO, uncertain data are described by probability distribution functions which are not known exactly and restricted in a functional ambiguity set constructed from available information and structured properties. The expected cost associated with the worst-case distribution is minimized, and the probability of constraint violations can be controlled via robust chance constraints. In many cases, the DRO can be reformulated as a convex optimization problem, or solved iteratively via convex optimization. RO and DRO approaches are young and active research fields, and the challenge is to explore tractable reformulations with various kinds of uncertainties. SO is a relatively mature technique, and the current research is focusing on probabilistic modeling of uncertainty, chance constrained programming, multi-stage SO such as stochastic dual dynamic programming, as well as more efficient computational methods.

There are several ways to categorize robust optimization methods. According to how uncertainty is dealt with, they can be classified into static (single-stage) RO and dynamic (multi-stage) RO. According to how uncertainty is modeled, they can be divided into RO and DRO. In the latter category, the ambiguity set for probability distribution can be further classified into the moment based one and the divergence based one. We will shed light on each of them in this chapter. Specifically, RO will be discussed in Sects. C.1 and C.2, moment-based DRO will be presented in Sect. C.3, and divergence-based DRO, also called robust SO will be illuminated in Sect. C.4. In the operations research community, DRO and robust SO refer to the same thing: optimization problem with distributional uncertainty, and can be used interchangeably, although DRO is preferred by the majority of researchers. In this book, we intentionally distinguish them because the moment ambiguity set can be set up with little information and is more likely an RO; the divergence based set relies on an empirical distribution (may be inexact), so is more similar to an SO. In fact, the gap between SO and RO has been significantly narrowed by recent research progress in the sense of data-driven optimization.

C.1 Static Robust Optimization

For the purpose of clarity, we begin to explain the paradigm of static RO from LPs, the best known and most frequently used mathematical programming problem in engineering applications. It is relatively easy to derive tractable robust counterparts with various uncertainty sets. Nevertheless, most results can be readily generalized to robust conic programs. The general form of an LP with uncertain parameters can be written as follows:

$$\min_x \left\{ c^T x \;\middle|\; Ax \le b \right\} : (A, b, c) \in W \tag{C.1}$$

where x is the decision variable, A, b, c are coefficient matrices with compatible dimensions, and W denotes the set of all possible data realizations constructed from available information or historical data, or merely a rough estimation.

Without loss of generality, we can assume that the objective function and the constraint right-hand side in (C.1) are certain, and uncertainty only exists in coefficient matrix A. To see this, it is not difficult to observe that problem (C.1) can be written as an epigraph form

$$\min_{t,x,y} \left\{t \mid c^T x - t \le 0,\ Ax - by \le 0,\ y = 1\right\} : (A, b, c) \in W$$

By introducing additional scalar variables t and y, coefficients appearing in the objective function and constraint right-hand side are constants. With this transformation, it will be more convenient to define the feasible solution and the optimal solution to (C.1). Hereinafter, we neglect the uncertainty in cost coefficient vector c and constraint right-hand vector b without particular mention, and consider problem

$$\min_{x} \left\{c^T x \;\middle|\; Ax \le b\right\} : A \in W \tag{C.2}$$

Next we present solution concepts of static RO under uncertain data.

C.1.1 Basic Assumptions and Formulations

Basic assumptions and definitions in static RO [90] are summarized as follows.

Assumption C.1 *Vector x represents "here-and-now" decisions: they should be determined without knowing exact values of uncertain parameters.*

Assumption C.2 *Once the decisions are made, constraints must be feasible when the actual data is within the uncertainty set W, and may be either feasible or not when the actual data step outside the uncertainty set W.*

These assumptions bring about the definition for a feasible solution of (C.2).

Definition C.1 A vector x is called a robust feasible solution to (C.2) if the following condition holds:

$$Ax \le b,\ \forall A \in W \tag{C.3}$$

To prescribe an optimal solution, the worst-case criterion is widely accepted in RO studies, leading to the following definition:

Definition C.2 The robust optimal value of (C.2) is the minimum value of the objective function over all possible x that satisfies (C.3).

After we have agreed on the meanings of feasibility and optimality of (C.2), we can seek the optimal solution among all robust feasible solutions to the problem. Now, the robust counterpart (RC) of the uncertain LP (C.2) can be described as:

$$\begin{aligned} \min_x \quad & c^T x \\ \text{s.t.} \quad & a_i^T x \le b_i, \ \forall i, \ \forall A \in W \end{aligned} \tag{C.4}$$

where a_i^T is the i-th row of matrix A, and b_i is the i-th element of vector b. We have two observations on the formulation of robust constraints in (C.4).

Proposition C.1 *Robust feasible solutions of (C.4) remain the same if we replace W with the Cartesian product $\hat{W} = W_1 \times \cdots \times W_n$, where $W_i = \{a_i | \exists A \in W\}$ is the projection of W on the coefficient space of i-th row of A.*

This is called the constraint-wise property in static RO [90]. The reason is

$$a_i^T x \le b_i, \ \forall A \in W \ \Leftrightarrow \ \max_{A \in W} a_i^T x \le b_i \ \Leftrightarrow \ \max_{a_i \in W_i} a_i^T x \le b_i$$

As a result, problem (C.4) comes down to

$$\begin{aligned} \min_x \quad & c^T x \\ \text{s.t.} \quad & a_i^T x \le b_i, \forall a_i \in W_i, \ \forall i \end{aligned} \tag{C.5}$$

Proposition C.1 seems rather counter-intuitive. One may perceive that (C.4) will be less conservative with uncertainty set W since it is a subset of $\hat{W}$. In fact, later we will see that this intuition is true for adjustable robustness.

Proposition C.2 *Robust feasible solutions of (C.5) remain the same if we replace W_i with its convex hull* $\text{conv}(W_i)$.

To see this, let vector a_i^j, $j = 1, 2, \cdots$ be the extreme points of W_i, then any point $\bar{a}_i \in \text{conv}(W_i)$ can be expressed by $\bar{a}_i = \sum_j \lambda_j a_i^j$, where $\lambda_j \ge 0, \sum_j \lambda_j = 1$ are weight coefficients. If x is feasible for all extreme points a_i^j, i.e., $a_i^j x \le b_i, \forall j$, then

$$\bar{a}_i^T x = \sum_j \lambda_j a_i^j x \le \sum_j \lambda_j b_i = b_i$$

which indicates that the constraint remains intact for all uncertain parameters reside in $\text{conv}(W_i)$.

Combining Propositions C.1 and C.2, we can conclude that the robust counterpart of an uncertain LP with a certain objective remains intact even if sets W_i of uncertain data are extended to their closed convex hulls, and W to the Cartesian product of the resulting sets. In other words, we can make a further assumption on the uncertainty set without loss of generality.

Assumption C.3 *The uncertainty set W is the Cartesian product of closed and convex sets.*

C.1.2 Tractable Reformulations

The constraint-wise property enables us to analyze the robustness of each constraint $a_i^T x \le b_i$, $\forall a_i \in W_i$ separately. Without particular mention, we will omit the subscript i for brevity. To facilitate discussion, it is convenient to parameterize the uncertain vector as $a = \bar{a} + P\zeta$, where $\bar{a}$ is the nominal value of a, P is a constant matrix, ζ is a new variable that is uncertain. This section will focus on how to derive tractable reformulation for robust constraints in the form of

$$(\bar{a} + P\zeta)^T x \le b, \ \forall \zeta \in Z \tag{C.6}$$

where Z is the uncertainty set of variable ζ. For same reasons, we can assume that Z is closed and convex. A "computationally tractable" problem means that there are known solution algorithms which can solve the problem with polynomial running time in its input size even in the worst case. It has been shown in [90] that problem (C.5) is generally intractable even if each W_i is closed and convex. Nevertheless, tractability can be preserved for some special classes of uncertainty sets. Some well-known results are summarized in the following.

Condition (C.6) contains an infinite number of constraints due to the enumeration over set Z. Later we will see that for some particular uncertainty sets, the $\forall$ quantifier as well as the uncertain parameter ζ can be eliminated by using duality theory, and the resulting constraint in variable x is still convex.

1. Polyhedral Uncertainty Set

We start with a commonly used uncertainty set: a polyhedron

$$Z = \{\zeta \mid D\zeta + q \ge 0\} \tag{C.7}$$

where D and q are constant matrices with compatible dimensions.

To exclude the $\forall$ quantifier for variable ζ, we investigate the worst case of the left-hand side and require

$$\bar{a}^T x + \max_{\zeta \in Z} \left(P^T x\right)^T \zeta \le b \tag{C.8}$$

For a fixed x, the second term is the optimum of an LP in variable ζ. Duality theory of LP says that the following relation holds

$$\left(P^T x\right)^T \zeta \le q^T u,\ \forall \zeta \in Z,\ \forall u \in U \tag{C.9}$$

where u is the dual variable, and $U = \{u \mid D^T u + P^T x = 0,\ u \ge 0\}$ is the feasible region of the dual problem. Please be cautious on the sign of u. We actually replace u with $-u$ in the original dual LP. Therefore, a necessary condition to validate (C.6) is

$$\exists u \in U : \bar{a}^T x + q^T u \le b \tag{C.10}$$

It is also sufficient if the second term takes its minimum value over U, because strong duality always holds for LPs, i.e. $(P^T x)^T \zeta = q^T u$ is satisfied at the optimal solution. In this regard, (C.8) is equivalent to

$$\bar{a}^T x + \min_{u \in U} q^T u \le b \tag{C.11}$$

In fact, the "min" operator in (C.11) can be omitted in an RC optimization problem that minimizes the objective function, and thus renders polyhedral constraints, although (C.8) is not given in a closed form and seems non-convex.

In summary, the RC problem of an uncertain LP with polyhedral uncertainty

$$\begin{aligned} \min_x \ & c^T x \\ \text{s.t.} \ & (\bar{a}_i + P_i \zeta_i)^T x \le b_i,\ \forall \zeta_i \in Z_i,\ \forall i \\ & Z_i = \{\zeta_i \mid D_i \zeta_i + q_i \ge 0\},\ \forall i \end{aligned} \tag{C.12}$$

can be equivalently formulated as

$$\begin{aligned} \min_x \ & c^T x \\ \text{s.t.} \ & \bar{a}_i^T x + q_i^T u_i \le b_i,\ \forall i \\ & D_i^T u_i + P_i^T x = 0,\ u_i \ge 0,\ \forall i \end{aligned} \tag{C.13}$$

which is still an LP.

2. Cardinality Constrained Uncertainty Set

Cardinality constrained uncertainty set is a special class of polyhedral uncertainty set which incorporates a budget constraint and defined as follows:

$$Z(\Gamma) = \left\{ \zeta \,\middle|\, -1 \le \zeta_j \le 1,\ \forall j,\ \sum_j |\zeta_j| \le \Gamma \right\} \tag{C.14}$$

where Γ is called the budget of uncertainty [96]. Motivated by the fact that each entry ζ_j is unlikely to reach 1 or -1 at the same time, the budget constraint controls the total data deviation from their forecast values. In other words, the decision maker can achieve a compromise between the level of solution robustness and the optimal cost by adjusting the value of Γ, which should be less than the dimension of ζ, otherwise the budget constraint will be redundant.

Although the cardinality constrained uncertainty set $Z(\Gamma)$ is essentially a polyhedron, the number of its facets, or the number of linear constraints in (C.7), grows exponentially in the dimension of ζ, leading to a huge and dense coefficient matrix for the uncertainty set. To circumvent this difficulty, we can lift it into a higher dimensional space as follows by introducing auxiliary variables

$$Z(\Gamma) = \left\{ \zeta, \sigma \;\middle|\; -\sigma_j \le \zeta_j \le \sigma_j,\ \sigma_j \le 1,\ \forall j,\ \sum_j \sigma_j \le \Gamma \right\} \tag{C.15}$$

The first inequality naturally suggests $\sigma_j \ge 0$, $\forall j$. It is easy to see the equivalence of (C.14) and (C.15), and the numbers of variables and constraints in the latter one grow linearly in the dimension of ζ.

Following a similar paradigm, certifying constraint robustness with a cardinality constrained uncertainty set requires the optimal value function of the following LP in variables ζ and σ representing the uncertainty

$$\begin{aligned} \max_{\zeta,\sigma} \quad & (P^T x)^T \zeta \\ \text{s.t.} \quad & -\zeta_j - \sigma_j \le 0,\ \forall j : u_j^n \\ & \zeta_j - \sigma_j \le 0,\ \forall j : u_j^m \\ & \sigma_j \le 1,\ \forall j : u_j^b \\ & \sum_j \sigma_j \le \Gamma : u_r \end{aligned} \tag{C.16}$$

where u_j^n, u_j^m, u_j^b, $\forall j$, and u_r following a colon are the dual variables associated with each constraint. The dual problem of (C.16) is given by

$$\begin{aligned} \min_{u^n, u^m, u^b, u_r} \quad & u_r \Gamma + \sum_j u_j^b \\ \text{s.t.} \quad & u_j^m - u_j^n = (P^T x)_j,\ \forall j \\ & -u_j^m - u_j^n + u_j^b + u_r = 0,\ \forall j \\ & u_j^m,\ u_j^n,\ u_j^b \ge 0,\ \forall j,\ u_r \ge 0 \end{aligned} \tag{C.17}$$

In summary, the RC problem of an uncertain LP with cardinality constrained uncertainty

$$\begin{aligned} &\min_x \ c^T x \\ &\text{s.t. } (\bar{a}_i + P_i \zeta_i)^T x \le b_i, \ \forall \zeta_i \in Z_i(\Gamma_i), \ \forall i \end{aligned} \tag{C.18}$$

can be equivalently formulated as

$$\begin{aligned} &\min_x \ c^T x \\ &\text{s.t. } \bar{a}_i^T x + u_{ri}\Gamma_i + \sum_j u_{ij}^b \le b_i, \ \forall i \\ &u_{ij}^m - u_{ij}^n = (P_i^T x)_j, \ \forall i, \ \forall j \\ &- u_{ij}^m - u_{ij}^n + u_{ij}^b + u_{ir} = 0, \ \forall i, \ \forall j \\ &u_{ij}^m, \ u_{ij}^n, \ u_{ij}^b \ge 0, \ \forall i, \ \forall j, \ u_{ir} \ge 0, \ \forall i \end{aligned} \tag{C.19}$$

which is still an LP.

3. Several Other Uncertainty Sets

Equivalent convex formulations of the uncertain constraint (C.6) with some other uncertainty sets are summarized in Table C.1 [92]. These outcomes are derived using the similar method described previously.

Table C.1 includes three cases: the p-norm uncertainty, the conic uncertainty, and general convex uncertainty. In the p-norm case, the Hölder's inequality is used, i.e.:

$$\left(P^T x\right)^T \zeta \le \|P^T x\|_p \|\zeta\|_q \tag{C.20}$$

Table C.1 Equivalent convex formulations with different uncertainty sets

Uncertainty	Z	Robust reformulation	Tractability
Box	$\|\zeta\|_\infty \le 1$	$\bar{a}^T x + \|P^T x\|_1 \le b$	LP
Ellipsoidal	$\|\zeta\|_2 \le 1$	$\bar{a}^T x + \|P^T x\|_2 \le b$	LP
p-norm	$\|\zeta\|_p \le 1$	$\bar{a}^T x + \|P^T x\|_q \le b$	Convex program
Proper cone	$D\zeta + q \in K$	$\begin{cases} \bar{a}^T x + q^T u \le b \\ D^T u + P^T x = 0 \\ u \in K^* \end{cases}$	Conic LP
Convex constraints	$h_k(\zeta) \le 0, \forall k$	$\begin{cases} \bar{a}^T x + \sum_k \lambda_k h_k^* \left(\frac{u^k}{\lambda_k}\right) \le b \\ \sum_k u^k = P^T x \\ \lambda_k \ge 0, \forall k \end{cases}$	Convex program

where $\|\cdot\|_p$ and $\|\cdot\|_q$ with $p^{-1}+q^{-1}=1$ are a pair of dual norms. Since norm function of any order is convex [97], the resulting RC is a convex program. Moreover, if q is a positive rational number, the q-order cone constraints can be represented by a set of SOC inequalities [98], which is computationally more friendly. Box (∞-norm) and ellipsoidal (2-norm) uncertainty sets are special kinds of p-norm ones.

In the general conic case, conic duality theory [99] is used. K^* stands for the dual cone of K, and the polyhedral uncertainty is a special kind of this case when K is the nonnegative orthant.

In the general convex case, Fenchel duality, a basic theory in convex analysis, is needed. Notation h^* stands for the convex conjugate function, i.e. $h^*(x) = \sup_y x^T y - h(y)$. The detailed proof of RC reformulations and more examples can be found in [100].

The above analysis focuses on the situation in which problem functions are linear in decision variables, and problem data are affine in some uncertain parameters, such as the form $a = \bar{a} + P\zeta$. For robust quadratic optimization, robust semidefinite optimization, robust conic optimization, and robust discrete optimization, in which the optimization problem is nonlinear and discontinuous, please refer to [90] and [91]; for quadratic type uncertainty, please refer to [90, (in Sect. 1.4)] and [100].

C.1.3 Formulation Issues

To help practitioners build a well-defined and easy-to-solve robust optimization model, some important modeling issues and deeper insights are discussed in this section.

1. Choosing the Uncertainty Set

Since a robust solution remains feasible if the uncertain data does not step outside the uncertainty set, the level of robustness mainly depends on the shape and size of the uncertainty set. The more reliable, the higher the cost. One may wish to seek a trade-off between reliability and economy. This inspires the development of smaller uncertainty sets with a certain probability guarantee that the constraint violation is unlikely to happen. Such guarantees are usually described via a chance constraint

$$\Pr_\zeta\left[a(\zeta)^T x \le b\right] \ge 1-\varepsilon \tag{C.21}$$

For $\varepsilon = 0$, chance constraint (C.21) is protected in the traditional sense of RO. When $\varepsilon > 0$, it becomes challenging to derive tractable reformulation for (C.21), especially when the probability distribution of uncertain data is unclear or inaccurate. In fact, this issue is closely related to the DRO that will be discussed later on. Here we provide some simple results which help the decision maker choose the parameter of the uncertainty set.

It is revealed that if $\mathbb{E}[\zeta] = 0$, the components of ζ are independent, and the uncertainty set takes the form

$$Z = \{\zeta \mid \|\zeta\|_2 \le \Omega,\ \|\zeta\|_\infty \le 1\} \tag{C.22}$$

then chance constraint (C.21) holds with a probability of at least $1 - \exp(-\Omega^2/2)$ (see [90, Proposition 2.3.3]).

Moreover, if the uncertainty set takes the form

$$Z = \{\zeta \mid \|\zeta\|_1 \le \Gamma,\ \|\zeta\|_\infty \le 1\} \tag{C.23}$$

then chance constraint (C.21) holds with a probability of at least $1-\exp(-\Gamma^2/2L)$, where L is the dimension of ζ (see [90, Proposition 2.3.4], and [96]).

It is proposed to construct uncertainty sets based on the central limit theorem. If each component of ζ is independent and identically distributed with mean μ and variance σ^2, the uncertainty set can be built as [101]

$$Z = \left\{\zeta \;\middle|\; \left|\sum_{i=1}^{L} \zeta_i - L\mu\right| \le \rho\sqrt{L}\sigma,\ \|\zeta\|_\infty \le 1\right\} \tag{C.24}$$

where parameter ρ is used to control the probability guarantee. Variations of this formulation can take other distributional information into account, such as data correlation and long tail-effect. It is a special kind of polyhedral uncertainty, however, it is unbounded for $L > 1$, since the components can be arbitrarily large as long as their summation is relatively small. Unboundedness may prevent establishing tractable RCs.

Additional references are introduced in further reading.

2. How to Solve a Problem Without a Clear Tractable Reformulation?

The existence of a tractable reformulation for a static RO problem largely depends on the type of the uncertainty set. If the robust counterpart cannot be written as a tractable convex program, a smart remedy is to use an adaptive scenario generation procedure: first solve the problem with a smaller uncertainty set Z_S which is a subset of the original one Z, and the problem with Z_S has a known tractable reformulation. If the optimal solution x^* is robust against all scenarios in Z, it is also an optimal solution of the original problem. Otherwise, we have to identify a scenario $\zeta^* \in Z$ which leads to the most severe violation, which can be implemented by solving

$$\max \left\{(P^T x^*)^T \zeta \mid \zeta \in Z\right\} \tag{C.25}$$

where Z is a closed and convex set as validated in Assumption C.3, and then append a cutting plane

$$a(\zeta^*)^T x \le b \tag{C.26}$$

to the reformulation problem. Equation (C.26) removes x that will cause infeasibility in scenario ζ^*, so is called a feasibility cut. It is linear and does not alter tractability. Then the updated problem is solved again. According to Proposition C.2, the new solution x^* will be robust for uncertain data in the convex hull of $Z_S \cup \zeta^*$. The above procedure continues until robustness is certified over the original uncertainty set Z.

This simple approach often converges quickly in a few number of iterations. Its advantage is that tractability is preserved. When we choose $Z_S = \zeta^0$, where ζ^0 is the nominal scenario or forecast, it could be more efficient than using convex reformulations, because only LPs (whose sizes are almost equal to the problem without uncertainty, and grows slowly) and simple convex programs (C.25) are solved, see [102] for a comparison. This paradigm is an essential strategy for solving the adjustable RO problems in the next section.

3. How to Deal with Equality Constraints?

Although the theory of static RO is relatively mature, it encounters difficulties in dealing with equality constraints. For example, consider $x + a = 1$ where $a \in [0, 0.1]$ is uncertain. However, one can seldom find a solution that makes the equality hold true for multiple values of a. The problem remains if you write a equality into a pair of opposite inequalities. In fact, this issue is inevitable in the static setting. In addition, this limitation will lead to completely different robust counterpart formulations for originally equivalent deterministic problems.

Consider the inequality $ax \le 1$, which is equivalent to $ax + s = 1$, $s \ge 0$. Suppose a is uncertain and belongs to interval $[1, 2]$, their respective robust counterparts are given by

$$ax \le 1,\ \forall a \in [1,2] \tag{C.27}$$

and

$$ax + s = 1,\ \forall a \in [1,2],\ s \ge 0 \tag{C.28}$$

The feasible set for (C.27) is $x \le 1/2$, and is $x = 0$ for (C.28). By observing this difference, it is suggested that a static RO model should avoid using slack variables in constraints with uncertain parameters.

Sometimes, the optimization problem may contain state variables which can respond to parameter changes by adjusting their values. In such circumstance, equality constraint can be used to eliminate state variables. Nevertheless, such an action may lead to a problem that contains nonlinear uncertainties, which are challenging to solve. An example is taken from [92] to illustrate this issue. The constraints are

$$\begin{aligned} &\zeta_1 x_1 + x_2 + x_3 = 1 \\ &x_1 + x_2 + \zeta_2 x_3 \le 5 \end{aligned} \tag{C.29}$$

where ζ_1 and ζ_2 are uncertain.

If x_1 is a state variable and $\zeta_1 \neq 0$, substituting $x_1 = (1 - x_2 - x_3)/\zeta_1$ in the second inequality results in

$$\left(1 - \frac{1}{\zeta_1}\right) x_2 + \left(\zeta_2 - \frac{1}{\zeta_1}\right) x_3 \leq 5 - \frac{1}{\zeta_1}$$

in which the uncertainty becomes nonlinear in the coefficients.

If x_2 is a state variable, substituting $x_2 = 1 - \zeta_1 x_1 - x_3$ in the inequality yields

$$(1 - \zeta_1)x_1 + (\zeta_2 - 1)x_3 \leq 4$$

in which the uncertainty sustains linear in the coefficients.

If x_3 is a state variable, substituting $x_3 = 1 - \zeta_1 x_1 - x_2$ in the inequality gives

$$(1 - \zeta_1\zeta_2)x_1 + (1 - \zeta_2)x_2 \leq 5 - \zeta_2$$

in which the uncertainty is nonlinear in the coefficients.

In conclusion, in the case that x_2 is a state variable, the problem is easier from a computational perspective. It is important to note that the physical interpretation of variable elimination is to determine the adjustable variable with exact information on the uncertain data. If no adjustment is allowed in (C.29), the only robust feasible solution is $x_1 = x_3 = 0$, $x_2 = 1$, which is rather restrictive. The adjustable RO will be elaborated in detail in the next section.

4. Pareto Efficiency of the Robust Solution

The concept of Pareto efficiency in RO problems is proposed in [103]. If the optimal solution under the worst-case data realization is not unique, it is rational to compare their performances in non-worst-case scenarios: an alternative solution may give an improvement in the objective value for at least one data scenario without deteriorating the objective performances in all other scenarios. To present related concept tersely, we restrict the discussion on the following robust LP with objective uncertainty

$$\max\left\{p^T x \mid \text{s.t. } x \in X,\ \forall p \in W\right\} = \max_{x \in X}\left\{\min_{p \in W} p^T x\right\} \tag{C.30}$$

where $W = \{p \mid Dp \geq d\}$ is a polyhedral uncertainty set for the price vector p; $X = \{x \mid Ax \leq b\}$ is the feasible region which is independent of the uncertainty. More general cases are elaborated in [103]. We consider this form because it is easy to discuss related issues, although objective uncertainty can be moved into constraints.

For a given strategy x, the worst-case uncertainty is

$$\min\left\{p^T x \mid \text{s.t. } p \in W\right\} = \max\left\{d^T y \mid \text{s.t. } y \in Y\right\} \tag{C.31}$$

where $Y = \{y \mid D^T y = x,\ y \geq 0\}$ is the feasible set for dual variable y. Substituting (C.31) in (C.30) gives

$$\max \left\{ d^T y \mid \text{s.t. } D^T y = x,\ y \geq 0,\ x \in X \right\} \tag{C.32}$$

which is an LP. Its solution x is the robust optimal one to (C.30), and the worst-case price p can be found by solving the left-hand side LP in (C.31). Let z^{RO} be the optimal value of (C.32), and then the set of robust optimal solutions for (C.30) can be expressed via

$$X^{RO} = \left\{ x \mid x \in X : \exists y \in Y \text{ such that } y^T d \geq z^{RO} \right\} \tag{C.33}$$

If (C.32) has a unique optimal solution, X^{RO} is a singleton; otherwise, a Pareto optimal robust solution can be formally defined.

Definition C.3 ([103]) $x \in X^{RO}$ is a Pareto optimal solution for problem (C.30) if there is no other $\bar{x} \in X$ such that $p^T \bar{x} \geq p^T x$, $\forall p \in W$ and $\bar{p}^T \bar{x} > \bar{p}^T x$ for some $\bar{p} \in W$.

The terminology "Pareto optimal" is borrowed from multi-objective optimization theory: RO problem (C.30) is viewed as a multi-objective LP with infinitely many objectives, each of which corresponds to a particular $p \in W$. Some interesting problems are elaborated.

a. Pareto Efficiency Test

In general, it is not clear whether X^{RO} contains multiple solutions, at least before a solution $x \in X^{RO}$ is found. To test whether a given solution x is a robust optimal one or not, it is proposed to solve a new LP

$$\begin{aligned} \max_{y} \quad & \bar{p}^T y \\ \text{s.t.} \quad & y \in W^* \\ & x + y \in X \end{aligned} \tag{C.34}$$

where $\bar{p}$ is a relative interior of the polyhedral uncertainty set W, which is usually set to the nominal scenario, and $W^* = \{y \mid \exists \lambda : d^T \lambda \geq 0,\ D^T \lambda = y,\ \lambda \geq 0\}$ is the dual cone of W. Please refer to Sect. A.2.1 and Eq. (A.27) for the dual cone of a polyhedral set. Since $y = 0$, $\lambda = 0$ is always feasible in (C.34), the optimal value is either zero or strictly positive. In the former case, x is also a Pareto optimal solution; in the latter case, $\bar{x} = x + y^*$ dominates x and itself is Pareto optimal for any y^* that solves LP (C.34) [103]. The interpretation of (C.34) is clear: since $y \in W^*$, $y^T p$ must be non-negative for all $p \in W$. If we can find y that leads to a strict objective improvement for $\bar{p}$, then $x + y$ would be Pareto optimal.

In view of the above interpretation, it is a direct conclusion that for an arbitrary relative interior point $\bar{p} \in W$, the optimal solutions to the problem

$$\max \left\{ \bar{p}^T x \mid x \in X^{RO} \right\} \tag{C.35}$$

are Pareto optimal.

b. Characterizing the Set of Pareto Optimal Solutions
It is interesting to characterize the Pareto optimal solution set X^{PRO}.

After we get z^{RO} and X^{RO}, solve the following LP

$$\begin{aligned} \max_{x,y,\lambda} \ & \bar{p}^T y \\ \text{s.t.} \ & d^T \lambda \geq 0, \ D^T \lambda = y, \ \lambda \geq 0 \\ & x \in X^{RO}, \ x + y \in X \end{aligned} \tag{C.36}$$

and we can conclude $X^{PRO} = X^{RO}$ if and only if the optimal value of (C.36) is equal to 0 [103]. If this is true, the decision maker would not have to worry about Pareto efficiency, as any solution in X^{RO} is also Pareto optimal. More broadly, the set X^{PRO} is shown to be non-convex and is contained in the boundary of X^{RO}.

c. Optimization Over Pareto Optimal Solutions
In the case that X^{PRO} is not a singleton, one may consider to optimize a linear secondary objective over X^{PRO}, i.e.:

$$\max \left\{ r^T x \mid \text{s.t. } x \in X^{PRO} \right\} \tag{C.37}$$

It is demonstrated in [103] that if r lies in the relative interior of W, the decision maker can simply replace X^{PRO} with X^{RO} in (C.37) without altering the problem solution, due to the property revealed in (C.35). In more general cases, problem (C.37) can be formulated as an MILP [103]

$$\begin{aligned} \max_{x,\mu,\eta,z} \ & r^T x \\ \text{s.t.} \ & x \in X^{RO} \\ & \mu \leq M(1 - z) \\ & b - Ax \leq Mz \\ & DA^T \mu - d\eta \geq D\bar{p} \\ & \mu, \eta \geq 0, z \in \{0, 1\}^m \end{aligned} \tag{C.38}$$

where M is a sufficiently large number, m is the dimension of vector z. To show their equivalence, it is revealed that the feasible set of (C.38) depicts an optimal solution of (C.34) with a zero objective value [103]. In other words, the constraints of (C.38) contain the KKT optimality condition of (C.34). To see this, the binary vector z imposes the complementarity and slackness condition $\mu^T(b - Ax) = 0$, which ensures λ, μ, η are the optimal solution of the following primal-dual LP pair

$$\text{Primal}: \begin{aligned} \max_{\lambda} \ & \bar{p}^T D^T \lambda \\ \text{s.t.} \ & \lambda \geq 0 \\ & d^T \lambda \geq 0 \\ & AD^T \lambda \leq b - Ax \end{aligned} \qquad \text{Dual}: \begin{aligned} \min_{\mu,\eta} \ & \mu^T(b - Ax) \\ \text{s.t.} \ & \mu \geq 0 \\ & \eta \geq 0 \\ & DA^T \mu - d\eta \geq D\bar{p} \end{aligned}$$

The original variable y in (C.34) is eliminated via equality $D^T \lambda = y$ in the dual cone W^*. According to strong duality, the optimal value of the primal LP (C.34) is $\bar{p}^T D^T \lambda = \bar{p}^T y = 0$, and Pareto optimality is guaranteed.

In practice, Pareto inefficiency is not a contrived phenomenon, see various examples in [103] and power market examples in [104, 105].

5. On Max-Min and Min-Max Formulations

In many literatures, the robust counterpart problem of (C.2) is written as a min-max form

$$\text{Opt-1} = \min_{x} \max_{A \in W} \left\{ c^T x \mid \text{s.t. } Ax \leq b \right\} \tag{C.39}$$

which means x is determined before A takes a value in W, and the decision maker can foresee the worst consequence of deploying x brought by the perturbation of A. To make a prudent decision that is insensitive to data perturbation, the decision maker resorts to minimizing the maximal objective.

The max-min formulation

$$\text{Opt-2} = \max_{A \in W} \min_{x} \left\{ c^T x \mid \text{s.t. } Ax \leq b \right\} \tag{C.40}$$

has a different interpretation: the decision maker can first observe the realization of uncertainty, and then recovers the constraints by deploying a corrective action x as a response to the observed A. Certainly, this specific x may not be feasible for other $A \in W$. On the other hand, the uncertainty, like a rational player, can foresee the optimal action taken by the human decision maker, and select a strategy that will yield a maximal objective value even an optimal corrective action is deployed.

From the above analysis, the feasible region of x in (C.39) is a subset of that in (C.40), because (C.40) only accounts for a special scenario in W. As a result, their optimal values satisfy Opt-1 $\geq$ Opt-2.

Consider the following problem in which the uncertainty is not constraint-wise

$$\begin{aligned} \min_{x} \ & x_1 + x_2 \\ \text{s.t.} \ & x_1 \ge a_1, \ x_2 \ge a_2, \ \forall a \in W \end{aligned} \tag{C.41}$$

where $W = \{a \mid a \ge 0, \ \|a\|_2 \le 1\}$.

For the min-max formulation, since x should be feasible for all possible values of a, it is necessary to require $x_1 \ge 1$ and $x_2 \ge 1$, and Opt-1 $= 2$ for problem (C.41).

As for the max-min formulation, as x is determined in response to the value of a, it is clear that the optimal choice is $x_1 = a_1$ and $x_2 = a_2$, so the problem becomes

$$\begin{aligned} \max_{a} \ & a_1 + a_2 \\ \text{s.t.} \ & a_1^2 + a_2^2 \le 1 \end{aligned}$$

whose optimal value is Opt-2 $= \sqrt{2} <$ Opt-1.

As a short conclusion, static RO models discussed in this section are used to immunize against constraint violation or objective volatility caused by data perturbations, without jeopardizing computational tractability. General approaches involve reformulating the original uncertainty dependent constraints into deterministic convex ones without uncertain data, such that feasible solutions of the robust counterpart program remain feasible for all data realizations in the pre-specified uncertainty set, which interprets the meaning of robustness.

C.2 Adjustable Robust Optimization

Several reasons call for developing new decision-making mechanisms to overcome limitations of the static RO approach: (1) Equality constraints often give rise to infeasible robust counterpart problems in the static setting; (2) real-world decision-making process may involve multiple stages, in which some decisions indeed can be made after the uncertain data has been known or can be predicted accurately. Take power system operation for an example, the on-off status of generation units must be made several hours before real-time dispatch when the renewable power is unclear; however, the output of some units (called AGC units) can change in response to the real values of system demands and renewable generations. This section will be devoted to the adjustable robust optimization (ARO) with two stages, which leverages the adaptability in the second stage. We still focus our attention on the linear case.

C.2.1 Basic Assumptions and Formulations

The essential difference between static RO and ARO approaches stems from the manner of decision making.

Assumption C.4 *In an ARO problem, some variables are "here-and-now" decisions, whereas the rest are "wait-and-see" decisions: they can be made at a later moment according to the observed data.*

In analogy to the static case, the decision-making mechanism can be explained.

Assumption C.5 *Once the here-and-now decisions are made, there must be at least one valid wait-and-see decision which is able to recover constraints in response to the observed data realization, if the actual data is within the uncertainty set.*

In this regard, we can say here-and-now decisions are robust against the uncertainty, and wait-and-see decisions are adaptive to the uncertainty. These terminologies are borrowed from two-stage SO models. In fact, there is a close relation between two-stage SO and two-stage RO [106, 107].

Now we are ready to post the compact form of a linear ARO problem with an uncertain constraint right-hand side:

$$\min_{x \in X} \left\{ c^T x + \max_{w \in W} \min_{y(w) \in Y(x,w)} d^T y(w) \right\} \tag{C.42}$$

where x is the here-and-now decision variable (or the first-stage decision variable), and X is the feasible region of x; w is the uncertain parameter, and W is the uncertainty set, which has been discussed in the previous section; $y(w)$ is the wait-and-see decision variable (or second-stage decision variable), which can be adjusted according to the actual data of w, so it is represented as a function of w; Y is the feasible region of y given the values of x and w, because the here-and-now decision is not allowed to change in this stage, and the exact value of w is known. It has a polyhedral form

$$Y(x, w) = \{y \mid Ax + By + Cw \le b\} \tag{C.43}$$

where A, B, C, and b are constant matrices and vector with compatible dimensions. It is clear that both of the here-and-now decision x and the data uncertainty w can influence the feasible region Y in the second stage. We define $w = 0$ the nominal scenario and assume 0 is a relative interior of W. Otherwise, we can decompose the uncertainty as $w = w^0 + \Delta w$ and merge the constant term Cw^0 into the right-hand side as $b \to b - Cw^0$, where w^0 is the predicted or expected value of w, and Δw is the forecast error, which is the real uncertain parameter.

It should be pointed out that x, w, and y may contain discrete decision variables. Later we will see, integer variables in x and w do not significantly alter the solution algorithm of ARO. However, because integrality in y prevents the use of LP duality theory, the computation will be greatly challenged. Although we assume coefficient matrices are constants in (C.43), most results in this section can be generalized if matrix A is a linear function in w; the situation would be complicated in matrix B is uncertainty-dependent. The purpose for the specific form in (C.43) is that it is more dedicated to the problems considered in this book: uncertainties originates from renewable/load volatility can be modeled by term Cw in (C.43), and the coefficients representing component and network parameters are constants.

Assumption C.5 inspires the definition for a feasible solution of ARO (C.42).

Definition C.4 A first-stage decision x is called robust feasible in (C.42) if the feasible region $Y(x, w)$ is non-empty for all $w \in W$, and the set of robust feasible solutions are given by:

$$X_R = \{x \mid x \in X : \forall w \in W,\ Y(x, w) \neq \emptyset\} \tag{C.44}$$

Please be aware of the sequence in (C.44): x takes its value first, and then parameter w chooses a value in W before some $y \in Y$ does. The non-emptiness of Y is guaranteed by the selection of x for an arbitrary w. If we swap the latter two terms and write $Y(x, w) \neq \emptyset$, $\forall w \in W$, like the form in a static RO, it sometimes causes confusion that both x and y are here-and-now type decisions, the adaptiveness vanishes, and thus X_R may become empty if uncertainty appears in an equality constraint, as analyzed in the previous section.

The definition of an optimal solution depends on the decision maker's attitude towards the cost in the second stage. In (C.42), we adopt the following definition.

Definition C.5 (Min-Max Cost Criterion) An optimal solution of (C.42) is a pair of here-and-now decision $x \in X_R$ and wait-and-see decision $y(w^*)$ corresponding to the worst-case scenario $w^* \in W$, such that the total cost in scenario w^* is minimal, where the worst-case scenario w^* means that for the fixed x, the optimal second-stage cost is maximized over W.

Other criteria may give different robust formulations. For example, the minimum nominal cost formulation and min-max regret formulation.

Definition C.6 (Minimum Nominal Cost Criterion) An optimal solution under the minimum nominal cost criterion is a pair of here-and-now decision $x \in X_R$ and wait-and-see decision y^0 corresponding to the nominal scenario $w^0 = 0$, such that the total cost in scenario w^0 is minimal.

The minimum nominal cost criterion leads to the following robust formulation

$$\begin{aligned} \min \ & c^T x + d^T y \\ \text{s.t.} \ & x \in X_R \\ & y \in Y(x, w^0) \end{aligned} \tag{C.45}$$

where robustness is guaranteed by X_R.

To explain the concept of regret, the minimum perfect-information total cost is

$$C_P(w) = \min \left\{ c^T x + d^T y \mid x \in X,\ y \in Y(x, w) \right\}$$

where w is known to be the decision maker. For a fixed first-stage decision x, the maximum regret is defined as

$$\text{Reg}(x) = \max_{w \in W} \left\{ \min_{y \in Y(x,w)} \{c^T x + d^T y\} - C_P(w) \right\}$$

Definition C.7 (Min-Max Regret Criterion) An optimal solution under the min-max regret criterion is a pair of here-and-now decision x and wait-and-see decision $y(w)$, such that the worst-case regret under all possible scenarios $w \in W$ is minimized.

The min-max regret cost criterion leads to the following robust formulation

$$\min_{x \in X} \left\{ c^T x + \max_{w \in W} \left\{ \min_{y \in Y(x,w)} d^T y - \min_{x' \in X, y' \in Y(x,w)} \left\{ c^T x' + d^T y' \right\} \right\} \right\} \tag{C.46}$$

In an ARO problem, we can naturally assume that the uncertainty set is a polyhedron. To see this, if x is a robust solution under an uncertainty set consists of discrete scenarios, i.e., $W = \{w^1, w^2, \cdots w^S\}$, according to Definition C.4, there exist corresponding $\{y^1, y^2, \cdots y^S\}$ such that

$$\begin{aligned} By^1 &\le b - Ax - Cw^1 \\ By^2 &\le b - Ax - Cw^2 \\ &\vdots \\ By^S &\le b - Ax - Cw^S \end{aligned}$$

For non-negative weighting parameters $\lambda_1, \lambda_2, \cdots, \lambda_S \ge 0$, $\sum_{s=1}^{S} \lambda_s = 1$, we have

$$\sum_{s=1}^{S} \lambda_s (By^s) \le \sum_{s=1}^{S} \lambda_s (b - Ax - Cw^s)$$

or equivalently

$$B \sum_{s=1}^{S} \lambda_s y^s \le b - Ax - C \sum_{s=1}^{S} \lambda_s w^s$$

indicating that for any $w = \sum_{s=1}^{S} \lambda_s w^s \in \text{conv}(\{w^1, w^2, \cdots w^S\})$, the wait-and-see decision $y = \sum_{s=1}^{S} \lambda_s y^s$ can recover all constraints, and thus $Y(x, w) \neq \emptyset$. This property inspires the following proposition that is in analogy to Proposition C.2

Proposition C.3 *Suppose x is a robust feasible solution for a discrete uncertainty set $\{w^1, w^2, \cdots w^S\}$, then it remains robust feasible if we replace the uncertainty set with its convex hull.*

Proposition C.3 also implies that in order to ensure the robustness of x, it is sufficient to consider the extreme points of a bounded polytope. Suppose the vertices of the polyhedral uncertainty set are w^s, $s = 1, 2, \cdots, S$. Consider the following set

$$\Xi = \{x, y^1, y^2, \cdots, y^S \mid Ax + By^s \leq b - Cw^s,\ s = 1, 2, \cdots, S\} \tag{C.47}$$

Robust feasible region X_R is the projection of polyhedron Ξ on x-space, which is also a polyhedron (Theorem B.2.5 in [99]).

Proposition C.4 *If the uncertainty set has a finite number of extreme points, set X_R is a polytope.*

Despite the nice theoretical properties, it is still difficult to solve an ARO problem in its general form (C.42). There have been considerable efforts spent on developing different approximations and approaches to tackle the computational challenges. We leave the solution methods of ARO problems to the next subsection. Here we demonstrate the benefit from postponing some decisions to the second stage via a simple example taken from [90].

Consider an uncertain LP

$$\begin{aligned} \min_{x} \ & x_1 \\ \text{s.t. } & x_2 \geq 0.5\xi x_1 + 1 \quad (a_\xi) \\ & x_1 \geq (2 - \xi) x_2 \quad (b_\xi) \\ & x_1 \geq 0,\ x_2 \geq 0 \quad (c_\xi) \end{aligned}$$

where $\xi \in [0, \rho]$ is an uncertain parameter and ρ is a constant (level of uncertainty) which may take a value in the open interval (0, 1).

In a static setting, both x_1 and x_2 must be independent of ξ. When $\xi = \rho$, constraint (a_ξ) suggests $x_2 \geq 0.5\rho x_1 + 1$; when $\xi = 0$, constraint (b_ξ) indicates $x_1 \geq 2x_2$; as a result, we arrive at the conclusion $x_1 \geq \rho x_1 + 2$, so the optimal value in the static case satisfies

$$\text{Opt} \geq x_1 \geq \frac{2}{1 - \rho}$$

Thus the optimal value tends to infinity when ρ approaches 1.

Now consider the adjustable case, in which x_2 is a wait-and-see decision. Let $x_2 = 0.5\xi x_1 + 1$, (a_ξ) is always satisfied; substituting x_2 in constraint (b_ξ) yields:

$$x_1 \geq (2-\xi)(\frac{1}{2}\xi x_1 + 1), \ \forall \xi \in [0, \rho]$$

Substituting $x_1 = 4$ into above inequality we have

$$4 \geq 2(2-\xi)\xi + 2 - \xi, \forall \xi \in [0, \rho]$$

This inequality can be certified by the fact that $\xi \geq 0$ and $\xi(2-\xi) \leq 1, \forall \xi \in \mathbb{R}$, indicating that $x_1 = 4$ is a robust feasible solution. Therefore, the optimal value should be no greater than 4 in the adjustable case for any ρ. The difference of optimal values in two cases can go arbitrarily large, depending on the value of ρ.

C.2.2 Affine Policy Based Approximation Model

ARO problem (C.42) is difficult to solve because the functional dependence of the wait-and-see decision on w is arbitrary, and there lacks a closed-form formula to characterize the optimal solution function $y(w)$ or certify whether $Y(x, w)$ is empty or not. At this point, we consider to approximate the recourse function $y(w)$ using a simpler one, naturally, an affine function

$$y(w) = y^0 + Gw \tag{C.48}$$

where y^0 is the action in the second stage for the nominal scenario $w = 0$, and G is the gain matrix to be designed. (C.48) is called a linear decision rule or affine policy. It explicitly characterizes the wait-and-see decisions as an affine function in the revealed uncertain data. The rationality for employing an affine policy instead of other parametric ones is that it yields computationally tractable robust counterpart reformulations. This finding is firstly reported in [108].

To validate (C.44) under the linear decision rule, substituting (C.48) in (C.43)

$$Ax + By^0 + (BG + C)w \leq b, \ \forall w \in W \tag{C.49}$$

In (C.49), decision variables are x, y^0, and G, which should be made before w is known, and thus are here-and-now decisions. The wait-and-see decision (or the incremental part) is naturally determined from (C.48) without further optimization, and cost reduction is considered in the determination of gain matrix G. (C.49) is in form of (C.6), and hence its robust counterpart can be derived via the methods in Appendix C.1.2. Here we just provide the results of polyhedral uncertainty as an example.

Suppose the uncertainty set is described by

$$W = \{w \mid Sw \le h\}$$

If we assume that y^0 is the optimal second stage decision when $w = 0$, then we have

$$Ax + By^0 \le b$$

Furthermore, (C.49) must hold if

$$\max_{w \in W} (BG + C)_i w \le 0, \ \forall i \tag{C.50}$$

where $(\cdot)_i$ stands for the i-th row of the input matrix. According to LP duality theory,

$$\max_{w \in W} \ (BG + C)_i w = \min_{\Lambda_i \in \Pi_i} \ \Lambda_i h, \ \forall i \tag{C.51}$$

where Λ is a matrix consists of the dual variables, Λ_i is the i-th row of Λ and also the dual variable of the i-th LP in (C.51), and the set

$$\Pi_i = \{\Lambda_i \mid \Lambda_i \ge 0, \ \Lambda_i S = (BG + C)_i\}$$

is the feasible region of the i-th dual LP.

The minimization operator in the right-hand side of (C.51) can be omitted if the objective is to seek a minimum. Moreover, if we adopt the minimum nominal cost criterion, the ARO problem with a linear decision rule in the second stage can be formulated as an LP

$$\begin{aligned} \min \ & c^T x + d^T y^0 \\ \text{s.t.} \ & Ax + By^0 \le b, \Lambda h \le 0 \\ & \Lambda \ge 0, \ \Lambda S = BG + C \end{aligned} \tag{C.52}$$

In (C.52), decision variables are vectors x and y^0, gain matrix G and dual matrix Λ. The constraints actually constitute a lifted formulation for X_R in (C.44). If the min-max cost criterion is employed, the objective can be transformed into a linear inequality constraint with uncertainty via an epigraph form, whose robust form can be derived using similar procedures shown above.

Affine policy based method is attractive because it reduces the conservatism in the static RO approach by incorporating corrective actions, and sustains computational tractability. In theory, the affine assumption more or less restricts the adaptability in the recourse stage. Nevertheless, research work in [109–111] shows that linear decision rules are indeed optimal or near optimal for many practical problems.

For more information on other decision rules and their reformulations, please see [90, (Chapter 14.3)] for the quadratic decision rule, [112] for the extended linear decision rule, [113, 114] for the piecewise constant decision rule (finite adaptability), [115, 116] for the piecewise linear decision rule, and [117] for generalized decision rules. The methods in [113, 116] can be used to cope with integer wait-and-see decision variables. See also [92].

C.2.3 Algorithms for Fully Adjustable Models

Fully adjustable models are generally NP-hard [118]. To find the solution in Definition C.2, the model is decomposed into a master problem and a subproblem, which are solved iteratively, and a sequence of lower bound and upper bound of the optimal values are generated, until they get close enough to each other. To explain the algorithm for ARO problems, we discuss two instances.

1. Second-Stage Problem is an LP
Now we consider problem (C.42) without specific functional assumptions on the wait-and-see variables. We start from the second-stage LP with fixed x and w:

$$\begin{aligned} \min_{y} \ & d^T y \\ \text{s.t.} \ & By \le b - Ax - Cw : u \end{aligned} \tag{C.53}$$

where u is the dual variable, and the dual LP of (C.53) is

$$\begin{aligned} \max_{u} \ & u^T (b - Ax - Cw) \\ \text{s.t.} \ & B^T u = d, \ u \le 0 \end{aligned} \tag{C.54}$$

If the primal LP (C.53) has a finite optimum, the dual LP (C.54) is also feasible and has the same optimum; otherwise, if (C.53) is infeasible, then (C.54) will be unbounded. Sometimes, an improper choice of x indeed leads to an infeasible second-stage problem. To detect infeasibility, consider the following LP with slack variables

$$\begin{aligned} \min_{y,s} \ & 1^T s \\ \text{s.t.} \ & s \ge 0 \\ & By - Is \le b - Ax - Cw : u \end{aligned} \tag{C.55}$$

Its dual LP is

$$\begin{aligned} \max_u \quad & u^T(b - Ax - Cw) \\ \text{s.t.} \quad & B^T u = 0, \quad -1 \le u \le 0 \end{aligned} \tag{C.56}$$

(C.55) and (C.56) are always feasible and have the same finite optimums. If the optimal value is equal to 0, then LP (C.53) is feasible; otherwise, if the optimal value is strictly positive, then LP (C.53) is infeasible.

For notation brevity, define feasible sets for the dual variable

$$U_O = \{u \mid B^T u = d, \ u \le 0\}$$

$$U_F = \{u \mid B^T u = 0, \ -1 \le u \le 0\}$$

The former one is associated with the dual form (C.54) of the second-stage optimization problem (C.53); the latter one corresponds to the dual form (C.56) of the second-stage feasibility test problem (C.55).

Next, we proceed to the middle level with fixed x:

$$R(x) = \max_{w \in W} \min_{y \in Y(x,w)} d^T y \tag{C.57}$$

which is a linear max-min problem that identifies the worst-case uncertainty. If LP (C.53) is feasible for an arbitrarily given value of $w \in W$, then we conclude $x \in X_R$ defined in (C.44); otherwise, if LP (C.53) is infeasible for some $w \in W$, then $x \notin X_R$ and $R(x) = +\infty$.

To check whether $x \in X_R$ or not, we investigate the following problem

$$\begin{aligned} \max_w \quad & \min_{y,s} 1^T s \\ \text{s.t.} \quad & w \in W, \ s \ge 0 \\ & By - Is \le b - Ax - Cw : u \end{aligned} \tag{C.58}$$

It maximizes the minimum of (C.55) over all possible values of $w \in W$. Since the minimums of (C.55) and (C.56) are equal, problem (C.58) is equivalent to maximizing the optimal value of (C.56) over the uncertainty set W, leading to a bilinear program

$$\begin{aligned} r(x) = \max_{u,w} \quad & u^T(b - Ax - Cw) \\ \text{s.t.} \quad & w \in W, \ u \in U_F \end{aligned} \tag{C.59}$$

Because both W and U_F are bounded, (C.59) must have a finite optimum. Clearly, $0 \in U_F$, so $r(x)$ must be non-negative. In fact, if $r(x) = 0$, then $x \in X_R$; if

$r(x) > 0$, then $x \notin X_R$. With the duality transformation, the opposite optimization operators in (C.58) come down to a traditional NLP.

For similar reasons, by replacing the second-stage LP (C.53) with its dual LP (C.54), problem (C.57) is equivalent to the following bilinear program

$$\begin{aligned} r(x) = \max_{u,w} \quad & u^T(b - Ax - Cw) \\ \text{s.t.} \quad & w \in W, \ u \in U_O \end{aligned} \tag{C.60}$$

The fact that a linear max-min problem can be transformed as a bilinear program using LP duality is reported in [119]. Bilinear programs can be locally solved by general purpose NLP solvers, but the non-convexity prevents a global optimal solution from being found easily. In what follows, we introduce some methods that exploit specific features of the uncertainty set and are widely used by the research community. In view that (C.59) and (C.60) only differ in the dual feasibility set, we will use set U to refer either U_F or U_O in the unified solution method.

a. General Polytope

Suppose that the uncertainty set is described by

$$W = \{w \mid Sw \le h\}$$

An important feature in (C.59) and (C.60) is that the constraint set W and U are separated and there is no constraint that involves w and u simultaneously, so the bilinear program can be considered in the following format

$$\begin{aligned} \max_{u \in U} \ u^T(b - Ax) + \max_{w} \quad & (-u^T Cw) \\ \text{s.t.} \quad & Sw \le h : \xi \end{aligned} \tag{C.61}$$

The bilinear term $u^T Cw$ is non-convex. If we treat the second part $\max_{w \in W}$ $(-u^T Cw)$ as an LP in w where u is a parameter, whose KKT optimality condition is given by

$$\begin{aligned} & 0 \le \xi \perp h - Sw \ge 0 \\ & S^T \xi + C^T u = 0 \end{aligned} \tag{C.62}$$

The stationary point of LCP (C.62) gives the optimal primal and dual solutions simultaneously. As the uncertainty set is a bounded polyhedron, the optimal solution must be bounded, and strong duality holds, so we can replace $-u^T Cw$ in the objective with a linear term $h^T \xi$ and additional constraints in (C.62). Moreover, the complementarity and slackness condition in (C.62) can be linearized via the method in Appendix B.3.5. In summary, problem (C.61) can be solved via an equivalent MILP

$$\begin{aligned} \max_{u,w,\xi} \quad & u^T(b - Ax) + h^T\xi \\ \text{s.t.} \quad & u \in U, \ \theta \in \{0,1\}^m \\ & S^T\xi + C^T u = 0 \\ & 0 \le \xi \le M(1-\theta) \\ & 0 \le h - Sw \le M\theta \end{aligned} \tag{C.63}$$

where m is the dimension of θ, and M is a large enough constant. Compared with (C.61), non-convexity migrates from the objective function to the constraints with binary variables. The number of binary variables in (C.63) only depends on the number of constraints in set W, and is independent of the dimension of x.

Another heuristic method for bilinear programs in the form of (C.59) and (C.60) is the mountain climbing method in [120], which is summarized in Algorithm C.1

Algorithm C.1 Mountain climbing

1: Choose a convergence tolerance $\varepsilon > 0$, and an initial $w^* \in W$;

2: Solve the following LP with current w^*

$$R_1 = \max_{u \in U} \ u^T(b - Ax - Cw^*) \tag{C.64}$$

The optimal solution is u^* and the optimal value is R_1;

3: Solve the following LP with current u^*

$$R_2 = \max_{w \in W} \ (b - Ax - Cw)^T u^* \tag{C.65}$$

The optimal solution is w^* and the optimal value is R_2;

4: If $R_2 - R_1 \le \varepsilon$, report the optimal value R_2 as well as the optimal solution w^*, u^*, and terminate; otherwise, go to step 2.

The optimal solutions of LPs must be found at one of the vertices of its feasible region, hence $w^* \in \text{vert}(W)$ and $u^* \in \text{vert}(U)$ hold. As its name implies, the sequence of objective values generated by Algorithm C.1 is monotonically increasing, until a local maximum is found [120]. The convergence is guaranteed by the finiteness of vert(U) and vert(W). If we try multiple initial points that are chosen elaborately and pick up the best one among the returned results, the solution quality is often satisfactory. The key point is, these initial points should span along most directions in the w-subspace. For example, one may search the $2m$ points on the boundary of W in directions $\pm e_i^m$, $i = 1, 2, \cdots, m$, where m is the dimension of w, and e_i^m is the i-th column of an $m \times m$ identity matrix. As LPs can be solved very efficiently, Algorithm C.1 is especially suitable for the instances with very complicated U and W, and usually outperforms general NLP solvers for bilinear programs with disjoint constraints.

Algorithm C.1 is also valid if W is other convex set, say, an ellipsoid, and converges to a local optimum in a finite number of iterations for a given precision [121].

b. Cardinality Constrained Uncertainty Set

A continuous cardinality constrained uncertainty set in the form of (C.14) is a special class of the polyhedral case, see the transformation in (C.15). Therefore, the previous method can be applied, and the number of inequalities in the polyhedral form is $3m+1$, which is equal to the number of binary variables in MILP (C.63). As revealed in Proposition C.3, for a polyhedral uncertainty set, we can merely consider the extreme points.

Consider a discrete cardinality constrained uncertainty set

$$W = \left\{ w \left| \begin{array}{c} w_j = w_j^0 + w_j^+ z_j^+ - w_j^- z_j^-, \forall j \\ \exists\, z^+, z^- \in Z \end{array} \right. \right\} \tag{C.66a}$$

$$Z = \left\{ z^+, z^- \left| \begin{array}{c} z^+,\ z^- \in \{0, 1\}^m \\ z_j^+ + z_j^- \le 1,\ \forall j \\ 1^T (z^+ + z^-) \le \Gamma \end{array} \right. \right\} \tag{C.66b}$$

where the budget of uncertainty $\Gamma \le m$ is an integer. In (C.66a), each element w_j takes one of three possible values: w_j^0, $w_j^0 + w_j^+$, and $w_j^0 - w_j^-$, and at most Γ of the m elements w_j can take a value that is not equal to w_j^0. If the forecast error is symmetric, i.e., $w_j^+ = w_j^-$, then (C.66) is called symmetric as the nominal scenario locates at the center of W. We discuss this case separately because this representation allows to linearize the non-convexity in (C.59) and (C.60) with fewer binary variables.

Expanding the bilinear term $u^T C w$ in an element-wise form

$$u^T C w = u^T C w^0 + \sum_i \sum_j \left(c_{ij} w_j^+ u_i z_j^+ - c_{ij} w_j^- u_i z_j^- \right)$$

where c_{ij} is the element of matrix C. Let

$$v_{ij}^+ = u_i z_j^+,\ v_{ij}^- = u_i z_j^-,\ \forall i,\ \forall j$$

the bilinear term can be expressed via a linear function. The product involving a binary variable and a continuous variable can be linearized via the method illuminated in Appendix B.2.2.

In conclusion, bilinear subproblems (C.59) and (C.60) can be solved via MILP

$$\begin{aligned} \max \ & u^T(b - Ax) - u^T C w^0 - \sum_i \sum_j \left(c_{ij} w_j^+ v_{ij}^+ - c_{ij} w_j^- v_{ij}^- \right) \\ \text{s.t.} \ & u \in U, \ \{z^+, z^-\} \in Z \\ & 0 \le v_{ij}^+ - u_j \le M\left(1 - z_j^+\right), -M z_j^+ \le v_{ij}^+ \le 0, \ \forall i, \forall j \\ & 0 \le v_{ij}^- - u_j \le M(1 - z_j^-), -M z_j^- \le v_{ij}^- \le 0, \ \forall i, \forall j \end{aligned} \tag{C.67}$$

where $M = 1$ for problem (C.59) since $-1 \le u \le 0$, and M is a sufficiently large number for problem (C.60), because there is no clear bounds for the dual variable u. The number of binary variables in MILP (C.67) is $2m$, which is less than that in (C.63) if the uncertainty set is replaced by its convex hull. The number of additional continuous variables v_{ij}^+ and v_{ij}^- is also moderate since the matrix C is sparse.

Finally, we are ready to give the decomposition algorithm which is proposed in [122]. In light of Proposition C.3, it is sufficient to consider the extreme points $w^1, w^2, \cdots, w^S$ in the uncertainty set, inspiring the following epigraph formulation which is equivalent to (C.42)

$$\begin{aligned} \min_{x, y^s, \eta} \ & c^T x + \eta \\ \text{s.t.} \ & x \in X \\ & \eta \ge d^T y^s, \ \forall s \\ & Ax + By^s \le b - Cw^s, \forall s \end{aligned} \tag{C.68}$$

Recall (C.47), the last constraint is in fact a lifted formulation for X_R. For polytope and cardinality constrained uncertainty sets, the number of extreme points is finite, but may grow exponentially in the dimension of uncertainty. Actually, it is difficult and also unnecessary to enumerate every extreme point, because most of them actually provide redundant constraints. A smart method is to identify active scenarios which contribute binding constraints in X_R. This motivation has been widely used in complex optimization problems and formalized in Sect. C.1.3. The procedure of the adaptive scenario generation algorithm for ARO is summarized in Algorithm C.2.

Algorithm C.2 converges in a finite number of iterations, which is bounded by the number of extreme points of the uncertainty set. In practice, this algorithm often converges in a few iterations, because problems (C.59) and (C.60) always identify the most critical scenario that should be considered. This is why we name the algorithm "adaptive scenario generation." It is called "constraint-and-column generation algorithm" in [122], because the numbers of decision variables

Algorithm C.2 Adaptive scenario generation

1: Choose a tolerance $\varepsilon > 0$, set $LB = -\infty$, $UB = +\infty$, iteration index $k = 0$, and the critical scenario set $O = w^0$;

2: Solve the following master problem

$$\begin{aligned} \min_{x, y^s, \eta} \quad & c^T x + \eta \\ \text{s.t.} \quad & x \in X, \ \eta \geq d^T y^s, \ s = 0, \cdots, k \\ & Ax + By^s \leq b - Cw^s, \forall w^s \in O \end{aligned} \tag{C.69}$$

The optimal solution is x^{k+1}, η^{k+1}, and update $LB = c^T x^{k+1} + \eta^{k+1}$;

3: Solve bilinear feasibility testing problem (C.59) with x^{k+1}, the optimal solution is w^{k+1}, u^{k+1}; if the optimal value $r^{k+1} > 0$, update $O = O \cup w^{k+1}$, and add a scenario cut

$$\eta \geq d^T y^{k+1}, \ Ax + By^{k+1} \leq b - Cw^{k+1} \tag{C.70}$$

with a new variable y^{k+1} to the master problem (C.69), update $k \leftarrow k + 1$, and go to Step 2;

4: Solve bilinear optimality testing problem (C.60) with x^{k+1}, the optimal solution is w^{k+1}, u^{k+1}, and the optimal value is R^{k+1}; update $O = O \cup w^{k+1}$ and $UB = c^T x^{k+1} + R^{k+1}$, create scenario cut (C.70) with a new variable y^{k+1}.

5: If $UB - LB \leq \varepsilon$, report the optimal solution, terminate; otherwise, add the scenario cut in step 4 to the master problem (C.69), update $k \leftarrow k + 1$, and go to step 2;

(columns) and constraints increase simultaneously. Please note that the scenario cut streamlines the feasibility cut and optimality cut used in the existing literature.

Bilinear subproblems (C.59) and (C.60) can be solved by the methods discussed previously, according to the form of the uncertainty set. In Algorithm C.2, we utilize w to create scenario cuts, which are also called primal cuts. In fact, the optimal dual variable u of (C.59) and (C.60) provides sensitivity information, and can be used to construct dual cuts, which is a single inequality in the first-stage variable x. See Benders decomposition algorithm in [118]. Since scenario cuts are much tighter than Benders cuts, Algorithm C.2 is the most prevalent method for solving ARO problems.

If matrix A is uncertainty-dependent, the scenario constraints in the master problem (C.69) become $A(w^s)x + By^s \leq b - Cw^s$, $\forall w^s$, where $A(w^s)$ a constant matrix associated with scenario s; the objective function of bilinear subproblems changes to $u^T[b - A(w)x - Cw]$, where x is given in the subproblem. If A can be expressed as a linear function in w, the problem structure remains the same, and previous methods are still valid. Even if the second-stage problem is an SOCP, the adaptive scenario generation framework remains applicable, and the key procedure is to solve a max-min SOCP. Such a model originates from the robust operation of a power distribution network with uncertain generation and demand. By dualizing the inner-most SOCP, the max-min SOCP is cast as a bi-convex program, which can be globally or locally solved via an MISOCP or the mountain climbing method. An example can be found in Sect. 5.4.

Recently, the duality theory of fully-adjustable robust optimization problem has been proposed in [123]. It has been shown that this kind of problem is self-dual, i.e., the dual problem remains an ARO. However, solving the dual problem may enjoy better efficiency. An extended CCG algorithm which always produces a feasible fist-stage decision (if one exists) is proposed in [124].

2. Second-Stage Problem is an MILP

Now we consider the case in which some of the wait-and-see decisions are discrete. As what can be observed from the previous case, the most important task in solving an ARO problem is to validate feasibility and optimality, which can boil down to solving a linear max-min problem. When the wait-and-see decisions are continuous and the second-stage problem is linear, LP duality theory is applied such that the linear max-min problem is cast as a traditional bilinear program. However, discrete variables appearing in the second stage make the recourse problem a mixed-integer linear max-min problem with a non-convex inner level, preventing the use of LP duality theory. As a result, validating feasibility and optimality becomes more challenging.

The compact form of an ARO problem with integer wait-and-see decisions can be written as

$$\min_{x \in X} \left\{ c^T x + \max_{w \in W} \min_{y,z \in Y(x,w)} d^T y + g^T z \right\} \tag{C.71}$$

where z is binary and depends on the exact value of w; the feasible region

$$Y(x, w) = \left\{ y, z \;\middle|\; \begin{array}{c} By + Gz \leq b - Ax - Cw \\ y \in \mathbb{R}^{m_1}, \ z \in \Phi \end{array} \right\}$$

where feasible set $\Phi = \{z | z \in \mathbb{B}^{m_2}, Tz \leq v\}$; m_1 and m_2 are dimensions of y and z; T and v are constant coefficients; all coefficient matrices have compatible dimensions. We assume that the uncertainty set W can be represented by a finite number of extreme points. This kind of problem is studied in [125]. A nested constraint-and-column generation algorithm is proposed.

Different from the mainstream idea that directly solves a linear max-min program as a bilinear program, the mixed-integer max-min program in (C.71) is expanded to a tri-level problem

$$\begin{aligned} \max_{w \in W} \min_{z \in \Phi} g^T z + \min_{y} \ & d^T y \\ \text{s.t.} \ & By \leq b - Ax - Cw - Gz \end{aligned} \tag{C.72}$$

For the ease of discussion, we assume all feasible sets are bounded, because decision variables of practical problems have physical bounds. By replacing the innermost LP in variable y with its dual LP, problem (C.72) becomes

$$\max_{w \in W} \left\{ \min_{z \in \Phi} \left\{ g^T z + \max_{u \in U} u^T (b - Ax - Cw - Gz) \right\} \right\} \tag{C.73}$$

where u is the dual variable, and set $U = \{u \mid u \leq 0,\ B^T u = d\}$. Because both w and z are expressed via binary variables, bilinear terms $u^T Cw$ and $u^T Gz$ have linear representations by using the method in Appendix B.2.2. Since Φ has a countable number of elements, problem (C.73) (in its linearized version) has the same form as ARO problem (C.42), and can be solved by Algorithm C.2. More exactly, write (C.73) into an epigraph form by enumerating all possible elements $z \in \Phi$, then perform Algorithm C.2 and identify binding elements. In this way, the minimization operator in the middle level is eliminated.

The nested adaptive scenario generation algorithm for ARO problem (C.71) with mixed-integer recourses is summarized in Algorithm C.3. Because both W and Φ are finite sets with countable elements, Algorithm C.3 converges in a finite number of iterations. Notice that we do not distinguish feasibility and optimality subproblems in the above algorithm due to their similarities. One can also introduce slack here-and-now variables in the second stage and penalty terms in the objective function, such that the recourse problem is always feasible. It should be pointed out that Algorithm C.3 incorporates double loops, and an MILP should be solved in each iteration in the inner loop, so we'd better not expect too much on its efficiency. Nonetheless, it is the first systematic method to solve an ARO problem with integer variables in the second stage. Another concept which should be clarified is that although the second-stage discrete variable z is treated as scenario and enumerated on the fly when solving problem (C.73) in step 3 (the inner loop), it is a decision variable of the master problem (C.74) in the outer loop.

As a short conclusion, to overcome the limitation of traditional static RO approaches which require all decisions should be made without exact information on the underlying uncertainty, ARO employs a two-stage decision-making framework and allows a subset of decision variables to be made after the uncertain data are revealed. Under some special decision rules, computational tractability can be preserved. In fully adjustable cases, the ARO problem can be solved by a decomposition algorithm. The subproblem comes down to a (mixed-integer) linear max-min problem, which is generally challenging to solve. We introduce MILP reformulations for special classes of uncertainty sets, which are compatible with commercial solvers, and help solve an engineering optimization problem in a systematic way.

Algorithm C.3 Nested adaptive scenario generation

1: Choose a tolerance $\varepsilon > 0$, set $LB = -\infty$, $UB = +\infty$, iteration index $k = 0$, and the critical scenario set $O = w^0$;

2: Solve the following master problem

$$
\begin{aligned}
\min_{x,y,z,\eta} \quad & c^T x + \eta \\
\text{s.t.} \quad & x \in X \\
& \eta \geq d^T y^s + g^T z^s, \ z^s \in \Phi, \ s = 0, \cdots, k \\
& Ax + By^s + Gz^s \leq b - Cw^s, \ \forall w^s \in O
\end{aligned}
\tag{C.74}
$$

The optimal solution is x^{k+1}, η^{k+1}, and update $LB = c^T x^{k+1} + \eta^{k+1}$;

3: Solve problem (C.73) with x^{k+1}, the optimal solution is $(z^{k+1}, w^{k+1}, u^{k+1})$, and optimal value is R^{k+1}; update $O = O \cup w^{k+1}$, $UB = \min\{UB, c^T x^{k+1} + R^{k+1}\}$, create new variables (y^{k+1}, z^{k+1}) and scenario cuts

$$
\begin{aligned}
& \eta \geq d^T y^{k+1} + g^T z^{k+1}, \ z^{k+1} \in \Phi \\
& Ax + By^{k+1} + Gz^{k+1} \leq b - Cw^{k+1}
\end{aligned}
\tag{C.75}
$$

4: If $UB - LB \leq \varepsilon$, terminate and report the optimal solution and optimal value; otherwise, add scenario cuts (C.75) to the master problem (C.74), update $k \leftarrow k + 1$, and go to step 2;

C.3 Distributionally Robust Optimization

Static and adjustable RO models presented in Sects. C.1 and C.2 do not rely on specifying probability distributions of the uncertain data, which are used in SO approaches for generating scenarios, evaluating probability of constraint violation, or deriving analytic solutions for some specific problems. Instead, RO design principle aims to cope with the worst-case scenario in a pre-defined uncertainty set in the space of uncertain variables, which is a salient distinction between these two approaches. If the exact probability distribution is precisely known, optimal solutions to SO models would be less conservative than the robust ones from the statistical perspective. However, the optimal solution to SO models could have poor statistical performances if the actual distribution is not identical to the designated one [126]. As for the RO approach, as it hedges against the worst-case scenario, which rarely happens in reality, the robust strategy could be conservative thus suboptimal in most cases.

A method which aims to build a bridge connecting SO and RO approaches is the DRO, whose optimal solutions are designed for the worst-case probability distribution within a family of candidate distributions, which are described by statistic information, such as moments, and structure properties, including symmetry, unimodality, and so on. This approach is generally less conservative than the traditional RO because dispersion effect of uncertainty is taken into account, i.e., the probability of an extreme event is low. Meanwhile, the statistic performance of the solution is less sensitive to the perturbation in probability distributions than that of an SO model, as it hedges against the worst distribution. Publications on this method have been proliferating rapidly in the past few years. This section only sheds light on some most representative methods which have been used in energy system studies.

C.3.1 Static Distributionally Robust Optimization

In analogy with the terminology used in Sect. C.1, "static" means that all decision variables are here-and-now type. Theoretical outcomes in this part mainly come from [127]. A static DRO problem can be formulated as

$$\begin{aligned} \min_x \ & c^T x \\ \text{s.t. } & x \in X \\ & \Pr\left(a_i(\xi)^T x \le b_i(\xi),\ i = 1, \cdots, m\right) \ge 1 - \varepsilon,\ \forall f(\xi) \in \mathcal{P} \end{aligned} \tag{C.76}$$

where x is the decision variable, X is a closed and convex set that is independent of the uncertain parameter, c is a deterministic vector, and ξ is the uncertain data, whose probability density function $f(\xi)$ is not known exactly, and belongs to $\mathcal{P}$, a set comprised of candidate distributions. Robust chance constraint in (C.76) requires a finite number of linear inequalities depending on ξ to be met with a probability of at least $1 - \varepsilon$, regardless of the true probability density function of ξ. We assume uncertain coefficients a_i and b_i are linear functions in ξ, i.e.

$$a_i(\xi) = a_i^0 + \sum_{j=1}^{k} a_i^j \xi_j$$

$$b_i(\xi) = b_i^0 + \sum_{j=1}^{k} b_i^j \xi_j$$

where a_i^0, a_i^j are constant vectors and b_i^0, b_i^j are constant scalars. Define

$$y_i^j(x) = (a_i^j)^T x - b_i^j,\ \forall i,\ \forall j$$

the chance constraint in (C.76) can be expressed via

$$\Pr\left(y_i^0(x) + y_i(x)^T \xi \le 0,\ i = 1, \cdots, m\right) \ge 1 - \varepsilon,\ \forall f(\xi) \in \mathcal{P} \tag{C.77}$$

where vector $y_i(x) = [y_i^1(x), \cdots, y_i^k(x)]^T$ is affine in x. Since the objective is certain and constraint violation is bounded by a small probability, problem (C.76) is also called a robust chance-constrained program.

Chance constraints can be transformed into tractable ones that are convex in variable x only for a few special cases. For example, if ξ follows a Gaussian distribution, $\varepsilon \le 0.5$, and $m = 1$, then the individual chance constraint without distribution uncertainty is equivalent to a single SOC constraint [128]. For $m > 1$, joint chance constraints form convex feasible region when the right-hand side terms $b_i(\xi)$ are uncertain and follow a log-concave distribution [127, 129], while coefficients $a_i, i = 1, \cdots, m$ are deterministic.

Constraint (C.77) is even more challenging at first sight: not only the random vector ξ, but also the probability distribution function $f(\xi)$ itself is uncertain. Because in many practical situations, probability distribution must be estimated from enough historical data, which may not be available at hand. Typically, one may only have access to some statistical indicators about $f(\xi)$, e.g. its mean value, covariance, and support set. Using a specific $f(\xi) \in \mathcal{P}$ may lead to over-optimistic solutions which fail to satisfy the probability guarantee under the true distribution.

Similar to the paradigm in static RO, a prudent way to immunize a chance constraint against uncertain probability distribution is to investigate the situation in the worst case, inspiring the following distributionally robust chance constraint, which is equivalent to (C.77)

$$\inf_{f(\xi)\in\mathcal{P}} \Pr\left(y_i^0(x) + y_i(x)^T\xi \le 0,\ i = 1, \cdots, m\right) \ge 1 - \varepsilon \tag{C.78}$$

Clearly, if x satisfies (C.78), the probability of constraint violation is upper bounded by ε for the true probability distribution of ξ.

This section introduces convex optimization models for approximating robust chance constraints under uncertain probability distributions, whose first- and second-order moments as well as the support set (or equivalently the feasible region) of random variable are known. More precisely, we let $\mathbb{E}_P(\xi) = \mu \in \mathbb{R}^k$ be the mean value and $\mathbb{E}_P((\xi - \mu)(\xi - \mu)^T) = \Sigma \in \mathbb{S}_{++}^k$ be the covariance matrix of random variable ξ under the true distribution P. We define the moment matrix

$$\Omega = \begin{bmatrix} \Sigma + \mu\mu^T & \mu \\ \mu^T & 1 \end{bmatrix}$$

for ease of notation.

To help readers understand the fundamental ideas in DRO, we briefly introduce the worst-case expectation problem, which will be used throughout this section. Recall that $\mathcal{P}$ represents the set of all probability distributions on $\mathbb{R}^k$ with mean vector μ and covariance matrix $\Sigma \succ 0$, the problem is formulated by

$$\theta_P^m = \sup_{f(\xi)\in\mathcal{P}} \mathbb{E}\left[(g(\xi))^+\right]$$

where $g : \mathbb{R}^k \to \mathbb{R}$ is a function of ξ; $(g(\xi))^+$ means the maximum between 0 and $g(\xi)$. Write the problem into an integral format

$$\begin{aligned}
\theta_P^m = & \sup_{f(\xi)\in\mathcal{P}} \int_{\xi\in\mathbb{R}^k} \max\{0, g(\xi)\} f(\xi)\mathrm{d}\xi \\
& \text{s.t. } f(\xi) \ge 0, \ \forall \xi \in \mathbb{R}^k \\
& \int_{\xi\in\mathbb{R}^k} f(\xi)\mathrm{d}\xi = 1 : \ \lambda_0 \\
& \int_{\xi\in\mathbb{R}^k} \xi f(\xi)\mathrm{d}\xi = \mu : \ \lambda \\
& \int_{\xi\in\mathbb{R}^k} \xi\xi^T f(\xi)\mathrm{d}\xi = \Sigma + \mu\mu^T : \ \Lambda
\end{aligned} \tag{C.79}$$

In problem (C.79), the decision variables are the values of $f(\xi)$ over all possible $\xi \in \mathbb{R}^k$, so there are infinitely many decision variables, and problem (C.79) is an infinite-dimensional LP. The former two constraints enforce $f(\xi)$ to be a valid distribution function; the latter two ensure consistent first- and second-order moments. The optimal solution gives the worst-case distribution. However, it is difficult to solve (C.79) in its primal form. We now associate dual variables $\lambda_0 \in \mathbb{R}$, $\lambda \in \mathbb{R}^k$, and $\Lambda \in \mathbb{S}^k$ with each integral constraint, and the dual problem of (C.79) can be constructed following the duality theory of conic LP, which is given by

$$\begin{aligned}
\theta_D^m = & \inf_{\lambda_0,\lambda,\Lambda} \ \lambda_0 + \mu^T\lambda + \mathrm{tr}[\Lambda^T(\Sigma + \mu\mu^T)] \\
& \text{s.t. } \lambda_0 + \xi^T\lambda + \mathrm{tr}[\Lambda^T(\xi\xi^T)] \\
& \ge \max\{0, g(\xi)\}, \forall \xi \in \mathbb{R}^k
\end{aligned} \tag{C.80}$$

To understand this dual form in (C.80), we can image a discrete version of (C.79), in which $\xi_1, \cdots, \xi_n$ are sampled scenarios of the uncertain parameter, and their associated probabilities $f(\xi_1), \cdots, f(\xi_n)$ are decision variables of (C.79). Moreover, if we replace the integral arithmetic in the constraints with the summation arithmetic, (C.79) comes down to a traditional LP, and its dual is also an LP, where the constraint becomes

$$\lambda_0 + \xi_i^T\lambda + \mathrm{tr}\left[\Lambda^T(\xi_i\xi_i^T)\right] \ge \max\{0, g(\xi_i)\}, \ i = 1, \cdots, n$$

Let $n \to +\infty$ and ξ spread over $\mathbb{R}^k$, we can get the dual problem (C.80).

Unlike the primal problem (C.79) that has infinite decision variables, the dual problem (C.80) has finite variables and an infinite number of constraints. In fact, we are optimizing over the coefficients of a polynomial in ξ. Because $\Sigma \succ 0$, Slater condition is met, and thus strong duality holds (this conclusion can be found in many other literatures, such as [130]), i.e., $\theta_P^m = \theta_D^m$. In the following, we will eliminate ξ and reduce the constraint into convex ones in dual variables λ_0, λ, and Λ. Recall the definition of matrix Ω, the compact form of problem (C.80) can be expressed as

$$\begin{aligned} \inf_{M \in \mathbb{S}^{k+1}} \quad & \operatorname{tr}[\Omega^T M] \\ \text{s.t.} \quad & \left[\xi^T \ 1\right] M \left[\xi^T \ 1\right]^T \geq 0, \ \forall \xi \in \mathbb{R}^k \\ & \left[\xi^T \ 1\right] M \left[\xi^T \ 1\right]^T \geq g(\xi), \ \forall \xi \in \mathbb{R}^k \end{aligned} \tag{C.81}$$

where the matrix decision variable is

$$M = \begin{bmatrix} \Lambda & \dfrac{\lambda}{2} \\ \dfrac{\lambda^T}{2} & \lambda_0 \end{bmatrix}$$

and the first constraint is equivalent to an LMI $M \succeq 0$.

A special case of the worst-case expectation problem is

$$\theta_P^m = \sup_{f(\xi) \in \mathcal{P}} \Pr[\xi \in S] \tag{C.82}$$

which quantifies the maximum probability of the event $\xi \in S$, where S is a Borel measurable set. This problem has a close relationship with generalized probability inequalities discussed in [130] and the generalized moments problem studied in [131]. By defining an indicator function as

$$\mathbb{I}_S(\xi) = \begin{cases} 1 & \text{if } \xi \in S \\ 0 & \text{otherwise} \end{cases}$$

The dual problem of (C.82) can be written as

$$\begin{aligned} \inf_{M \in \mathbb{S}^{k+1}} \quad & \operatorname{tr}[\Omega^T M] \\ \text{s.t.} \quad & M \succeq 0, \ \left[\xi^T \ 1\right] M \left[\xi^T \ 1\right] \geq 1, \ \forall \xi \in S \end{aligned} \tag{C.83}$$

which is a special case of (C.81) when $g(\xi) = \mathbb{I}_S(\xi)$.

Next we present how to formulate a robust chance constraint (C.78) as convex constraints that can be recognized by convex optimization solvers.

1. Individual Chance Constraints

Consider a single robust chance constraint

$$\inf_{f(\xi) \in \mathcal{P}} \Pr\left(y^0(x) + y(x)^T \xi \leq 0\right) \geq 1 - \varepsilon \tag{C.84}$$

The feasible set in x is denoted by X_R^S.

To eliminate the optimization over function $f(\xi)$, we leverage the concept of conditional value-at-risk (CVaR) introduced by [132]. For a given loss function $L(\xi)$ and tolerance $\varepsilon \in (0, 1)$, the CVaR at level ε is defined as

$$\mathrm{CVaR}(L(\xi), \varepsilon) = \inf_{\beta \in \mathbb{R}} \beta + \frac{1}{\varepsilon} \mathbb{E}_{f(\xi)} \left([L(\xi) - \beta]^+ \right) \tag{C.85}$$

where the expectation is taken over a given probability distribution $f(\xi)$. CVaR is the conditional expectation of loss greater than the $(1 - \varepsilon)$-quantile of the loss distribution. Indeed, condition

$$\Pr[L(\xi) \le \mathrm{CVaR}(L(\xi), \varepsilon)] \ge 1 - \varepsilon$$

holds regardless of the probability distribution and loss function $L(\xi)$ [127]. Therefore, to certify $\Pr(L(\xi) \le 0) \ge 1-\varepsilon$, a sufficient condition without probability evaluation is $\mathrm{CVaR}(L(\xi), \varepsilon) \le 0$, or more precisely:

$$\begin{aligned} &\sup_{f(\xi) \in \mathcal{P}} \mathrm{CVaR}\left(y^0(x) + y(x)^T \xi, \varepsilon \right) \le 0 \\ \Longrightarrow \ &\inf_{f(\xi) \in \mathcal{P}} \Pr\left(y^0(x) + y(x)^T \xi \le 0 \right) \ge 1 - \varepsilon \end{aligned} \tag{C.86}$$

According to (C.85), above worst-case CVaR can be expressed by

$$\begin{aligned} &\sup_{f(\xi) \in \mathcal{P}} \mathrm{CVaR}\left(y^0(x) + y(x)^T \xi, \varepsilon \right) \\ &= \sup_{f(\xi) \in \mathcal{P}} \inf_{\beta \in \mathbb{R}} \left\{ \beta + \frac{1}{\varepsilon} \mathbb{E}_{f(\xi)} \left(\left[y^0(x) + y(x)^T \xi - \beta \right]^+ \right) \right\} \\ &= \inf_{\beta \in \mathbb{R}} \left\{ \beta + \frac{1}{\varepsilon} \sup_{f(\xi) \in \mathcal{P}} \mathbb{E}_{f(\xi)} \left(\left[y^0(x) + y(x)^T \xi - \beta \right]^+ \right) \right\} \end{aligned} \tag{C.87}$$

The maximization and minimization operators are interchangeable because of the saddle point theorem in [133]. Recall previous analysis; the worst-case expectation can be computed from problem

$$\begin{aligned} &\inf_{\beta, M \in \mathbb{S}^{k+1}} \ \mathrm{tr}[\Omega^T M] \\ &\text{s.t. } M \succeq 0, \\ &\left[\xi^T \ 1 \right] M \left[\xi^T \ 1 \right] \ge y^0(x) + y(x)^T \xi - \beta, \ \forall \xi \in \mathbb{R}^k \end{aligned}$$

The semi-infinite constraint has a matrix quadratic form

$$\begin{bmatrix} \xi \\ 1 \end{bmatrix}^T \left(M - \begin{bmatrix} 0 & \dfrac{y(x)}{2} \\ \dfrac{y(x)^T}{2} & y^0(x) - \beta \end{bmatrix} \right) \begin{bmatrix} \xi \\ 1 \end{bmatrix} \geq 0, \ \forall \xi \in \mathbb{R}^k$$

which is equivalent to

$$M - \begin{bmatrix} 0 & \dfrac{y(x)}{2} \\ \dfrac{y(x)^T}{2} & y^0(x) - \beta \end{bmatrix} \succeq 0$$

As a result, the worst-case CVaR can be calculated from an SDP

$$\begin{aligned} & \sup_{f(\xi) \in \mathcal{P}} \ \text{CVaR}\left(y^0(x) + y(x)^T \xi, \varepsilon \right) \\ & = \inf_{\beta, M} \ \beta + \frac{1}{\varepsilon} \text{tr}(\Omega^T M) \\ & \text{s.t. } \ M \succeq 0 \\ & M \succeq \begin{bmatrix} 0 & \dfrac{y(x)}{2} \\ \dfrac{y(x)^T}{2} & y^0(x) - \beta \end{bmatrix} \end{aligned} \tag{C.88}$$

It is shown that the indicator $\Rightarrow$ in (C.86) is in fact an equivalence $\Leftrightarrow$ [127] in static DRO. In conclusion, robust chance constraint (C.84) can be written as a convex set in variable x, β, and M as follows:

$$X_R^S = \left\{ x \ \middle| \ \begin{array}{l} \exists \beta \in \mathbb{R}, \ M \succeq 0 \text{ such that} \\ \beta + \dfrac{1}{\varepsilon} \text{tr}(\Omega^T M) \leq 0 \\ M \succeq \begin{bmatrix} 0 & \dfrac{y(x)}{2} \\ \dfrac{y(x)^T}{2} & y^0(x) - \beta \end{bmatrix} \end{array} \right\} \tag{C.89}$$

2. Joint Chance Constraints

Now consider the joint robust chance constraints

$$\inf_{f(\xi) \in \mathcal{P}} \Pr\left(y_i^0(x) + y_i(x)^T \xi \leq 0, i = 1, \cdots, m \right) \geq 1 - \varepsilon \tag{C.90}$$

The feasible set in x is denoted by X_R^J.

Let α be the vector of strictly positive scaling parameters, and $\mathcal{A} = \{\alpha \mid \alpha > 0\}$. It is clear that constraint

$$\inf_{f(\xi)\in\mathcal{P}} \Pr\left[\max_{i=1,\cdots,m}\left\{\alpha_i\left(y_i^0(x) + y_i(x)^T\xi\right)\right\} \le 0\right] \ge 1-\varepsilon \tag{C.91}$$

imposes the same feasible region in variable x as (C.90). Nonetheless, it turns out that parameter α_i can be co-optimized to improve the quality of the convex approximation for X_R^J. (C.91) is a single robust chance constraint, and can be conservatively approximated by a worst-case CVaR constraint

$$\sup_{f(\xi)\in\mathcal{P}} \text{CVaR}\left[\max_{i=1,\cdots,m}\left\{\alpha_i\left(y_i^0(x) + y_i(x)^T\xi\right)\right\}, \varepsilon\right] \le 0 \tag{C.92}$$

It defines a feasible region in variable x with auxiliary parameter $\alpha \in \mathcal{A}$, which is denoted by $X_R^J(\alpha)$. Clearly, $X_R^J(\alpha) \subseteq X_R^J$, $\forall \alpha \in \mathcal{A}$. Unlike (C.88), condition (C.92) is α-dependent.

By observing the fact that

$$\begin{aligned}
&[\xi^T\ 1]\, M\, [\xi^T\ 1] \ge \max_{i=1,\cdots,m}\left\{\alpha_i\left(y_i^0(x) + y_i(x)^T\xi\right)\right\} - \beta,\ \forall \xi \in \mathbb{R}^k \\
\Longleftrightarrow\ &[\xi^T\ 1]\, M\, [\xi^T\ 1] \ge \alpha_i\left(y_i^0(x) + y(x)_i^T\xi\right) - \beta,\ \forall \xi \in \mathbb{R}^k,\ i = 1,\cdots,m \\
\Longleftrightarrow\ &M - \begin{bmatrix} 0 & \dfrac{\alpha_i y_i(x)}{2} \\ \dfrac{\alpha_i y_i(x)^T}{2} & \alpha_i y_i^0(x) - \beta \end{bmatrix} \succeq 0,\ i = 1,\cdots,m
\end{aligned}$$

and employing the optimization formulation of the worst-case expectation problem, the worst-case CVaR in (C.92) can be calculated by

$$\begin{aligned}
J(x,\alpha) &= \sup_{f(\xi)\in\mathcal{P}} \text{CVaR}\left[\max_{i=1,\cdots,m}\left\{\alpha_i\left(y_i^0(x) + y_i(x)^T\xi\right)\right\}, \varepsilon\right] \\
&= \inf_{\beta\in\mathbb{R}}\left\{\beta + \frac{1}{\varepsilon}\sup_{f(\xi)\in\mathcal{P}} \mathbb{E}_{f(\xi)}\left(\left[\max_{i=1,\cdots,m}\left\{\alpha_i\left(y_i^0(x) + y_i(x)^T\xi\right)\right\} - \beta\right]^+\right)\right\} \\
&= \inf_{\beta,M}\left\{\beta + \frac{1}{\varepsilon}\text{tr}(\Omega^T M)\ \middle|\ \text{s.t. } M \succeq 0,\ M \succeq \begin{bmatrix} 0 & \dfrac{\alpha_i y_i(x)}{2} \\ \dfrac{\alpha_i y_i(x)^T}{2} & y_i^0(x) - \beta \end{bmatrix}, \forall i\right\}
\end{aligned} \tag{C.93}$$

In conclusion, for any fixed $\alpha \in \mathcal{A}$, the worst-case CVaR constraint (C.92) can be written as a convex set in variables x, β, and M as follows:

$$X_R^J(\alpha) = \left\{ x \;\middle|\; \begin{array}{l} \exists \beta \in \mathbb{R},\ M \succeq 0 \text{ such that} \\ \beta + \dfrac{1}{\varepsilon}\text{tr}(\Omega^T M) \le 0 \\ M \succeq \begin{bmatrix} 0 & \dfrac{\alpha_i y_i(x)}{2} \\ \dfrac{\alpha_i y_i(x)^T}{2} & y_i^0(x) - \beta \end{bmatrix}, \forall i \end{array} \right\} \tag{C.94}$$

Moreover, it is revealed in [127] that the union $\bigcup_{\alpha \in \mathcal{A}} X_R^J(\alpha)$ gives an exact description of X_R^J, which indicates that the original robust chance constrained program

$$\min_x \left\{ c^T x \;\middle|\; \text{s.t. } x \in X \cap X_R^J \right\} \tag{C.95}$$

and the worst-case CVaR formulation

$$\min_{x,\alpha} \left\{ c^T x \;\middle|\; \text{s.t. } x \in X \cap X_R^J(\alpha),\ \alpha \in \mathcal{A} \right\}$$

or equivalently

$$\min_{x,\alpha} \left\{ c^T x \;\middle|\; \text{s.t. } x \in X,\ \alpha \in \mathcal{A},\ J(x,\alpha) \le 0 \right\} \tag{C.96}$$

have the same optimal value. The constraints of (C.96) contain bilinear matrix inequalities, which means that if either x or α is fixed, $J(x,\alpha) \le 0$ in (C.96) can come down to LMIs, however, when both x and α are variables, the constraint is non-convex, making problem (C.96) difficult to solve. In view of the biconvex feature [121], a sequential convex optimization procedure is presented to find an approximated solution.

Algorithm C.4 Sequential convex optimization procedure

1: Choose a convergence tolerance $\varepsilon > 0$; Let the iteration counter $k = 1$, $x^0 \in X \cap X_R^J(\alpha)$ be a feasible solution for some α and $f^0 = c^T x^0$;

2: Solve the following subproblem with input x^{k-1}

$$\min_\alpha \ \{J(x,\alpha) \mid \text{s.t. } \alpha \ge \delta \mathbf{1}\} \tag{C.97}$$

where $\mathbf{1}$ denotes the all-one vector with a compatible dimension, and $\delta > 0$ is a small constant; the worst-case CVaR functional is defined in (C.93). The optimal solution is α^k;

3: Solve the following master problem with input α^k

$$\min_x \ \left\{ c^T x \mid \text{s.t. } x \in X,\ J(x,\alpha^k) \le 0 \right\} \tag{C.98}$$

The optimal solution is x^k and the optimal value is f^k;

4: If $|f^k - f^{k-1}|/|f^{k-1}| \le \varepsilon$, terminate and report the optimal solution x^k; otherwise, update $k \leftarrow k+1$, and go to step 2.

The main idea of this algorithm is to identify the best feasible region $X_R^J(\alpha)$ through successively solving the subproblem (C.97), and therefore improving the objective value. The performance of Algorithm C.4 is intuitively explained below.

Because parameter α is optimized in the subproblem (C.97) given the value x^k, there must be $J(x^k, \alpha^{k+1}) \le J(x^k, \alpha^k) \le 0$, $\forall k$, demonstrating that x^k is a feasible solution of the master problem (C.98) in iteration $k+1$; therefore, the optimal values of (C.98) in two consecutive iterations satisfy $c^T x^{k+1} \le c^T x^k$, as the objective evaluated at the optimal solution x^{k+1} in iteration $k+1$ deserves a value no greater than that is incurred at any feasible solution. In this regard, the optimal value sequence f^k, $k = 1, 2, \cdots$ is monotonically decreasing. If X is bounded, the optimal solution sequence x^k is also bounded, and the optimal value converges. Algorithm C.4 does not necessarily find the global optimum of problem (C.96). Nevertheless, it is desired by practical problems due to its robustness since it involves only convex optimization.

In many practical applications, the uncertain data ξ is known to be within a strict subset of $\mathbb{R}^k$, which is called the support set. We briefly outline how to incorporate the support set in the distributionally robust chance constraints. We assume the support set Ξ is the intersection of a finite number of ellipsoids, i.e.

$$\Xi = \left\{ \xi \in \mathbb{R}^k \;\middle|\; \xi^T W_i \xi \le 1,\ i = 1, \cdots, l \right\} \tag{C.99}$$

where $W_i \in \mathbb{S}_+^k$, $i = 1, \cdots, l$, and we have $\Pr(\xi \in \Xi) = 1$. Let $\mathcal{P}_\Xi$ be the set of all candidate probability distributions supported on Ξ which have identical first- and second-order moments.

Consider the worst-case expectation problem (C.79). If we replace $\mathcal{P}$ with $\mathcal{P}_\Xi$, the constraints of the dual problem (C.80) become

$$\left[\xi^T \ 1\right] M \left[\xi^T \ 1\right]^T \ge 0, \qquad \forall \xi \in \Xi \tag{C.100}$$

$$\left[\xi^T \ 1\right] M \left[\xi^T \ 1\right]^T \ge g(\xi), \quad \forall \xi \in \Xi \tag{C.101}$$

According to (C.99), $1 - \xi^T W_i \xi$ must be non-negative if and only if $\xi \in \Xi$, and hence a sufficient condition for (C.100) is the existence of constants $\tau_i \ge 0$, $i = 1, \cdots, l$, such that

$$\left[\xi^T \ 1\right] M \left[\xi^T \ 1\right]^T - \sum_{i=1}^{l} \tau_i \left(1 - \xi^T W_i \xi\right) \ge 0 \tag{C.102}$$

Under this condition, as long as $\xi \in \Xi$, we have

$$\left[\xi^T \ 1\right] M \left[\xi^T \ 1\right]^T \ge \sum_{i=1}^{l} \tau_i \left(1 - \xi^T W_i \xi\right) \ge 0$$

Arrange (C.102) as a matrix quadratic form

$$[\xi^T \; 1]\left(M - \sum_{i=1}^{l} \tau_i \begin{bmatrix} -W_i & \mathbf{0} \\ \mathbf{0}^T & 1 \end{bmatrix}\right)\begin{bmatrix} \xi \\ 1 \end{bmatrix} \geq 0, \; \forall \xi \in \mathbb{R}^k$$

As a result, (C.100) can be reduced to an LMI in variables M and τ

$$M - \sum_{i=1}^{l} \tau_i \begin{bmatrix} -W_i & \mathbf{0} \\ \mathbf{0}^T & 1 \end{bmatrix} \succeq 0 \tag{C.103}$$

For similar reasons, by letting $g(\xi) = y^0(x) + y(x)^T\xi - \beta$, (C.101) can be conservatively approximated by the following LMI

$$M - \sum_{i=1}^{l} \tau_i \begin{bmatrix} -W_i & \mathbf{0} \\ \mathbf{0}^T & 1 \end{bmatrix} \succeq \begin{bmatrix} 0 & \frac{1}{2}y(x) \\ \frac{1}{2}y(x)^T & y^0(x) - \beta \end{bmatrix} \tag{C.104}$$

In fact, (C.103) and (C.104) are special cases of S-Lemma. Based upon these outcomes, most formulations in this section can be extended to consider the bounded support set Ξ in the form of (C.99). For polyhedral and some special classes of convex support sets, one may utilize the nonlinear Farkas lemma (Lemma 2.2 in [127]) to derive tractable reformulations.

C.3.2 *Adjustable Distributionally Robust Optimization*

As explained in Appendix C.1, the traditional static RO encounters difficulties in dealing with equality constraints. This plight remains in the DRO approach following a static setting. Consider $x + \xi = 1$ where $\xi \in [0, 0.1]$ is uncertain, while its mean and variance are known. For any given x^*, the worst-case probability $\inf_{f(\xi)\in\mathcal{P}} \Pr[x^* + \xi = 1] = 0$, because one can always find a feasible probability distribution function $f(\xi)$ that satisfies the first- and second-order moment constraints, whereas $f(1 - x^*) = 0$.

To vanquish this difficulty, it is necessary to incorporate wait-and-see decisions. A simple remedy is to impose an affine recourse policy without involving optimization in the second stage, giving rise to an affine-adjustable RO with distributional uncertainty and linear decision rule, which can be solved by the method in Appendix C.3.1.

This section aims to investigate the following adjustable DRO with completely flexible wait-and-see decisions

$$\min_{x \in X} \left\{ c^T x + \sup_{f(w) \in \mathcal{P}} \mathbb{E}_{f(w)} Q(x, w) \right\} \tag{C.105}$$

where x is the first-stage (here-and-now) decision, and X is its feasible set; the uncertain parameter is denoted by w; the probability distribution $f(w)$ belongs to the Chebyshev ambiguity set (whose first- and second-order moments are known)

$$\mathcal{P} = \left\{ f(w) \left| \begin{array}{c} f(w) \geq 0, \ \forall w \in W \\ \displaystyle\int_{w \in W} f(w) \mathrm{d}w = 1 \\ \displaystyle\int_{w \in W} w f(w) \mathrm{d}w = \mu \\ \displaystyle\int_{w \in W} w w^T f(w) \mathrm{d}w = \Theta \end{array} \right. \right\} \tag{C.106}$$

supported on $W = \{w \mid (w - \mu)^T Q(w - \mu) \leq \Gamma\}$, where matrix $\Theta = \Sigma + \mu\mu^T$ represents the second-order moment; μ is the mean value and Σ is the covariance matrix. The expectation in (C.105) is taken over the worst-case $f(w)$ in $\mathcal{P}$, and the second-stage problem under fixed x and w is an LP

$$Q(x, w) = \min_{y \in Y(x, w)} d^T y \tag{C.107}$$

$Q(x, w)$ is its optimal value function under fixed x and w. The feasible set of the second-stage problem is

$$Y(x, w) = \{y \mid By \leq b - Ax - Cw\}$$

Matrices A, B, C and vectors b, c, d are constant coefficients in the model. We assume that the second-stage problem is always feasible, i.e., $\forall x \in X, \forall w \in W$: $Y(x, w) \neq \emptyset$ and is bounded, and thus $Q(x, w)$ has a finite optimal value. This can be implemented by introducing wait-and-see type slack variables and adding penalties in the objective of (C.107).

The difference between problems (C.42) and (C.105) stems from the descriptions of uncertainty and the criteria in the objective function: more information of the dispersion effect, such as the covariance matrix, is taken into account in the latter one, and the objective function in (C.105) is an expectation reflecting the statistical behavior of the second-stage cost, rather than the one in (C.42) which is associated with only a single worst-case scenario, and leaves the performances in all other scenarios un-optimized. Because the probability distribution is uncertain, it is prudent to investigate the worst-case outcome in which the expected cost of the second stage is maximized. This formulation is advantageous in several ways: first, the requirement on the exact probability distribution is not necessary, and the optimal solution is insensitive to the family of distributions with common mean and covariance; second, the dispersion of the uncertainty is also taken into account, which helps reduce model conservatism: since the variance is fixed, a scenario that leaves far away from the forecast would have a low probability; finally, it is often important to tackle the tail effect, which indicates that the occurrence of a rare event may induce heavy losses in spite of its low probability. Such phenomenon is naturally taken into account in (C.105). In what follows, we outline the method proposed in [134] to solve the adjustable DRO problem (C.105). A slight modification is that an ellipsoid support set is considered.

1. The Worst-Case Expectation Problem

We consider the following worst-case expectation problem with a fixed x

$$\sup_{f(w)\in\mathcal{P}} \mathbb{E}_{f(w)} Q(x,w) \tag{C.108}$$

According to the discussions for problem (C.79), the dual problem of (C.108) is

$$\begin{aligned} \min_{H,h,h_0} \quad & \operatorname{tr}(H^T\Theta) + \mu^T h + h_0 \\ \text{s.t.} \quad & w^T H w + h^T w + h_0 \ge Q(x,w), \ \forall w \in W \end{aligned} \tag{C.109}$$

where H, h, h_0 are dual variables. Nevertheless, the optimal value function $Q(x, w)$ is not given in a closed form. From the LP duality theory

$$Q(x,w) = \max_{u\in U} u^T(b - Ax - Cw)$$

where u is the dual variable of LP (C.107), and its feasible set is given by

$$U = \{u \mid B^T u = d, \ u \le 0\}$$

Because we have assumed that $Q(x, w)$ is bounded, the optimal solution of the dual problem can be found at one of the extreme points of U, i.e.,

$$\exists u^* \in \operatorname{vert}(U): \ Q(x,w) = (b - Ax - Cw)^T u^* \tag{C.110}$$

where $\operatorname{vert}(U) = \{u^1, u^2, \cdots, u^{N_E}\}$ stands for the vertices of polyhedron U, and $N_E = |\operatorname{vert}(U)|$ is the cardinality of $\operatorname{vert}(U)$. In view of this, the constraint of (C.109) can be expressed as

$$w^T H w + h^T w + h_0 \ge (b - Ax - Cw)^T u^i, \ \forall w \in W, \ i = 1, \cdots, N_E$$

Recall the definition of W; a certification for above condition is

$$\begin{aligned} & w^T H w + h^T w + h_0 - (b - Ax - Cw)^T u^i \\ & \quad \geq \lambda[\Gamma - (w-\mu)^T Q (w-\mu)] \geq 0, \ \forall w \in \mathbb{R}^k, \ i = 1, \cdots, N_E \end{aligned}$$

which has the following compact matrix form

$$\begin{bmatrix} w \\ 1 \end{bmatrix}^T M^i \begin{bmatrix} w \\ 1 \end{bmatrix} \geq 0, \forall w \in \mathbb{R}^k, \ i = 1, \cdots, N_E \tag{C.111}$$

where

$$M^i = \begin{bmatrix} H + \lambda Q & \dfrac{h - C^T u^i}{2} - \lambda Q \mu \\ \dfrac{h^T - (u^i)^T C}{2} - \lambda \mu^T Q & h_0 - (b - Ax)^T u^i - \lambda(\Gamma - \mu^T Q \mu) \end{bmatrix} \tag{C.112}$$

and (C.111) simply reduces to $M^i \succeq 0, i = 1, \cdots, N_E$.

Finally, problem (C.109) comes down to the following SDP

$$\begin{aligned} \min_{H,h,h_0,\lambda} \quad & \text{tr}(H^T \Theta) + \mu^T h + h_0 \\ \text{s.t.} \quad & M^i(H, h, h_0, \lambda) \succeq 0, \ i = 1, \cdots, N_E \\ & \lambda \in \mathbb{R}^+ \end{aligned} \tag{C.113}$$

where $M^i(H, h, h_0, \lambda)$ is defined in (C.112). The above results can be readily extended if the support set is the intersection of ellipsoids.

2. Adaptive Constraint Generation Algorithm

Due to the positive semi-definiteness of the covariance matrix Σ, the duality gap between problems (C.108) and (C.109) is zero [134], and hence we can replace the worst-case expectation in (C.105) with its dual form, yielding

$$\begin{aligned} \min \quad & c^T x + \text{tr}(H^T \Theta) + \mu^T h + h_0 \\ \text{s.t.} \quad & M^i(H, h, h_0, \lambda) \succeq 0, \ i = 1, \cdots, N_E \\ & x \in X, \ \lambda \in \mathbb{R}^+ \end{aligned} \tag{C.114}$$

Problem (C.114) is an SDP. However, the number of vertices in set U ($|\text{vert}(U)|$) may increase exponentially in the dimension of U. It is non-trivial to enumerate all of them. However, because of weak duality, only the one which is optimal in the dual problem provides an active constraint, as shown in (C.110), and the rest are redundant inequalities. To identify the critical vertex in (C.110), we solve problem (C.114) in iterations: in the master problem, a subset of vert(U) is used to formulate a relaxation, then check whether the following constraint

$$w^T Hw + h^T w + h_0 \geq (b - Ax - Cw)^T u, \ \forall w \in W, \ \forall u \in U \tag{C.115}$$

is fulfilled. If yes, the relaxation is exact and the optimal solution is found; otherwise, find a new vertex of U at which constraint (C.115) is violated, and then add a cut to the master problem so as to tighten the relaxation, till constraint (C.115) is satisfied. The flowchart is summarized in Algorithm C.5.

Algorithm C.5 terminates in a finite number of iterations which is bounded by $|\text{vert}(U)|$. Actually, it will converge within a few iterations, because the subproblem (C.117) in step 3 always identifies the most critical vertex in vert(U). It is worth mentioning that the subproblem (C.117) is a non-convex program. Despite that it can be solved by general NLP solvers, we suggest three approaches with different computational complexity and optimality guarantees.

1. If the support set $W = \mathbb{R}^k$, it can be verified that matrix M_i becomes

$$\begin{bmatrix} H & \dfrac{h + C^T u^i}{2} \\ \dfrac{(h + C^T u^i)^T}{2} & h_0 - (b - Ax)^T u^i \end{bmatrix} \succeq 0$$

Then there must be $H \succeq 0$, and non-convexity appears in the bilinear term $u^T Cw$. In such circumstance, problem (C.117) can be solved via a mountain

---

Algorithm C.5 Adaptive constraint generation

1: Choose a convergence tolerance $\epsilon > 0$ and an initial vertex set $V_E \subseteq \text{vert}(U)$.
2: Solve the following master problem

$$\begin{aligned} \min \ & c^T x + \text{tr}(H^T \Theta) + \mu^T h + h_0 \\ \text{s.t.} \ & M^i(H, h, h_0, \lambda) \succeq 0, \ \forall u^i \in V_E \\ & x \in X, \ \lambda \in \mathbb{R}^+ \end{aligned} \tag{C.116}$$

The optimal value is R^*, and the optimal solution is (x^*, H, h, h_0).
3: Solve the following sub-problem with obtained (x^*, H, h, h_0)

$$\begin{aligned} \min_{w,u} \ & w^T Hw + h^T w + h_0 - (b - Ax^* - Cw)^T u \\ \text{s.t.} \ & w \in W, \ u \in U \end{aligned} \tag{C.117}$$

The optimal value is r^*, and the optimal solution is $u^*$and w^*.
4: If $r^* \geq -\varepsilon$, terminate and report the optimal solution $x^*$and the optimal value R^*; otherwise, $V_E = V_E \cup u^*$, add an LMI cut $M(H, h, h_0, \lambda) \succeq 0$ associated with the current u^* to the master problem (C.116), and go to step 2.

---

climbing method similar to Algorithm C.1 (but here the mountain is actually a pit because the objective is to be minimized).

2. In the case that W is an ellipsoid, the above iterative approach is still applicable; however, the w-subproblem in which w is to be optimized may become non-convex because H may be indefinite. Since u is fixed in the w-subproblem, non-convex term $w^T H w$ can be decomposed as the difference of two convex functions as $w^T(H + \alpha I)w - \alpha w^T w$, where α is a constant such that $H + \alpha I$ is positive-definite, and the w-subproblem can be solved by the convex-concave procedure elaborated in [135], or any existing NLP solver.
3. As a non-convex QP, problem (C.117) can be globally solved by the MILP method presented in Appendix A.4. This method could be time consuming with the growth in problem sizes.

C.4 Data-Driven Robust Stochastic Program

Most classical SO methods assume that the probability distribution of uncertain factors is exactly known, which is an input of the problem. However, such information heavily relies on historical data, and may not be available at hand or accurate enough. Using an inaccurate distribution in a classical SO model could lead to biased results. To cope with ambiguous probability distributions, a natural way is to consider a set of possible candidates derived from available data, instead of a single distribution, just as the moment-inspired ambiguity set used in DRO. In this section, we investigate some useful SO models with distributional uncertainty described by divergence ambiguity sets, which is referred to as robust SO. When the distribution is discrete, the distributional uncertainty is interpreted by the perturbation of probability value associated with each scenario; when the distribution is continuous, the distance of two density functions should be specified first. In this section, we consider Φ-divergence and Wasserstein metric based ambiguity sets.

C.4.1 Robust Chance Constrained Stochastic Program

We introduce robust chance-constrained stochastic programs with distributional robustness. The ambiguous PDF is modeled based on ϕ-divergence, and the optimal solution provides constraint feasibility guarantee with desired probability even in the worst-case distribution. In short, the underlying problem possesses the following features:

1. The PDF is continuous and the constraint violation probability is a functional.
2. Uncertain parameters do not explicitly appear in the objective function.

The main results of this section come from [136].

1. Problem Formulation
In a traditional chance-constrained stochastic linear program, the decision maker seeks a cost-minimum solution at which some certain constraints can be met with a given probability, yielding:

$$\begin{aligned} \min \quad & c^T x \\ \text{s.t.} \quad & \Pr[C(x,\xi)] \geq 1-\alpha \\ & x \in X \end{aligned} \tag{C.118}$$

where x is the vector of decision variables; ξ is the vector of uncertain parameters, and the exact (joint) probability distribution is apparent to the decision maker; vector c represents the cost coefficients; X is a polyhedron that is independent of ξ; α is the risk level or the maximum allowed probability of constraint violation; $C(x,\xi)$ collects all uncertainty dependent constraints, whose general form is given by

$$C(x,\xi) = \{\xi \mid \exists y : A(\xi)x + B(\xi)y \leq b(\xi)\} \tag{C.119}$$

where A, B, b are constant coefficient matrices that may contain uncertain parameters; y is a recourse action that can be made after ξ is known. In the presence of y, we call (C.118) a two-stage problem; otherwise, it is a single-stage problem if y is null. We don't consider the cost of recourse actions in the objective function in its current form. In case of need, we can add the second-stage cost $d^T y(\xi)$ in the objective function, and ξ is a specific scenario which $y(\xi)$ corresponds to; for instance, robust optimization may consider a max-min cost scenario or a max-min regret scenario; traditional SO often tackles the expected second-stage cost $\mathbb{E}[d^T y(\xi)]$. We leave it to the end of this section to discuss how to deal with the second-stage cost in the form of worst-case expectation like (C.108), and show that the problem can be convexified under some technical assumptions.

In the chance constraint, for a given x, the probability of constraint satisfaction can be evaluated for a particular probability distribution of ξ. Traditional studies on chance-constrained programs often assume that the distribution of ξ is perfectly known. However, this assumption can be very strong because it requires a lot of historical data. Moreover, the optimal solution may be sensitive to the true distribution and thus highly suboptimal in practice. To overcome these difficulties, a prudent method is to consider a set of probability distributions belonging to a pre-specified ambiguity set D, and require that the chance constraint should be satisfied under all possible distributions in D, resulting in the following robust chance-constrained programming problem:

$$\begin{aligned} \min \quad & c^T x \\ \text{s.t.} \quad & \inf_{f(\xi)\in D} \Pr[C(x,\xi)] \geq 1-\alpha \\ & x \in X \end{aligned} \tag{C.120}$$

where $f(\xi)$ is the probability density function of random variable ξ.

The ambiguity set D in (C.120) which includes distributional information can be constructed in a data-driven fashion, such as the moment based ones used in Appendix C.3. Please see [137] for more information on establishing D based on moment data and other structural properties, such as symmetry and unimodality. The tractability of (C.120) largely depends on the form of D. For example: if D is built on the mean value and covariance matrix (which is called a Chebyshev ambiguity set), a single robust chance constraint can be reformulated as an LMI and a set of joint robust chance constraints can be approximated by BMIs [127]; probability of constraint violation under more general moment based ambiguity sets can be evacuated by solving conic optimization problems [137].

A shortcoming of moment description is that it does not provide a direct measure on the distance between the candidate PDFs in D and a reference distribution. Two PDFs with the same moments may differ a lot in other aspects. Furthermore, the worst-case distribution corresponding to a Chebyshev ambiguity set always puts more weights away from the mean value, subject to the variance. As such, the long-tail effect is a source of conservatism. In this section, we consider the confidence set built around a reference distribution. The motivation is: the decision maker may have some knowledge on what distribution the uncertainty follows, although such a distribution could be inexact, and the true density function would not deviate far away from it.

To describe distributional ambiguity in terms of a PDF, the first problem is how to characterize the distance between two functions. One common measure on the distance between density functions is the ϕ-divergence, which is defined as [138]

$$D_\phi(f \| f_0) = \int_\Omega \phi\left(\frac{f(\xi)}{f_0(\xi)}\right) f_0(\xi) d\xi \tag{C.121}$$

where f and f_0 stand for the particular density function and the estimated one (or the reference distribution), respectively; function ϕ satisfies:

(C1) $\phi(1) = 0$

(C2) $0\phi(x/0) = \begin{cases} x \lim_{p \to +\infty} \phi(p)/p & \text{if } x > 0 \\ 0 & \text{if } x = 0 \end{cases}$

(C3) $\phi(x) = +\infty$ for $x < 0$

(C4) $\phi(x)$ is a convex function on $\mathbb{R}^+$

It is proposed in [138] that the ambiguity set can be built as:

$$D = \{P : D_\phi(f \| f_0) \le d, \, f = \mathrm{d}P/\mathrm{d}\xi\} \tag{C.122}$$

where the tolerance d can be adjusted by the decision maker according to their attitudes towards risks. The ambiguity set in (C.122) can be denoted as D_ϕ, without causing confusion with the definition of ϕ-divergence $D_\phi(f \| f_0)$. Compared to the moment-based ambiguity sets, especially the Chebyshev ambiguity set, where only the first- and second-order moments are involved, the density based description captures the overall profile of the ambiguous distribution, so may hopefully provide less conservative solutions. However, it hardly guarantees consistent moments. Which one is better depends on data availability: if we are more confident on the reference distribution, (C.122) may be better; otherwise, if we only have limited statistic information such as mean and variance, then the moment-based ones are more straightforward.

Many commonly seen divergence measures are special cases of ϕ-divergence, coinciding with a particular choice of function ϕ. Some examples are given in Table C.2 [139]. In what follows, we will use the KL-divergence. According to its corresponding function ϕ, the KL-divergence is given by

$$D_\phi(f \| f_0) = \int_\Omega \log\left(\frac{f(\xi)}{f_0(\xi)}\right) f(\xi) d\xi \tag{C.123}$$

Before presenting the main results in [136], the definition of conjugate duality is given. For a univariate function $g : \mathbb{R} \to \mathbb{R} \cup \{+\infty\}$, its conjugate function $g^* : \mathbb{R} \to \mathbb{R} \cup \{+\infty\}$ is defined as

$$g^*(t) = \sup_{x \in \mathbb{R}} \{tx - g(x)\}$$

For a valid function ϕ for ϕ-divergence satisfying (C1)–(C4), its conjugate function ϕ^* is convex, nondecreasing, and the following condition holds [136]

$$\phi^*(x) \geq x \tag{C.124}$$

Besides, if ϕ^* is a finite constant on a closed interval $[a, b]$, then it is a finite constant on the interval $(-\infty, b]$.

Table C.2 Instances of ϕ-divergences

Divergence	Function $\phi(x)$
KL-divergence	$x \log x - x + 1$
Reverse KL-divergence	$-\log x$
Hellinger distance	$(\sqrt{x} - 1)^2$
Variation distance	$\lvert x - 1 \rvert$
J-divergence	$(x - 1)\log x$
χ^2 divergence	$(x - 1)^2$
α-divergence	$\begin{cases} \frac{4}{1-\alpha^2}\left(1 - x^{(1+\alpha)/2}\right) & \text{If } \alpha \neq \pm 1 \\ x \ln x & \text{If } \alpha = 1 \\ -\ln x & \text{If } \alpha = -1 \end{cases}$

2. Equivalent Formulation

It is revealed in [136] that when the confidence set D is constructed based on ϕ-divergence, robust chance constrained program (C.120) can be easily transformed into a traditional chance-constrained program (C.118) at the reference distribution by calibrating the confidence tolerance α.

Theorem C.1 ([136]) *Let $\mathbb{P}_0$ be the cumulative distribution function generated by density function f_0, then the robust chance constraint*

$$\inf_{\mathbb{P}(\xi)\in\{D_\phi(f\|f_0)\le d\}} \Pr[C(x,\xi)] \ge 1-\alpha \tag{C.125}$$

constructed based on ϕ-divergence is equivalent to a traditional chance constraint

$$\Pr_0[C(x,\xi)] \ge 1-\alpha'_+ \tag{C.126}$$

where $\Pr_0$ *means that the probability is evaluated at the reference distribution* $\mathbb{P}_0$, $\alpha'_+ = \max\{\alpha', 0\}$, *and α' can be computed by*

$$\alpha' = 1 - \inf_{z\in Z}\left\{\frac{\phi^*(z_0+z) - z_0 - \alpha z + d}{\phi^*(z_0+z) - \phi^*(z_0)}\right\}$$

where

$$Z = \left\{ z \left| \begin{array}{l} z>0,\ z_0+\pi z \le l_\phi \\ \underline{m}(\phi^*) \le z+z_0 \le \overline{m}(\phi^*) \end{array} \right. \right\}$$

In the above formula, constants $l_\phi = \lim_{x\to+\infty}\phi(x)/x$, $\overline{m}(\phi^*) = \inf\{m : \phi^*(m) = +\infty\}$, $\underline{m}(\phi^*) = \sup\{m : \phi^* \text{ is a finite constant on } (-\infty, m]\}$, *Table C.3 summarizes the values of these parameters for typical ϕ-divergence measures, and*

$$\pi = \begin{cases} -\infty & \text{if } Leb\{[f_0=0]\}=0 \\ 0 & \text{if } Leb\,\{[f_0=0]\}>0 \text{ and } Leb\{[f_0=0]\backslash C(x,\xi)\}=0 \\ 1 & \text{otherwise} \end{cases}$$

where $Leb\{\cdot\}$ is the Lebesgue measure on $\mathbb{R}^{Dim(\xi)}$.

Table C.3 Values of l_ϕ, $\underline{m}(\phi^*)$, and $\overline{m}(\phi^*)$ for ϕ-divergences

ϕ-Divergence	l_ϕ	$\underline{m}(\phi^*)$	$\overline{m}(\phi^*)$
KL-divergence	$+\infty$	$-\infty$	$+\infty$
Hellinger distance	1	$-\infty$	1
Variation distance	1	-1	1
J-divergence	$+\infty$	$-\infty$	$+\infty$
χ^2 divergence	$+\infty$	-2	$+\infty$

Table C.4 Values of α' for some ϕ-divergences

ϕ-Divergence	α'
χ^2 divergence	$\alpha' = \alpha - \dfrac{\sqrt{d^2 + 4d(\alpha - \alpha^2)} - (1 - 2\alpha)d}{2d + 2}$
Variation distance	$\alpha' = \alpha - \dfrac{1}{2}d$
KL-divergence	$\alpha' = 1 - \inf_{x \in (0,1)} \left\{ \dfrac{e^{-d} x^{1-\alpha} - 1}{x - 1} \right\}$

The values of α' for the Variation distance and the χ^2 divergence have analytical expressions; for the KL divergence, α' can be computed from one-dimensional line search. Results are shown in Table C.4.

For the KL divergence, calculating α' entails solving $\inf_{x \in (0,1)} h(x)$ where

$$h(x) = \frac{e^{-d} x^{1-\alpha} - 1}{x - 1}$$

Its first-order derivative is given by

$$h'(x) = \frac{1 - \alpha e^{-d} x^{1-\alpha} - (1 - \alpha) e^{-d} x^{-\alpha}}{(x - 1)^2}, \quad \forall x \in (0, 1)$$

To claim the convexity of $h(x)$, we need to show that $h'(x)$ is an increasing function in $x \in (0, 1)$. To this end, first notice that the denominator $(x - 1)^2$ is a decreasing function in x on the open interval $(0, 1)$; then we can show the numerator is an increasing function in x, because its first-order derivative gives

$$(1 - \alpha e^{-d} x^{1-\alpha} - (1 - \alpha) e^{-d} x^{-\alpha})'_x = \alpha(1 - \alpha) e^{-d} (x^{-\alpha - 1} - x^{-\alpha}) > 0, \ \forall x \in (0, 1)$$

Hence $h'(x)$ is monotonically increasing, and $h(x)$ is a convex function in x. Moreover, because $h'(x)$ is continuous in $(0, 1)$, and $\lim_{x \to 0^+} h'(x) = -\infty$, $\lim_{x \to 1^-} h'(x) = +\infty$, there must be some $x^* \in [\delta, 1 - \delta]$ such that $h'(x^*) = 0$, i.e., the infimum of $h(x)$ is attainable. The minimum of $h(x)$ can be calculated by solving a nonlinear equation $h'(x) = 0$ via Newton's method, or a derivative-free line search, such as the golden section search algorithm. Either scheme is computationally inexpensive.

Finally, we discuss the connection between the modified tolerance α' and its original value α. Because a set of distributions are considered in (C.125), the threshold in (C.126) should be greater than the original one, i.e., $1 - \alpha' \geq 1 - \alpha$ must hold. To see this, recall inequality (C.124) of conjugate function, we have

$$\alpha \phi^*(z_0 + z) + (1 - \alpha)\phi(z_0) \geq \alpha(z_0 + z) + (1 - \alpha) z_0$$

The right-hand side gives $\alpha z + z_0$; in the ambiguity set (C.123), d is strictly positive, therefore

$$\alpha\phi^*(z_0 + z) + (1 - \alpha)\phi(z_0) \geq \alpha z + z_0 - d$$

which gives

$$\phi^*(z_0 + z) - z_0 - az + d \geq (1 - \alpha)(\phi(z_0 + z) - \phi(z_0))$$

Recall the expression of α' in Theorem C.1, we arrive at

$$1 - \alpha' = \frac{\phi^*(z_0 + z) - z_0 - az + d}{\phi(z_0 + z) - \phi(z_0)} \geq 1 - \alpha$$

which is the desired conclusion.

Theorem C.1 concludes that the complexity of handling a robust chance constraint is almost the same as that of tackling a traditional chance constraint associated with the reference distribution $\mathbb{P}_0$, except for the efforts on computing α'. If $\mathbb{P}_0$ belongs to the family of log-concave distributions, then the chance constraint is convex. As a special case, if $\mathbb{P}_0$ is the Gaussian distribution or a uniform distribution on ellipsoidal support, a single chance constraint can boil down to a second-order cone [140]. For more general cases, the chance constraint is non-convex in x. In such circumstance, we will use risk based reformulation and the sampling average approximation (SAA) approach.

3. Risk and SAA Based Reformulation

Owing to the different descriptions on dispersion ambiguity and presence of the wait-and-see decision y, unlike DRO problem (C.76) with static robust chance constraint (C.78) which can be transformed into an SDP, constraint (C.125) is treated in a different way, as demonstrated in Theorem C.1: it comes down to a traditional chance constraint (C.126) while the dispersion ambiguity is taken into account by a modification in the confidence level. The remaining task is to express (C.126) as a solver-compatible form.

(1) Loss Function

For given x and ξ, constraints in $C(x, \xi)$ cannot be met if no y satisfying $A(\xi)x + B(\xi)y \leq b(\xi)$ exists. To quantify the constraint violation under scenario $\boldsymbol{\xi}$ and first-stage decision x, define the following loss function $L(x, \xi)$

$$\begin{aligned} L(x, \xi) = \min_{y,\sigma} \ & \sigma \\ \text{s.t.} \ & A(\xi)x + B(\xi)y \leq b(\xi) + \sigma\mathbf{1} \end{aligned} \tag{C.127}$$

where $\mathbf{1}$ is an all-one vector with compatible dimension. If $L(x, \xi) \geq 0$, the minimum of slackness σ under the joint efforts of the recourse action y is defined as the loss; otherwise, demands are satisfiable after the uncertain parameter is known.

As we assume $C(x, \xi)$ is a bounded polytope, problem (C.127) is always feasible and bounded below. Therefore, the loss function $L(\boldsymbol{x}, \boldsymbol{\xi})$ is well-defined, and the chance constraint (C.126) can be written as

$$\text{Pr}_0[L(x, \xi) \leq 0] \geq 1 - \alpha'_{+} \tag{C.128}$$

In this way, the joint chance constraints are consolidated into a single one, just like what has been done in (C.90) and (C.91).

(2) VaR Based Reformulation: An MILP

For a given probability tolerance β and a first-stage decision x, the β-VaR for loss function $L(x, \xi)$ under the reference distribution PDF $\mathbb{P}_0$ is defined as

$$\beta\text{-VaR}(x) = \min \left\{ a \in \mathbb{R} \middle| \int_{L(x,\xi) \leq a} f_0(\xi) \mathrm{d}\xi \geq \beta \right\} \tag{C.129}$$

which interprets the threshold a such that the loss is no greater than a will hold with a probability no less than β. According to (C.129), an equivalent expression of chance constraint (C.128) is

$$(1 - \alpha'_{+})\text{-VaR}(x) \leq 0 \tag{C.130}$$

So that probability evaluation is obviated. Furthermore, if SAA is used, (C.128) and (C.130) indicate that the scenarios which will lead to $L(x, \xi) > 0$ account for a fraction of α_{1+} among all sampled data.

Let $\xi_1, \xi_2, \cdots, \xi_q$ be q scenarios sampled from random variable $\boldsymbol{\xi}$. We use q binary variables $z_1, z_2, \cdots, z_q$ to identify possible infeasibility: $z_k = 1$ implies that constraints cannot be satisfied in scenario ξ_k. To this end, let M be a large enough constant, consider inequality

$$A(\xi_k)x + B(\xi_k)y_k \leq b(\xi_k) + Mz_k \tag{C.131}$$

In (C.131), if $z_k = 0$, recourse action y_k will recover all constraints in scenario ξ_k, and thus $C(x, \xi_k)$ is non-empty; otherwise, if no such a recourse action y_k exists, then constraint violation will take place. To reconcile infeasibility, $z_k = 1$ so that (C.131) becomes redundant, and there is actually no constraint for scenario $\boldsymbol{\xi}_k$. The fraction of sampled scenarios which will incur inevitable constraint violations is counted by $\sum_{k=1}^{q} z_k/q$. So we can write out the following MILP reformulation for robust chance-constrained program (C.120) based on VaR and SAA

$$\begin{aligned} \min \ & c^T x \\ & \text{s.t. } x \in X \\ & A(\xi_k)x + B(\xi_k)y_k \leq b(\xi_k) + Mz_k, \ k = 1, \cdots, q \\ & \sum_{k=1}^{q} z_k \leq q\alpha'_{+}, \ z_k \in \{0, 1\}, \ k = 1, \cdots, q \end{aligned} \tag{C.132}$$

In MILP (C.132), constraint violation can happen in at most $q\alpha_{1+}$ out of q scenarios in the reference distribution, according to Theorem C.1, and the reliability requirement (C.125) under all possible distributions in ambiguity set D_ϕ can be guaranteed by the selection of α'_+. Improved MILP formulations of chance constraints which do not rely on the specific big-M parameter are comprehensively studied in [141], and some structure properties of the feasible region are revealed.

(3) CVaR Based Reformulation: An LP

The number of binary variables in MILP (C.132) is equal to the number of sampled scenarios. To guarantee the accuracy of SAA, a large number of scenarios are required, preventing MILP (C.132) from being solved efficiently. To ameliorate this plight, we provide a conservative LP approximation for problem (C.120) based on the properties of CVaR revealed in [132].

The β-CVaR for the loss function $L(x,\xi)$ is defined as

$$\beta\text{-CVaR}(x) = \frac{1}{1-\beta}\int_{L(x,\xi)\geq\beta\text{-VaR}(x)} L(x,\xi)f(\boldsymbol{\xi})d\xi \tag{C.133}$$

which interprets the conditional expectation of loss that is no less than β-VaR; therefore, relation

$$\beta\text{-VaR} \leq \beta\text{-CVaR} \tag{C.134}$$

always holds, and a conservative approximation of constraint (C.130) is

$$(1-\alpha'_+)\text{-CVaR}(x) \leq 0 \tag{C.135}$$

Inequality (C.135) is a sufficient condition for (C.130) and (C.128). This conservative replacement is apposite to the spirit of robust optimization. In what follows, we will reformulate (C.135) in a solver-compatible form.

According to [132], the left-hand side of (C.135) is equal to the optimum of the following minimization problem

$$\min_\gamma \left\{\gamma + \frac{1}{\alpha'_+}\int_{\xi\in\mathbb{R}^K} \max\{L(x,\xi)-\gamma,0\}f(\xi)\mathrm{d}\xi\right\} \tag{C.136}$$

By performing SAA, the integral in (C.136) renders a summation over discrete sampled scenarios $\xi_1, \xi_2, \cdots, \xi_q$, resulting in

$$\min_\gamma \left\{\gamma + \frac{1}{q\alpha'_+}\sum_{k=1}^{q} \max\{L(x,\xi_k)-\gamma,0\}\right\} \tag{C.137}$$

By introducing auxiliary variable s_k, the feasible region defined by (C.135) can be expressed via

$$\begin{gathered}
\exists \gamma \in \mathbb{R}, s_k \in \mathbb{R}^+, \ \sigma_k \in \mathbb{R}, \ k = 1, \cdots, q \\
\sigma_k - \gamma \le s_k, \ k = 1, \cdots, q \\
A(\xi_k)x + B(\xi_k)y_k \le b(\xi_k) + \sigma_k \mathbf{1}, \ k = 1, \cdots, q \\
\gamma + \frac{1}{q\alpha'_+} \sum_{k=1}^{q} s_k \le 0
\end{gathered}$$

Now we can write out the conservative LP reformulation for robust chance constrained program (C.120) based on CVaR and SAA

$$\begin{aligned}
\min_{x,y,s,\gamma} \quad & c^T x \\
\text{s.t.} \quad & x \in X, \ \gamma + \frac{1}{q\alpha'_+} \sum_{k=1}^{q} s_k \le 0, \ s_k \ge 0, \ k = 1, \cdots, q \\
& A(\xi_k)x + B(\xi_k)y_k - b(\xi_k) \le (\gamma + s_k)\mathbf{1}, \ k = 1, \cdots, q
\end{aligned} \tag{C.138}$$

where σ_k is eliminated.

According to (C.134), condition (C.135) guarantees (C.130) as well as (C.128), so chance constraint in (C.125) holds with a probability no less (usually higher) than $1 - \alpha$, regardless of the true distributions in confidence set D_ϕ. Since (C.134) is usually a strict inequality, this fact will introduce some extent of conservatism in the CVaR based LP model (C.138).

Relations among different mathematical models discussed in this section are summarized in Fig. C.1.

4. Considering Second-Stage Cost

Finally, we elaborate how to solve problem (C.120) with a second-stage cost in the sense of worst-case expectation, i.e.

$$\begin{aligned}
\min_{x} \quad & \left\{ c^T x + \max_{P(\xi) \in D_{KL}} \mathbb{E}_P[Q(x, \xi)] \right\} \\
\text{s.t.} \quad & x \in X \\
& \sup_{P(\xi) \in D'} \Pr[C(x, \xi)] \ge 1 - \alpha
\end{aligned} \tag{C.139}$$

where $Q(x, \xi)$ is the optimal value function of the second-stage problem

$$\begin{aligned}
Q(x, \xi) = \min \quad & q^T y \\
\text{s.t.} \quad & B(\xi)y \le b(\xi) - A(\xi)x
\end{aligned}$$

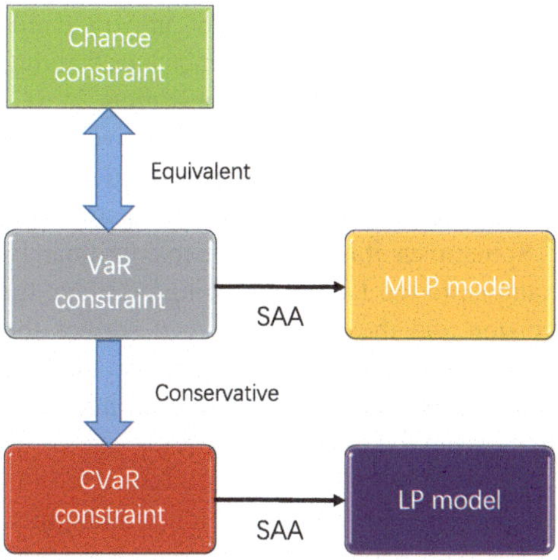

Fig. C.1 Relations of the models discussed in this section

which is an LP for a fixed first-stage decision x and a given parameter ξ;

$$D_{KL} = \{P(\xi) \mid D_\phi^{KL}(f \| f_0) \le d_{KL}(\alpha^*),\ f = \mathrm{d}P/\mathrm{d}\xi\}$$

is the KL-divergence based ambiguity set, and d_{KL} is an α-dependent threshold which determines the size of the ambiguity set, and α^* reflects the confidence level: the real distribution is contained in D_{KL} with a probability no less than α^*. For discrete distributions, the KL-divergence measure has the form of

$$D_\phi^{KL}(f \parallel f_0) = \sum_s \rho_s \log \frac{\rho_s}{\rho_s^0}$$

In either case, there are infinitely many PDFs satisfying the inequality in the ambiguity set D_{KL} when $d_{KL} > 0$. Otherwise, when $d_{KL} = 0$, the ambiguity set D_{KL} becomes a singleton, and the model (C.139) degenerates to a traditional SO problem. In practice, the user can specify the value of d_{KL} according to the attitude towards risks. Nevertheless, the proper value of d_{KL} can be obtained from probability theory. Intuitively, the more historical data we possess, the closer the reference PDF f_0 leaves from the true one, and the smaller d_{KL} should be set.

Suppose we have totally M samples with equal probabilities to fit in N bins, and there are $M_1, M_2, \cdots, M_N$ samples fall into each bin, then the discrete reference PDF for the histogram is $\{\pi_1, \cdots, \pi_N\}$, where $\pi_i = M_i/M$, $i = 1, \cdots, N$. Let $\pi_1^r, \cdots, \pi_N^r$ be the real probability of each bin, according to the discussions in [138], random variable $2M \sum_{i=1}^N \pi_i^r \log(\pi_i^r/\pi_i)$ follows χ^2 distribution with $N-1$ degrees of freedom. Therefore, the confidence threshold can be calculated from

$$d_{KL}(\alpha^*) = \frac{1}{2M} \chi^2_{N-1,\alpha^*}$$

where χ^2_{N-1,α^*} stands for the α^* upper quantile of χ^2 distribution with $N-1$ degrees of freedom. For other divergence based ambiguity sets, please see more discussions in [138]. Robust chance constraints in (C.139) are tackled using the method presented previously, and the objective function will be treated independently. The ambiguity sets in the objective function and chance constraints could be the same one or different ones, and thus are distinguished by D_{KL} and D'.

Sometimes, it is imperative to coordinately optimize the costs in both stages. For example, in the facility planning problem, the first stage represents the investment decision and the second stage describes the operation management. If we only optimize the first-stage cost, then the facilities with lower investment costs will be preferred, but they may suffer from higher operating costs, and not be the optimal choice from the long-term aspect.

To solve (C.139), we need a tractable reformulation for the worst-case expectation problem under KL-divergence ambiguity set

$$\max_{P(\xi)\in D_{KL}} \quad \mathbb{E}_P[Q(x,\xi)] \tag{C.140}$$

under fixed x. It is proved in [138, 142] that problem (C.140) is equivalent to

$$\min_{\alpha\geq 0} \alpha \log \mathbb{E}_{P_0}[\mathrm{e}^{Q(x,\xi)/\alpha}] + \alpha d_{KL} \tag{C.141}$$

where α is the dual variable. Formulation (C.141) has two advantages: first, the expectation is evaluated associated with the reference distribution P_0, which is much easier than optimizing over the ambiguity set D_{KL}; second, the maximum operator switches to a minimum operator, which is consistent with the objective function of the decision making problem. We will use SAA to express the expectation, giving rise to a discrete version of problem (C.141). In fact, in discrete cases, (C.141) can be derived from (C.140) using Lagrange duality. The following interpretation is given in [143].

Denote by $\xi_1, \cdots, \xi_s$ the representative scenarios in the discrete distribution; their corresponding probabilities in the reference PDF and the actual PDF are given by $P_0 = \{p_1^0, \cdots, p_s^0\}$ and $P = \{p_1, \cdots, p_s\}$, respectively. Then problem (C.140) can be written in a discrete form as

$$\begin{aligned} \max_{p} \quad & \sum_{i=1}^{s} p_i Q(x,\xi_i) \\ \text{s.t.} \quad & \sum_{i=1}^{s} p_i \log\left(\frac{p_i}{p_i^0}\right) \leq d_{KL} \\ & p \geq 0,\ 1^T p = 1 \end{aligned} \tag{C.142}$$

where vector $p = [p_1, \cdots, p_s]^T$ is the decision variable. According to Lagrange duality theory, the objective function of the dual problem is

$$g(\alpha, \mu) = \alpha d_{KL} + \mu + \sum_{i=1}^{s} \max_{p_i \geq 0} p_i \left(Q(x, \xi_i) - \mu - \alpha \log \left(\frac{p_i}{p_i^0} \right) \right) \tag{C.143}$$

where μ is the dual variable associated with equality constraint $1^T p = 1$, and α with the KL-divergence inequality. Substituting $t_i = p_i / p_i^0$ into (C.143) and eliminating p_i, we get

$$g(\alpha, \mu) = \alpha d_{KL} + \mu + \sum_{i=1}^{s} \max_{t_i \geq 0} \; p_i^0 t_i \left(Q(x, \xi_i) - \mu - \alpha \log t_i \right)$$

Calculating the first-order derivative of $t_i(Q(x, \xi_i) - \mu - \alpha \log t_i)$ with respect to t_i, the optimal solution is

$$t_i = \mathrm{e}^{\frac{Q(x,\xi_i)-\mu-\alpha}{\alpha}} > 0$$

and the maximum is

$$\alpha \mathrm{e}^{\frac{Q(x,\xi_i)-\mu-\alpha}{\alpha}}$$

As a result, the dual objective reduces to

$$g(\alpha, \mu) = \alpha d_{KL} + \mu + \alpha \sum_{i=1}^{s} p_i^0 \mathrm{e}^{\frac{Q(x,\xi_i)-\mu-\alpha}{\alpha}} \tag{C.144}$$

and the dual problem of (C.142) can be rewritten as

$$\min_{\alpha \geq 0, \mu} \; g(\alpha, \mu) \tag{C.145}$$

The optimal solution μ^* must satisfy $\partial g / \partial \mu = 0$, yielding

$$\sum_{i=1}^{s} p_i^0 \mathrm{e}^{\frac{Q(x,\xi_i)-\mu^*-\alpha}{\alpha}} = 1$$

or

$$\mu^* = \alpha \log \sum_{i=1}^{s} p_i^0 \, \mathrm{e}^{Q(x,\xi_i)/\alpha} - \alpha$$

Substituting the above relations into $g(\alpha, \mu)$ results in the following dual problem

$$\min_{\alpha \geq 0} \left\{ \alpha d_{KL} + \alpha \log \sum_{i=1}^{s} p_i^0 \, \mathrm{e}^{Q(x,\xi_i)/\alpha} \right\} \tag{C.146}$$

which is a discrete form of (C.141).

In (C.139), replacing the inner problem (C.140) with its Lagrangian dual form (C.146), we can obtain an equivalent mathematical program

$$\begin{aligned} \min \ & \left\{ c^T x + \alpha d_{KL} + \alpha \log \sum_{i=1}^{s} p_i^0 \, \mathrm{e}^{\theta_i/\alpha} \right\} \\ \text{s.t.} \ & x \in X, \ \alpha \geq 0, \ \theta_i = q^T y_i, \ \forall i \\ & A(\xi_i)x + B(\xi_i)y_i \leq b(\xi_i), \ \forall i \\ & \text{Cons-RCC} \end{aligned} \tag{C.147}$$

where Cons-RCC stands for the LP based formulation of robust chance constraints, so the constraints in problem (C.147) are all linear, and the only nonlinearity rests in the last term of the objective function. In what follows, we will show it is actually a convex function in θ_i and α.

In the first step, we claim that the following function is convex [97, p. 87, in Example 3.14]

$$h_1(\theta) = \log \left(\sum_{i=1}^{s} \mathrm{e}^{\theta_i} \right)$$

Since the composition with an affine mapping preserves convexity [97, Sect. 3.2.2], a new function

$$h_2(\theta) = h_1(A\theta + b)$$

remains convex under linear mapping $\theta \to A\theta + b$. Let A be an identity matrix, and

$$b = \begin{bmatrix} \log p_1^0 \\ \vdots \\ \log p_s^0 \end{bmatrix}$$

then we have

$$h_2(\theta) = \log \left(\sum_{i=1}^{s} p_i^0 \mathrm{e}^{\theta_i} \right)$$

is a convex function; at last, function

$$h_3(\alpha, \theta) = \alpha h_2(\theta/\alpha)$$

is the perspective of $h_2(\theta)$, so is also convex [97, p. 89, Sect. 3.2.6].

In view of this convex structure, (C.147) essentially gives rise to a convex program, and the local minimum is also the global one. However, according to our experiments, general purpose NLP solvers still have difficulty to solve (C.147). Therefore, we employ the outer approximation method [144, 145]. The motivation is to solve the epigraph form of (C.147), in which nonlinearity is moved into the constraints; then linearize the feasible region with an increasing number of cutting planes generated in an iteration algorithm, until certain convergence criterion is met. In this way, the hard problem (C.147) can be solved via a sequence of LPs. The outer approximation algorithm is outlined in Algorithm C.6. Because (C.147) is a convex program, the cutting planes will not remove any feasible point, and Algorithm C.6 finds the global optimal solution in finite steps, regardless of the initial point. But for sure, the number of iterations is affected by the quality of initial guess. A proper initiation could be obtained by solving a traditional SO problem without considering distribution uncertainty.

The motivation of Algorithm C.6 is illustrated in Fig. C.2. The original objective function is nonlinear but convex. In the epigraph form (C.148), we generate a set of linear cuts (C.149) dynamically according to the optimal solution found in

---

Algorithm C.6 Outer approximation

1: Choose an initial point (θ^1, α^1) and convergence tolerance $\epsilon > 0$, the initial objective value is $R^1 = 0$, and iteration index $k = 1$.

2: Solve the following master problem which is an LP

$$\begin{aligned}
\min_{\alpha,\theta,\gamma,x} \quad & c^T x + \alpha d_{KL} + \gamma \\
\text{s.t.} \quad & h_3(\alpha^j, \theta^j) + \nabla h_3(\alpha^j, \theta^j) \begin{bmatrix} \alpha - \alpha^j \\ \theta - \theta^j \end{bmatrix} \le \gamma, \ j = 1, \cdots, k \\
& x \in X, \ \alpha \ge 0, \ \theta_i = q^T y_i, \ \forall i \\
& A(\xi_i)x + B(\xi_i)y_i \le b(\xi_i), \ \forall i \\
& \text{Cons-RCC}
\end{aligned} \tag{C.148}$$

The optimal value is R^{k+1}, and the optimal solution is $(x^{k+1}, \theta^{k+1}, \alpha^{k+1})$.

3: If $R^{k+1} - R^k \le \varepsilon$, terminate and report the optimal solution $(x^{k+1}, \theta^{k+1}, \alpha^{k+1})$; otherwise, update $k \leftarrow k + 1$, calculate the gradient ∇h_3 at the obtained solution (α^k, θ^k), add the following cut to problem (C.148), and go to step 2.

$$h_3(\alpha^k, \theta^k) + \nabla h_3(\alpha^k, \theta^k) \begin{bmatrix} \alpha - \alpha^k \\ \theta - \theta^k \end{bmatrix} \le \gamma \tag{C.149}$$

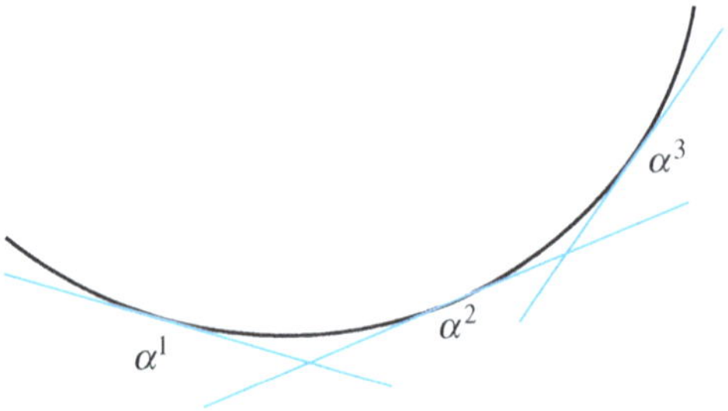

Fig. C.2 Illustration of the outer approximation algorithm

step 2, then the convex region can be approximated with arbitrarily high accuracy around the optimal solution. The convergence of the very basic version of outer approximation method has been analyzed in [146, 147]. In fact, Algorithm C.6 is very efficient to solve problem (C.147), because problem (C.148) is an LP, the objective function is smooth, and the algorithm often converges in a few number of iterations.

C.4.2 Stochastic Program with Discrete Distributions

In ARO discussed in Appendix C.2, the uncertain parameter is assumed to reside in the so-called uncertainty set. Every element in this set is treated equally, so the scenario in the worst case must be one of the extreme points of the uncertainty set, which is the main source of conservatism in the traditional RO paradigm. In contrast, in the classic two-stage SO, uncertain parameter ξ is modeled through a certain probability distribution P, and the expected cost is minimized, giving rise to

$$\begin{aligned} \min \ & c^T x + \mathbb{E}_P[Q(x,\xi)] \\ \text{s.t.} \ & x \in X \end{aligned} \tag{C.150}$$

where the bounded polyhedron X is the feasible region of first-stage decision x, ξ is the uncertain parameter, and $Q(x,\xi)$ is the optimal value function of the second-stage problem, which is an LP for fixed x and ξ

$$\begin{aligned} Q(x,\xi) = \min \ & q^T y \\ \text{s.t.} \ & B(\xi)y \le b(\xi) - A(\xi)x \end{aligned} \tag{C.151}$$

where q is the cost coefficients, $A(\xi)$, $B(\xi)$, and $b(\xi)$ are constant matrices affected by uncertain data, $y(\xi)$ is the second-stage decision, which is the reaction to the realization of uncertainty.

Since the true PDF of ξ is difficult to obtain in some circumstances, in this section, we do not require perfect knowledge on the probability distribution $\mathbb{P}$ of random variable ξ, and let it be ambiguous around a reference distribution and reside in an ambiguity set D, which can be constructed from limited historical data. We take all possible distributions in the ambiguity set into consideration, so as to minimize the expected cost in the worst-case distribution, resulting in the following model

$$\begin{aligned} \min \ & c^T x + \max_{P(\xi)\in D} \mathbb{E}_P[Q(x,\xi)] \\ \text{s.t.} \ & x \in X \end{aligned} \tag{C.152}$$

Compared with (C.120), constraint violation is not allowed in problem (C.152), and the second-stage expected cost in the worst-case distribution is considered. It is a particular case of (C.139) without chance constraints. Specifically, we will utilize discrete distributions in this section. This formulation enjoys several benefits. One is the easy exposition of the density function. In previous sections, the candidate in the moment or divergence based ambiguity sets is not given in an analytical form, and vanishes during the dual transformation. As a result, we don't have clear knowledge on the worst-case distribution. For discrete distributions, the density function is a vector of real entries associated with the probability of each representative scenario. We can easily construct the ambiguity set and optimize an expectation over discrete distributions. The other originates from the computational perspective, which can be seen later. The main results in this section come from [148, 149].

1. Modeling the Confidence Set

For a given set of historical data with M elements, which can be regarded as M samples of the random variable, we can draw a histogram with K bins as an estimation of the reference distribution. Suppose that the numbers of samples fall in each bin is $M_1, M_2, \cdots, M_K$, where $\sum_{i=1}^K M_i = M$, then the reference (empirical) distribution of the uncertain data is given by $\mathbb{P}_0 = [p_1^0, \cdots, p_K^0]$, where $p_i^0 = M_i/M$, $i = 1, \cdots, K$. Since the data may not be enough to fit a PDF with high accuracy, the actual distribution should be close to but might be different from its reference. It is proposed in [148] to construct the ambiguity set using statistical inference corresponding to a given tolerance. Two types of ambiguity sets are suggested based on L_1 norm and L_∞ norm

$$D_1 = \left\{ \mathbb{P} \in \mathbb{R}_+^K \middle| \|\mathbb{P} - \mathbb{P}_0\|_1 \le \theta \right\} = \left\{ p \in \Delta_K \middle| \sum_{i=1}^K \left| p_i - p_i^0 \right| \le \theta \right\} \tag{C.153}$$

$$D_\infty = \left\{ \mathbb{P} \in \mathbb{R}_+^K \middle| \|\mathbb{P} - \mathbb{P}_0\|_\infty \le \theta \right\} = \left\{ p \in \Delta_K \middle| \max_{1\le i\le K} \left| p_i - p_i^0 \right| \le \theta \right\} \tag{C.154}$$

where $\Delta_K = \{p \in [0,1]^K : \mathbf{1}^T p = 1\}$. These two ambiguity sets can be easily expressed by polyhedral sets as follows:

$$D_1 = \left\{ p \in \Delta_K \left| \begin{array}{l} \exists t \in \mathbb{R}_+^K : \sum_{k=1}^{K} t_k \le \theta \\ t_k \ge p_k - p_k^0, \ k = 1, \cdots, K \\ t_k \ge p_k^0 - p_k, \ k = 1, \cdots, K \end{array} \right. \right\} \tag{C.155}$$

$$D_\infty = \left\{ p \in \Delta_K \left| \begin{array}{l} \theta \ge p_k - p_k^0, \ k = 1, \cdots, K \\ \theta \ge p_k^0 - p_k, \ k = 1, \cdots, K \end{array} \right. \right\} \tag{C.156}$$

where $p = [p_1, \cdots, p_K]^T$ is the variable in the ambiguity set; $t = [t_1, \cdots, t_K]^T$ is the lifting (auxiliary) variable in D_1; parameter θ reflects decision maker's confidence level on the distance between the reference distribution and the true one. Apparently, the more historical data we utilize, the smaller their distance will be. Provided with M observations and K bins, the quantitative relation between the value of θ and the number of samples is given by [148]

$$\Pr\{\|\mathbb{P} - \mathbb{P}_0\|_1 \le \theta\} \ge 1 - 2K\mathrm{e}^{-2M\theta/K} \tag{C.157}$$

$$\Pr\{\|\mathbb{P} - \mathbb{P}_0\|_\infty \le \theta\} \ge 1 - 2K\mathrm{e}^{-2M\theta} \tag{C.158}$$

According to (C.157) and (C.158), if we want to maintain (C.153) and (C.154) with a confidence level of β, parameter θ should be selected as

$$\text{For } D_1 : \ \theta_1 = \frac{K}{2M} \ln \frac{2K}{1-\beta} \tag{C.159}$$

$$\text{For } D_\infty : \ \theta_\infty = \frac{1}{2M} \ln \frac{2K}{1-\beta} \tag{C.160}$$

As the size of sampled data approaches infinity, θ_1 and θ_∞ decrease to 0, and the reference distribution converges to the true one. Accordingly, problem (C.152) becomes a traditional two-stage SO.

2. CCG Based Decomposition Algorithm

Let ξ^k denote the representative scenario of the k-th bin, p_k be the corresponding probability, and $P = [p_1, \cdots, p_K]$ belongs to the ambiguity set in form of (C.153) or (C.154), then problem (C.152) can be written as

$$
\begin{aligned}
&\min c^T x + \max_{\mathbb{P}} \sum_{k=1}^{K} p_k \min q^T y^k \\
&\text{s.t. } x \in X, \ \mathbb{P} \in D \\
&A(\xi^k)x + B(\xi^k)y^k \le b(\xi^k), \forall k
\end{aligned}
\tag{C.161}
$$

Problem (C.161) has a min-max-min structure and can be solved by the Benders decomposition method [148] or the CCG method [149]. The latter one will be introduced in the rest of this section. It decomposes problem (C.161) into a lower bounding master problem and an upper bounding subproblem, which are solved iteratively until the gap between the upper bound and lower bound gets smaller than a convergence tolerance. The basic idea has been explained in Appendix C.2.3. As we can see in [149], the second-stage problem can be a broader class of convex programs, such as an SOCP.

(1) Subproblem

For a given first-stage decision x, the subproblem aims to find the worst-case distribution, which comes down to a max-min program shown below

$$
\max_{\mathbb{P} \in D} \sum_{k=1}^{K} p_k \min_{y^k \in Y_k(x)} q^T y^k \tag{C.162}
$$

where

$$
Y_k = \{y^k \mid B(\xi^k)y^k \le b(\xi^k) - A(\xi^k)x\}, \ \forall k \tag{C.163}
$$

Problem (C.162) has some unique features that facilitate the computation:

1. Feasible sets Y_k are decoupled.
2. The probability variables p_k do not affect feasible sets Y_k.
3. The ambiguity set D and feasible sets Y_k are decoupled.

Although (C.162) seems nonlinear due to the production of scalar variable p_k and vector variable y^k in the objective function, as we can see in the following discussion, it is equivalent to an LP or can be decomposed into several LPs, and thus can be solved efficiently.

An Equivalent LP

Because $p_k \ge 0$, we can exchange the summation operator and the minimization operator, and problem (C.162) can be written as

$$
\max_{\mathbb{P} \in D} \min_{y^k \in Y_k(x)} \sum_{k=1}^{K} p_k q^T y^k \tag{C.164}
$$

For the inner minimization problem, p_k is constant, so it is an LP, whose dual problem is

$$\begin{aligned} &\max_{\mu^k} \sum_{k=1}^{K} \left(b(\xi^k) - A(\xi^k)x\right)^T \mu^k \\ &\text{s.t. } \mu_k \le 0,\ B^T(\xi^k)\mu^k = p_k q,\ \forall k \end{aligned}$$

where μ^k are dual variables. Substituting it into (C.164), and combining two maximization operators, we obtain

$$\begin{aligned} &\max_{p_k, \mu^k} \sum_{k=1}^{K} \left(b(\xi^k) - A(\xi^k)x\right)^T \mu^k \\ &\text{s.t. } \mu_k \le 0,\ B^T(\xi^k)\mu^k = p_k q,\ \forall k \\ &\quad (p_1, \cdots, p_k) \in D \end{aligned} \tag{C.165}$$

Since D is polyhedral, problem (C.165) is in fact an LP. The optimal solution offers the worst-case distribution $[p_1^*, \cdots, p_K^*]$, which will be used to generate cuts in the master problem. The recourse actions y^k in each scenario will be provided by the optimal solution of the master problem.

Despite the fact that LP is acknowledged as the most tractable mathematical programming problem, however, when K is extremely large, it is still challenging to solve (C.165) or even store it in a computer. Nevertheless, the separability of feasible regions allows solving (C.162) in a decomposition manner.

A Decomposition Method

As mentioned above, p_k has no impact on Y_k, which are decoupled; moreover, because p_k is a scalar in the objective function of each inner minimization problem, it does not affect the optimal solution y^k. In view of this convenience, problem (C.162) can be decomposed into $K + 1$ smaller LPs, and can be solved in parallel. To this end, for each ξ^k, solve the following LP:

$$h_k^* = \min_{y^k \in Y_k(x)} q^T y^k,\ k = 1, \cdots, K$$

The optimal value is h_k^*; after obtaining optimal values $(h_1^*, \cdots, h_K^*)$ of the K LPs, we can retrieve the worst-case distribution through solving an additional LP

$$\max_{\mathbb{P} \in D} \sum_{k=1}^{K} p_k h_k^*$$

In fact, if the second-stage problem is a conic program (in [149], it is an SOCP), the above discussions are still valid, as long as the strong duality holds.

It is interesting to notice that in the ARO problem in Sect. C.2.3, the subproblem comes down to a non-convex bilinear program after dualizing the inner minimization problem, and is generally NP-hard; in this section, the subproblem actually gives rise to LPs, whose complexity is polynomial in problem sizes. The reason accounting for this difference is that the uncertain parameter in (C.161) is expressed by sampled scenarios and thus is constant; the distributional uncertainty appearing in the objective function does not influence the constraints of the second stage problem, and thus the linear max-min problem (C.164) reduces to an LP after a dual transformation.

(2) The CCG Algorithm
The motivation of CCG algorithm has been thoroughly discussed in Appendix C.2.3. In this section, for a fixed x, the optimal value of subproblem (C.162) is denoted by $Q(x)$, and $c^T x + Q(x)$ gives an upper bound of the optimal solution of (C.161), because the first-stage variable is un-optimized. Then a set of new variables and optimality cuts are generated and added into master problem. If the subproblem is infeasible in some scenario, then a set of feasibility cuts are assigned to the master problem. The master problem starts from a subset of D, which is updated by including the worst-case distribution identified by the subproblem. Forasmuch, the master problem is a relax version of the original problem (C.161), and provides a lower bound on the optimal value. The flowchart of the CCG procedure for problem (C.161) is given in Algorithm C.7. This algorithm will terminate in a finite number of iterations, as the confidence set D has finite extreme points.

C.4.3 Formulations Based on Wasserstein Metric

Up to now, the KL-divergence based ambiguity set based formulations have received plenty of research, because it enjoys some convenience when deriving the robust counterpart. For example, it has already known in Sect. C.4.1 that robust chance constraints under KL-divergence ambiguity set can reduce to a traditional chance constraints under the empirical distribution with a rescaled confidence level, and the worst-case expectation problem under KL-divergence ambiguity set is equivalent to a convex program. However, according to its definition, KL-divergence ambiguity set may encounter theoretical difficulty to represent confidence sets for continuous distribution [150], because the empirical distribution calibrated from finite data must be discrete, and any distribution in the KL-divergence ambiguity set must assign positive probability mass to each sampled scenario. As a continuous distribution has a density function, it must reside outside the KL-divergence ambiguity set regardless of the sampled scenarios. In contrast, Wasserstein metric based ambiguity sets contain both discrete and continuous distributions. It offers an explicit confidence level for the unknown distribution belonging to the set, and enables the decision maker more informative guidance to control the model conservativeness. This section introduces state-of-the-art results in robust SO with Wasserstein metric

Algorithm C.7 CCG based decomposition algorithm

1: Choose a convergence tolerance $\varepsilon > 0$, and an initial probability vector $p^0 \in D$; Set LB $= -\infty$, UB $= +\infty$, and iteration index $s = 0$.

2: Solve the master problem

$$\begin{aligned} \min_{x,\eta,y^{k,m}} \quad & c^T x + \eta \\ \text{s.t.} \quad & x \in X, \ \ \eta \geq \sum_{k=1}^{K} p_k^m q^T y^{k,m}, \ m \in \text{Opt}\{0, 1, \cdots, s\} \\ & A(\xi^k)x + B(\xi^k)y^{k,m} \leq b(\xi^k), \ m \in \text{Opt}\{0, 1, \cdots, s\}, \ \forall k \\ & A(\xi^k)x + B(\xi^k)y^{k,m} \leq b(\xi^k), \ m \in \text{Fea}\{0, 1, \cdots, s\}, \ k \in I(s) \end{aligned} \tag{C.166}$$

where Opt$\{*\}$/Fea$\{*\}$ selects the iterations in which an optimality (feasibility) cut is generated; $I(s)$ depicts the index of scenarios in which the second-stage problem is infeasible in iteration s. The optimal solution is (x^*, η^*); update LB $= c^T x^* + \eta^*$;

3: Solve subproblem (C.162) with current x^*. If there exists some ξ^k such that $Y_k(x^*) = \emptyset$, then generate new variable $y^{k,s}$, update $I(s)$, and add the following feasibility cut to the master problem

$$A(\xi^k)x + B(\xi^k)y^{k,s} \leq b(\xi^k), \ k \in I(s) \tag{C.167}$$

Otherwise, if $Y_k(x^*) \neq \emptyset, \forall k$, subproblem (C.162) can be solved. The optimal solution is p^{s+1}, and the optimal value is $Q(x^*)$; update UB $= \min\{\text{UB}, c^T x^* + Q(x^*)\}$, create new variables $(y^{1,s+1}, \cdots, y^{k,s+1})$, and add the following optimality cut to the master problem

$$\begin{aligned} & \eta \geq \sum_{k=1}^{K} p_k^{s+1} q^T y^{k,s+1} \\ & A(\xi^k)x + B(\xi^k)y^{k,s+1} \leq b(\xi^k), \ \forall k \end{aligned} \tag{C.168}$$

4: If UB$-$LB$< \varepsilon$, terminate and report the optimal first-stage solution x^* as well as the worst-case distribution p^{s+1}; otherwise, update $s \leftarrow s + 1$, and go to step 2.

based ambiguity sets. The most critical problem is the robust counterparts of the worst-case expectation problem and robust chance constraints, which will be discussed respectively. They can be embedded in single- and two-stage robust SO problems without substantial barriers. The materials in this section mainly come from [150].

1. Wasserstein Metric Based Ambiguity Set

Let Ξ be the support set of multi-dimensional random variable $\xi \in \mathbb{R}^m$. $M(\Xi)$ represent all probability distributions $\mathbb{Q}$ supported on Ξ, and $\mathbb{E}_{\mathbb{Q}}[\|\xi\|] = \int_{\Xi} \|\xi\| \mathbb{Q}(\mathrm{d}\xi) < \infty$, where $\|\cdot\|$ stands for an arbitrary norm on $\mathbb{R}^m$.

Definition C.8 Wasserstein metric $d_W : M(\Xi) \times M(\Xi) \to \mathbb{R}_+$ is defined as

$$d_W(\mathbb{Q}, \mathbb{Q}_0) = \inf \left(\int_{\Xi^2} \left\| \xi - \xi^0 \right\| \Pi(\mathrm{d}\xi, \mathrm{d}\xi^0) \middle| \begin{array}{l} \Pi \text{ is a joint distribution of } \xi \text{ and} \\ \xi^0 \text{ with marginals } \mathbb{Q} \text{ and } \mathbb{Q}_0 \end{array} \right)$$

for two probability distributions $\mathbb{Q}, \mathbb{Q}_0 \in M(\Xi)$.

As a special case, for two discrete distributions, Wasserstein metric is given by

$$d_W(\mathbb{Q}, \mathbb{Q}_0) = \inf_{\pi \geq 0} \left(\sum_i \sum_j \pi_{ij} \left\| \xi_j - \xi_i^0 \right\| \middle| \begin{array}{l} \sum_j \pi_{ij} = p_i^0, \ \forall i \\ \sum_i \pi_{ij} = p_j, \ \forall j \end{array} \right) \tag{C.169}$$

where p_i^0 and p_j denote the probability of representative scenario ξ_i^0 and ξ_j.

In either case, the decision variable Π (or π_{ij}) represents the probability mass transported from ξ_i^0 to ξ_j, therefore, the Wasserstein metric can be viewed as the minimal cost of a transportation plan, where the distance $\|\xi_j - \xi_i^0\|$ encodes the transportation cost of unit mass.

Sometimes, the Wasserstein metric can be represented in the dual form

$$d_W(\mathbb{Q}, \mathbb{Q}_0) = \sup_{f \in L} \left(\int_\Xi f(\xi)\mathbb{Q}(\mathrm{d}\xi) - \int_\Xi f(\xi)\mathbb{Q}_0(\mathrm{d}\xi) \right) \tag{C.170}$$

where $L = \{f : |f(\xi) - f(\xi^0)| \leq \|\xi - \xi^0\|, \forall \xi, \xi^0 \in \Xi\}$ (Theorem 3.2, [150], which was firstly discovered by Kantorovich and Rubinstein [151] for distributions with a bounded support).

With the above definition, the Wasserstein ambiguity set is the ball of radius ϵ centered at the empirical distribution $\mathbb{Q}_0$

$$D_W = \{\mathbb{Q} \in M(\Xi) : d_W(\mathbb{Q}, \mathbb{Q}_0) \leq \epsilon\} \tag{C.171}$$

where $\mathbb{Q}_0$ is constructed with N independent data samples

$$\mathbb{Q}_0 = \frac{1}{N} \sum_{i=1}^{N} \delta_{\xi_i^0}$$

where $\delta_{\xi_i^0}$ stands for Dirac distribution concentrating unit mass at ξ_i^0.

Particularly, we require the unknown distribution $\mathbb{Q}$ follow a light tail assumption, i.e., there exists $a > 1$ such that

$$\int_\Xi \mathrm{e}^{\|\xi\|^a} \mathbb{Q}(\mathrm{d}\xi) < \infty$$

This assumption indicates that the tail of distribution $\mathbb{Q}$ decays at an exponential rate. If Ξ is bounded and compact, this assumption trivially holds. Under this assumption, modern measure concentration theory provides the following finite sample guarantee for the unknown distribution belonging to Wasserstein ambiguity set

$$\Pr\left[d_W(\mathbb{Q}, \mathbb{Q}_0) \geq \epsilon\right] \leq \begin{cases} c_1 \mathrm{e}^{-c_2 N \epsilon^{\max\{m,2\}}} & \text{if } \epsilon \leq 1 \\ c_1 \mathrm{e}^{-c_2 N \epsilon^a} & \text{if } \epsilon > 1 \end{cases} \tag{C.172}$$

where c_1, c_2 are positive constants depending on a, A, and m and $m \neq 2$.

Equation (C.172) provides a priori estimate of the confidence level for $\mathbb{Q} \notin D_W$. On the other hand, we can utilize (C.172) to select parameter ϵ of the Wasserstein ambiguity set such that D_W contains the uncertain distribution $\mathbb{Q}$ with probability $1 - \beta$ for some prescribed β. This requires solving ϵ from the right-hand side of (C.172) with a given left-hand side β, resulting in

$$\epsilon = \begin{cases} \left(\dfrac{\ln(c_1 \beta^{-1})}{c_2 N}\right)^{1/\max\{m,2\}} & \text{if } N \geq \dfrac{\ln(c_1 \beta^{-1})}{c_2} \\ \left(\dfrac{\ln(c_1 \beta^{-1})}{c_2 N}\right)^{1/a} & \text{if } N < \dfrac{\ln(c_1 \beta^{-1})}{c_2} \end{cases} \tag{C.173}$$

Wasserstein ambiguity set with above radius can be regarded as a confidence set for the unknown distribution $\mathbb{Q}$ as in statistical testing.

2. Worst-Case Expectation Problem

A robust SO problem under Wasserstein metric naturally requests to minimize the worst-case expected cost:

$$\inf_{x \in X} \sup_{\mathbb{Q} \in D_W} \mathbb{E}_{\mathbb{Q}}[h(x, \xi)] \tag{C.174}$$

We demonstrate how to solve the core problem: the worst-case expectation

$$\sup_{\mathbb{Q} \in D_W} \mathbb{E}_{\mathbb{Q}}[l(\xi)] \tag{C.175}$$

where $l(\xi) = \max_{1 \leq k \leq K} l_k(\xi)$ is the payoff function, consisting of the point-wise maximum of K elementary functions. For notation brevity, the dependence on x is suppressed and will be recovered later on when necessary. We further assume that the support set Ξ is closed and convex, and specific $l(\xi)$ will be discussed.

Problem (C.175) renders an infinite-dimensional optimization problem for continuous distribution. Nonetheless, the inspiring work in [150] shows that (C.175) can be reformulated as a finite-dimensional convex program for various payoff functions. To see this, expand the worst-case expectation as

$$\sup_{\mathbb{Q}\in D_W} \mathbb{E}_{\mathbb{Q}}[l(\xi)] = \begin{cases} \sup\limits_{\Pi} \displaystyle\int_{\Xi} l(\xi)\mathbb{Q}(d\xi) \\ \text{s.t.} \displaystyle\int_{\Xi^2} \left\| \xi - \xi^0 \right\| \Pi(d\xi, d\xi^0) \le \epsilon \\ \qquad \Pi \text{ is a joint distribution of } \xi \\ \qquad \xi^0 \text{ with marginals } \mathbb{Q} \text{ and } \mathbb{Q}_0 \end{cases}$$

According to the law of total probability, Π can be decomposed as the marginal distribution $\mathbb{Q}_0$ of ξ^0 and the conditional distributions $\mathbb{Q}_i$ of ξ given $\xi^0 = \xi_i^0$:

$$\Pi = \frac{1}{N}\sum_{i=1}^{N} \delta_{\xi_i^0} \otimes \mathbb{Q}_i$$

and the worst-case expectation evolves into a generalized moment problem in conditional distributions $\mathbb{Q}_i$, $i \le N$

$$\sup_{\mathbb{Q}\in D_W} \mathbb{E}_{\mathbb{Q}}[l(\xi)] = \begin{cases} \sup\limits_{\mathbb{Q}_i\in M(\Xi)} \displaystyle\frac{1}{N}\sum_{i=1}^{N}\int_{\Xi} l(\xi)\mathbb{Q}_i(d\xi) \\ \text{s.t.} \displaystyle\frac{1}{N}\sum_{i=1}^{N}\int_{\Xi} \left\| \xi - \xi_i^0 \right\| \mathbb{Q}_i(d\xi) \le \epsilon \end{cases}$$

Using standard Lagrangian duality, we obtain

$$\begin{aligned} \sup_{\mathbb{Q}\in D_W} \mathbb{E}_{\mathbb{Q}}[l(\xi)] &= \sup_{\mathbb{Q}_i\in M(\Xi)} \inf_{\lambda\ge 0} \frac{1}{N}\sum_{i=1}^{N}\int_{\Xi} l(\xi)\mathbb{Q}_i(d\xi) \\ &\quad + \lambda\left(\epsilon - \frac{1}{N}\sum_{i=1}^{N}\int_{\Xi} \left\| \xi - \xi_i^0 \right\| \mathbb{Q}_i(d\xi)\right) \\ &\le \inf_{\lambda\ge 0} \sup_{\mathbb{Q}_i\in M(\Xi)} \lambda\epsilon + \frac{1}{N}\sum_{i=1}^{N}\int_{\Xi} \left(l(\xi) - \lambda\left\| \xi - \xi_i^0 \right\|\right) \mathbb{Q}_i(d\xi) \\ &= \inf_{\lambda\ge 0} \lambda\epsilon + \frac{1}{N}\sum_{i=1}^{N} \sup_{\xi\in\Xi} \left(l(\xi) - \lambda\left\| \xi - \xi_i^0 \right\|\right) \end{aligned}$$

Decision variables λ and ξ have finite dimensions. The last problem can be reformulated as

$$\begin{aligned}
\inf_{\lambda, s_i} \ & \lambda\epsilon + \frac{1}{N}\sum_{i=1}^{N} s_i \\
\text{s.t.} \ & \sup_{\xi \in \Xi} \left(l_k(\xi) - \lambda \left\| \xi - \xi_i^0 \right\| \right) \le s_i \\
& i = 1, \cdots, N, \ k = 1, \cdots, K \\
& \lambda \ge 0
\end{aligned} \tag{C.176}$$

From the definition of dual norm, we know $\lambda \left\| \xi - \xi_i^0 \right\| = \max_{\|z_{ik}\|_* \le \lambda} \langle z_{ik}, \xi - \xi_i^0 \rangle$, so the constraints give rise to

$$\begin{aligned}
\sup_{\xi \in \Xi} \left(l_k(\xi) - \lambda \left\| \xi - \xi_i^0 \right\| \right) &= \sup_{\xi \in \Xi} \left(l_k(\xi) - \max_{\|z_{ik}\|_* \le \lambda} \langle z_{ik}, \xi - \xi_i^0 \rangle \right) \\
&= \sup_{\xi \in \Xi} \min_{\|z_{ik}\|_* \le \lambda} l_k(\xi) - \langle z_{ik}, \xi - \xi_i^0 \rangle \\
&\le \min_{\|z_{ik}\|_* \le \lambda} \sup_{\xi \in \Xi} l_k(\xi) - \langle z_{ik}, \xi - \xi_i^0 \rangle
\end{aligned}$$

Substituting it into problem (C.176) leads to a more restricted feasible set and a larger objective value, yielding

$$\begin{aligned}
\inf_{\lambda, s_i} \ & \lambda\epsilon + \frac{1}{N}\sum_{i=1}^{N} s_i \\
\text{s.t.} \ & \min_{\|z_{ik}\|_* \le \lambda} \sup_{\xi \in \Xi} l_k(\xi) - \langle z_{ik}, \xi - \xi_i^0 \rangle \le s_i \\
& i = 1, \cdots, N, \ k = 1, \cdots, K \\
& \lambda \ge 0
\end{aligned} \tag{C.177}$$

The constraints of (C.177) trivially suggest the feasible set of λ is $\lambda \ge \|z_{ik}\|_*$, and the min operator in constraints can be omitted because it is in compliance with the objective function. Therefore, we arrive at

$$\begin{aligned}
\inf_{\lambda, s_i} \ & \lambda\epsilon + \frac{1}{N}\sum_{i=1}^{N} s_i \\
\text{s.t.} \ & \sup_{\xi \in \Xi} \ (l_k(\xi) - \langle z_{ik}, \xi \rangle) + \langle z_{ik}, \xi_i^0 \rangle \le s_i, \ \lambda \ge \|z_{ik}\|_* \\
& i = 1, \cdots, N, \ k = 1, \cdots K
\end{aligned} \tag{C.178}$$

It is proved in [150] that problems (C.175) and (C.178) are actually equivalent. Next, we will derive the concrete forms of (C.178) under specific payoff function $l(\xi)$ and uncertainty set Ξ. Unlike [150] which relies on conjugate functions in convex analysis, we mainly exploit LP duality theory, which is more friendly to readers with engineering background.

Case 1 Convex PWL payoff function $l(\xi) = \max_{1 \le k \le K} \{a_k^T \xi + b_k\}$ and bounded polyhedral uncertainty set $\Xi = \{\xi \in \mathbb{R}^m : C\xi \le d\}$. The key point is the supremum regarding ξ in the following constraint

$$\sup_{\xi \in \Xi} \left(a_k^T \xi - \langle z_{ik}, \xi \rangle \right) + b_k + \langle z_{ik}, \xi_i^0 \rangle \le s_i$$

For each k, the supremum is an LP

$$\begin{aligned} \max \ & (a_k - z_{ik})^T \xi \\ \text{s.t.} \ & C\xi \le d \end{aligned}$$

Its dual LP reads

$$\begin{aligned} \min \ & d^T \gamma_{ik} \\ \text{s.t.} \ & C^T \gamma_{ik} = a_k - z_{ik} \\ & \gamma_{ik} \ge 0 \end{aligned}$$

Therefore, $z_{ik} = a_k - C^T \gamma_{ik}$. Because of strong duality, we can replace the supremum by the objective of the dual LP, which gives rise to:

$$d^T \gamma_{ik} + b_k + \langle a_k - C^T \gamma_{ik}, \xi_i^0 \rangle \le s_i, \ i = 1, \cdots, N, \ k = 1, \cdots, K$$

Arrange all constraints together, we obtain a convex program which is equivalent to problem (C.178) in Case 1:

$$\begin{aligned} \inf_{\lambda, s_i} \ & \lambda \epsilon + \frac{1}{N} \sum_{i=1}^{N} s_i \\ \text{s.t.} \ & b_k + a_k^T \xi_i^0 + \gamma_{ik}^T (d - C^T \xi_i^0) \le s_i, \ i = 1, \cdots, N, \ k = 1, \cdots, K \\ & \lambda \ge \| a_k - C^T \gamma_{ik} \|_*, \ \gamma_{ik} \ge 0, \ i = 1, \cdots, N, \ k = 1, \cdots, K \end{aligned} \tag{C.179}$$

In the absence of distributional uncertainty, or $\epsilon = 0$ which implies that Wasserstein ambiguity set D_W is a singleton, λ can take any non-negative value without changing the objective function. Because all sampled scenarios must belong to the support set, i.e. $d - C^T \xi_i^0 \ge 0, \forall i$ holds, so there must be $\gamma_{ik} = 0$

at the optimal solution, leading to an optimal value of $\sum_{i=1}^{N} s_i/N$, where $s_i = \max_{1 \le k \le K}\{a_k^T \xi_i^0 + b_k\}$, which represents the sample average of the payoff function under the empirical distribution.

Case 2 Concave PWL payoff function $l(\xi) = \min_{1 \le k \le K}\{a_k^T \xi + b_k\}$ and bounded polyhedral uncertainty set $\Xi = \{\xi \in \mathbb{R}^m : C\xi \le d\}$. In such circumstance, the supremum regarding ξ in the constraint becomes

$$\max_{\xi \in \Xi} \left\{ -z_i^T \xi + \min_{1 \le k \le L} \left\{ a_k^T \xi + b_k \right\} \right\}$$

which is equivalent to an LP

$$\begin{aligned} \max \quad & -z_i^T \xi + \tau_i \\ \text{s.t.} \quad & A\xi + b \ge \tau_i \mathbf{1} \\ & C\xi \le d \end{aligned}$$

where the k-th row of A is a_k^T; the k-th entry of b is b_k; $\mathbf{1}$ is all-one vector with a compatible dimension. Its dual LP reads

$$\begin{aligned} \min \quad & b^T \theta_i + d^T \gamma_i \\ \text{s.t.} \quad & -A^T \theta_i + C^T \gamma_i = -z_i \\ & \mathbf{1}^T \theta_i = 1, \ \theta_i \ge 0, \ \gamma_i \ge 0 \end{aligned}$$

Therefore, $z_i = A^T \theta_i - C^T \gamma_i$. Because of strong duality, we can replace the supremum by the objective of the dual LP, which gives rise to:

$$b^T \theta_i + d^T \gamma_i + \langle A^T \theta_i - C^T \gamma_i, \xi_i^0 \rangle \le s_i, \ i = 1, \cdots, N$$

Arrange all constraints together, we obtain a convex program which is equivalent to problem (C.178) in Case 2:

$$\begin{aligned} \inf_{\lambda, s_i} \quad & \lambda \epsilon + \frac{1}{N} \sum_{i=1}^{N} s_i \\ \text{s.t.} \quad & \theta_i^T (b + A\xi_i^0) + \gamma_i^T (d - C\xi_i^0) \le s_i, \ i = 1, \cdots, N \\ & \lambda \ge \| A^T \theta_i - C^T \gamma_i \|_*, \ i = 1, \cdots, N \\ & \gamma_i \ge 0, \ \theta_i \ge 0, \ \mathbf{1}^T \theta_i = 1, \ i = 1, \cdots, N \end{aligned} \tag{C.180}$$

There will be no k index for the constraints, because it is packaged in A and b.

An analogous analysis shows that if $\epsilon = 0$, there must be $\gamma_i = 0$ and

$$s_i = \min\{\theta_i^T(b + A\xi_i^0) : \theta_i \geq 0,\ \mathbf{1}^T\theta_i = 1\} = \min_{1\leq k\leq K}\{a_k^T\xi + b_k\}$$

implying $\sum_{i=1}^N s_i/N$ is the sample average of the payoff function under the empirical distribution.

Now we focus our attention on the min-max problem (C.174) which frequently arises in two-stage robust SO, which entails evaluation of the expected recourse cost from an LP parameterized in ξ. We investigate two cases depending on where ξ appears.

Case 3 Uncertain cost coefficients: $l(\xi) = \min_y\{y^T Q\xi : Wy \geq h - Ax\}$ where x is the first-stage decision variable, y represents the recourse action, and the feasible region is always non-empty. In this case, the supremum regarding ξ in the constraint becomes

$$\begin{aligned}&\max_{\xi\in\Xi}\left\{-z_i^T\xi + \min_y\left\{y^T Q\xi : Wy \geq h - Ax\right\}\right\}\\&= \min_y\left\{\max_{\xi\in\Xi}\left\{\left(Q^T y - z_i\right)^T\xi\right\} : Wy \geq h - Ax\right\}\end{aligned}$$

Replace the inner LP with its dual, we get an equivalent LP

$$\begin{aligned}\min_{\gamma_i, y_i}\ & d^T\gamma_i\\ \text{s.t. }\ & C^T\gamma_i = Q^T y_i - z_i,\ \gamma_i \geq 0\\ & Wy_i \geq h - Ax\end{aligned}$$

Here we associated variable y with a subscript i to highlight its dependence on the value of ξ. Therefore, $z_i = Q^T y_i - C^T\gamma_i$, and we can replace the supremum by the objective of the dual LP, which gives rise to:

$$d^T\gamma_i + \left\langle Q^T y_i - C^T\gamma_i, \xi_i^0\right\rangle \leq s_i,\ i = 1, \cdots, N$$

Arrange all constraints together, we obtain a convex program which is equivalent to problem (C.178) in Case 3:

$$\begin{aligned}\inf_{\lambda, s_i}\ & \lambda\epsilon + \frac{1}{N}\sum_{i=1}^N s_i\\ \text{s.t. }\ & y_i^T Q\xi_i^0 + \gamma_i^T(d - C^T\xi_i^0) \leq s_i,\ i = 1, \cdots, N\\ & \lambda \geq \|Q^T y_i - C^T\gamma_i\|_*,\ \gamma_i \geq 0,\ i = 1, \cdots, N\\ & Wy_i \geq h - Ax,\ i = 1, \cdots, N\end{aligned} \tag{C.181}$$

Without distributional uncertainty, $\epsilon = 0$, λ can be arbitrary nonnegative value; for similar reason, we have $\gamma_i = 0$ and $s_i = y_i^T Q\xi_i^0$ at optimum. So problem (C.181) is equivalent to the SAA problem under the empirical distribution

$$\min_{y_i} \left\{ \frac{1}{N} \sum_{i=1}^{N} y_i^T Q\xi_i^0 : W y_i \geq h - Ax \right\}$$

Case 4 Uncertain constraint right-hand side:

$$\begin{aligned} l(\xi) &= \min_{y} \{q^T y : Wy \geq H\xi + h - Ax\} \\ &= \max_{\theta} \left\{\theta^T (H\xi + h - Ax) : W^T\theta = q, \theta \geq 0\right\} \\ &= \max_{k} v_k^T (H\xi + h - Ax) = \max_{k} \left\{v_k^T H\xi + v_k^T (h - Ax)\right\} \end{aligned}$$

where v_k is the vertices of polyhedron $\{\theta : W^T\theta = q, \theta \geq 0\}$. In this way, $l(\xi)$ is expressed as a convex PWL function. Applying the result in Case 1, we obtain a convex program which is equivalent to problem (C.178) in Case 4:

$$\begin{aligned} \inf_{\lambda, s_i} \ & \lambda\epsilon + \frac{1}{N}\sum_{i=1}^{N} s_i \\ \text{s.t.} \ & v_k^T (h - Ax) + v_k^T H\xi_i^0 + \gamma_{ik}^T (d - C^T\xi_i^0) \leq s_i, \ i = 1, \cdots, N, \ \forall k \\ & \lambda \geq \|H^T v_k - C^T \gamma_{ik}\|_*, \ \gamma_{ik} \geq 0, \ i = 1, \cdots, N, \ \forall k \end{aligned} \tag{C.182}$$

For similar reason, without distributional uncertainty, we have $\gamma_{ik} = 0$ and $s_i = v_k^T (h - Ax) + v_k^T H\xi_i^0 = q^T y_i$ at optimum, where the last equality is because of strong duality. So problem (C.181) is equivalent to the SAA problem under the empirical distribution

$$\min_{y_i} \left\{ \frac{1}{N} \sum_{i=1}^{N} q^T y_i : W y_i \geq H\xi + h - Ax \right\}$$

The following discussions are devoted to the computational tractability.

- If the 1-norm or ∞-norm is used to define Wasserstein metric, their dual norms are ∞-norm and 1-norm respectively, then problems (C.179)–(C.182) reduce to LPs whose sizes grow with the number N of sampled data. If the Euclidean norm is used, the resulting problems will be SOCP.
- For Case 1, Case 2 and Case 3, the remaining equivalent LPs scale polynomially and can be therefore readily solved. As for Case 4, the number of vertices may grow exponential in the problem size. However, one can adopt a decomposition

algorithm similar to CCG which iteratively identifies critical vertices without enumerating all of them.

- The computational complexity of all equivalent convex programs is independent of the size of the Wasserstein ambiguity set.
- It is shown in [150] that the worst-case expectation can also be computed from the following problem

$$\begin{aligned}
\sup_{\alpha_{ik}, q_{ik}} \quad & \frac{1}{N} \sum_{i=1}^{N} \sum_{k=1}^{K} \alpha_{ij} l_k \left(\xi_i^0 - \frac{q_{ik}}{\alpha_{ik}} \right) \\
\text{s.t.} \quad & \frac{1}{N} \sum_{i=1}^{N} \sum_{k=1}^{K} \| q_{ik} \| \le \epsilon \\
& \alpha_{ik} \ge 0, \forall i, \forall k, \ \sum_{k=1}^{K} \alpha_{ik} = 1, \forall i \\
& \xi_i^0 - \frac{q_{ik}}{\alpha_{ik}} \in \Xi, \ \forall i, \forall k
\end{aligned} \tag{C.183}$$

Non-convex term arises from the fraction q_{ik}/α_{ik}. In fact, problem (C.183) is convex following the definition of extended perspective function [150]. Moreover, if $[\alpha_{ik}(r), q_{ik}(r)]_{r \in \mathbb{N}}$ is a sequence of feasible solutions and the corresponding objective values converge to the supremum of (C.183), then the discrete distribution

$$\mathbb{Q}_r = \frac{1}{N} \sum_{i=1}^{N} \sum_{k=1}^{K} \alpha_{ik}(r) \delta_{\xi_{ik}(r)}, \ \xi_{ik}(r) = \xi_i^0 - \frac{q_{ik}(r)}{\alpha_{ik}(r)}$$

approaches the worst-case distribution in D_W [150].

3. Static Robust Chance Constraints

Another important issue in SO is chance constraint. Here we discuss robust joint chance constraints in the following form

$$\inf_{\mathbb{Q} \in D_W} \Pr[a(x)^T \xi_i \le b_i(x), i = 1, \cdots, I] \ge 1 - \beta \tag{C.184}$$

where x is the decision variable; the chance constraint involves I inequalities with uncertain parameter ξ_i supported on set $\Xi_i \subseteq \mathbb{R}^n$ for each i. The joint probability distribution $\mathbb{Q}$ belongs to the Wasserstein ambiguity set. $a(x) \in \mathbb{R}^n$ and $b(x) \in \mathbb{R}$ are affine mappings of x, where $a(x) = \eta x + (1 - \eta)\mathbf{1}$, $\eta \in \{0, 1\}$, and $b_i(x) = B_i^T x + b_i^0$. When $\eta = 1$ ($\eta = 0$), (C.184) involves the left-hand (right-hand) uncertainty. $\Xi = \prod_i \Xi_i$ is the support set of $\xi = [\xi_1^T, \cdots, \xi_I^T]^T$. The robust chance constraint (C.184) requires that all inequalities be met for all possible distributions in Wasserstein ambiguity set D_W with a probability of at least $1 - \beta$, where $\beta \in (0, 1)$

denotes a prescribed risk tolerance. The feasible region stipulated by (C.184) is X. We will introduce the main results from [152] while avoiding rigorous mathematical proofs.

Assumption C.6 *The support set Ξ is an $n \times I$-dimensional vector space, and the distance metric in Wasserstein ambiguity set is $d(\xi, \zeta) = \|\xi - \zeta\|$.*

Theorem C.2 ([152]) *Under Assumption (C.6), $X = Z_1 \cup Z_2$, where*

$$Z_1 = \left\{ x \in \mathbb{R}^n \;\middle|\; \begin{array}{l} \epsilon v - \beta\gamma \le \dfrac{1}{N}\displaystyle\sum_{j=1}^{N} z_j \\ z_j + \gamma \le \max\left\{b_i(x) - a(x)^T \zeta_i^j, 0\right\} \\ i = 1, \cdots I,\ j = 1, \cdots, N \\ z_j \le 0,\ j = 1, \cdots, N \\ \|a(x)\|_* \le v,\ \gamma \ge 0 \end{array} \right\} \tag{C.185}$$

where ϵ is the radius of the Wasserstein ambiguity set, N is the number of sampled scenarios in the empirical distribution, and

$$Z_2 = \{x \in \mathbb{R}^n \mid a(x) = 0,\ b_i(x) \ge 0,\ i = 1, \cdots I\} \tag{C.186}$$

In Theorem C.2, Z_2 is trivial: If $\eta = 1$, then $Z_2 = \{x \in \mathbb{R}^n \mid x = 0,\ b_i \ge 0,\ \forall i\}$; If $\eta = 0$, then $Z_2 = \emptyset$. Z_1 can be reformulated as an MILP compatible form if it is bounded. By linearizing the second constraint, we have

$$Z_1 = \left\{ x \in \mathbb{R}^n \;\middle|\; \begin{array}{l} \epsilon v - \beta\gamma \le \dfrac{1}{N}\displaystyle\sum_{j=1}^{N} z_j \\ z_j + \gamma \le s_{ij},\ \forall i, \forall j \\ b_i(x) - a(x)^T \zeta_i^j \le s_{ij} \le M_{ij} y_{ij},\ \forall i, \forall j \\ s_{ij} \le b_i(x) - a(x)^T \zeta_i^j + M_{ij}(1 - y_{ij}),\ \forall i, \forall j \\ \|a(x)\|_* \le v,\ \gamma \ge 0,\ z_j \le 0,\ \forall j \\ s_{ij} \ge 0,\ y_{ij} \in \{0, 1\},\ \forall i,\ \forall j \end{array} \right\} \tag{C.187}$$

where $\forall i$ and $\forall j$ are short for $i = 1, \cdots, I$ and $j = 1, \cdots, N$, respectively;

$$M_{ij} \ge \max_{x \in Z_1} \left| b_i(x) - a(x)^T \zeta_i^j \right|$$

It is easy to see that if $b_i(x) - a(x)^T \zeta_i^j < 0$, then $y_{ij} = 0$ (otherwise $s_{ij} \le b_i(x) - a(x)^T \zeta_i^j < 0$), hence $s_{ij} = 0 = \max\{b_i(x) - a(x)^T \zeta_i^j, 0\}$. If $b_i(x) - a(x)^T \zeta_i^j > 0$, then $y_{ij} = 1$ (otherwise $b_i(x) - a(x)^T \zeta_i^j \le M_{ij} y_{ij} = 0$), hence $s_{ij} = b_i(x) - a(x)^T \zeta_i^j = \max\{b_i(x) - a(x)^T \zeta_i^j, 0\}$. If $b_i(x) - a(x)^T \zeta_i^j = 0$, then we have $s_{ij} = 0$ regardless of the value of y_{ij}. In conclusion, (C.185) and (C.187) are equivalent.

For the right-hand uncertainty in which $\eta = 0$, $a(x) = \mathbf{1}$, $X = Z_1$ because $Z_2 = \emptyset$. Moreover, variable v in (C.187) is equal to 1 if 1-norm is used in Wasserstein ambiguity set D_W, indicating $v \ge \|\mathbf{1}\|_\infty = 1$ in Z_1.

In (C.187), a total number of $I \times N$ binary variables are introduced to linearize the $\max\{a, b\}$ function, making the problem challenging to solve. An inner approximation of Z is to simply replace $\max\{b_i(x) - a(x)^T \zeta_i^j, 0\}$ with its first input, yielding a parameter-free approximation

$$Z = \left\{ x \in \mathbb{R}^n \;\middle|\; \begin{aligned} & \epsilon v - \beta \gamma \le \frac{1}{N} \sum_{j=1}^{N} z_j \\ & z_j + \gamma \le b_i(x) - a(x)^T \zeta_i^j, \ \forall i, \forall j \\ & z_j \le 0, \ \forall j, \ \|a(x)\|_* \le v, \ \gamma \ge 0 \end{aligned} \right\} \tag{C.188}$$

This formulation can be derived from CVaR model, and enjoys better computational tractability.

4. Adaptive Robust Chance Constraints

Robust chance constraint program with Wasserstein metric is studied in [153] in a different but more general form. The problem is as follows:

$$\begin{aligned} \min_{x \in X} \ & c^T x \\ \text{s.t.} \ & \inf_{\mathbb{Q} \in D_W} \Pr[F(x, \xi) \le 0] \ge 1 - \beta \end{aligned} \tag{C.189}$$

where X is a bounded polyhedron, $F : \mathbb{R}^n \times \Xi \to \mathbb{R}$ is a scalar function that is convex in x for every ξ. This formulation is general enough to capture joint chance constraints. To see this, suppose F contains K individual constraints, then F can be defined as the component-wise maximum as in (C.91).

Here we develop a technique to solve two-stage problems where $F(x, \xi)$ is the optimal value of another LP parameterized in x and ξ. More precisely, we consider

$$\begin{aligned} \min_{x \in X} \ & c_1^T x \\ \text{s.t.} \ & \sup_{\mathbb{Q} \in D_W} \Pr[f(x, \xi) \ge c_2^T \xi] \le \beta \end{aligned} \tag{C.190a}$$

$$
\begin{aligned}
f(x,\xi) = \min \ & c_3^T y \\
\text{s.t.} \ & Ax + By + C\xi \le d
\end{aligned}
\tag{C.190b}
$$

where in (C.190a), the robust chance constraint can be regarded a risk limiting requirement, and the threshold value depends on uncertain parameter ξ. We assume LP (C.190b) is always feasible (relatively complete recourse) and has finite optimum. Second-stage cost can be considered in the objective function of (C.190a) in form of worst-case expectation which has been discussed in previous sections and is omitted here for the sake of brevity. Here we focus on coping with second-stage LP in robust chance constraint.

Define loss function

$$
g(x,\xi) = f(x,\xi) - c_2^T \xi \tag{C.191}
$$

Recall the relation between chance constraint and CVaR discussed in Sect. C.3.1, a sufficient condition of robust chance constraint in (C.190a) is CVaR$(g(x,\xi),\beta) \le 0$, $\forall \mathbb{Q} \in D_W$, or equivalently

$$
\sup_{\mathbb{Q}\in D_W} \inf_{\gamma\in\mathbb{R}} \beta\gamma + \mathbb{E}_{\mathbb{Q}}(\max\{g(x,\xi)-\gamma,0\}) \le 0 \tag{C.192}
$$

According to [153], constraint (C.192) can be conservatively approximated by

$$
\epsilon L + \inf_{\gamma\in\mathbb{R}} \left\{ \beta\gamma + \frac{1}{N}\sum_{i=1}^{N} \max\{g(x,\xi^i)-\gamma,0\} \right\} \le 0 \tag{C.193}
$$

where ϵ is the parameter in Wasserstein ambiguity set D_W, L is a constant satisfying $g(x,\xi) \le L\|\xi\|_1$, and ξ^i, $i=1,\cdots,N$ are samples of uncertain data. Substituting (C.190b) and (C.191) into (C.193), we obtain an LP that is equivalent to problem (C.190)

$$
\begin{aligned}
\min \ & c_1^T x \\
\text{s.t.} \ & x \in X,\ \epsilon L + \beta\gamma + \frac{1}{N}\sum_{i=1}^{N} s_i \le 0 \\
& s_i \ge 0,\ s_i \ge c_3^T y^i - c_2^T \xi^i - \gamma,\ i=1,\cdots,N \\
& Ax + By^i + C\xi^i \le d,\ i=1,\cdots,N
\end{aligned}
\tag{C.194}
$$

where y^i is the second-stage decision associated with ξ^i. This formulation could be very conservative due to three reasons. First, worst-case distribution is considered; second, CVaR constraint (C.192) is a pessimistic approximation of chance constraints; finally, sampling constraint (C.193) is a pessimistic approximation of (C.192).

More discussions on robust chance constraints with Wasserstein metric under various settings can be found in [153].

5. Use of Forecast Data

Wasserstein metric enjoys many advantages, such as finite-sample performance guarantee and existence of tractable reformulation. However, moment information is not used, especially the first-order moment reflecting the prediction, which can be updated with time rolling on, so the worst-case distribution generally has a mean value different from the forecast (if available). To incorporate forecast data, we propose the following Wasserstein ambiguity set with fixed-mean

$$D_W^M = \left\{ \mathbb{Q} \in D_W \middle| \mathbb{E}_{\mathbb{Q}}[\xi] = \hat{\xi} \right\} \tag{C.195}$$

and the worst-case expectation problem can be expressed as

$$\sup_{\mathbb{Q} \in D_W^M} \quad \mathbb{E}_{\mathbb{Q}}[l(\xi)] \tag{C.196a}$$

$$= \sup_{f^n(\xi)} \quad \frac{1}{N} \sum_{n=1}^{N} \int_{\Xi} l(\xi) f^n(\xi) \mathrm{d}\xi \tag{C.196b}$$

$$\text{s.t.} \quad \frac{1}{N} \sum_{n=1}^{N} \int_{\Xi} \|\xi - \xi^n\|_p f^n(\xi) \mathrm{d}\xi \le \epsilon : \lambda \tag{C.196c}$$

$$\int_{\Xi} f^n(\xi) \mathrm{d}\xi = 1 : \theta_n, \ n = 1, \cdots, N \tag{C.196d}$$

$$\frac{1}{N} \sum_{n=1}^{N} \int_{\Xi} \xi f^n(\xi) \mathrm{d}\xi = \hat{\xi} : \rho \tag{C.196e}$$

where $l(\xi)$ is a loss function similar to that in (C.175), $f^n(\xi)$ is the conditional density function under historical data sample ξ^n, dual variables λ, θ_n, and ρ are listed following a colon. Similar to the discussions for problem (C.79), the dual problem of (C.196) is

$$\min_{\lambda \ge 0, \theta_n, \rho} \quad (\lambda \epsilon + \rho^T \hat{\xi}) N + \sum_{n=1}^{N} \theta_n \tag{C.197a}$$

$$\text{s.t.} \ \theta_n + \lambda \|\xi - \xi^n\|_p + \rho^T \xi \ge l(\xi), \forall \xi \in \Xi, \ \forall n \tag{C.197b}$$

For $p = 2$, polyhedral Ξ and PWL $l(\xi)$, constraint (C.197b) can be transformed into the intersection of PSD cones, and problem (C.197) gives rise to an SDP;

some examples can be found in Sect. C.3.1. If Ξ is described by a single quadratic constraint, constraint (C.197b) can be reformulated by using the well-known S-Lemma, which has been discussed in Sect. A.2.4, and problem (C.197) still comes down to an SDP. For $p = 1$ or $p = +\infty$, polyhedral Ξ and PWL $l(\xi)$, constraint (C.197b) can be transformed into a polyhedron using duality theory, and problem (C.197) gives rise to an LP. Because the ambiguity set is more restrictive, problem (C.197) would be less conservative than problem (C.178) in which the mean value of uncertain data is free.

A Wasserstein-moment metric with variance is exploited in [154] and applied to wind power dispatch. Nevertheless, the ambiguity set neglects first-order moment and considers second-order moment. This formulation is useful when little historical data is available at hand.

As a short conclusion, distributionally robust optimization and data-driven robust stochastic optimization leverage statistical information on the uncertain data and overcome the conservatism of traditional robust optimization approaches which are built upon the worst-case scenario. The core issue is the equivalent convex reformulation of the worst-case expectation problem or the robust chance constraint over the uncertain probability distribution restricted in the ambiguity set. Optimization over a moment based ambiguity set can be formulated as a semi-infinite LP, whose dual problem gives rise to SDPs, and hence can be readily solved. When additional structure property is taken into account, such as unimodality, more sophisticated treatment is needed. As for the robust stochastic programming, tractable reformulation of the worst-case expectation and robust chance constraints is the central issue. Robust chance constraint under a ϕ-divergence based ambiguity set is equivalent to traditional chance constraint under the empirical distribution but with a modified confidence level, and it can be transformed into an MILP or approximated by LP based on risk theory under the help of sampling average approximation technique, so does a robust chance constraint under a Wasserstein metric based ambiguity set, following somewhat different expressions. The worst-case expectation under ϕ-divergence based ambiguity set boils down to a convex program with linear constraints and a nonlinear objective function, which can be efficiently solved via outer approximation algorithm. The worst-case expectation under Wasserstein ambiguity set comes down to a conic program which is convex and readily solvable. Unlike the max-min problem in traditional robust optimization method identifying the worst-case scenario which the decision maker wishes to avoid, the worst-case expectation problem in distributionally robust optimization and robust stochastic programming is solved in its dual form, whose solution is less intuitive to the decision maker; moreover, it may not be easy to recover the primal optimal solution, i.e., the worst-case probability. The worst-case distribution in the robust chance constrained stochastic programming is discussed in [136, 150]; the worst-case discrete distribution in a two-stage stochastic program with min-max expectation can be computed via a polynomial complexity algorithm. Nonetheless, from a practical perspective, what the human decision makers actually need to deploy is merely the here-and-now decision, and the worst probability distribution is usually not very important, since corrective actions can be postponed to a later stage when the uncertain data have been observed or can be predicted with high accuracy.

C.5 Further Reading

Uncertainty is ubiquitous in real-life decision-making problems, and the decision maker usually has limited information and statistic data on the uncertain factors, which makes robust optimization very attractive in practice, as it is tailored to the available information at hand, and often gives rise to computationally tractable reformulations. Although the original idea can date back to [155] in the 1970s, it is during the past two decades that the fundamental theory of robust optimization has been systematically developed. This research field is even more active during the past 5 years. This chapter aims to help beginners get an overview on this method and understand how to apply robust optimization in practice. We provide basic models and tractable reformulations, called the robust counterparts, for various robust optimization models under different assumptions on the uncertainty and decision-making manner. Basic theory of robust optimization is provided in [90, 91]. Comprehensive surveys can be found in [92, 94]. Here we shed more light on several important topics in robust optimization.

Uncertainty sets play a decisive role on the performance of a robust solution. A larger set could protect the system against a higher level of uncertainty, and increase the cost as well. However, the probability that uncertain data take their worst-case values is usually small. The decision-maker needs to make a trade-off between reliability and economy. Ambiguous chance constraints and their approximations are discussed in Chapter 2 of [90], based on which the parameter in the uncertainty set can be selected. It is proposed in [156] to construct uncertainty sets from historical data and statistical tests. The connection of uncertainty sets and coherent risk measures is revealed in [157]. It is shown that the distortion risk measure leads to a polyhedral uncertainty set. Specifically, the connection of CVaR and uncertainty sets is discussed in [158]. A reverse correspondence is reported in [159], demonstrating that robust optimization could generalize the concepts of risk measures. A data-driven approach is proposed in [156] to construct uncertainty sets for robust optimization based on statistical hypothesis tests. The counterpart problems are shown to be tractable, and optimal solutions satisfy constraints with finite-sample probabilistic guarantee.

Distributionally robust optimization integrates statistic information, worst-case expectation, and robust probability guarantee in a holistic optimization framework, in which the uncertainty is modeled via an ambiguous probability distribution. The choice of ambiguity sets for candidate distributions affects not only the model conservatism, but also the existence of tractable reformulations. Various ambiguity sets have been proposed in the literature, which can be roughly classified into two categories:

1. Moment ambiguity sets. All PDFs share the same moment data, usually the first- and second-order moments, and structured properties, such as symmetry and unimodality. For example, Markov ambiguity set contains all distributions with the same mean and support, and the worst-case expectation is shown to be equivalent to LPs [160]. Chebyshev ambiguity set is composed of all distributions with known expectation and covariance matrix, and usually leads to SDP counterparts [127, 161, 162]; the Gauss ambiguity set contains all unimodal distributions in the Chebyshev ambiguity set, and also gives rise to SDP reformulations [163].
2. Divergence ambiguity sets. All PDFs are close to a reference distribution in terms of a specified measure. For example, the Wasserstein ambiguity quantifies the divergence via Wasserstein metric [150, 152, 164]; the ϕ-divergence ambiguity [138, 165] characterizes the divergence of two probability density functions through the distance of special non-negative weights (for discrete distributions) or integrals (for continuous distributions).

More information on the types of ambiguity sets and reformulations of their distributionally robust counterparts can be found in [137]. According to the latest research progress, the moment based distributionally robust optimization is relatively mature and has been widely adopted in engineering, because the semi-infinite LP formulation and its dual for the worst-case expectation problem offer a systematic approach to analyze the impact of uncertain distributions. However, when more complicated ambiguity sets are involved, such as the Gauss ambiguity set, deriving a tractable reformulation needs more sophisticated approaches. The study on the latter category, which directly imposes uncertainty on the distributions is attracting growing attentions in the past 2 or 3 years, because it makes full use of historical data, which can better capture the unique feature of uncertain factors under investigation.

Data-driven robust stochastic programming, conceptually the same as distributionally robust optimization but preferred by some researchers, has been studied using ϕ-divergence in [166, 167] and Wasserstein metric in [150, 152, 153, 164, 168–170], because a tractable counterpart problem can be derived under such ambiguity sets.

Many decision-making problems in engineering and finance often require that a certain risk measure associated with random variables should be limited below a threshold. However, the probability distribution of random variables is not exactly known; therefore, the risk limiting constraint must be able to withstand perturbations of distribution in a reasonable range. This entails a tractable reformulation of a risk measure under distributional uncertainty. This problem has been comprehensively discussed in [164]. In more recent publications, CVaR under moment ambiguity set with unimodality is studied in [171]; VaR and CVaR under moment ambiguity set are discussed in [172]; distortion risk measure under Wasserstein ambiguity set is considered in [173].

In multi-stage decision making, causality is a pivotal issue for practical implementation, which means that the wait-and-see decisions in the current stage cannot depend on the information of uncertainty in future stages. For example, in a unit commitment problem with 24 periods, the wind power output is observed period-by-period. It is shown in [174] that the two-stage robust model in [118] offers non-causal dispatch strategies, which are in fact not robust. A multi-stage causal unit commitment model is suggested in [174, 175] based on affine policy. Causality is put to effect by imposing block diagonal constraints on the gain matrix of affine policy. Causality is also called non-anticipativity in some literature, such as [176], which is attracting attention from practitioners [177, 178].

For some other interesting topics on robust optimization, such as the connection with stochastic optimization, connection with risk theory, and applications in engineering problems other than those in power systems, readers can refer to [94]. Nonlinear issues have been addressed in [100, 179]. Optimization models with uncertain SOC and SDP constraints are discussed in [180, 181]. The connection among robust optimization, data utilization, and machine learning has been reviewed in [182].

Appendix D
Equilibrium Problems

Life is not a game. Still, in this life, we choose the games we live to play.

The concept of an equilibrium describes a state that the system has no incentive to change. These incentives can be profit-driven in the case of competitive markets or a reflection of physical laws such as energy flow equations. In this sense, equilibrium encompasses broader concepts than the solution of a game. Equilibrium is a fundamental notation appearing in various disciplines in economics and engineering. Identifying the equilibria allows eligible authorities to predict the system state at a future time or design reasonable policies for regulating a system or a market. This is not saying that an equilibrium state must appear sooner or later, partly because decision makers in reality have only limited rationality and information. Nevertheless, the awareness of such an equilibrium could be helpful for system design and operation. In this chapter, we restrict our attention in the field of game theory, which entails simultaneously solving multiple interactive optimization problems. We review the notions of some quintessential equilibrium problems and show how they can be solved via traditional optimization methods. These problems can be roughly categorized into two classes: the first one contains only one level: all players must make a decision simultaneously, which is referred to as a Nash-type game; the second one has two levels: decisions are made sequentially by two groups of players, called the leaders and the followers. This category is widely known as Stackelberg-type games, or multi-leader-follower games, or equilibrium programs with equilibrium constraints (EPEC). Unlike a traditional mathematical programming problem where the decision maker is unique, in an equilibrium problem or a game, multiple decision makers seek optimums of individual optimization problems parameterized in the optimal solutions of others.

General notations used throughout this chapter are defined as follows. Specific symbols are explained in the individual sections. In the game theoretic language, a decision maker is called a player. Vector $x = (x_1, \cdots, x_n)$ refers to the joint decisions of all upper-level players or the so-called leaders in a bilevel setting, where x_i stands for the decisions of leader i; $x_{-i} = (x_1, \cdots, x_{i-1}, x_{i+1}, \cdots, x_n)$ refers to the rivals' actions for leader i. Similarly, $y = (y_1, \cdots, y_m)$ refers to the

W. Wei, J. Wang, *Modeling and Optimization of Interdependent Energy Infrastructures*, https://doi.org/10.1007/978-3-030-25958-7

joint decisions of all lower-level players or the so-called followers, where y_j stands for the decisions of follower j; $y_{-j} = (y_1, \cdots, y_{j-1}, y_{j+1}, \cdots, y_m)$ refers to the rivals' actions for follower j. λ and μ are Lagrangian dual multipliers associated with inequality and equality constraints.

D.1 Standard Nash Equilibrium Problem

After J. F. Nash published his work on the equilibrium of n-person non-cooperative games in the early 1950s [183, 184], game theory quickly became a new branch of operational research. Nash equilibrium problem (NEP) captures the interactive behaviors of strategic players, in which each player's utility depends on the actions of other players. During decades of wonderful research, a variety of new concepts and algorithms of Nash equilibriums have been proposed and applied to almost every area of knowledge. This section just reviews some basic concepts and the most prevalent best-response algorithms.

D.1.1 Formulation and Optimality Condition

In a standard n-person non-cooperative game, each player minimizes his payoff function $f_i(x_i, x_{-i})$ which depends on all players' actions. The strategy set $X_i = \{x_i \in \mathbb{R}^{k_i} \mid g_i(x_i) \le 0\}$ of player i is independent of x_{-i}. The joint strategy set of the game is the Cartesian product of X_i, i.e., $X = \prod_{i=1}^{n} X_i$, and $X_{-i} = \prod_{j \neq i} X_j$. Roughly speaking, the non-cooperative game is a collection of coupled optimization problems, where player i chooses $x_i \in X_i$ that minimizes his payoff $f_i(x_i, x_{-i})$ given his rivals' strategies x_{-i}, or mathematically

$$\left.\begin{aligned} \min_{x_i} \quad & f_i(x_i, x_{-i}) \\ \text{s.t.} \quad & g_i(x_i) \le 0 : \lambda_i \end{aligned}\right\}, \quad i = 1, \cdots, n \tag{D.1}$$

In the problem of player i, the decision variable is x_i, and x_{-i} is regarded as parameters; λ_i is the dual variable.

The Nash equilibrium consists of a strategy profile such that every player's strategy constitutes the best response to all other players' strategies, or in other words, no player can further reduce his payoff by changing his action unilaterally. Therefore, the Nash equilibrium is a stable state which can sustain spontaneously. The mathematical definition is formally given below.

Definition D.1 A strategy vector $x^* \in X$ is a Nash equilibrium if the condition

$$f_i(x_i^*, x_{-i}^*) \le f_i(x_i, x_{-i}^*), \ \forall x_i \in X_i \tag{D.2}$$

holds for all players.

Condition (D.2) naturally interprets the fact that at a Nash equilibrium, if any player choose an alternative strategy, his payoff may grow, which is undesired. To depict a Nash equilibrium, a usual approach is the fixed-point of best-response mapping. Let $B_i(x_{-i})$ be the set of optimal strategies of player i given the strategies x_{-i} of others, then set $B(x) = \prod_{i=1}^{n} B_i(x_{-i})$ is the best-response mapping of the game. It is clear that x^* is a Nash equilibrium if and only if $x^* \in B(x^*)$, i.e., x^* is a fixed point of $B(x)$. This fact establishes the foundation for analyzing Nash equilibria using the well-developed fixed-point theory. However, conducting the fixed-point analysis usually requires the best-response mapping $B(x)$ in a closed form. Moreover, to declare the existence and uniqueness of a Nash equilibrium, the mapping should be contractive [185]. These strong assumptions inevitably limit the applicability of fixed-point method. For example, in many instances, the best-response mapping $B(x)$ is neither contractive nor continuous, but Nash equilibria may still exist.

Another way to characterize the Nash equilibrium is the KKT system approach. Generally speaking, in a standard Nash game, each player is facing an NLP parameterized in the rivals' strategies. If we consolidate the KKT optimality conditions of all these NLPs in (D.1), we get the following KKT system

$$\left.\begin{aligned} &\nabla_{x_i} f_i(x_i, x_{-i}) + \lambda_i^T \nabla_{x_i} g_i(x_i) = 0 \\ &\lambda_i \geq 0,\ g(x_i) \leq 0,\ \lambda_i^T g_i(x_i) = 0 \end{aligned}\right\} \ i = 1, \cdots, n \tag{D.3}$$

If x^* is a Nash equilibrium that satisfies (D.2), and any standard constraint qualification holds for every player's problem in (D.1), then x^* must be a stationary point of the concentrated KKT system (D.3) [186]; and vice versa: if all problems in (D.1) meet a standard constraint qualification, and a point x^* together with a proper vector of dual multipliers $\lambda = (\lambda_1, \cdots, \lambda_n)$ solves KKT system (D.3), then x^* is also a Nash equilibrium that satisfies (D.2).

Problem (D.3) is an NCP and is the optimality condition of Nash equilibrium. It is a natural attempt to retrieve an equilibrium by solving NCP (D.3) without deploying an iterative algorithm, which may suffer from divergence. To obviate the computational challenges brought by the complementarity and slackness constraints in KKT system (D.3), a merit function approach and an interior-point method are comprehensively discussed in [186].

D.1.2 Variational Inequality Formulation

An alternative perspective to study the NEP is to formulate it as a variational inequality (VI) problem. This approach is pursued in [187]. The advantage of variational inequality approach is that it permits an easy access to existence and uniqueness results without the best-response mapping. From a computational point of view, it naturally leads to easily implementable algorithms along with provable convergence performances.

Given a closed and convex set $X \in \mathbb{R}^n$ and a mapping $F : X \to \mathbb{R}^n$, a variational inequality problem, denoted by VI(X, F), is to determine a point $x^* \in X$ satisfying [188]

$$(x - x^*)^T F(x^*) \geq 0, \; \forall x \in X \tag{D.4}$$

To see the connection between a VI problem and a traditional convex optimization problem that seeks a minimum of a convex function $f(x)$ over a convex set X, let us assume that the optimal solution is x^*, then the feasible region must not lie in the half space where $f(x)$ decreases; geometrically, the line segment connecting any $x \in X$ with x^* must form an acute angle with the gradient of f at x^*, which can be mathematically described as $(x - x^*)^T \nabla f(x^*) \geq 0, \; \forall x \in X$. This condition can be concisely expressed by VI(X, ∇f) [189].

However, when the Jacobian matrix of F is not symmetric, F cannot be written as the gradient of another scalar function, and hence the variational inequality problem encompasses broader classes of problems than traditional mathematical programs. For example, when $X = \mathbb{R}^n$, problem (D.4) degenerates into a system of equations $F(x^*) = 0$; when $X = \mathbb{R}^n_+$, problem (D.4) comes down to an NCP $0 \leq x^* \perp F(x^*) \geq 0$.

To see the latter case, $x^* \geq 0$ because it belongs to X; if any element of $F(x^*)$ is negative, say the first element $[F(x^*)]_1 < 0$, we let $x_1 = x_1^* + 1$, and $x_i = x_i^*$, $i = 2, \cdots$, then $(x - x^*)^T F(x^*) = [F(x^*)]_1 < 0$, which is contradictive to (D.4). Hence $F(x^*) \geq 0$ must hold. Let $x = 0$ in (D.4), we have $(x^*)^T F(x^*) \leq 0$. Because $x^* \geq 0$ and $F(x^*) \geq 0$, there must be $(x^*)^T F(x^*) = 0$, resulting in the target NCP.

The monotonicity of F plays a central role in the theoretical analysis of VI problems, just like the role of convexity in mathematical programming. It has a close relationship with the Jacobian matrix ∇F [187, 190]:

$$\begin{array}{lcl} F(x) \text{ is monotone on } X & \Leftrightarrow & \nabla F(x) \succeq 0, \forall x \in X \\ F(x) \text{ is strictly monotone on } X & \Leftarrow & \nabla F(x) \succ 0, \forall x \in X \\ F(x) \text{ is strongly monotone on } X & \Leftrightarrow & \nabla F(x) - c_m I \succeq 0, \forall x \in X \end{array}$$

where c_m is a strictly positive constant. As a correspondence to convexity, a differentiable function f is convex (strictly convex, strongly convex) on X if and only if ∇f is monotone (strictly monotone, strongly monotone) on X.

Conceptually, monotonicity (convexity) is the weakest, since the matrix $\nabla F(x)$ can have zero eigenvalues; strict monotonicity (strict convexity) is stronger, as all eigenvalues of matrix $\nabla F(x)$ are strictly positive; strong monotonicity (strong convexity) is the strongest, because the smallest eigenvalue of matrix $\nabla F(x)$ should be greater than a given positive number. Intuitively, a strong convex function must be more convex than a given convex quadratic function; for example, $f(x) = x^2$ is strongly convex on $\mathbb{R}$; $f(x) = 1/x$ is convex on $\mathbb{R}_{++}$ and strongly convex on $(0, 1]$.

To formulate an NEP as a VI problem and establish the existence and uniqueness result of Nash equilibria, we list some assumptions on the convexity and smoothness of each player's problem.

Assumption D.1 ([187])

(1) The strategy set X_i is non-empty, closed, and convex;
(2) Function $f_i(x_i, x_{-i})$ is convex in $x_i \in X_i$ for fixed $x_{-i} \in X_{-i}$;
(3) Function $f_i(x_i, x_{-i})$ is continuously differentiable in $x_i \in X_i$ for fixed $x_{-i} \in X_{-i}$;
(4) Function $f_i(x_i, x_{-i})$ is twice continuously differentiable in $x \in X$ with bounded second derivatives.

Proposition D.1 ([187]) *In a standard NEP NE(X, f), where $f = (f_1, \cdots, f_n)$, if conditions (1)–(3) in Assumption D.1 are met, then the game is equivalent to a variational inequality problem VI(X, F) with*

$$X = X_1 \times \cdots \times X_n$$

and

$$F(x) = (\nabla_{x_1} f_1(x), \cdots, \nabla_{x_n} f_n(x))$$

In the VI problem corresponding to a traditional mathematical program, the Jacobian matrix ∇F is symmetric, because it is the Hessian matrix of a scalar function. However, in Proposition D.1, the Jacobian matrix ∇F for an NEP is generally non-symmetric. Building upon the VI reformulation, the standard results on solution properties of VI problems [188] can be extended to standard NEPs.

Proposition D.2 *Given an NEP NE(X, f), all conditions in Assumption D.1 are met, then we have the following statements:*

(1) If $F(x)$ is strictly monotone, then the game has at most one Nash equilibrium.
(2) If $F(x)$ is strongly monotone, then the game has a unique Nash equilibrium.

Some sufficient guarantees for $F(x)$ to be (strictly, strongly) monotone are given in [187]. It should be pointed out that the equilibrium concept in the sense of Proposition D.2 is termed the pure-strategy Nash equilibrium, so as to distinguish it from the mixed-strategy Nash equilibrium which will appear later on.

D.1.3 Best Response Algorithms

A major benefit of the VI reformulation is that it leads to easily implementable solution algorithms. Here we list two of them. Readers who are interested in the proofs on their performances can consult [187].

1. Algorithms for Strongly Convex Cases
The first algorithm is a totally asynchronous-iterative one, in which players may update their strategies with different frequencies. Let $T = \{0, 1, 2, \cdots\}$ be the indices of iteration steps, and $T_i \subseteq T$ be the set of steps in which player i updates his own strategy x_i. The notation x_i^k implies that at step $k \notin T_i$, x_i^k remains unchanged. Let $t_j^i(k)$ be the latest step at which the strategy of player j is received by player i at step k. Therefore, if player i updates his strategy at step k, he uses the following strategy profile offered by other players:

$$x_{-i}^{t^i(k)} = \left(x_1^{t_1^i(k)}, \cdots, x_{i-1}^{t_{i-1}^i(k)}, x_{i+1}^{t_{i+1}^i(k)}, \cdots, x_n^{t_n^i(k)}\right) \tag{D.5}$$

Using above definitions, the totally asynchronous-iterative algorithm is summarized in Algorithm D.1. Some technique conditions for which the schedules T_i and $t_j^i(k)$ should satisfy in order to be implementable in practice are discussed in [191, 192], which are assumed to be satisfied without particular mention.

Algorithm D.1 Asynchronous best-response algorithm

1: Choose a convergence tolerance $\varepsilon > 0$ and a feasible initial point $x^0 \in X$; the iteration index is $k = 0$;
2: For player $i = 1, \cdots, n$, update the strategy x_i^{k+1} as

$$x_i^{k+1} = \begin{cases} x_i^* \in \arg\min_{x_i} \left\{ f_i\left(x_i, x_{-i}^{t^i(k)}\right) \,\middle|\, x_i \in X_i \right\} & \text{if } k \in T_i \\ x_i^n & \text{otherwise} \end{cases} \tag{D.6}$$

3: If $\|x^{k+1} - x^k\|_2 \leq \varepsilon$, terminate and report x^{k+1} as the Nash equilibrium; otherwise, update $k \leftarrow k + 1$, and go to step 2.

A sufficient condition which guarantees the convergence of Algorithm D.1 is provided in [187]. Roughly speaking, Algorithm D.1 would converge if $f_i(x)$ is strongly convex in x_i. However, this is a strong assumption, which cannot be satisfied even if there is only one point where the partial Hessian matrix $\nabla_{x_i}^2 f_i(x)$ of player i is singular.

Algorithm D.1 reduces to some classic algorithms by enforcing a special updating procedure, i.e., a particular selection of T_i and $t_j^i(k)$. For example, if players update their strategies simultaneously (sequentially), Algorithm D.1 becomes the Jacobi (Gauss-Seidel) type iterative scheme. Interestingly, the asynchronous best-response algorithm is robust against data missing or delay, and is guaranteed to find the unique Nash equilibrium. This feature greatly relaxes the requirement on data synchronization and simplifies the design of communication systems, and makes this class of algorithm very appealing in distributed system operations.

2. Algorithms for Convex Cases

To relax the strong monotonicity assumption on $F(x)$, the second algorithm has been proposed in [187], which only uses the monotonicity property and is summarized below. Algorithm D.2 converges to a Nash equilibrium, if each player's optimization problem is convex (or $F(x)$ is monotone), which significantly improves its applicability.

Algorithm D.2 Proximal decomposition algorithm

1: Given $\{\rho_n\}_{n=0}^{\infty}$, $\varepsilon > 0$, and $\tau > 0$, choose a feasible initial point $x^0 \in X$;

2: Find an equilibrium z^0 of the following NEP using Algorithm D.1

$$\left.\begin{aligned} \min_{x_i} \quad & f_i(x_i, x_{-i}) + \tau \|x_i - x_i^0\|_2^2 \\ \text{s.t.} \quad & x_i \in X_i \end{aligned}\right\}, i = 1, \cdots, n \tag{D.7}$$

3: If $\|z^0 - x^0\|_2 \le \varepsilon$, terminate and report x^0 as the Nash equilibrium; otherwise, update $x^0 \leftarrow (1 - \rho_n)x^0 + \rho_n z^0$, and go to step 2.

Algorithm D.2 is a double-loop method: the inner loop identifies a Nash equilibrium of the regularized game (D.7) with x^0 being a parameter, which is updated in each iteration, and the outer loop updates x^0 by selecting a new point along the line connecting x^0 and z^0. Notice that in step 2, as long as τ is large enough, the Hessian matrix $\nabla_{x_i}^2 f_i(x) + 2\tau I$ must be positive definite, and thus the best-response algorithm applied to (D.7) is guaranteed to converge to the unique Nash equilibrium. See [187] for more details about parameter selection.

The penalty term $\tau \|x_i - x_i^0\|^2$ limits the change of optimal strategies in two consecutive iterations, and can be interpreted as a damping factor that attenuates possible oscillations during the computation. It is worth mentioning that the penalty parameter τ significantly impacts the convergence rate of Algorithm D.2 and should be carefully selected. If it is too small, the damping effect of the penalty term is limited, and the oscillation may still take place; if it is too large, the increment of x in each step is very small, and Algorithm D.2 may suffer from a slow convergence rate. The optimal value of τ is problem-dependent. There is not a universal way to determine its best value.

Recently, single-loop distributed algorithms for monotone Nash games are proposed in [193], which authors believe to be promising in practical applications. In these two schemes, the regularization parameter is updated at once after each iteration is completed, rather than when the regularized problem is approximately solved, and players can select their parameter independently.

D.1.4 Nash Equilibrium of Matrix Games

As explained before, not all Nash games have an equilibrium, especially when the strategy set and the payoff function are non-convex or discrete. To widen the equilibrium notion and reveal deeper insights on the behaviors of players in such instances, it is instructive to revisit some simple games, called the matrix game, which is the primary research object of game theorists.

The bimatrix game refers to a matrix game involving two players P1 and P2. The numbers of possible strategies of P1 and P2 are m and n, respectively. $A = \{a_{ij}\} \in \mathbb{M}^{m \times n}$ is the payoff matrix of P1: when P1 chooses strategy i and P2 selects strategy j, the payoff of P1 is a_{ij}. The payoff matrix $B \in \mathbb{M}^{m \times n}$ of P2 can be defined in the same way. In a matrix game, each player is interested to determine a probability distribution of his actions, such that his expected payoff is minimized. Let x_i (y_j) be the probability that P1 (P2) will use strategy i (j), vectors $x = [x_1, \cdots, x_m]^T$ and $y = [y_1, \cdots, y_n]^T$ are called mixed strategies, clearly,

$$
\begin{aligned}
&x \geq 0, \ \sum_{i=1}^{m} x_i = 1 \quad \text{or} \quad x \in \Delta_m \\
&y \geq 0, \ \sum_{j=1}^{n} y_j = 1 \quad \text{or} \quad y \in \Delta_n
\end{aligned}
\tag{D.8}
$$

where Δ_m and Δ_n are simplex slices in $\mathbb{R}^m$ and $\mathbb{R}^n$.

1. Two-Person Zero-Sum Games

The zero-sum game represents a totally competitive situation: P1's gain is P2's loss, so the sum of their payoff matrices is $A + B = 0$, as its name suggests. Such type of game has been well studied in vast literature since von Neumann found the famous Minimax Theorem in 1928. The game is revisited from a mathematical programming perspective in [194]. The proposed linear programming method is especially powerful for instances with a high-dimensional payoff matrix. Next, we briefly introduce this method.

Let us begin with a payoff matrix $A = \{a_{ij}\}$, $a_{ij} > 0, \forall i, j$ with strictly positive entries (otherwise, we can add a constant to every entry, such that the smallest entry becomes positive, and the equilibrium strategy remains the same). The expected payoff of P1 is given by

$$
V_A = \sum_{i=1}^{m} \sum_{j=1}^{n} x_i a_{ij} y_j = x^T A y \tag{D.9}
$$

which must be positive because of the element-wise positivity assumption on A.

Since $B = -A$ and minimizing $-x^T Ay$ is equivalent to maximizing $x^T Ay$, the two-person zero-sum game has a min-max form as

$$\min_{x \in \Delta_m} \max_{y \in \Delta_n} \; x^T Ay \tag{D.10}$$

or

$$\max_{y \in \Delta_n} \min_{x \in \Delta_m} \; x^T Ay \tag{D.11}$$

The solution to the two-person zero-sum matrix game (D.10) or (D.11) is called a mixed-strategy Nash equilibrium, or the saddle point of a min-max problem. It satisfies

$$(x^*)^T Ay^* \le x^T Ay^*, \; \forall x \in \Delta_m$$

$$(x^*)^T Ay^* \ge y^T Ax^*, \; \forall y \in \Delta_n$$

To solve this game, consider (D.10) in the following format

$$\min_x \{v_1(x) \mid x \in \Delta_m\} \tag{D.12}$$

where $v_1(x)$ is the optimal value function of the problem faced by P2 with the fixed strategy x of P1

$$v_1(x) = \max_y \left\{ x^T Ay \,\middle|\, y \in \Delta_n \right\}$$

In view of the feasible region defined in (D.8), $v_1(x)$ is equal to the maximal element of vector $x^T A$, which is strictly positive, and the inequality

$$A^T x \le \mathbf{1}^n v_1(x)$$

holds. Furthermore, introducing a normalized vector $\bar{x} = x/v_1(x)$, we have

$$\bar{x} \ge 0, \quad A^T \bar{x} \le \mathbf{1}^n$$

$$v_1(x) = (\bar{x}^T \mathbf{1}^m)^{-1}$$

Taking these relations into account, problem (D.12) becomes

$$\begin{aligned} \min_{\bar{x}} \quad & (\bar{x}^T \mathbf{1}^m)^{-1} \\ \text{s.t.} \quad & A^T \bar{x} \le \mathbf{1}^n \\ & \bar{x} \ge 0 \end{aligned} \tag{D.13}$$

Because the objective is strictly positive and monotonic, the optimal solution of (D.13) keeps unchanged if we choose to maximize $\bar{x}^T \mathbf{1}^m$ under the same constraints, giving rise to the following LP

$$\begin{aligned} \max_{\bar{x}} \quad & \bar{x}^T \mathbf{1}^m \\ \text{s.t.} \quad & A^T \bar{x} \le \mathbf{1}^n \\ & \bar{x} \ge 0 \end{aligned} \tag{D.14}$$

Let $\bar{x}^*$ and $\bar{v}_1^*$ be the optimal solution and optimal value of LP (D.14). According to the analysis of variable transformation, the optimal expected payoff v_1^* and the optimal mixed strategy x^* of P1 in this game are given by

$$v_1^* = 1/\bar{v}_1^*, \quad x^* = \bar{x}^*/\bar{v}_1^* \tag{D.15}$$

Consider (D.11) in the same way, we obtain the following LP for P2:

$$\begin{aligned} \min_{\bar{y}} \quad & \bar{y}^T \mathbf{1}^n \\ \text{s.t.} \quad & A\bar{y} \ge \mathbf{1}^m \\ & \bar{y} \ge 0 \end{aligned} \tag{D.16}$$

Denote by $\bar{y}^*$ and $\bar{v}_2^*$ the optimal solution and optimal value of LP (D.16), and then the optimal expected payoff v_2^* and the optimal mixed strategy y^* of P2 in this game can be posed as

$$v_2^* = 1/\bar{v}_2^*, \quad y^* = \bar{y}^*/\bar{v}_2^* \tag{D.17}$$

In summary, the mixed-strategy Nash equilibrium of two-person zero-sum matrix game (D.10) is (x^*, y^*), and the payoff of P1 is v_1^*. Interestingly, we notice that problems (D.14) and (D.16) constitute a pair of dual LPs, implying that their optimal values are equal, and the optimal solution y^* in (D.17) also solves the inner LP of (D.10). This observation leads to two important conclusions:

1. The Nash equilibrium of a two-person zero-sum matrix game, or the saddle point, can be computed by solving a pair of dual LPs. In fact, if one player's strategy, say x^*, have been obtained from (D.14) and (D.15), the rival's strategy can be retrieved by solving (D.10) with $x = x^*$.
2. The decision sequence of a two-person zero-sum game is interchangeable without influencing the saddle point.

2. General Bimatrix Games

In more general two-person matrix games, the sum of payoff matrices is not equal to zero, and each player wishes to minimize its own expected payoff taking the other player's strategy as given. In the setting of mixed strategies, players are selecting the probability distribution among available strategies rather than a single action (the pure strategy), and the respective optimization problems are as follows:

$$\begin{aligned} \min_{x} \ & x^T A y \\ \text{s.t.} \ & x^T \mathbf{1}^m = 1 : \lambda \\ & x \geq 0 \end{aligned} \tag{D.18}$$

$$\begin{aligned} \min_{y} \ & x^T B y \\ \text{s.t.} \ & y^T \mathbf{1}^n = 1 : \gamma \\ & y \geq 0 \end{aligned} \tag{D.19}$$

The pair of probability distributions (x^*, y^*) is called a mixed-strategy Nash equilibrium if

$$\begin{aligned} (x^*)^T A y^* &\leq x^T A y^*, \ \forall x \in \Delta_m \\ (x^*)^T B y^* &\leq (x^*)^T B y, \ \forall y \in \Delta_n \end{aligned}$$

Unlike the zero-sum case, there is not an equivalent LP that can extract the Nash equilibrium. Performing the KKT system approach, we write out the KKT condition for (D.18)

$$\begin{aligned} Ay - \lambda \mathbf{1}^m - \mu = 0 \\ 0 \leq \mu \perp x \geq 0 \\ x^T \mathbf{1}^m = 1 \end{aligned}$$

where μ is the dual variable associated with the non-negative constraint, and can be eliminated from the first equality. Concentrating the KKT conditions of LPs (D.18) and (D.19) gives

$$\begin{aligned} 0 \leq Ay - \lambda \mathbf{1}^m \perp x \geq 0, \ x^T \mathbf{1}^m = 1 \\ 0 \leq B^T x - \gamma \mathbf{1}^n \perp y \geq 0, \ y^T \mathbf{1}^n = 1 \end{aligned} \tag{D.20}$$

Complementarity condition (D.20) can be solved by setting $\lambda = \gamma = 1$ and omitting equality constraints, and recovering them at a later normalization step, i.e., we first solve

$$\begin{aligned} 0 \leq Ay - \mathbf{1}^m \perp x \geq 0 \\ 0 \leq B^T x - \mathbf{1}^n \perp y \geq 0 \end{aligned} \tag{D.21}$$

Suppose that the solution is $(\bar{x}, \bar{y})$, then the Nash equilibrium is

$$\begin{aligned} x^* &= \bar{x}/\bar{x}^T \mathbf{1}^m \\ y^* &= \bar{y}/\bar{y}^T \mathbf{1}^n \end{aligned} \tag{D.22}$$

and the corresponding multipliers are derived from (D.20) as

$$\begin{aligned} \lambda^* &= (x^*)^T A y^* \\ \gamma^* &= (x^*)^T B y^* \end{aligned} \tag{D.23}$$

On the other hand, if (x^*, y^*) is a Nash equilibrium and solves (D.20) with multipliers (λ^*, γ^*), we can observe that $(x^*/\gamma^*, y^*/\lambda^*)$ solves (D.21), therefore

$$\begin{aligned} \bar{x} &= \frac{x^*}{(x^*)^T B y^*} \\ \bar{y} &= \frac{y^*}{(x^*)^T A y^*} \end{aligned} \tag{D.24}$$

Now we can see that identifying the mixed-strategy Nash equilibrium of a bimatrix game entails solving KKT system (D.20) or (D.21), which is called a linear complementarity problem (LCP). A classical algorithm for LCP is the Lemke's method [195, 196]. Another systematic way to solve an LCP is to reformulate it as an MILP using the method described in Appendix B.3.5. Nonetheless, there are more tailored MILP models for LCPs, which will be detailed in Sect. D.4.2.

Unlike the pure-strategy Nash equilibrium, whose existence relies on some assumptions on convexity, the mixed-strategy Nash equilibrium for matrix games, which is the discrete probability distribution among available actions, always exists [184]. If a game with two players has no pure-strategy Nash equilibrium, and each player can choose actions from a finite strategy set, we can then calculate the payoff matrices as well as the mixed-strategy Nash equilibrium, which informs the likelihood that the player will adopt each corresponding pure strategy.

D.1.5 Potential Games

Despite that a direct certification of the existence and uniqueness of a pure-strategy Nash equilibrium for a general game model is non-trivial, when the game possesses some special structures, such a certification becomes axiomatic. One of these guarantees is the existence of an exact potential function, and the associated problem is known as the potential game [197]. Four types of potential games are listed in [197], categorized by the type of the potential function. Other extensions of the potential game have been studied as well. For a complete introduction, we recommend [198].

Definition D.2 (Exact Potential Game) A game is an exact potential game if there is a potential function $U(x)$ such that:

$$\begin{gathered} f_i(x_i, x_{-i}) - f_i(y_i, x_{-i}) = U(x_i, x_{-i}) - U(y_i, x_{-i}) \\ \forall x_i, y_i \in X_i, \ \forall x_{-i} \in X_{-i}, \ i = 1, \cdots, n \end{gathered} \tag{D.25}$$

In an exact potential game, the change in the utility/payoff of any single player due to the unilateral strategy deviation leads to the same amount of change in the potential function. Among various variations of potential games which are defined by relaxing the strict equality (D.25), the exact potential game is the most fundamental one and has attracted the majority of research interests. Throughout this section, the term potential game means the exact one without particular mention.

The condition for a game being a potential game and the method for constructing the potential function are given in the following proposition.

Proposition D.3 ([197]) *Suppose the payoff functions f_i, $i = 1, \cdots, n$ in a game are twice continuously differentiable, then a potential function exists if and only if*

$$\frac{\partial^2 f_i}{\partial x_i \partial x_j} = \frac{\partial^2 f_j}{\partial x_i \partial x_j}, \quad \forall i, j = 1, \cdots, n \tag{D.26}$$

and the potential function can be constructed as

$$U(v) - U(z) = \sum_{i=1}^{n} \int_0^1 (x_i'(t))^T \frac{\partial f_i}{\partial x_i}(x(t)) dt \tag{D.27}$$

where $x(t) : [0, 1] \to X$ is a continuously differentiable path in X connecting strategy profile v and a fixed strategy profile z, such that $x(0) = z$, $x(1) = v$.

To obtain (D.27), first, a direct consequence of (D.25) is

$$\frac{\partial f_i}{\partial x_i} = \frac{\partial U}{\partial x_i}, \quad i = 1, \cdots, n \tag{D.28}$$

For any smooth curve $C(t) : [0, 1] \to X$ and any function U with a continuous gradient ∇U, the gradient theorem in calculus tells us

$$U(C_{end}) - U(C_{start}) = \int_C \nabla U(s) ds$$

where vector s represents points along the integral trajectory C parameterized in a scalar variable. Introducing $s = x(t)$: when $t = 0$, $s = C_{start} = z$; when $t = 1$, $s = C_{end} = v$. By the chain rule, $ds = x'(t)dt$, and hence we get

$$U(v) - U(z) = \int_0^1 (x'(t))^T \nabla U(x(t)) dt$$
$$= \sum_{i=1}^{n} \int_0^1 (x_i'(t))^T \frac{\partial U}{\partial x_i}(x(t)) dt$$

Then, if U is a potential function, substituting (D.28) into the above equation gives Eq. (D.27).

In summary, for a standard NEP with continuous payoff functions, we can check whether it is a potential game, and further construct its potential function, if the right-hand side of (D.27) has a closed-form expression. Nevertheless, in some particular cases, the potential function can be observed without calculating an integral.

1. The payoff functions of the game can be decomposed as

$$f_i(x_i, x_{-i}) = p_i(x_i) + Q(x), \ \forall i = 1, \cdots, n$$

   where the first term only depends on x_i, and the second term that couples all players' strategies and appears in every utility function is identical. In such circumstance, the potential function is instantly posed as

$$U(x) = Q(x) + \sum_{i=1}^{n} p_i(x_i)$$

   which can be verified through its definition in (D.25).
2. The payoff functions of the game can be decomposed as

$$f_i(x_i, x_{-i}) = p_i(x_{-i}) + Q(x), \ \forall i = 1, \cdots, n$$

   where the first term only depends on the joint actions of opponents x_{-i}, and the second term is common and identical to all players. In such circumstance, the potential function is $Q(x)$. This is easy to understand because x_{-i} is constant in the decision-making problem of player i and thus the first term $p_i(x_{-i})$ can be omitted from the objective function.
3. The payoff function of each player has a form of

$$f(x_i, x_{-i}) = \left(a + b \sum_{j=1}^{n} x_j \right) x_i + c_i(x_i), \ i = 1, \cdots, n$$

Obviously,

$$\frac{\partial^2 f_i}{\partial x_i \partial x_j} = \frac{\partial^2 f_j}{\partial x_i \partial x_j} = b$$

therefore, a potential function exists and is given by

$$U(x) = a \sum_{i=1}^{n} x_i + \frac{b}{2} \sum_{i=1}^{n} \sum_{j \neq i} x_i x_j + b \sum_{i=1}^{n} x_i^2 + \sum_{i=1}^{n} c_i(x_i)$$

The potential function provides a convenient way to analyze the Nash equilibria of potential games, since the function coincides with incentives of all players.

Proposition D.4 ([198]) *If game G_1 is a potential game with potential function $U(x)$; G_2 is another game with the same number of players and their payoff functions are $F_1(x_1, x_{-1}) = \cdots = F_n(x_n, x_{-n}) = U(x)$. Then G_1 and G_2 have the same set of Nash equilibria.*

This is easy to understand because an equilibrium of G_1 satisfies

$$f_i(x_i^*, x_{-i}^*) \le f_i(x_i, x_{-i}^*), \forall x_i \in X_i, \ i = 1, \cdots, n$$

By the definition of potential function (D.25), this gives

$$U(x_i^*, x_{-i}^*) \le U(x_i, x_{-i}^*), \forall x_i \in X_i, \ i = 1, \cdots, n \tag{D.29}$$

So any equilibrium of G_1 is an equilibrium of G_2. Similarly, the reverse holds, too.

In Proposition D.4, the identical interest game G_2 is actually an optimization problem. The potential function builds a bridge between an NEP and a mathematical programming problem. Let $X = X_1 \times \cdots \times X_n$, (D.29) can be written as

$$U(x^*) \le U(x), \forall x \in X \tag{D.30}$$

On this account, we have

Proposition D.5 ([197]) *Every minimizer of the potential function $U(x)$ in X is a (pure-strategy) Nash equilibrium of the potential game.*

Proposition D.5 is very useful. It reveals the fact that computing a Nash equilibrium of a potential game is equivalent to solving a traditional mathematical program. Meanwhile, the existence and uniqueness results of Nash equilibrium for potential games can be understood from the solution property of NLPs.

Proposition D.6 ([198]) *Every potential game with a continuous potential function $U(x)$ and a compact strategy space X has at least one (pure-strategy) Nash equilibrium. If $U(x)$ is strictly convex, then the Nash equilibrium is unique.*

Propositions D.5–D.6 make no reference on the convexity of individual payoff functions of players. Moreover, if the potential function $U(x)$ is non-convex and has multiple local minimums, then each local optimizer corresponds to a local Nash equilibrium where X_i in (D.29) is replaced with the intersection of X_i with a neighborhood region of x_i^*.

D.2 Generalized Nash Equilibrium Problem

In the above developments for standard NEPs, we have assumed that the strategy sets are decoupled: the available strategies of each player do not depend on other players' choices. However, there are indeed many practical cases where the strategy sets are interactive. For example, when players consume a common resource, the total consumption should not exceed the inventory quantity. The generalized Nash equilibrium problem (GNEP), invented in [199], relaxes the strategy independence assumption in classic NEPs and allows the feasible set of each player's actions to depend on the rivals' strategies. For a comprehensive review, we recommend [200].

D.2.1 Formulation and Optimality Condition

Denote by $X_i(x_{-i})$ the strategy set of player i when others select x_{-i}. In a GNEP, given the value of x_{-i}, each player i determines a strategy $x_i \in X_i(x_{-i})$ which minimizes a payoff function $f_i(x_i, x_{-i})$. In this regard, a GNEP with n players is the joint solution of n coupled optimization problems

$$\left.\begin{aligned} \min_{x_i} \quad & f_i(x_i, x_{-i}) \\ \text{s.t.} \quad & x_i \in X_i(x_{-i}) \end{aligned}\right\}, \quad i = 1, \cdots, n \tag{D.31}$$

In (D.31), correlation takes place not only in the objective function, but also in the constraints.

Definition D.3 A generalized Nash equilibrium (GNE), or the solution of a GNEP, is a feasible point x^* such that

$$f(x_i^*, x_{-i}^*) \le f(x_i, x_{-i}^*), \ \forall x_i \in X_i(x_{-i}) \tag{D.32}$$

holds for all players.

In its full generality, the GNEP is much more difficult than an NEP due to the variability of strategy sets. In this section, we restrict our attention to a particular class of GNEP: the so-called GNEP with shared convex constraints. In such a problem, the strategy sets can be expressed as

$$X_i(x_{-i}) = \{x_i \mid x_i \in Q_i,\ g(x_i, x_{-i}) \le 0\},\ i = 1, \cdots, n \tag{D.33}$$

where Q_i is a closed and convex set which involves only x_i; $g(x_i, x_{-i}) \le 0$ represents the shared constraints. They consist of a set of convex inequalities coupling all players' strategies and are identical in $X_i(x_{-i})$, $i = 1, \cdots, n$. Sometimes, Q_i and $g(x_i, x_{-i}) \le 0$ are also mentioned as local and global constraints, respectively.

In the absence of shared constraints, the GNEP reduces to a standard NEP. Define the feasible set of strategy profile $x = (x_1, \cdots, x_n)$ in a GNEP

$$X = \left\{ x \;\middle|\; x \in \prod_{i=1}^{n} Q_i,\ g(x) \le 0 \right\} \tag{D.34}$$

It is easy to see that $X_i(x_{-i})$ is a slice of X. A geometric interpretation of (D.33) is illustrated in Fig. D.1. It is seen that the choice of x_1 influences the feasible interval $X_2(x_1)$ of Player 2.

We make some assumptions on the smoothness and convexity for a GNEP with shared constraints.

Assumption D.2

(1) Strategy set Q_i of each player is nonempty, closed, and convex.
(2) Payoff function $f_i(x_i, x_{-i})$ of each player is twice continuously differentiable in x and convex in x_i for every fixed x_{-i}.
(3) Functions $g(x) = (g_1(x), \cdots, g_m(x))$ are differentiable and convex in x.

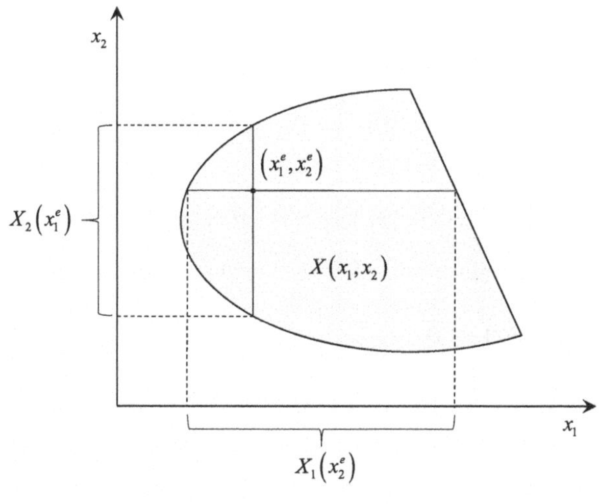

Fig. D.1 Relations of X and the individual strategy sets

In analogy with the NEP, concatenating the KKT optimality condition of each optimization problem in (D.31) gives us what is called the KKT condition of the GNEP. For notation brevity, we omit local constraints ($Q_i = \mathbb{R}^{n_i}$) and assume that $X_i(x_{-i})$ contains only global constraints. Write out the KKT condition of GNEP (D.31)

$$\left.\begin{aligned} \nabla_{x_i} f_i(x_i, x_{-i}) + \lambda_i^T \nabla_{x_i} g(x) = 0 \\ \lambda_i \ge 0,\ g(x) \le 0,\ \lambda_i^T g(x) = 0 \end{aligned}\right\} i = 1, \cdots, n \tag{D.35}$$

where λ_i is the Lagrange multiplier vector associated with the global constraints in the i-th player's problem.

Proposition D.7 ([200])

(1) Let $\bar{x} = (\bar{x}_1, \cdots, \bar{x}_n)$ be the equilibrium of a GNEP, then a multiplier vector $\bar{\lambda} = (\bar{\lambda}_1, \cdots, \bar{\lambda}_n)$ exists, such that the pair $(\bar{x}, \bar{\lambda})$ solves KKT system (D.35).
(2) If $(\bar{x}, \bar{\lambda})$ solves KKT system (D.35), and Assumption D.2 holds, then $\bar{x}$ is an equilibrium of GNEP (D.31) with shared convex constraints.

However, in contrast to an NEP, the solutions of an GNEP may be non-isolated and constitute a low dimensional manifold, because $g(x)$ is a common constraint shared by all, and the Jacobian of the KKT system may appear to be singular. A meticulous explanation is provided in [200]. We give a graphic interpretation for this phenomenon.

Consider a GNEP with two players:

$$\text{Player 1:} \quad \begin{cases} \max\limits_{x_1} \ x_1 \\ \quad \text{s.t. } x_1 \in X_1(x_2) \end{cases}$$

$$\text{Player 2:} \quad \begin{cases} \max\limits_{x_2} \ x_2 \\ \quad \text{s.t. } x_2 \in X_2(x_1) \end{cases}$$

where

$$X_1(x_2) = \{x_1 \mid x_1 \ge 0,\ g(x_1, x_2) \le 0\}$$
$$X_2(x_1) = \{x_2 \mid x_2 \ge 0,\ g(x_1, x_2) \le 0\}$$

and the global constraint set is

$$\left\{ x \,\middle|\, \begin{aligned} g_1 = 2x_1 + x_2 \le 0 \\ g_2 = x_1 + 2x_2 \le 0 \end{aligned} \right\}$$

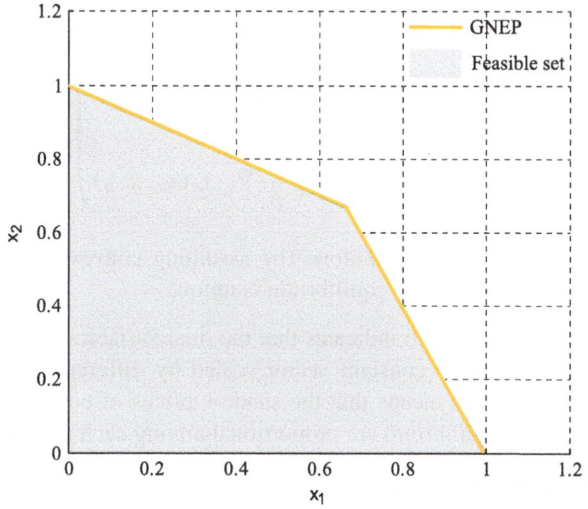

Fig. D.2 Illustration of the equilibria of a simple GNEP

The feasible set X of the strategy profile is plotted in Fig. D.2. It can be verified that any point on the line segments

$$L_1 = \left\{ (x_1, x_2) \;\middle|\; 0 \le x_1 \le \frac{2}{3},\ \frac{2}{3} \le x_2 \le 1,\ x_1 + 2x_2 = 2 \right\}$$

and

$$L_2 = \left\{ (x_1, x_2) \;\middle|\; \frac{2}{3} \le x_1 \le 1,\ 0 \le x_2 \le \frac{2}{3},\ 2x_1 + x_2 = 2 \right\}$$

is an equilibrium point that satisfies Definition D.3.

To refine a meaningful equilibrium from the infinitely many candidates, it is proposed to impose additional conditions on the Lagrange multipliers associated with shared constraints [201]. The outcome is called a restricted Nash equilibrium. Two special cases are discussed here.

1. Normalized Nash Equilibrium

The normalized Nash equilibrium is firstly introduced in [202]. It incorporates a cone constraint on the dual multipliers

$$\lambda_i = \beta_i \lambda_0,\ \beta_i > 0,\ i = 1, \cdots, n \tag{D.36}$$

where $\lambda_0 \in \mathbb{R}^m_+$. Solving KKT system (D.35) with constraint (D.36) gives an equilibrium solution. It is shown that for any given $\beta \in \mathbb{R}^n_{++}$, a normalized Nash equilibrium exists as long as the game is feasible. Moreover, if the mapping

$$F(\beta) : \mathbb{R}^n \to \mathbb{R}^n = \begin{pmatrix} \dfrac{1}{\beta_1} \nabla_{x_1} f_1(x_1, x_{-1}) \\ \vdots \\ \dfrac{1}{\beta_n} \nabla_{x_n} f_n(x_n, x_{-n}) \end{pmatrix}$$

parameterized in β is strictly monotone (by assuming convexity of payoff functions), then the normalized Nash equilibrium is unique.

The relation given in (D.36) indicates that the dual variables λ_i associated with the shared constraints are a constant vector scaled by different scalars. From an economic perspective, this means that the shadow prices of common resources at any normalized Nash equilibrium are proportional among each player.

2. Variational Equilibrium

Recall the variational inequality formulation for the NEP in Proposition D.1, a GNEP with shared convex constraints can be treated in the same way: Let $F(x) = (\nabla_{x_1} f_1(x), \cdots, \nabla_{x_n} f_n(x))$ be a mapping, and the feasible region X is defined in (D.34), then every solution of variational inequality problem VI(X, F) gives an equilibrium solution of the GNEP, which is called the variational equilibrium (VE).

However, unlike an NEP and its associated VI problem which have the same solutions, not all equilibria of the GNEP are preserved when it is passed to a corresponding VI problem; see [203, 204] for examples and further details. In fact, a solution x^* of a GNEP is a VE if and only if it solves KKT system (D.35) with the following constraints on the Lagrange dual multipliers [187, 200, 203]:

$$\lambda_1 = \cdots = \lambda_n = \lambda_0 \in \mathbb{R}^m_+ \tag{D.37}$$

implying that all players perceive the same shadow prices of common resource at a VE. The VI approach has two important implications. First, it allows us to analyze a GNEP using well-developed VI theory, such as conditions which could guarantee the existence and uniqueness of the equilibrium point; second, condition (D.37) gives an interesting economic interpretation of the VE, and inspires pricing-based distributed algorithms to compute an equilibrium solution, which will be discussed in the next section.

The concept of potential game for NEPs directly applies to GNEPs. If a GNEP with shared convex constraints possesses a potential function $U(x)$ which satisfies (D.25), an equilibrium can be retrieved from a mathematical program which minimizes the potential function over the feasible set X defined in (D.34). To reveal the connection of the optimal solution and the VE, we omit constraints in the local strategy sets Q_i, $i = 1, \cdots, n$ for notation simplicity, and write out the mathematical program as follows:

$$\begin{aligned} \min_{x} \quad & U(x) \\ \text{s.t.} \quad & g(x) \le 0 \end{aligned} \tag{D.38}$$

whose KKT optimality condition is given by

$$\begin{aligned} \nabla_x U(x) + \lambda^T \nabla_x g(x) = 0 \\ \lambda \ge 0,\ g(x) \le 0,\ \lambda^T g(x) = 0 \end{aligned} \tag{D.39}$$

The first equality can be decomposed into n sub-equations

$$\nabla_{x_i} U(x) + \lambda^T \nabla_{x_i} g(x) = 0,\ i = 1, \cdots, n \tag{D.40}$$

Recall (D.28), $\nabla_{x_i} U(x) = \nabla_{x_i} f_i(x_i, x_{-i})$, substituting it into (D.39) we have

$$\begin{aligned} \nabla_{x_i} f_i(x_i, x_{-i}) + \lambda^T \nabla_{x_i} g(x) = 0,\ i = 1, \cdots, n \\ \lambda \ge 0,\ g(x) \le 0,\ \lambda^T g(x) = 0 \end{aligned}$$

which is exactly KKT system (D.35) with identical shadow price constraint (D.37). In this regard, we can see

Proposition D.8 *Optimizing the potential function of a GNEP with shared convex constraints gives a variational equilibrium.*

Consider the example shown in Fig. D.2 again, $(2/3, 2/3)$ is the unique VE of the GNEP, which is plotted in Fig. D.3. The corresponding dual variables of global constraints $g_1 \le 0$ and $g_2 \le 0$ are $(1/3, 1/3)$.

D.2.2 Best-Response Algorithm

The presence of shared constraints wrecks the Cartesian structure of $\prod_{i=1}^{n} Q_i$ in a standard Nash game, and prevents a direct application of the best response methods presented in Appendix D.1.3 to solve an NEP. Moreover, even if an equilibrium can be found, it may depend on the initial point as well as the optimization sequence, because solutions of a GNEP are non-isolated. To illustrate this pitfall, take Fig. D.3 for an example. Suppose we pick up an arbitrary point $x_0 \in X$ as the initial value. If we first maximize x_1 (x_2), the point moves to B (A), and then in the second step, x_2 (x_1) does not change, because it is already an equilibrium solution in the sense of Definition D.3. In view of this, fixed-point iteration may give any outcome on the line segments connecting $(2/3, 2/3)$ and $(0, 1)/(1, 0)$, depending on the initiation.

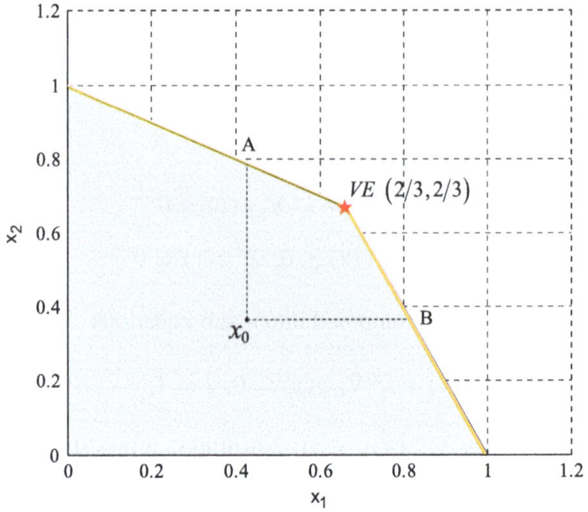

Fig. D.3 Illustration of the variational equilibrium

This section introduces the distributed algorithms proposed in [187] which identify a VE of GNEP (D.31) with shared convex constraints. Motivated by the Lagrange decomposition framework, we can rewrite problem (D.31) in a more convenient form. Consider finding a pair (x, λ), where x is the equilibrium of the following standard NEP $\mathcal{G}(\lambda)$ with a given vector λ of Lagrange multipliers

$$\mathcal{G}(\lambda): \quad \left\{ \begin{array}{ll} \min\limits_{x_i} & f_i(x_i, x_{-i}) + \lambda^T g(x) \\ \text{s.t.} & x_i \in Q_i \end{array} \right\}, \ i = 1, \cdots, n \tag{D.41}$$

and furthermore, a complementarity constraint

$$0 \leq \lambda \perp - g(x) \geq 0 \tag{D.42}$$

Problem (D.41)–(D.42) has a clear economic interpretation: suppose the shared constraints represent the availability of some common resources, vector λ can be viewed as the prices paid by players for consuming these resources. Actually, when a resource is adequate, the inequality constraint is not binding and the Lagrange dual multiplier is zero; the dual multiplier or shadow price is positive only if a resource becomes scarce, indicated by a binding inequality constraint. This relation has been imposed in constraint (D.42).

The KKT conditions of $\mathcal{G}(\lambda)$ (D.41) in conjunction with condition (D.42) turn out to be the VE condition of GNEP (D.31). In view of this connection, a VE can be found by solving (D.41)–(D.42) in a distributed manner based on previous algorithms developed for NEPs. Likewise, we discuss strongly convex cases and convex cases separately, due to their different convergence guarantees.

1. Algorithms for Strongly Convex Cases
Suppose that the game $\mathcal{G}(\lambda)$ in (D.41) is strongly convex and has a unique Nash equilibrium $x(\lambda)$ for any given $\lambda \geq 0$. This uniqueness condition allows defining the map

$$\Phi(\lambda) : \lambda \to -g(x(\lambda)) \tag{D.43}$$

which quantifies the negative violation of the shared constraints at $x(\lambda)$. Based on (D.41)–(D.42), the distributed algorithm is provided as follows.

Algorithm D.3 Projection with variable step sizes

1: Choose an initial price vector $\lambda^0 \geq 0$. The iteration index is $k = 0$.
2: Given λ^k, find the unique equilibrium $x(\lambda^k)$ of $\mathcal{G}(\lambda^k)$ using Algorithm D.1.
3: If $0 \leq \lambda^k \perp \Phi(\lambda^k) \geq 0$ is satisfied, terminate and report $x(\lambda^k)$ as the VE; otherwise, choose $\tau_k > 0$, and update the price vector according to

$$\lambda^{k+1} = \left[\lambda^k - \tau_k \Phi(\lambda^k)\right]^+$$

set $k \leftarrow k + 1$, and go to step 2.

Algorithm D.3 is a double-loop method. The range of parameter τ_n and convergence proof have been thoroughly discussed in [187] based on the monotonicity of the mapping $F + \nabla g(x)\lambda$, where $F = (\nabla_{x_i} f_i)_{i=1}^n$, and $\nabla g(x)$ is a matrix whose i-th column is equal to ∇g_i.

2. Algorithms for Convex Cases
Now we consider the case in which the VI associated with problem (D.41)–(D.42) is merely monotone (at least one problem in (D.41) is not strongly convex). In such circumstance, the convergence of Algorithm D.3 is no longer guaranteed. This is not only because Algorithm D.1 for the inner loop game $\mathcal{G}(\lambda)$ may not converge, but also because the outer loop has to be complicated. To circumvent this difficulty, we try to convexify the game using regularization terms as what has been done in Algorithm D.2. To this end, we have to explore an optimization reformulation for the complementarity constraint (D.42), which is given by

$$\lambda \in \arg\min_{\bar{\lambda}} \left\{ -\bar{\lambda}^T g(x) \,\middle|\, \bar{\lambda} \geq 0 \right\}$$

Then, consider the following ordinary NEP with $n+1$ players in which the last player controls the price vector λ:

$$\begin{aligned} &\min_{x_i} \left\{ f_i(x_i, x_{-i}) + \lambda^T g(x) \,\middle|\, x_i \in Q_i \right\}, \ i = 1, \cdots, n \\ &\min_{\lambda} \left\{ -\lambda^T g(x) \,\middle|\, \lambda \geq 0 \right\} \end{aligned} \tag{D.44}$$

where the last player solves an LP in variable λ parameterized in x. At the equilibrium, $g(x) \leq 0$ is implicitly satisfied. To see this, because Q_i is bounded, problems of the first n players must have a finite optimum for arbitrary λ. If $g(x) \nleq 0$, the last problem has an infinite optimum, imposing a large penalty on the constraint that is violated, and thus the first n players will alter their strategies accordingly. Whenever $g(x) \leq 0$ is met, the last LP must have a zero minimum, which satisfies (D.42). In summary, this extended game (D.44) has the same equilibria as problem (D.41)–(D.42). Since the strategy sets of (D.44) have a Cartesian structure, Algorithm D.2 can be applied to find an equilibrium.

The convergence of Algorithm D.4 is guaranteed under a sufficiently large τ. More quantitative discussions on parameter selection and convergence conditions can be found in [187]. In practice, the value of τ should be carefully chosen to achieve satisfactory computational performances.

In NEPs and GNEPs, players make simultaneous decisions. In real-life decision making problems, there are many situations in which players can move sequentially. In the rest of this chapter, we consider three kinds of bilevel games, in which the upper-level (lower-level) players are called leaders (followers), and leaders make decisions prior to follower's. The simplest one is the Stackelberg game, or the single-leader-single-follower game, or just the bilevel program; Stackelberg game can be generalized by incorporating multiple players in the upper and lower levels. Players at the same level make decisions simultaneously, whereas followers' actions are subject to leaders' movements, forming an NEP parameterized in the leaders'

Algorithm D.4 Regularization method for general convex cases

1: Given $\{\rho_n\}_{n=0}^{\infty}$, $\varepsilon > 0$, and $\tau > 0$, choose a feasible initial point $x^0 \in X$ and an initial price vector λ^0; the iteration index is $k = 0$.

2: Given $z^k = (x^k, \lambda^k)$, find a Nash equilibrium $z^{k+1} = (x^{k+1}, \lambda^{k+1})$ of the following regularized NEP using Algorithm D.1

$$\begin{aligned} &\min_{x_i} \left\{ f_i(x_i, x_{-i}) + \lambda^T g(x) + \tau \left\| x_i - x_i^k \right\|_2^2 \,\middle|\, x_i \in Q_i \right\}, \ i = 1, \cdots, n \\ &\min_{\lambda} \left\{ -\lambda^T g(x) + \tau \left\| \lambda - \lambda^k \right\|_2^2 \,\middle|\, \lambda \geq 0 \right\} \end{aligned}$$

3: If $\|z^{k+1} - z^k\|_2 \leq \varepsilon$, terminate and report x^{k+1} as the variational equilibrium; otherwise, update $k \leftarrow k+1$, $z^k \leftarrow (1-\rho_k) z^{k-1} + \rho_k z^k$, and go to step 2.

---

decisions. When there is only one leader, the problem is called a mathematical program with equilibrium constraints (MPEC); when there are multiple leaders, the problem is referred to as an equilibrium program with equilibrium constraints (EPEC). It is essentially a bilevel GNEP among the leaders.

D.3 Bilevel Programs

Bilevel program is a special mathematical program with another optimization problems nested in the constraints. The main problem is called the upper-level problem, and the decision maker is the leader; the one nested in constraints is called the lower-level problem, and the decision maker is the follower. In game theory, a bilevel program is usually referred to as the Stackelberg game, which arises in many economic and engineering design problems.

D.3.1 Bilevel Programs with a Convex Lower Level

1. Mathematic Model and Single-Level Equivalence

A bilevel program is the most basic instance of bilevel games. The leader moves first and chooses a decision x; then the follower selects its strategy y solving the lower-level problem parameterized in x

$$\begin{aligned} \min_y \quad & f(x, y) \\ \text{s.t.} \quad & g(x, y) \le 0 : \lambda \\ & h(x, y) = 0 : \mu \end{aligned} \tag{D.45}$$

where λ and μ following the colon are dual variables associated with inequality and equality constraints, respectively. We assume that problem (D.45) is convex and the KKT condition is necessary and sufficient for a global optimum

$$\text{Cons-KKT} = \left\{ (x, y, \lambda, \mu) \;\middle|\; \begin{array}{c} \nabla_y f(x, y) + \lambda^T \nabla_y g(x, y) + \mu^T \nabla_y h(x, y) = 0 \\ 0 \le \lambda \perp - g(x, y) \ge 0, \; h(x, y) = 0 \end{array} \right\} \tag{D.46}$$

The set of optimal solutions of problem (D.45) is denoted by $S(x)$. If (D.45) is strictly convex, the optimal solution is unique, and $S(x)$ reduces to a singleton.

When the leader minimizes its payoff function $F(x, y)$, the best response $y(x) \in S(x)$ is taken into account. The leader's problem is formally described as

$$\begin{aligned} \min_{x,\bar{y}} \quad & F(x, \bar{y}) \\ \text{s.t.} \quad & x \in X \\ & \bar{y} \in S(x) \end{aligned} \tag{D.47}$$

Notice that although $\bar{y}$ acts as a decision variable of the leader, it is actually controlled by the follower through the best response mapping $S(x)$. When the leader makes decisions, it will take the response from the follower into account. When $S(x)$ is a singleton, qualifier $\in$ reduces to $=$; otherwise, if $S(x)$ contains more than one elements, (D.47) assumes that the follower will choose the one which is preferred by the leader. Therefore, (D.47) is called an optimistic equivalence. On the contrary, the pessimistic equivalence assumes that the follower will choose the one which is unfavorable for the leader, which is more difficult to solve. As for the optimistic case, replacing $\bar{y} \in S(x)$ with KKT condition (D.46) leads to the NLP formulation of the bilevel program, or more exactly, a mathematical program with complementarity constraints (MPCC)

$$\begin{aligned} \min_{x,\bar{y},\lambda,\mu} \quad & F(x, \bar{y}) \\ \text{s.t.} \quad & x \in X, \ (x, \bar{y}, \lambda, \mu) \in \text{Cons-KKT} \end{aligned} \tag{D.48}$$

Although the lower-level problem (D.45) is convex, the best reaction map of the follower characterized by Cons-KKT is non-convex, so a bilevel program is intrinsically non-convex and generally difficult to solve.

2. Why Bilevel Programs Are Difficult to Solve?
Two difficulties prevent an MPCC from being solved reliably and efficiently.

1. The feasible region of (D.47) is non-convex: even if objective functions and constraints of the leader and the follower are linear, the complementarity and slackness condition in (D.46) is still non-convex. An NLP solver only finds a local solution for non-convex problems, if succeeds, and global optimality can hardly be guaranteed.
2. Despite its non-convexity, the failure to meet ordinary constraint qualifications creates another barrier for solving an MPCC. NLP algorithms generally stop when a stationary point of the KKT conditions is found; however, due to the presence of the complementarity and slackness condition, the dual multipliers may not be well-defined because of the violation of standard constraint qualifications. Therefore, NLP solvers may fail to find a local optimum without particular treatment on the complementarity constraints. To see how constraint qualifications are violated, consider the following simplest linear complementarity constraint

$$x \ge 0, \ y \ge 0, \ x^T y = 0, \ x \in \mathbb{R}^5, \ y \in \mathbb{R}^5$$

The Jacobian matrix of the active constraints at point $(\bar{x}, \bar{y})$ is

$$J = \begin{bmatrix} e_{\bar{x}} & 0 \\ 0 & e_{\bar{y}} \\ \bar{y} & \bar{x} \end{bmatrix}$$

where $e_{\bar{x}}$ and $e_{\bar{y}}$ are zero-one matrices corresponding to the active constraints $x_i = 0, i \in I$, $y_j = 0, j \in J$, where $I \bigcup J = \{1, 2, 3, 4, 5\}$, and $I \bigcap J$ is not necessarily empty. Suppose that $I = \{1, 2, 4\}$ and $J = \{3, 5\}$, then

$$J = \begin{bmatrix} 1 & 0 & 0 & 0 & 0 & 0 & 0 & 0 & 0 & 0 \\ 0 & 1 & 0 & 0 & 0 & 0 & 0 & 0 & 0 & 0 \\ 0 & 0 & 0 & 1 & 0 & 0 & 0 & 0 & 0 & 0 \\ 0 & 0 & 0 & 0 & 0 & 0 & 0 & 1 & 0 & 0 \\ 0 & 0 & 0 & 0 & 0 & 0 & 0 & 0 & 0 & 1 \\ \bar{y}_1 & \bar{y}_2 & \bar{y}_3 & \bar{y}_4 & \bar{y}_5 & \bar{x}_1 & \bar{x}_2 & \bar{x}_3 & \bar{x}_4 & \bar{x}_5 \end{bmatrix}$$

Since $\bar{x}_1 = \bar{x}_2 = \bar{x}_4 = 0$ and $\bar{y}_3 = \bar{y}_5 = 0$, it is apparent that the row vectors of J are linearly dependent at point $(\bar{x}, \bar{y})$. The same applies to any $(\bar{x}, \bar{y})$ regardless of the indices I and J of active constraints, because whenever $y_j > 0$, complementarity will enforce $x_i = 0, i = j$, creating a binding inequality in $x \geq 0$ and a row in matrix J whose i-th element is 1; whenever $x_i > 0$, complementarity will enforce $y_j = 0, j = i$, creating a binding inequality in $y \geq 0$ and a row in matrix J whose $(i + 5)$-th element is 1. Therefore, the last row of J can be represented by a linear combination of the other rows.

The above discussion and conclusion on linear complementarity constraints also apply to the nonlinear case, because the Jacobian matrix J has the same structure. In this regard, the above difficulty is an intrinsic phenomenon in MPCCs.

Proposition D.9 *Complementarity and slackness conditions violate the linear independent constraint qualification at any feasible solution.*

From a geometric perspective, the feasible region of complementarity constraints consists of slices like $x_i = 0$, $y_j = 0$; there is no strictly feasible point and the Slater's condition does not hold. In conclusion, general purpose NLP solvers are not numerically reliable for solving MPCCs, although they were once used to carry out such tasks.

3. Methods for Solving MPCCs

In view of the limitations of standard NLP algorithms, new constraint qualifications are proposed to define stationary solutions so as to solve MPCCs through conventional NLP methods, such as the Bouligand-, Clarke-, Mordukhovich-, weakly-, and Strongly-stationary constraint qualifications. See [205–208] for further information.

Through some proper transformation, MPCCs can be solved via standard NLP algorithms as well. Several approaches are available for this task.

a. Regularization Method [209–211]
In this approach, the non-negativity and complementarity requirements

$$x \geq 0, \ y \geq 0, \ xy = 0 \tag{D.49}$$

are approximated by

$$x \geq 0, \ y \geq 0, \ xy \leq \varepsilon \tag{D.50}$$

Please note that $xy \geq 0$ is a natural result of non-negativity requirements on x and y. When $\varepsilon = 0$, (D.50) is equivalent to (D.49); when $\varepsilon > 0$, (D.50) defines a larger feasible region than (D.49), so this approach is sometimes called a relaxation method. The smaller ε is, the closer any feasible point (x, y) is to achieve complementarity. If x and y are vectors with non-negative elements, $x^T y = 0$ is the same as $x_i y_i = 0$, $i = 1, 2, \cdots$. The same procedure can be applied if x and y are replaced by nonlinear functions.

Since Slater's condition holds for the feasible set defined by (D.50) with $\epsilon > 0$, NLP solvers can be used to solve related optimization problem. In a regularization procedure for solving an MPCC, the relaxation (D.50) is applied with gradually decreased value of ε for implementation issues. If the initial value of ε is too small, the solver may be numerically unstable and fail to find a feasible solution.

b. Penalization Method [212–214]
In this approach, the complementarity condition $xy = 0$ is removed from the set of constraints; instead, an associated penalty term xy/ε is added to the objective function to create an extra cost whenever complementarity is not satisfied. Since x and y are non-negative, as indicated by (D.49), the penalty term would never take a negative value. In this way, the feasible region becomes much simpler.

In a penalization procedure for solving an MPCC, a sequence of NLPs are solved iteratively with gradually decreased value of ε, and the violation of complementarity condition gradually approaches to 0 as iterations proceed. If ε is initiated too small, the penalty coefficient $1/\varepsilon$ is very large which may cause an ill-conditioned problem and numeric instability. One advantage of this iterative procedure is that the optimal solution in iteration k can be used as the initial guess in iteration $k + 1$, since the feasible region does not change, and the solution in every iteration is feasible in the next one. A downside of this approach is that the NLP solver generally identifies a local optimum. In consequence, a smaller ϵ may not necessarily lead to a solution that gets closer to the feasible region.

c. Smoothing Method [215, 216]
This approach employs the perturbed Fischer-Burmeister function

$$\phi(x, y, \varepsilon) = x + y - \sqrt{x^2 + y^2 + \varepsilon} \tag{D.51}$$

which is firstly introduced in [217] for LCPs, and shown particularly useful in SQP methods for solving MPCCs in [218]. Clearly, when $\varepsilon = 0$, the function ϕ reduces to the standard Fischer-Burmeister function

$$\phi(x, y, 0) = 0 \iff x \geq 0, \ y \geq 0, \ xy = 0 \tag{D.52}$$

$\phi(x, y, 0)$ is not smooth at the origin $(0, 0)$. When $\varepsilon > 0$, the function ϕ satisfies

$$\phi(x, y, \varepsilon) = 0 \iff x \geq 0, \ y \geq 0, \ xy = \varepsilon/2 \tag{D.53}$$

and is smooth in x and y.

In view of this, complementarity and slackness condition (D.49) can be replaced by $\phi(x, y, \varepsilon) = 0$ and further embedded in NLP models. When ε tends to 0, (D.49) is enforced approximately.

d. Sequential Quadratic Programming (SQP) [217–219]
SQP is a general purpose NLP method. In each iteration of SQP, the quadratic functions in complementarity constraints are approximated by a linear one, and the nonlinear objective function is replaced with their second-order Taylor series, constituting a quadratic program with linear constraints (maybe in conjunction with trust region bounds). At the optimal solution, nonlinear constraints are linearized, the objective function is approximated again, and then the SQP algorithm proceeds to the next iteration.

When applied to an MPCC, the SQP method is often capable of finding a local optimal solution, without a sequence of user-specified ε_k approaching to 0, probably because the SQP solver itself is endowed with some softening ability, e.g., when a quadratic program encounters numeric issues, the SQP solver SNOPT automatically relaxes some hard constraints and penalizes violations in the objective function.

The aforementioned classical methods are discussed in [220], and numeric experiences are reported in [221].

e. MINLP Methods
Due to the wide applications in various engineering disciplines, solution methods of MPCCs continue to be an active research area. Recall that the complementarity constraints in form of $g(x) \geq 0$, $h(x) \geq 0$, $g(x)h(x) = 0$ is equivalent to

$$0 \leq g(x) \leq Mz, \ 0 \leq h(x) \leq M(1 - z)$$

where z is a binary variable, M is a sufficiently large constant. Therefore, an MPCC can be converted to a mixed integer nonlinear program (MINLP). MINLP removes the numeric difficulty in MPCC; however, the computation complexity remains. If all functions in (D.48) are linear, or there are only a few complementarity constraints, the resulting MILP or MINLP model may be solved within reasonable time; otherwise, the branch-and-bound algorithm could offer upper and lower bounds on the optimal value.

f. Convex Relaxation/Approximation Methods
If all functions in MPCC (D.48) are linear, it is a non-convex QCQP in which non-convexity originates from the complementarity constraints. When the problem scale is large, the MILP method may be time-consuming. Inspired by the success of convex relaxation methods in non-convex QCQPs, there have been increasing interests for developing convex relaxation methods for MPCCs. An SDP relaxation method is proposed in [222], which is embedded in a branch-and-bound algorithm to solve the MPCC. For the MPCC derived from a bilevel polynomial program, it is proposed to solve a sequence of SDPs with increasing problem sizes, so as to solve the original problem globally [223, 224]. Convex relaxation methods have been applied to power market problems in [225, 226]. Numerical experiments show that the combination of MILP and SDP relaxation can greatly reduce the computation time. Nonetheless, please bear in mind that in the SDP relaxation model, the decision variable is a matrix with a dimension of $n \times n$, so solving the SDP model may still be a challenging task, although it is convex.

Recently, a DC programming approach is proposed in [227] to solve LPCC in the penalized version. In this approach, the quadratic penalty term is decomposed into the difference of two convex quadratic functions, and the concave part is then linearized. Computational performances reported in [227] are very promising.

D.3.2 Special Bilevel Programs

Although general bilevel programs are difficult, there are special cases which can be solved relatively easily. One of such classes of programs is the linear bilevel program, in which objective functions are linear and constraints are polyhedra. The linear max-min problem is a special case of the linear bilevel program, in which the leader and the follower have completely opposite targets. Furthermore, two special market models are studied.

1. Linear Bilevel Program
A linear bilevel program can be written as

$$\begin{aligned} &\max_x c^T x + d^T y(x) \\ &\text{s.t.} Cx \le d \\ &y(x) \in \arg\min_y f^T y \\ &\text{s.t.}\;\; By \le b - Ax \end{aligned} \tag{D.54}$$

In problem (D.54), the follower makes a decision y after the leader deploys its action x, which influences the feasible region of y. Meanwhile, the leader can predict the follower's optimal response $y(x)$, and choose a strategy that finally optimizes $c^T x + d^T y(x)$. Other matrices and vectors are constant coefficients.

Given the upper level decision x, the follower is facing an LP, whose KKT optimality condition is given by

$$
\begin{aligned}
& B^T u = f \\
& 0 \geq u \perp Ax + By - b \leq 0
\end{aligned}
$$

The last constraint is equivalent to the following linear constraints

$$
\begin{aligned}
-M(1-z) &\leq u \leq 0 \\
-Mz &\leq Ax + By - b \leq 0
\end{aligned}
$$

where z is a vector consisting of binary variables, and M is a large enough constant.

In problem (D.54), replacing follower's LP with its KKT condition gives rise to an MILP

$$
\begin{aligned}
& \max_{x,y,u,z} \ c^T x + d^T y \\
& \text{s.t. } Cx \leq d, \ B^T u = f \\
& - M(1-z) \leq u \leq 0 \\
& - Mz \leq Ax + By - b \leq 0
\end{aligned}
\tag{D.55}
$$

If the number of complementarity constraints is moderate, MILP (D.55) can be often solved efficiently, despite its NP-hard complexity in the worst-case. Since MILP solvers and computation hardware keep improving nowadays, it is always worthy of bearing this technique in mind. Please also be aware that the big-M parameter notably impacts the performance of solving MILP (D.55). A heuristic method to determine such a parameter in linear bilevel programs is proposed in [228]. This method firstly solves two LPs and generates a feasible solution of the equivalent MPCC; then solves a regularized version of the MPCC model using NLP solvers and identifies a local optimal solution near the obtained feasible point; finally, the big-M parameter and the binary variables are initiated according to the local optimal solution. In this way, no manually-supplied parameter is needed, and the MILP model is properly strengthened.

Another optimality certification of follower's LP is the following primal-dual optimality condition

$$
\begin{aligned}
& B^T u = f, \ u \leq 0, \ Ax + By \leq b \\
& u^T (b - Ax) = f^T y
\end{aligned}
$$

The first line summarizes feasible regions of the primal and dual variables. The last equation enforces equal values on the optimums of the primal and the dual problems, which is known as the strong duality condition.

Replacing follower's LP with the primal-dual optimality condition gives an NLP:

$$\begin{aligned} \max_{x,y,u} \quad & c^T x + d^T y \\ \text{s.t.} \quad & Cx \le d, \ Ax + By \le b \\ & u \le 0, \ B^T u = f \\ & u^T(b - Ax) = f^T y \end{aligned} \tag{D.56}$$

The following discussion is divided into two categories based on the type of variable x.

(a) x is **continuous**. In such a general situation, there is no effective way to solve problem (D.56), due to the last bilinear equality. Notice the fact that $f^T y \ge u^T(b - Ax)$ always holds on the feasible region because of the weak duality, the last constraint can be relaxed and penalized in the objective function, resulting in a bilinear program over a polyhedron [229–231]

$$\begin{aligned} \max_{x,y,u} \quad & c^T x + d^T y - \sigma[f^T y - u^T(b - Ax)] \\ \text{s.t.} \quad & Cx \le d, \ Ax + By \le b \\ & u \le 0, \ B^T u = f \end{aligned} \tag{D.57}$$

where $\sigma > 0$ is a penalty parameter. In problem (D.57), the constraints on u and (x, y) are decoupled, so this problem can be solved by Algorithm C.1 (mountain climbing) in Appendix C.2.3, if global optimality is not mandatory.

In some problems, the upper-level decision influences the lower-level cost function, and has no impact on the feasible region in the lower level. For example, the tax rate design or a retail market pricing belongs to such category. The same procedure can be performed to solve this kind of bilevel problem. We recommend the MILP model, because in the penalized model, both $f^T y$ and $u^T Ax$ are non-convex. A tailored retail market model will be introduced later.

(b) x is **binary**. In such circumstance, the bilinear term $u^T Ax = \sum_{ij} A_{ij} u_i x_j$ can be linearized by replacing $u_i x_j$ with a new continuous variable v_{ij} together with auxiliary linear inequalities enforcing $v_{ij} = u_i x_j$. In this way, the last inequality translates into

$$\begin{gathered} u^T b - \sum_{ij} A_{ij} v_{ij} = f^T y \\ u_i^l x_j \le v_{ij} \le 0, \ u_i^l(1 - x_j) \le u_i - v_{ij} \le 0, \ \forall i, j \end{gathered}$$

where u_i^l is a proper bound that does not discard the original optimal solution. As we can see, a bilevel linear program with binary upper-level variables is not necessarily harder than all continuous instances. This formulation is very useful to model interdiction problems in which x mimics attack strategy.

(c) x can be **discretized**. Even if x is continuous, we can approximate it via binary expansion

$$x_i = x_i^l + \Delta_i \sum_{k=0}^{K} 2^k z_{ik}, \quad z_{ik} \in \{0, 1\}$$

where x_i^l (x_i^m) is the lower (upper) bound of x_i, and $\Delta_i = (x_i^m - x_i^l)/2^{K+1}$ is the step size. With this transformation, the bilinear term $u^T Ax$ becomes

$$\sum_{ij} A_{ij} u_i x_i^l + \sum_{ij} A_{ij} u_i \Delta_j \sum_{k=0}^{K} 2^k z_{jk}$$

The first term is linear, and $u_i z_{jk}$ in the second term can be linearized in a similar way. However, this entails introducing continuous variable with respect to indices i, k, and k. A low-complexity linearization method is suggested in [232]. It re-orders the summations in the second term as

$$\sum_j \sum_{k=0}^{K} \Delta_j 2^k z_{jk} \sum_i A_{ij} u_i$$

which can be linearized through defining an auxiliary continuous variable $v_{jk} = z_{jk} \sum_i A_{ij} u_i$ and stipulating

$$-M z_{jk} \le v_{jk} \le M z_{jk}, \quad -M(1 - z_{jk}) \le \sum_i A_{ij} u_i - v_{jk} \le M(1 - z_{jk})$$

where M is a large enough constant.

The core idea behind this trick is to treat $u^T A$ as a whole vector which has the same dimension as x, because for bilinear form $x^T v = \sum_i x_i v_i$, the dimension of summation is one, while for $x^T Q v = \sum_{ij} Q_{ij} x_i v_j$, the dimension of summation is two. This observation inspires us to conform vector dimensions while deploying such linearization.

2. Linear Max-Min Problem

A linear max-min problem is a special case of the linear bilevel program, which can be written as

$$\begin{aligned}
&\max_x \ c^T x + d^T y(x) \\
&\text{s.t. } x \in X \\
&y(x) \in \arg\min_y \ c^T x + d^T y \\
&\text{s.t. } y \in Y, \ By \le b - Ax
\end{aligned} \tag{D.58}$$

In problem (D.58), the follower seeks an objective that is completely opposite to that of the leader. This kind of problem frequently arises in robust optimization and has been discussed in Appendix C.2.3 from the computational perspective. Here we revisit it from a game theoretical point of view.

Problem (D.58) can be expressed as a two-person zero-sum game

$$\max_{x \in X} \min_{y \in Y} \left\{ c^T x + d^T y \,\middle|\, Ax + By \le b \right\} \tag{D.59}$$

However, the coupled constraints make it different from a saddle point problem in the sense of a Nash game or a matrix game. Indeed, it is a Stackelberg game. Let us investigate the interchangeability of the max and min operators (decision sequence). We have already shown in Appendix D.1.4 that swapping the order of max and min operators in a two-person zero-sum matrix game does not influence the equilibrium. However, this is not the case of (D.59) [233], because

$$\begin{aligned}
&\max_{x \in X} \min_{y \in Y} \{c^T x + d^T y \mid Ax + By \le b\} \\
&= \max_{x \in X} \left\{ c^T x + \min_{y \in Y} \{d^T y \mid By \le b - Ax\} \right\} \\
&\ge \max_{x \in X} \left\{ c^T x + \min_{y \in Y} \ d^T y \right\} \\
&= \max_{x \in X} \ c^T x + \min_{y \in Y} \ d^T y \\
&= \min_{y \in Y} \left\{ d^T y + \max_{x \in X} \ c^T x \right\} \\
&\ge \min_{y \in Y} \left\{ d^T y + \max_{x \in X} \{c^T x \mid Ax \le b - By\} \right\} \\
&= \min_{y \in Y} \max_{x \in X} \{c^T x + d^T y \mid Ax + By \le b\}
\end{aligned}$$

In fact, strict inequality usually holds in the third and sixth line. This result implies that owing to the presence of strategy coupling, the leader rests in a superior status, which is different from the Nash game in which players possess the same positions.

To solve linear max-min problem (D.59), there is no doubt that the aforementioned MILP transformation for general linear bilevel programs gives a possible mean for this task. Nevertheless, the special structure of (D.59) allows several alternatives which are more dedicated and effective. To this end, we will transform it into an equivalent optimization problem using LP duality theory. For the ease of notation, we merge polytope Y into the coupled constraint, and the dual of lower-level LP in (D.58) (or the inner LP in (D.59)) reads

$$\max_{u} \ \{u^T(b - Ax) \mid u \in U\}$$

where $U = \{u \mid B^T u = d,\ u \leq 0\}$ is the feasible region of dual variable u. As strong duality always holds for LPs, we have $d^T y = u^T(b - Ax)$. Substituting it into (D.58) we obtain

$$\begin{aligned} \max \ & c^T x + u^T b - u^T Ax \\ \text{s.t.} \ & x \in X,\ u \in U \end{aligned} \tag{D.60}$$

Problem (D.60) is a bilinear program due to the product term $u^T Ax$ in variables u and x. Several methods for solving such a problem locally or globally have been set forth in Appendix C.2.3, as a fundamental methodology in robust optimization. Although variable y of the follower does not appear in (D.60), it can be easily recovered from the lower level of (D.58) with the obtained leader's strategy x.

3. A Retail Market Problem

In a retail market, a retailer releases the prices of some goods; according to the retail prices, the customer decides on the optimal purchasing strategy subject to the demands on each goods as well as production constraints; finally, the retailer produces or trades with a higher level market to manage the inventory, and delivers the goods to customers. This retail market can be modeled through a bilevel program. In the upper level

$$\max_{x,z} \ x^T D_C y(x) - p^T D_M z \tag{D.61a}$$

$$\text{s.t.} \ Ax \leq a \tag{D.61b}$$

$$B_1 y(x) + B_2 z \leq b \tag{D.61c}$$

(D.61a)–(D.61c) form retailer's problem, where vector x denotes the prices of goods released by the retailor; vector $y(x)$ stands for the amounts of goods purchased by the customer, which is determined from an optimal production planning problem; p is the production cost or the price in the higher level market; z represents the production/purchase strategy of the retailer. Other matrices and vectors are constant coefficients. The first term in objective function (D.61a) is the income paid by the customer, and the second term is the payoff of the retailer. The objective function is the total profit to be maximized. Because there is no competition and the retailer has full market power, to avoid unfair retail prices, we assume that both sides have reached certain agreements on the pricing policy, which is modeled through constraint (D.61b). It includes simple lower and upper bounds as well as other bilateral contract, such as the restriction on the average price over a certain period or the price correlation among multiple goods. The inventory dynamics and other technique constraints are depicted by constraint (D.61c).

Given the retail prices, customers solve the optimal production planning problem in the lower level

$$\begin{aligned} \min_{y} \quad & x^T D_C y \\ \text{s.t.} \quad & F y \geq f \end{aligned} \tag{D.62}$$

and determine the optimal purchasing strategy. The objective function in (D.62) is the total cost of customers, where the price vector x is constant coefficient; constraints capture the demands and all other technique requirements in the production process.

Bilevel program (D.61)–(D.62) are not linear, although (D.62) is indeed an LP, because of the bilinear term $x^T D_C y$ in (D.61a), where both x and y are variables (the retailer controls y indirectly through prices). The KKT condition of LP (D.62) reads

$$\begin{aligned} D_C^T x = F^T u \\ 0 \leq u \perp F y - f \geq 0 \end{aligned}$$

where u is the dual variable. The complementarity constraints can be linearized via binary variables, which has been clarified in Appendix B.3.5. Furthermore, strong duality gives

$$x^T D_C y = f^T u$$

The right-hand side is linear in u. Therefore, problem (D.61)–(D.62) and the following MILP

$$\begin{aligned} \max_{x,y,u,v,z} \quad & f^T u - p^T D_M z \\ \text{s.t.} \quad & Ax \leq a, \ B_1 y + B_2 z \leq b \\ & v \in \mathbb{B}^{N_f}, \ D_C^T x = F^T u \\ & 0 \leq u \leq M(1 - v) \\ & 0 \leq F y - f \leq M v \end{aligned} \tag{D.63}$$

have the same optimal solution in primal variables, where N_f is the dimension of f.

We can learn from this example that when the problem exhibits a certain structure, the non-convexity can be eliminated without introducing additional dimensions of complexity. In problem (D.61), the price is a primal variable quoted by a decision maker, and is equal to the knock-down price. This scheme is called pay-as-bid. Next, we give an example of a marginal pricing market where the price is determined by the dual variables of a market clearing problem.

4. A Wholesale Market Problem

In a wholesale market, a provider bids its offering prices to a market organizer. The organizer collects information on available resources and the bidding of the provider, and then clears the market by scheduling the production in the most economic way. The provider is paid at the marginal cost. This problem can be modeled by a bilevel program

$$\max_{\beta} \; \lambda(\beta)^T p(\beta) - f(p(\beta)) \tag{D.64}$$

where β is the offering price vector of the provider, $p(\beta)$ is the quantity of goods ordered by the market organizer, function $f(p) = \sum_i f_i(p_i)$, where $f_i(p_i)$ is a univariate convex function representing the production cost, and $\lambda(\beta)$ is the marginal prices of each kind of goods. Both of them depend on the value of β, and are determined from the market clearing problem in the lower level

$$\min_{p,u} \; \beta^T p + c^T u \tag{D.65a}$$

$$\text{s.t.} \; p_n \le p \le p_m : \eta_n, \eta_m \tag{D.65b}$$

$$p + Fu = d : \lambda \tag{D.65c}$$

$$Au \le a : \xi \tag{D.65d}$$

where u includes all other variables, such as the amount of each kind of goods collected from other providers or produced locally, the system operating variable, and so on; c is the coefficient including prices of goods offered by other providers, and the production cost if the organizer wishes to produce the goods by itself. Objective function (D.65a) represents the total cost in the market to be minimized. Constraint (D.65b) defines offering limits of the upper-level provider; constraint (D.65c) is the system-wide production-demand balancing condition of each goods, the dual variable λ at the optimal solution gives the marginal cost of each goods; (D.65d) imposes constraints which the system operation must obey, such as network flow and inventory dynamics.

In the provider's problem (D.64), the offering price β is not restricted by finite upper bounds pricing policies (but such a policy can certainly be modeled), because the competition appears in the lower level: if β is not reasonable, the market organizer would resort to other providers or count on its own production capability.

Compared with the situation in a retail market, problems (D.64)–(D.65) are even more complicated: the dual variable λ appears in the objective function of the provider, and the term $\lambda^T p$ is non-convex. In the following, we reveal that it can be exactly expressed as a linear function in the primal and dual variables via (somehow tricky) algebraic transformations.

KKT conditions of the market clearing LP (D.65) are summarized as follows:

$$\beta = \lambda + \eta_n + \eta_m \tag{D.66a}$$

$$\eta_n^T (p - p_n) = 0 \tag{D.66b}$$

$$\eta_m^T (p_m - p) = 0 \tag{D.66c}$$

$$c = A^T \xi + F^T \lambda \tag{D.66d}$$

$$\xi^T (Au - a) = 0 \tag{D.66e}$$

$$\eta_n \ge 0, \; \eta_m \le 0, \; \xi \le 0 \tag{D.66f}$$

$$\text{(D.65b)} - \text{(D.65d)} \tag{D.66g}$$

According to (D.66a),

$$\beta^T p = \lambda^T p + \eta_n^T p + \eta_m^T p \tag{D.67a}$$

From (D.66b) and (D.66c) we have

$$\eta_n^T p = \eta_n^T p_n, \; \eta_m^T p = \eta_m^T p_m, \tag{D.67b}$$

Substituting (D.67b) in (D.67a) renders

$$\lambda^T p = \beta^T p - \eta_n^T p_n - \eta_m^T p_m \tag{D.67c}$$

Furthermore, strong duality of LP implies the following equality

$$\beta^T p + c^T u = \eta_n^T p_n + \eta_m^T p_m + d^T \lambda + a^T \xi$$

or

$$\beta^T p - \eta_n^T p_n - \eta_m^T p_m = d^T \lambda + a^T \xi - c^T u \tag{D.67d}$$

Substituting (D.67d) in (D.67c) results in

$$\lambda^T p = d^T \lambda + a^T \xi - c^T u \tag{D.67e}$$

The right-hand side is a linear expression for $\lambda^T p$ in primal variable u and dual variables λ and ξ.

Combining the KKT condition (D.66) and (D.67e) gives an MPCC which is equivalent to the bilevel wholesale market problem (D.64)–(D.65)

$$\begin{aligned} \max \quad & d^T \lambda + a^T \xi - c^T u - f(p(\beta)) \\ \text{s.t.} \quad & \text{(D.66a)} - \text{(D.66g)} \end{aligned} \tag{D.68}$$

Because complementarity conditions (D.66b), (D.66c), (D.66e) can be linearized, and convex function $f(p)$ can be approximated by PWL functions, MPCC (D.68) can be recast as an MILP.

D.3.3 Bilevel Mixed-Integer Program

Although LP can tackle many economic problems and market activities in real life, there are indeed even more decision-making problems which are beyond the reach of LP, for example, power market clearing considering unit commitment [234]. KKT optimality condition or strong duality from LP theory does not apply to discrete optimization problems due to their intrinsic non-convexity. Furthermore, this is no computationally viable approach to express the optimality condition of a general discrete program in closed form, making a bilevel mixed-integer program much more challenging to solve than a bilevel linear program. Some traditional algorithms either rely on enumerative branch-and-bound strategies based on a weak relaxation or depend on complicated operations that are problem-specific. To our knowledge, the reformulation and decomposition algorithm proposed in [235] is the first approach that can solve general bilevel mixed-integer programs in a systematic way, and will be introduced in this section.

The bilevel mixed-integer program has the following form

$$\begin{aligned} \min \ & f^T x + g^T y + h^T z \\ & \text{s.t. } Ax \le b, \ x \in \mathbb{R}^{m_c} \times \mathbb{B}^{m_d} \\ & (y, z) \in \arg\max \ \ w^T y + v^T z \\ & \text{s.t. } \ Py + Nz \le r - Kx \\ & y \in \mathbb{R}^{n_c}, \ z \in \mathbb{B}^{n_d} \end{aligned} \tag{D.69}$$

where x is the upper-level decision variable and appears in constraints of the lower-level problem; y and z represent lower-level continuous decision variable and discrete decision variable, respectively. We do not distinguish upper-level continuous variable and discrete variable because they have little impact on the exposition of the algorithm, unlike the ones appeared in the lower level. If the lower-level has multiple solutions, the follower chooses the onc in favor of the leader. In the current form, the upper-level constraints are independent of lower-level variables. Nevertheless, coupling constraints in the upper level can be easily incorporated [236].

In this section, we assume that the relatively complete recourse property in [235] holds, i.e., for any feasible pair (x, z), the feasible set for lower-level continuous variable y is non-empty. Under this premise, the optimal solution exists. This assumption is mild because we can add slack variables in the lower-level constraints and penalize constraint violation in the lower-level objective function. For instances in which the relatively complete recourse property is missing, please refer to the remedy in [236].

To eliminate $\in$ qualifier in (D.69), we duplicate decision variables and constraints of the lower-level problem and set up an equivalent formulation:

$$\begin{aligned} &\min\ f^T x + g^T y^0 + h^T z^0 \\ &\text{s.t. } Ax \le b,\ x \in \mathbb{R}^{m_c} \times \mathbb{B}^{m_d} \\ &Kx + Py^0 + Nz^0 \le r \\ &w^T y^0 + v^T z^0 \ge \max\ \ w^T y + v^T z \\ &\text{s.t. } \ Py + Nz \le r - Kx \\ &y \in \mathbb{R}^{n_c},\ z \in \mathbb{B}^{n_d} \end{aligned} \tag{D.70}$$

In this formulation, the leader controls its original variable x as well as replicated variables y^0 and z^0. Conceptually, the leader will use (y^0, z^0) to anticipate the response of follower and its impact on his objective function. Clearly, if the lower-level problem has a unique optimal solution, it must be equal to (y^0, z^0). It is worth mentioning that although more variables and constraints are incorporated in (D.70), this formulation is actually an informative and convenient expression for algorithm development, as $\ge$ would be more friendly to general purpose mathematical programming solvers.

Up to now, the obstacle of solving (D.70) remains: discrete variable z in the lower level, which prevents the use of optimality condition of LP. To overcome this difficulty, we treat y and z separately and restructure the lower-level problem as:

$$w^T y^0 + v^T z^0 \ge \max_{z \in Z}\ v^T z + \max_y \{w^T y | Py \le r - Kx - Nz\} \tag{D.71}$$

where Z represents the set consisting of all possible values of z. Despite the large cardinality of Z, the second optimization is a pure LP, and can be replaced with its KKT condition, resulting in:

$$\begin{aligned} &w^T y^0 + v^T z^0 \ge \max_{z \in Z}\ v^T z + w^T y \\ &\text{s.t. } P^T \pi = w \\ &0 \le \pi \perp r - Kx - Nz - Py \ge 0 \end{aligned} \tag{D.72}$$

The complementarity constraints can be linearized via the method in Sect. B.3.5. Then, by enumerating z^j over Z with associated variables (y^j, π^j), we arrive at an MPCC that is equivalent to problem (D.70)

$$\begin{aligned}
\min \ & f^T x + g^T y^0 + h^T z^0 \\
\text{s.t.} \ & Ax \le b, \ x \in \mathbb{R}^{m_c} \times \mathbb{B}^{m_d} \\
& Kx + Py^0 + Nz^0 \le r, \ P^T \pi^j = w, \ \forall j \\
& 0 \le \pi^j \perp r - Kx - Nz^j - Py^j \ge 0, \ \forall j \\
& w^T y^0 + v^T z^0 \ge w^T y^j + v^T z^j, \ \forall j
\end{aligned} \tag{D.73}$$

Without particular mention, (D.73) is compatible with MILP solvers.

Except for the KKT optimality condition, another popular approach entails applying primal-dual condition for the LP regarding the lower-level continuous variable y. Following this line, rewrite this LP in (D.71) by strong duality, we obtain

$$\begin{aligned}
w^T y^0 + v^T z^0 \ge \max_{z \in Z} \ v^T z + \min \ & \pi^T (r - Kx - Nz) \\
\text{s.t.} \ & P^T \pi = w, \ \pi \ge 0
\end{aligned} \tag{D.74}$$

In (D.74), if all variables in x are binary, the bilinear terms $\pi^T Kx$ and $\pi^T Nz$ from the leader's point of view can be linearized via the method in Sect. B.2.2. The min operator in the right-hand side can be omitted because the upper-level objective function is to be minimized, giving rise to

$$\begin{aligned}
\min \ & f^T x + g^T y^0 + h^T z^0 \\
\text{s.t.} \ & Ax \le b, \ x \in \mathbb{R}^{m_c} \times \mathbb{B}^{m_d} \\
& Kx + Py^0 + Nz^0 \le r \\
& w^T y^0 + v^T z^0 \ge v^T z^j + (r - Kx - Nz)^T \pi^j, \ \forall j \\
& P^T \pi^j = w, \ \pi^j \ge 0, \ \forall j
\end{aligned} \tag{D.75}$$

Clearly, (D.75) has fewer constraints compared to (D.73). Nevertheless, whenever x contains continuous variables, linearizing $\pi^T Kx$ would incur more binary variables.

One may think that it is hopeless to solve above enumeration forms (D.73) and (D.75) due to the large cardinality of Z. In a way similar to the CCG algorithm for solving robust optimization, we can start with a subset of Z and solve relaxed version of problem (D.73), until the lower bound and upper bound of optimal value converge. The flowchart is shown in Algorithm D.5

Because Z has finite elements, Algorithm D.5 must terminate in a finite number of iterations, which is bounded by the cardinality of Z. When it converges, LB equals to UB without a positive gap.

To see this, suppose that in iteration l_1, (x^*, y^{0*}, z^{0*}) is obtained in step 2 with LB < UB, and z^* is produced in step 4. Particularly, we assume that z^* was previously derived in some iteration $l_0 < l_1$. Then, in step 5, new variables and cuts associated with $z^* = z^{l_1+1}$ will be generated and augmented with the master problem. As those variables and constraints already exist after iteration l_0, the augmentation is essentially redundant, and the optimal value of master problem in iteration $l_1 + 1$ remains the same as that in iteration l_1, so does LB. Consequently, in iteration $l_1 + 1$

Algorithm D.5 CCG algorithm for bilevel MILP

1: Set LB $= -\infty$, UB $= +\infty$, and $l = 0$;

2: Solve the following master problem

$$\begin{aligned} \min \quad & f^T x + g^T y^0 + h^T z^0 \\ \text{s.t.} \quad & Ax \le b, \ x \in \mathbb{R}^{m_c} \times \mathbb{B}^{m_d} \\ & Kx + Py^0 + Nz^0 \le r, \ P^T \pi^j = w, \ \forall j \le l \\ & 0 \le \pi^j \perp r - Kx - Nz^j - Py^j \ge 0, \ \forall j \le l \\ & w^T y^0 + v^T z^0 \ge w^T y^j + v^T z^j, \ \forall j \le l \end{aligned} \tag{D.76}$$

The optimal solution is $(x^*, y^{0*}, z^{0*}, y^{1*}, \cdots, y^{l*}, \pi^{1*}, \cdots, \pi^{l*})$, and the optimal value is v^*. Update lower bound LB $= v^*$.

3: Solve the following lower-level MILP with obtained x^*

$$\begin{aligned} \theta(x^*) = \max \quad & w^T y + v^T z \\ \text{s.t.} \quad & Py + Nz \le r - Kx^* \\ & y \in \mathbb{R}^{n_c}, \ z \in \mathbb{B}^{n_d} \end{aligned} \tag{D.77}$$

The optimal value is $\theta(x^*)$.

4: Solve an additional MILP to refine a solution that is favor of the leader

$$\begin{aligned} \Theta(x^*) = \min \quad & g^T y + h^T z \\ \text{s.t.} \quad & w^T y + v^T z \ge \theta(x^*) \\ & Py + Nz \le r - Kx^* \\ & y \in \mathbb{R}^{n_c}, \ z \in \mathbb{B}^{n_d} \end{aligned} \tag{D.78}$$

The optimal solution is (y^*, z^*), and the optimal value is $\Theta(x^*)$. Update upper bound UB $= \min\{\text{UB}, f^T x^* + \Theta(x^*)\}$.

5: If UB − LB = 0, terminate and report optimal solution; otherwise, set $z^{l+1} = z^*$, create new variables (y^{l+1}, π^{l+1}), adding the following cuts to master problem

$$w^T y^0 + v^T z^0 \ge w^T y^{l+1} + v^T z^{l+1}$$

$$0 \le \pi^{l+1} \perp r - Kx - Nz^{l+1} - Py^{l+1} \ge 0, \ P^T \pi^{l+1} = w$$

Update $l \leftarrow l + 1$, and go to step 2.

$$
\begin{aligned}
\text{LB} &= f^T x^* + g^T y^{0*} + h^T z^{0*} \\
&= f^T x^* + \min\ g^T y^0 + h^T z^0 \\
&\quad \text{s.t. } P y^0 + N z^0 \le r - K x^*,\ P^T \pi^j = w,\ \forall j \le l_1 + 1 \\
&\quad 0 \le \pi^j \perp r - K x^* - N z^j - P y^j \ge 0,\ \forall j \le l_1 + 1 \\
&\quad w^T y^0 + v^T z^0 \ge w^T y^j + v^T z^j,\ \forall j \le l_1 + 1 \\
&\ge f^T x^* + \min\ g^T y^0 + h^T z^0 \\
&\quad \text{s.t. } P y^0 + N z^0 \le r - K x^*,\ P^T \pi^j = w,\ j = l_1 + 1 \\
&\quad 0 \le \pi^j \perp r - K x^* - N z^j - P y^j \ge 0,\ j = l_1 + 1 \\
&\quad w^T y^0 + v^T z^0 \ge w^T y^j + v^T z^j,\ j = l_1 + 1 \\
&\ge f^T x^* + \min\ g^T y^0 + h^T z^0 \\
&\quad \text{s.t. } P y^0 + N z^0 \le r - K x^* \\
&\quad w^T y^0 + v^T z^0 \ge \theta(x^*) \\
&= f^T x^* + \Theta(x^*)
\end{aligned}
$$

The second $\ge$ follows from the fact that z^{l_1+1} is the optimal solution to problem (D.77) and KKT condition in constraints warrants that $v^T z^{l_1+1} + w^T y^{l_1+1} = \theta(x^*)$. In the next iteration, the algorithm terminates since LB $\ge$ UB.

It should be pointed out that although a large amount of variables and constraints are generated in step 5, in practice, Algorithm D.5 often converges to an optimal solution within a small number of iterations that could be drastically smaller than the cardinality of Z, because the most critical scenarios in Z can be discovered from problem (D.77).

It is suggested in [235] that the master problem could be tightened by introducing variables $(\hat{y}, \hat{\pi})$ representing the primal and dual variables of lower-level problem corresponding to (x, z^0) and augmenting the following constraints

$$
\begin{gathered}
w^T y^0 + v^T z^0 \ge w^T \hat{y} + v^T z^0 \\
0 \le \hat{\pi} \perp r - K x - N z^0 - P \hat{y} \ge 0,\ P^T \hat{\pi} = w
\end{gathered}
$$

It is believed that such constraints include some useful information that is parametric not only to x but also to z^0, and is not available from any fixed samples $z^1, \cdots, z^l$. It is also pointed out that for instance with pure integer variables in the lower-level problem, this strategy is generally ineffective.

D.4 Mathematical Programs with Equilibrium Constraints

A mathematical program with equilibrium constraints (MPEC) is an extension of the bilevel program by incorporating multiple followers competing with each other, resulting in a GNEP in the lower level. In this regard, an MPEC is a single-leader-multi-follower Stackelberg game. In a broader sense, MPEC is an optimization problem with variational inequalities. MPECs are difficult to solve because of the complementarity constraints.

D.4.1 Mathematical Formulation

In an MPEC, the leader deploys its action x prior to the followers; then each follower selects its optimal decision y_j taking the decision of the leader x and rivals' strategies y_{-j} as given. The MPEC can be formulated in two levels:

$$\text{Leader:} \quad \left\{ \begin{aligned} \min_{x,\bar{y},\bar{\lambda},\bar{\mu}} \quad & F(x, \bar{y}, \bar{\lambda}, \bar{\mu}) \\ \text{s.t.} \quad & G(x, \bar{y}) \le 0 \\ & (\bar{y}, \bar{\lambda}, \bar{\mu}) \in S(x) \end{aligned} \right. \tag{D.79a}$$

$$\text{Followers:} \quad \left\{ \begin{aligned} \min_{y_j,\lambda_j,\mu_j} \quad & f_j(x, y_j, y_{-j}) \\ \text{s.t.} \quad & g_j(x, y_j) \le 0 : \mu_j \\ & h(x, y) \le 0 : \lambda_j \end{aligned} \right\}, \ \forall j \tag{D.79b}$$

In (D.79a), the leader minimizes its payoff function F which depends on the choice of its own x, the decisions of the followers y, and the dual variables λ and μ from the lower level, because these dual variables may represent the prices of goods determined by the lower-level market clearing model. Constraints include inequalities and equalities (as a pair of opposite inequalities), as well as the optimality condition of the lower-level problem. In (D.79b), x is treated as a parameter, and the competition among followers comes down to a GNEP with shared convex constraints: the payoff function $f_j(x, y_j, y_{-j})$ of follower j is assumed to be convex in y_j; inequality $g_j(x, y_j) \le 0$ defines a local constraint of follower j which is convex in y_j and does not involve y_{-j}; inequality $h(x, y) \le 0$ is the shared constraint which is convex in y. Since each follower's problem is convex, the KKT condition is both necessary and sufficient for optimality. We assume that the set of GNEPs $S(x)$ is always non-empty.

The GNEP encompasses several special cases in the lower level. If the global constraint is absent, it degenerates into an NEP; moreover, if the objective functions of followers are also decoupled, the lower level reduces to independent convex optimization programs.

By replacing the lower-level GNEP with its KKT condition (D.35), the MPEC (D.79) becomes an MPCC, which can be solved by some suitable methods explained before. As the lower level GNEP usually possesses infinitely many equilibria, the outcome found by the MPCC reformulation is the favorite one from the leader's perspective. We can also require the Lagrange multipliers for the shared constraints should be equal, so as to restrict the GNEP to VEs. If the followers' problems are linear, the primal-dual optimality condition is an alternative choice in addition to the KKT condition, as it often involves fewer constraints. Nevertheless, the strong duality may introduce products involving primal and dual variables, such as those in (D.56) and (D.57), which remain non-convex and require special treatments.

D.4.2 Linear Complementarity Problem

A linear complementarity problem (LCP) requires finding a feasible solution subject to the following constraints

$$0 \le x \perp Px + q \ge 0 \tag{D.80}$$

where P is a square matrix; q is a vector. Their dimensions are compatible with x.

LCP is a special case of MPCC without an objective function. This type of problem frequently arises in various disciplines including market equilibrium analysis, computational mechanics, game theory, and mathematical programming. The theory of LCPs is a well-developed field. Detailed discussions can be found in [237]. In general, an LCP is NP-hard, although it is polynomially solvable for some special cases. One situation is when the matrix P is positive semidefinite. In such circumstance, problem (D.80) can be solved via the following convex quadratic program

$$\begin{aligned} \min \quad & x^T P x + q^T x \\ \text{s.t.} \quad & x \ge 0, \; Px + q \ge 0 \end{aligned} \tag{D.81}$$

Equation (D.81) is a CQP which is readily solvable. Its optimum must be non-negative according to the constraints. If the optimal value of (D.81) is 0, then its optimal solution also solves LCP (D.80); otherwise, if the optimal value is strictly positive, LCP (D.80) is infeasible. In fact, this conclusion holds no matter whether P is positive semidefinite or not. However, if P is indefinite, identifying the global optimum of a non-convex QP (D.81) is also NP-hard, and thus does not facilitate solving the LCP.

There is a large body of literature discussing algorithms for solving LCPs. One of the most representative ones is the Lemke's pivoting method developed in [238], and another emblematic one is the interior-point method proposed in [239]. One drawback of the former method is its exponentially growing worst-case complexity, which makes it less efficient for large problems. The latter approach runs in polynomial time, but it requires the positive semidefiniteness of P, which is a strong assumption and limits its application. In this section, we will not present comprehensive reviews on the algorithms for LCP. We will introduce MILP formulations for problem (D.80) devised in [240, 241]. They make no reference on any special structure of matrix P. More importantly, they offer an option to access the solutions of practical problems in a systematic way.

Recall the MILP formulation techniques presented in Appendix B.3.5, it is easy to see that problem (D.80) can be equivalently expressed as linear constraints with additional binary variable z as follows:

$$0 \le x \le Mz,\ 0 \le Px + q \le M(1 - z) \tag{D.82}$$

Integrality of z maintains the element-wise complementarity of x and $Px + q$: at most one of x_i and $(Px + q)_i$ can be strictly positive. Formulation (D.82) entails a manually specified parameter M, which is not instantly available at hand. On the one hand, it must be big enough to preserve all extreme points of (D.80). On the other hand, it is expected to be as small as possible from a computational perspective, otherwise, the continuous relaxation of (D.82) would be very loose. In this regard, (D.82) is too cursory, although it might work well.

To circumvent above difficulty, it is proposed in [240] to solve a bilinear program without a big-M parameter

$$\begin{aligned} \min_{x,z}\ & z^T(Px + q) + (\mathbf{1} - z)^T x \\ \text{s.t.}\ & x \ge 0,\ Px + q \ge 0,\ z \text{ binary} \end{aligned} \tag{D.83}$$

If (D.80) has a solution x^*, the optimal value of (D.83) is 0: for $x_i^* > 0$, we have $z_i^* = 1$ and $(Px^* + q)_i = 0$; for $(Px^* + q)_i > 0$, we have $z_i^* = 0$ and $x_i^* = 0$. The optimal solution is consistent with the feasible solution of (D.82). The objective can be linearized by introducing auxiliary variables $w_{ij} = z_i x_j$, $\forall i, j$. However, applying normal integer formulation techniques in Appendix B.2.2 on variable w_{ij} again needs the upper bound of x_i, another interpretation of the big-M parameter.

A parameter-free MILP formulation is suggested in [240]. To understand the basic idea, recall the fact that $(1 - z_i)x_i = 0$; if we impose $x_i = w_{ii} = x_i z_i$, $i = 1, 2, \cdots$ in the constraint, $(\mathbf{1} - z)^T x$ in the objective can be omitted. Furthermore, multiplying both sides of $\sum_j P_{kj} x_j + q_k \ge 0$, $k = 1, 2, \cdots$ with z_i gives $\sum_j P_{kj} w_{ij} + q_k z_i \ge 0, \forall i, k$. Since $z_i \in \{0, 1\}$, $\sum_j P_{kj} x_j + q_k \ge \sum_j P_{kj} w_{ij} + q_k z_i$, $\forall i, k$ and $0 \le w_{ij} \le x_j$, $\forall i, j$ naturally hold. Collecting up these valid inequalities, we obtain an MILP

$$\begin{aligned} \min_{x,z,w} \quad & q^T z + \sum_i \sum_j P_{ij} w_{ij} \\ \text{s.t.} \quad & \sum_j P_{kj} x_j + q_k \geq \sum_j P_{kj} w_{ij} + q_k z_i \geq 0, \ \forall i, k \\ & 0 \leq w_{ij} \leq x_j, \ \forall i, j, \ w_{jj} = x_j, \ \forall j, \ z \text{ binary} \end{aligned} \tag{D.84}$$

Instead of enforcing every $\sum_j P_{kj} w_{ij} + q_k z_i$ being at 0, we relax them as inequalities and minimize their summation. More valid inequalities can be added in (D.84) by exploiting linear cuts of z. It is proved in [240] that relation $w_{ij} = z_i x_j, \forall i, j$ is implicitly guaranteed at the optimal solution of (D.84). In view of this, MILP (D.84) is equivalent to LCP (D.80) in the following sense: (D.80) has a solution if and only if (D.84) has an optimal value equal to zero, and the optimal solution to (D.84) incurring a zero objective value is a solution of LCP (D.80). MILP (D.84) is superior compared with (D.82) and big-M linearization based MILP formulation of MINLP (D.83) because it is parameter-free and gives tighter continuous relaxation. Nevertheless, the number of constraints in (D.84) is significantly larger than that in formulation (D.82). This method has been further analyzed in [241] and extended to binary-constrained mixed LCPs.

Another parameter-free MILP formulation is suggested in [242], which takes the form of

$$\begin{aligned} \max_{\alpha, y, z} \quad & \alpha \\ \text{s.t.} \quad & 0 \leq (Py)_i + q_i \alpha \leq 1 - z_i, \ \forall i \\ & 0 \leq y_i \leq z_i, \ z_i \in \{0, 1\}, \ \forall i \\ & 0 \leq \alpha \leq 1 \end{aligned} \tag{D.85}$$

Since $\alpha = 0$, $y = 0$, $z = 0$ is always feasible, MILP (D.85) is feasible and has an optimum no greater than 1. By observing the constraints, we can conclude that if MILP (D.85) has a feasible solution with $\bar{\alpha} > 0$, then $x = y/\bar{\alpha}$ solves problem (D.80). If the optimal solution $\bar{\alpha} = 0$, then problem (D.80) has no solution; otherwise, suppose $\bar{x}$ solves (D.80), and let $\bar{\alpha}^{-1} = \max\{\bar{x}_i, (P\bar{x})_i + q_i, i = 1, \cdots\}$, then for any $0 < \alpha \leq \bar{\alpha}$, $\bar{y} = \alpha \bar{x}$ is feasible in (D.85). As a result, the optimal solution should be no less than $\bar{\alpha}$, rather than 0. Compared with formulation (D.82), the big-M parameter is adaptively scaled by optimizing α.

Because (D.85) works with an intermediate variable y, when LCP (D.80) should be jointly solved with other conditions on x, formulation (D.85) is not advantageous, because non-convex variable transformation $x = y/\alpha$ must be appended to link both parts.

Robust solutions of LCPs with uncertain P and q are discussed in [243]. It is found that when $P \succeq 0$, robust solutions can be extracted from an SOCP under some mild assumptions on the uncertainty set; otherwise, the more general problem with uncertainty can be reduced to a deterministic non-convex QCQP. This technique is particularly useful in uncertain traffic equilibrium problems and uncertain Nash-Cournot games. Uncertain VI problems and MPCCs can be tackled in the similar vein after some proper transformations.

It is shown in [244] that a linear bilevel program or its equivalent MPEC can be globally solved via a sequential LCP method. A hybrid enumerative method is suggested which substantially reduces the effort for searching a solution of the LCP or certifying that the LCP has no solution. When the LCP is easy to solve, this approach is attractive.

Several extensions of LCP, including the discretely-constrained mixed LCP, discretely-constrained Nash-Cournot game, discretely-constrained MPEC, and logic constrained equilibrium problem as well as their applications in energy markets and traffic system equilibrium have been investigated in [245–248]. In a word, due to its wide applications, LCP is still an active research field, and MILP remains an appealing method for solving LCPs for practical problems.

D.4.3 Linear Programs with Complementarity Constraints

A linear program with complementarity constraints (LPCC) entails solving a linear optimization problem with linear complementarity constraints. It is a special case of MPCC if all functions in the problem are linear, and a generalization of LCP by incorporating an objective function to be optimized. An LPCC has the following form

$$\begin{aligned} \max_{x,y} \quad & c^T x + d^T y \\ \text{s.t.} \quad & Ax + By \geq f \\ & 0 \leq y \perp q + Nx + My \geq 0 \end{aligned} \tag{D.86}$$

A standard approach for solving (D.86) is to linearize the complementarity constraint by introducing a binary vector z and solve the following MILP

$$\begin{aligned} \max_{x,y,z} \quad & c^T x + d^T y \\ \text{s.t.} \quad & Ax + By \geq f \\ & 0 \leq q + Nx + My \leq Mz \\ & 0 \leq y \leq M(1 - z) \\ & z \in \{0, 1\}^m \end{aligned} \tag{D.87}$$

If both of x and y are bounded variables, we can readily derive the proper value of M in each inequality; otherwise, finding high quality bounds is nontrivial even if they do exist. The method in [228] can be used to determine proper bounds of M, if the NLP solver can successfully find local solutions of the bounding problems.

Using an arbitrarily large value may solve the problem correctly. Nevertheless, parameter-free method is still of great theoretical interest. A smart Benders decomposition algorithm is proposed in [249] to solve (D.87) without requiring the value of M. The completely positive programming method developed in [250] can also be used to solve (D.87). For more theory and algorithm for LPCC, please see [227, 251–257] and references therein. Interesting connections among conic QPCCs, QCQPs, and completely positive programs are revealed in [258].

D.5 Equilibrium Programs with Equilibrium Constraints

An equilibrium program with equilibrium constraints (EPEC) is the most general extension of the bilevel program. It incorporates multiple leaders and multiple followers competing with each other in the upper level and the lower level, respectively, resulting in two GNEPs in both levels. In this regard, an EPEC is a multi-leader-follower Stackelberg game.

D.5.1 Mathematical Model

In an EPEC, each leader i deploys an action x_i prior to the followers while taking movements of other leaders x_{-i} into account and anticipating the best responses $y(x)$ from the followers; then each follower selects its optimal decision y_j by taking the strategies of leaders x and rivals' actions y_{-j} as given. The EPEC can be formulated in two levels

$$\text{Leaders:} \quad \left\{ \begin{aligned} \min_{x_i, \bar{y}, \bar{\lambda}, \bar{\mu}} \quad & F_i(x_i, x_{-i}, \bar{y}, \bar{\lambda}, \bar{\mu}) \\ \text{s.t.} \quad & G_i(x_i) \le 0 \\ & (\bar{y}, \bar{\lambda}, \bar{\mu}) \in S(x_i, x_{-i}) \end{aligned} \right\}, \ \forall i \tag{D.88a}$$

$$\text{Followers:} \quad \left\{ \begin{aligned} \min_{y_j} \quad & f_j(x, y_j, y_{-j}) \\ \text{s.t.} \quad & g_j(x, y_j) \le 0 : \mu_j \\ & h(x, y) \le 0 : \lambda_j \end{aligned} \right\}, \ \forall j \tag{D.88b}$$

In (D.88a), each leader minimizes its payoff function F_i which depends on its own choice x_i, the decisions of followers y, dual variables λ and μ are parameterized

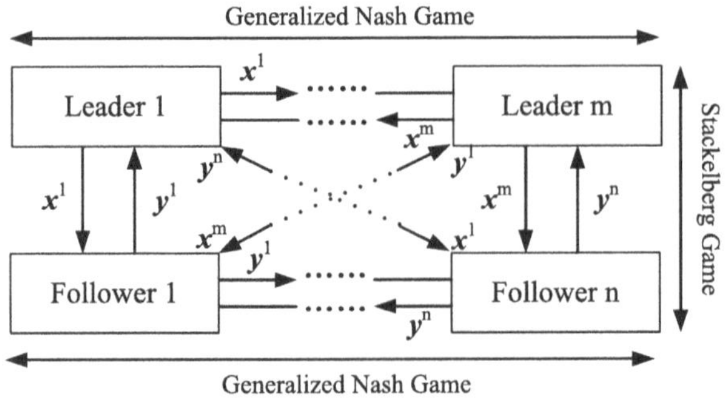

Fig. D.4 The structure of an EPEC

in competitors' strategies x_{-i}. Tuple $(\bar{y}, \bar{\lambda}, \bar{\mu})$ in the upper level is restricted by the optimality condition of the lower-level problem. Although the inequality constraints of leaders are decoupled, and we do not explicitly consider global constraints in the upper-level GNEP, the leaders' strategy sets as well as their payoff functions are still correlated through the best reaction map $S(x_i, x_{-i})$, and hence (D.88a) itself is a GNEP, which is non-convex. The followers' problem (D.88b) is a GNEP with shared constraints, which is the same as the situation in an MPEC. The same convexity assumptions are made in (D.88b). The structure of EPEC (D.88) is depicted in Fig. D.4. The equilibrium solution of EPEC (D.88) is defined as the GNE among leaders' MPECs. It is common knowledge that EPECs often have no pure strategy equilibrium due to the intrinsic non-convexity of MPECs.

D.5.2 Methods for Solving an EPEC

An EPEC can be viewed as a set of coupled MPEC problems: leader i is facing an MPEC composed of problem i in (D.88a) together with all followers' problems in (D.88b), which is parameterized in x_{-i}. By replacing lower-level GNEP with its KKT optimality conditions, it can be imaged that the GNEP among leaders have non-convex constraints which inherit the tough properties of complementarity constraints. Thus, solving an EPEC is usually extremely challenging. To the best of our knowledge, systematic algorithms of EPEC are firstly developed in dissertations [259–261]. The primary application of such an equilibrium model is found in energy market problems, see [262] for an excellent introduction.

Unlike the NEP and GNEP discussed in Sects. D.1 and D.2, where the strategy sets are convex or jointly convex, because the lower-level problems are replaced with KKT optimality conditions and the MPEC for the leader is intrinsically non-convex, provable existence and uniqueness guarantees for the solution to EPECs are non-trivial. There is sustainable attempt on the analysis of EPEC solution properties. For example, the existence of a unique equilibrium for certain EPEC instances is discussed in [263, 264], and in [265] for a nodal price based power market model. However, the existence and uniqueness of solution are only guaranteed under restrictive conditions. Counterexamples have been given in [266] to demonstrate that there is no general result for the existence of solutions to EPECs due to their non-convexity. The non-uniqueness issue is studied in [260, 267]. It is shown that even in the simplest instances, local uniqueness of the EPEC equilibrium solution may not be guaranteed, and a manifold of equilibria may exist. This can be understood because EPEC is a generalization of GNEP, whose solution property is illustrated in Sect. D.2.1. When the payoff functions possess special structures, say, a potential function exists, then the existence of a global equilibrium can be investigated using the theory of potential games [268–270]. In summary, the theory of EPEC solutions is much more complicated than the single-level NEP and GNEP.

This section reviews several representative algorithms which are widely used in literature. The former two are generic and seen in [259, 260]; the third one is motivated by the convenience brought by the property of potential games, and reported in [270]; at last, a pricing game in a competitive market, which appears to be non-convex at first sight, is presented to show the hidden convexity in such a special equilibrium model.

1. Best Response Algorithm

Since the equilibrium of an EPEC is a GNEP among leaders' MPEC problems, the most intuitive strategy for identifying an equilibrium solution is the best response algorithm. In some literature, it is also called diagonalization method or sequential MPEC method. This approach can be further categorized into Jacobian type and Gauss-Seidel type method, according to the information used when players update their strategies.

To explain the algorithmic details, denote by MPEC(i) the problem of leader i: the upper level is problem i in (D.88a), and the lower level is the GNEP described in (D.88b) given all leaders' strategies. Let x_i^k be the strategy of leader i in iteration k, and $x^k = (x_1^k, \cdots, x_m^k)$ the strategy profile of leaders. The Gauss-Seidel type algorithm proceeds as follows [271]:

Algorithm D.6 Best-response (Diagonalization) algorithm for EPEC

1: Choose an initial strategy profile x^0 for leaders, set convergence tolerance $\varepsilon > 0$, an allowed number of iterations K, and the iteration index $k = 0$;
2: Let $x^{k+1} = x^k$. Loop for players $i = 1, \cdots, m$:

  a. Solve MPEC(i) for leader i given x_{-i}^{k+1}.
  b. Replace x_i^{k+1} with the optimal strategy of leader i just obtained.

3: If $\|x^{k+1} - x^k\|_2 \le \varepsilon$, the upper level converges; solve lower-level GNEP (D.88b) with $x^* = x^{k+1}$ using the algorithms elaborated in Sect. D.2.2, and the equilibrium among followers is y^*. Report (x^*, y^*) and terminate.
4: If $k = K$, report failure of convergence and quit.
5: Update $k \leftarrow k + 1$, and go to step 2.

Without an executable criterion to judge the existence and uniqueness of solution, possible outcomes of Algorithm D.6 are discussed in three situations.

1. There is no equilibrium. Algorithm D.6 does not converge. In such circumstance, one may turn to seeking a mixed-strategy Nash equilibrium, which always exists. Examples are given in [271]: if there are two leaders, we can list possible strategy combinations and solve the lower-level GNEP among followers, then compute respective payoffs of the two leaders, and then build a bimatrix game, whose mixed-strategy Nash equilibrium can be calculated from solving an LCP, as explained in Sect. D.1.4.
2. There is a unique equilibrium, or there are multiple equilibria. Algorithm D.6 may converge or not, and which equilibrium will be found (if it converges) depends on the initial strategy profile offered in step 1.
3. Algorithm D.6 may converge to a local equilibrium in the sense of [265], if each MPEC is solved by a local NLP method which does not guarantee global optimality. The true equilibrium can be found only if each leader's MPEC can be globally solved. The MILP reformulation (if possible) offers one plausible way for this task.

2. KKT System Method

To tackle the divergence issue in the best response algorithm, it is proposed to apply the KKT condition to each leader's MPEC and solve the resulting KKT systems simultaneously [261, 272]. The solution turns out to be a strong stationary equilibrium point of EPEC (D.88). There is no convergence issue in this approach, since no iteration is deployed. However, special attention should be paid to some potential problems mentioned below.

1. Since the EPEC is essentially a GNEP among leaders, the concentrated KKT system may have non-isolated solutions. To refine a meaningful outcome, we can manually specify a secondary objective function, which is optimized subject to the KKT system.
2. The embedded (twice) application of KKT condition for the lower-level problems and upper-level problems inevitably introduces extensive complementarity and slackness conditions, which greatly challenges solving the concentrated KKT system. In this regard, scalability may be a main bottleneck for this approach. If the lower-level GNEP is linear, it may be better to use primal-dual optimality condition first for followers, and then KKT condition for leaders.
3. Because each leader's MPEC is non-convex, a stationary point of the KKT condition is not necessarily an optimal solution of the leader; as a result, the solution of the concentrated KKT system may not be an equilibrium of the EPEC. To validate the result, one can conduct the best-response method initiated at the candidate solution with a slight perturbation.

3. Potential MPEC Method
When the upper-level problems among leaders admit a potential function satisfying (D.25), the EPEC can be reformulated as an MPEC, and the relations of their solutions are revealed by comparing the KKT condition of the normalized Nash stationary points of the EPEC and the KKT condition of the associated MPEC [270].

For example, if the leaders' objectives are given by

$$F_i(x_i, x_{-i}, y) = F_i^S(x_i) + H(x, y)$$

or in other words, the payoff function $F_i(x_i, x_{-i}, y)$ can be decomposed as the sum of two parts: the first one $F_i^S(x_i)$ only depends on the local variable x_i, and the second one $H(x, y)$ is common to all leaders. In such circumstance, the potential function can be expressed as

$$U(x, y) = H(x, y) + \sum_{i=1}^{m} F_i^S(x_i)$$

Please see Sect. D.1.5 for the condition under which a potential function exists and special instances in which a potential function can be easily found.

Suppose that leaders' local constraints are given by $x_i \in X_i$ which is independent of x_{-i} and y, and the best reaction map of followers with fixed x is given by $(\bar{y}, \bar{\lambda}, \bar{\mu}) \in S(x)$. Clearly, the solution of MPEC

$$\begin{aligned} \min_{x, \bar{y}, \bar{\lambda}, \bar{\mu}} \quad & U(x, \bar{y}) \\ \text{s.t.} \quad & x_i \in X_i, \ i = 1, \cdots, m \\ & (\bar{y}, \bar{\lambda}, \bar{\mu}) \in S(x) \end{aligned}$$

must be an equilibrium solution of the original EPEC.

This approach leverages the property of potential games and is superior over the previous two methods (if a potential function exists): KKT condition is applied only once to the lower level problems, and the equilibrium can be retrieved by solving MPEC only once.

4. A Pricing Game in a Competitive Market
We consider an EPEC taken from the examples in [273], which models a strategic pricing game in a competitive market. The hidden convexity in this EPEC is revealed. For ease of exposition, we study the case with two leaders and one follower. The results can be extended to the situation where more than two leaders exist. The pricing game with two leaders can be formulated by the following EPEC

$$\text{Leader 1:} \quad \max_{x_1} \left\{ y^T(x_1, x_2) A_1 x_1 \;\middle|\; B_1 x_1 \le b_1 \right\} \tag{D.89a}$$

$$\text{Leader 2:} \quad \max_{x_2} \left\{ y^T(x_1, x_2) A_2 x_2 \;\middle|\; B_2 x_2 \le b_2 \right\} \tag{D.89b}$$

$$\text{Follower:} \quad \max_{y} \left\{ f(y) - y^T A_1 x_1 - y^T A_2 x_2 \;\middle|\; Cy = d \right\} \tag{D.89c}$$

In (D.89a) and (D.89b), two leaders announce their offering prices x_1 and x_2, respectively, subject to some certain pricing policy described in their corresponding constraints. The follower then decides how many goods should be purchased from each leader, according to the optimal solution of problem (D.89c), where the profit of the follower

$$f(y) = -\frac{1}{2} y^T Q y + c^T y$$

is a strongly concave quadratic function, i.e. $Q \succ 0$, and matrix C has full rank in its rows. Each player in the market wishes to maximize his own profit. The utilities of leaders are the payments from trading with the follower; the profit of follower is the revenue minus the purchasing cost.

At first sight, EPEC (D.89) is non-convex, not only because the leaders' objective functions are bilinear, but also because the best response mapping is generally non-convex. In light of the strong convexity of (D.89c), the following KKT condition:

$$c - Qy - A_1 x_1 - A_2 x_2 - C^T \lambda = 0$$
$$Cy - d = 0$$

is necessary and sufficient for a global optimum. Because constraints in (D.89c) are all equalities, there is no complementarity and slackness condition. Solve this set of linear equations, we can obtain the optimal solution y in a closed form. To this end, substituting

$$y = Q^{-1}(c - A_1 x_1 - A_2 x_2 - C^T \lambda)$$

into the second equation, we have

$$\lambda = M\left[N(c - A_1 x_1 - A_2 x_2) - d\right]$$

where

$$M = \left[C Q^{-1} C^T \right]^{-1}, \quad N = C Q^{-1}$$

Moreover, eliminating λ in the expression of y gives the best reaction map

$$y = r + D_1 x_1 + D_2 x_2 \tag{D.90}$$

where

$$r = Q^{-1}c + N^T Md - N^T MNc$$
$$D_1 = N^T MNA_1 - Q^{-1}A_1$$
$$D_2 = N^T MNA_2 - Q^{-1}A_2$$

Substituting (D.90) into the objective functions of leaders, EPEC (D.89) reduces to a standard Nash game

$$\text{Leader 1:} \quad \max_{x_1} \ \{\theta_1(x_1, x_2) \mid B_1 x_1 \le b_1\}$$
$$\text{Leader 2:} \quad \max_{x_2} \ \{\theta_2(x_1, x_2) \mid B_2 x_2 \le b_2\}$$

where

$$\theta_1(x_1, x_2) = r^T A_1 x_1 + x_1^T D_1^T A_1 x_1 + x_2^T D_2^T A_1 x_1$$
$$\theta_2(x_1, x_2) = r^T A_2 x_2 + x_2^T D_2^T A_2 x_2 + x_1^T D_1^T A_2 x_2$$

The partial Hessian matrix of $\theta_1(x_1, x_2)$ can be calculated as

$$\nabla^2_{x_1}\theta_1(x_1, x_2) = 2A_1^T(N^T MN - Q^{-1})A_1$$

As $Q \succ 0$, its inverse matrix $Q^{-1} \succ 0$; denote by $Q^{-1/2}$ the square root of Q^{-1}, and

$$P_J = I - Q^{-1/2}C^T(CQ^{-1}C^T)^{-1}CQ^{-1/2}$$

It is easy to check that P_J is a projection matrix, which is symmetric and idempotent, i.e., $P_J = P_J^2 = P_J^3 = \cdots$. Moreover, it can be verified that the Hessian matrix $\nabla^2_{x_1}\theta_1(x_1, x_2)$ can be expressed via

$$\nabla^2_{x_1}\theta_1(x_1, x_2) = 2A_1^T(N^T MN - Q^{-1})A_1 = -2A_1^T Q^{-1/2} P_J Q^{-1/2} A_1$$

For any vector z with a proper dimension,

$$\begin{aligned} z^T \nabla^2_{x_1}\theta_1(x_1, x_2) z &= -2z^T\left(A_1^T Q^{-1/2} P_J Q^{-1/2} A_1\right) z \\ &= -2z^T A_1^T Q^{-1/2} P_J^T P_J Q^{-1/2} A_1 z \\ &= -2(P_J Q^{-1/2} A_1 z)^T (P_J Q^{-1/2} A_1 z) \le 0 \end{aligned}$$

We can see that $\nabla^2_{x_1}\theta_1(x_1, x_2) \preceq 0$. The similar analysis also applies to $\nabla^2_{x_2}\theta_2(x_1, x_2)$. Therefore, the problems of leaders are actually convex programs, and a pure-strategy Nash equilibrium exists.

D.6 Conclusions and Further Reading

Equilibrium problems entail solving interactive optimization problems simultaneously, and serve as the foundation for modeling competitive behaviors among strategic decision makers, and analyzing the stable outcome of a game. This chapter provides an overview on two kinds of equilibrium problems that frequently arise in various economic and engineering applications.

One-level equilibrium problems, including the NEP and GNEP, are introduced first. The existence of equilibrium can be ensured under some convexity and monotonicity assumptions. Distributed methods for solving one-level games are presented. When each player solves a strictly convex optimization problem, distributed algorithms converge with provable guarantee, and thus are preferred, whereas the KKT system renders nonlinear equations and is relatively difficult to solve. To address incomplete information and uncertainty in player's decision making, a robust optimization based game model is proposed in [274], which is distribution-free and relaxes Harsanyi's assumptions on Bayesian games. Particularly, the robust Nash equilibrium of a bimatrix game with uncertain payoffs can be characterized via the solution of a second-order cone complementarity problem [275], and more general cases involving n players and continuous payoffs are discussed in [276]. Distributional uncertainty is tackled in [277], in which the mixed-strategy Nash equilibrium of a distributionally robust chance-constrained game is studied. A generalized Nash game arises when the strategy sets of players are coupled. Due to practical interest from a variety of engineering disciplines, the solution method for GNEPs is still an active research area. The volume of articles is growing quickly in recent years, say, [278–286], to name just a few. GNEPs with uncertainties are studied in [287, 288].

Bilevel equilibrium problems, including the bilevel program, MPEC, and EPEC, are investigated. These problems are intrinsically hard to solve, due to the non-convexity induced by the best reaction map of followers, and solution properties have been revealed for specific instances under restrictive assumptions. We recommend [271, 289, 290] for theoretical foundations and energy market applications of bilevel equilibrium models, and [291] for an up-to-date survey. The theories on bilevel programs and MPEC are relatively mature. Recent research efforts have been spent on new constraint qualifications and optimality conditions, for example, the work in [292–295]. The MILP reformulation is preferred by most power system applications, because the ability of MILP solvers keep improving, and a global optimal solution can be found. Stochastic MPEC is proposed in [296] to model uncertainty using probability distributions. Algorithms are developed in [297–300], and a literature review can be found in [301]. Owing to the inherent hardness, discussions on EPEC models are limited to special cases, such as those with shared P-matrix linear complementarity constraints [302], power market models [302–304], those with convex quadratic objectives and linear constraints [302], and Markov game models [305]. Methods for solving EPEC are based on relaxing or regularizing complementarity constraints [305, 306], as well as evolutionary algorithms [307]. Robust equilibria of EPEC are discussed in [308]. An interesting connection between the bilevel program and the GNEP has been revealed in [309], establishing a new look on these game models.

We believe that the equilibrium programming models will become an imperative tool for designing and analyzing interconnected energy systems and related markets, in view of the physical interdependence of heterogeneous energy flows and strategic interactions among different network operators.

References

1. Ben-Tal, A., Nemirovski, A.: Lectures on Modern Convex Optimization: Analysis, Algorithms, and Engineering applications. SIAM, Philadelphia (2001). http://www2.isye.gatech.edu/~nemirovs/Lect_ModConvOpt.pdf
2. Boyd, S., Vandenberghe, L.: Convex Optimization. Cambridge University Press, New York (2004)
3. Boyd, S., Vandenberghe, L.: Convex Optimization Solution Manual (2006). http://egrcc.github.io/docs/math/cvxbook-solutions.pdf
4. Hiriart-Urruty, J.B., Seeger, A.: A variational approach to copositive matrices. SIAM Rev. **52**(4), 593–629 (2010)
5. Berman, A., Shaked-Monderer, N.: Completely Positive Matrices. World Scientific, Singapore (2003)
6. Dantzig, G.: Linear Programming and Extensions. Princeton University Press, Princeton (2016)
7. Bertsimas, D., Tsitsiklis, J.N.: Introduction to Linear Optimization. Athena Scientific, Belmont (1997)
8. Vanderbei, R.J.: Linear Programming. Springer, New York (2015)
9. Lobo, M.S., Vandenberghe, L., Boyd, S., Lebret, H.: Applications of second-order cone programming. Linear Algebra Appl. **284**(1–3), 193–228 (1998)
10. Ben-Tal, A., Nemirovski, A.: On polyhedral approximations of the second-order cone. Math. Oper. Res. **26**(2), 193–205 (2001)
11. Burer, S., Saxena, A.: The MILP Road to MIQCP. In: Lee, J., Leyffer, S. (eds.) Mixed Integer Nonlinear Programming. The IMA Volumes in Mathematics and its Applications, vol. 154. Springer, New York (2012)
12. Sherali, H.D., Adams, W.P.: A Reformulation-Linearization Technique for Solving Discrete and Continuous Nonconvex Problems. Springer, Berlin (1999)
13. McCormick, G.P.: Computability of global solutions to factorable nonconvex programs: part I - convex underestimating problems. Math. Program. **10**(1), 147–175 (1976)
14. Saxena, A., Bonami, P., Lee, J.: Convex relaxations of non-convex mixed integer quadratically constrained programs: extended formulations. Math. Program. **124**(1), 383–411 (2010)
15. Lipp, T., Boyd, S.: Variations and extension of the convex-concave procedure. Optim. Eng. **17**(2), 263–287 (2016)

16. Yang, W., Zhao, J., Xu, F.: An efficient method for convex constrained rank minimization problems based on DC programming. Math. Probl. Eng. **2016**, 7473041 (2016)
17. Vandenberghe, L., Boyd, S.: Semidefinite programming. SIAM Rev. **38**(1), 49–95 (1996)
18. Han, S.P., Mangasarian, O.L.: Exact penalty functions in nonlinear programming. Math. Program. **17**(1), 251–269 (1979)
19. Di Pillo, G., Grippo, L.: Exact penalty functions in constrained optimization. SIAM J. Control Optim. **27**(6), 1333–1360 (1989)
20. Bomze, I.M., Dür, M., de Klerk, E., Roos, C., Quist, A.J., Terlaky, T.: On copositive programming and standard quadratic optimization problems. J. Global Optim. **18**(4), 301–320 (2000)
21. Burer, S.: On the copositive representation of binary and continuous nonconvex quadratic programs. Math. Program. **120**(2), 479–495 (2009)
22. Povh, J.: Towards the Optimum by Semidefinite and Copositive Programming: New Approach to Approximate Hard Optimization Problems. VDM Verlag (2009)
23. Parrilo, P.: Structured Semidefinite Programs and Semi-algebraic Geometry Methods in Robustness and Optimization. Ph.D. Thesis, California Institute of Technology (2000)
24. Dong, H., Luo, Y.: Compact disjunctive approximations to nonconvex quadratically constrained programs. arXiv:1811.08122v1 (2018)
25. Hu, J., Mitchell, J.E., Pang, J.S.: An LPCC approach to nonconvex quadratic programs. Math. Program. **133**(1), 243–277 (2012)
26. Giannessi, F., Tomasin, E.: Nonconvex quadratic programs, linear complementarity problems, and integer linear programs. In: Conti, R., Ruberti, A. (eds.) Fifth Conference on Optimization Techniques (Rome 1973), Part I. Lecture Notes in Computer Science, vol. 3, pp. 437–449. Springer, Berlin (1973)
27. Chen, J., Burer, S.: Globally solving nonconvex quadratic programming problems via completely positive programming. Math. Prog. Comput. **4**(1), 33–52 (2012)
28. Gondzio, J., Yildirim, E.A.: Global solutions of nonconvex standard quadratic programs via mixed integer linear programming reformulations. arXiv preprint arXiv:1810.02307 (2018)
29. Rockafellar, R.T.: Convex analysis. Princeton University Press, Princeton (2015)
30. Bertsekas, D.P., Nedi, A., Ozdaglar, A.E.: Convex Analysis and Optimization. Athena Scientific, Belmont (2003)
31. Tuy, H.: Convex Analysis and Global Optimization, 2nd edn. Springer, Berlin (2016)
32. Boyd, S., Kim, S.J., Vandenberghe, L., Hassibi, A.: A tutorial on geometric programming. Optim. Eng. **8**(1), 67–127 (2007)
33. Dür, M.: Copositive programming - a survey. In: Diehl, M., Glineur, F., Jarlebring, E., Michiels, W. (eds.) Recent Advances in Optimization and its Applications in Engineering. Springer, Berlin (2010)
34. Gijben, L.: On approximations, complexity, and applications for copositive programming. Rijksuniversiteit Groningen (2015)
35. Burer, S.: Copositive programming. In: Handbook on Semidefinite, Conic and Polynomial Optimization, pp. 201–218. Springer, New York (2012)
36. Lasserre, J.B.: New approximations for the cone of copositive matrices and its dual. Math. Program. **144**(1), 265–276 (2014)
37. Bomze, I.M.: Copositive relaxation beats Lagrangian dual bounds in quadratically and linearly constrained quadratic optimization problems. SIAM J. Optim. **25**(3), 1249–1275 (2015)
38. Berman, A., Dur, M., Shaked-Monderer, N.: Open problems in the theory of completely positive and copositive matrices. Electron. J. Linear Algebra **29**(1), 46–58 (2015)
39. Grant, M., Boyd, S., Ye, Y.: Disciplined convex programming. In: Global Optimization: From Theory to Implementation, pp. 155–210. Springer, Berlin (2006)
40. Palomar, D.P., Eldar, Y.C.: Convex Optimization in Signal Processing and Communications. Cambridge University Press, Cambridge (2010)
41. Luo, Z.Q., Ma, W.K., So, A.M.C., Ye, Y.: Semidefinite relaxation of quadratic optimization problems. IEEE Signal Proc. Mag. **27**(3), 20–34 (2010)

42. Maher, A.: Semidefinite programming: methods and algorithms for energy management. Working Paper https://tel.archives-ouvertes.fr/tel-00881025/document.
43. Taylor, J.A.: Convex Optimization of Power Systems. Cambridge University Press, Cambridge (2015)
44. Bao, X., Sahinidis, N.V., Tawarmalani, M.: Semidefinite relaxations for quadratically constrained quadratic programming: a review and comparisons. Math. Program. **129**(1), 129–157 (2011)
45. Anstreicher, K.M.: On convex relaxations for quadratically constrained quadratic programming. Math. Program. **136**(2), 233–251 (2012)
46. Bose, S., Gayme, D.F., Chandy, K.M., Low, S.H.: Quadratically constrained quadratic programs on acyclic graphs with application to power flow. IEEE Trans. Control Network Syst. **2**(3), 278–287 (2015)
47. Xu, Z., Hong, M., Luo, Z.Q.: Semidefinite approximation for mixed binary quadratically constrained quadratic programs. SIAM J. Optim. **24**(3), 1265–1293 (2014)
48. Billionnet, A., Elloumi, S., Lambert, A.: Exact quadratic convex reformulations of mixed-integer quadratically constrained problems. Math. Program. **158**(1), 235–266 (2016)
49. Burer, S., Kim, S., Kojima, M.: Faster, but weaker, relaxations for quadratically constrained quadratic programs. Comput. Optim. Appl. **59**(1–2), 27–45 (2014)
50. Jiang, R., Li, D.: Convex relaxations with second-order cone constraints for nonconvex quadratically constrained quadratic programming. Working Paper, arXiv:1608.02096 (2016)
51. Shtern, S., Ben-Tal, A.: Computational methods for solving nonconvex block-separable constrained quadratic problems. SIAM J. Optim. **26**(2), 1174–1206 (2016)
52. Ben-Tal, A., Den Hertog, D.: Hidden conic quadratic representation of some nonconvex quadratic optimization problems. Math. Program. **143**(1), 1–29 (2014)
53. Burer, S., Monteiro, R.D.C.: Local minima and convergence in low-rank semidefinite programming. Math. Program. **103**(3), 427–444 (2005)
54. Burer, S., Choi, C.: Computational enhancements in low-rank semidefinite programming. Optim. Methods Softw. **21**(3), 493–512 (2006)
55. Journée, M., Bach, F., Absil, P., Sepulchre, R.: Low-rank optimization on the cone of positive semidefinite matrices. SIAM J. Optim. **20**(5), 2327–2351 (2010)
56. Madani, R., Sojoudi, S., Fazelnia, G., Lavaei, J.: Finding low-rank solutions of sparse linear matrix inequalities using convex optimization. SIAM J. Optim. **27**(2), 725–758 (2017)
57. Lemon, A., So, A.M.C., Ye, Y.: Low-rank semidefinite programming: Theory and applications. Found. Trends Optim. **2**(1–2), 1–156 (2016)
58. Parrilo, P.A.: Semidefinite programming relaxations for semialgebraic problems. Math. Program. **96**(2), 293–320 (2003)
59. Papachristodoulou, A., Anderson, J., Valmorbida, G., Prajna, S., Seiler, P., Parrilo, P.: SOSTOOLS version 3.00 sum-of-squares optimization toolbox for MATLAB. User's guide, https://arxiv.org/abs/1310.4716 (2013)
60. Lasserre, J.B.: Moments, positive polynomials and their applications. Imperial College Press, London (2009)
61. De Klerk, E., Laurent, M.: On the Lasserre hierarchy of semidefinite programming relaxations of convex polynomial optimization problems. SIAM J. Optim. **21**(3), 824–832 (2011)
62. Lasserre, J.B., Toh, K.C., Yang, S.: A bounded degree SOS hierarchy for polynomial optimization. EUR. J. Comput. Optim. **5**(1), 87–117 (2017)
63. Jeyakumar, V., Lasserre, J.B., Li, G.: On polynomial optimization over non-compact semi-algebraic sets. J. Optim. Theory Appl. **163**(3), 707–718 (2014)
64. Nie, J.: Optimality conditions and finite convergence of Lasserre's hierarchy. Math. Program. **146**(1), 97–121 (2014)
65. Chen, D.S., Batson, R.G., Dang, Y.: Applied Integer Programming: Modeling and Solution. Wiley, Hoboken (2011)
66. Beale, E., Forrest, J.: Global optimization using special ordered sets. Math. Program. **10**(1), 52–69 (1976)

67. Vielma, J.P., Nemhauser, G.L.: Modeling disjunctive constraints with a logarithmic number of binary variables and constraints. Math. Program. **128**(1), 49–72 (2011)
68. Vielma, J.P., Ahmed, S., Nemhauser, G.: Mixed-integer models for nonseparable piecewise-linear optimization: unifying framework and extensions. Oper. Res. **58**(2), 303–315 (2010)
69. Silva, T.L., Camponogara, E.: A computational analysis of multidimensional piecewise-linear models with applications to oil production optimization. Eur. J. Oper. Res. **232**(3), 630–642 (2014)
70. Bernreuther, M.: Solving mixed-integer programming problems using piecewise linearization methods. Tech. Rep. 01/897074, University of Konstanz (2017)
71. Frenzen, C.L., Sasao, T., Butler, J.T.: On the number of segments needed in a piecewise linear approximation. J. Comput. Appl. Math. **234**(2), 437–446 (2010)
72. Tsai, J., Lin, M.: An improved framework for solving NLIPs with signomial terms in the objective or constraints to global optimality. Comput. Chem. Eng. **53**(53), 44–54 (2013)
73. Zhan, Y., Zheng, Q., Tseng, C., Pasiliao, E.: An accelerated extended cutting plane approach with piecewise linear approximations for signomial geometric programming. J. Global Optim. **70**(6), 1–21 (2018)
74. Fortuny-Amat, J., McCarl, B.: A representation and economic interpretation of a two-level programming problem. J. Oper. Res. Soc. **32**(9), 783–792 (1981)
75. FICO Xpress Optimization Suite: MIP formulations and linearizations. Quick reference. https://www.artelys.com/uploads/pdfs/Xpress/mipformref-1.pdf
76. Jünger, M., Liebling, T.M., Naddef, D., Nemhauser, G.L., Pulleyblank, W., Reinelt, G., Rinaldi, G., Wolsey, L.: 50 years of integer programming 1958–2008: from the early years to the state-of-the-art. Springer, New York (2010)
77. Vielma, J.P.: Mixed integer linear programming formulation techniques. SIAM Rev. **57**(1), 3–57 (2015)
78. Misener, R., Floudas, C.A.: Piecewise-linear approximations of multidimensional functions. J. Optim. Theory App. **145**(1), 120–147 (2010)
79. Rovatti, R., D'Ambrosio, C., Lodi, A., Martelloa, S.: Optimistic MILP modeling of non-linear optimization problems. Eur. J. Oper. Res. **239**(1), 32–45 (2014)
80. Sridhar, S., Linderoth, J., Luedtke, J.: Locally ideal formulations for piecewise linear functions with indicator variables. Oper. Res. Lett. **41**(6), 627–632 (2013)
81. Gupte, A., Ahmed, S., Cheon, M.S., Dey, S.: Solving mixed integer bilinear problems using MILP formulations. SIAM J. Optim. **23**(2), 721–744 (2013)
82. Kolodziej, S., Castro, P.M., Grossmann, I.E.: Global optimization of bilinear programs with a multiparametric disaggregation technique. J. Global Optim. **57**(4), 1039–1063 (2013)
83. Bonami, P., Lodi, A., Tramontani, A., Wieseet, S.: On mathematical programming with indicator constraints. Math. Program. **151**(1), 191–223 (2015)
84. Trespalacios, F., Grossmann, I.E.: Improved big-M reformulation for generalized disjunctive programs. Comput. Chem. Eng. **76**, 98–103 (2015)
85. Vielma J.P.: Embedding formulations and complexity for unions of polyhedra. Manage. Sci. **64**(10), 4721–4734 (2018)
86. Basu, A., Martin, K., Ryan, C.T., Wang, G.: Mixed-integer linear representability, disjunctions, and variable elimination. In: Proceeding International Conference on Integer Programming and Combinatorial Optimization, pp. 75–85. Springer, Cham (2017)
87. Vielma, J.P., Dunning, I., Huchette, J., & Lubin, M.: Extended formulations in mixed integer conic quadratic programming. Math. Program. Comput. **9**(3), 369–418 (2017)
88. Morán, R., Dey, S.S., Vielma, J.P.: A strong dual for conic mixed-integer programs. SIAM J. Optim. **22**(3), 1136–1150 (2012)
89. Huchette, J.A.: Advanced mixed-integer programming formulations: methodology, computation, and application. Doctoral Dissertation, MIT (2018)
90. Ben-Tal, A., El Ghaoui, L., Nemirovski, A.: Robust Optimization. Princeton University Press, Princeton (2009)
91. Bertsimas, D., Brown, D.B., Caramanis, C.: Theory and applications of robust optimization. SIAM Rev. **53**(3), 464–501 (2011)

92. Gorissen, B.L., Yanıkoğlu, İ., den Hertog, D.: A practical guide to robust optimization. Omega **53**, 124–137 (2015)
93. Ben-Tal, A., Nemirovski, A.: Selected topics in robust convex optimization. Math. Program. **112**(1), 125–158 (2008)
94. Gabrel, V., Murat, C., Thiele, A.: Recent advances in robust optimization: an overview. Eur. J. Oper. Res. **235**(3), 471–483 (2014)
95. Yanıkoğlu, İ, Gorissen, B.L., den Hertog, D.: A survey of adjustable robust optimization. Eur. J. Oper. Res. **277**, 799–813 (2018)
96. Bertsimas, D., Sim, M.: The price of robustness. Oper. Res. **52**(1), 35–53 (2004)
97. Boyd, S., Vandenberghe, L.: Convex Optimization. Cambridge University Press, New York (2004)
98. Alizadeh, F., Goldfarb, D.: Second-order cone programming. Math. Program. **95**(1), 3–51 (2003)
99. Ben-Tal, A., Nemirovski, A.: Lectures on Modern Convex Optimization: Analysis, Algorithms, and Engineering Applications. SIAM, Philadelphia (2001). http://www2.isye.gatech.edu/~nemirovs/Lect_ModConvOpt.pdf
100. Ben-Tal, A., den Hertog, D., Vial, J.-Ph.: Deriving robust counterparts of nonlinear uncertain inequalities. Math. Program. **149**(1), 265–299 (2015)
101. Bandi, C., Bertsimas, D.: Tractable stochastic analysis in high dimensions via robust optimization. Math. Program. **134**(1), 23–70 (2012)
102. Bertsimas, D., Dunning, I., Lubin, M.: Reformulation versus cutting-planes for robust optimization. Comput. Manag. Sci. 13(2), 195–217 (2016)
103. Iancu, D.A., Trichakis, N.: Pareto efficiency in robust optimization. Manage. Sci. **60**(1), 130–147 (2013)
104. Wei, W., Liu, F., Mei, S.: Offering non-dominated strategies under uncertain market prices. IEEE Trans. Power Syst. **30**(5), 2820–2821 (2015)
105. Wei, W., Liu, F., Mei, S.: Energy pricing and dispatch for smart grid retailers under demand response and market price uncertainty. IEEE Trans. Smart Grid **6**(3), 1364–1374 (2015)
106. Bertsimas, D., Goyal, V.: On the power of robust solutions in two-stage stochastic and adaptive optimization problems. Math. Oper. Res. **35**(2), 284–305 (2010)
107. Chen, X., Sim, M., Sun, P.: A robust optimization perspective on stochastic programming. Oper. Res. **55**(6), 1058–1071 (2007)
108. Ben-Tal, A., Goryashko, A., Guslitzer, E., Nemirovski, A.: Adjustable robust solutions of uncertain linear programs. Math. Program. **99**(2), 351–376 (2004)
109. Bertsimas, D., Iancu, D.A., Parrilo, P.A.: Optimality of affine policies in multistage robust optimization. Math. Oper. Res. 35(2), 363–394 (2010)
110. Bertsimas, D., Goyal, V.: On the power and limitations of affine policies in two-stage adaptive optimization. Math. Program. **134**(2), 491–531 (2012)
111. Bertsimas, D., Bidkhori, H.: On the performance of affine policies for two-stage adaptive optimization: a geometric perspective. Math. Program. **153**(2), 577–594 (2015)
112. Chen, X., Zhang, Y.: Uncertain linear programs: extended affinely adjustable robust counterparts. Oper. Res. **57**(6), 1469–1482 (2009)
113. Bertsimas, D., Caramanis, C.: Finite adaptability in multistage linear optimization. IEEE Trans. Autom. Control **55**(12), 2751–2766 (2010)
114. Bertsimas, D., Goyal, V., Sun, X.A.: A geometric characterization of the power of finite adaptability in multistage stochastic and adaptive optimization. Math. Oper. Res. **36**(1), 24–54 (2011)
115. Chen, X., Sim, M., Sun, P., Zhang, J.: A linear decision-based approximation approach to stochastic programming. Oper. Res. **56**(2), 344–357 (2008)
116. Bertsimas, D., Georghiou, A.: Design of near optimal decision rules in multistage adaptive mixed-integer optimization. Oper. Res. **63**(3), 610–627 (2015)
117. Georghiou, A., Wiesemann, W., Kuhn, D.: Generalized decision rule approximations for stochastic programming via liftings. Math. Program. **152**(1), 301–338 (2015)

118. Bertsimas, D., Litvinov, E., Sun, X.A., Zhao, J., Zheng, T.: Adaptive robust optimization for the security constrained unit commitment problem. IEEE Trans. Power Syst. **28**(1), 52–63 (2013)
119. Falk, J.E.: A linear max-min problem. Math. Program. **5**(1), 169–188 (1973)
120. Konno, H.: A cutting plane algorithm for solving bilinear programs. Math. Program. **11**(1), 14–27 (1976)
121. Korski, J., Pfeuffer, F., Klamroth, K.: Biconvex sets and optimization with biconvex functions: a survey and extensions. Math. Methods Oper. Res. **66**(3), 373–407 (2007)
122. Zeng, B., Zhao, L.: Solving two-stage robust optimization problems using a column-and-constraint generation method. Oper. Res. Lett. **41**(5), 457–461 (2013)
123. Bertsimas, D., de Ruiter, F.: Duality in two-stage adaptive linear optimization: faster computation and stronger bounds. INFORMS J. Comput. **28**(3), 500–511 (2016)
124. Bertsimas, D., Shtern, S.: A scalable algorithm for two-stage adaptive linear optimization. arXiv preprint arXiv:1807.02812 (2018)
125. Zhao, L., Zeng, B.: An exact algorithm for two-stage robust optimization with mixed integer recourse problems. Working Paper (2012), http://www.optimization-online.org/DB_FILE/2012/01/3310.pdf
126. Bertsimas, D., Thiele, A.: A robust optimization approach to inventory theory. Oper. Res. **54**(1), 150–168 (2006)
127. Zymler, S., Kuhn, D., Rustem, B.: Distributionally robust joint chance constraints with second-order moment information. Math. Program. **137**(1), 167–198 (2013)
128. Alizadeh, F., Goldfarb, D.: Second-order cone programming. Math. Program. **95**(1), 3–51 (2003)
129. Prékopa, A.: On probabilistic constrained programming. In: Proceedings of the Princeton Symposium on Mathematical Programming, pp. 113–138. Princeton University Press, Princeton (1970)
130. Bertsimas, D., Popescu, I.: Optimal inequalities in probability theory: a convex optimization approach. SIAM J. Optim. **15**(3), 780–804 (2005)
131. Lasserre, J.B.: Moments, Positive Polynomials and Their Applications. Imperial College Press, London (2009)
132. Rockafellar, R.T., Uryasev, S.: Optimization of conditional value-at-risk. J. Risk **2**, 21–24 (2002)
133. Shapiro, A., Kleywegt, A.J.: Minimax analysis of stochastic problems. Optim. Method. Softw. **17**(3), 523–542 (2002)
134. Bertsimas, D., Doan, X.V., Natarajan, K., Teo, C.P.: Models for minimax stochastic linear optimization problems with risk aversion. Math. Oper. Res. **35**(3), 580–602 (2010)
135. Lipp, T., Boyd, S.: Variations and extension of the convex-concave procedure. Optim. Eng. **17**(2), 263–287 (2016)
136. Jiang, R., Guan, Y.: Data-driven chance constrained stochastic program. Math. Program. **158**(1) 291–327 (2016)
137. Hanasusanto, G.A., Roitch, V., Kuhn, D., Wiesemann, W.: A distributionally robust perspective on uncertainty quantification and chance constrained programming. Math. Program. **151**(1), 35–62 (2015)
138. Ben-Tal, A., Den Hertog, D., De Waegenaere, A., Melenberg, B., Rennen, G.: Robust solutions of optimization problems affected by uncertain probabilities. Manag. Sci. **59**(2), 341–357 (2013)
139. Liese, F., Vajda, I.: On divergences and informations in statistics and information theory. IEEE Trans. Inform. Theory **52**(10), 4394–4412 (2006)
140. Calafiore, G.C., El Ghaoui, L.: On distributionally robust chance-constrained linear programs. J. Optim. Theory Appl. **130**(1), 1–22 (2006)
141. Ahmed, S., Xie, W.: Relaxations and approximations of chance constraints under finite distributions. Math. Program. **170**(1), 43–65 (2018)
142. Hu, Z., Hong, L.J.: Kullback–Leibler divergence constrained distributionally robust optimization. http://www.optimization-online.org/DB_FILE/2012/11/3677.pdf

143. Chen, Y., Guo, Q., Sun, H., Li, Z., Wu, W., Li, Z.: A distributionally robust optimization model for unit commitment based on Kullback–Leibler divergence. IEEE Trans. Power Syst. **33**(5), 5147–5160 (2018)
144. Duran, M.A., Grossmann, I.E.: An outer-approximation algorithm for a class of mixed-integer nonlinear programs. Math. Program. **36**(3), 307–339 (1986)
145. Fletcher, R., Leyffer, S.: Solving mixed integer nonlinear programs by outer approximation. Math. Program. **66**(1–3), 327–349 (1994)
146. Kelley, J.E.: The cutting-plane method for solving convex programs. J. Soc. Ind. Appl. Math. **8**(4), 703–712 (1960)
147. Hogan, W.W.: Applications of a general convergence theory for outer approximation algorithms. Math. Program. **5**(1), 151–168 (1973)
148. Zhao, C., Guan, Y.: Data-driven stochastic unit commitment for integrating wind generation. IEEE Trans. Power Syst. **31**(4), 2587–2596 (2016)
149. Ding, T., Yang, Q., Yang, Y., Li, C., Bie, Z.: A data-driven stochastic reactive power optimization considering uncertainties in active distribution networks and decomposition method. IEEE Trans. Smart Grid **9**(5), 4994–5004 (2018)
150. Esfahani, P.M., Kuhn, D.: Data-driven distributionally robust optimization using the Wasserstein metric: performance guarantees and tractable reformulations. Math. Program. **171**(1–2), 115–166 (2018)
151. Kantorovich, L.V., Rubinshtein, G.S.: On a space of totally additive functions. Vestn. Leningr. Univ. **13**, 52–59 (1958)
152. Xie, W.: On distributionally robust chance constrained program with Wasserstein distance. arXiv:1806.07418 (2018)
153. Hota, A.R., Cherukuri, A., Lygeros, J.: Data-driven chance constrained optimization under Wasserstein ambiguity sets. arXiv preprint arXiv:1805.06729 (2018)
154. Yao, L., Wang, X., Duan, C., Wu, X., Zhang, W.: Risk-based distributionally robust energy and reserve dispatch with Wasserstein-moment metric. In: IEEE Power and Energy Society General Meeting (2018)
155. Soyster, A.L.: Convex programming with set-inclusive constraints and applications to inexact linear programming. Oper. Res. **21**(5), 1154–1157 (1973)
156. Bertsimas, D., Gupta, V., Kallus, N.: Data-driven robust optimization. Math. Program. **167**(2), 235–292 (2018)
157. Bertsimas, D., David, B.: Constructing uncertainty sets for robust linear optimization. Oper. Res. **57**(6), 1483–1495 (2009)
158. Chen, W., Sim, M., Sun, J., Teo, C.P.: From CVaR to uncertainty set: implications in joint chance-constrained optimization. Oper. Res. **58**(2), 470–485 (2010)
159. Natarajan, K., Pachamanova, D., Sim, M.: Constructing risk measures from uncertainty sets. Oper. Res. **57**(5), 1129–1141 (2009)
160. Wiesemann, W., Kuhn, D., Sim, M.: Distributionally robust convex optimization. Oper. Res. **62**(6), 1358–1376 (2014)
161. Delage, E., Ye, Y.: Distributionally robust optimization under moment uncertainty with application to data-driven problems. Oper. Res. **58**(3), 596–612 (2010)
162. Vandenberghe, L., Boyd, S., Comanor, K.: Generalized Chebyshev bounds via semidefinite programming. SIAM Rev. **49**(1), 52–64 (2007)
163. Van Parys, B.P.G., Goulart, P.J., Kuhn, D.: Generalized Gauss inequalities via semidefinite programming. Math. Program. **156**(1), 271–302 (2016)
164. Postek, K., den Hertog, D., Melenberg, B.: Computationally tractable counterparts of distributionally robust constraints on risk measures. SIAM Rev. **58**(4), 603–650 (2016)
165. Shapiro, A.: Distributionally robust stochastic programming. SIAM J. Optim. **27**(4), 2258–2275 (2017)
166. Bayraksan, G., Love, D.K.: Data-driven stochastic programming using phi-divergences. In: The Operations Research Revolution, pp. 1–19, INFORMS, Catonsville (2015)
167. Love, D., Bayraksan, G.: Phi-divergence constrained ambiguous stochastic programs for data-driven optimization. Working Paper. http://www.optimization-online.org/DB_FILE/2016/03/5350.pdf (2015)

168. Zhao, C., Guan, Y.: Data-driven risk-averse stochastic optimization with Wasserstein metric. Oper. Res. Lett. **46**(2), 262–267 (2018)
169. Hanasusanto, G.A., Kuhn, D.: Conic programming reformulations of two-stage distributionally robust linear programs over Wasserstein balls. Oper. Res. **66**(3), 849–869 (2018)
170. Luo, F., Mehrotra, S.: Decomposition algorithm for distributionally robust optimization using Wasserstein metric. arXiv preprint arXiv:1704.03920 (2017)
171. Li, B., Jiang, R., Mathieu, J.L.: Ambiguous risk constraints with moment and unimodality information. Math. Program. **173**, 151–192 (2019)
172. Lotfi, S., Zenios, S.A.: Equivalence of robust VaR and CVaR optimization. Working Paper 16-03. The Wharton Financial Institutions Center (2016)
173. Pichler, A., Xu, H.: Quantitative stability analysis for minimax distributionally robust risk optimization. Math. Program. 1–31 (2019)
174. Lorca, A., Sun, X.A., Litvinov, E., Zheng, T.: Multistage adaptive robust optimization for the unit commitment problem. Oper. Res. **64**(1), 32–51 (2016)
175. Lorca, A., Sun, X.A.: Multistage robust unit commitment with dynamic uncertainty sets and energy storage. IEEE Trans. Power Syst. **32**(3), 1678–1688 (2017)
176. Gupta, V., Grossmann, I.E.: Solution strategies for multistage stochastic programming with endogenous uncertainties. Comput. Chem. Eng. **35**(11), 2235–2247 (2011)
177. Ding, T., Hu, Y., Bie, Z.: Multi-stage stochastic programming with nonanticipativity constraints for expansion of combined power and natural gas systems. IEEE Trans. Power Syst. **33**(1), 317–328 (2018)
178. Zhai, Q., Li, X., Lei, X., Guan, X.: Transmission constrained UC with wind power: an all-scenario-feasible MILP formulation with strong nonanticipativity. IEEE Trans. Power Syst. **32**(3), 1805–1817 (2017)
179. Ben-Tal, A., Brekelmans, R., Den Hertog, D., Vial, J.P.: Globalized robust optimization for nonlinear uncertain inequalities. INFORMS J. Comput. **29**(2), 350–366 (2017)
180. El Ghaoui, L., Oustry, F., Lebret, H.: Robust solutions to uncertain semidefinite programs. SIAM J. Optim. **9**(1), 33–52 (1998)
181. Zhen, J., de Ruiter, F., den Hertog, D.: Robust optimization for models with uncertain SOC and SDP constraints. http://www.optimization-online.org/DB_FILE/2017/12/6371.pdf (2017)
182. Ning, C., You, F.: Optimization under uncertainty in the era of big data and deep learning: when machine learning meets mathematical programming. Comput. Chem. Eng. **125**(9), 434–448 (2019)
183. Nash, J.: Equilibrium points in n-person games. Proc. Natl. Acad. Sci. **36**(1), 48–49 (1950)
184. Nash, J.: Non-cooperative games. Ann. Math. **54**(2), 286–295 (1951)
185. Granas, A., Dugundji, J.: Fixed Point Theory. Springer, New York (2003)
186. Dreves, A., Facchinei, F., Kanzow, C., Sagratella, S.: On the solution of the KKT conditions of generalized Nash equilibrium problems. SIAM J. Optim. **21**(3), 1082–1108 (2011)
187. Scutari, G., Palomar, D.P., Facchinei. F., Pang, J.S.: Monotone games for cognitive radio systems. In: Distributed Decision Making and Control, pp. 83–112. Springer, London (2012)
188. Facchinei, F., Pang, J.S.: Finite-Dimensional Variational Inequalities and Complementarity Problem. Springer, New York (2003)
189. Scutari, G., Palomar, D.P., Facchinei, F., Pang, J.S.: Convex optimization, game theory, and variational inequality theory. IEEE Signal Process. Mag. **27**(3), 35–49 (2010)
190. Ortega, J.M., Rheinboldt, W.C.: Iterative Solution of Nonlinear Equations in Several Variables. SIAM, Philadelphia (1987)
191. Scutari, G., Palomar, D.P., Barbarossa, S.: Asynchronous iterative water-filling for Gaussian frequency-selective interference channels. IEEE Trans. Inf. Theory **54**(7), 2868–2878 (2008)
192. Bertsekas, D.P., Tsitsiklis, J.N.: Parallel and Distributed Computation: Numerical Methods, 2nd edn. Athena Scientific Press, Belmont (1989)
193. Kannan, A., Shanbhag, U.V.: Distributed computation of equilibria in monotone Nash games via iterative regularization techniques. SIAM J. Optim. **22**(4), 1177–1205 (2012)

194. Dantzig, G.B.: Linear Programming and Extensions. Princeton University Press, Princeton (1963)
195. Lemke, C.E., Howson, J.T.: Equilibrium points of bimatrix games. SIAM J. Appl. Math. **12**(2), 413–423 (1964)
196. Lemke, C.E.: Bimatrix equilbrium points and mathematical programming, Manage. Sci. **11**(7), 681–689 (1965)
197. Monderer, D., Shapley, L.S.: Potential games. Games Econ. Behav. **14**(1), 124–143 (1996)
198. Lã, Q.D., Chew, Y.H., Soong, B.H.: Potential Game Theory. Springer, Switzerland (2016)
199. Harker, P.: Generalized Nash games and quasi-variational inequalities. Eur. J. Oper. Res. **54**(1), 81–94 (1991)
200. Facchinei, F., Kanzow, C.: Generalized Nash equilibrium problems. Quart. J. Oper. Res. **5**(3), 173–210 (2007)
201. Fukushima, M.: Restricted generalized Nash equilibria and controlled penalty algorithm. Comput. Manag. Sci. **8**(3), 201–218 (2011)
202. Rosen, J.B.: Existence and uniqueness of equilibrium points for concave N-person games. Econometrica **33**(3), 520–534 (1965)
203. Facchinei, F., Fischer, A., Piccialli, V.: On generalized Nash games and variational inequalities. Oper. Res. Lett. **35**(2), 159–164 (2007)
204. Facchinei, F., Sagratella, S.: On the computation of all solutions of jointly convex generalized Nash equilibrium problems. Optim. Lett. **5**(3), 531–547 (2011)
205. Schwartz, A.: Mathematical programs with complementarity constraints: theory, methods and applications. Ph.D. Thesis, Institute of Applied Mathematics and Statistics, University of Wrzburg (2011)
206. Guo, L., Lin, G.H.: Notes on some constraint qualifications for mathematical programs with equilibrium constraints. J. Optim. Theory Appl. **156**(3), 600–616 (2013)
207. Ye, J.: Constraint qualifications and KKT conditions for bilevel programming problems. Math. Oper. Res. **31**(4), 811–824 (2006)
208. Guo, L., Lin, G.H., Jane, J.Y.: Second-order optimality conditions for mathematical programs with equilibrium constraints. J. Optim. Theory Appl. **158**(1), 33–64 (2013)
209. Scholtes, S.: Convergence properties of a regularization scheme for mathematical programs with complementarity constraints. SIAM J. Optim. **11**(4), 918–936 (2001)
210. Ralph, D., Wright, S.J.: Some properties of regularization and penalization schemes for MPECs. Optim. Method. Softw. **19**(5), 527–556 (2004)
211. Hoheisel, T., Kanzow, C., Schwartz, A.: Theoretical and numerical comparison of relaxation methods for mathematical programs with complementarity constraints. Math. Program. **137**(1), 257–288 (2013)
212. Hu, X., Ralph, D.: Convergence of a penalty method for mathematical programming with complementarity constraints. J. Optim. Theory Appl. **123**(2), 365–390 (2004)
213. Lin, G.H., Fukushima, M.: Some exact penalty results for nonlinear programs and mathematical programs with equilibrium constraints. J. Optim. Theory Appl. **118**(1), 67–80 (2003)
214. Leyffer, S., López-Calva, G., Nocedal, J.: Interior methods for mathematical programs with complementarity constraints. SIAM J. Optim. **17**(1), 52–77 (2006)
215. Facchinei, F., Jiang, H., Qi, L.: A smoothing method for mathematical programs with equilibrium constraints. Math. Program. **85**(1), 107–134 (1999)
216. Fukushima, M., Pang, J.S.: Convergence of a smoothing continuation method for mathematical progams with complementarity constraints. In: Ill-posed Variational Problems and Regularization Techniques, pp. 99–110. Springer, Berlin (1999)
217. Kanzow, C.: Some noninterior continuation methods for linear complementarity problems. SIAM J. Matrix Anal. Appl. **17**(4), 851–868 (1996)
218. Fukushima, M., Luo, Z.Q., Pang, J.S.: A globally convergent sequential quadratic programming algorithm for mathematical programs with linear complementarity constraints. Comput. Optim. Appl. **10**(1), 5–34 (1998)
219. Fletcher, R., Leyffer, S., Ralph, D., Scholtes, S.: Local convergence of SQP methods for mathematical programs with equilibrium constraints. SIAM J. Optim. **17**(1), 259–286 (2006)

220. Ralph, D.: Mathematical programs with complementarity constraints in traffic and telecommunications networks. Philos. Trans. R. Soc. A **366**(1872), 1973–1987 (2008)
221. Fletcher, R., Leyffer, S.: Solving mathematical programs with complementarity constraints as nonlinear programs. Optim. Methods Softw. **19**(1), 15–40 (2004)
222. Fampa, M.H.C., Melo, W.A.X., Maculan, N.: Semidefinite relaxation for linear programs with equilibrium constraints. Int. Trans. Oper. Res. **20**(2), 201–212 (2013)
223. Jeyakumar, V., Lasserre, J.B., Li, G., Pham, T.S.: Convergent semidefinite programming relaxations for global bilevel polynomial optimization problems. SIAM J. Optim. **26**(1), 753–780 (2016)
224. Nie, J., Wang, L., Ye, J.: Bilevel polynomial programs and semidefinite relaxation methods. SIAM J. Optim. **27**(3), 1728–1757 (2017)
225. Fampa, M., Pimentel, W.: Linear programing relaxations for a strategic pricing problem in electricity markets. Int. Trans. Oper. Res. **24**(1–2), 159–172 (2017)
226. Ghamkhari, M., Sadeghi-Mobarakeh, A., Mohsenian-Rad, H.: Strategic bidding for producers in nodal electricity markets: a convex relaxation approach. IEEE Trans. Power Syst. **32**(3), 2324–2336 (2017)
227. Jara-Moroni, F., Pang, J.S., Wächter, A.: A study of the difference-of-convex approach for solving linear programs with complementarity constraints. Math. Program. **169**(1), 221–254 (2018)
228. Pineda, S., Bylling, H., Morales, J.M.: Efficiently solving linear bilevel programming problems using off-the-shelf optimization software. Optim. Eng. **19**(1), 187–211 (2018)
229. White, D.J., Anandalingam, G.: A penalty function approach for solving bi-level linear programs. J. Global Optim. **3**(4), 397–419 (1993)
230. Lv, Y., Hu, T., Wang, G., Wan, Z.: A penalty function method based on Kuhn-Tucker condition for solving linear bilevel programming. Appl. Math. Comput. **188**(1), 808–813 (2007)
231. Wan, Z., Wang, G., Lv, Y.: A dual-relax penalty function approach for solving nonlinear bilevel programming with linear lower level problem. Acta Math. Sci. **31**(2), 652–660 (2011)
232. Zare, M., Borrero, J., Zeng, B., Prokopyev, O.: A note on linearized reformulations for a class of bilevel linear integer problems. Ann. Oper. Res. 1–19 (2017)
233. Falk, J.: A linear max-min problem. Math. Program. **5**(1), 169–188 (1973)
234. Haghighat, H., Zeng, B.: Bi-level mixed integer transmission planning. IEEE Trans. Power Syst. **33**(6), 7309–7312 (2018)
235. Zeng, B., An, Y.: Solving bilevel mixed integer program by reformulations and decomposition. Optimization online (2014). http://www.optimization-online.org/DB_FILE/2014/07/4455.pdf
236. Yue, D., Gao, J., Zeng, B., You, F.: A projection-based reformulation and decomposition algorithm for global optimization of a class of mixed integer bilevel linear programs. J. Global Optim. **73**, 27–57 (2018)
237. Cottle, R.W., Pang, J.S., Stone, R.E.: The Linear Complementarity Problem. Society for Industrial and Applied Mathematics, Philadelphia (2009)
238. Lemke, C.E.: Bimatrix equilibrium points and mathematical programming. Manag. Sci. **11**(7), 681–689 (1965)
239. Kojima, M., Mizuno, S., Yoshise, A.: A polynomial time algorithm for a class of linear complementarity problems. Math. Program. **44**(1), 1–26 (1988)
240. Sherali, H.D., Krishnamurthy, R.S., Al-Khayyal, F.A.: Enumeration approach for linear complementarity problems based on a reformulation-linearization technique. J. Optim. Theory Appl. **99**(2), 481–507 (1998)
241. Fomeni, F.D., Gabriel, S.A., Anjos, M.F.: An RLT approach for solving the binary-constrained mixed linear complementarity problem. Tech. Rep. G-2015-60, GERAD-HEC Montréal (2015)
242. Pardalos, P.M., Rosen, J.B.: Global optimization approach to the linear complementarity problem. SIAM J. Sci. Stat. Comput. **9**(2), 341–353 (1988)
243. Xie, Y., Shanbhag, U.V.: On robust solutions to uncertain linear complementarity problems and their variants. SIAM J. Optim. **26**(4), 2120–2159 (2016)

244. Júdice, J.J., Faustino, A.M.: A sequential LCP method for bilevel linear programming. Ann. Oper. Res. **34**(1), 89–106 (1992)
245. Gabriel, S.A., Conejo, A.J., Ruiz, C., Siddiqui, S.: Solving discretely constrained, mixed linear complementarity problems with applications in energy. Comput. Oper. Res. **40**(5), 1339–1350 (2013)
246. Gabriel, S.A., Siddiqui, S.A., Conejo, A.J., Ruiz, C.: Solving discretely-constrained Nash-Cournot games with an application to power markets. Netw. Spat. Econ. **13**(3), 307–326 (2013)
247. Gabriel, S.A., Leuthold, F.U.: Solving discretely-constrained MPEC problems with applications in electric power markets. Energy Econ. **32**(1), 3–14 (2010)
248. Fomeni, F.D., Gabriel, S.A., Anjos, M.F.: Applications of logic constrained equilibria to traffic networks and to power systems with storage. Journal of the Operational Research Society **70**, 310–325 (2016). https://doi.org/10.13140/RG.2.2.23671.424072016
249. Hu, J., Mitchell, J.E., Pang, J.S., Bennett, K.P., Kunapuli, G.: On the global solution of linear programs with linear complementarity constraints. SIAM J. Optim. **19**(1), 445–471 (2008)
250. Chen, J., Burer, S.: Globally solving nonconvex quadratic programming problems via completely positive programming. Math. Prog. Comput. **4**(1), 33–52 (2012)
251. Hu, J., Mitchell, J.E., Pang, J.S., Yu, B.: On linear programs with linear complementarity constraints. J. Global Optim. **53**(1), 29–51 (2012)
252. Bai, L., Mitchell, J.E., Pang, J.S.: On convex quadratic programs with linear complementarity constraints. Comput. Optim. Appl. **54**(3), 517–554 (2013)
253. Fang, H., Leyffer, S., Munson, T.: A pivoting algorithm for linear programming with linear complementarity constraints. Optim. Methods Softw. **27**(1), 89–114 (2012)
254. Júdice, J.J.: Algorithms for linear programming with linear complementarity constraints. Top **20**(1), 4–25 (2012)
255. Watson, L.T., Billups, S.C., Mitchell, J.E., Easterling, D.R.: A globally convergent probability-one homotopy for linear programs with linear complementarity constraints. SIAM J. Optim. **23**(2), 1167–1188 (2013)
256. de Farias, I.R., Kozyreff, E., Zhao, M.: Branch-and-cut for complementarity-constrained optimization. Math. Program. Comput. **6**(4), 365–403 (2014)
257. Yu, B., Mitchell, J.E., Pang, J.S.: Solving linear programs with complementarity constraints using branch-and-cut. arXiv preprint arXiv:1802.02941 (2018)
258. Bai, L., Mitchell, J.E., Pang, J.S.: On conic QPCCs, conic QCQPs and completely positive programs. Math. Program. **159**(1–2), 109–136 (2016)
259. Hu, X.: Mathematical programs with complementarity constraints and game theory models in electricity markets. Doctor Dissertation, University of Melbourne (2003)
260. Ehrenmann, A.: Equilibrium problems with equilibrium constraints and their application to electricity markets. Doctor Dissertation, University of Cambridge (2004)
261. Su, C.L.: Equilibrium Problems with Equilibrium Constraints: Stationarities, Algorithms, and Applications. Doctor Dissertation, Stanford University (2005)
262. Gabriel, S.A., Conejo, A.J., Fuller, J.D., Ruiz, C.: Complementarity Modeling in Energy Markets. Springer, New York (2012)
263. Sherali, H.D.: A multiple leader Stackelberg model and analysis. Oper. Res. **32**(2), 390–404 (1984)
264. DeMiguel, V., Xu, H.: A stochastic multiple-leader Stackelberg model: analysis, computation, and application. Oper. Res. **57**(5), 1220–1235 (2009)
265. Hu, X., Ralph, D.: Using EPECs to model bilevel games in restructured electricity markets with locational prices. Oper. Res. **55**(5), 809–827 (2007)
266. Pang, J.S., Fukushima, M.: Quasi-variational inequalities, generalized Nash equilibria, and multi-leader-follower games. Comput. Manag. Sci. **2**(1), 21–56 (2005)
267. Ehrenmann, A.: Manifolds of multi-leader Cournot equilibria. Oper. Res. Lett. **32**(2), 121–125 (2004)
268. Kulkarni, A.A., Shanbhag, U.V.: A shared-constraint approach to multi-leader multi-follower games. Set-Valued Var. Anal. **22**(4), 691–720 (2014)

269. Kulkarni, A.A., Shanbhag, U.V.: An existence result for hierarchical Stackelberg v/s Stackelberg games. IEEE Trans. Autom. Control **60**(12), 3379–3384 (2015)
270. Guo, L., Lin, G.H., Zhang, D., Zhu, D.: An MPEC reformulation of an EPEC model for electricity markets. Oper. Res. Lett. **43**(3), 262–267 (2015)
271. Gabriel, S.A., Conejo, A.J., Fuller, J.D., Ruiz, C.: Complementarity Modeling in Energy Markets. Springer, New York (2012)
272. Leyffer, S., Munson, T.: Solving multi-leader-common-follower games. Optim. Methods Softw. **25**(4), 601–623 (2010)
273. Hu, M., Fukushima, M.: Variational inequality formulation of a class of multi-leader-follower games. J. Optim. Theory Appl. **151**(3), 455–473 (2011)
274. Aghassi, M., Bertsimas, D.: Robust game theory. Math. Program. **107**(1), 231–273 (2006)
275. Hayashi, S., Yamashita, N., Fukushima, M.: Robust Nash equilibria and second-order cone complementarity problems. J. Nonlinear Convex Anal. **6**(2), 283–296 (2005)
276. Nishimura, R., Hayashi, S., Fukushima, M.: Robust Nash equilibria in N-person non-cooperative games: uniqueness and reformulation. Pac. J. Optim. **5**(2), 237–259 (2009)
277. Singh, V.V., Jouini, O., Lisser, A.: Distributionally robust chance-constrained games: existence and characterization of Nash equilibrium. Optim. Lett. **11**(7), 1385–1405 (2017)
278. Facchinei, F., Fischer, A., Piccialli, V.: Generalized Nash equilibrium problems and Newton methods. Math. Program. **117**(1), 163–194 (2009)
279. Facchinei, F., Kanzow, C.: Penalty methods for the solution of generalized Nash equilibrium problems. SIAM J. Optim. **20**(5), 2228–2253 (2010)
280. von Heusinger, A., Kanzow, C.: Optimization reformulations of the generalized Nash equilibrium problem using Nikaido-Isoda-type functions. Comput. Optim. Appl. **43**(3), 353–377 (2009)
281. Nabetani, K., Tseng, P., Fukushima, M.: Parametrized variational inequality approaches to generalized Nash equilibrium problems with shared constraints. Comput. Optim. Appl. **48**(3), 423–452 (2011)
282. Kubota, K., Fukushima, M.: Gap function approach to the generalized Nash equilibrium problem. J. Optim. Theory Appl. **144**(3), 511–531 (2010)
283. von Heusinger, A., Kanzow, C.: Relaxation methods for generalized Nash equilibrium problems with inexact line search. J. Optim. Theory Appl. **143**(1), 159–183 (2009)
284. Schiro, D.A., Pang, J.S., Shanbhag, U.V.: On the solution of affine generalized Nash equilibrium problems with shared constraints by Lemke's method. Math. Program. **142**(1), 1–46 (2013)
285. Dreves, A., Sudermann-Merx, N.: Solving linear generalized Nash equilibrium problems numerically. Optim. Method. Softw. **31**(5), 1036–1063 (2016)
286. Kanzow, C., Steck, D.: Augmented Lagrangian methods for the solution of generalized Nash equilibrium problems. SIAM J. Optim. **26**(4), 2034–2058 (2016)
287. Luna, J.P., Sagastizábal C., Solodov, M.: An approximation scheme for a class of risk-averse stochastic equilibrium problems. Math. Program. **157**(2), 451–481 (2016)
288. Yu, C.K., van der Schaar, M., Sayed, A.H.: Distributed learning for stochastic generalized Nash equilibrium problems. IEEE Trans. Signal Process. **65**(15), 3893–3908 (2017)
289. Dempe, S.: Foundations of Bilevel Programming. Kluwer Academic Publishers, New York (2002)
290. Dempe, S., Kalashnikov, V., Pérez-Valdés, G.A., Kalashnykova, N.: Bilevel Programming Problems: Theory, Algorithms and Applications to Energy Networks. Springer, Berlin (2015)
291. Pozo, D., Sauma, E., Contreras, J.: Basic theoretical foundations and insights on bilevel models and their applications to power systems. Ann. Oper. Res. **254**(1–2), 303–334 (2017)
292. Chieu, N.H., Lee, G.M.: Constraint qualifications for mathematical programs with equilibrium constraints and their local preservation property. J. Optim. Theory Appl. **163**(3), 755–776 (2014)
293. Guo, L., Lin, G.H., Jane, J.Y.: Second-order optimality conditions for mathematical programs with equilibrium constraints. J. Optim. Theory Appl. **158**(1), 33–64 (2013)

294. Jane, J.Y., Zhang, J.: Enhanced Karush-Kuhn-Tucker conditions for mathematical programs with equilibrium constraints. J. Optim. Theory Appl. **163**(3), 777–794 (2014)
295. Gfrerer, H., Ye, J.J.: New constraint qualifications for mathematical programs with equilibrium constraints via variational analysis. SIAM J. Optim. **27**(2), 842–865 (2017)
296. Patriksson, M., Wynter, L.: Stochastic mathematical programs with equilibrium constraints. Oper Res. Lett. **25**(4), 159–167 (1999)
297. Lin, G., Fukushima, M.: A class of stochastic mathematical programs with complementarity constraints: reformulations and algorithms. J. Ind. Manag. Optim. **1**(1), 99–122 (2005)
298. Lin, G., Chen, X., Fukushima, M.: New restricted NCP functions and their applications to stochastic NCP and stochastic MPEC. Optimization **56**(5–6), 641–653 (2007)
299. Lin, G., Chen, X., Fukushima, M.: Solving stochastic mathematical programs with equilibrium constraints via approximation and smoothing implicit programming with penalization. Math. Program. **116**(1), 343–368 (2009)
300. Shapiro, A., Xu, H.: Stochastic mathematical programs with equilibrium constraints, modelling and sample average approximation. Optimization **57**(3), 395–418 (2008)
301. Lin, G.H., Fukushima, M.: Stochastic equilibrium problems and stochastic mathematical programs with equilibrium constraints: a survey. Pac. J. Optim. **6**(3), 455–482 (2010)
302. Hu, M.: Studies on multi-leader-follower games and related issues. Doctor Dissertation, Kyoto University, Kyoto (2012)
303. Aussel, D., Bendotti, P., Pištěk, M.: Nash equilibrium in a pay-as-bid electricity market: part 1-existence and characterization. Optimization **66**(6), 1013–1025 (2017)
304. Aussel, D., Bendotti, P., Pištěk, M.: Nash equilibrium in a pay-as-bid electricity market: part 2-best response of a producer. Optimization **66**(6), 1027–1053 (2017)
305. Solis, C.U., Clempner, J.B., Poznyak, A.S.: Modeling multi-leader-follower noncooperative Stackelberg games. Cybernet. Syst. **47**(8), 650–673 (2016)
306. Steffensen, S., Bittner, M.: Relaxation approach for equilibrium problems with equilibrium constraints. Comput. Oper. Res. **41**, 333–345 (2014)
307. Sinha, A., Malo, P., Frantsev, A., Deb, K.: Finding optimal strategies in a multi-period multi-leader-follower Stackelberg game using an evolutionary algorithm. Comput. Oper. Res. **41**, 374–385 (2014)
308. Hu, M., Fukushima, M.: Existence, uniqueness, and computation of robust Nash equilibria in a class of multi-leader-follower games. SIAM J. Optim. **23**(2), 894–916 (2013)
309. Lampariello, L., Sagratella, S.: A bridge between bilevel programs and Nash games. J. Optim. Theory Appl. **174**(2), 613–635 (2017)

Index

W. Wei, J. Wang, *Modeling and Optimization of Interdependent Energy Infrastructures*, https://doi.org/10.1007/978-3-030-25958-7

D

T

The manufacturer's authorised representative in the EU is Springer Nature Customer Service Centre GmbH, Europaplatz 3, 69115 Heidelberg, Germany. If you have any concerns regarding our products, please contact ProductSafety@springernature.com

Printed and bound by CPI Group (UK) Ltd, Croydon, CR0 4YY
02/10/2026
02227132-0012